Christoph Reigber
Hermann Lühr
Peter Schwintzer
Jens Wickert

# Earth Observation with CHAMP

Results from Three Years in Orbit

Christoph Reigber
Hermann Lühr
Peter Schwintzer
Jens Wickert
(Editors)

# Earth Observation with CHAMP

## Results from Three Years in Orbit

With 349 Figures, 96 in Color

Springer

**Prof. Dr. Dr. Christoph Reigber**
GeoForschungsZentrum Potsdam
Telegrafenberg, 14473 Potsdam
e-mail: reigber@gfz-potsdam.de

**Prof. Dr. Hermann Lühr**
GeoForschungsZentrum Potsdam
Telegrafenberg, 14473 Potsdam
e-mail: hluehr@gfz-potsdam.de

**Dr. Peter Schwintzer**
GeoForschungsZentrum Potsdam
Telegrafenberg, 14473 Potsdam
e-mail: psch@gfz-potsdam.de

**Dr. Jens Wickert**
GeoForschungsZentrum Potsdam
Telegrafenberg, 14473 Potsdam
e-mail: wickert@gfz-potsdam.de

Library of Congress Control Number: 2004109780

ISBN 3-540-22804-7 **Springer Berlin Heidelberg New York**



**Springer is a part of Springer Science+Business Media**
springeronline.com

Printed in Germany



Cover design: Achim Helm and Erich Kirchner
Production: Luisa Tonarelli
Typesetting: Katrin Weiße
Printing: Mercedes Druck, Berlin
Binding: Stein + Lehmann, Berlin

Printed on acid-free paper 32/3141/LT – 5 4 3 2 1 0

# PREFACE

The German geoscience satellite CHAMP (*CHA*llenging *M*inisatellite *P*ayload) was injected into its near circular, almost polar and low altitude (450 km) orbit from the cosmodrome Plesetsk in Russia on July 15, 2000. After a nine month commissioning period during which all spacecraft systems and instruments were checked out, calibrated and validated, the satellite is delivering an almost uninterrupted flow of science data since May 2001. Since that date, all science data have been made available to more than 300 selected co-investigator teams around the globe.

The scientific goals of the CHAMP mission are to gain a better understanding of dynamic processes taking place in the Earth's interior and in the space near Earth. These goals can be achieved by improved observation of the Earth's gravity and magnetic fields and their time variability with high-performance on-board instrumentation and by exploring the structure of the Earth's atmosphere and ionosphere through radio occultation measurements.

CHAMP carries a Global Positioning System (GPS) receiver, a high-precision accelerometer and a laser retro-reflector for the orbit, gravity and occultation tasks, and vector and scalar magnetometers and an ion drift meter for the magnetic field observations. Two pairs of star sensors provide the orientation for the vector instruments. This instrumentation, partially flown for the first time, is innovative in its combination for scientific and operational use

From September 1-4, 2003, a *CHAMP Science Meeting* was held at the GeoForschungsZentrum Potsdam, Germany. 207 scientists from 20 countries attended the meeting which stimulated a broad discussion within the international science community, especially among the CHAMP data and product users, on the exploitation and application of CHAMP data. The outcome of this meeting, organized along the main mission objectives, is largely mirrored in the main chapters of this book.

## Gravity

The gravity field related chapter clearly demonstrates that with CHAMP a new era of global Earth gravity field modeling and analysis has begun. The meanwhile released CHAMP-derived gravity models with their unprecedented accuracy enhance applications in geosciences, such as global to regional interpretation of gravity signatures with respect to the Earth interior, and ocean circulation studies using the CHAMP geoid. The challenges of detecting environmentally induced gravitational temporal field variations from space are discussed in several contributions. Stimulated by the high quality and continuity of the generated CHAMP orbit and gravity products, advanced and novel methods of data reduction, precise orbit determination and gravity field modeling as well as gravity model evaluation procedures are maturing, showing most convincing results.

**Magnetics**

Concerning the magnetic field research chapter, it is shown that detailed models of the field distribution, going down to a spatial half-wavelength resolution of 250 km, could be derived. Besides this precise mapping of the geomagnetic field, subtle details of the observed signals can be utilised for remote sensing purposes. This innovative approach opens a whole range of new aspects that can be addressed with the available measurements. In a number of articles information has been deduced from the magnetic field measurements on the conductivity in the Earth's interior, on electric current systems in the upper atmosphere and even on the flow patters of ocean water.

**Atmosphere/Ionosphere**

CHAMP pioneers an innovative technique for remote sensing of the Earth's neutral atmosphere and ionosphere on global scale, the GPS radio occultation method. This technique exploits the accuracy potential of the GPS for providing precise information on the vertical distribution of the main atmospheric parameters as temperature, water vapor and electron density. Various articles demonstrate the efficient and fast generation of high quality CHAMP occultation data products. For the first time a near-real-time analysis is demonstrated, which fulfils the operational requirements for the assimilation into global weather forecast systems. As shown in this chapter, measurements from CHAMP stimulate a multitude of research studies, leading to a new quality of the data analysis in the lower troposphere region and supporting the preparation of future GPS occultation missions.

We would like to thank all the participants of the CHAMP Science Meeting for their contributions to the realization of this book. In particular, we acknowledge the reviewers for their intensive work in improving the accepted contributions. Special thanks are due to Mrs. Katrin Weisse of GFZ Potsdam for her unresting efforts in collecting, formatting and editing the papers. Lastly, the financial and operational support by the Federal Ministry of Education and Research (BMBF) through its GEOTECHNOLOGIE programme, the GeoForschungsZentrum Potsdam (GFZ) and the German Aerospace Center (DLR) through their space-related programmes is gratefully acknowledged, without which neither the CHAMP satellite nor its operation would have been feasible.

Potsdam, August 2004

Christoph Reigber
Hermann Lühr
Peter Schwintzer
Jens Wickert

# Table of Contents


## I Orbit and Earth Gravity Field 1

Ice mass balance and Antarctic gravity change: Satellite and terrestrial perspectives 3
*Erik R. Ivins, Eric Rignot, Xiaoping Wu, Thomas S. James, Gino Casassa*

Gravity model TUM-2Sp based on the energy balance approach and kinematic CHAMP orbits 13
*Lóránt Földváry, Dražen Švehla, Christian Gerlach, Martin Wermuth, Thomas Gruber, Reiner Rummel, Markus Rothacher, Björn Frommknecht, Thomas Peters, Peter Steigenberger*

On the contribution of CHAMP to temporal gravity field variation studies 19
*Zhang Qiang and Philip Moore*

Earth gravity field and seasonal variability from CHAMP 25
*Christoph Reigber, Horst Jochmann, Johann Wünsch, Svetozar Petrovic, Peter Schwintzer, Franz Barthelmes, Karl-Hans Neumayer, Rolf König, Christoph Förste, Georges Balmino, Richard Biancale, Jean-Michel Lemoine, Sylvain Loyer, Felix Perosanz*

Comparison of superconducting gravimeter and CHAMP satellite derived temporal gravity variations 31
*Jürgen Neumeyer, Peter Schwintzer, Franz Barthelmes, Olaf Dierks, Yuichi Imanishi, Corinna Kroner, Bruno Meurers, He-Ping Sun, Heikki Virtanen*

Improvements in Arctic gravity and geoid from CHAMP and GRACE: An evaluation 37
*David McAdoo, Carl Wagner, Seymour Laxon*

Evaluation of gravity data by EIGEN-2 (CHAMP-only) model in China 47
*Yang Lu and Hongling Shi*

Energy balance relations for validation of gravity field models and orbit determinations applied to the CHAMP mission 53
*Anno Löcher and Karl Heinz Ilk*

Evaluation of terrestrial gravity data by independent global gravity field models 59
*Markus Roland and Heiner Denker*

Recent developments in CHAMP orbit determination at GFZ 65
*Rolf König, Grzegorz Michalak, Karl Hans Neumayer, Roland Schmidt, Sheng Yuan Zhu, Heribert Meixner, Christoph Reigber*

On calibrating the CHAMP on-board accelerometer and attitude quaternion processing 71
*Karl Hans Neumayer, Grzegorz Michalak, Rolf König*

Evaluation of the CHAMP accelerometer on two years of mission 77
*Félix Perosanz, Richard Biancale, Jean Michel Lemoine, Nicole Vales, Sylvain Loyer, Sean Bruinsma*

A new method to detect and estimate CHAMP clock bias change and cycle slip 83
*Bibo Peng, Bin Wu, Jun Li, Houze Hsu*

Comparison of different stochastic orbit modeling techniques 89
*Adrian Jäggi, Heike Bock, Urs Hugentobler, Gerhard Beutler*

Determination of non-conservative accelerations from orbit analysis 95
*Jose van den Ijssel, Pieter Visser, Roger Haagmans*

CHAMP and resonances 101
*Robert H. Gooding, Carl A. Wagner, Jaroslav Klokočník, Jan Kostelecký, Christoph Reigber*

CHAMP gravity field solutions and geophysical constraint studies 108
*Shin-Chan Han, C.K. Shum, Christopher Jekeli, Alexander Braun, Yiqun Chen, and Chung-Yen Kuo*

Application of Eigenvalue decomposition in the parallel computation of a CHAMP 100x100 gravity field 115
*Mark B. Hinga, Steve R. Poole, Byron D. Tapley*

Time-variable gravity seen by satellite missions: On its sampling and its parametrization 121
*Martin Wiehl and Reinhard Dietrich*

Gravity field recovery by analysis of short arcs of CHAMP 127
*Karl Heinz Ilk, Torsten Mayer-Gürr, Martin Feuchtinger*

Statistical assessment of CHAMP data and models using the energy balance approach 133
*Jürgen Kusche and Jasper van Loon*

Multiscale geopotential solutions from CHAMP orbits and accelerometry 139
*Martin J. Fengler, Willi Freeden, Jürgen Kusche*

Multiscale modeling from EIGEN-1S, EIGEN-2, EIGEN-GRACE01S, UCPH2002_0.5, EGM96 145
*Martin J. Fengler, Willi Freeden, Martin Gutting*

A comparison of various procedures for global gravity field recovery from CHAMP orbits 151
*Torsten Mayer-Gürr, Martin Feuchtinger, Jürgen Kusche*

Precise orbit determination for CHAMP using an efficient kinematic and reduced-dynamic procedure 157
*Heike Bock, Urs Hugentobler, Adrian Jäggi, Gerhard Beutler*

On bias and scale and thrust factors for CHAMP accelerometry 163
*Zhang Qiang and Philip Moore*

CHAMP accelerometer preprocessing at GeoForschungsZentrum Potsdam 169
*Christoph Förste and Sunchan Choi*

CHAMP clock characterization revisited 175
*Rolf König, Grzegorz Michalak, Ludwig Grunwaldt, Christoph Reigber*

How Baltic Sea water mass variations mask the postglacial rebound signal in CHAMP and GRACE gravity field solutions 181
*Martin Wiehl, Reinhard Dietrich, Andreas Lehmann*

The impact of the new CHAMP and GRACE gravity models on the measurement of the general relativistic Lense-Thirring effect 187
*Lorenzo Iorio*

Recovery of isostatic topography over North America from topographic and CHAMP gravity correlations 193
*Laramie V. Potts, C.K. Shum, Ralph von Frese, Shin-Chan Han, Rainer Mautz*

Dynamic topography as reflected in the global gravity field 199
*Mikhail K. Kaban, Peter Schwintzer, Christoph Reigber*

Impact of the CHAMP mission on estimating the mean sea surface 205
*Verena Seufer, Jens Schröter, Manfred Wenzel, Wolfgang Keller*

Improved estimates of the oceanic circulation using the CHAMP geoid 211
*Gennady Kivman, Sergey Danilov, Bernadette Fritzsch, Sven Harig, Christian Reick, Jens Schröter, Verena Seufer, Dmitry Sidorenko Joanna Staneva, Manfred Wenzel*

Contemporary changes in the geoid about Greenland: Predictions relevant to gravity space missions 217
*Kevin Fleming, Zdeněk Martinec, Jan Hagedoorn, Detlef Wolf*

Mantle viscosity and S-wave to density conversion profiles using CHAMP geoid data 223
*Gabriele Marquart and Radboud Koop*

Regional geoid undulations from CHAMP, represented by locally supported basis functions 230
*Rainer Mautz, Burkhard Schaffrin, C. K. Shum, Shin-Chan Han*

## II Earth Magnetic Field 237

Ionospheric plasma effects for geomagnetic LEO missions at mid- and low-latitudes 239
*Matthias Förster, Martin Rother, Hermann Lühr*

Interpretation of CHAMP crustal field anomaly maps using Geographical Information System (GIS) technique 249
*Kumar Hemant, Stefan Maus, Volker Haak*

Magnetic crustal thickness in Greenland from CHAMP and Ørsted data 255
*Cathrine Fox Maule, Michael E. Purucker, Nils Olsen*

CHAMP magnetic anomalies of the Antarctic crust 261
*Hyung Rae Kim, Luis R. Gaya-Piqué, Ralph R. B. von Frese, Patrick T. Taylor, Jeong Woo Kim*

Magnetic petrology database for interpretation satellite magnetic anomalies 267
*Katherine A Nazarova*

Balloon geomagnetic survey at stratospheric altitudes 273
*Katherine A. Nazarova, Yuri Tsvetkov, James Heirtzler, Terence Sabaka*

Effect of varying crustal thickness on CHAMP geopotential data 279
*Patrick T. Taylor, Károly I. Kis, Ralph R. B. von Frese, Juha V. Korhonen, Géza Wittmann, Hyung Rae Kim, Laramie V. Potts*

Reliability of CHAMP anomaly continuations 287
*Ralph R.B. von Frese, Hyung Rae Kim, Patrick T. Taylor, Mohammad F. Asgharzadeh*

Introducing POMME, the POtsdam Magnetic Model of the Earth 293
*Stefan Maus, Hermann Lühr, Georgios Balasis, Martin Rother, Mioara Mandea*

Alternative parameterisations of the external magnetic field and its induced counterpart for 2001 and 2002 using Ørsted, Champ and observatory data 299
*Vincent Lesur, Susan Macmillan, Alan Thomson*

New insight into secular variation between MAGSAT and CHAMP/ØRSTED 305
*Ingo Wardinski and Richard Holme*

Time structure of the 1991 magnetic jerk in the core-mantle boundary zone by inverting global magnetic data supported by satellite measurements 311
*Ludwig Ballani, Ingo Wardinski, Dietrich Stromeyer, Hans Greiner-Mai*

Use of CHAMP magnetic data to improve the Antarctic geomagnetic reference model 317
*Luis R. Gaya-Piqué, Angelo De Santis, Joan Miquel Torta*

Secular variation of the geomagnetic field from satellite data 323
*Vadim P. Golovkov, Tatiana I. Zvereva, Tatiana A. Chernova*

The spectrum of the magnetic secular variation 329
*Richard Holme and Nils Olsen*

Geomagnetic induction modeling based on CHAMP magnetic vector data 335
*Heather McCreadie and Zdeněk Martinec*

Electromagnetic induction by *Sq* ionospheric currents in a heterogeneous Earth: Modeling using ground-based and satellite measurements 341
*Jakub Velímský and Mark E. Everett*

Wavelet analysis of CHAMP flux gate magnetometer data 347
*Georgios Balasis, Stefan Maus, Hermann Lühr, Martin Rother*

Modelling the ocean effect of geomagnetic storms at ground and satellite altitude 353
*Alexei Kuvshinov, Nils Olsen*

3-D modelling of the magnetic fields due to ocean tidal flow 359
*Alexei Kuvshinov, Nils Olsen*

The enhancement of the thermospheric density during the Sept. 25-26, 2001 magnetic storm 366
*Huixin Liu, Hermann Lühr, Wolfgang Köhler*

On the modelling of field-aligned currents from magnetic observations by polar orbiting satellites 371
*Peter Stauning, Freddy Christiansen, Jürgen Watermann*

The low-altitude cusp: Multi-point observations during the February 2002 SIRCUS campaign 375
*Jurgen Watermann, Hermann Lühr, Kristian Schlegel, Peter Stauning, Jeffrey P. Thayer, Freddy Christiansen, Patrick T. Newell*

Detection of intense fine-scale field-aligned current structures in the Cusp region 381
*Peter Stauning, Freddy Christiansen, Jürgen Watermann*

A comparative study of geomagnetic Pi2 pulsations observed by CHAMP and on the ground 389
*Peter R Sutcliffe and Hermann Lühr*

ULF wave magnetic measurements by CHAMP satellite and SEGMA ground magnetometer array: Case study of July 6, 2002 395
*Massimo Vellante, Hermann Lühr, Tielong Zhang, Viktor Wesztergom, Umberto Villante, Marcello De Lauretis, Andrea Piancatelli, Martin Rother, Konrad Schwingenschuh, Wolfgang Koren, Werner Magnes*

Classes of the equatorial electrojet 401
*Heather McCreadie*

The ESPERIA project: A mission to investigate the near-Earth space 407
*Vittorio Sgrigna, Rodolfo Console, Livio Conti, Arkady Moiseev Galper, Valeria Malvezzi, Michel Parrot, Piergiorgio Picozza, Renato Scrimaglio, Piero Spillantini, David Zilpimiani*

Status of the CHAMP ME data processing 413
*Martin Rother, Sungchan Choi, Wolfgang Mai, Hermann Lühr, David Cooke*

## III Neutral Atmosphere and Ionosphere 419

Atmospheric and ocean sensing with GNSS 421
*Thomas P. Yunck and George A. Hajj*

Amplitude variations in CHAMP radio occultation signal as an indicator of the ionospheric activity 431
*Alexander Pavelyev, Jens Wickert, Christoph Reigber, Torsten Schmidt, Yuei-An Liou, Chen-Young Huang, Stanislav Matyugov, Dmitrii Pavelyev*

About the potential of GPS radio occultation measurements for exploring the ionosphere 441
*Norbert Jakowski, Konstantin Tsybulya, Stanimir M. Stankov, Andreas Wehrenpfennig*

Validation of GPS ionospheric radio occultation results onboard CHAMP by vertical sounding observations in Europe 447
*Norbert Jakowski, Konstantin Tsybulya, Jens Mielich, Anna Belehaki, David Altadill, Jean-Claude Jodogne, and Bruno Zolesi*

Ionospheric tomography with GPS data from CHAMP and SAC-C 453
*Miquel García-Fernández, Angela Aragón, Manuel Hernandez-Pajares, Jose Miguel Juan, Jaume Sanz, Victor Rios*

Topside plasma scale height modelling based on CHAMP measurements: First results 459
*Stanimir M. Stankov and Norbert Jakowski*

Differential code bias of GPS receivers in low Earth orbit: An assessment for CHAMP and SAC-C 465
*Stefan Heise, Claudia Stolle, Stefan Schlüter, Norbert Jakowski*

Ionosphere/plasmasphere imaging based on GPS navigation measurements from CHAMP and SAC-C 471
*Stefan Heise, Norbert Jakowski, David Cooke*

Three-dimensional monitoring of the polar ionosphere with ground- and space-based GPS 477
*Claudia Stolle, Stefan Schlüter, Christoph Jacobi, Norbert Jakowski, Stefan Heise, Armin Raabe*

Comparison of electron density profiles from CHAMP data with NeQuick model 483
*Norbert Jakowski, Konstantin Tsybulya, Sandro M. Radicella, Marta Cueto, Miguel Herraiz*

Model for short-term atmospheric density variations 489
*Mark Zijlstra, Stephan Theil, Silvia Scheithauer*

Atmospheric profiling with CHAMP: Status of the operational data analysis, validation of the recent data products and future prospects 495
*Jens Wickert, Torsten Schmidt, Georg Beyerle, Grzegorz Michalak, Rolf König, Julia Kaschenz, Christoph Reigber*

Simulated temperature and water vapor retrieval from bending angles and refractivity measurements using an optimal estimation approach 501
*Axel von Engeln and Gerald Nedoluha*

An analysis of the lower tropospheric refractivity bias by heuristic sliding spectral methods 507
*Georg Beyerle, Jens Wickert, Torsten Schmidt, Rolf König, Christoph Reigber*

Diffractive integrals for bistatic remote sensing using GPS signals 513
*Alexander Pavelyev, Jens Wickert, Yuei-An Liou*

Canonical transform methods for analysis of radio occultations 519
*Michael E. Gorbunov and Kent B. Lauritsen*

GPS radio occultation with CHAMP: Comparison of atmospheric profiles from GFZ Potsdam and IGAM Graz 525
*Jens Wickert, Andreas Gobiet, Georg Beyerle, Andrea K. Steiner, Gottfried Kirchengast, Ulrich Foelsche, Torsten Schmidt*

Evaluation of stratospheric radio occultation retrieval using data from CHAMP, MIPAS, GOMOS, and ECMWF analysis fields 531
*Andreas Gobiet, Gottfried Kirchengast, Jens Wickert, Christian Retscher, Ding-Yi Wang, Alain Hauchecorne*

Derivation of the water vapor content from the GNSS radio occultation observations 537
*Francesco Vespe, Jens Wickert, Catia Benedetto, Rosa Pacione*

Processing of CHAMP radio occultation data using GRAS SAF software 543
*Georg Bergeton Larsen, Kent Bækgaard Lauritsen, Frans Rubek, Martin Bjært Sørensen*

Gravity wave "portrait" reconstructed by radio holographic analysis of the amplitude of GPS radio occultation signals 549
*Yuei-An Liou, Jens Wickert, Alexander Pavelyev, Christoph Reigber, Torsten Schmidt, Chen-Young Huang, Shen Yan*

Global analysis of stratospheric gravity wave activity using CHAMP radio occultation temperatures 555
*Christoph Jacobi, Madineni Venkat Ratnam, Gerd Tetzlaff*

Tropical tropopause characteristics from CHAMP 561
*Torsten Schmidt, Jens Wickert, Georg Beyerle, Christoph Reigber*

Comparisons of MIPAS/ENVISAT and GPS-RO/CHAMP temperatures 567
*Ding-Yi Wang, Jens~Wickert, Gabriele P. Stiller, Thomas von Clarmann, Georg Beyerle, Torsten Schmidt, Manuel López-Puertas, Bernd Funke, Sergio Gil-López, Norbert Glatthor, Udo Grabowski, Michael Höpfner, Sylvia Kellmann, Michael Kiefer, Andrea Linden, Gizaw Mengistu Tsidu, Mathias Milz, Tilman Steck, Herbert Fischer*

Comparison of GPS/SAC-C and MIPAS/ENVISAT temperature profiles and its possible implementation for EOS MLS observations 573
*Jonathan H. Jiang, Ding-Yi Wang, Larry L. Roman, Chi O. Ao, Michael J. Schwartz, Gabriele P. Stiller, Thomas von Clarmann, Manuel López-Puertas, Bernd Funke, Sergio Gil-López, Norbert Glatthor, Udo Grabowski, Michael Höpfner, Sylvia Kellmann, Michael Kiefer, Andrea Linden, Gizaw Mengistu Tsidu, Mathias Milz, Tilman Steck, Herbert Fischer*

Structure and variability of the tropopause obtained from CHAMP radio occultation temperature profiles 579
*Madineni Venkat Ratnam, Gerd Tetzlaff, Christoph Jacobi*

An assessment of an ionospheric GPS data assimilation process 585
*Matthew Angling*

The Continuous Wavelet Transform, a valuable analysis tool to detect atmospheric and ionospheric signatures in GPS radio occultation phase delay data 591
*Achim Helm, Georg Beyerle, Stefan Heise, Torsten Schmidt, Jens Wickert*

The CHAMP atmospheric processing system for radio occultation measurements 597
*Torsten Schmidt, Jens Wickert, Georg Beyerle, Rolf König, Roman Galas, Christoph Reigber*

Potential contribution of CHAMP occultation to pressure field improvement for gravity recovery 603
*Shengjie Ge and C. K. Shum*

Analysis of gravity wave variability from SAC-C and CHAMP occultation profiles between June 2001 and March 2003 609
*Alejandro de la Torre, Toshitaka Tsuda, Ho Fang Tsai, George Hajj, Jens Wickert*

The CHAMPCLIM project: An overview 615
*Ulrich Foelsche, Andreas Gobiet, Armin Löscher, Gottfried Kirchengast, Andrea K. Steiner, Jens Wickert, and Torsten Schmidt*

**Author Index** **621**

**Keyword Index** **625**

# Orbit and Earth Gravity Field

Instruments, Orbit Determination and Gravity Field Recovery

# Ice Mass Balance and Antarctic Gravity Change: Satellite and Terrestrial Perspectives

Erik R. Ivins[1], Eric Rignot[1], Xiaoping Wu[1], Thomas S. James[2], and Gino Casassa[3]

[1] Jet Propulsion Lab, Caltech, Pasadena, USA *eri@fryxell.jpl.nasa.gov*
[2] Geological Survey of Canada, Sidney, BC, Canada
[3] Centro de Estudios Cientificos, Climate Change & Glaciol. Lab., Valdiva, Chile

**Summary.** Recent advances in the spatial and temporal retrieval of land-based cryospheric change information south of 42.5° allow fairly robust construction of forward model predictions of the time-rate of change in gravity. A map-view prediction is presented for the time-rate of change in geoid, $d\mathsf{N}/dt$ that might be retrieved from the currently orbiting gravity space craft, CHAMP (Challenging Mini-Satellite Payload for Geophysical Research and Application) and/or GRACE (Gravity Recovery and Climate Experiment). Complementary computation of the surface gravity change, $d\delta g/dt$, is also presented. The latter can be recovered from terrestrial absolute gravity measurements. Also, the computed rate of change Stokes coefficients for degree and order $\ell$, $m$ 1-12 may be used as reliable estimates of the Southern Hemisphere cryospheric change contribution to the global low-degree harmonic variability recorded in multidecadal satellite laser ranging (SLR) data sets.

**Key words:** CHAMP, GRACE, sealevel rise, cryosphere

## 1 Introduction

With the launching of the CHAMP and GRACE gravity satellite missions scientists are now provided with a new opportunity to decipher interannual and interdecadal hydrological changes that may have wide-sweeping societal implications about changes in the storage and replenishment of large freshwater reservoirs across the globe [1]. The Antarctic ice sheet stores a water mass having the potential to raise equivalent sealevel by 57 meters. The stability of this mass to wasting is quite sensitive to year-to-year changes in oceanic thermal structure [2]. The level of accuracy provided by the new gravity missions [3] implies that stand-alone satellite gravity data sets might be brought to bear on the question of Antarctic ice mass change. The main goal of this paper is to employ very recent estimates of ice mass balance to investigate the implications for measurement of the time-dependent gravity south of 42.5°.

## 2 Time-varying Stokes Coefficients

A main goal of non-tidal time-varying gravity field determination is to isolate perturbations to the external gravitational potential on a rigid Earth:

$$\mathcal{T}(r,\theta,\lambda,t) = \frac{GM}{r} \sum_{\ell=1}^{L} \sum_{m=0}^{\ell} \sum_{j=1}^{2} \{\frac{R_e}{r}\}^{\ell} \,\dot{\bar{C}}_{\ell mj}(t)\, \bar{Y}_{\ell mj}(\theta,\lambda) \tag{1}$$

where $r, \theta, \lambda, t$ are radial position, colatitude, longitude and time, respectively, and $GM$ is the product of the universal gravitational constant and planetary mass. The Earth's surface gravity is $g$, and mean radius, $R_e$. Here $\bar{Y}_{\ell mj}$ are normalized real surface harmonics [4] and $\bar{C}_{\ell mj}$ are the Stokes coefficients with the symbol $\dot{\{\ \}}$ for time differentiation. We call these later parameters the time-varying Stokes coefficients and consider physics in which these are primarily driven by surface density changes $\dot{\sigma}(\theta, \lambda, t)$ at $r = R_e$. Variations that conserve mass allow the sum (1) to begin at $\ell = 1$.

### 2.1 The 27-year Record from the Lageos Class SLR Measurements

Since the launch of Starlette and Lageos I in the mid-1970's continuous monitoring of $\dot{\bar{C}}_{\ell 01}$ for $\ell = 2, 3, 4, 5, 6, 7$ has been maintained. Since about 1992, when a constellation of similar small passive satellites were in orbit, these have been increasingly reliable observations. There are mass balance coefficient relations that link these observations to total nonsteric sealevel change [4] [5] [6]. However, there exists a large number of poorly modeled sources, aside from glacial isostatic adjustment (GIA) which drives a large component of the secular part of the signal [5]. A main inference derived from the even and odd-chained secular zonal signal, after correcting for GIA, is that Antarctica may be contributing to equivalent nonsteric sealevel rise rate (ESLRR) of about $\dot{\xi}_A$ = -0.05 to +0.6 mm/yr [5] [6], but with a number of critical caveats and rather indeterminable error estimates. The situation for Antarctica, however, has now taken a turn for the better as we now discuss.

### 2.2 12-year Record of Mass Balance Monitoring

Radar and laser altimetry, on-ice GPS, speckle tracking of ice flow by remote sensing, ocean temperature and salinity measurements, passive microwave monitoring from space, ice core data, and grounding line migrations determined from InSAR, now provide a wealth of new data from which the mass balance of the principal ice drainage basins of Antarctica may be obtained [7]. In some parts of East Antarctica the formal errors are comparable to the drainage basin imbalance estimate. However, these form a relatively small portion of the total, and are sufficiently small in amplitude that they do

**Table 1.** Model ESLRR (equivalent sealevel rise rates) in mm/yr.

| $\dot{\xi}_A$ | $\dot{\xi}_{NZ}$ | $\dot{\xi}_{SSA}$ | $\dot{\xi}_{AP}$ |
|---|---|---|---|
| 0.2410 | 0.0002 | 0.1050 | 0.0809 |

not degrade a quantitative estimation of satellite retrieved gravity changes south of the Antarctic Circle. With new Shuttle Radar Topography Mission (SRTM) data now available for elevation control, revised estimates for the Northern and Southern Patagonian icefields show that the ice mass loss to the oceans during the period 1995-2000 is substantial [8], occurring at rates that rival those of the entire Antarctic continent. Estimates are also available for glaciers of the Southern Alps of New Zealand [9].

The greatest uncertainty in computing southern hemispheric long wavelength gravity changes due to continental cryospheric imbalance is a lack of sound observational constraint on the Antarctic Peninsula region.

### 2.3 Space and Terrestrial Fields from Cryospheric Imbalances

It is straightforward to convert volumetric ice imbalance into gravity and crustal motion for elastic Earth models. First, we note that a set of surface density rate coefficients, $\dot{\bar{\sigma}}_{\ell mj}$, may be calculated in an expansion similar to (1) and related to the Stokes rate coefficients as:

$$\dot{\bar{\sigma}}_{\ell mj} = \frac{M}{2\pi R_e^2}(\ell + \frac{1}{2})\,\dot{\bar{C}}_{\ell mj}\,. \quad (2)$$

With the direct loading of the ocean (of total rise rate $\dot{\xi}$) accounted for, these coefficients are computed from a series of spherical caps with rates of ice height changes $\dot{D}_i$. Assuming conservation of mass, the density coefficients are

$$\dot{\bar{\sigma}}_{\ell mj} = \rho_{\rm ice} \sum_{i=1}^{\mathcal{I}_{\rm disk}} \tilde{w}_{\ell mj,i}\dot{D}_i - \frac{\rho_{\rm w}}{2\ell+1}\,\dot{\xi}\bar{a}_{\ell mj} \quad (3)$$

where $\rho_{\rm w}$ and $\rho_{\rm ice}$ are densities of ocean water and continental ice, and $\bar{a}_{\ell mj}$ are normed ocean coefficients [4]. The total number of caps is $\mathcal{I}_{\rm disk}$ and $\tilde{w}_{\ell mj,i}$ are disk coefficients (eq. 30b in [10]). The rate of change in the geoid [4] is:

$$\frac{d\mathsf{N}}{dt} = R_e \sum_{\ell=2}^{L}\sum_{m=0}^{\ell}\sum_{j=1}^{2} \dot{\bar{C}}_{\ell mj}(t)\,(1+k'_\ell)\,\bar{Y}_{\ell mj}(\theta,\lambda) \quad (4)$$

and the time-rate of change in surface gravity [10] is

$$\frac{d\delta g}{dt} = -g\sum_{\ell=2}^{L}\sum_{m=0}^{\ell}\sum_{j=1}^{2}[\frac{1}{2} + (\ell+1)k'_\ell + 2h'_\ell]\dot{\bar{C}}_{\ell mj}(t)\,\bar{Y}_{\ell mj}(\theta,\lambda) \quad (5)$$

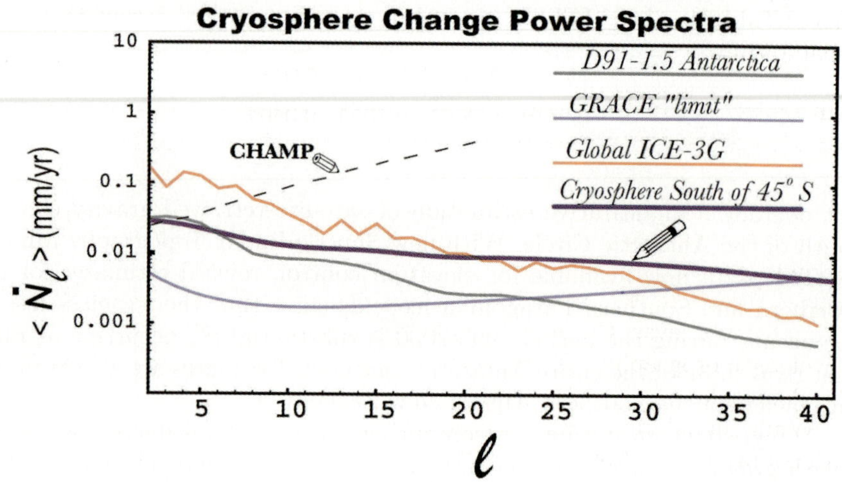


**Fig. 1.** Power spectra for cryosphere change and GIA models. GRACE annual recovery error estimate [1] and CHAMP sensitivity estimate for three years [14] correspond to theoretical limits.

wherein elastic deformation is accounted for by the surface potential and radial displacement Love numbers, $k'_\ell$ and $h'_\ell$, respectively. Geoid rates may be constructed using the stand-alone satellite data. The surface rate of change is measured using absolute gravity instruments placed on bedrock [11] [12].

**Integration of Present-day Cryospheric Information** Using the cap-grid structure for ice covered regions, each of the terms $\tilde{w}_{\ell mj,i}\dot{D}_i$ in (3) is created with grid-size varying from small sizes in the Cordillera Darwin (Tierra del Fuego), Patagonia and the Antarctic Peninsula (radii $\alpha_i \simeq 0.05°$) to large caps in East Antarctica (radii $\alpha_i \simeq 1.28°$). Each of these were treated, in essence, as cap harmonics. For our purposes, neither spatial nor spectral smoothing was required and the series were truncated at $L = 256$, as were the ocean function coefficients. The total ESLRR associated with land-ocean mass transfer for four regions south of 42.5° S is given in Table 1. The total rate of change in sealevel, $\dot{\xi}$, in (3) is assumed to be derived from the net sum of ice mass changes in the model.

**Degree Power Spectra** The detectability of the time-dependent field generated by changes in the continental cryosphere may be scrutinized via the secular geoid rate power spectrum [13]. Such plots (Figure 1) are diagnostic of the ability for satellite data to recover the time-dependent field associated with the large-scale ice mass changes. In Figure 1 the combined mass exchange of all four regions (*A*, *AP*, *SSA* and New Zealand; *NZ*) are assumed. For comparison, the spectra for both global and Antarctic GIA are also shown (see [13]). With the exception of Antarctica, GIA has a fairly reliable time-dependent gravity field.

**Antarctic Peninsula Scenario** The status of glacial and ice sheet balance has been greatly improved during the last few years in all but one region: the Antarctic Peninsula. Data-based models of glacier discharge that scale seasonal temperature changes to elevation and precipitation, suggest that the Peninsula is in negative mass balance [15] [16] [17]. The evidence, however, is somewhat anecdotal. (For example, the observed changes in adjacent ice shelves [18]). There is not enough data for constructing a complete balance estimate, and even the net surface mass balance is complicated by short wavelength orography [19]. We incorporate a scaled model [16] to account for the Peninsula.

While model inputs for Patagonia are likely significant at a $2{-}\sigma$ level [8], uncertainty of the Antarctic continent struggles at the $1{-}\sigma$ level [7], a situation possibly exacerbated by sub-decadal variability [20]. The estimate also considers reanalysis with newly acquired airborne laser altimetry and radio echo sounding data [21]. In spite of its large uncertainty, the new estimate is a great improvement over the simple 'scenarios' assumed previously [5] [10].

## 3 Terrestrial and Space Rate Predictions

### 3.1 Zonal Field

Observations of the amplitudes of the secular zonal harmonic rates are roughly $\mid \dot{J}_\ell \mid \sim 0.3 \longleftrightarrow 3.0 \times 10^{-11}\mathrm{yr}^{-1}$ for $2 \leq \ell \leq 6$ [6]. Table 2 gives ice change predictions for the very low order zonal gravity field rate coefficients. Clearly, these predictions are of an amplitude comparable to those derived from long-term SLR observations. For the first time, then, it is more definitively show that interdecadal cryospheric change in the high latitude Southern Hemisphere is a major contributor to the overall $\dot{J}_\ell$-sealevel budget. This is true, even if the '$AP$' scenario is ignored (the column $\dot{J}_\ell^{-AP}$ in Table 2 is for coefficients computed with the net imbalance of the Peninsula set to zero).

### 3.2 Non-zonal and Surface Gravity Change

The non-zonal coefficient amplitude is strong enough that CHAMP, GRACE and follow-on gravity missions could be employed to 'monitor' the cryospheric balance state in the Southern Hemisphere. For such analyses, we provide the lowest 12 degree/order normed rate coefficients in Table 3 for the forward model that includes the '$AP$' scenario and provide a map of the corresponding surface gravity rate change, $d\delta g/dt$ in Figure 2. Measurement of the surface

**Table 2.** Zonal rate coefficients $\dot{J}_\ell \equiv -\sqrt{2\ell+1}\dot{C}_{\ell 0}$ (units: $10^{-11}\mathrm{yr}^{-1}$)

| $\ell$ | $\dot{J}_\ell$ | $\dot{J}_\ell^{-AP}$ | $\ell$ | $\dot{J}_\ell$ | $\dot{J}_\ell^{-AP}$ |
|---|---|---|---|---|---|
| 1 | -2.034 | -1.587 | 7 | -0.170 | -0.218 |
| 2 | 1.211 | 0.929 | 8 | 0.055 | 0.163 |
| 3 | -1.040 | -0.780 | 9 | 0.081 | -0.064 |
| 4 | 0.755 | 0.559 | 10 | -0.266 | 0.104 |
| 5 | -0.495 | -0.381 | 11 | 0.478 | 0.321 |
| 6 | 0.298 | 0.270 | 12 | -0.677 | -0.542 |

gravity change may be important for separation of solid Earth GIA signatures from current ice mass change related gravity [22] [10]. The future for terrestrial gravity measurements may, indeed, be a bright one, due to recent technological advances that allow such instruments greater portability and lower power levels for continuous operation [23]. The predicted rates of gravity change in West Antarctica and Patagonia exceed annual change amplitude of 1.5 $\mu$gals over wavelengths of several hundred km.

### 3.3 Geoid Change

Continued improvements in the processing of CHAMP and GRACE data may soon mean that secular changes in the geoid shall be determined at, or below, the 1 mm level at half wavelengths, $\Delta\lambda \simeq \pi R_e/\ell$, of $\Delta\lambda \sim 1600$ km ($\ell \sim 12$) [24]. Figure 3 shows in mapview the computed annual geoid changes anticipated for cryospheric land/ocean mass transfer south of 42.5° using recent mass balance estimates [7] [8] [21]. Although preliminary analysis of CHAMP and GRACE data do not yet show this level of sensitivity, the large signal ($\sim$ -1 mm/yr) contoured over West Antarctica has $600 \leq \Delta\lambda \leq 2000$ km, and thus, there is potential for this signal to be detected.

## 4 Conclusions

The goal of the present paper is to clarify the current status of forward modeling predictions for interdecadal to multidecadal time-scale cryospheric change south of 42.5° S at high spatial resolution and to offer predictive maps for estimating both space and terrestrial-based gravity changes. Largely due to the application of space and airborne monitoring systems and the emergence of InSAR and GPS, great strides have been made in determining ice mass balance since about 1990. Folding such information into the various forward/inverse architectures for solving for global water mass fluxes is important in two distinct areas of gravity research: (1) analysis of mass flux budgets from the zonal field monitoring provided by SLR; and (2) for the future analyses of ice mass budgets in which high resolution data may be able to provide an additional tool for deciphering the GIA component of secular Antarctic gravity change.

**Table 3.** Non-zonal rate coefficients for ice mass south of 42.5° (units: $10^{-11}\text{yr}^{-1}$)

| $\ell$ | $m$ | $\dot{\bar{C}}_{\ell m}$ | $\dot{\bar{S}}_{\ell m}$ | $\ell$ | $m$ | $\dot{\bar{C}}_{\ell m}$ | $\dot{\bar{S}}_{\ell m}$ |
|---|---|---|---|---|---|---|---|
| 1 | 1 | -0.08773 | 0.50261 | 9 | 4 | 0.01458 | 0.05543 |
| 2 | 1 | 0.02346 | -0.43268 | 9 | 5 | -0.07292 | -0.00113 |
| 2 | 2 | 0.08552 | 0.03383 | 9 | 6 | 0.01382 | -0.05086 |
| 3 | 1 | -0.01104 | 0.48439 | 9 | 7 | 0.02130 | 0.01109 |
| 3 | 2 | -0.14452 | -0.05936 | 9 | 8 | -0.00569 | 0.00524 |
| 3 | 3 | 0.03178 | -0.03324 | 9 | 9 | -0.00065 | -0.00137 |
| 4 | 1 | -0.01045 | -0.46513 | 10 | 1 | -0.04409 | -0.19464 |
| 4 | 2 | 0.19257 | 0.05242 | 10 | 2 | 0.14171 | -0.10706 |
| 4 | 3 | -0.06405 | 0.05614 | 10 | 3 | 0.05232 | 0.02614 |
| 4 | 4 | -0.00828 | -0.02516 | 10 | 4 | -0.00072 | -0.02467 |
| 5 | 1 | 0.03937 | 0.40150 | 10 | 5 | 0.06542 | 0.00484 |
| 5 | 2 | -0.19575 | -0.02886 | 10 | 6 | -0.01848 | 0.06052 |
| 5 | 3 | 0.07950 | -0.07598 | 10 | 7 | -0.03325 | -0.01775 |
| 5 | 4 | 0.01583 | 0.04669 | 10 | 8 | 0.01150 | -0.01090 |
| 5 | 5 | -0.01305 | -0.00067 | 10 | 9 | 0.00222 | 0.00465 |
| 6 | 1 | -0.06038 | -0.33111 | 10 | 10 | -0.00106 | 0.00031 |
| 6 | 2 | 0.18137 | -0.00915 | 11 | 1 | 0.02961 | 0.15748 |
| 6 | 3 | -0.07874 | 0.08100 | 11 | 2 | -0.15805 | 0.09909 |
| 6 | 4 | -0.02673 | -0.06773 | 11 | 3 | -0.07236 | -0.02612 |
| 6 | 5 | 0.03113 | -0.00083 | 11 | 4 | -0.01162 | -0.00789 |
| 6 | 6 | -0.00193 | 0.00743 | 11 | 5 | -0.04577 | -0.00992 |
| 7 | 1 | 0.07038 | 0.27696 | 11 | 6 | 0.02180 | -0.06062 |
| 7 | 2 | -0.15689 | 0.05078 | 11 | 7 | 0.04319 | 0.02398 |
| 7 | 3 | 0.05696 | -0.07326 | 11 | 8 | -0.01948 | 0.01825 |
| 7 | 4 | 0.02943 | 0.07988 | 11 | 9 | -0.00525 | -0.01015 |
| 7 | 5 | -0.05086 | 0.00189 | 11 | 10 | 0.00302 | -0.00091 |
| 7 | 6 | 0.00488 | -0.01935 | 11 | 11 | -0.00006 | 0.00059 |
| 7 | 7 | 0.00414 | 0.00241 | 12 | 1 | -0.01830 | -0.10427 |
| 8 | 1 | -0.06880 | -0.24268 | 12 | 2 | 0.17152 | -0.08579 |
| 8 | 2 | 0.13730 | -0.08460 | 12 | 3 | 0.07896 | 0.03729 |
| 8 | 3 | -0.02101 | 0.05541 | 12 | 4 | 0.01921 | 0.03356 |
| 8 | 4 | -0.02493 | -0.07597 | 12 | 5 | 0.01933 | 0.01545 |
| 8 | 5 | 0.06719 | -0.00099 | 12 | 6 | -0.02387 | 0.04968 |
| 8 | 6 | -0.00916 | 0.03514 | 12 | 7 | -0.04760 | -0.02821 |
| 8 | 7 | -0.01151 | -0.00576 | 12 | 8 | 0.02790 | -0.02559 |
| 8 | 8 | 0.00206 | -0.00153 | 12 | 9 | 0.00924 | 0.01796 |
| 9 | 1 | 0.05850 | 0.21989 | 12 | 10 | -0.00731 | 0.00189 |
| 9 | 2 | -0.13236 | 0.10379 | 12 | 11 | 0.00015 | -0.00203 |
| 9 | 3 | -0.01913 | -0.03720 | 12 | 12 | 0.00045 | 0.00024 |

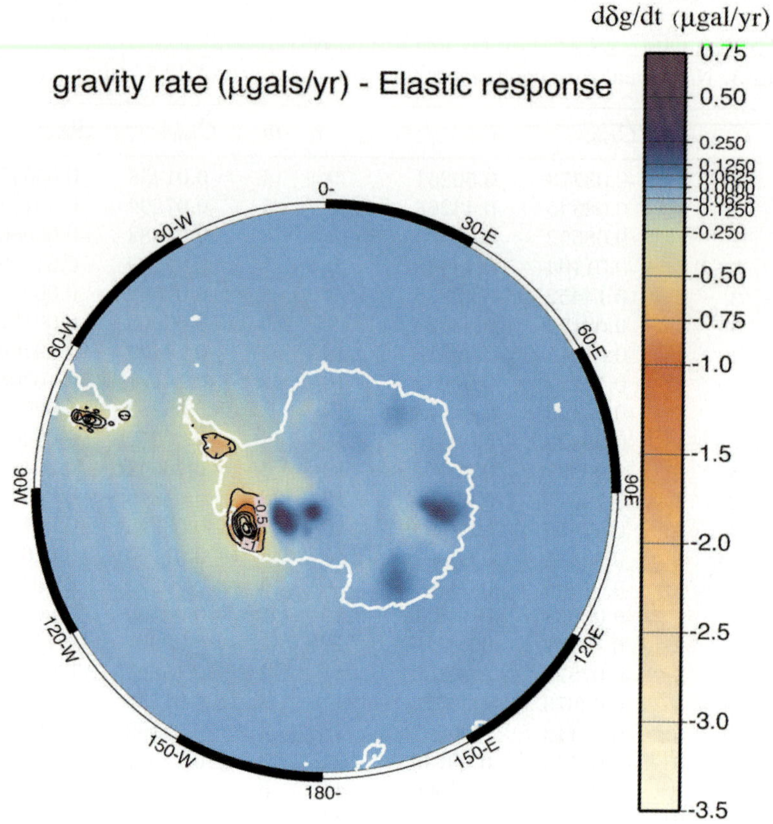


**Fig. 2.** Surface gravity change for absolute and relative terrestrial gravity measurements. Contours of negative $d\delta g/dt$ are at 0.5 $\mu$gal/yr intervals.

## References

1. Wahr J, Molenaar M, Bryan F (1998) Time-variability of the Earth's gravity field: hydrological and oceanic effects and their possible detection using GRACE. J Geophys Res *103*: 30,231–30,229.
2. Jacobs SS, Giulivi CF, Mele PA (2002) Freshening of the Ross Sea during the late 20th Century. Science *297*: 386–389.
3. Reigber C, Balmino G, Schwintzer P, Biancale R, Bode A, Lemoine JM, Konig R, Loyer S, Neumayer H, Marty JC, Barthelmes F, Perosanz F, Zhu SY (2003) Global gravity field recovery using solely GPS tracking and accelerometer data from CHAMP. Space Sci Rev *108*: 55–66.
4. Lambeck K (1980) The Earth's Variable Rotation: Geophysical Causes and Consequences. Cambridge U. Press, Cambridge New York Sydney, pp. 449.

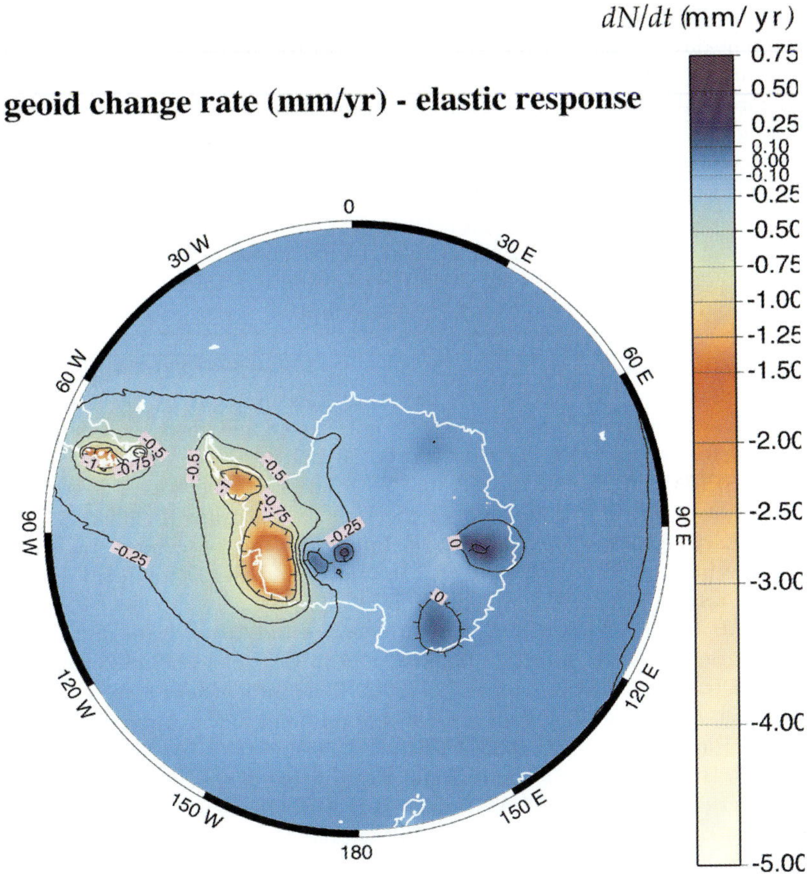


**Fig. 3.** Geoid changes caused by glacial imbalances, associated ocean filling and elastic response.


5. James TS, Ivins ER (1997) Global geodetic signatures of the Antarctic ice sheet. J Geophys Res *102*: 605–633.
6. Cox CM, SM Klosko BF Chao (2001) Changes in ice-mass balance inferred from time variations of the geopotental observed through SLR and DORIS tracking. In: Sideris M (ed) Gravity, geoid and geodynamics 2000: IAG symposium series *123*, Springer, Berlin Heidelberg New York: 353–360.
7. Rignot E, Thomas RH (2002) Mass balance of polar ice sheets. Science *297*: 1502–1506.
8. Rignot E, Rivera A, Casassa G (2003) Contribution of the Patagonia icefields of South America to sealevel rise. Science *302*: 434–437.
9. Dyurgerov MB, Meier MF (1997) Mass balance of mountain and subpolar glaciers: a new global assessment for 1961-1990. Arctic Alp Res *29*: 379–391.

10. James TS, Ivins ER (1998) Predictions of Antarctic crustal motions driven by present-day ice sheet evolution and by isostatic memory of the Last Glacial Maximum. J Geophys Res *103*: 4993–5017.
11. Lambert A, Laird JO, Courtier N, Bower DR (1994) Absolute gravimetry applied to postglacial rebound studies: progress in Laurentia. In: Schutz BE, Anderson A, Froidevaux C, Parke M (ed) Gravimetry and Space Techniques Applied to Geodynamics and Ocean Dynamics, Geophysical Monograph *82*, IUGG vol 17, AGU, Washington DC 1–7.
12. Amalvict M, Hinderer J, Luck B (2001) First absolute gravity measurements at the French station Dumont d'Urville (Antarctica). In: Sideras M (ed), Gravity, geoid and geodynamics 2000: IAG symposia series *123*, Springer, Berlin Heidelberg New York: 373-377.
13. Ivins ER, Wu X, Raymond CA, Yoder CF, James TS (2001) Temporal geoid of a rebounding Antarctica and potential measurement by the GRACE and GOCE satellites. In: Sideris M (ed) Gravity, geoid and geodynamics 2000: IAG symposium series *123*, Springer, Berlin Heidelberg New York: 361–366.
14. Visser PNAM, Rummel R, Balmino G, Sünkel H, Johanessen J, Aguirre M, Woodworth PL, Le Provost C, Tischering CC, Sabadini R (2002) The European Earth Explorer Mission: GOCE: Impact for the geosciences. In: Mitrovica JX, Vermerrsen B (ed) Ice Sheets, Sea Level and the Dynamics Earth, Geodynamics Series *29*, AGU Washington DC: 95–107.
15. Fox AJ, Cooper APR (1998) Climate-change indicators from archival aerial photography of the Antarctic Peninsula. Ann Glaciology *27*: 636–642.
16. Ivins ER Raymond CA, James TS (2000) The influence of 5000 year-old and younger glacial variability on present-day rebound in the Antarctic Peninsula. Earth Planets Space *52*: 1023–1029.
17. Rau F, Braun M (2002) The regional distribution of the dry-snow zone on the Antarctic Peninsula north of 70 degrees S Ann Glaciology *34*: 95-100.
18. Shepherd A, Wingham D, Payne T, Skvarca P (2003) Larsen Ice Shelf has progressively thinned. Science *302*: 856–859.
19. Turner J, Lachlan-Cope TA, Marshall GJ, Morris EM, Mulvaney R, Winter W (2002) Spatial variability of Antarctic Peninsula net surface mass balance. J Geophys Res *107(D13)*: 4173, doi:10.1029/2001JD000755.
20. Joughin I, Rignot E, Rosanova CE, Lucchitta BK, Bohlander J (2003) Timing of recent accelerations of Pine Island Glacier. Antarctic Geophys Res Lett *30*: art. no. 1706.
21. Rignot E, Thomas RH, Kanagaratnam P, Casassa G, Frederick E, Gogineni S, Krabill W, Rivera A, Russell R, Sonntag J, Swift R, Yungel J (2004) Improved estimation of the mass balance of glaciers draining into the Amundsen Sea sector of West Antarctica from the CECS/NASA 2002 Campaign. Ann Glaciology *39*, (in press).
22. Wahr J, Han D, Trupin A (1995) Predictions of vertical uplift caused by changing polar ice volumes on a viscoelastic Earth. Geophys Res Lett *22*: 977–901.
23. Vitouchkine AL, Faller JE (2002) Measurement results with a small cam-driven absolute gravimeter. Metrologia *39*: 465–469.
24. Schrama EJO (2003) Error characteristics estimated from CHAMP, GRACE and GOCE derived geoids and from satellite altimetry derived mean dynamic topography. Space Sci Rev *108*: 179–193.

# Gravity Model TUM-2Sp Based on the Energy Balance Approach and Kinematic CHAMP Orbits

Lóránt Földváry*, Dražen Švehla, Christian Gerlach, Martin Wermuth, Thomas Gruber, Reiner Rummel, Markus Rothacher, Björn Frommknecht, Thomas Peters, Peter Steigenberger

Institut für Astronomische und Physikalische Geodäsie, Technische Universität München, *foeldvary@bv.tum.de*

**Summary.** We have used one year of CHAMP data for deriving a gravity field model based on the energy balance approach. In order to avoid the use of any a priori gravity information, purely kinematic orbits have been computed from GPS measurements only. Subsequently velocities have been derived from these kinematic positions by two different methods, namely smoothing splines and Newton-Gregory interpolation. Using the principle of energy conservation, the satellite's positions and velocities are transformed into gravitational potential. CHAMP onboard micro-accelerometry is used to correct for surface forces. For spherical harmonic analysis the so-called direct approach has been implemented using the full normal equation matrix. The model, called TUM2Sp, was found to be a more accurate gravity field than EIGEN-2 model.

**Key words:** gravity field, energy integral, kinematic orbit

## 1 Introduction

Due to the BlackJack GPS receivers onboard of the CHAMP satellite [Reigber et al. (1999)], continuous satellite tracking became feasible. Simultaneously, using the micro-accelerometer onboard, the non-gravitational accelerations acting on the satellite are measured. These two instruments enable the application of the energy integral for gravity recovery from the CHAMP satellite [cf. Jekeli (1999), or Visser et al. (2003)],

$$E = \frac{1}{2}\dot{\boldsymbol{x}}^2 - V - \frac{1}{2}(\boldsymbol{\omega} \times \boldsymbol{x})^2 - \int_{\boldsymbol{x}} \boldsymbol{a}_e \cdot \mathrm{d}\boldsymbol{x} - \int_{\boldsymbol{x}} \boldsymbol{a}_{ng} \cdot \mathrm{d}\boldsymbol{x} \qquad (1)$$

The left hand side of equation (1) is the observed energy along the path of the satellite, and should be a constant, the Hamiltonian in a conservative force field. The Hamiltonian is the sum of the gravitational potential, $V$, and

* on leave from MTA-BME Research Group for Physical Geodesy and Geodynamics, Department of Geodesy and Surveying, Budapest University of Technology and Economics

kinetic energy, $\frac{1}{2}\dot{\boldsymbol{x}}^2$ . Employing a rotating Earth-fixed coordinate system the rotation in it also consumes energy, which is accounted for by the centrifugal term of equation (1) ($\frac{1}{2}(\boldsymbol{\omega} \times \boldsymbol{x})^2$ ). Further energy variations occur due to non-conservative forces. Variations due to external gravitational forces (i.e. direct, solid Earth, pole and ocean tides) are included in the acceleration $\boldsymbol{a_e}$, while non-gravitational forces acting on the satellite are contained in the $\boldsymbol{a_{ng}}$ vector.

The directly observed variables by CHAMP are the positions, $\boldsymbol{x}$, and the non-gravitational accelerations, $\boldsymbol{a_{ng}}$. The velocities, $\dot{\boldsymbol{x}}$, were derived from positions, while the remaining input data are from other sources (Earth rotation parameters from IERS, planetary ephemerides from JPL, named DE405, and ocean tide model from UT-CSR).

## 2 Method

As shown in Gerlach et al. (2003) the disturbing potential derived from (reduced-)dynamic orbits by the energy integral method is strongly influenced by the gravity model used for orbit determination. Therefore we stick to a purely kinematic solution in the present study (for kinematic POD see Švehla and Rothacher (2003a) and Švehla and Rothacher (2003b)). The flowchart in Figure 1 illustrates the determination of the gravity field using the energy integral, starting with a kinematic orbit.

In case of a kinematic orbit we face the problem that only positions and no velocities are determined. As we need velocities, compare equation 1, these must be derived from the kinematic positions numerically. Assuming the kinematic orbit being noisier than the (reduced-)dynamic one, in our first solution we attempted to reduce the noise of the kinematic positions by applying a smoothing on the position residuals (i.e. kinematic minus reduced-dynamic positions). Unavoidably the smoothing in the high frequencies also affected the gravity signal part to the same degree. This method was applied for the

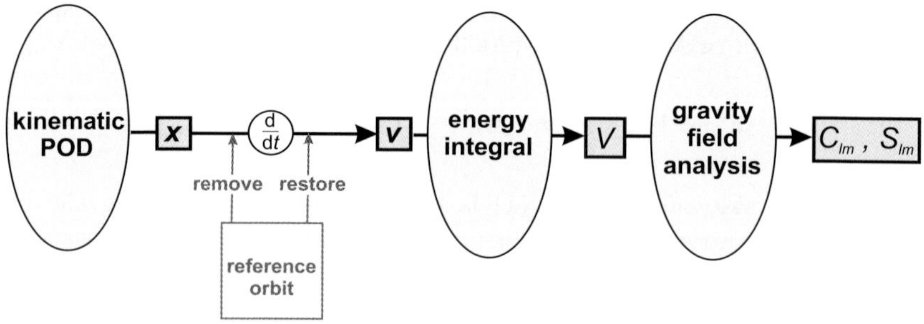


**Fig. 1.** Flowchart of the performed CHAMP gravity inversion in the present study.

TUM-1S model [Gerlach et al. (2003)]. In this case, in a remove-restore solution, smoothing cubic splines were fitted to the position residuals, then the derivative was taken analytically, and added to the reduced-dynamic velocities. TUM-1S made use of half a year of CHAMP data, which has been extended for this study to one year. In addition, some minor improvements in modelling have been introduced. The extended solution is referred as 'cs44', and processing steps of it are displayed on the flowchart (Figure 1) with the reference orbit step included.

Since for derivation of the kinematic velocities in remove-restore way reduced-dynamic orbits were used (which are known to be dependent on a priori gravity field), this solution seems to be affected by a priori gravity information. Prohibiting any possible dependence on the a priori gravity field, kinematic velocities have now been derived in a reference-field-free manner. For this we implicitly assume that the measurements are free of systematic errors, so the one year of data exhibits random distribution of the noise (white noise). For this solution we are approximating the kinematic positions by a simple interpolation method, namely Newton-Gregory interpolation, and the derivatives are computed analytically. The interpolation is applied on the pure kinematic positions. This solution leads to the 'cs45' coefficient set, which is named TUM-2Sp officially now. The processing sequence is illustrated in flowchart (Figure 1) without the remove-restore step for velocity determination.

## 3 Gravity Recovery

The differences of the aforementioned three estimations for a CHAMP-only gravity field are summarized in Table 1.

The pseudo-observable for the spherical harmonic analysis is the gravitational potential of the Earth derived from equation (1). The unknown potential coefficients, $C_{nm}$ and $S_{nm}$, were determined (from the spherical harmonic expansion of the potential) by least squares adjustment. The observations have been weighted equally.

The spherical harmonic coefficients were solved up to degree and order 100. As a consequence of the sampling of 30 seconds of the kinematic CHAMP orbit, which corresponds approximately to gravity information up to degree and order 90, and the numerical differentiation to obtain kinematic velocities,

**Table 1.** Gravity models discussed in this study.

| data set | length of data | kinematic velocity | reference orbit |
|---|---|---|---|
| TUM-1S | 1/2 year | smoothing splines | reduced-dynamic |
| cs44 | 1 year | smoothing splines | reduced-dynamic |
| cs45 | 1 year | Newton-Gregory interpolation | none |

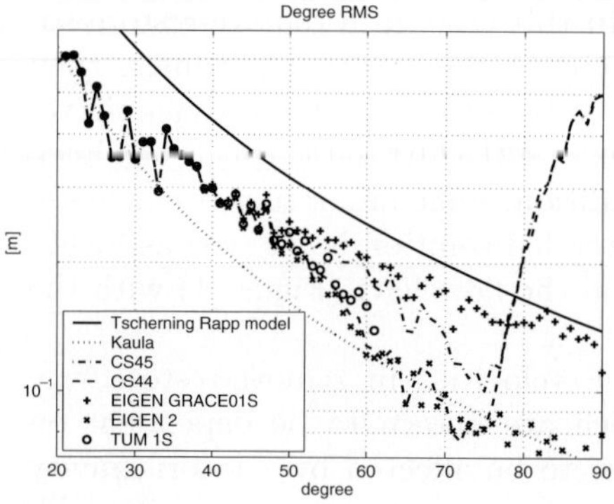


**Fig. 2.** Degree RMS of the geoid height.

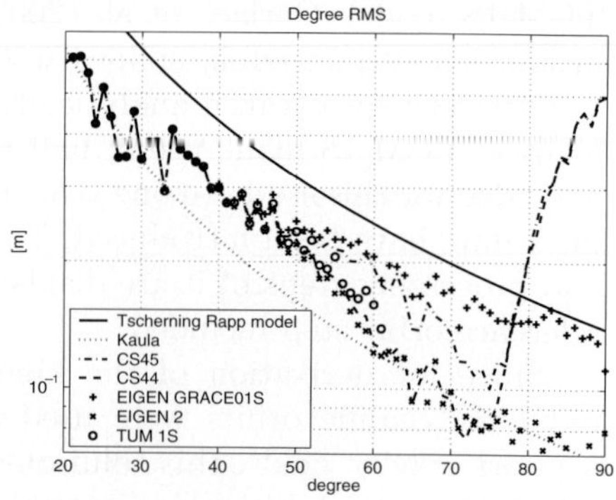


**Fig. 3.** RMS differences of geoid height differences at certain distances.

**Table 2.** Comparison between geoid heights from GPS/levelling and global potential models (low-pass filter above degree/order 60) in [cm].

| data set | EIGEN-2 | TUM-1S | CS44 | CS45 | EIGEN-GRACE01S |
|---|---|---|---|---|---|
| USA (5168 points) | 60.2 | 64.1 | 56.4 | 47.1 | 41.5 |
| Europe (180 points) | 59.3 | 56.4 | 55.7 | 33.1 | 19.4 |
| Australia (197 points) | 67.4 | 63.3 | 63.8 | 52.7 | 50.3 |
| Japan (837 points) | 69.5 | 65.5 | 66.7 | 54.8 | 51.5 |

we found disturbing potential signal up to about degree and order 60. In the higher degrees the noise is strongly dominating (cf. Figure 2). According to this, the presented sets of coefficients were truncated at degree 60.

## 4 Results

We ended up with three different gravity fields described in Table 1. These models were compared to GPS/leveling data over different regions (USA, Australia, Europe, Japan). The RMS values of the derived geoid height differences are listed in Table 2. The Table shows that EIGEN-GRACE01S model [Reigber et al. (2003b)] provides always the best solution, and cs45 model gets close to that, while the other 3 models are on similar level with a best performance of the cs44 model. EIGEN-2 [Reigber et al. (2003a)] is only superior to TUM-1S over the USA.

Table 3 shows the RMS differences of gravity anomalies over land and ocean and the Arctic based on gravimetric and altimetric data sets (NIMA [Lemoine et al. (1998)] and AGP [Kenyon and Forsberg (2002)]) with respect to the five models considered in Table 2.

**Table 3.** Comparison between gravity anomalies from terrestrial/altimetric data and the considered set of five models (low-pass filter at degree/order 60) in [mGal].

| data set | EIGEN-2 | TUM-1S | CS44 | CS45 | EIGEN-GRACE01S |
|---|---|---|---|---|---|
| Land (gravimetry) | 13.92 | 13.78 | 13.72 | 13.27 | 13.25 |
| Ocean (altimetry) | 6.61 | 6.58 | 6.50 | 6.12 | 5.87 |
| Arctic (gravimetry) | 14.67 | 14.55 | 14.54 | 14.17 | 14.14 |

Figure 3 shows RMS values of the differences of geoid height differences arranged according to various distances [Gruber (2001)]. The figure can be interpreted as a kind of resolution dependency of the geoid errors, even though there is no theoretically exact correspondence to degrees of spherical harmonics. This figure makes use only of the US data. In Figure 3 the EIGEN-2 is superior to TUM-1S - compare also Table 2. In other regions TUM-1S has slightly lower RMS values than EIGEN-2 (Table 2 and Table 3). CS44 set is slightly better than EIGEN-2 model, due to the extended one-year data set. The cs45 set performed better than EIGEN-2 model at all the degrees, and exhibits characteristic similarities with the EIGEN-GRACE01S model.

## 5 Conclusion

Based on the tests in the previous section we conclude that (1) EIGEN-2, TUM-1S and cs44 show similar error characteristics (cf. Figure 3); (2) solution cs45 and the first GRACE model (EIGEN-GRACE01S) show no error increase with decreasing distance (Figure 3). This suggest that (a) the white noise characteristic of the CHAMP kinematic orbits leads to an improved spatial resolution of the gravity model; (b) the gravity model used for reduced-dynamic orbit determination affects the gravity estimation when one makes use of the reduced-dynamic orbit as a reference for kinematic velocity estimation – therefore (reduced-)dynamic orbits should not be involved in gravity modelling.

As for the accuracy of the models, we should keep in mind that EIGEN-2, TUM-1S, cs44, cs45 and EIGEN-GRACE01S models are all significant improvement above pre-CHAMP models. This proves that with CHAMP we enter into a new era of gravity field modelling. The sequence EIGEN-2, TUM-1S, cs44, cs45, EIGEN-GRACE01S corresponds in increasing order with this accuracy. The TUM-1S and the EIGEN-2 models seem to be on a similar accuracy level, according to the results of the various tests; cs44 is slightly better due to the extension of the processed data to one year. The accuracy has significantly improved for cs45 due to change of the processing method from smoothing to interpolation without any use of a reference orbit. EIGEN-GRACE01S model reflects the refinements of low-low SST over high-low SST.

Some of the above mentioned models are available via internet. Visit *http://step.iapg.verm.tu-muenchen.de/iapg/forschen.html* for the TUM-1S and the TUM-2Sp models (the latter is referred as the cs45 set in this paper).

*Acknowledgement.* This work is supported by the German Ministry of Education and Research, Geotechnology program, Grant No. 03F0329A. The Australian GPS-levelling data have been provided by Geoscience Australia. The Japanese GPS-levelling data have been provided by the Geodetic Division of the Geographical Survey Institute, Tsukuba. Their contributions are gratefully acknowledged.

## References

Jekeli C (1999): The determination of gravitational potential differences from satellite-to-satellite tracking. Celestial Mechanics And Dynamical Astronomy *75*: 85–101.

Gerlach C, Földváry L, Švehla D, Gruber T, Wermuth M, Sneeuw N, Frommknecht B, Oberndorfer H, Peters T, Rothacher M, Rummel R, Steigenberger P (2003): A CHAMP-only gravity field model from kinematic orbit using the energy integral. Geophys Res Lett *30*: 20, 2037, doi:10.1029/2003GL018025.

Gruber T (2001): High resolution gravity field modeling with full variance-covariance matrices, J Geodesy *75(9/10)*: 505–514.

Kenyon S, Forsberg R (2003): Arctic Gravity Project. Web Site: http://earth-info.nima.mil/GandG/agp.

Lemoine FG, Kenyon SC, Factor JK, Trimmer RG, Pavlis NK, Chinn DS, Cox CM, Klosko SM, Luthcke SB, Torrence MH, Wang YM, Williamson RG, Pavlis EC, Rapp RH, Olson TR (1998): The development of the joint NASA GSFC and the National Imagery and Mapping Agency (NIMA) geopotential model EGM96; NASA Technical Paper NASA/TP-1998-206861, Greenbelt, Maryland, USA.

Reigber C, Schwintzer P, Lühr H (1999): The CHAMP geopotential mission. In: Proceedings of the 2nd Joint Meeting of the International Gravity and the International Geoid Commission (ed. Marson I, Sünkel H), Trieste 7-12 September, 1998. Boll Geofis Teor Appli *40(3-4)*: 285–289.

Reigber C, Schwintzer P, Neumayer KH, Barthelmes F, König R, Förste C, Balmino G, Biancale R, Lemoine JL, Loyer S, Bruinsma S, Perosanz F, Fayard T (2003): The CHAMP-only EIGEN-2 earth gravity field model, Adv Space Res *31(8)*: 1883–1888.

Reigber C, Schmidt R, Flechtner F, König R, Meyer U, Neumayer KH, Schwintzer P, Zhu SY (2003): First EIGEN gravity field model based on GRACE mission data only, in preparation for J Geodyn.

Švehla D, Rothacher M (2003a): Kinematic and reduced-dynamic precise orbit determination of low earth orbiters, Adv Geosciences *1*: 47–56.

Švehla D, Rothacher M (2003b): Kinematic precise orbit determination for gravity field determination, submitted to the Proceedings of the IUGG General Assembly 2003, June 30 July 11 2003, Sapporo, Japan, Springer Verlag, IAG *126*.

Visser PNAM, Sneeuw N, Gerlach C (2003): Energy integral method for gravity field determination from satellite orbit coordinates. J Geodesy *77*: 207–216.

# On the Contribution of CHAMP to Temporal Gravity Field Variation Studies

Zhang Qiang and Philip Moore

School of Civil Engineering and Geosciences, Newcastle University, UK
*Qiang.Zhang@ncl.ac.uk*

**Summary.** This study investigates the effect of temporal gravity field variability on CHAMP and other geodetic satellites. The sensitivity of these satellites to the lower order and degree harmonics is presented along with the dominant tidal periodicities. Lumped harmonics from analyses of CHAMP are discussed with a singular value decomposition identifying the dominant combinations. Temporal variability in lower order and degree harmonics are presented in studies with and without CHAMP. These results are compared against those derived from geophysical data.

**Key words:** Gravity, Lumped harmonics, Temporal variability

## 1 Sensitivity analyses of CHAMP and other geodetic satellites

The sensitivity of a particular satellite to the lower degree and order geopotential coefficients can be quantified by using analytical orbit theory (Kaula, 1966). This sensitivity is a combination of the orbital inclination and the orbital height given the accentuation of gravity field effects with altitude. For example, on consideration of the even and odd zonal harmonics

$$d\sigma^e/dt = \Sigma\sigma_l{}^e, \sigma = \Omega, \omega, M \tag{1}$$

$$d\sigma^o/dt = \Sigma\sigma_l{}^o cos\omega, \sigma = \omega, e \tag{2}$$

where the superscript, e or o, denotes even or odd zonals respectively. Utilising mean orbital elements and rates of change of the angular arguments the sensitivity of five geodetic satellites, Lageos I and II, Starlette, Ajisai and Stella, as well as CHAMP, are summarised in Tables 1 and 2. The tables illustrate the effect of a small change in the zonal harmonics, $J_n$ on the respective Keplerian elements. The tables show the insensitivity of the Lageos I and II satellites ($ht \approx 6000km$) to the higher degree harmonics. In contrast, Ajisai ($ht \approx 1200km$), Stella and Starlette ($ht \approx 800km$) and CHAMP ($ht \approx 400 - 450km$) are sensitive to degree 30 and beyond. It is the different sensitivities that facilitates the separation of the lower degree and order temporal variability from multisatellite analyses.

Another important consideration for temporal field studies are the periodicities of the dominant solid earth and ocean tides. In particular, any

| | Starlette | | Ajisai | | Stella | | LageosI | | LageosII | | Champ | |
|---|---|---|---|---|---|---|---|---|---|---|---|---|
| deg | $u^e$ | $\Omega^e$ | $u^e$ | $\Omega^e$ | $u^e$ | $\Omega^e$ | $u^e$ | $\Omega^e$ | $u^e$ | $\Omega^e$ | $u^e$ | $\Omega^e$ |
| 2 | 49.4 | -47.8 | 38.0 | -37.7 | -72.6 | 12.1 | -6.6 | 4.2 | 6.0 | -7.7 | -94.8 | -4.5 |
| 4 | 61.1 | -1.9 | 42.8 | -1.9 | -43.3 | 16.9 | 0.5 | 1.5 | 3.9 | -0.6 | -76.2 | -7.4 |
| 6 | -11.1 | 26.5 | -5.7 | 15.9 | -22.3 | 17.8 | 0.6 | 0.3 | 0.1 | 0.5 | -60.9 | -9.4 |
| 8 | -34.0 | 7.1 | -19.1 | 4.2 | -7.2 | 16.2 | 0.3 | 0.0 | -0.3 | 0.1 | -52.1 | -10.6 |
| 10 | -2.7 | -13.8 | -2.0 | -6.0 | 2.7 | 13.3 | 0.0 | 0.0 | 0.0 | 0.0 | -33.1 | -11.3 |
| 12 | 19.7 | -7.4 | -7.7 | -3.2 | 8.5 | 10.0 | 0.0 | 0.0 | -0.1 | 0.0 | -35.1 | -11.6 |
| 14 | 6.4 | 5.9 | 2.7 | 1.9 | 61.0 | 6.7 | 0.0 | 0.0 | -0.1 | 0.0 | -28.3 | -11.5 |
| 16 | -9.6 | 5.7 | -2.8 | 1.8 | 11.3 | 3.9 | 0.0 | 0.0 | 0.0 | 0.0 | -22.5 | -11.2 |
| 18 | -6.1 | -2.0 | -1.7 | -0.4 | 10.1 | 1.7 | 0.0 | 0.0 | 0.0 | 0.0 | -17.6 | -10.6 |
| 20 | 4.0 | -3.8 | 0.8 | -0.9 | 8.2 | 0.1 | 0.0 | 0.0 | 0.0 | 0.0 | -13.4 | -10.0 |
| 22 | 4.5 | 0.2 | 1.0 | 0.0 | 6.1 | -0.9 | 0.0 | 0.0 | 0.0 | 0.0 | -9.9 | -9.3 |
| 24 | -1.2 | 2.3 | -0.2 | 0.4 | 4.1 | -1.4 | 0.0 | 0.0 | 0.0 | 0.0 | -5.1 | -8.5 |
| 26 | -2.9 | 0.5 | -0.4 | 0.1 | 2.3 | -1.5 | 0.0 | 0.0 | 0.0 | 0.0 | -4.7 | -7.7 |
| 28 | 0.0 | -1.2 | 0.0 | -0.1 | 1.0 | -1.5 | 0.0 | 0.0 | 0.0 | 0.0 | -2.8 | -6.9 |
| 30 | 1.7 | -0.6 | 0.1 | -0.1 | 0.0 | -1.2 | 0.0 | 0.0 | 0.0 | 0.0 | -1.1 | -6.1 |

**Table 1.** Sensitivity of satellite orbits to even zonal variations (per unit $\Delta Jn$ x 1.0d-11) (milliarcseconds/year); $u^e = M^e + \omega^e$

mis-modelling of tidal constituent giving rise to a signal at the annual or semi-annual frequency will be aliased into those periods and possibility misinterpreted as mass redistribution. Table 3 presents the theoretical amplitudes and periodicities due to the second degree solid earth tide, with $k2 = 0.3$. Most of the dominant tidal amplitudes are long-periodic but apart from the annual and semi-annual zonal solar tide, should not aliase recovery of signatures at the seasonal to annual frequencies.

## 2 CHAMP Normal Equations: Singular Value Decomposition

Utilising the precise positioning from GFZ rapid science orbits for CHAMP and SLR tracking to LageosI/II, Starlette, Stella and Ajisai we have recovered the gravity field up to degree and order 10 over 15 day arcs along with the other parameters of Table 4. In all orbit computations the a priori gravity field was GGM01C with ocean tidal model CSR3.0. For CHAMP, k0(L) denotes local accelerometer bias, k1(G) global accelerometer scale factors, th(L) local thruster accelerations with L indicating a daily solution and G a 15day solution. For other satellites, Cd denotes daily drag coefficients, Cr a global solar radiation coefficient and atacc along track accelerations every ($5day$) for Lageos.

Normal equations for the gravity field component over each 15day period were combined with weights according to perceived accuracy of $J_2$ namely,

| | Starlette | | Ajisai | | Stella | | LageosI | | LageosII | | Champ | |
|---|---|---|---|---|---|---|---|---|---|---|---|---|
| deg | $e^o$ | $\omega^0$ | $e^o$ | $\omega^0$ | $e^o$ | $\omega^0$ | $e^o$ | $\omega^0$ | $e^o$ | $\omega^0$ | $e^o$ | $\omega^0$ |
| 3 | -13.3 | 325.8 | -9.7 | 4847.8 | 15.5 | -377.6 | 0.6 | -72.6 | -1.1 | 40.7 | 22.1 | -2757.3 |
| 5 | -13.8 | 337.5 | -9.1 | 4571.6 | 12.0 | -291.2 | -0.2 | 19.0 | -0.6 | 21.7 | 23.6 | -2950.4 |
| 7 | 7.0 | -173.3 | 3.7 | -1851.8 | 6.5 | -158.1 | -0.1 | 15.7 | 0.0 | -1.0 | 21.9 | -2735.9 |
| 9 | 10.8 | -265.2 | 5.4 | -2717.2 | 1.7 | -41.3 | -0.1 | 5.2 | 0.1 | -2.0 | 19.3 | -2410.7 |
| 11 | -1.7 | 42.1 | -0.5 | 246.4 | -1.7 | 43.2 | -0.0 | 1.0 | 0.0 | -0.2 | 16.5 | -2063.1 |
| 13 | -6.9 | 172.2 | -2.6 | 1309.9 | -3.8 | 94.4 | 0.0 | 0.0 | 0.0 | 0.1 | 13.8 | -1727.3 |
| 15 | -0.8 | 18.0 | -0.4 | 193.8 | -4.7 | 117.2 | 0.0 | -0.1 | 0.0 | 0.0 | 11.3 | -1417.9 |
| 17 | 3.9 | -98.4 | 1.1 | -544.2 | -4.7 | 118.8 | 0.0 | 0.0 | 0.0 | 0.0 | 9.1 | -1140.6 |
| 19 | 1.4 | -35.6 | 0.4 | -214.0 | -4.2 | 106.4 | 0.0 | 0.0 | 0.0 | 0.0 | 7.2 | -897.2 |
| 21 | -1.9 | 49.4 | -0.4 | 192.3 | -3.4 | 86.5 | 0.0 | 0.0 | 0.0 | 0.0 | 5.5 | -686.9 |
| 23 | -1.3 | 33.7 | -0.3 | 140.0 | -2.5 | 63.8 | 0.0 | 0.0 | 0.0 | 0.0 | 4.1 | -507.9 |
| 25 | 0.8 | -20.8 | 0.1 | -52.4 | -1.6 | 42.1 | 0.0 | 0.0 | 0.0 | 0.0 | 2.9 | -357.7 |
| 27 | 1.0 | -25.3 | 0.2 | -74.2 | -0.9 | 23.2 | 0.0 | 0.0 | 0.0 | 0.0 | 1.9 | -233.4 |
| 29 | -0.2 | 5.9 | -0.0 | 5.8 | -0.3 | 8.5 | 0.0 | 0.0 | 0.0 | 0.0 | 1.1 | -132.1 |
| 31 | -0.6 | 16.6 | -0.1 | 34.1 | 0.1 | -2.1 | 0.0 | 0.0 | 0.0 | 0.0 | 0.4 | -51.0 |

**Table 2.** Sensitivity of satellite orbits to odd zonal variations (per unit $\Delta Jn$ x 1.0d-11) (milliarcseconds/year)

Lageos I (1.2), LageosII (1.0), Starlette (0.8), Ajisai (0.8), Stella (0.6) and CHAMP (0.1). On eliminating the contribution of the local parameters the geopotential normal equations $N\underline{x} = \underline{b}$ can be written using Singular Value Decomposition (SVD) as $N = QWP^T$, $W\underline{x}' = \underline{b}'$ where $\underline{x}' = P^T\underline{x}$ and $\underline{b}' = Q^{-1}\underline{b}$. $W$ is a diagonal weight matrix with covariance for $\underline{x}'$ being $C = W^{-1}$. SVD identifies the dominant linear combinations for each 15day arc. Thus, for example, for Lageos 1 (MJD 50864-50879) the dominant combination in a 4x4 gravity field recovery was the lumped harmonic

$$J_2' = 1.000J_2 + 0.369J_4 - 0.068J_3 - 0.036C_{2,1} + \ldots\ldots \quad (3)$$

The $1\sigma$ sd was $\sigma_{J_2'} = 2.6e - 11$. Equation (3) is to be compared with the theoretical value from $d\Omega/dt$ in Table 1, namely

$$J_2' = 1.000J_2 + 0.371J_4 + 0.0795J_6 + \ldots\ldots \quad (4)$$

In contrast, solving for a 10 by 10 field from CHAMP the $46^{th}$ ranking lumped harmonic ($1\sigma$ sd=3.8e-11) for MJD 50249-50264 is

$$J_2' = 1.000J_2 + 1.613J_4 + 2.023J_6 + 2.285J_8 + 2.433J_{10} + 1.622C_{4,3} - 1.619C_{2,2} \quad (5)$$

which can be compared against the theoretical value of $d\Omega/dt$ in Table 1, namely

$$J_2' = 1.000J_2 + 1.630J_4 + 2.060J_6 + 2.336J_8 + 2.490J_{10} + \ldots\ldots \quad (6)$$

| Tide | Theoretical Amplitude (arc secs) $\Delta I$ | $\Delta\omega$ | $\Delta\Omega$ | Period (days) |
|---|---|---|---|---|
| 055.545 | 0.0000 | 1.0934 | 11.3904 | 6798.375 |
| 056.554 Sa | 0.0000 | 0.0106 | 0.1101 | 365.260 |
| 057.555 Ssa | 0.0000 | 0.0328 | 0.3415 | 182.621 |
| 145.555 O1 | 0.0084 | 0.1742 | -0.0330 | 13.470 |
| 163.555 P1 | 0.0443 | 0.7821 | -0.1417 | 153.586 |
| 165.555 K1 | 0.8394 | -0.0399 | 0.9475 | 966.000 |
| 165.565 | -0.1324 | 0.4678 | -0.2434 | 1125.996 |
| 166.554 $\chi$1 | -0.0041 | -0.1374 | 0.0286 | 587.344 |
| 255.555 M2 | -0.4210 | 0.1419 | 0.6008 | 13.285 |
| 272.556 T2 | -0.0833 | 0.1804 | 0.0826 | 97.239 |
| 273.555 S2 | -1.9390 | 5.6909 | 1.5701 | 132.517 |
| 274.554 | 0.0254 | -0.1165 | -0.0107 | 207.968 |
| 275.545 | -0.0229 | 0.2260 | -0.0192 | 450.961 |
| 275.555 K2 | 1.9153 | -20.2475 | 1.9255 | 483.000 |
| 275.565 | -0.6131 | 6.9751 | -0.7337 | 519.940 |
| 275.575 | -0.0720 | 0.8871 | -0.1023 | 562.998 |
| 276.554 | -0.0473 | -1.5431 | 0.4381 | 1498.387 |

**Table 3.** CHAMP tidal amplitudes and periodicities

| Sat | data | data period | Parameters | Arcs |
|---|---|---|---|---|
| Ajisai | SLR | 50859-52424 | $\underline{x}$, $\dot{\underline{x}}$, Cd, Cr | 5d |
| CHAMP | x,y,z | 52049-52829 | $\underline{x}$, $\dot{\underline{x}}$, k0(L), k1(G), th(L) | 1d |
| Lageos1 | SLR | 50864-52424 | $\underline{x}$, $\dot{\underline{x}}$, Cr, atacc | 15d |
| Lageos2 | SLR | 50864-52424 | $\underline{x}$, $\dot{\underline{x}}$, Cr, atacc | 15d |
| Starllete | SLR | 50904-52424 | $\underline{x}$, $\dot{\underline{x}}$, Cd, Cr | 5d |
| Stella | SLR | 50904-52424 | $\underline{x}$, $\dot{\underline{x}}$, Cd, Cr | 5d |

**Table 4.** Satellite data for gravity field variability

The SVD shows that the dominant combinations are long-periodic for the 5 geodetic satellites while the dominant lumped harmonics for CHAMP involve the sectorial harmonics.

## 3 The temporal gravity field: satellite solutions and geophysical data

Temporal gravity field solutions were recovered every 15day utilising SLR for the five geodetic satellites with and without CHAMP and from CHAMP by itself with applied constraints of 0.6e-10 to each amplitude. The amplitude and phase of the annual signal for a field to degree and order 4 (recovered

| order | combinations |
|---|---|
| 1 | S2,2; S4,2; S6,2; S8,2 |
| 2 | C2,2; C4,2; C6,2; C8,2 |
| 3 | Ist order Cs and Ss |
| 4 | S4,4; S6,4; S8,4; S10,4 |
| 5 | C4,4; C6,4; C8,4; C10,4 |
| 6 | Ist order Cs and Ss |
| 7 | S3,2; S5,2; S7,2; S9,2 |
| 8 | C3,2; C5,2; C7,2; C9,2 |
| 9 | S6,6; S8,6; S10,6 |
| 10 | C6,6; C8,6; C10,6 |

**Table 5.** Dominant CHAMP combinations in order of significance from SVD: 10 x 10 field, MJD 52409-52424

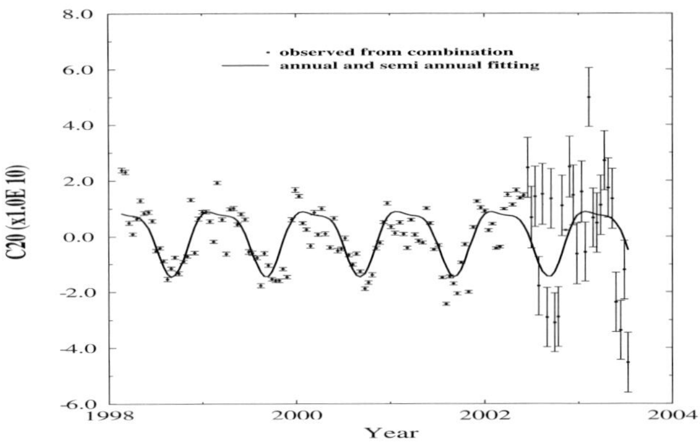


**Fig. 1.** Annual and semi-annual variation in $C_{2,0}$

from a 6 by 6 solution) are compared against geophysical data in Table 6. The latter comprised annual and semi-annual variations inferred from CDAS-1 atmospheric pressure data from Jan 1989 to Mar 2002; ocean mass redistribution from the Ocean Circulation and Climate Model (OCCAM) (Webb et al., 1998) for Jan 1992 to Apr 1996 and land hydrology from VIC (Nijssen et al., 2001) for 1980 to 1993. The CHAMP only results required the application of the constraint to avoid excessively amplitudes. The disparity between CHAMP and the SLR results is not unacceptable although the results can be overinterpreted. For, example Fig.1 plots the 15day SLR solutions from 1998 to mid 2002 followed by the CHAMP only solution to mid 2003. The satellite data has been fitted by annual and semi-annual sinusoids. The strength of the multisatellite solution prior to 2002 is evident in the consistency of fit and the low error bars. For the CHAMP only solution, the current methodolgy gives rise to both larger variability and larger error bars.

| n,m | $C_{n,m}$ | | | | $S_{n,m}$ | | | |
|---|---|---|---|---|---|---|---|---|
| | Geophy | SLR | CHAMP | SLR&CH | Geophy | SLR | CHAMP | SLR&CH |
| 2,0 | 1.19/294 | 0.70/291 | 0.30/316 | 0.66/288 | | | | |
| 2,1 | 0.28/321 | 0.37/334 | 0.75/ 32 | 0.45/ 9 | 0.49/351 | 0.71/ 5 | 0.59/ 97 | 0.15/307 |
| 2,2 | 0.28/152 | 0.32/ 15 | 0.43/269 | 0.25/ 95 | 0.25/ 37 | 0.72/ 43 | 0.60/ 69 | 0.36/ 96 |
| 3,0 | 0.98/204 | 1.79/286 | 1.43/312 | 1.04/295 | | | | |
| 3,1 | 0.36/309 | 2.52/239 | 2.21/152 | 0.58/ 89 | 0.80/314 | 0.98/ 33 | 1.45/321 | 0.80/277 |
| 3,2 | 0.29/147 | 0.12/ 30 | 0.21/308 | 0.51/ 7 | 0.53/298 | 0.11/ 18 | 1.00/279 | 0.36/283 |
| 3,3 | 0.40/ 91 | 1.02/277 | 0.32/153 | 0.71/333 | 0.38/256 | 0.18/257 | 0.79/259 | 1.51/251 |
| 4,0 | 0.01/174 | 0.35/ 17 | 0.25/305 | 0.17/ 84 | | | | |
| 4,1 | 0.27/255 | 0.50/156 | 0.31/160 | 1.14/177 | 0.38/256 | 0.16/116 | 0.15/113 | 0.16/ 39 |
| 4,2 | 0.22/134 | 0.10/104 | 0.73/221 | 0.22/155 | 0.68/298 | 0.78/272 | 0.70/228 | 0.43/242 |
| 4,3 | 0.20/ 61 | 0.96/166 | 0.65/100 | 0.23/190 | 0.46/311 | 0.49/192 | 0.32/ 24 | 0.56/142 |
| 4,4 | 0.55/ 91 | 0.86/149 | 0.70/ 81 | 0.67/105 | 0.18/307 | 0.11/ 42 | 0.05/355 | 0.38/227 |

**Table 6.** Annual variation (amp/phase) in normalized hramonics from mass distribution of atmosphere (CDAS-1), ocean (OCCAM) and hydrology (VIC); amplitude A (*1.e-10) and phase P (deg) defined by $Acos(2\pi(t - t_0)/365.25 + P)$

With the method adopted in this study, there is no evidence as yet that CHAMP can facilitate temporal gravity field recovery due to both the sensitivity to higher degree and order harmonics and the possibility that gravity field signal is absorbed within the solution vector for CHAMP in Table 4. Other possibilities include the use of geophysical data to constrain the harmonics but care must be exercised as global hydrology is poorly defined.

*Acknowledgement.* The authors would like to thank the U.K. Natural Environment Research Council (Grant No. NER/A/0000/00612) for financing this study and for GFZ for supplying the data and technical support.

## References

1. Kaula WM (1966) *Theory of Satellite Geodesy.* Blaisdell, Waltham, Massachusetts.
2. Nijssen B, Schnur R and Lettenmaier DP (2001) Global retrospective estimation of soil moisture using the Variable Infiltration Capacity land surface model, 1980-93. Journal of Climate *14*: 1790–1808.
3. Webb DJ, de Cuevas BA and Coward AC (1998) The first run of the OCCAM global ocean model. Internal Document No. 14, Southampton Oceanography Centre, Southampton, UK.

# Earth Gravity Field and Seasonal Variability from CHAMP

Christoph Reigber[1], Horst Jochmann[1], Johann Wünsch[1], Svetozar Petrovic[1], Peter Schwintzer[1], Franz Barthelmes[1], Karl-Hans Neumayer[1], Rolf König[1], Christoph Förste[1], Georges Balmino[2], Richard Biancale[2], Jean-Michel Lemoine[2], Sylvain Loyer[2], Felix Perosanz[3]

[1] GeoForschungsZentrum Potsdam (GFZ), Department 1 'Geodesy and Remote Sensing', Telegrafenberg A 17, 14473 Potsdam, Germany, *reigber@gfz-potdam.de*
[2] CNES/GRGS, 18 avenue Edouard Belin, 31400 Toulouse, France,
[3] Noveltis, 2 avenue de Europe, 31520 Ramenville, France

**Summary.** GPS-CHAMP satellite-to-satellite and accelerometry data covering 2.5 years of the CHAMP mission period were exploited to generate the global gravity field model EIGEN-3p revealing considerable improvements in both accuracy and resolution with respect to the previous model EIGEN-2. For the year 2001, CHAMP and satellite laser ranging data of four satellites were combined to recover largest scale monthly gravity field variations that are subsequently analyzed for the annually varying constituents. The temporal gravity field variations observed by CHAMP and the SLR satellites are compared in the spectral and spatial domain with geophysically (atmosphere, ocean, hydrology) predicted gravity variations that do not reflect the large observed scattering in the monthly solutions but are of comparable size and distribution on the annual time scale.

**Key words:** CHAMP satellite gravity mission, global gravity field model, temporal gravity field variations

## 1 The CHAMP-only mean Earth gravity field model EIGEN-3p

EIGEN-3p (European Improved Gravity model of the Earth by New techniques, issue 3 preliminary) is the global gravity field model following EIGEN-2 (Reigber et al. 2003a) and, like EIGEN-2, is computed solely from CHAMP's gravitational orbit perturbations. The about half a year's worth of GPS-CHAMP satellite-to-satellite tracking and accelerometer data that were exploited for the construction of EIGEN-2 was expanded to about 2.5 years out of the period July 2000 through June 2003. In addition modifications in the data processing and solution procedure took place: (1) The bias parameters for the three accelerometer axes are in the newly processed two years of CHAMP data solved once per orbit revolution (along- and across- track axes) and twice per revolution for the malfunctioning radial channel, instead of only daily in the EIGEN-2 processing. (2) The stabilization of the overall normal equation system by introducing zero pseudo-observation for the spherical harmonic gravitational coefficients with weights increasing with

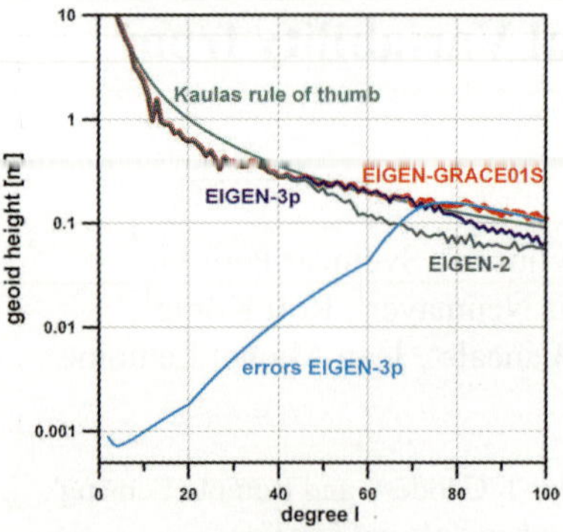


**Fig. 1.** Signal/error amplitudes per degree in terms of geoid heights.

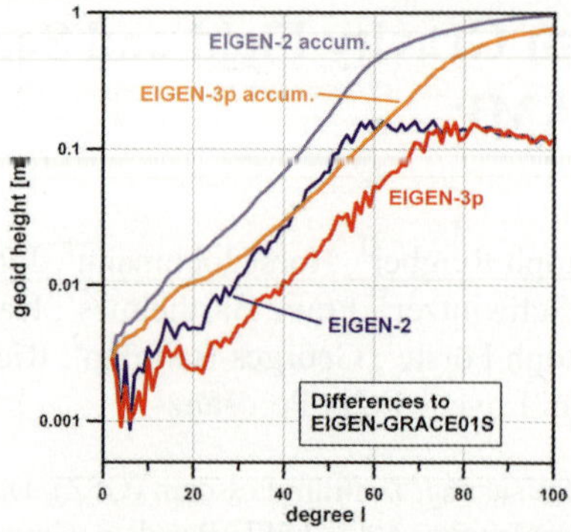


**Fig. 2.** CHAMP minus GRACE models' difference amplitudes per degree and (accum.) as a function of maximum degree in terms of geoid heights.

increasing degree starts at degree 65 in the EIGEN-3p solution instead at degree 25 in EIGEN-2.

The positive effect of the enlarged database and processing modifications with respect to EIGEN-2 are obvious from Figures 1 and 2, where the EIGEN-2 and EIGEN-3p solutions are compared with the higher resolving GRACE gravity field model EIGEN-GRACE01S (Reigber et al. 2003b). Figure 1 gives the signal amplitudes per degree in terms of geoid undulations for the three models and for Kaula's 'rule of thumb' (Kaula 1966). It can be seen that the EIGEN-3p solution has got full power up to almost degree/order 65, corresponding to a half wavelength spatial resolution of $\lambda/2 = 300$ km, compared to only 450 km of EIGEN-2.

The CHAMP minus GRACE models' difference degree amplitudes as a function of maximum degree, shown in Figure 2, demonstrate that the EIGEN-3p model recovers the geoid with an average accuracy of $\pm 1$ cm accumulated over all terms up to degree/order 20 (EIGEN-2 up to degree/order 14), or $\lambda/2 = 1000$ km spatially, and $\pm 10$ cm up to degree/order 55 (EIGEN-2 up to degree/order 44) or $\lambda/2 = 350$ km. In spite of the overall superior performance of EIGEN-3p, a degradation in accuracy over the polar areas has been reported for EIGEN-3p compared to previous solutions (D. Mc Adoo, C. Wagner, personal communication) that is very likely arising from the too weak solution's stabilization. EIGEN-3p will be followed by EIGEN-3 as soon as the reprocessing of all data with the new processing scheme is completed and an optimum stabilization procedure has been found.

## 2 Large Scale Temporal Gravity Field Variations

First attempts to resolve large scale monthly gravity field variations solely from CHAMP observations revealed an unrealistically large scattering in the low degree/order spherical harmonic gravitational coefficients when computed as

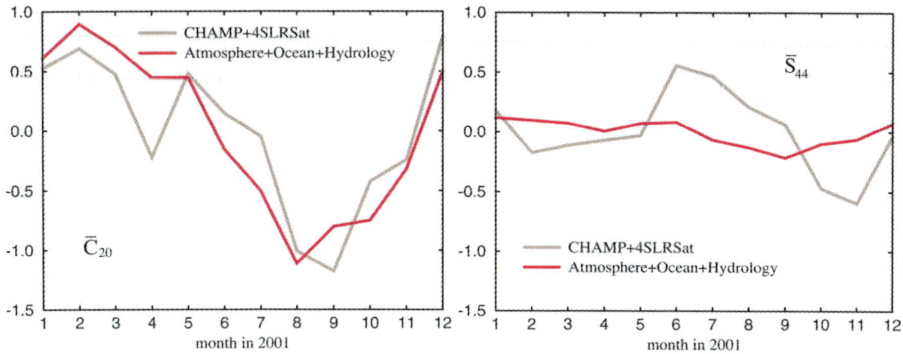


**Fig. 3.** Observed (CHAMP+4SLRsat) and geophysically predicted (atmosphere+ocean+ hydrology) spherical harmonic coefficients' time series (example for $\overline{C}_{20}$ and $\overline{S}_{44}$) in terms of geoid heights, unit: mm.

monthly time series. Therefore, additional normal equation systems have been combined with the CHAMP system which were generated from LAGEOS-1 and -2, STARLETTE and STELLA laser tracking data covering the year 2001. The 2001 CHAMP+4SLRsat combined normal equation system then was solved yielding a full gravity field model solution where the coefficients from degree/order 2 through 4 were resolved at monthly intervals in order to investigate seasonal gravity field variations.

The gravity field variations at monthly intervals for the year 2001 were also predicted from geophysical models, i.e. from atmospheric pressure fields over the continents (ECMWF), an ocean circulation model (Hirose et al. 2001) including atmospheric forcing and a continental hydrologic model (Milly, Shmakin 2002). The resulting fields were then individually and in summation expanded in spherical harmonics up to degree/order 4 for comparison with the CHAMP+4SLRsat observed coefficients. Figure 3 gives two typical examples of coefficients time series: the agreement 'observed vs. predicted' is quite good as far as $\overline{C}_{20}$ is

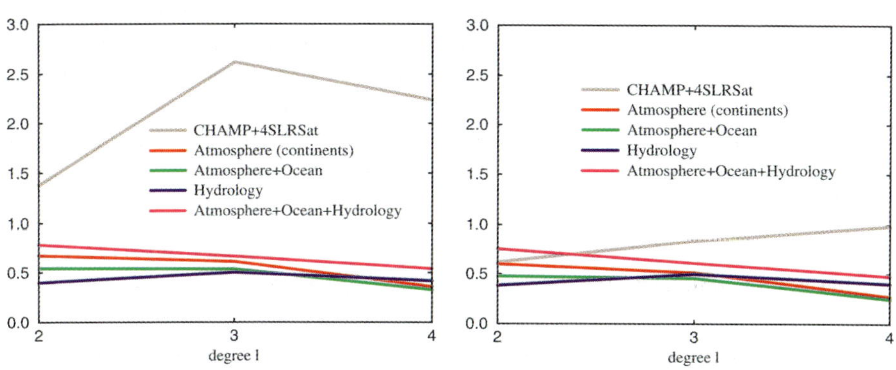


**Fig. 4.** Signal amplitudes per degree in terms of geoid height variability over the year 2001: monthly (left) and annual fit (right), unit: mm.

concerned whereas the $\overline{S}_{44}$ series reveal large discrepancies. The coefficients' time series were then used to determine annual amplitudes and phases for both the CHAMP + 4SLRsat solution and the geophysical model predictions. Figure 4 shows in terms of geoid amplitudes per degree the root mean square (rms) of the coefficients' monthly variations about the mean and the variations of the annual fit. The reduction in power by a factor 3 to 5 for the CHAMP+4SLRsat solution is due to the considerably large scattering of the monthly values compared to the smooth curves obtained from the geophysical models, where the annual constituent is dominating (cf. Figure 3).

Table 1 gives the adjusted annual amplitudes and phases per coefficient from the CHAMP + 4SLRsat solution and for comparison the solution by Cheng et al. (2002), derived from laser tracking data of six satellites over the years 2000 and 2001, as well as the sum of the annual variations predicted by the geophysical models.

Figure 5 shows the geoid contribution, synthesized from the CHAMP+4SLRsat observed and geophysically predicted annual coefficients' amplitudes and phases (degree 2 to 4), at t = 3 months, the sine-terms, and at t ≡ 0, the cosine-terms, repectively.

| degree n | order m | $\overline{C}_{nm}$ Amp (0.01mm) | $\overline{C}_{nm}$ Phase (deg) | $\overline{S}_{nm}$ Amp (0.01mm) | $\overline{S}_{nm}$ Phase (deg) |
|---|---|---|---|---|---|
| 2 | 0 | 69 79 (90) | 54 31 (57) | - | - |
| 2 | 1 | 39 28 (22) | 22 313 (350) | 6 33 (51) | 86 327 (348) |
| 2 | 2 | 28 25 (9) | 181 132 (195) | 22 50 (25) | 274 305 (325) |
| 3 | 0 | 31 32 (53) | 311 357 (122) | - | - |
| 3 | 1 | 25 55 (16) | 51 37 (352) | 15 40 (49) | 21 52 (13) |
| 3 | 2 | 89 98 (22) | 221 175 (181) | 40 38 (28) | 53 44 (47) |
| 3 | 3 | 49 85 (24) | 242 286 (269) | 39 158 (21) | 93 30 (71) |
| 4 | 0 | 14 54 (30) | 263 128 (108) | - | - |
| 4 | 1 | 18 4 (6) | 192 346 (77) | 59 55 (18) | 334 357 (34) |
| 4 | 2 | 56 64 (24) | 348 170 (205) | 107 42 (35) | 34 49 (56) |
| 4 | 3 | 7 15 (9) | 246 292 (281) | 30 118 (19) | 346 49 (19) |
| 4 | 4 | 45 73 (34) | 254 250 (261) | 29 45 (12) | 163 17 (64) |

The phase convention is cos(ft-φ), where t is time past Jan. 1, f is the annual frequency, and φ is the phase.

**Table 1.** Amplitude/phase annual variation of the spherical harmonic coefficients in terms of geoid heights, unit: 0.01 mm (left column: CHAMP+4SLRsat 2001 observed, middle column in grey: 6SLRsat 2001/2002 observed, Cheng et al. (2002), in parenthesis: 2001 predicted from atmosphere+ocean+hydrology.

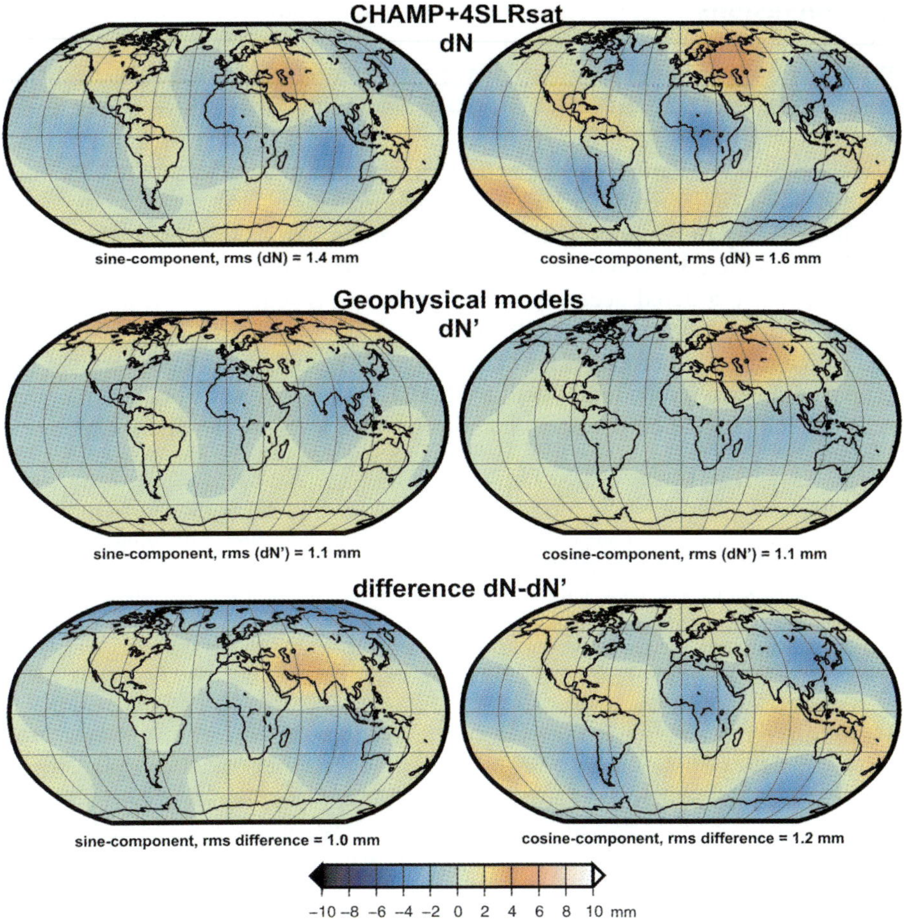


**Fig. 5.** Sine (left) and cosine (right) components of the annual variation in the geoid (in mm) from spherical harmonic terms of degree 2 to 4 in Table 1, as determined from CHAMP+4SLRsat observations (top) and predicted by atmosphere, ocean and continental hydrology modelling (middle), and the difference between observed minus predicted fields (bottom). Root mean square (rms) values are computed using cosine of latitude as weight.

The comparisons in Table 1 and Figure 5 show agreements and disagreements in both the spectral and spatial domain, which must be attributed to the CHAMP+4SLRsat observed time series as well as to the geophysical models where especially the hydrologic contributions are quite uncertain. As the results obtained so far shall only give a first impression, a thorough investigation and discussion of the quality of CHAMP resolved temporal gravity variations is left for further studies.

## 3 Conclusions

With the single-satellite mission CHAMP, the Earth's gravity field is resolved with a 10 cm geoid accuracy up to degree/order 55 of a spherical harmonic expansion of the geopotential, corresponding to a spatial resolution of 350 km (half wavelength) at the Earth's surface. The 1 cm geoid accuracy threshold is obtained at a resolution of degree/order 20, i.e. 1000 km spatially.

The recovery of temporal gravity field variations due to sub-seasonal and seasonal climatologic mass transports in the atmosphere, oceans and continental hydrology, where a geoid accuracy on the mm-level is required, is restricted with CHAMP to the longest wavelengths (degree/order 4, half wavelength 5000 km) and a time resolution of longer than three months. With these limitations, CHAMP cannot compete with the dual-satellite GRACE mission's results providing higher accuracy and resolution thanks to the ultra-precise low-low ranging instrumentation.

*Acknowledgements.* We thank Chris Milly for computing and providing the 2001 hydrological data from his model. The support of the CHAMP exploitation phase within the GEOTECHNOLOGIEN geoscientific R+D programme (grant FKZ 50EP9587) on the German side and by the French Space Agency CNES on the French side are gratefully acknowledged.

## References

Cheng MK, Gunter B, Ries JC, Chambers DP, Tapley BD (2002) Temporal Variation in the Earth's Gravity Field from SLR and CHAMP GPS Data. in: I.N. Tziavos (ed.), Gravity and Geoid, 3rd Meeting of the International Gravity and geoid Commission (IGGC), Univ. of Thessaloniki, Greece: 424-431.

Hirose N, Fukumori I, Zlotnicki V, Ponte RM (2001) High-frequency barotrophic response to atmospheric disturbances: Sensitivity to forcing, topography, and friction. J Geophys Res *106*: 30,987-30,995.

Kaula W (1966) Theory of Satellite Geodesy. Blaisdale Press, Waltham, Mass.

Milly PCD, Shmakin AB (2002) Global modeling of land water and energy balances. Part I: The land dynamics (LaD) model. Journal of Hydrometeorology *3(3)*: 283-299.

Reigber Ch, Schwintzer P, Neumayer K-H, Barthelmes F, König R, Förste Ch, Balmino G, Biancale R, Lemoine J-M, Loyer S, Bruinsma S, Perosanz F, Fayard T (2003a) The CHAMP-only Earth Gravity Field Model EIGEN-2. Adv Space Res *31(8)*: 1883-1888.

Reigber Ch, Schmidt R, Flechtner F, König R, Meyer U, Neumayer K-H, Schwintzer P, Zhu SY (2003b) First GFZ GRACE gravity field model EIGEN-GRACE01S. http://op.gfz-potdam.de/grace/results.

# Comparison of Superconducting Gravimeter and CHAMP Satellite Derived Temporal Gravity Variations

Jürgen Neumeyer[1], Peter Schwintzer[1], Franz Barthelmes[1], Olaf Dierks[1], Yuichi Imanishi[2], Corinna Kroner[3], Bruno Meurers[4], He-Ping Sun[5], and Heikki Virtanen[6]

[1] GeoForschungsZentrum Potsdam, Dept. Geodesy and Remote Sensing, Germany (neum@gfz-potsdam.de)
[2] Ocean Research Institute Tokyo, Japan
[3] Institute of Geosciences, FSU Jena, Germany
[4] Institute of Meteorology and Geophysics, University of Vienna, Austria
[5] Institute of Geodesy and Geophysics Wuhan, China
[6] Finnish Geodetic Institute Masala, Finland

**Summary.** The operational Superconducting Gravimeter (SG) network can play an important role for validation of satellite-derived temporal gravity field variations. A comparison shows a quite good agreement between SG and CHAMP results within their estimated error bars. It could be proved that the SG-derived temporal gravity variations are representative for a large area within the μgal accuracy, if the local gravity effects are removed.. The long-periodic tidal waves are well determined by ground measurements, therefore they can be applied as a reference for validation. For further validation, field SG measurements should be carried out in representative areas with large gravity variations (e.g. Amazon area).

**Key words:** gravity field, CHAMP, Superconducting Gravimeter, gravity variation

## 1 Introduction

One objective of the new-generation satellite gravity missions CHAMP and GRACE is the recovery of temporal Earth gravity field variations. For the CHAMP mission the gravity resolution is about 1 μgal at a half wavelength spatial resolution of $\lambda/2$ = 5000 km with a temporal resolution of 1 month. For the GRACE mission a largely increased resolution is expected.

Of fundamental interest is the combination of satellite-based and surface gravity measurements. Because CHAMP's temporal resolution ranges from 1 month to years, surface gravity measurements must have a long-term stability, which only Superconducting Gravimeters (SG) fulfil.

On the Earth surface high-precision gravity measurements are carried out with Superconducting Gravimeters forming the SG network of the Global Geodynamic Project (GGP). These measurements have a gravity resolution in ngal range and a linear drift of some μgal per year.

When we compare satellite-derived with ground-measured gravity variations, we must ensure to the best knowledge that after preprocessing and reduction of known gravity effects both data sets represent the same sources of gravitation and spatial resolution. Therefore all local gravity effects must be removed from the ground measurements.

## 2 CHAMP recovery of temporal gravity field variations

The CHAMP data processing is performed with an extended EIGEN-2 Earth gravity field model [Reigber et al. 2003]. The input data for this model are:

- CHAMP-GPS satellite-to-satellite tracking and accelerometer data,
- Satellite Laser Ranging data of Lageos-1, and -2, Starlette and Stella.

Normal equation systems were generated to compute the EIGEN-2CHAMP+4SLR solution for:

- Spherical harmonic coefficients of the gravitational potential complete to degree/order 120,
- Coefficients up to degree/order 4 at 30d (monthly) intervals.

The time related gravity reduction for the CHAMP gravity field solution is performed for the following effects:

- Atmospheric pressure (direct attraction and loading term): 6-hourly data (ECMWF) are used to model temporal variations in the gravitational potential.
- Ocean tides and ocean loading: The FES 2000 ocean model is applied for semidiurnal to long-periodic constituents.
- Pole tide: The IERS polar motion series are used.
- Earth tides according to the IERS conventions [McCarthy 2000].

From the resulting 12 sets of spherical harmonic coefficients of degree 2 through 4 ($\lambda/2$ = 5000 km) monthly gravity variations have been calculated for the selected SG positions in a time span from Dec. 2000 to Dec. 2001 with an estimated standard deviation of 1 μgal.

## 3 Superconducting Gravimeter derived gravity variations

The SG gravity data are reduced for the same gravity effects as the CHAMP solution. Additionally local and instrumental effects are removed.

In a first step the raw SG gravity data are corrected for spikes and offsets. Then the linear instrumental drift is removed These preprocessed data are reduced for the following gravity effects:

- Atmospheric pressure: By using the local air pressure a single admittance coefficient is calculated, which is used for reduction of the air-pressure-induced gravity effect (attraction and loading term),
- Ocean loading calculated with the FES 95 model [Francis & Mazzega 1990],
- Pole tide calculated with IERS polar motion series and a gravimetric factor of 1.16 [Torge 1989],

- Earth tides calculated with analysed tidal parameters based on the Wahr-Dehant Earth tide model [Wenzel 1996],
- Local groundwater-level-induced gravity effect: By using water table measurements a single admittance coefficient is determined for reduction of this effect [Neumeyer et al. 1999].

After these gravity reductions we got the SG gravity variations. For comparing with CHAMP, monthly means of the gravity variations were calculated.

## 4 Comparison results

The present SG network comprises 20 stations in operation. Here, six SG stations have been selected: Sutherland / South Africa ($\phi$ =-32.381 deg, $\lambda$=20.811 deg, h=1791 m), Vienna / Austria ($\phi$=48.25 deg, $\lambda$=16.358 deg, h=192 m), Moxa / Germany ($\phi$=50.645 deg, $\lambda$=11.616 deg, h=455 m), Metsahovi / Finland ($\phi$=60.217 deg, $\lambda$=24.396 deg, h=56 m), Wuhan / China ($\phi$=30.516 deg, $\lambda$=114.49 deg, h=80 m), Matsushiro / Japan ($\phi$=36.543 deg, $\lambda$=138.207 deg, h=406 m).

For the time period from Dec. 2001 to Dec. 2002 the gravity data of these sites were processed according to the above processing procedure. For comparison monthly averages of the gravity variations are used. The assigned CHAMP values are taken from the monthly global gravity field solutions for the coordinates of the selected SG sites. Gravity variations due to Earth and ocean tides, pole tide and atmosphere are reduced in both SG and CHAMP derived series, the remaining time variable effects should be due to continental large scale hydrology.

Figure 1 shows the comparison result. Each station is represented by one box with two graphs. The upper part displays the SG measured gravity variations (grey) and the monthly averages (black). The lower part shows the CHAMP derived gravity variations (black) and again the monthly averages of the SG gravity variations (grey). For a the better visualisation the curves are created about their mean values.

The comparison shows a reasonable agreement in the trend behaviour of the gravity variations for all stations: For the stations with large variations (Metsahovi, Wuhan and Matsushiro) the CHAMP solution follows quite well the SG result too. Concluding we note a significant result of the comparison. The CHAMP and the SG are measuring approximately the same gravity variations within their estimated error bars. The GRACE result with an expected improved spatial and temporal resolution will further improve the significance of the comparison.

On the other hand the result proves that the Superconducting Gravimeter derived gravity variations, although being point measurements, are representative for a large area within the µgal accuracy. Neighbouring stations show coherent signals [Crossley & Hinderer 2002]. This is shown on two examples:

1. The sites Moxa and Vienna which are 435 km apart show nearly the same gravity variability. In particular there is also a good agreement in time periods smaller than 1 month (Figure 2).

2. The distance between Wuhan and Matsushiro is about 2300 km. The gravity

variations also show a similar behaviour concerning the longer period variations.

The large scale validity of SG derived gravity variations requires the reduction of local gravity effects from the SG series. Therefore, beside precise air pressure measurements water table, soil moisture, rainfall, snow loading etc. measurements should be carried out at SG sites. For comparison purposes those SG sites are only useful to be employed, where local gravity effects can be well monitored and modelled.

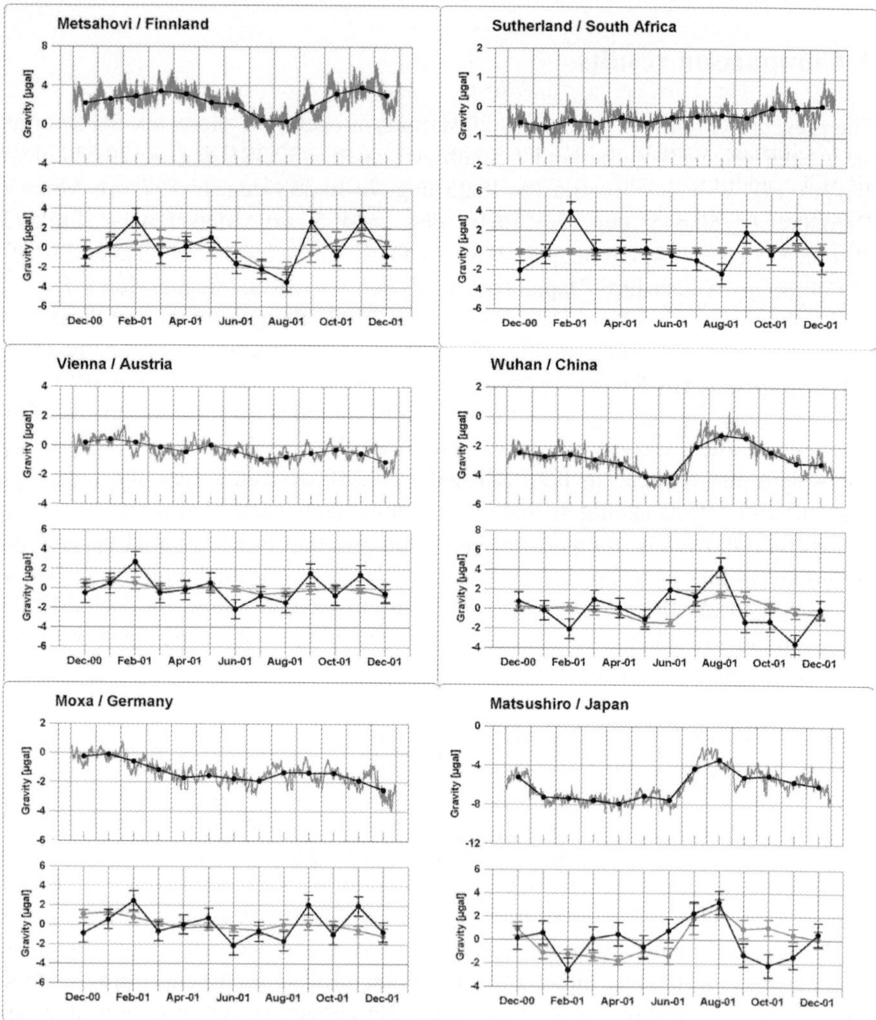


**Fig. 1.** Gravity variations at Superconducting Gravimeter sites:
Upper panel per box: - SG gravity variations (grey)
- SG monthly mean of gravity variations (black)
Lower panel per box: - CHAMP monthly gravity field solution (black)
- SG monthly mean of gravity variations (grey)

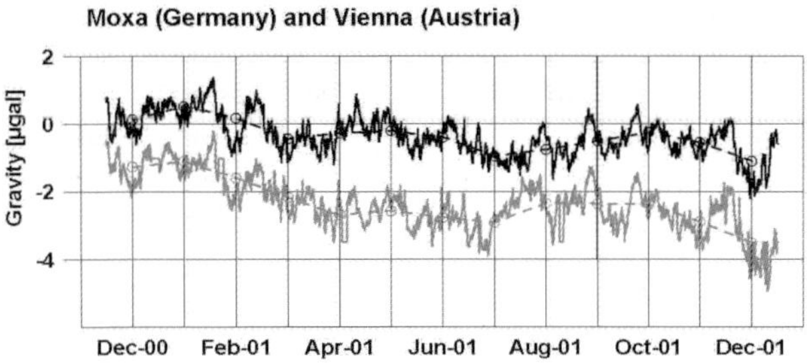


**Fig. 2.** SG gravity variation of Vienna (black) and Moxa (grey).

For further validation of satellite-based gravity variations, especially for GRACE, additional surface gravity measurements with Field Superconducting Gravimeters should be carried out in areas with large gravity variations and negligible local gravity effects considering the spatial resolution of the satellite measurements, for instance in the Amazon area where seasonal gravity changes can be observed in the order of some 10 µgal.

## 5 Long-periodic Earth tides as reference

For Earth tidal-forces-induced gravity variations the relation between satellite and surface gravity variation measurements can be expressed in a first approximation by the body Love numbers h and k and the gravimetric factor δ. On the Earth surface the gravimeter is measuring beside the gravitational attraction (mass) the gravity effect due to elastic deformation (vertical surface shift) and the deformation potential (mass redistribution due to the vertical surface shift).

The gravimetric factor is a function of the Love numbers h and k. For the spherical harmonic expansion degree l = 2 we have

$$\delta 2 = 1 + h2 - 3/2 \cdot k2 \qquad (1)$$

With the nominal Love numbers $h2$ = 0.614 for elastic deformation and $k2$ = 0.302 for deformation potential. The gravimetric factor is δ2= 1.16.

The satellite is not sensitive to the vertical surface shift. Therefore h2 = 0 and the gravimetric factor for the satellite measurements can be expressed by

$$\delta 2_S = 1 - 3/2 \cdot k2 \qquad (2)$$

With $k2$ = 0.302 one obtains $\delta 2_S$ = 0.547. The ratio between $\delta 2_S$ and δ2 is rδ = 0.471. Accordingly the gravity signal from the satellite should be smaller than the gravimeter signal by this factor rδ. This assumption is valid for the tidal-forces-induced gravity variations. Therefore the SG gravity data have been reduced by the factor Rδ for the CHAMP comparison of tidal waves.

The long periodic tidal waves STA (period = 121.75 days), SSA (period = 182.62 days) and SA (period = 365.63 days) are well determined by ground measurements. They have the same source for SG and CHAMP temporal gravity field variations. Therefore they can be applied as reference for validation. For non-tidal

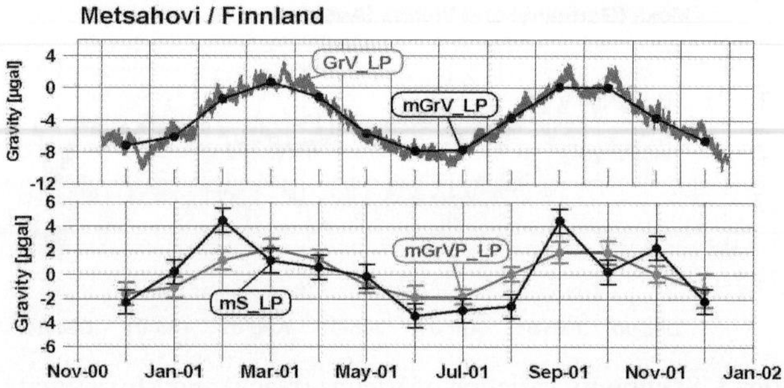


**Fig. 3.** Gravity variations for Metsahovi station including Earth tidal STA, SSA and SA.

forces induced elastic deformation the load Love numbers must be applied.

Figure 3 shows an example of Metsahovi station. For this purpose the SG data have been processed according to the above processing procedure but without reduction of the long-periodic tidal waves STA, SSA and SA (GrV_LP). From this data set monthly averages (mGrV_LP) are calculated. For comparing with CHAMP the mass and deformation potential terms must be considered. Therefore the monthly averages are divided by the factor $r\delta$ (mGrVP_LP). The station Metsahovi has been selected for this simulation because it has the largest long-periodic tidal amplitudes of the six selected SG stations.

Additionally the long-periodic gravity part caused by the tidal waves STA, SSA and SA is added to the CHAMP result (mS_LP). This simulation shows that for further evaluation studies the well resolved long-periodic Earth tides can be used as reference for comparison. Then, in the corresponding satellite based (monthly) gravity solution these tidal constituents shall not be reduced.

## References

Crossley D, Hinderer J (2002) GGP Ground Truth for Satellite Missions. Marees Terrestres Bulletin d'Informations, Bruxelles *136*: 10,735-10,742.

Francis O, Mazzega P (1990) Global charts of ocean tides loading effects. J Geophys Res *95(C7)*: 11411-11424.

McCarthy DD (2000) IERS conventions 2000, International Earth Rotation Service.

Neumeyer J, Barthelmes F, Wolf D (1999) Estimates of environmental effects in Superconducting Gravimeter data. Marees Terrestres d'Informations, Bruxelles *131*: 10,153-10,159.

Reigber Ch, Schwintzer P, Neumayer K-H, Barthelmes F, König R, Förste Ch, Balmino G, Biancale R, Lemoine J-M, Loyer S, Bruinsma S, Perosanz F, Fayard T (2003) The CHAMP-only Earth Gravity Field Model EIGEN-2. Adv Space Res *31(8)*: 1883-1888.

Torge W (1989) Gravimetry. de Gruyter, Berlin, New York.

Wenzel H-G (1996) Earth tide data processing package ETERNA 3.3. Marees Terrestres Bulletin d'Informations, Bruxelles *124*: 9425-9439.

# Improvements in Arctic Gravity and Geoid from CHAMP and GRACE: An Evaluation

David McAdoo[1], Carl Wagner[1] and Seymour Laxon[2]

[1] ORAD / Laboratory for Satellite Altimetry, National Oceanic and Atmospheric Administration (NOAA), Silver Spring, MD USA 21094 *dave.mcadoo@noaa.gov*

[2] Center for Polar Observation and Modeling, University College London, Gower Street, London, WC1E 6BT, UK

**Summary.** The near-polar CHAMP and GRACE satellites are now acquiring vitally important new information on the geoid and gravity field of the polar regions. This investigation demonstrates that CHAMP and GRACE data are dramatically reducing the large gaps in our knowledge of the Arctic region, constraining the long wavelength geopotential (>300 km) and beginning to yield the high accuracy marine geoid which is needed for Arctic oceanographic and sea ice studies. Using a detailed Arctic surface gravity field and an independent altimetric gravity field as benchmarks we have evaluated the intermediate-to-long wavelength (> 300km) component of seven CHAMP and two GRACE satellite-only gravity models such as the GFZ EIGEN, the NASA PGS and UT/CSR. We evaluate, spectrally, the errors in - and differences between - these satellite-only models in the Arctic at wavelengths from 300 to 2500 km. The GRACE models accurately resolve Arctic gravity to *full* wavelengths as short as 500 km while the CHAMP models do so to full wavelengths as short as 1000 km. However the CHAMP models continue to show improved resolution as more and better (e.g. lower elevation) data are incorporated. The best CHAMP models agree well with the detailed Arctic ARC-GP model to an rms (error of commission) of better than 2.06 mGal (gravity)and 31 cm (geoid) for all wavelengths (full) longer than 1100 km. GRACE-only geoids are precise to 40 cm or better (all wavelengths) over large areas of the Arctic. CHAMP and GRACE-based geoids could have the accuracy required to detect (together with altimetry) the poorly known dynamic topography of the Arctic Ocean. As an example, a GRACE/detailed-gravity hybrid geoid is presented.

**Key words:** Arctic, CHAMP, geoid, GRACE, gravity, sea ice

## 1 Introduction

An accurate Arctic gravity field - and geoid - is critically needed by Arctic geologists, geophysicists, geodesists and particularly by oceanographers and cryosphere scientists. Studies of Arctic ocean dynamics and sea ice should benefit substantially from new geoids derived using CHAMP (CHAllenging Minisatellite Payload), GRACE plus surface data. GRACE and CHAMP will provide the long wavelength portion of an accurate marine geoid to which recent and forthcoming altimeter data can be referred. Such a geoid is vital, for example, if we are to use ICESat altimeter data for precise estimation of sea ice elevation and thickness (Farrell et al., 2003). Similarly, this geoid will be needed for applying ENVISAT and CryoSat altimetry to estimation of dynamic topography and ice elevation. The near-polar orbit of the GRACE and CHAMP missions enables them to reduce long-wavelength gravity and geoid errors in the Arctic by at least as much as is possible globally.

**1b. Detailed Gravity** A decade ago gravity fields of the Arctic were filled with large data voids. Since 1993 great progress has been made in detailed gravity mapping of this ice-covered, inaccessible region. A gravity field covering the ocean between 60°N and 81.5°N (Fig. 1a) was derived using ERS-1 altimeter data (Laxon and McAdoo, 1994, 1998). Recently, the international Arctic Gravity Project (ARC-GP; Forsberg, Kenyon et al., 2002) gravity field was derived from airborne, surface and submarine data and was released. This ARC-GP field (Fig. 1b) fills a number of the large gaps in data coverage that existed heretofore. Despite this progress, gaps in gravity data coverage (Childers et al., 2001) remain particularly in the northernmost Arctic Ocean. To fill in these gaps and accurately underpin the detailed gravity fields, the accurate long (>300 km) wavelength geopotential information from the GRACE and CHAMP satellite missions is essential. Only by combining the accurate, long-wavelength information from GRACE and CHAMP with detailed surface gravity will it be possible to compute the required high-accuracy marine geoid. Note the strong similarity between the detailed gravity fields in Fig. 1a (south of 81.5°N, ERS' northern limit) and Fig. 1b even though the input gravity data are largely independent. The gravitational expression of seafloor details such as continental shelf edges, the Nansen-Gakkel ridge and the Chukchi Borderland can be seen clearly in both figures. This similarity or coherence between the ARC-GP and ERS fields validates the ARC-GP. Therefore we will use the more extensive ARC-GP field as our "benchmark" for evaluating GRACE and CHAMP gravity.

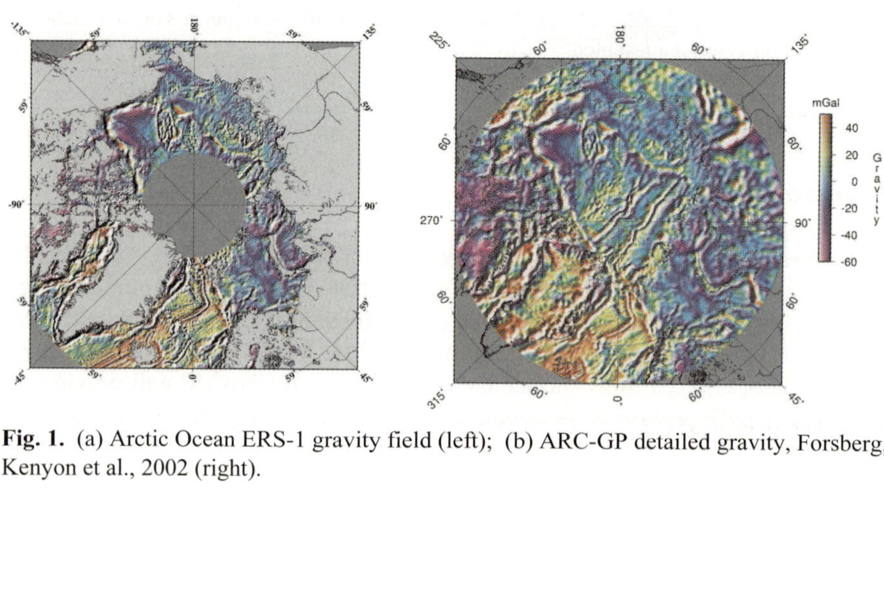


**Fig. 1.** (a) Arctic Ocean ERS-1 gravity field (left); (b) ARC-GP detailed gravity, Forsberg, Kenyon et al., 2002 (right).

## 2 GRACE and CHAMP satellite-only Arctic gravity

Fig. 2 displays a plot of the Gravity Recovery And Climate Experiment (GRACE) GGM01S satellite-only gravity model (Tapley et al., 2003) and has been evaluated here in the Arctic for spherical harmonic coefficients complete to degree and order, n, m, =120. This model was based upon a preliminary analysis of 111 days of micro/s level range-rate data between the two GRACE sub-satellites. Although the attenuation of short wavelength gravity signals at GRACE's 480 km current elevation prevents detection of anomalies with wavelengths shorter than 300 km, one can easily see a correspondence between the GRACE gravity and the detailed ARC-GP field in Fig. 1b. Gravity anomalies associated with the continental shelf break, the Nansen-Gakkel ridge and the northernmost Mid-Atlantic (Knipovich) ridge are evident in both Fig. 1b and Fig. 2. Even though the full GRACE GGM01S gravity field displays some artificial north-south, sectorial striping with an approximate wavelength of 350 km at lower latitudes (< 65°N), this GRACE model exhibits little or no such spurious striping in the Arctic. GFZ's GRACE-only EIGEN-GRACE01S model (Reigber et al., 2003a) from 39 days of tracking yields Arctic gravity very similar to that of GGM01S in Fig. 2.

Two recent CHAMP-only gravity models, PGS7772 (Lemoine et al., 2003) and EIGEN2ee (Reigber et al., 2003b) are evaluated over the Arctic and shown in Figs. 3a and 3b, respectively, below. South of 85°N one can see good similarity between the two CHAMP models as well as between each of the two CHAMP models and the GRACE model (Fig. 2). Although one can see some spurious north-south banding in both Fig. 3a and 3b, real geophysical signals with full wavelengths as short as 500 km are clearly evident. Examples include the pro-

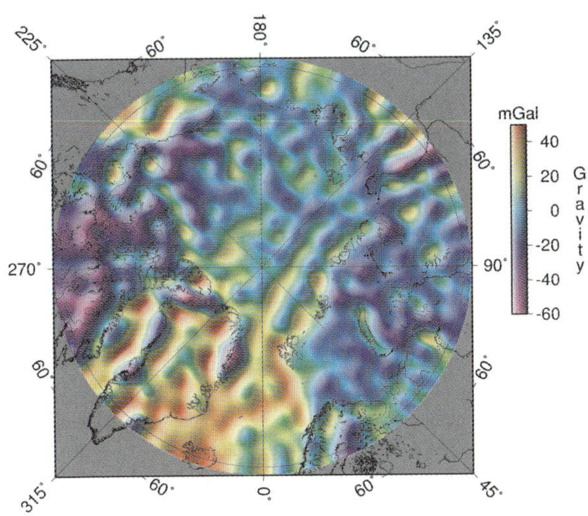


**Fig. 2.** GRACE GGM01S (120x120) gravity over the Arctic (Tapley et al., 2003).

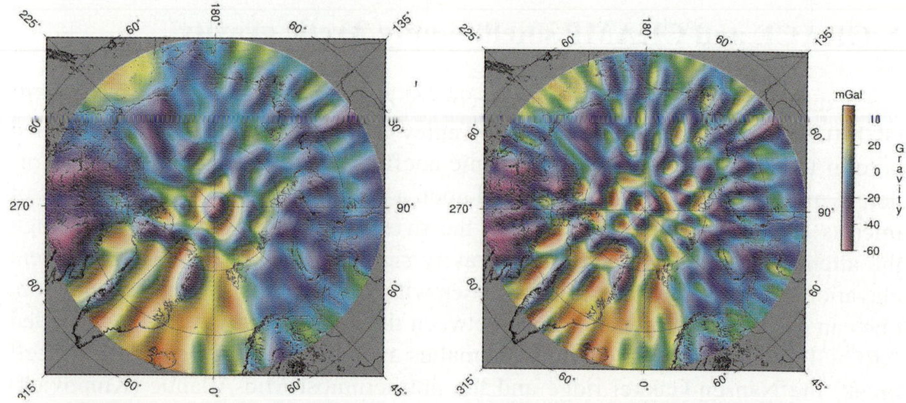


**Fig. 3.** CHAMP-only Arctic gravity: (a) [LEFT] PGS7772 (100x100) model (Lemoine et al., 2003), (b) [RIGHT] EIGEN-2ee (120x120 plus some terms to 140) model (Reigber et al., 2003b).

nounced gravity low off the northeastern coast of Greenland, the north-south trending, positive, shelf-break anomaly at ~228°E along the western edge of the Canadian Archipelago (e.g. Banks Island). Note that north of 85°N the CHAMP models in Fig. 3 show ring-like anomalies that differ substantially from "reality" (i.e., Fig. 1b or 2). These poor results north of 85°N are likely due to CHAMP having a lower tracking precision than GRACE or to the inclination of CHAMP's orbit which is 87° as opposed to GRACE's more nearly polar, 89°, inclination.

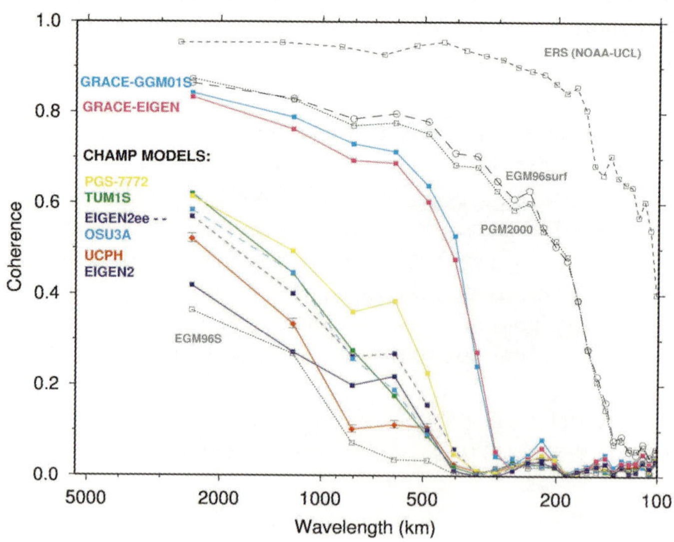


**Fig. 4.** Squared coherency between the ARC-GP Arctic gravity and various GRACE and CHAMP satellite only-models (in colors) plus coherency (in gray) between ARC-GP and OTHER gravity models (see text).

To assess the similarity between our benchmark ARC-GP field (Fig. 1b) and the new GRACE-only and CHAMP-only models we have computed the squared coherency (or the coherency spectrum) between each of these satellite models and the ARC-GP. Coherency, C, can be interpreted as the correlation coefficient as a function of wavelength. Signal-to-noise (SNR) equals C/(1 – C) so SNR = 1 when C = 0.5. Among satellite-only models prior to CHAMP, e.g. EGM96S (Lemoine et al., 1998) and GRIM5S1 (Biancale et al., 2000) we have found that EGM96S (70x70) produces the highest coherency (see Fig. 4) and, in turn, all CHAMP models shown produce higher coherency than EGM96S. The CHAMP models shown include PGS7772 (100x100), TUM1S (60x60/140x140, Gerlach et al., 2003), Eigen2ee, OSU3A (70x70, Han et al., 2004), UCPH (90x90, Howe et al., 2003) and EIGEN-2 (Reigber et al., 2003c). The coherency curves indicate that CHAMP models shown all do a significantly better job of resolving Arctic gravity than any pre-existing satellite-only model. Moreover CHAMP models confidently resolve full wavelengths as short as 1000 km and are improving. Of the CHAMP models, PGS7772 yields the highest coherencies by a slight margin over TUM1S, EIGEN2ee and OSU3A. Coherencies for the two GRACE models in Fig. 4, GGM01S (Tapley et al., 2003) and EIGEN-GRACE01S (Reigber et al., 2003a) are very similar and are each substantially higher than those of the CHAMP models. The PGM2000 and EGM96surf models (360x360; Lemoine et al., 1998; Pavlis, pers comm) are based wholly (EGMsurf) or in part (PGM) on detailed surface gravity data some of which were included in ARC-GP. The NOAA-UCL model (Laxon and McAdoo, 1998; Fig. 1a) was derived from re-tracked ERS-1 data.

## 3 Low-pass filtered gravity

Short-wavelength, surface gravity anomalies are attenuated at satellite elevation (>450 km for CHAMP and GRACE) and hence are difficult (or impossible) to detect with satellites. This difficulty is referred to as "omission error". In order to minimize the effects of this omission error in our comparisons we have low-pass filtered the data using a gaussian filter with a full width of 550km, which is equivalent to a 2.5° radius. Before filtering the data are projected from geographical to rectangular map coordinates. In Fig. 5 the detailed ARC-GP (cf. Fig. 1b) is shown after filtering on the left (5a) and the identically filtered GGM01S is shown on the right (5b). Note the striking similarity of these two models after filtering. The correlation coefficient between them is 0.987! The EIGEN GRACE01S (Reigber et al., 2003a) yields a nearly identical correlation coefficient of 0.986.

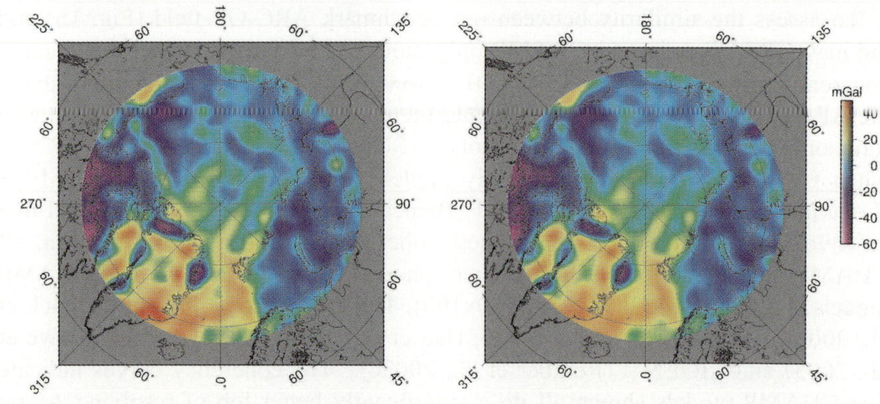


**Fig. 5.** (a) ARC-GP low-pass filtered with a 2.5° gaussian (left); (b) low-pass filtered GRACE GGM01S (right).

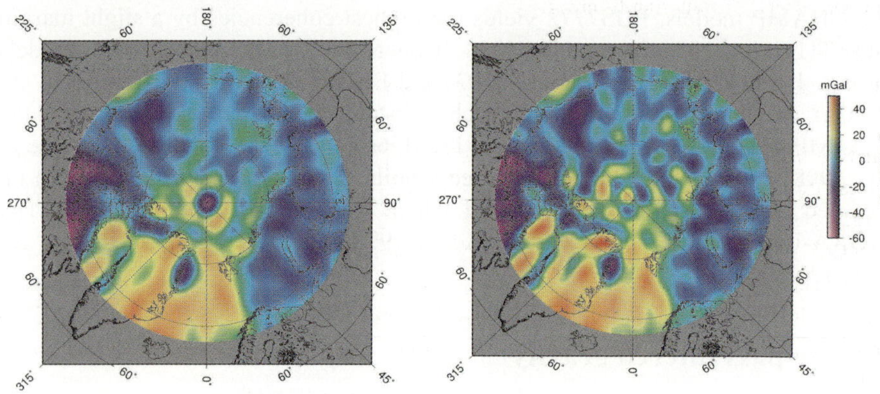


**Fig. 6.** (a) CHAMP Eigen2ee low-pass filtered with a 2.5° gaussian (left); (b) low-pass filtered CHAMP PGS-7772 (right).

In Fig. 6, a gaussian filter identical to that used on ARC-GP and GRACE models (cf. Fig. 5) was applied to the two CHAMP models Eigen2ee and PGS-7772 (cf. Fig. 3a,b). These two filtered CHAMP models show excellent qualitative agreement with our benchmark ARC-GP model (Fig. 5a) except within about 6° of about the north pole where disagreement is clear as ring-like artifacts discussed above persist. The correlation coefficients between the filtered Eigen2ee and ARC-GP as well as between PGS-7772 and ARC-GP are 0.862 and 0.909 respectively. We have done the same filtering to other CHAMP models and the corresponding correlation coefficients are shown in Table 1. The residuals or differences between filtered ARC-GP and the various CHAMP and GRACE models are also shown in Table 1.

| GRAVITY RMS Difference | (mGal) | GEOID RMS Differences (m) | Correlation Coef w.r.t ARC-GP |
|---|---|---|---|
| GRACE (GGM01S) | 2.34 | 0.35 | 0.987 |
| GRACE (EIGEN-GRA) | 2.47 | 0.36 | 0.986 |
| CHAMP (PGS-7772) | 5.69 | 0.59 | 0.909 |
| CHAMP (TUM-1S) | 6.29 | 0.664 | 0.886 |
| CHAMP (OSU3A) | 6.43 | 0.681 | 0.881 |
| CHAMP (EIGEN-2ee) | 6.86 | 0.696 | 0.862 |
| CHAMP (EIGEN-2) | 8.02 | 0.778 | 0.817 |
| CHAMP (UCPH) | 8.31 | 0.993 | 0.807 |
| vs other global satellite model: | | | |
| Multi-Sat(EGM96S) | 8.24 | 0.995 | 0.780 |

**Table 1.** RMS Differences and correlation coefficients between ARC-GP and satellite models after 2.5° gaussian filtering.

## 4 Geoids combining GRACE and surface data

In Fig. 7a note the GRACE GGM01S (Tapley et al., 2003) geoid which has been constructed using all s.h. coefficients complete to degree and order 120. Plotted in Fig. 7b is a hybrid geoid which combines GRACE data at long wavelengths and ArcGP at short ones using a remove-restore method as follows. First the GRACE/ggm01s and and the ArcGP geoids were both low- pass filtered (with a 3.5° radius gaussian) . Then the filtered ARC-GP is subtracted from the ArcGP itself to get residuals which then are added the back to the low pass-filtered GRACE.

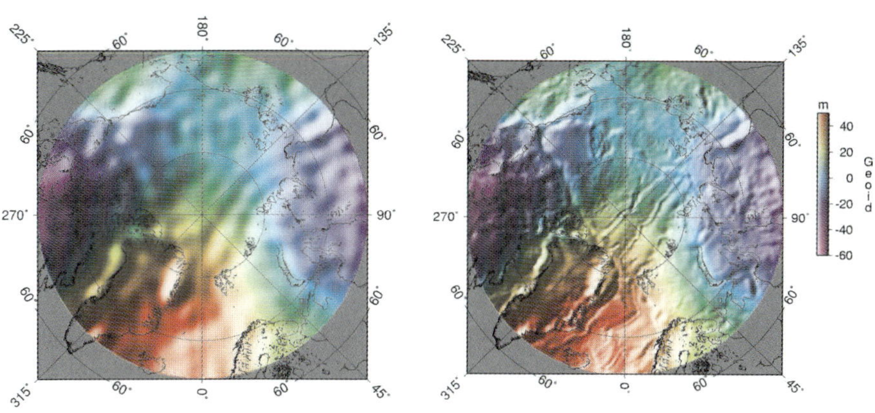


**Fig. 7.** (a) GRACE GGM01S 120x120 geoid (left), and (b) a hybrid geoid combining GRACE and ARC-GP models (right, see text).

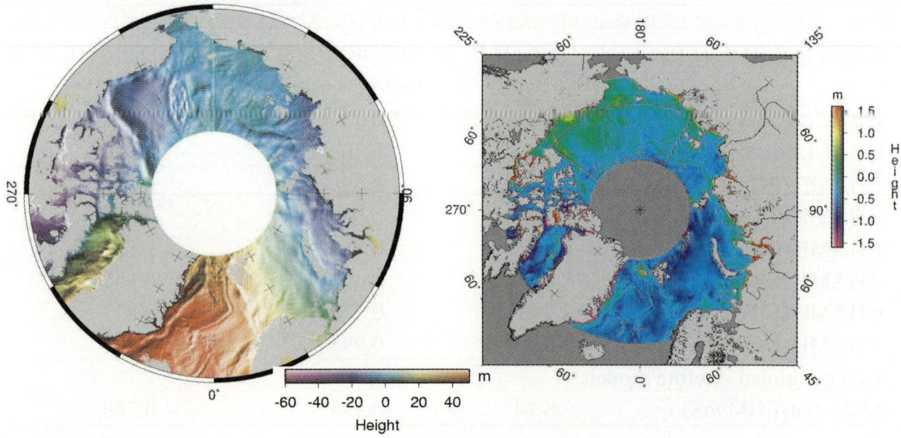


**Fig. 8.** (a) Left: ERS-1, four-year mean sea surface (MSS), (b) Right: ERS MSS minus hybrid geoid (Fig. 7b); Note the windshield-shaped test region F extending from 150E to 230E (see text).

As a preliminary demonstration of the utility of such a "Hybrid" Arctic geoid we have subtracted it from a mean sea surface (Fig. 8a) constructed from 4 years ( '94-'97') of re-tracked ERS-1 data (Peacock and Laxon, 2003) . The difference between this mean sea surface and the hybrid geoid is plotted in Fig. 8b. Does this mean sea surface (MSS) minus geoid represent dynamic topography in the Arctic Ocean? Clearly there are some residual, fine scale geoid artifacts which result however some evidence of dynamic topography, e.g Beaufort Gyre effects may be seen in Fig. 8b. There are few conventional hydrographic observation in the Arctic with which to judge Fig. 8b. In test region F (Fig. 8b) the rms difference signal is less than 19 cm. The corresponding difference between the ERS MSS and the ARC-GP geoid is 27 cm and the difference between the ERS MSS and GRACE geoid is 36 cm. We appear to be on the threshold of detecting dynamic topography of the Arctic Ocean. However the Arctic Ocean's dynamic signal is thought to be more subtle (Maslowski, 2000), more transient and of a smaller spatial scale than that of other ocean basins. So we will need to compute a more precise, detailed, GRACE-based hybrid geoid if we are to confidently detect dynamic ocean topography in the Arctic using satellite altimetry from new and future missions.

## 5 Conclusions

The new CHAMP and GRACE gravity models are yielding substantial, scientifically important improvements in our understanding of the Arctic geopotential. Our analyses show that CHAMP and GRACE accurately resolve Arctic gravity to *full* wavelengths as short as, or shorter than, 1000 km and 500 km respectively. Moreover CHAMP models continue to show improvement in resolution – a process which should continue as the three-and-half years worth of current CHAMP observations, plus years of future observations, are analyzed. Both CHAMP and GRACE are reducing long-wavelength gravity errors in the Arctic by at least as much as they are globally. The precision of the EIGEN-GRACE and GGM01S GRACE models are nearly identical notwithstanding the larger amount of data in GGM01S.

As an example of how these new satellite missions will benefit Arctic oceanographic and sea ice studies we have presented a preliminary GRACE-ArcGP hybrid geoid. Future GRACE- and CHAMP-based geoids should have the accuracy needed to detect - with altimetry - poorly known dynamic topography of the Arctic Ocean.

*Acknowledgements.* We wish to thank S.C. Han, Frank Lemoine, Nikos Pavlis, Christoph Reigber, Peter Schwintzer, C.K. Shum for kindly, and in a most timely way, providing us with their various gravity model results. The views, opinions, and findings contained in this report are those of the author(s) and should not be construed as an official National Oceanic and Atmospheric Administration or U.S. Government position, policy, or decision.

## References

Biancale R, Balmino G, Lemoine, J-M, Marty J-C, Moynot B, Barlier F, Exertier P, Laurain O, Gegout P, Schwintzer P, Reigber Ch, Bode A, König R, Massmann F-H, Raimondo J-C, Schmidt R, Zhu SY (2000) A New global Earth's Gravity Field Model from Satellite Orbit Perturbations: GRIM5-S1. Geophys Res Lett *27*: 3611-3614.

Childers V, McAdoo DC, Brozena, J, Laxon, S (2001) New gravity data in the Arctic Ocean: Comparison of airborne and ERS gravity. J Geophys Res *105* (B5).

Farrell SL, Laxon SW, McAdoo DC (2003) Sea Ice Thickness Determination Using Laser Altimetry. Arctic Climate System Study (ACSYS), Final Science Conference "Progress in Understanding the Arctic Climate System", submitted.

Forsberg R and Kenyon SC and Arctic Gravity Project (ARC-GP) working group under the International Gravity and Geoid Commission, International Association of Geodesy (2002) See http://earth-info.nima.mil/GandG/agp/ and Eos Trans. AGU *80(46)*, Fall Meeting Suppl., Abstract F992, 1999.

Gerlach C, Foldvary L, Svehla D, Gruber Th, Wermuth M, Sneeuw N, Frommknecht, B, Oberndorfer H, Peters Th, Rummel R (2003) A CHAMP-only gravity field from kinematic orbits using the energy integral. Geophys Res Lett *30(20)*: 2037.

Han S-C, Shum CK, Jekeli C, Braun A, Chen Y, Kuo C (2004) CHAMP Gravity Field Solutions and Geophysical Constraint Studies. This issue.

Howe E, Stenseng L, Tscherning CC (2003) CHAMP Gravity Field Model UCPH2003. Geophys. Res. Abstracts (EGS/AGU) v. 5, 09582, also http://www.gfy.ku.dk/~eva/en/sagrada.php

Laxon SW and McAdoo DC (1994) Arctic Ocean gravity field derived from ERS-1 satellite altimetry. Science *265*: 621- 624.

Laxon SW and DC McAdoo (1998) Satellites provide new insights into polar geophysics. Eos Trans. AGU *79(6)*, 69: 72-73.

Lemoine FG, Kenyon SC, Factor JK, Trimmer RG, Pavlis NK, Chinn DS, Cox CM, Klosko SM, Luthcke SB, Torrence MH, Wang, YM, Williamson RG, Pavlis EC, Rapp RH,

Olson TR (1998) The development of the joint NASA GSFC and the National Imagery and Mapping Agency (NIMA) geopotential model EGM96. NASA Technical Paper NASA/TP-1998-206861, Goddard Space Flight Center, Greenbelt.

Maslowski W, Newton B, Schlosser P, Semtner AJ and Martinson DG (2000) Modeling Recent Climate Variability in the Arctic Ocean. Geophys Res Lett *27(22)*: 3743-3746.

Peacock NR and Laxon SW (2003) Sea surface height determination in the Arctic Ocean from ERS altimetry. J Geophys Res, in press.

Reigber Ch, Schmidt R, Flechtner F, König R, Meyer U, Neumayer KH, Schwintzer P, Zhu SY (2003a) First EIGEN Gravity Field Model based on GRACE Mission Data Only. (in preparation for Geophys Res Lett).

Reigber Ch, Schwintzer P, Stubenvoll R, König R, Neumayer H, Förste Ch, Barthelmes F, Balmino G, Biancale R, Lemoine J-M, Loyer S, Bruinsma S, Perosanz F, Fayard T (2003b) CHAMP Global Gravity Field Recovery - Status in Satellite-Only and Combined Mean Field Solution. IUGG 2003 (abstract).

Reigber Ch, Schwintzer P, Neumayer K-H, Barthelmes F, König R, Förste Ch, Balmino G, Biancale R, Lemoine J-M, Loyer S, Bruinsma S, Perosanz F, Fayard T (2003c) The CHAMP-only Earth Gravity Field Model EIGEN-2. Adv Space Res *31(8)*: 1883-1888 (doi: 10.1016/S0273--1177(03)00162-5).

Tapley B et al. (2003) The GRACE Gravity Model 01 (GGM01) released on July 21, 2003. This model was based upon a preliminary analysis of 111 days of in-flight data gathered during the commissioning phase of the GRACE mission (2003) see http://www.csr.utexas.edu/grace/gravity/

# Evaluation of Gravity Data by EIGEN-2 (CHAMP-only) Model in China

Yang Lu and Hongling Shi

Institute of Geodesy & Geophysics, Chinese Academy of Sciences, 54 Xudong Road, Wuhan, 430077, P.R. China, *luyang@asch.whigg.ac.cn*

**Summary.** This paper discussed consistency of long wavelength components of gravity field between the EGM96 and CHAMP gravity models and compared the long wavelength components of surface gravity data with ones from the models of CHAMP within the same spatial resolution, based on the 2-D Gaussian low-pass filter in China and its vicinity. The results show that the models of EGM96 and CHAMP are consistent up to about degree 35, while above this degrees the EIGEN-2 may have inferior estimates. The evaluation of the terrestrial gravity data for China and its vicinities by comparisons with the models of CHAMP has confirmed the existence of larger errors and systemic discrepancy.

**Key words:** Satellite gravity field mission, Gravity anomaly, Gaussian filter.

## 1 Introduction

Up to the present, we have three main ways to observe gravity field. One is directly observing on the earth surface, including terrestrial gravity measurement, airborne gravity measurement and shipboard gravity measurement. Another is radar satellite altimetry. Third way is satellite gravity field missions such as CHAMP, GRACE and GOCE mission. In the three measurements, only the terrestrial measurement can provided the full wavelength information of gravity field for points. However the gravity data recovery from satellite altimetry, the presently best space resolution is 2', included about 98.8% information of gravity field according to the study by Tscherning-Rapp spectrum characteristic model for the gravity field (see Tscherning 1974). The gravity satellite mission can provide the medium and long wavelength information of gravity field. The space resolution of gravity field from the CHAMP satellite come to a head about 220 km, only include about 38% information of gravity field. It means that the gravity data from multifarious source has the different resolution, and including multiform wavelength band of gravity field. Therefore calibration and evaluation of gravity data should be transacted under the same dimensional scale. In this paper, we use the 2-D Gaussian low-pass filter to distill the common components of gravity data sets, namely long wavelength components, for comparing and evaluating the gravity data from multifarious source, including the satellite gravity data, satellite altimetry data and terrestrial gravity data in China and its vicinity.

## 2 Filtering Method

As well known, a grid gravity field with a certain space resolution is as a digitalization image, which is similarly to the two dimensions discrete image function. That so, we can dispose the gravity field data according to the method of the image processing. The 2-D Gaussian low-pass filter will be used to withdraw the long wavelength component of the gravity field. Usually, the Gaussian distribution model definition is writ (see Zhang 1998)

$$G(x,y) = \frac{1}{2\pi\sigma^2} e^{-\frac{x^2+y^2}{2\sigma^2}} \tag{1}$$

Here σ is standard deviation.

The 2-D Gaussian filtering made one two-dimension function (original image) *f(x, y)* convolute with the Gaussian function *g(x, y)*, then get the purpose of restraining the noise and smooth. The convolution result is *h(x, y)*. In the frequency domain, according to the convolution theorem we have

$$H(u,v) = G(u,v)F(u,v) \tag{2}$$

Where *G(u, v), H(u, v)* and *F(u, v)* is Fourier transform functions of *g(x, y), h(x, y)* and *f (x, y)*, respectively. As the linear system theory, the *G(u, v)* is a transition function. According to the Fourier transform we can know the Gaussian function after Fourier transformed will be still the Gaussian function with the same form, then using Fourier inversion we have

$$h(x,y) = F^{-1}[G(u,v)F(u,v)] \tag{3}$$

## 3 Short descriptions of used data

The test area is over the China mainland, China Sea and its vicinities, and western Pacific within the pale of the area ($0° \le \varphi \le 55°N$, $70° \le \lambda \le 140°E$), with three types data areas i.e. mainland area, marine shipboard measurement area and overlapping marine area with altimetry derived gravity datasets (see Fig. 1). All data should been transformed to the reference ellipsoid with a=6378136.46, 1/f=298.25765 as well as EIGEN-2 model used (see Reigber 2003), and arranged in 30'×30' grid.

### 3.1 Satellite gravity data EIGEN-2 and EIGEN-1S model

EIGEN-2, the CHAMP-only Earth Gravity Field Model, derived from altogether six months of CHAMP data, made by GFZ according to the CHAMP GPS satellite to satellite and

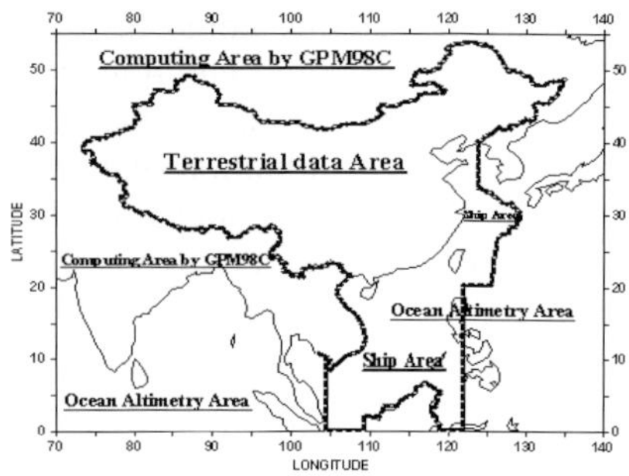


**Fig. 1.** Distribution of datasets in test area.

accelerometer data. The EIGEN-2 has been developed to degree 140. The accuracy of EIGEN-2 is well below 10 cm and 0.5 mGal in terms of geoid heights and gravity anomalies, respectively, at λ/2=550 km (http://op.gfz-potsdam.de/champ/results/ index_RESULTS.html, see Reigber 2003).

EIGEN-1S, the satellite-only Earth Gravity Field Model, to degree 119 including 88 days of CHAMP data was made by GFZ. The solution has got full power only up to about degree/order 35 (http://op.gfz-potsdam.de/champ/results/index_RESULTS.html). The accuracy of EIGEN-1S is ±2.5 mGal in terms of the gravity anomaly, at λ/2=550 km (see Reigber 2001, 2002).

### 3.2 Terrestrial data sets and marine satellite altimetry data sets

Terrestrial gravity data include ground gravity data in China mainland, shipboard gravity data on China Sea and its vicinity with 30'×30' grid, and altimetry derived gravity data sets in West Pacific, include KMS02 (see Andersen 1998, 2001), Sandwell's V.9.2 (see Sandwell 1997), and GMGA97 (see Hwang 1998).

## 4 Comparison and evaluation of gravity data

Before evaluating terrestrial gravity data we compare the satellite gravity models EIGEN-1S, -2 and EGM96 in order to examine or look the differences of CHAMP models at in a comprehensive way. One possibility to evaluate the models is the computation of anomaly difference grids with a varying maximum degree of the expansion. The Table 1 and Table 2 show gravity anomaly differences between EGM96 with EIGEN-2 and EGM-1S in the test area for a varying maximum degree of the expansion, respectively. The statistics shows that the RMS difference between the EIGEN-2 and EGM96 model miss 1.0 mGal (Table 1), while EIGEN-1S exceeds 1.0 mGal (Table 2) at degree 40. The agreement of the two models is quite good up to degree 35(less 0.8 mGal), and thus three models should be useful for the detection of long wavelength errors in the terrestrial gravity data. However, we would rather EIGEN-2 model than others.

| $L_{max}$ | Num. | Mean | RMS | Min. | Max. |
|---|---|---|---|---|---|
| 5 | 5516 | -0.000 | 0.004 | -0.010 | 0.010 |
| 10 | | 0.002 | 0.026 | -0.070 | 0.050 |
| 15 | | -0.012 | 0.106 | -0.210 | 0.190 |
| 20 | | 0.000 | 0.191 | -0.620 | 0.530 |
| 25 | | -0.016 | 0.293 | -0.850 | 0.680 |
| 30 | | 0.005 | 0.397 | -1.150 | 1.110 |
| 35 | | 0.011 | 0.703 | -2.20 | 1.760 |
| 40 | | 0.031 | 0.956 | -2.780 | 2.850 |
| 45 | | -0.060 | 1.982 | -7.100 | 6.700 |

**Table 1.** Gravity anomaly differences between EIGEN-2 and EGM96 in the test area for a varying maximum degree of the expansion. Units are mGal.

| $L_{max}$ | Num. | Mean | RMS | Min. | Max. |
|---|---|---|---|---|---|
| 5 | 5516 | 0.001 | 0.004 | -0.010 | 0.010 |
| 10 | | 0.005 | 0.042 | -0.100 | 0.100 |
| 15 | | -0.010 | 0.096 | -0.200 | 0.200 |
| 20 | | -0.002 | 0.253 | -0.600 | 0.600 |
| 25 | | -0.025 | 0.382 | -0.800 | 1.100 |
| 30 | | -0.003 | 0.543 | -1.600 | 1.700 |
| 35 | | -0.012 | 0.806 | -2.800 | 2.600 |
| 40 | | -0.025 | 1.118 | -3.700 | 3.200 |
| 45 | | -0.137 | 2.761 | -10.900 | 9.000 |

**Table 2.** Gravity anomaly differences between EIGEN-1 and EGM96 in the test area for a varying maximum degree of the expansion. Units are mGal.

| Data type | Num | Mean | RMS | Min | Max |
|---|---|---|---|---|---|
| EIGEN-2—GMGA97 | 4357 | -0.03 | 2.12 | -15.9 | 12.1 |
| EIGEN-2—V.9.2 | 4357 | -0.08 | 2.68 | -12.5 | 32.6 |
| EIGEN-2—Kms02 | 4357 | -0.08 | 2.87 | -17.5 | 33.6 |
| Average | | -0.06 | 2.56 | -15.3 | 26.1 |

**Table 3.** Evaluations of the altimeter derived gravity anomalies, with a spatial wavelength about 1100 km, by EIGEN-2 model in the western Pacific area. Units are mGal.

For evaluating the terrestrial data we compute the gravity anomalies from EIGEN-2 up to degree 140 on 30' grid in the test area (Fig. 2.a). The original gravity is plot on Fig. 2 b, c, d, and e. Using 2-D Gaussian low-pass filter we distill the common components from the gravity datasets. In this way, we have the long wavelength component of datasets with a space domain wavelength roughly 10° (see Fig. 3). From Fig. 2 we can see obviously different from EIGEN-2 and terrestrial datasets. However, after filtering the long wavelength components of gravity from EIGEN-2 are primitively similar to ones of the terrestrial and altimetry data (comparable Fig. 3.a with Fig. 3.b, c, d, e). In other words, we have accomplished successfully the common components of gravity field from multifarious source by low-pass filtering.

Fig. 4 shows the residuals of long wavelength gravity between EIGEN-2 and ground and shipboard in the test area, respectively. We can see that the plus biggish

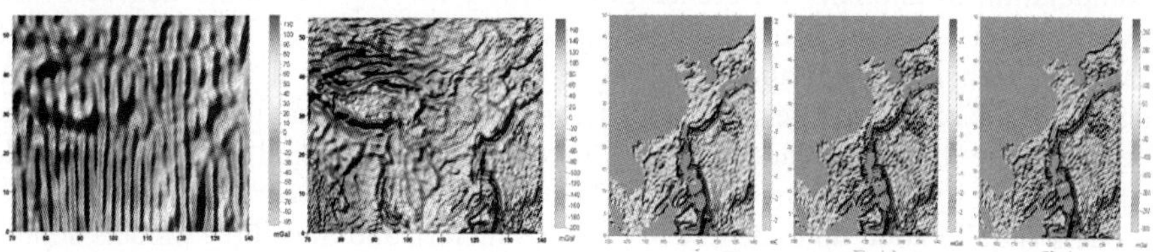

Fig. 2.a. Fig. 2.b. Fig. 2.c. Fig. 2.d. Fig. 2.e.

**Fig. 2.** The gravity anomalies data sets from multifarious source in the test area. Fig. 2.a is anomaly computed from EIGEN-2 up to degree 140. Fig. 2.b is the ground- shipboard dataset. Fig. 2.c, d, e are the data of KMS02, GMGA97 and V.9.2, respectively.

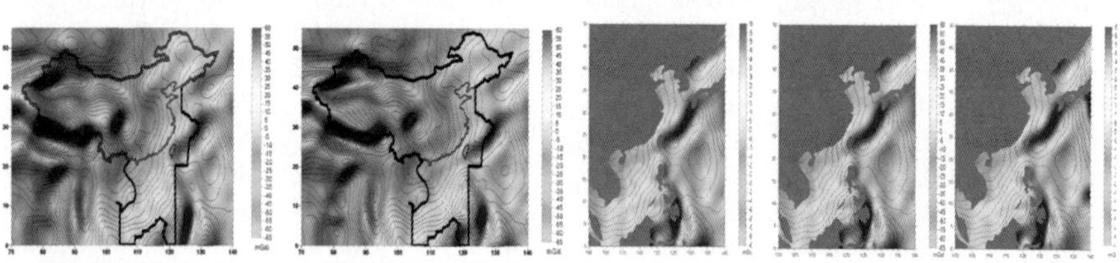

Fig. 3.a. Fig. 3.b. Fig. 3.c. Fig. 3.d. Fig. 3.e.

**Fig. 3.** Long wavelength gravity anomalies after low-pass filtering by 2D Gaussian filter Fig. 3. a, b, c, d, e are long wavelength components of the EIGEN-2, the ground-shipboard, and KMS02, GMGA97 and V9.2, respectively.

residuals present to the northwestern China and its vicinities, and Qinghai-Tibet Plateau (*75°≤λ≤107°E, 26°≤φ≤35°N*). The negative biggish residual was located on the Chaidamu basin of China (*81°≤λ≤87°E, 35°≤φ≤38°N*). The statistics shows that the RMS difference between the EIGEN-2 and the ground data has a stupendous value about ±8.7 mGal. The mean value of the differences is largish about 7.0 mGal, which should be predicated an existence of the biggish systemic discrepancy for ground data. On marine of test area with the shipboard data, the biggish long wavelength residuals present to the Southern China Sea and its vicinities. The statistics shows that the RMS difference between the EIGEN-2 and shipboard data is about ±4.19 mGal. The mean value of the differences is 2.68 mGal, which should be implied systemic discrepancy for the shipboard data. On the West Pacific, the residual differences between EIGEN-2 and altimetry data are plot in Fig. 4.b, c, d. The complicated area is mainly centralized nearby offshore with Philippines, Indonesia and Malaysia, nearby Ryukyu, and South China Sea. In these region there are more islands and reef, which influence the altimetry accuracy. The statistics shows that the RMS of the long wavelength differences between EIGEN-2 and altimetry data exceeds ±2.0 mGal (average is ±2.56 mgal). The mean difference is less than -0.08 mGal (average -0.06 mgal, see Table 3), which means system error from altimetry data is acceptable in the West Pacific.

## 5 Conclusions

The comparisons of gravity models showed that the global gravity field models EIGEN-2, EIGEN-1S and EGM96 are consistent up to about degree 35, while above this degrees the EIGEN-2 may have inferior estimates due to the limitation of CHAMP resolution. That imply three models should be useful for detection of long wavelength errors in the terrestrial data. The evaluation of the gravity data for China and its vicinities by comparisons of long wavelength with the model EIGEN-2 has confirmed the existence of larger errors and systemic discrepancy. The majority of biggish differences converged in the Northwestern and the Qinghai-Tibet plateau of China. It was found apparent inconsistencies between

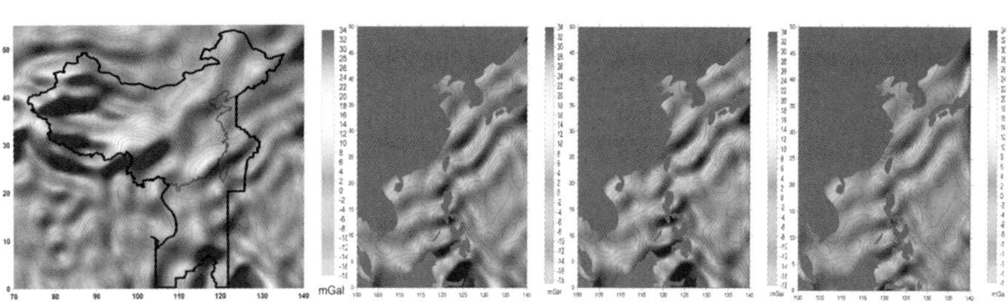

Fig. 4.a. Fig. 4.b. Fig. 4.c. Fig. 4.d.

**Fig. 4.** Long wavelength anomaly residuals with the spatial wavelength about 550 km. Fig. 4.a, b, c, d are long wavelength residuals between EIGE-2 and ground-shipboard, and KMS02, GMGA97 and V9.2, respectively.

shipboard data and EIGEN-2 on the China Sea and its vicinities, and between altimetry derived datasets and EIGEN-2 model on the Western Pacific. The reasons for the revealed errors may be traced back to lacking or poor quality gravity data and substitute errors, especially in the Western mainland of China, while on the ocean we suspect datum inconsistencies or errors from the shipboard surveying and the satellite altimetry, which one is satellite altimetry errors, the other is recovery errors of gravity anomalies from the altimetry data. Otherwise, it is not allow neglect that possible reference system inconsistencies or inaccurate coordination of gravity stations used. Certainly, it is difficult to avoid these infections. Therefore, we expect by dint of the high accuracy global gravity field models derived from the satellite gravity missions such as CHAMP, GRACE and GOCE to significantly improve the gravity field in terms of a homogeneous and high accuracy and resolution.

*Acknowledgement.* Thanks are due to C.C. Tscherning and R. Forsberg for providing GRAVSOFT, and to D.T. Sandwell, C. Hwang, O.B. Andersen, and P. Knudsen for providing the altimetry gravity datasets. This work is supported in part by Key Project of Chinese Academy of Sciences (KZCX2- SW- T1, KZCX3- SW-132).

## References

Andersen OB, Knudsen P (1998) Globe marine gravity field from the ERS-1 and GEOSAT geodetic mission altimetry. J Geophys Res *103(C4)*: 8129-8137.

Andersen OB, Knudsen P, Trimmer R (2001) THE KMS2001 global mean sea surface and gravity field. Proc. IAG2001 Scientific Assembly, Budapest, Hungary [CD-ROM]

Hwang C, Kao EC, Parsons BE (1998) Global derivation of marine gravity anomalies from SEASAT, GEOSAT, ERS-1 and TOPEX/POSEIDON altimeter data. Geophy J Int *134*: 449–459.

Reigber C, Balmino G, Schwintzer P, Biancale R, Bode A, Lemoine JM, König R, Loyer S, Neumayer H, Marty JC, Barthelmes F, Perosanz F, and Zhu SY (2002) A high quality global gravity field model from CHAMP GPS tracking data and Accelerometry (EIGEN-1S). Geophys Res Lett *29(14)*: 371-374.

Reigber Ch, Schwintzer P, König R, Neumayer KH, Bode A, Barthelmes F, Förste Ch, Balmino G., Biancale R, Lemoine J M, Loyer S, Perosanz F (2001) Earth Gravity Field Solutions from Several Months of CHAMP Satellite Data. EOS Trans. AGU, Fall Meet. Suppl. *82(47)*: G4IC-02.

Reigber Ch, Schwintzer P, Neumayer K H, Barthelmes F, König R, Förste Ch, Balmino G., Biancale R, Lemoine J M, Loyer S, Bruinsma S, Perosanz F, Fayard T (2003) The CHAMP-only Earth Gravity Field Model EIGEN-2. Adv Space Res *31(8)*: 1883-1888.

Sandwell DT, Smith WHF (1997) Marine gravity anomaly from GEOSAT and ERS-1 satellite altimetry. J Geophys Res *102(B5)*: 10,039-10,054.

Tscherning CC and Rapp RH (1974) Closed covariance expressions for gravity anomalies, geoid undulations, and deflections of the vertical implied by anomaly degree variance models. Ohio State University, Columbus, Ohio, OSU Rep. No. 208.

Zhang YJ (1998) Image engineering I: Image processing and image analysis. Qinghua University publishing Company, Beijing.

# Energy Balance Relations for Validation of Gravity Field Models and Orbit Determinations Applied to the CHAMP Mission

Anno Löcher and Karl Heinz Ilk

Institute of Theoretical Geodesy, University Bonn, Germany, *ilk@theor.geod.uni-bonn.de*

**Summary.** An extended Jacobi integral describes the energy balance of the motion of a satellite referred to a terrestrial reference frame along its orbit. In addition to its classical form inertia forces and non-conservative force function contributions are included. If force function models and observed satellite orbits are consistent the energy balance sums up to a constant. Deviations from it can be caused either by orbit errors or by insufficiencies in the force function models. Therefore, the energy integral offers itself as a validation tool for consistency checks of force functions and orbit determination results. A basic question is the separation of the various sources of inconsistency. In this paper the theoretical foundation of the validation procedure is presented. It is shown that the validation method can be used to detect deficiencies in the orbit modeling and in the gravity field recovery results. Examples are presented how to separate the various causes of inconsistency. Applications to the results of the CHAMP mission demonstrate the procedure.

**Key words:** Gravity field validation, Satellite orbit validation, Energy balance, Jacobi integral, CHAMP, Satellite-to-satellite-tracking.

## 1 Introduction

The satellite missions CHAMP, GRACE and in future GOCE will provide unprecedented views of the Earth's gravity field and its changes with time. The gravity field recovery results based on three-years observations of CHAMP and the first results derived from satellite-to-satellite tracking of GRACE seem to confirm the expectations in precision and consistency of the gravity field. A basic problem related to these high-precise gravity field results is a proper validation because no comparable information exists. The frequently applied validation and verification procedures based on comparisons with existing models in well-determined regions, with altimetric geoids, with GPS-levelling results or based on orbit predictions may not fulfil the demands of a rigorous validation. All these procedures are susceptible for a broad spectrum of additional error sources based on the processing procedures and may, therefore, camouflage inconsistencies of kinematically observed orbits and the dynamical modelling of the satellite's motion. A kind of "absolute" criterion for the proof of consistency of observed orbit and the dynamical model of satellite's motion is the energy integral along the orbit, which has been applied in satellite geodesy already in 1969 by Reigber (1969). If the various energy constituents do not sum up to a constant then either the orbit is incorrect or the force function models are wrong. An advantage is the point-wise application along the orbit, avoiding accumulations of errors or instability effects. The size of the constant is only of secondary importance, while the structure of the deviations from the constant may give hints to specific force function or orbit determination deficiencies. This is based on the fact that the energy exchange relations caused by the various force function components show typical properties which may open the possibility to separate the different sources of inconsistency. A sophisticated validation procedure is important not only to control the success of global gravity field recovery result; it can be used as well for checking the possibility of a focus to regional gravity field features. The inhomogeneous structure of the gravity field will produce inhomogeneous signals in the satellite's motion or in the high-low satellite-to-satellite connections of CHAMP and the GPS-satellites. A consequence is that regions with rough gravity field features are not adequately modelled by a series of spherical harmonics. The validation procedure can be used to check whether regional improvements of the gravity field may be successful – this can be applied a-priori and a-posteriori.

## 2 Energy integral along a satellite orbit

It is well-known that the classical energy balance which describes the exchange of kinetic and potential energy is valid only in case of conservative forces. The energy function $\hat{E}(t)$ reads, referred to an inertial system:

$$\hat{E}(t) = \frac{1}{2}\frac{\mathbf{P}^2}{M} - MV(\mathbf{R}) + \hat{E}_0 = const\,, \qquad (1)$$

with the linear momentum (mass $M$, velocity $\mathbf{v}$) $\mathbf{P} = M\mathbf{v}$, a gravitational potential $V(\mathbf{R})$ fixed with respect to the inertial system and the energy constant $\hat{E}_0$. In case of a uniformly rotating Earth (rotation vector $\mathbf{\Omega}$) the gravitational force function is not conservative, and therefore, the Jacobi integral has to be applied instead of the simple energy balance relation along the orbits (e.g. Löcher, 2003):

$$\hat{E}'(t) = \frac{1}{2}\frac{\mathbf{P}'^2}{M} - \mathbf{P}'\cdot(\mathbf{\Omega}\times\mathbf{R}') - MV(\mathbf{R}') + \hat{E}'_0 = const\,, \qquad (2)$$

with the potential energy $\hat{E}'_{pot}(t) := -MV'(\mathbf{R}')$ (quantities which refer to the rotating reference system are marked by an apostrophe). The Jacobi integral has to be extended if the rotation of the Earth is considered time variable and non-conservative disturbing forces as tidal forces, Earth tides etc. and surface forces as e.g. air drag and solar radiation pressure have to be included in the balance relation. The same holds if the sum of all surface forces are measured in-situ by accelerometers. From the inertia forces, only the centrifugal force and the Eulerian force show an effect on the energy function, while the Coriolis force does not as one easily verifies. The energy contribution of the centrifugal force $\mathbf{Z}'$ reads, (Ilk and Löcher, 2003),

$$\hat{E}'_Z = -\int \mathbf{Z}'\cdot\dot{\mathbf{R}}'dt = -\frac{1}{2}\cdot(\mathbf{\Omega}\times\mathbf{R}')^2\,, \qquad (3)$$

and the energy contribution of the Eulerian force $\mathbf{E}'$,

$$\hat{E}'_E\big|_{t_0}^{t} = -\int_{t_0}^{t} \mathbf{E}' \cdot \dot{\mathbf{R}}\, dt = M \int_{t_0}^{t} \dot{\mathbf{\Omega}} \cdot \left(\mathbf{R} \times \dot{\mathbf{R}}\right) dt \,. \tag{4}$$

The contribution of an arbitrary non-conservative force $\mathbf{K}'$ to the energy function $\hat{E}'_K\big|_{t_0}^{t}$ by the work $A'\big|_{t_0}^{t}$ performed by the satellite along the orbit reads:

$$\hat{E}'_K\big|_{t_0}^{t} = -A'\big|_{t_0}^{t} = -\int_{t_0}^{t} \mathbf{K}' \cdot d\mathbf{R}' = -\int_{t_0}^{t} \mathbf{K}' \cdot \left(\frac{\mathbf{P}'}{M} - \mathbf{\Omega} \times \mathbf{R}'\right) dt \,. \tag{5}$$

With these formulae the extended (specific) energy function for CHAMP reads:

$$J_{sat} := E'(t) + E'_{akz}\big|_{t_0}^{t} + E'_{M,S}\big|_{t_0}^{t} + E'_E\big|_{t_0}^{t} + E'_{ET}\big|_{t_0}^{t} + E'_{OT}\big|_{t_0}^{t} = const \,, \tag{6}$$

with the sum of kinetic, centrifugal and potential energy $E'(t)$, the surface energy $E'_{akz}\big|_{t_0}^{t}$, the tidal energy of Moon and Sun $E'_{M,S}\big|_{t_0}^{t}$, the Euler energy $E'_E\big|_{t_0}^{t}$, the Earth tides energy $E'_{ET}\big|_{t_0}^{t}$ and the ocean tides energy $E'_{OT}\big|_{t_0}^{t}$.

## 3 Energy contributions for CHAMP along PSO arcs

In this paragraph the extended Jacobi integral due to (6) shall be determined along an arc of the so-called Post-processed Science Orbits (PSO) of CHAMP. The three-dimensional accelerometer data sets as well as the observations and orbits at different production levels are provided by the CHAMP Information System and Data Centre (ISDC). The transformations between the terrestrial and celestial reference frames follow the conventions published by McCarthy (1996). For the computation of the tidal energy caused by Moon and Sun the numerical ephemeris DE405 of the Jet Propulsion Laboratory (JPL) have been used.

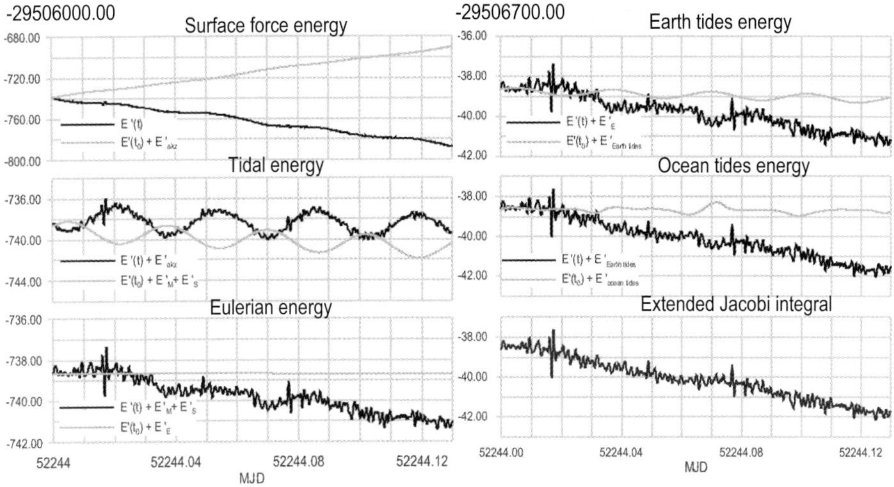


**Fig. 1.** Selected energy contributions ( $m^2/s^2$ ) to the extended Jacobi integral for a two-revolution PSO arc of CHAMP (1.12.2001).

The accelerometer measurements to determine the surface force energy for the CHAMP orbit have been processed according to the rules of the CHAMP data format (Förste et al., 2001). The energy contributions caused by the Earth tides as well as by the ocean tides have been based on the models as published by McCarthy (1996). As gravity field model the recent satellite-only gravity field solution EIGEN-2 has been used for this test (Reigber et al., 2003). The energy contributions to the extended Jacobi integral in case of a two-revolution arc are shown in Fig. 1 to demonstrate the sizes of these various effects. The kinetic and the potential energy as well as the centrifugal energy are not shown here; they are the largest by far. In case of consistency of orbit and force function the bottom graph at the right hand side should be a constant. The secular trend shown in this graph could be caused by an insufficient calibration of the accelerometer data. If a linear trend has been removed then a deviation from a constant in the size of 0,26 $m^2/s^2$ remains, which corresponds to approximately 3 *cm* in the positions.

Especially the gravity field will have an important influence on the energy constant. Fig. 2 shows the inconsistency effects of Eigen-2 plotted along the subsatellite tracks for a 10-days arc. Similar effects occur 6 months later; the differences are shown in Fig. 3. Only orbit-dependent effects are visible while field-dependent ones cancel out. These examples clearly demonstrates the possibility to separate

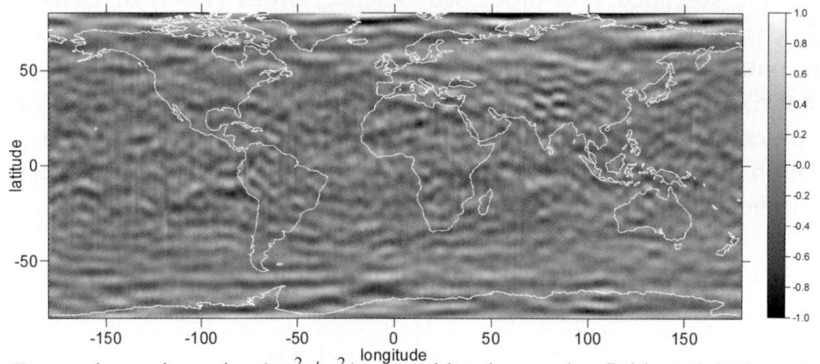


**Fig. 2.** Energy inconsistencies ( $m^2/s^2$ ) caused by the gravity field model Eigen-2, plotted along the subsatellite tracks of CHAMP for a period of 10 days.

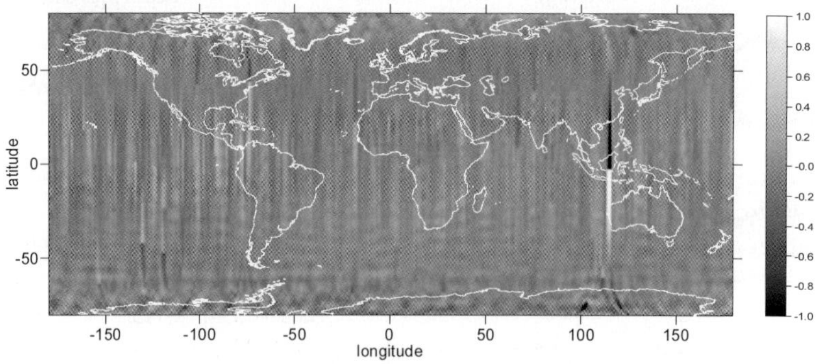


**Fig. 3.** Energy inconsistency differences ( $m^2/s^2$ ) for a period of 10 days with a time lag of 6 months referred to the situation shown in Fig. 2.

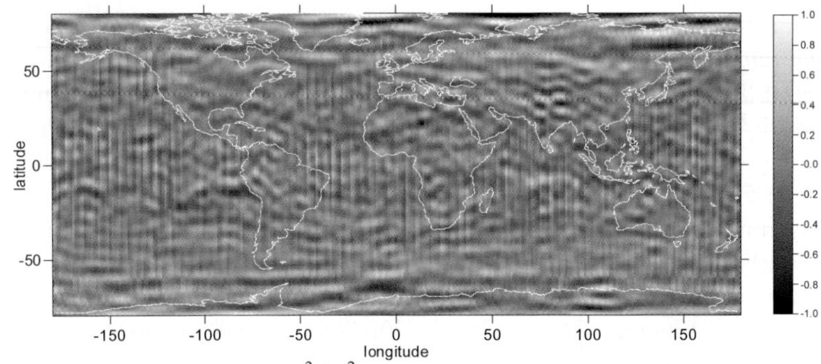


**Fig. 4.** Energy inconsistencies ( $m^2 / s^2$ ) as shown in Fig. 2, but without ocean tides, plotted along the subsatellite tracks for a 10 days arc of CHAMP.

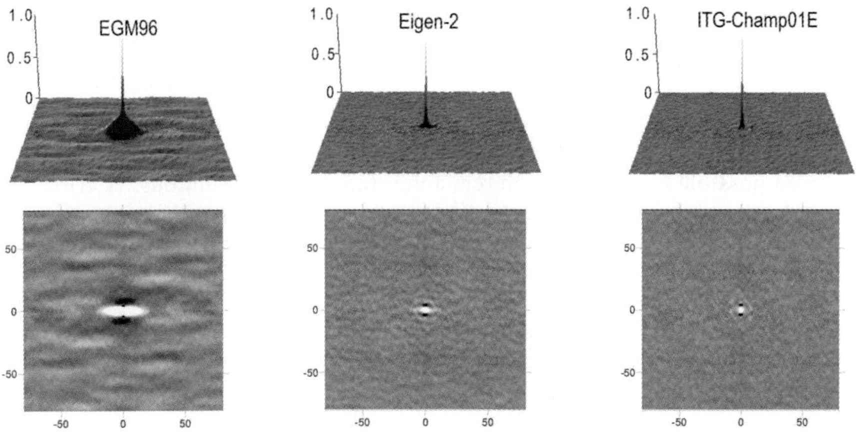


**Fig. 5.** Correlograms of the energy inconsistencies caused by the gravity field models EGM96, Eigen-2, and ITG-01s for kinematic orbits of CHAMP (60 days).

force function and orbit determination inconsistency effects. Even certain field function constituents leave a specific pattern in the deviation of the energy integral from a constant as Fig. 4 demonstrates: in this example the ocean tides are skipped in formula (6). Typical ocean related inconsistency effects can be observed.

## 4 Application to kinematic orbits of CHAMP

PSO are based on a dynamical gravity field model and, therefore, are rather smooth. Kinematic orbits are much more critical, because they do not contain any information on dynamical models and are based exclusively on satellite observations. The present tests are based on a pure kinematic orbit determination of CHAMP provided by M. Rothacher and D. Svehla from the FESG of the Technical University Munich (Svehla and Rothacher, 2003). The critical point of the application of the energy validation approach to kinematical orbits are the velocities of CHAMP. They have to be derived from the ephemeris of the kinematically determined satellite positions. In our approach, the kinematical positions with a sampling rate of 30 sec have been filtered twice, based on the variance-covariance matrices of the positions. Those positions with an rms larger than 8 *cm* are sorted out a-priori. Then the velocities are determined based on an interpolation polynomial of degree 6. Those velocities which result in an rms for the kinetic energy larger than $5\,m^2/s^2$ are deleted as well. The energy inconsistencies are superimposed by noise which hampers the detection of systematic effects. But correlograms clearly uncover systematic effects which show the inconsistencies of orbits and field functions (Fig. 5). The gravity field EGM96 shows large inconsistencies while in case of Eigen-2 (Reigber et al., 2003) only small inconsistencies can be detected. In case of the new gravity field model ITG-Champ01E, derived by the Institute of Theoretical Geodesy, University Bonn (Ilk et al., 2004), the consistency of field function and orbits seems to be realized up to a high degree.

## 5 Conclusions

The results of this investigation clearly demonstrate that the energy integral is a sensitive tool for consistency validation of orbits and force functions. It seems that it is even possible to separate different force function contributions. A critical aspect is the determination of the velocity based on kinematically derived positions.

*Acknowledgements:* We are grateful for providing the data by the ISDC of the GeoForschungsZentrum Potsdam (GFZ) and for the kinematical orbits by D. Svehla and M. Rothacher from the FESG at the Technical University Munich. The support by BMBF and DFG within the frame of the Geotechnologien-Programm is gratefully acknowledged.

## References

Förste Ch, Schwintzer P, Reigber Ch (2001) The CHAMP Data Format. Internal publication, GFZ, http://op.gfz-potsdam.de/champ/docs_CHAMP.

Ilk KH, Mayer-Gürr T, Feuchtinger M (2004) Gravity Field Recovery by Analysis of Short Arcs of CHAMP. This issue.

Ilk KH, Löcher A (2003) The Use of Energy Balance Relations for Validation of Gravity Field Models and Orbit Determination Results. presented at the Gen. Ass. of the IUGG 2003, Sapporo, Japan.

Löcher A (2003) Untersuchungen zur Energieerhaltung bei der Satellitenbewegung am Beispiel der CHAMP-Mission. Diploma thesis at the University Bonn.

McCarthy DD (ed), (1996) IERS Conventions 1996. Central Bureau of IERS – Observatoire de Paris, Paris.

Reigber Ch (1969) Zur Bestimmung des Gravitationsfeldes der Erde aus Satellitenbeobachtungen. DGK, Reihe C, Heft Nr. 137, München.

Reigber Ch, Schwintzer P, Neumayer K-H, Barthelmes F, König R, Förste Ch, Balmino G, Biancale R, Lemoine J-M, Loyer S, Bruinsma S, Perosanz F, Fayard T (2003) The CHAMP-only Earth Gravity Field Model Eigen-2. Adv Space Res *31(8):* 1883-1888.

Svehla D, Rothacher M (2003) Kinematic Precise Orbit Determination for Gravity Field Determination. presented at the Gen. Ass. of the IUGG 2003, Sapporo, Japan.

# Evaluation of Terrestrial Gravity Data by Independent Global Gravity Field Models

Markus Roland and Heiner Denker

Institut für Erdmessung, Universität Hannover, Schneiderberg 50, 30167 Hannover, Germany, *roland@ife.uni-hannover.de*

**Summary.** The new CHAMP and GRACE global gravity field models provide a significantly improved long wavelength gravity spectrum. These satellite-only models are therefore a good basis for studying long wavelength errors of the terrestrial gravity data, as they can be considered as a completely independent source of information.

In this contribution, the models from the CHAMP and GRACE mission as well as EGM96 are used for the evaluation of a terrestrial gravity data set for Europe. The differences are examined both geographically and spectrally. Different techniques are applied for the evaluation, including spherical harmonic expansions, degree variances and the multiresolution analysis based on spherical wavelets. All techniques confirm the existence of small long wavelength errors in the terrestrial gravity data. The reason for such errors may be various, e.g., lacking or poor quality gravity data in some regions, or effects of datum inconsistencies.

**Key words:** gravity anomaly errors, degree variances, spherical harmonics, multiresolution analysis

## 1 Introduction

Within the framework of the European Geoid Project, the Institut für Erdmessung (IfE) has collected about 3 million terrestrial gravity data and about 700 million terrain data for Europe and the surrounding marine areas for the computation of the geoid model EGG97, cf. [3]. The original detailed $1' \times 1.5'$ gravity grids include residual terrain reductions (RTM). For this study, these data were averaged in $90' \times 90'$ blocks to filter out high frequency effects. As the individual gravity data sources are coming from different national agencies, it is likely that different standards were used for the data processing. Therefore, small systematic errors may exist in some of the sources. Possible systematic error sources affecting terrestrial gravity data were studied in detail in [5], with the largest components coming from inconsistencies in the gravity and (horizontal and vertical) position reference systems. Moreover, a study on the effect of such datum inconsistencies on European geoid computations is given in [2].

The new satellite gravity field missions CHAMP and GRACE have improved the knowledge of the long wavelength spectrum of the gravity field significantly during the last years. They can therefore be used to identify systematic errors in the terrestrial gravity data at long wavelengths. In the following, the models EIGEN-1S and EIGEN-2 from the CHAMP mission (cf. [6] and [8]), EIGEN-GRACE01S and GGM01S from the GRACE mission (cf. [7] and [1]), and EGM96 as the standard high-degree model are used for the evaluation of the terrestrial data. From all models, $90' \times 90'$ mean gravity anomaly grids were computed up to degree and order 120 for the area covered by the EGG97 data set. The evaluation techniques include spherical harmonic expansions, degree variances, and the multiresolution analysis (MRA) with spherical wavelets, allowing the localization of features in the frequency and space domain.

## 2 Evaluation Techniques

Within the evaluation process, gravity anomaly differences between the terrestrial data and the global models as well as between different global models were computed and analysed. The first technique uses *RMS degree variances*, which were computed from spherical harmonic expansions of the difference grids (up to $\ell_{\max} = 120$). As the grids are only defined for a small part of the Earth's surface, a scaling of the RMS degree variances by a factor of $\frac{4\pi}{A}$ was done, where $A$ is the area size on the unit sphere. This leads to degree variances that have comparable magnitudes as those derived from planar approximation power spectral densities (PSDs), for details cf. [9]. For a geographical view of the difference patterns, i.e. in the space domain, a *spherical harmonic synthesis with varying maximum degree* $\ell_{\max}$ was performed.

Another technique, which allows a combined spectral and spatial evaluation, is the *multiresolution analysis* (MRA). Spherical wavelet functions $\Psi$ were used. They only depend on the spherical distance and are defined by a Legendre series, see [4]. For the MRA, the mother wavelet is contracted and dilated in fixed steps, resulting in discrete scales $j$ with corresponding wavelet functions $\Psi_j$. The Legendre coefficients for each function determine the spectral behavior. The range of degrees, where the coefficients are unequal zero, is the range of the spectral localization for the scale $j$. Table 1 shows the spectral bands localized by the smoothed Shannon wavelet (SSW), which is used in this study.

## 3 Evaluation Results

Fig. 1 shows the RMS degree variances for 4 selected difference data sets as well as the errors for 2 of the involved global models. All other cases are not shown here due to space restrictions. The differences between the terrestrial data and the CHAMP and GRACE models show a significant increase around degree 45 and 105, respectively, where the satellite-only models do

**Table 1.** Frequency localization of the smoothed Shannon wavelet

| scale $j$ | spectral band | | |
|---|---|---|---|
| 0 | 0 | – | 1 |
| 1 | | 2 | |
| 2 | 3 | – | 4 |
| 3 | 5 | – | 8 |
| 4 | 9 | – | 16 |
| 5 | 17 | – | 32 |
| 6 | 33 | – | 64 |
| 7 | 65 | – | 128 |

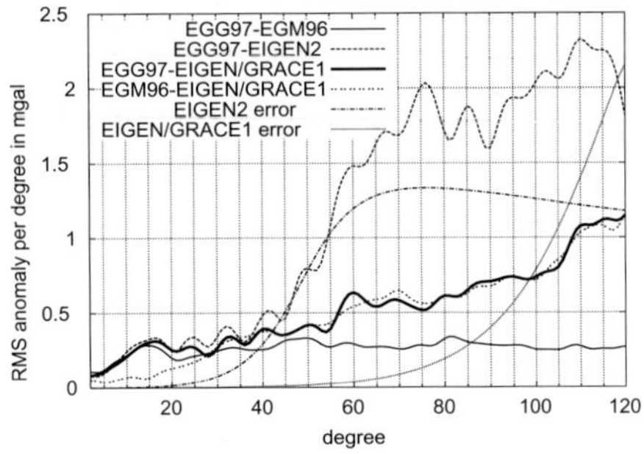


**Fig. 1.** RMS anomaly degree variances.

not have the full power due to attenuation effects. Considering also the errors of the data sets involved, one can find that the differences are exceeding the corresponding error estimates by a factor of two up to about degree 35 for the CHAMP models and up to about degree 90 for the GRACE models, respectively. In these comparisons, the EIGEN-2 model is performing slightly better than the EIGEN-1S model, while the two GRACE models (EIGEN-GRACE01S and GGM01S) show very similar results. It should also be noted that the agreement of the terrestrial data with the EGM96 model is superior to the other models at higher degrees. However, this is probably due to the fact, that terrestrial gravity data is already included in the EGM96 model. At present, the GRACE models should be the best source of information to determine long wavelength errors of the terrestrial data, allowing spherical harmonic expansions up to about degree 90.

A geographical view of the difference patterns, derived from the spherical harmonic expansions up to a fixed maximum degree, is given for some selected cases in Figs. 2-5. While the agreement of the two global models EIGEN-2 and EGM96 is very good up to degree 20, the differences between the EGG97 data and the EIGEN-2 model show a RMS value of 0.8 mgal and maximum values up to $\pm 3$ mgal in the Mediterranean Sea, Eastern Europe, and Greenland. All these areas are known for their weak data quality. On the other hand, the agreement between the terrestrial data and EIGEN-2 is significantly better over the continental parts of central, western and northern Europe. These regions are covered by high quality gravity data, and only small differences, possibly resulting from datum inconsistencies, are found. All these features are even more pronounced in the comparisons with the highly accurate EIGEN-GRACE01S model. The corresponding differences up to spherical harmonic degree 50 are shown in Figs. 4 and 5. In the differences between the EGG97 data and the EIGEN-GRACE01S model, the problem areas show up more clearly with the largest values being about $\pm 8$ mgal in the eastern part of the Mediterranean Sea and Greenland. Again, the discrepancies over central Europe, where high quality data are available, are very small and below 1-2 mgal.

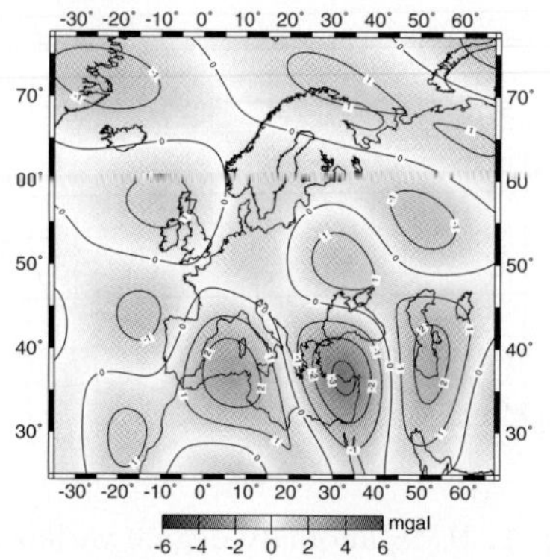


**Fig. 2.** Differences between EGG97 and EIGEN-2 up to $\ell_{\max} = 20$.

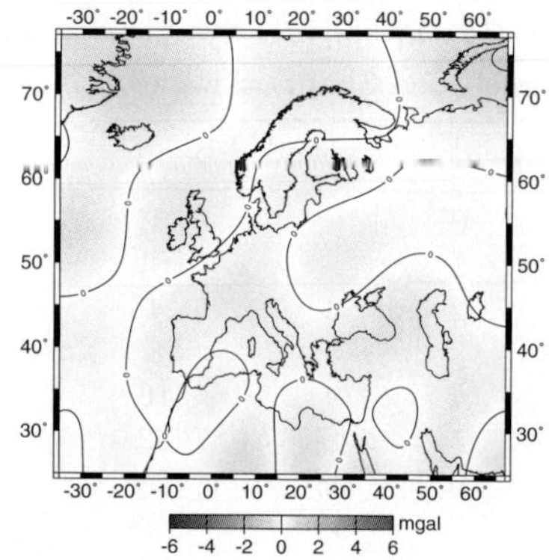


**Fig. 3.** Differences between EGM96 and EIGEN-2 up to $\ell_{\max} = 20$.

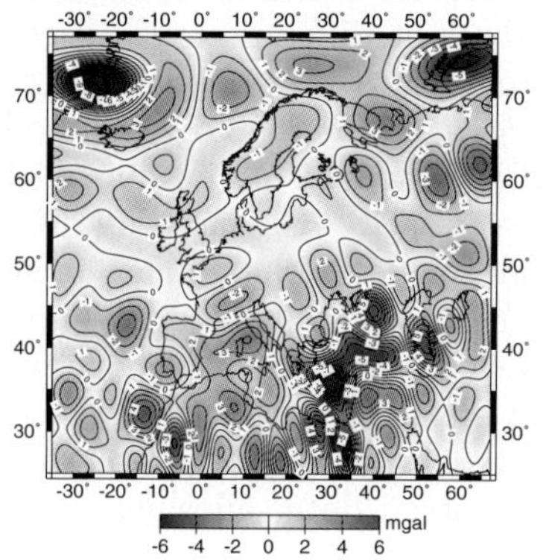


**Fig. 4.** Differences between EGG97 and EIGEN-GRACE01S up to $\ell_{\max} = 50$.

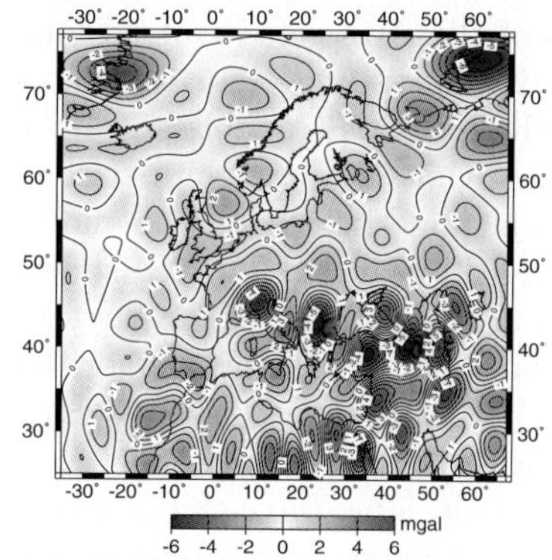


**Fig. 5.** Differences between EGM96 and EIGEN-GRACE01S up to $\ell_{\max} = 50$.

Over northern France, a small positive offset exists, while in southern France and over Switzerland a small negative offset can be seen. These features may be due to vertical and horizontal datum problems, e.g., the French and Swiss national height systems are offset with respect to UELN by several dm. Moreover, also the differences between the EGM96 and EIGEN-GRACE01S models up to degree 50 attain larger values, especially over the Alps and Southeast Europe, where the data used in the EGM96 development are probably weak.

Considering the results of the spherical harmonic analysis, the most interesting features of the MRA should reside at the scales 4 to 6 of the SSW (cf. Table 1). The results of the MRA are illustrated exemplarily for scales 4 and 5, again using the differences EGG97/EIGEN-2 and EGM96/EIGEN-2 (see Figs. 6-9). Moreover, an analysis with the GRACE models was done. The results showed a very good agreement with the EIGEN-2 analysis for scales up to 5. Due to space restrictions, no details are provided on the GRACE results. At scale 4 (see Figs. 6 and 7), representing the spectral band from $\ell = 9$

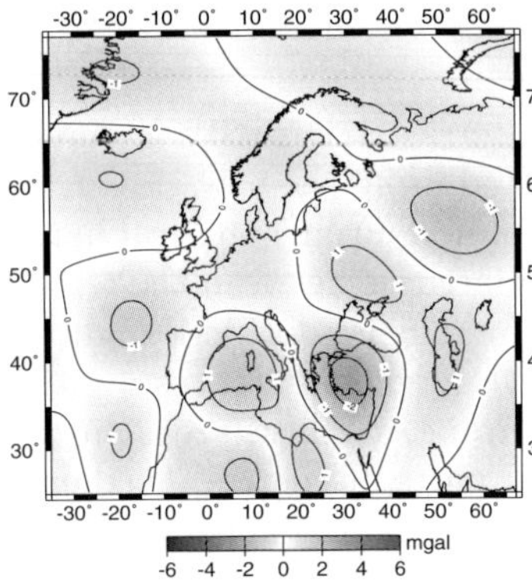


**Fig. 6.** Differences between EGG97 and EIGEN-2 from MRA, SSW, scale 4.

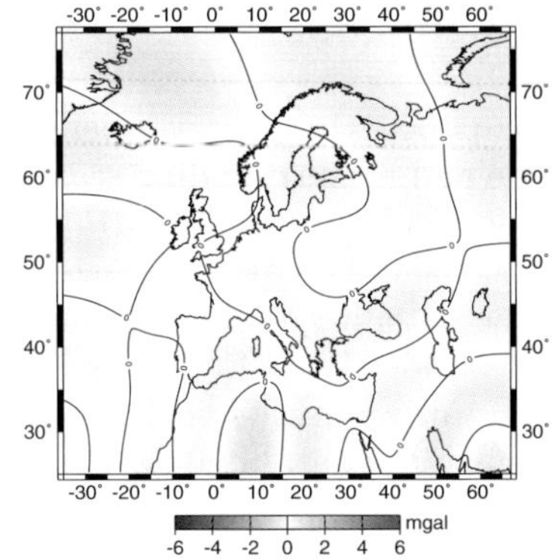


**Fig. 7.** Differences between EGM96 and EIGEN-2 from MRA, SSW, scale 4.

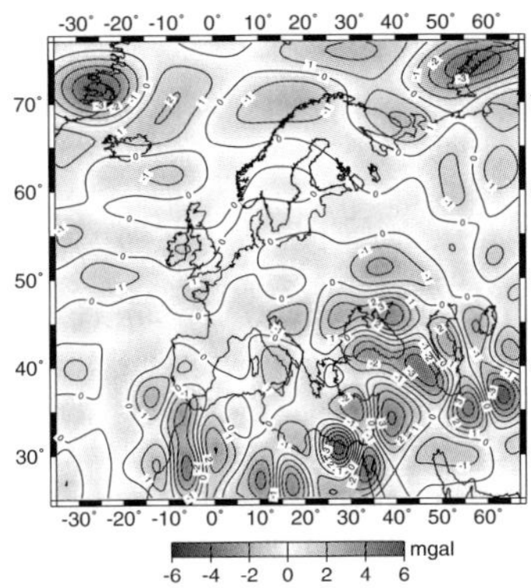


**Fig. 8.** Differences between EGG97 and EIGEN-2 from MRA, SSW, scale 5.

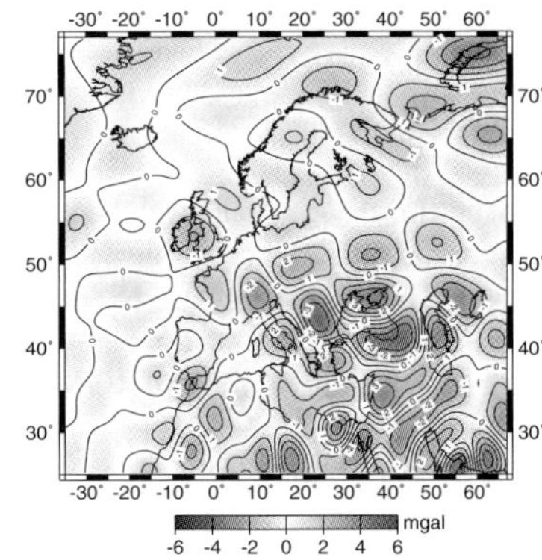


**Fig. 9.** Differences between EGM96 and EIGEN-2 from MRA, SSW, scale 5.

to 16, the differences between EGG97 and EIGEN-2 reach extreme values of $\pm 3$ mgal and a RMS of about 0.6 mgal. The largest differences are located in the Mediterranean Sea and in the Russian plains west of the Ural. Again, the differences between the global models EGM96 and EIGEN-2 are very smooth with a RMS around 0.2 mgal, and they do not show patterns like the EGG97 comparisons. At scale 5 ($\ell = 17$ to 32) the results are more diverse (see Figs. 8 and 9). The differences between EGM96 and EIGEN-2 show a larger RMS around 1.1 mgal, and the maximum values reach about $\pm 4$ mgal. As the analysis with the GRACE model showed similar features, this indicates areas with weak data sets in the EGM96 development (e.g., Southeast Europe). In the EGG97 comparison, small differences are found for most parts of continental Europe. Only in Southeast Europe, Africa, and at the extreme northern latitudes, larger differences up to about $\pm 5$ mgal are found. Again these are the areas with lacking or poor quality gravity data.

## 4 Conclusions

In this paper, the long wavelength quality of terrestrial gravity anomalies in Europe and its surrounding seas was investigated using the new global gravity field models from the CHAMP and GRACE mission. From spherical harmonic expansions and degree variances significant differences between the terrestrial and the satellite data were found up to degree 35 for CHAMP and up to degree 90 for GRACE. The problem areas with larger differences are the the Mediterranean Sea, Greenland and Eastern Europe, all these areas are known for their weak data quality. In the continental parts of central, northern and western Europe, covered by high quality gravity data, the agreement is much better. Here, only some small effects from datum inconsistencies may exist. The multiresolution analysis (MRA) confirmed the findings from the spherical harmonic analysis.

## References

1. Center for Space Research, The University of Texas at Austin (2003) GRACE Gravity Model. `http://www.csr.utexas.edu/grace/gravity/`.
2. Denker H (2001) On the effect of datum inconsistencies of gravity and position on European geoid computations. In: Proc IAG Scientific Assembly, Budapest, Hungary, 2-7 Sep, on CD-ROM.
3. Denker H and Torge W (1998) The European Gravimetric Quasigeoid EGG97 – An IAG supported continental enterprise. In: Forsberg R, Feissel M, and Dietrich R, eds, Geodesy on the Move – Gravity, Geoid, Geodynamics, and Antarctica, IAG Symposia, *119*, Springer Verlag: 249–254.
4. Freeden W, Grevens T, and Schreiner M (1998) Constructive Approximation on the Sphere with Application to Geomathematics. Clarendon Press, Oxford.
5. Heck B (1990) An evaluation of some systematic error sources affecting terrestrial gravity anomalies. Bulletin Géodésique *64*: 88–108.
6. Reigber C, Balmino G, Schwintzer P, Biancale R, Bode A, Lemoine J-M, König R, Loyer S, Neumayer H, Marty J-C, Barthelmes F, Perosanz F, and Zhu SY (2002) A high quality global gravity field model from CHAMP GPS tracking data and Accelerometry (EIGEN-1S). Geoph Res Lett *29(14)*: doi:10.1029/2002GL015064.
7. Reigber C, Schmidt R, Flechtner F, König R, Meyer U, Neumayer K-H, Schwintzer P, and Zhu SY (2003) First EIGEN Gravity Field Model based on GRACE Mission Data Only. Geoph Res Lett (submitted).
8. Reigber C, Schwintzer P, Neumayer K-H, Barthelmes F, König R, Förste C, Balmino G, Biancale R, Lemoine J-M, Loyer S, Bruinsma S, Perosanz F, and Fayard T (2003) The CHAMP-only Earth Gravity Field Model EIGEN-2. Adv Space Res *31(8)*: doi:10.1016/S0273–1177(03)00162-5.
9. Roland M and Denker H (2003) Evaluation of Terrestrial Gravity Data by New Global Gravity Field Models. In: Tziavos IN, ed, Gravity and Geoid – 3rd Meeting of the International Gravity and Geoid Commission, Thessaloniki, Greece, August 26-30, 2002, Publishing Ziti: 256–261.

# Recent Developments in CHAMP Orbit Determination at GFZ

Rolf König, Grzegorz Michalak, Karl Hans Neumayer, Roland Schmidt, Sheng Yuan Zhu, Heribert Meixner, and Christoph Reigber

GeoForschungsZentrum Potsdam (GFZ), Dept. 1 'Geodesy and Remote Sensing', Telegrafenberg, 14473 Potsdam, Germany, *koenigr@gfz-potsdam.de*

**Summary.** The GeoForschungsZentrum Potsdam (GFZ) operationally provides CHAMP orbit products for various purposes. Here the rapid and ultra-rapid orbits are highlighted. Significant developments in Precise Orbit Determination (POD) for Low Earth Orbiters (LEOs), in particular SAC-C and GRACE besides CHAMP, are described. GFZ also started to generate CHAMP-like rapid orbits for SAC-C with good accuracy. Furtheron improved LEO orbit accuracies are demonstrated by simultaneous orbit solutions of the GPS satellites and one or more LEOs in an integrated approach.

**Key words:** CHAMP, SAC-C, GRACE, GPS, Precise Orbit Determination

## 1 Introduction

The GeoForschungsZentrum Potsdam (GFZ) runs a science data system in the CHAMP mission [1] ground segment [2] that produces standard orbit products servicing e.g. the mission objectives i.e. gravity field and magnetic field evaluation and atmospheric and ionospheric applications. Standard orbits are firstly CHAMP orbit predictions (the PDO products) for ground net antenna steering, SLR ground station pointing, mission planning and data preprocessing. In [3] procedures and results are given. The second type of standard orbit is the Rapid Science Orbit (RSO) as input to magnetic field and atmospheric and ionospheric processors. Goals are laid down in [4], achievements are reported in [5]. The third type is the Ultra-rapid Science Orbit (USO) being of use for the evaluation of radio occultations for numerical weather prediction application. Finally the Post-processed Science Orbit (PSO) results in the course of gravity field recovery work as e.g. in [6].

In the beginning of the mission, POD of CHAMP was a challenge, see the articles refering to POD in the proceedings of the first CHAMP science meeting in 2002 [7]. Meanwhile CHAMP orbit accuracies are in the few centimeter range reported by various groups e.g. in this issue, and also shown in the sequel. In the following, focus is put on the RSO and USO for CHAMP: the RSO is mainly viewed in its historical evolution in terms of accuracy, the USO being accurate to the centimeters from its invention, is mainly viewed in its latency which by intention should be very small. Further on a RSO type of orbit for the SAC-C satellite is introduced and characterized in terms of accuracy. Finally an example of the new and promising approach of the integrated adjustment of various types of observations from a multitude of high and low altitude satellites [8], i.e. the GPS satellites, CHAMP, SAC-C and the GRACE satellites, is presented with emphasis on LEO orbit improvement.

## 2 CHAMP Rapid and Ultra-rapid Orbits

The RSO is generated with a latency of 17 hours on a day by day basis. It includes CHAMP 30-s ephemerides of two 14-h arcs overlapping by two hours covering the time between 10 pm the day before the previous day and 0 am the current day and one GPS 1-d arc with 300-s ephemerides and clocks for the previous day. The CHAMP orbit accuracy is measured by SLR measurements. Fig. 1 displays on its left side the development of the accuracy from the beginning of the production of the RSOs in March 2001, with slightly more than 20 cm RMS of SLR residuals down to the 4 cm level nowadays. The improvement is due to newer gravity field models updated by more CHAMP data and some tuning of the adjustment process. It should be noted that the SLR RMS values are reported as is with no outlier rejection applied. Therefore the moving mean of the RMS values indicated by the thick line in the left part of Fig. 1 gives rather a pessimistic assessment of orbit accuracy. It should also be noted that the SLR RMS gives a measure of position accuracy of the orbit as the SLR observations from ground to CHAMP cover all axis directions if the sample is large. Therefore the 1-D accuracy of the RSO or the accuracy per coordinate can be said to be slightly above 2 cm, i.e. 4 cm divided by $\sqrt{3}$, nowadays.

The GPS orbits obey an accuracy of 8 cm per coordinate assessed from comparisons with the IGS [9] rapid orbit (IGR). The right hand side of Fig. 1 gives the 1-D differences (RMS of the position differences divided by $\sqrt{3}$) and a moving mean of the point values indicated by the thick line.

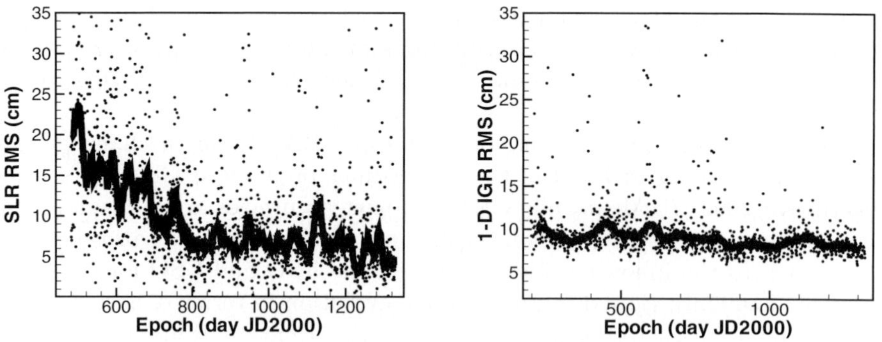


**Fig. 1.** CHAMP and GPS RSO Orbit Accuracies

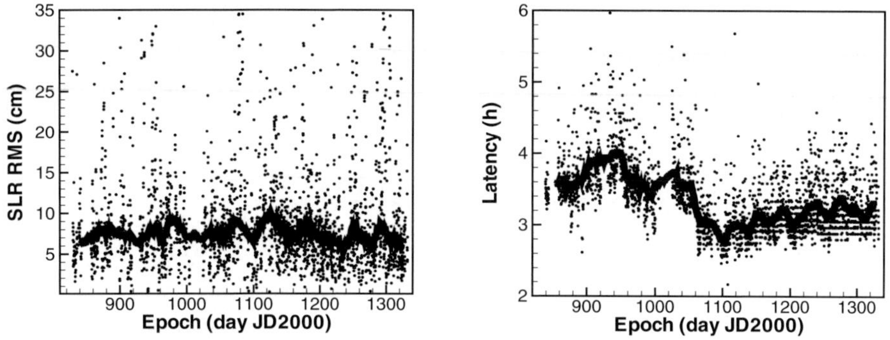


**Fig. 2.** CHAMP USO Orbit Accuracy and Latency

The USO is generated eight times per day at three hour intervals. As in case of the RSO, the GPS orbits contain 300-s ephemerides and clocks spanning one day, the CHAMP orbits contain 30-s ephemerides covering the last 14 hours of the GPS arc. Accuracies of the GPS orbits are 8 cm per coordinate as in the case of the RSO. The accuracies of the CHAMP orbits are slightly worse than in the RSO case. They size at 5 cm per coordinate as derived from the SLR residuals displayed on the left hand side of Fig. 2.

The USO was invented as a fast available orbit product in April 2002 as input to the rapid processing of radio occultation measurements [10]. Its latency, i.e. the difference in time between its availability for the users and its last time tagged ephemeride (covered by observations, not predicted), lies at 3 hours as displayed on the right hand side in Fig. 2 by the thick line representing the moving mean of all latencies ranging at the 3 hours level since some months.

## 3 SAC-C Orbits

The generation of SAC-C orbits started in July 2003 for particular periods of the SAC-C mission [11] on demand by the GFZ occultation processing group. This resulted in occultation products similar in quality to CHAMP occultation products [12]. Fig. 3 shows two quality measures for the assessment of the SAC-C orbit accuracies. Unfortunately there are no SLR observations from ground to SAC-C that could be used as an absolute measure of position accuracy. Instead we try an assessment by substitute means as the comparison of our orbits to the orbits produced by JPL [13] shown on the left hand side of Fig. 3. The mean difference in position is 12 cm with no bias detectable.

On the right hand side of Fig. 3 the standard deviations of the initial position parameters show globally 7 cm per coordinate axis. Interpreting these results with the experiences from CHAMP, the accuracy of the SAC-C orbits may assessed to lie below 6 cm per coordinate.

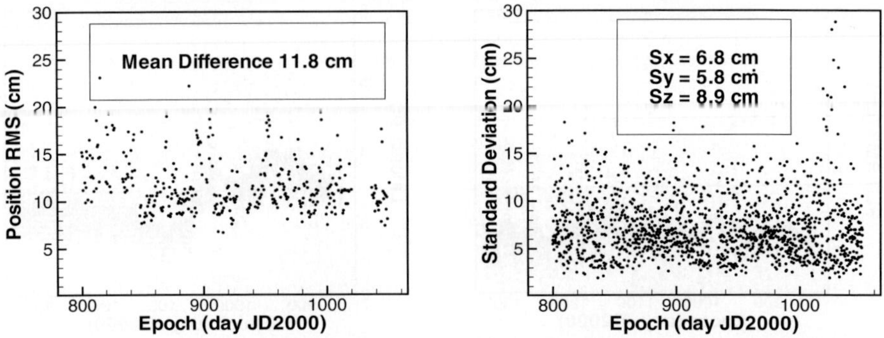


**Fig. 3.** SAC-C Orbit Accuracies

# 4 Integrated Orbit Adjustment for CHAMP, SAC-C, GRACE and the GPS Satellites

Integrated POD means processing of GPS code and phase observations from ground stations and from space-borne receivers together with all or some of all available mission data other than GPS. In this chapter we mix GPS ground data with CHAMP, SAC-C and GRACE data with focus on effects on POD of the LEO satellites. The integrated adjustment of GPS ground and space-borne observations can also be called 1-step solution in contrast to the conventional procedure where the GPS orbit and clocks are firstly solved, then fixed and then the LEO orbit is derived based on the space-borne GPS data in a second step (2-step solution).

In [8] it is shown that the integrated POD of the GPS and GRACE satellites yields, besides improved GPS ephemerides and Earth reference frame parameters, more accurate GRACE orbits. Table 1 compiles results for the CHAMP and SAC-C satellites where two 1.5-day arcs beginning of May 2002 demonstrate the improvement on behalf of the orbital fits of GPS and SLR observations. The SLR measurements are downweighted in the POD process in order to gain an independent quality parameter.

**Table 1.** Improved LEO orbits by the integrated adjustment

| Observation Type | RMS of Orbital Fit | | | | | | | |
|---|---|---|---|---|---|---|---|---|
| | CHAMP | | | | SAC-C | | | |
| | 2-step | | 1-step | | 2-step | | 1-step | |
| | (cm) | n | (cm) | n | (cm) | n | (cm) | n |
| Code | 73.89 | 60368 | 72.28 | 60367 | 124.00 | 54231 | 122.48 | 54230 |
| Phase | 2.55 | 60368 | 0.73 | 60367 | 2.91 | 54231 | 1.42 | 54230 |
| SLR | 5.97 | 264 | 5.04 | 264 | - | - | - | - |

Table 2 compiles results for the integrated POD of the GPS constellation with CHAMP plus one or more LEOs, the GRACE satellites and/or the SAC-C satellite. Table 2 seems to show that adding more LEOs does not improve

**Table 2.** CHAMP orbit quality by the integrated adjustment with additional LEOs

| Observation Type | RMS of Orbital Fit CHAMP (cm) | n | SAC-C (cm) | n | GRACE (cm) | n |
|---|---|---|---|---|---|---|
| *GPS + CHAMP + SAC-C:* | | | | | | |
| Code | 72.33 | 60369 | 122.47 | 54231 | | |
| Phase | 0.74 | 60369 | 1.42 | 54231 | | |
| SLR | 5.19 | 264 | - | - | | |
| *GPS + CHAMP + GRACE:* | | | | | | |
| Code | 72.30 | 60369 | | | 47.70 | 129151 |
| Phase | 0.74 | 60369 | | | 0.61 | 129151 |
| SLR | 5.26 | 264 | | | 4.77 | 297 |
| *GPS + CHAMP + SAC-C + GRACE:* | | | | | | |
| Code | 72.30 | 60369 | 122.46 | 54231 | 47.73 | 129152 |
| Phase | 0.74 | 60369 | 1.43 | 54231 | 0.61 | 129152 |
| SLR | 5.24 | 264 | - | - | 4.70 | 297 |

individual LEO orbit accuracies. However the distribution of the residuals of the space-borne observations is skew. The reason could come from the unique weighting of the ground and space-borne GPS observations where the number of ground observations is more than 10 times the number of the space-borne observations. Therefore further analyses need to be carried out.

## 5 Conclusions

GFZ operationally generates Rapid and Ultra-rapid Science Orbits within the CHAMP ground segment: the RSOs and the USOs. The CHAMP RSO accuracies have greatly improved over time to the 2 cm range nowadays. The USO accuracy is slightly worse however the orbits are delivered eight times a day three hours later than the latest observation in the orbit. GFZ also generates RSO-type SAC-C orbits with approximately 6 cm accuracy for occultation data processing. The integrated adjustment of ground and space-borne observation allows a considerable enhancement of LEO orbit accuracies. Additional LEOs in the integrated case seem not to increase individual LEO orbit quality.

## References

[1] Reigber Ch, Schwintzer P, Lühr H (1999) The CHAMP geopotential mission. Boll Geof Teor Appl *40*: 285–289.

[2] CHAMP - GFZ's Challenging Minisatellite Payload for Geophysical Research and Application (CHAMP) Home Page. http://op.gfz-potsdam.de/champ/index_CHAMP.html, cited December 2003.

[3] Schmidt R, Baustert G, König R, Reigber Ch (2003) Orbit Predictions for CHAMP - Development and Status. In: Reigber C, Lühr H, Schwintzer P (Eds), First CHAMP Mission Results for Gravity, Magnetic and Atmospheric Studies, Springer, Berlin Heidelberg: 104–111.

[4] König R, Zhu SY, Reigber Ch, Neumayer K-H, Meixner H, Galas R, Baustert G, Schwintzer P (2002) CHAMP Rapid Orbit Determination for GPS Atmospheric Limb Sounding. Adv Space Res *30*: 289–293.

[5] Michalak G, Baustert G, König R, Reigber Ch (2003) CHAMP Rapid Science Orbit Determination - Status and Future Prospects. In: Reigber C, Lühr H, Schwintzer P (Eds), First CHAMP Mission Results for Gravity, Magnetic and Atmospheric Studies, Springer, Berlin Heidelberg: 98–103.

[6] Reigber Ch, Jochmann H, Wünsch J, Petrovic S, Schwintzer P, Barthelmes F, Neumayer KH, König R, Förste C, Balmino G, Biancale R, Lemoine J-M, Loyer S, Perosanz F (2004) Earth Gravity Field and Seasonal Variability from CHAMP. This issue.

[7] Reigber C, Lühr H, Schwintzer P, eds (2003) First CHAMP Mission Results for Gravity, Magnetic and Atmospheric Studies. Springer, Berlin Heidelberg.

[8] Zhu SY, Reigber Ch, König R (2004) Integrated adjustment of CHAMP, GRACE and GPS Data. J Geodesy, accepted.

[9] IGS Home Page. http://igscb.jpl.nasa.gov, cited December 2003.

[10] Schmidt T, Wickert J, Beyerle G, König R, Galas R, Reigber Ch (2004) The CHAMP Atmospheric Processing System for Radio Occultation Measurements. This issue.

[11] SAC-C Home Page. http://www.conae.gov.ar/satelites/sac-c.html, cited December 2003.

[12] Wickert J, Schmidt T, Beyerle G, Michalak G, König R, Kaschenz J, Reigber Ch (2004) Atmospheric profiling with CHAMP: Status of the operational data analysis, validation of the recent data products and future prosspects. This issue.

[13] JPL Home Page. http://www.jpl.nasa.gov, cited December 2003.

# On Calibrating the CHAMP On-Board Accelerometer and Attitude Quaternion Processing

Karl Hans Neumayer, Grzegorz Michalak, and Rolf König

GeoForschungsZentrum Potsdam (GFZ), Dept. 1 'Geodesy and Remote Sensing', Telegrafenberg, 14473 Potsdam, Germany, *neumayer@gfz-potsdam.de*

**Summary.** The Rapid Science Orbit (RSO) is introduced into CHAMP gravity field processing as an auxilliary tool in order to facilitate the screening and to establish pre-estimated dynamical parameters, especially accelerometer calibration factors. It is furthermore shown how so-called pseudo-attitude angles help to edit the quaternions delivered by the CHAMP on-board star tracker cameras.

**Key words:** CHAMP, Rapid Science Orbit, accelerometer, attitude

## 1 Introduction

Gravity field processing at GFZ is presently done in 1.5 day batches [5]. For every one of those batches, first the so-called *screening runs* are performed, then the accumulation of arc-wise normal equations comprising the gravity field parameters follows. The purpose of the screening runs is to produce good orbits that are then plugged into the device that accumulates normal equations for each 1.5 day arc. This paper concentrates on some aspects of the separation of GPS data editing and the simultaneous establishment of *kinematic* parameters (like ambiguities, receiver clock offsets) and cycle slip search on the one hand from the dynamical parameter adjustment, especially the accelerometer calibration factors on the other hand. We also want to highlight some features of attitude data editing for the screening process.

## 2 Some remarks on the screening process

The arc-wise accumulation of normal equations is preceded by so-called *screening runs*. The solve-for parameters in these runs do not comprehend the gravity field expansion coefficients. The adjustment method applied is the so-called 2-step process [4], this is basically satellite-to-satellite tracking (SST) with separate, consecutive runs, first for the adjustment of sender ephemeris, and then that of the receiver satellite. The measurement types used are space-borne GPS code and phase observations only. The purpose of the screening runs is two-fold. On the one hand, there is the viewpoint of GPS-SST data editing. The orbit adjustment program EPOSOC of GFZ is capable of detecting and removing outliers, it can find cycle slips and estimate values for ambiguities and receiver clock offsets. On the other hand, there is the aspect of dynamical parameter adjustment, with a special focus on the acclerometer calibration factors. Conventionally, both tasks are accomplished in the screening process at the same time in one sweep in an interleaving manner. Dynamical parameter adjustment occurs together with the data editing. For arcs as short as 1.5 days this is perfectly legitimate. In the context of recent experiments, however, it has been discussed to extend the lengths of the individual screening arc to 4.5 days. Here unfavourable interactions between the two tasks are to be expected. Therefore, a new method had to be adopted.

## 3 Preprocessing via the RSO

The two tasks, data screening and parameter estimation, may effectively be de-correlated by using the so-called *Rapid Science Orbit* (RSO) [4] as an auxilliary tool. It is available to the external user via the CHAMP Info System and Data Center (ISDC) [1]. The RSO is reduced dynamically; typically with a considerable number of empirical parameters estimated. From the analysis of down-weighted SLR measurements we know that it has an accuracy of some 4-5 cm [2].

The method employed is presented in Fig. 1. The left side of the diagram describes the GPS-SST data cleaning process and the estimation of kinematic parameters. OBS denotes the types of observations involved, EST denotes the parameters to be estimated.

For the time interval of interest, the GPS sender arcs (1.5 day arc lengths, 30 s time steps) are merged with a CHAMP trajectory which is glued together from 14 hour snippets of the CHAMP RSO. The orbits of the GPS satellites, their clock offsets and the CHAMP orbit enter the adjustment process as fixed quantities. As a consequence, all dynamic orbit parameters drop out of the picture. Only the ambiguities and the receiver clock offsets remain. This step produces edited GPS-SST data where (hopefully) all cycle slips are found and marked and all outliers are removed. Furthermore, on the data records, entries for sender and receiver clocks as well as for ambiguity values are added.

The right side of Fig. 1 shows the data flow for the dynamic part of the preparatory dynamical orbit adjustment. The dynamical parametrization is just like that of the conventional screening. The GPS-SST observation data, however, are replaced by the CHAMP RSO orbit, the spacecraft positions being pseudo-observations. Again, the observed orbit here is glued together from 14 hour CHAMP trajectory segments. This step produces adjusted dynamical parameters including accelerometer calibration factors.

Both steps are, due to the role of the RSO, independent. Via the data editing part, clean GPS-SST data files can be produced for any period of interest

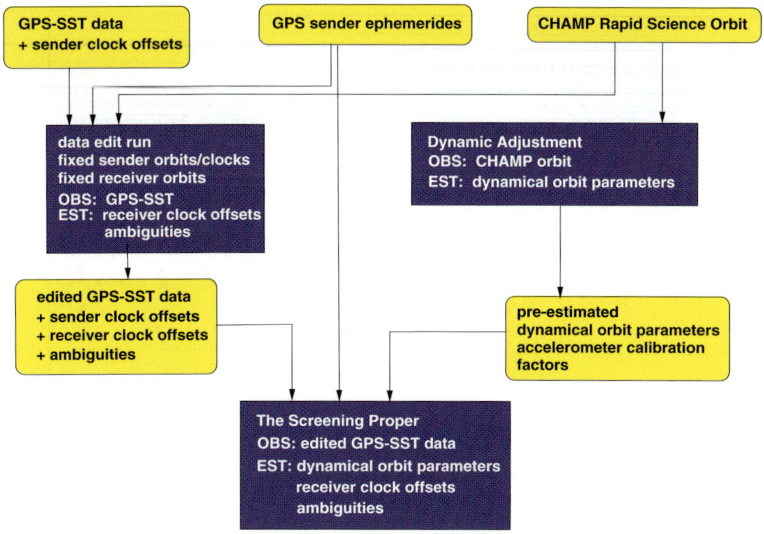


**Fig. 1.** The preprocessing procedure adopting the Rapid Science Orbit.

independently from the dynamic part and stored somewhere for subsequent use. Likewise, adjusted dynamical parameters may be produced without the need to wait for the GPS-SST data to be ready.

In the lower part of Fig. 1 then comes the screening proper. The parameter estimates from the right side of the diagram are taken together with the pre-screened GPS-SST data produced on the left side of the diagram. From this point on, conventional screening continues.

If, for a certain period of interest, the RSO is not available, the processor is capable of substituting a CHAMP kinematic orbit. An extension to the case of GRACE is possible, where the role of the RSO is taken over by the Level-1B product provided by JPL for both GRACE satellites [6]. However, these auxilliary capabilities will not be discussed here.

## 4 Adjusted accelerometer calibration factors

In order to illustrate one of the effects of introducing the RSO as an auxiliary tool, we show in Fig. 2 a plot of the along-track scaling factors of the accelerometer. The time horizon shown is the whole month of April, 2003. Most realistic and closest to one are the factors when CHAMP's trajectory is adjusted to the RSO, whereas for the conventional screening, when GPS data are used, the result is often strange and far away from 1. Moreover, the factors change wildly between individual arcs. If, however, arc-wise normal equations are accumulated from the same arcs for the whole month, and the accelerometer calibration factors are solved for together with the gravity field

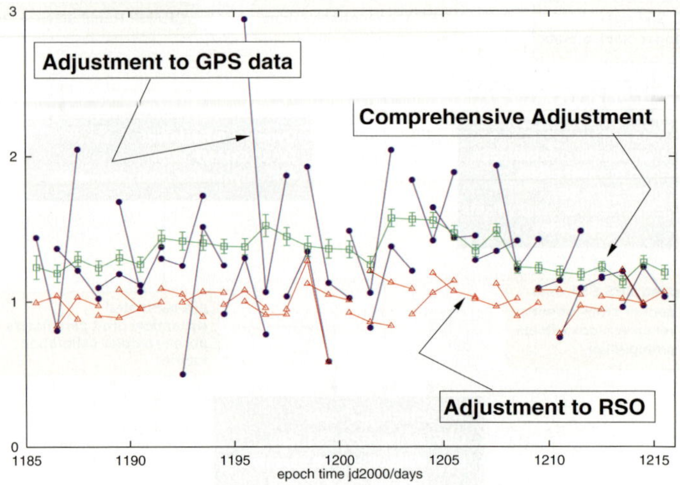


**Fig. 2.** Along-track calibration factors.

expansion coefficients, the appearance of the scaling factors is smooth again, but they do not settle around 1, but rather at a value between 1.2 and 1.3.

## 5 Some aspects of attitude quaternion treatment

A given time series is only then fit for being interpolated if it is smooth enough. It is furthermore only then fit for outlier removal if its individual undistorted members are small enough, and if the outliers are visibly larger than typical members of the time series.

The attitude quaternion time series, as delivered by the star tracker cameras, has neither of both properties: As it describes the rotation from the inertial reference frame to the satellite body frame, it is a macroscopic quantity, and most certainly not small. And it is not smooth, for quaternion time series are known to have jumps and cusps.

Much better candidates for gap filling and outlier removal are the yaw, pitch and roll angle of the CHAMP spacecraft, i.e. the angles of the true attitude vs. a nominal one; nominal meaning in this context that the satellite body axes are aligned with certain nominal directions like the radial, the orbit normal or the along-track direction of the spacecraft motion. Yaw, pitch and roll are small, for they are controlled by the on-board software to remain within a margin of 2 to 3 degrees [3]. Furthermore, those angles are continuous, as the satellite attitude does not jump wildly. These angles oscillate with low frequency between their maximal values.

The idea for quaternion gap filling and outlier removal now goes like this: From the spacecraft true attitude (expressed in quaternions), the nominal

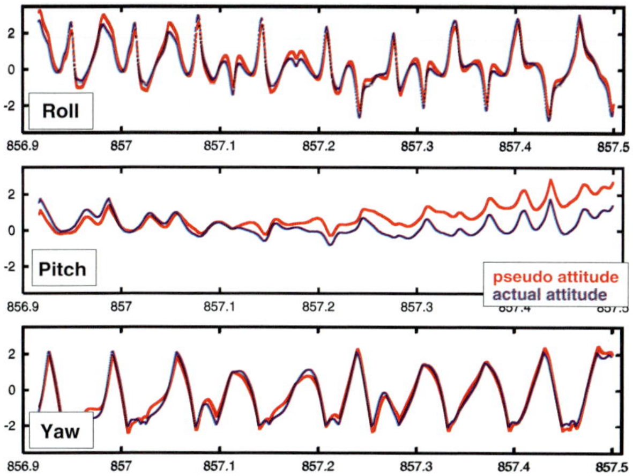


**Fig. 3.** True roll/pitch/yaw vs. pseudo-angles.

attitude is removed, and the small corrective attitude angles yaw, pitch and roll remain. From these small angles, outliers are eliminated and gaps are closed via brute-force linear interpolation. The corrected yaw, pitch and roll angles are then added back to the nominal attitude to give the "true" attitude with clean data and gaps closed.

This approach has one drawback: The nominal attitude is derived from the orbit state vector. It is necessary to have a time series for position and velocity, from which the nominal attitude matrix can be inferred via trivial geometrical operations (cross product, scalar product).

In the case of CHAMP, however, this can be avoided. In nominal orientation, the last column of the matrix transforming from the inertial reference frame to the satellite body frame is the negative radial direction of CHAMP's orbit. It is now easy to show that, if an ellipse-like curve described by a small number of parameters is fitted through the time series of those last matrix columns, a pseudo-trajectory results that - at least for the derivation of the nominal attitude - is as good as CHAMP's trajectory itself. It is therefore not necessary to pre-compute a CHAMP orbit to derive a nominal attitude matrix. Fig. 3 shows the comparison of the true attitude angles vs. the ones derived from the above-mentioned pseudo-trajectory. The only visible deviation occurs in the pitch axis.

Now to the gap filling: Fig. 4 shows the following experiment: Into a given quaternion time series, unrealistically large artificial random gaps have been introduced. Then those gaps have been closed via the described procedure with the brute-force interpolation of the pseudo-angles. The original gap-free yaw angle is finally plotted against the reconstructed yaw. Despite the fact that there are some distortions - when the gaps are too large - the angle time series exhibit a nearly 100% match.

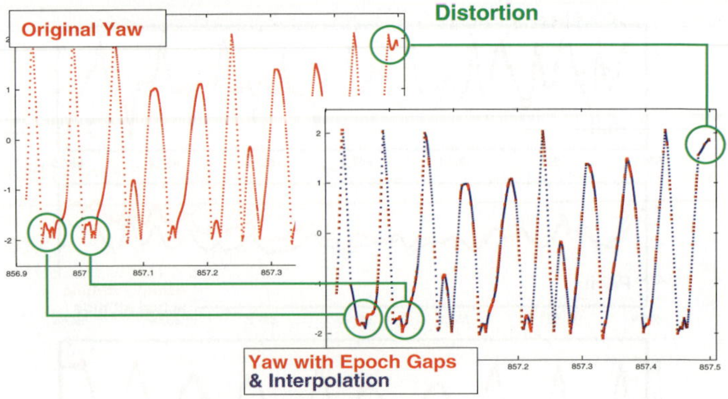


**Fig. 4.** Interpolated yaw angle.

## 6 Conclusions

Introducing the rapid science orbit (RSO) as an auxilliary tool in the screening process admits a separation of GPS data editing on the one hand from a preliminary adjustment of dynamical parameters on the other hand. One of the visible benefits are more realistic starting values for the along-track scaling factors of the accelerometer already in the screening process prior to the gravity field recovery.

A central tool for attitude quaternion editing is the introduction of so-called pseudo-attitude angles that closely mimick yaw, pitch and roll. This can be achieved using the attitude quaternions entirely alone, no preliminary CHAMP orbit is necessary.

## References

[1] *http://isdc.gfz-potsdam.de*

[2] König R, Michalak G, Neumayer KH, Schmidt R, Zhu SY, Meixner H and Reigber Ch (2004) Recent Developments in CHAMP Orbit Determination at GFZ. This issue.

[3] *http://op.gfz-potsdam.de/champ/systems/index_SYSTEMS.html*

[4] Michalak G, Baustert G, König R, Reigber Ch (2003) CHAMP Rapid Science Orbit Determination - Status and Future Prospects. In: Reigber C, Lühr H, Schwintzer P (eds), First CHAMP Mission Results for Gravity, Magnetic and Atmospheric Studies, Springer, Berlin Heidelberg: 98–103

[5] Reigber Ch, Jochmann H, Wünsch J, Petrovic S, Schwintzer P, Barthelmes F, Neumayer KH, König R, Förste C, Balmino G, Biancale R, Lemoine J-M, Loyer S, Perosanz F (2004) Earth Gravity Field and Seasonal Variability from CHAMP. This issue.

[6] Case K, Kruizinga G and Wu S (2002) GRACE Level 1B Data Product User Handbook. JPL Publication D-22027.

# Evaluation of the CHAMP Accelerometer on Two Years of Mission

Félix Perosanz[1], Richard Biancale[1], Jean Michel Lemoine[1], Nicole Vales[1], Sylvain Loyer[2], Sean Bruinsma[1]

[1] CNES/GRGS, 18 avenue Edouard Belin, 31400 Toulouse, France, *Felix.Perosanz@cnes.fr*
[2] Noveltis, 2 avenue de l'Europe 31520 Ramonville, France

**Summary.** This paper presents a synthesis of the CNES/GRGS routine activities regarding the analysis and calibration of the CHAMP accelerometer data. The impact of instrument operations and out of specification temperature variation on bias and scale factor estimation is demonstrated. Precise dynamic solutions of orbits and calibration parameters have been obtained from the processing of two years of data. The results presented here could be obtained thanks to the high performances of the accelerometer and GPS tracking.

**Key words:** CHAMP, accelerometer, calibration, orbit, dynamic, GPS, SLR

## 1 Introduction

Measuring accelerations variations of few $10^{-9}$ $ms^{-2}$ in space remains a challenge. A full understanding of the instrument and of its environment as well as a continuous surveying of its output is needed. CNES/GRGS is involved in the analysis and calibration of the accelerometer since the beginning of the mission. A continuous effort is done to improve the calibration strategy in order to estimate high accuracy dynamic orbit solutions and reach the scientific objectives of the gravity mission of CHAMP.

This presentation focuses first on the accelerometer signal status and analysis. Then the calibration definition and estimation strategy is given. Bias and scale factors solutions based on two years of data processing are presented. An alternative definition of the bias, which reduces its correlation with the scale factor, is also proposed.

## 2 Measurement status

### 2.1 Instrument analysis

The main concept of the Star accelerometer is reminded in Figure 1. For a full description refer to [Perret et al. 2001]. The polarization voltage $V_{proof}$ is applied to the charged proof mass which levitates in the electrostatic cage of the Sensor Unit (SU). Voltages from 6 pairs of electrodes maintain the proof mass at the centre of

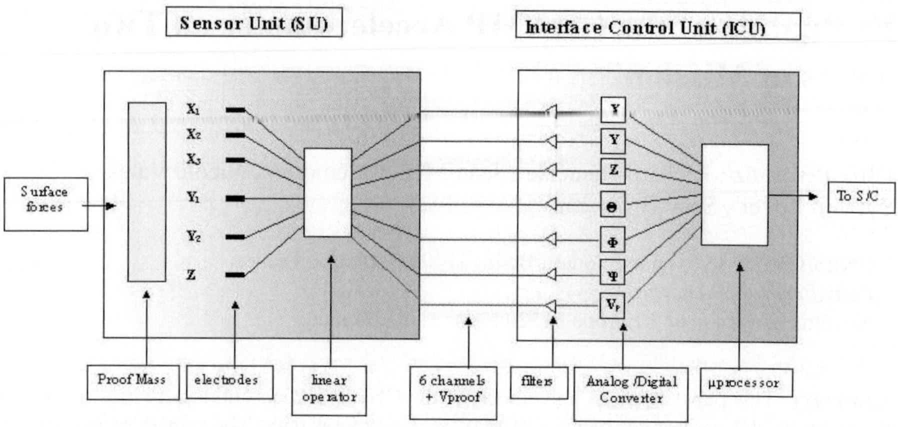


**Fig. 1.** Accelerometer block diagram.

the electrostatic cage. These voltages are proportional to the surface forces acting on the satellite and provide the linear and angular acceleration vectors through a linear combination. The 6 components of these vectors as well as the $V_{proof}$ voltage are then filtered and converted into the Interface and Control Unit (ICU) and finally delivered to the Space Craft (S/C). $V_{proof}$ is hardware controlled to be constant. However both the SU and the ICU introduce in the transfer function, biases and scale factors which are sensitive to instrumental operations (on/off, reboot, ICU calibration test…) and environment (temperature, satellite attitude and manoeuvres…). Finally the 7 measurement channels (6 acceleration and $V_{proof}$) maybe affected by SU and ICU calibration parameters variations.

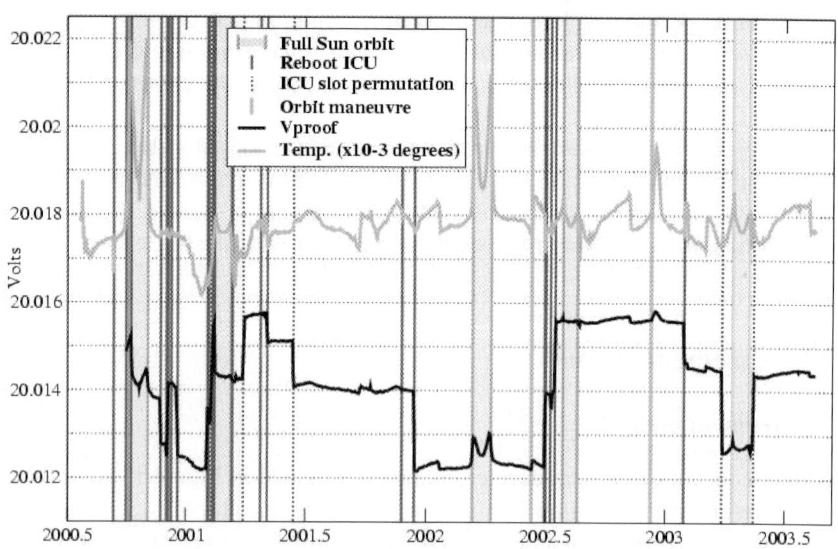


**Fig. 2.** Vproof voltage variations from May 2000 to May 2003.

### 2.2 Surveying the polarization voltage $V_{proof}$

We have represented in Figure 2 the temporal variations of $V_{proof}$ since the beginning of the mission. Several ICU operations like reboots, and slot permutation (nominal to spare) as well as the on board temperature (scaled by $10^{-3}$) are also mentioned. $V_{proof}$ voltage at the output of the accelerometer is obviously not constant and clearly correlated with the instrument operations and out of specification temperature changes which have occurred during full sun orbit periods. Similar variations of the calibration parameters of the acceleration measurement channels may be expected. Nevertheless no secular drift of $V_{proof}$ is observed.

### 2.3 Signal jumps

Signal "jumps" appear on both linear and angular acceleration channels since the beginning of the mission. These discontinuities are too fast (less than one second) and too large (few $10^{-8}$ m.s$^{-2}$ and few $10^{-7}$ rad.s$^{-2}$) to have a dynamic origin. Even if no clear instrumental explanation can yet be given, series of jumps seem to be concentrated after ICU operations and are correlated with satellite eclipse events (very few jumps during full sun orbit). Furthermore, even if the sign changes from one jump to the other, the accumulated signal systematically results in a drift. For example the sum of the observed jumps of the tangential linear acceleration channel gives, after 4 months, a variation of $10^{-6}$ m.s$^{-2}$ which is one order of magnitude larger than the bias variation observed on the 3 years of mission (see section 4.1). One explanation could be that after each jump the signal slowly comes backs to its original level so the secular instrumental bias is not affected.

## 3 Data processing

### 3.1 Calibration equation

The calibration equation links the surface acceleration (SF) acting on the proof mass with the accelerometer output (ACC) as it is delivered to the scientific community. It can be defined by :

$$\gamma_{ACC} = Scale_{SU+ICU} \cdot (\gamma_{SF} + Bias_{SU+ICU}) \quad (1)$$

with:

$$Bias_{SU+ICU} = Bias_{SU} + Bias_{ICU} \cdot Scale_{SU} \quad (1a)$$

$$Scale_{SU+ICU} = Scale_{SU} \cdot Scale_{ICU} \quad (1b)$$

The very conception of the instrument makes its a priori on ground calibration impossible. The accelerometer bias and scale factor have to be a posteriori estimated through dynamic orbit adjustment [Perret et al. 2001]. It is clear from equations (1a&b) that the SU and ICU contribution cannot be separated and only $Bias_{SU+ICU}$ and $Scale_{SU+ICU}$ will be estimated. In addition, calibration parameters are sensitive to on board operations and out of specification temperature variations as seen in chapter 2.2. Furthermore, the accelerometer itself is specified to be stable in a frequency bandwidth limited to $10^{-1}$-$10^{-4}$ Hz [Grunwaldt 1998].

Several strategies have been evaluated to estimate the global accelerometer calibration parameters. We propose to solve for one bias and one scale factor per satellite revolution using the initial values given in Table 1.

| | R | T | N |
|---|---|---|---|
| A priori BIAS ($m/s^2$) | -0.84 e-4 | 0.3555 e-5 | 0.43 e-4 |
| A priori SCALE | 0.83 | 0.83 | 0.83 |

**Table 1.** Initial bias and scale factor values for the period May 2001 - May2003.

### 3.2 Data processing

Calibration parameters are estimated through a classical least-squares adjustment in which the residuals between GPS Satellite to Satellite Tracking (SST) observations and a modeled orbit are minimized. The dynamic approach consists in estimating CHAMP's positions from the numerical integration of, on one hand modeled gravitational accelerations, and on the other hand, measured surface accelerations. We processed zero-difference pseudo-range and phase GPS data with the CNES/GRGS GINS POD software. GPS satellites orbits and clocks at a 30 second sampling were fixed to a priori solutions. More than 500 one-day arcs have been computed from May 2001 to May 2003. Down-weighted SLR observations were also included in the processing in order to get an objective checking of the orbit quality. The global RMS of phase and SLR observations residuals is respectively, 1.0 and 4.6 centimetres.

## 4 Calibration results

### 4.1 Solutions

Figure (3a) and (3b) represent the one per revolution bias and scale factor solution from the 500 arcs. Continuity constraints of $10^{-7}$ $ms^{-2}$ and 5% have been applied. All the solutions are remarkably stable (even if significant variations at the scale of a few weeks clearly show up) except for the radial bias that is obviously drifting. Its deviation exceeds $10^{-6}$ $ms^{-2}$ in two years while the expected physical signal (lift acceleration) should not vary by more than $10^{-8}$ $ms^{-2}$. Even if most of the effects of the X3 electrode failure have been corrected in level-2 accelerometer data [Perosanz et al. 2003], this remaining drift clearly belongs to this SU anomaly. Nevertheless, estimating new calibration parameters every satellite revolution compensates for this effect.

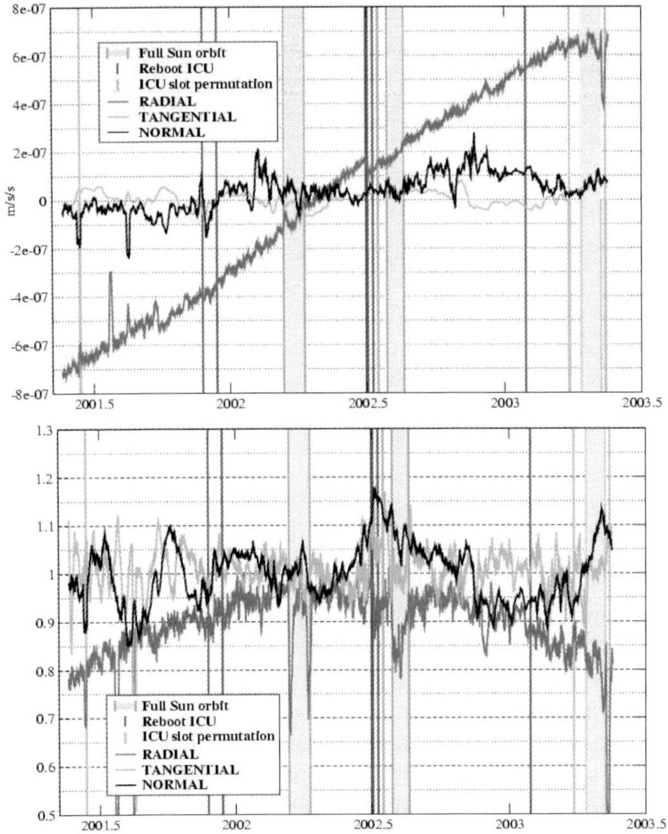


(a)

(b)

**Fig. 3.** Bias (a) and scale factor (b) solution from May 2001 to May 2003.

## 4.2 Bias and scale correlation

It can be anticipated from equation (1) that bias and scale solutions are correlated. This is even more critical when calibration parameters are applied to data that contain a systematic signal like the tangential linear acceleration (atmospheric drag). In such case, scale factor variations applied to the mean signal level has the same signature as bias variations. We could verify that the two sets of solutions were clearly correlated.

One alternative is to change the definition of calibration parameters and include in the bias the mean level of the considered signal. Figure 4 compares the classical "raw solution" to this new "scale correction solution" for the tangential bias. The apparent noise of the series is reduced so that a new signal signature seems to appear. During the full sun orbit period, for example, the series of biases seems to be correlated with the on board temperature changes.

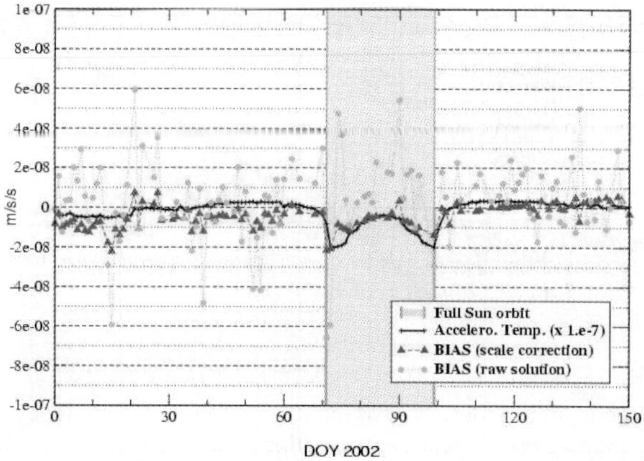


**Fig. 4.** Correlated and un-correlated tangential bias.

## 5 Conclusions

Our evaluation is based on more than two years of data analysis. The STAR accelerometer on board CHAMP is still working remarkably well with high performances and continuous data availability. High quality dynamic orbit are computed on a routine basis at CNES/GRGS.

However, a very careful calibration procedure is needed and calibration parameters must be estimated every satellite revolution in order to take into account the drift of the Radial signal and the sensitivity to satellite environment and operations (ICU reboots, out of specification temperature variations…).

In addition, the sets of tangential calibration parameters are clearly correlated. Including the mean signal in the definition of the biases, helps to decorrelate them from the scale factors.

*Acknowledgments.* We gratefully acknowledge GFZ and AIUB teams for providing GPS satellites orbits and clocks solutions.

## References

Grunwaldt L (1998) CHAMP accelerometer specification. GFZ note CH-GFZ-SP-0022.

Perret A, Biancale R, Camus AL, Lemoine JM, Fayard T, Loyer S, Perosanz F, Sarrailh M (2001) STAR commissioning phase calibration validation activities by CNES. Vol 1& 2, DSO/ED/TU/EI 65-2001, CNES, France.

Perosanz F, Biancale R, Loyer S, Lemoine JM, Perret A, Touboul P, Foulon B, Pradels G, Grunwald L, Fayard T, Vales N, Sarrailh M (2003). On board evaluation of the STAR accelerometer, in: Reigber C, Lühr H and Schwintzer P (Eds): First CHAMP Mission Results for Gravity, Magnetic and Atmospheric Studies, Springer-Verlag, ISBN 3-540-00206-5, 11-18.

# A New Method to Detect and Estimate CHAMP Clock Bias Change and Cycle Slip

Bibo Peng, Bin Wu, Jun Li, Houze Hsu

Institute of Geodesy and Geophysics, Chinese Academy of Sciences, Xudong Road 174, Wuchang, Hubei, P.R. China, *bobby@asch.whigg.ac.cn*

**Summary:** A new method to estimate CHAMP satellite-borne GPS receiver clock bias change is presented in this paper, according to the following approach. The difference between two neighboring epoch phase observations includes real CHAMP-GPS distance change, CHAMP clock receiver bias change, cycle slip and other corrections. Real CHAMP-GPS distance change can be obtained at a relatively high precision from a dynamic orbit. As a result, the CHAMP GPS receiver clock bias and cycle slip can also be determined at a high precision.

**Key words:** CHAMP, clock bias, cycle slip

## 1 Introduction

The observed distance of CHAMP satellite-borne GPS receiver is affected by several main sources such as GPS satellite clock bias, CHAMP GPS receiver clock bias, ionosphere, multi-path, and additional cycle slips for carrier phase distance(D. Svehla et al. 2003). For orbit determination in a high precision, all effects from these sources should be correctly detected and removed. Multi-path effect on the observed distance is in centimeters, and can be reduced by special designed antenna and GPS receiver. Ionosphere effect on the observed distance reaches meters, and can be decreased to a few centimeters by using dual-frequency observing data. GPS satellite clock bias may be hundreds microseconds, about tens of kilometers when converted to distance, but it can be obtained from the SP3 ephemeris file in a precision of centimeters. For CHAMP GPS receiver clock bias, it may have an effect of hundreds meters on the observing distance, and must be estimated and removed by using differential data during the process of orbit determination.

In this paper, we proposed a new strategy to detect CHAMP GPS receiver clock bias changes and cycle slips.

## 2 Theory

The observation equation for CHAMP zero differenced carrier phase measurements between CHAMP GPS receiver and GPS satellite can be written in an ionosphere-free linear combination as (Blewitt, 1990):

$$\rho_i = \frac{1}{f_1^2 - f_2^2}\left(f_1^2 L_{1i} - f_2^2 L_{2i}\right) - \frac{1}{f_1^2 - f_2^2}\left(f_1^2 \lambda_1 N_{1i} - f_2^2 \lambda_2 N_{2i}\right) - ct_{ri} + ct_{si} - \delta_{reli} - \delta_{rmassi} - \delta_{smassi} \tag{1}$$

where:

| | |
|---|---|
| $i$ | epoch index |
| $\rho$ | geometric distance |
| $c$ | light speed |
| $t_r$, $t_s$ | receiver clock bias and GPS clock bias |
| $f_1$, $f_2$ | GPS signal frequency |
| $L_1$, $L_2$ | phase measurements |
| $N_1$, $N_2$ | ambiguity |
| $\delta_{rel}$ | relativistic correction |
| $\delta_{rmass}$, $\delta_{smass}$ | mass center correction. |

All right terms of Equation (1) can be obtained, except receiver clock bias and ambiguity $N_1$, $N_2$. The geometric distance change between two neighboring epochs can be written as:

$$\Delta\rho = \frac{1}{f_1^2 - f_2^2}\left(f_1^2 \Delta L_1 - f_2^2 \Delta L_2\right) - \frac{1}{f_1^2 - f_2^2}\left(f_1^2 \lambda_1 \Delta N_1 - f_2^2 \lambda_2 \Delta N_2\right) - c\Delta t_r + c\Delta t_s - \Delta\delta_{rel} - \Delta\delta_{rmass} - \Delta\delta_{smass} \tag{2}$$

Δ denotes the change between two neighboring epochs. From the Equation (2), one can recognize that the change of the phase observations between two neighboring epochs includes real CHAMP-GPS distance change, CHAMP clock receiver bias change, cycle slip and other corrections. Theoretically, if geometric distance change can be obtained, the receiver clock bias change and cycle slip also can be detected and determined, because CHAMP clock bias change affects all GPS measurements at one observing epoch, while cycle slip only affects single measurement.

Fortunately, we can use a dynamic method to integrate the orbit to simulate the geometric distance change (Mader, 1986). However, at first, we must assume that the observing orbit and the integrated orbit are different. Then, we should find out the condition, under which the geometric distance change can be replaced by the integrated value.

The observing distance and the integrated one can be expressed as GPS earth center vector subtracting by CHAMP earth center vector:

$$\vec{\rho}_o = \vec{R} - \vec{r}_o \tag{3}$$

$$\vec{\rho}_c = \vec{R} - \vec{r}_c \tag{4}$$

where $c$ denotes integrated vector, o denotes observing vector. R represents GPS satellite earth center vector. The changes between two neighboring epoch 1 and 2 are

$$\Delta\vec{\rho}_o = \vec{R}_2 - \vec{r}_{o2} - (\vec{R}_1 - \vec{r}_{o1}) \\ = \vec{R}_2 - \vec{R}_1 - (\vec{r}_{o2} - \vec{r}_{o1}) \quad (5)$$

$$\Delta\vec{\rho}_c = \vec{R}_2 - \vec{r}_{c2} - (\vec{R}_1 - \vec{r}_{c1}) \\ = \vec{R}_2 - \vec{R}_1 - (\vec{r}_{c2} - \vec{r}_{c1}) \quad (6)$$

or

$$\Delta\vec{\rho}_o = \vec{R}_2 - \vec{R}_1 - \Delta\vec{r}_o \quad (7)$$

$$\Delta\vec{\rho}_c = \vec{R}_2 - \vec{R}_1 - \Delta\vec{r}_c \quad (8)$$

respectively.

The CHAMP earth center vector at the second epoch can be simply given by

$$\vec{r}_2 = \vec{r}_1 + \dot{\vec{r}}_1 t + \tfrac{1}{2}\vec{f}_1 t^2 \quad (9)$$

and, consequently

$$\Delta\vec{r} = \dot{\vec{r}}_1 t + \tfrac{1}{2}\vec{f}_1 t^2 \quad (10)$$

Of course, the actual acceleration is not a constant. We adopt this equation, because it helps us to understand easily that the vector change is not sensitive to the first epoch position vector but to the velocity.

For short time orbit integration, it's not necessary to consider non-conservative forces. This means that the force acting on CHAMP depends on position vector only. The force is directly proportional to $1/r^2$, so that it is not sensitive to the first position vector. Since it's easier to obtain velocity vector at high precision by kinematic method from code and phase observing data, from equation (10), we can draw the conclusion that $\Delta\vec{r}_c$ can be replaced by $\Delta\vec{r}_o$ under the condition of position vector roughly determined and velocity vector determined at high precision, or in other words, the geometric distance change can be obtained from dynamic integration.

## 3 Strategy

At each epoch, the approximate CHAMP position vector can be calculated from a kinematic method. The position vector change between two neighboring epochs is obtained from dynamic integration. From equation (2), the combination of CHAMP clock bias change and cycle slip effect reads

$$c\Delta t_r + \frac{1}{f_1^2 - f_2^2}(f_1^2\lambda_1\Delta N_1 - f_2^2\lambda_2\Delta N_2) = \\ \frac{1}{f_1^2 - f_2^2}(f_1^2\Delta L_1 - f_2^2\Delta L_2) - \Delta\rho + c\Delta t_s - \Delta\delta_{rel} - \Delta\delta_{rmass} - \Delta\delta_{smass} \quad (11)$$

At one observing epoch, we can calculate CHAMP clock bias change and cycle slip effect from the above equation for each GPS satellite measurement. If the number of observations is more than 3, we could decide which observation has no cycle slip, because cycle slip only affects one GPS observation. If there is no cycle slip in the observations, the right terms of equation (11) will be only the clock bias change. Therefore, we can estimate clock bias change for the epoch, and the cycle slip effects by subtracting the clock bias change for the cycle slip observation. Moreover, even in the case of one or two observations at one epoch, we could obtain clock bias change, under the condition of no cycle slip in the observation.

## 4 Data

GPS-SST data and GPS SP3 ephemeris for one day on Jan 3, 2002 was chosen to calculate receiver clock bias change and cycle slip. The first epoch state vector was chosen from GPZ ISDC CHAMP precise science orbit. In order to compare the results, the GPZ ISDC CHAMP precise science orbit was chosen to simulate the real champ orbit, and to estimate receiver clock bias change and cycle slip.

## 5 Results

Table 1 shows the results by using the above method and data. The interval between two neighboring epochs is 10 seconds. Results of the column $c\Delta t_r$ (GFZ) are calculated from the GFZ precise science orbit and the pre-processed observation data with no cycle slip effects. Thus, only CHAMP clock bias changes are remained in the column.

For each epoch, there are differences in several centimeters between different GPS measurements. These differences exist not only in our results, but also in results from the GFZ precise science orbit. The reason is that the GPS clock bias we used is taken from IGS SP3 ephemeris, and the precision is at a level of centimeters. If there is better GPS clock bias data, we assure that the differences could be decreased.

Although there exist effects of GPS clock bias, most of our results are still very close to those from GFZ precise science orbit. It indicates that our method is effective in CHAMP clock bias change determination.

For epoch 2000, the value calculated from GPS 29 measurement is obviously larger than other values, which differ from each other in centimeters. Then, from this fact, it is inferred that there exists a cycle slip in the measurement, or that, at least, there is a spike.

In Table 2, we chose 4 different initial position values to test the effect of initial position error on CHAMP clock bias change. The No.1 initial position was taken from GFZ precise science orbit; the No.2, the No.3 and the No.4 were added 1, 10 and 100 meters in the three directions, respectively. From Tab. 2, we could find that the position error at 1-meter in the three directions only causes millimeters effects on clock bias change, while the position error at 10-meters may cause centimeters effects.

| Epoch index | GPS PRN | $c\Delta t_r+\Delta r_{slip}$(m) | $c\Delta t_r$ (GFZ) (m) |
|---|---|---|---|
| 2 | 4 | 16.067 | 16.075 |
| | 5 | 16.126 | 16.145 |
| | 20 | 16.097 | 16.105 |
| | 29 | 16.088 | 16.116 |
| | 30 | 16.069 | 16.078 |
| 10 | 4 | 23.974 | 23.987 |
| | 5 | 24.004 | 24.016 |
| | 7 | 23.983 | 23.990 |
| | 20 | 23.982 | 23.994 |
| | 29 | 23.990 | 24.010 |
| | 30 | 23.987 | 23.998 |
| 50 | 20 | -27.121 | -27.120 |
| | 25 | -27.107 | -27.090 |
| 100 | 1 | -11.981 | -11.983 |
| | 11 | -11.985 | -11.979 |
| | 13 | -11.967 | -11.987 |
| | 20 | -11.963 | -11.978 |
| 1000 | 3 | 25.446 | 25.462 |
| | 14 | 25.454 | 25.446 |
| | 15 | 25.438 | 25.450 |
| | 17 | 25.439 | 25.473 |
| | 29 | 25.447 | 25.458 |
| 2000 | 14 | 21.261 | 21.309 |
| | 20 | 21.385 | 21.333 |
| | 25 | 21.310 | 21.316 |
| | 29 | 21.741 | 21.310 |
| 5000 | 6 | -16.508 | |
| | 25 | -16.428 | -16.445 |
| | 30 | -16.473 | -16.568 |

**Table 1.** Clock bias change comparison with GFZ results.

| Initial vector | GPS PRN | | | | |
|---|---|---|---|---|---|
| | 4 | 5 | 20 | 29 | 30 |
| 1 | 16.067 | 16.126 | 16.097 | 16.088 | 16.069 |
| 2 | 16.064 | 16.124 | 16.094 | 16.082 | 16.064 |
| 3 | 16.042 | 16.118 | 16.065 | 16.029 | 16.022 |
| 4 | 15.823 | 16.047 | 15.783 | 15.503 | 15.607 |

**Table 2.** Initial vector effects on clock bias change.

## 6 Conclusion

A method to detect and estimate CHAMP clock bias change and cycle slip was developed on the basis of the combination of dynamic and kinematic methods. Results show that CHAMP clock bias change and cycle slip effects can at least be estimated in a precision of centimeters level. If GPS clock bias correction with higher precision is used, results could be better. Up to now, this method has been only used to detect and estimate CHAMP clock bias change and cycle slip, and we hope this method could be used for CHAMP precision orbit determination.

## References

Blewitt G (1990) An automatic editing algorithm for GPS data. Geophys Res Lett *17*: 199-202.

Mader GL (1986) Dynamic positioning using GPS carrier phase measurements. Manus. Geod *11*: 272-277.

Svehla D and Rothacher M (2003) Kinematic and reduced-dynamic precise orbit determination of low earth orbiters. Adv Geosciences *1*: 47 – 56.

# Comparison of Different Stochastic Orbit Modeling Techniques

Adrian Jäggi, Heike Bock, Urs Hugentobler, Gerhard Beutler

Astronomical Institute, University of Berne, Sidlerstr. 5, CH-3012 Berne, Switzerland *adrian.jaeggi@aiub.unibe.ch*

**Summary.** Reduced-dynamic orbit determination for spaceborne GPS receivers is a method promising highest accuracy of the estimated LEO trajectories. We compare the performance of different pseudo-stochastic orbit parametrizations (instantaneous velocity changes and piecewise constant accelerations) and probe the range between dynamic and heavily reduced dynamic orbits. Internal indicators like formal accuracies of orbit positions, comparisons with orbits computed at the Technical University of Munich (TUM), and validations with SLR measurements are used to assess the quality of the estimated orbits. For piecewise constant accelerations comparisons between the estimated and the measured accelerations from the STAR accelerometer allow for an additional and independent validation of the estimated orbits.

**Key words:** Spaceborne GPS Receivers, Reduced-Dynamic Orbit Determination, Pseudo-Stochastic Orbit Modeling, Accelerometer Data

## 1 Introduction

Since the launch of CHAMP on July 15, 2000, the uninterrupted GPS tracking technique of low Earth orbiters (LEOs) has proved to be a reliable method for high quality precise orbit determination (POD). This article focuses on results achieved with a pseudo-stochastic orbit modeling in reduced-dynamic LEO POD, because this technique plays a key role if highest precision is demanded (see [3]). However, the challenging low altitude in the case of CHAMP requires an efficient and flexible pseudo-stochastic orbit model due to the rather large number of parameters involved.

Our approach of LEO POD is based on undifferenced GPS phase tracking data, whereas the GPS satellite orbits and high-rate clock corrections are introduced as known. This leads to a very efficient procedure for estimating LEO orbital parameters, LEO receiver clock corrections, and real-valued phase ambiguities as the only unknowns in a least-squares adjustment. The orbital parameters estimated are the six osculating elements, three constant accelerations in radial, along-track and cross-track directions acting over the whole orbital arc, and the so-called pseudo-stochastic parameters, which are in this article either instantaneous velocity changes (pulses) or piecewise constant accelerations (see [1]). These parameters are called pseudo-stochastic as they are characterized by an expectation value of zero and an a priori variance, which constrains the estimates not allowing them to deviate too much from zero.

## 2 Internal orbit quality assessment

Pseudo-stochastic parameters define the degree of "strength" reduction of the dynamic laws by allowing for a stochastic component in the equations of motion. Fig. 1 (left) shows for day 141/01 the variation of the postfit RMS as a function of equal constraints in three orthogonal directions for both types of pseudo-stochastic parameters set up every 6 minutes using the gravity field model EIGEN-2 ([2]). The similar dependency signifies that both parametrizations may be considered to some extent as equivalent. In both cases looser constraints (heavily reduced dynamic orbits) obviously allow for a better fit, which, however, does not necessarily guarantee the best orbit quality.

It might be more instructive to analyze formal accuracies of the orbit positions rather than postfit RMS values to shed light on the orbit quality. Several differently constrained solutions are subsequently highlighted, based either on accelerations, which are labeled with lower case letters (a $\leftrightarrow 5 \cdot 10^{-8} m/s^2$, b $\leftrightarrow 1 \cdot 10^{-8} m/s^2$, c $\leftrightarrow 5 \cdot 10^{-9} m/s^2$, d $\leftrightarrow 1 \cdot 10^{-9} m/s^2$), or on pulses, which are labeled with capital letters (A $\leftrightarrow 1 \cdot 10^{-5} m/s$, B $\leftrightarrow 5 \cdot 10^{-6} m/s$, C $\leftrightarrow 1 \cdot 10^{-6} m/s$, D $\leftrightarrow 5 \cdot 10^{-7} m/s$). Fig. 1 (right) shows the formal accuracies of the orbit positions (3D) for all solutions based on accelerations for day 141/01. Apart from the less accurate positions at the arc boundaries, which are common to all solutions, some obvious differences between the individual solutions may be observed, like time intervals of worse formal accuracies, especially for the weakly constrained solution (a). Because only 8 satellites could be tracked simultaneously at that time, the number of good observations may be reduced significantly at certain epochs, which affects kinematic and heavily

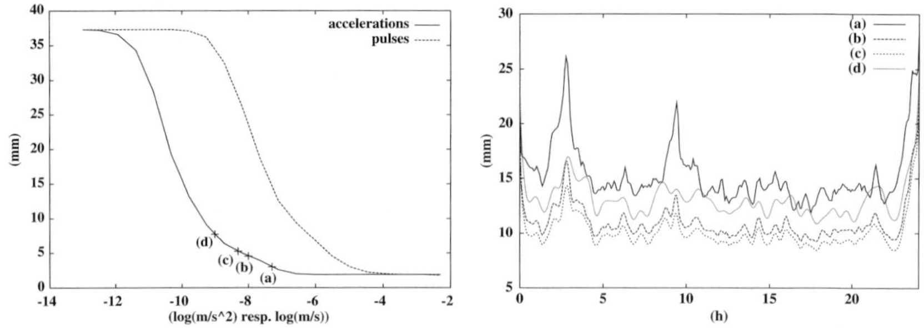


**Fig. 1.** Postfit RMS for day 141/01 as a function of differently constrained pseudo-stochastic parameters set up every 6 minutes using EIGEN-2 (left). Formal accuracies of orbit positions (3D) for differently constrained solutions (see text) (right).

reduced dynamic orbits like solution (a) very much. In contrast the "almost" dynamic solution (d) shows a very smooth (but also not optimal) accuracy curve which is barely affected by a poor geometry. The optimal choice may be found somewhere in-between depending on several factors, like the number of pseudo-stochastic parameters set up, the number of successfully tracked satellites, and the data quality. Detailed investigations for the CHAMP orbit comparison campaign (days 140/01 - 150/01) showed that solution (c) is close to the optimum when using the same constraints in three orthogonal directions (see [1]).

Fig. 2 (left) shows the formal accuracies of the orbit positions (3D) for the solutions (a), (b), and (c) for day 198/02 using the gravity field model EIGEN-2. Apart from the improved accuracy level (8.3 mm compared to 12.2 mm in Fig. 1 (right)) it is evident that even solution (a) shows a significantly improved performance compared to day 141/01 because it takes most benefit from the better tracking conditions (more than 8 satellites) of the CHAMP BlackJack receiver. This leads to an optimal constraining slightly favouring "more" kinematic orbits like solution (b). The differences between both solutions (c) and (b) (7.9 mm RMS of plain orbit differences) are small, however. This fact is illustrated by Fig. 2 (right) showing for a time window of about two revolution periods the actual differences in along-track direction with periodic deviations mostly below the 1.5 cm level. The solid curve in the same figure shows the differences between the solutions (B) and (b) in the same direction. Analyzing formal accuracies, plain orbit differences (0.9 mm RMS), and external comparisons (see next section) shows that the two solutions (B) and (b) must be considered as almost identical. But the sharp cusps every 6 minutes also illustrate the subtle differences between the two pseudo-stochastic parametrizations. Therefore the estimation of accelerations seems to be slightly preferable to get smooth (differentiable) orbits.

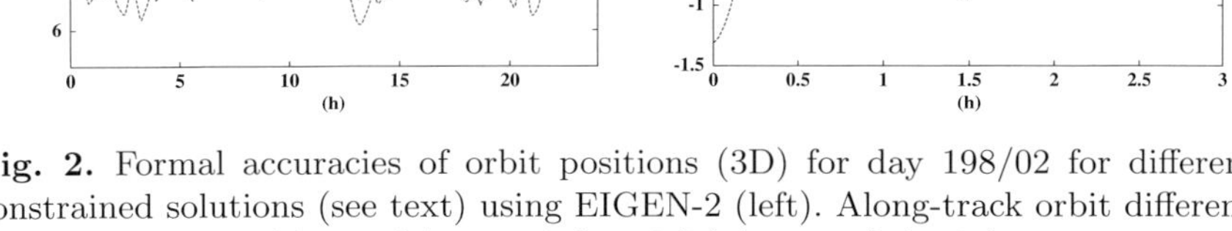


**Fig. 2.** Formal accuracies of orbit positions (3D) for day 198/02 for differently constrained solutions (see text) using EIGEN-2 (left). Along-track orbit differences between solutions (c) and (b) resp. (B) and (b) (see text) (right).

## 3 External orbit quality assessment

In order to assess the overall quality of the estimated orbits, detailed comparisons between our solutions and reduced-dynamic orbits computed at the Technical University of Munich (TUM) (see [3]) were carried out for a time interval of one week. Fig. 3 (left) shows the RMS of plain orbit differences for GPS week 1175 using pulses. The three left bars of each group represent the daily RMS of the comparison for the solutions (A), (B), and (C) w.r.t. (TUM). The three right bars ((A'), (B'), and (C')) represent the results w.r.t. (TUM) for solutions with pulses set up every 15 minutes instead of every 6 minutes. The right figure shows the analogue results for the solutions (a), (b), (c), (a'), (b'), and (c') (i.e., based on accelerations) w.r.t. (TUM).

The best agreement may be achieved for solutions (A) resp. (a) (2.29 cm resp. 2.35 cm mean RMS) implying that the solution (TUM) might be slightly "more" kinematic than our favourite solutions (B) resp. (b). The fact that the comparison favours the solutions (A) and (B) rather than (a) and (b) is simply because the solution (TUM) is generated with pulses as well (see [3]). It is, nevertheless, remarkable that solutions of similarly good quality may be obtained using accelerations with constraints and numbers of parameters varied over a broad range, whereas the solutions based on pulses tend to deviate more rapidly from the solution (TUM) when varying the parameter space in a similar way. The comparatively poor agreement for all solutions of day 195 w.r.t. (TUM) is caused by intentionally not taking into account the attitude information from the star sensors in order to visualize the impact on the orbits (approximately 1.3 cm on the RMS level).

SLR residuals were computed for our solutions for GPS Week 1175 as well. Fig. 4 (above) shows the daily SLR RMS for the different solutions. Most residuals (94%) are below the 6 cm limit without any significant SLR bias. An identical overall RMS of 3.43 cm is achieved for the two best solutions (b) and (B) using a screening threshold of 0.5 m. This RMS level could be

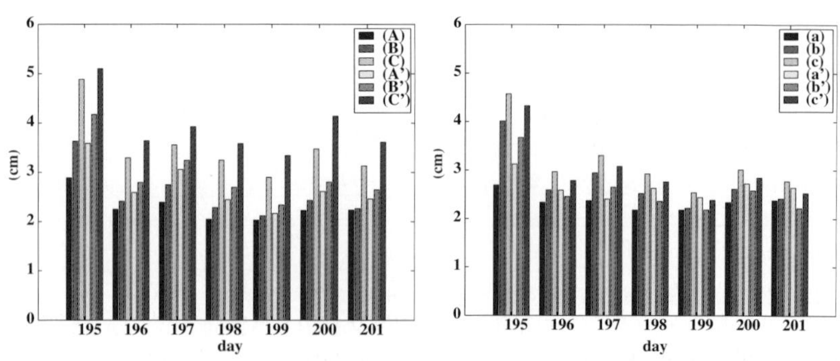


**Fig. 3.** Daily RMS of plain orbit differences for GPS week 1175 for different solutions (left: pulses, right: accelerations, see text) w.r.t. reduced-dynamic orbits from TUM.

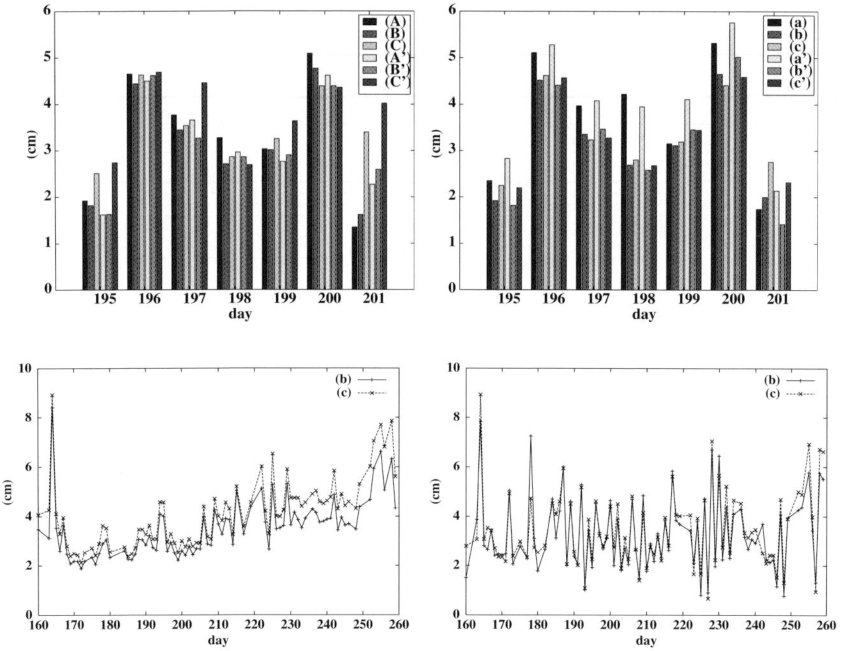


**Fig. 4.** Daily SLR RMS for GPS week 1175 for different solutions (above left: pulses, above right: accelerations, see text). Daily RMS of plain orbit differences for days 160/02 to 260/02 for solutions (b) and (c) w.r.t. (TUM) (below left). Daily SLR RMS for days 160/02 to 260/02 for solutions (b) and (c) (below right).

easily lowered to about 2.5 cm when applying a more restrictive screening procedure removing a few outliers.

Longer data series indicate that similar results may be achieved as for GPS week 1175. Fig. 4 (below) shows for the solutions (b) and (c) for about 100 days (data files having long observation gaps were ignored) the orbit differences w.r.t. the solution (TUM) (left) and the SLR residuals (right). Consistent with Fig. 3 (right) solution (b) agrees better with the solution (TUM) than solution (c) does. The SLR residuals also slightly favour solution (b) (3.30 cm mean of daily SLR RMS) over solution (c) (3.47 cm mean of daily SLR RMS). The observed drift in the comparison to the solution (TUM) might be due to small system inconsistencies (z-shift) which are greatly reduced after performing a Helmert transformation.

If a "good" gravity field model is available the estimated piecewise constant accelerations may be compared with the measured accelerations from the STAR accelerometer. Fig. 5 shows for a time interval of about three revolution periods the agreement (correlation: 94.8%) of the estimated piecewise constant accelerations in along-track direction with the measured accelerations (bias and scale applied) when the gravity field model EIGEN-2 is used.

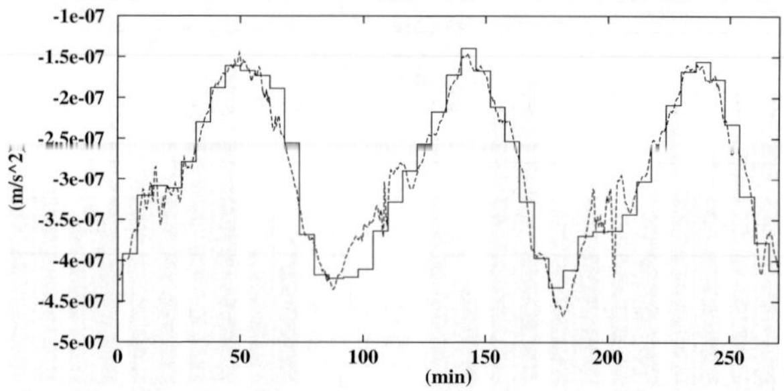


**Fig. 5.** Piecewise constant accelerations for day 198/02 every six minutes in along-track direction compared with accelerometer measurements (bias and scale applied) using the gravity field model EIGEN-2.

# 4 Summary

About 100 days of undifferenced GPS phase tracking data of the CHAMP satellite were processed to analyze the performance of different reduced-dynamic orbit parametrizations. SLR residuals proved that the orbits are accurate on an RMS level of about 3.5 cm, which was also supported by comparisons with reduced-dynamic orbits from the Technical University of Munich. Additionally, estimated piecewise constant accelerations were compared with accelerometer data showing a high correlation (94.8%) in along-track direction. Closer inspection of the agreement for the other directions and a refined piece-wise linear parametrization for the estimated accelerations could allow for a retrieval of accelerometer bias and scale parameters.

*Acknowledgement.* We are grateful to the Institute of Astronomical and Physical Geodesy at the Technical University of Munich for providing CHAMP orbits to support the presented comparisons.

# References

1. Jäggi A, Beutler G, Hugentobler U (2003) Efficient Stochastic Orbit Modeling Techniques using Least Squares Estimators. Presented at IUGG General Assembly, Sapporo, Japan, July 1-11, 2003.
2. Reigber Ch, Schwintzer P, Neumayer K-H, Barthelmes F, König R, Förste Ch, Balmino G, Biancale R, Lemoine J-M, Loyer S, Bruinsma S, Perosanz F, Fayard T (2003) The CHAMP-only Earth Gravity Field Model EIGEN-2. Adv Space Res *31(8)*: 1883-1888 (doi: 10.1016/S0273-1177(03)00162-5).
3. Švehla D, Rothacher M (2003) Kinematic and reduced-dynamic precise orbit determination of CHAMP satellite over one year using zero-differences. Presented at EGS-AGU-EGU Joint Assembly, Nice, France, 2003.

# Determination of Non-Conservative Accelerations from Orbit Analysis

Jose van den IJssel[1], Pieter Visser[1], and Roger Haagmans[2]

[1] Delft Institute for Earth-Oriented Space Research, Delft University of Technology, Kluyverweg 1, 2629 HS, Delft, The Netherlands, *jose@deos.tudelft.nl*

[2] ESA/ESTEC, Keplerlaan 1, 2200 AG, Noordwijk, The Netherlands

**Summary.** It is shown by means of an extensive simulation study as well as an experiment using real CHAMP data that it is feasible to accurately estimate non-conservative accelerations from precise GPS-based orbit perturbations. Assuming the availability of high-precision gravity field models, such as anticipated for GRACE and GOCE, an accuracy of better than 50 nm/s$^2$ seems possible for 30-seconds averaged accelerations. The remaining dominant error sources seem to be GPS receiver carrier-phase noise and GPS ephemeris errors.

**Key words:** CHAMP, GPS, precise orbit determination, non-conservative acceleration, accelerometer

## 1 Introduction

One of the key science instruments aboard CHAMP is the STAR accelerometer, which measures the non-conservative accelerations acting on the spacecraft in order to separate these from the gravitational ones when determining the gravity field from orbit perturbations. However, when a highly accurate gravity model and very precise GPS-based orbit determination are available, it is also possible to extract the non-conservative accelerations from the total accelerations. For satellites equipped with an accelerometer this could provide a good validation check for e.g. the obtained accelerometer calibration parameters and gravity models. For satellites without an accelerometer this could be used for e.g. atmospheric density modeling.

An extensive simulation study has been carried out in order to investigate the feasibility of determining non-conservative accelerations from orbit analysis. The CHAMP mission is used as the basis for this assessment study, which has the advantage that real accelerometer data are available. After a short description of the simulation scenario, the results of the simulation study are presented. Next, the results of experiments using real CHAMP data are shown and the paper concludes with a short summary of the results.

## 2 Simulation scenario

The non-conservative acceleration recovery experiments are largely based on our regular CHAMP precise orbit determination (POD) infrastructure [1].

The core of this infrastructure is the well-known GEODYN software package [5]. The orbit determination strategy is based on a reduced-dynamic approach and uses ionospheric-free triple differenced GPS phase measurements along with precise GPS orbits computed by the International GPS Service. For the assessment study 1 day was selected with a high level of atmospheric perturbations, 25 September 2001. For this day, triple differenced GPS phase measurements have been simulated between the CHAMP satellite, the existing GPS constellation and a network of 50 ground stations, with a data rate of 0.1 Hz. The CHAMP orbit has been simulated using real accelerometer observations for the non-conservative accelerations and state-of-the-art models for the conservative accelerations. This means that all non-conservative force models (drag, solar radiation and albedo) were switched off and replaced by CHAMP accelerometer observations, properly corrected for the advertised biases and scale factors. Full use was made of the observed along-track and cross-track accelerations, but the radial accelerations were put to zero because of the well-known electrode problems causing large biases and drifts in this direction [3]. Although the radial accelerations were put to zero in the simulation of GPS observations, in the recovery process constant empirical accelerations are estimated in all 3 directions, in order to have a realistic set of unknown parameters and the proper correlations.

# 3 Results

## 3.1 Simulations in an error-free environment

In order to assess the model error several recovery experiments are conducted based on error-free observations. The model error is caused by the fact that the simulated orbit is based on 10 seconds CHAMP accelerometer observations, whereas the empirical accelerations are accelerations averaged over their estimation time interval, which is in general larger than 10 seconds. Therefore it is expected that the model error will increase with longer estimation intervals for the empirical accelerations. Furthermore it is known that model errors usually increase with longer orbit arcs. The results shown in table 1 are in agreement with these expectations. The rms of the recovery error clearly decreases with smaller orbit arc and smaller estimation intervals. It can be concluded that for short arcs and estimation intervals the model error is very small, in the order of a few $nm/s^2$ ($10^{-9}$ $m/s^2$), and can be ignored.

In the recovery use is made of a weighted Bayesian least-squares estimator. The recovery tests showed that attention had to be paid to the effect of observation weighting in relation to constraining the range of the empirical accelerations. Figure 1 shows this sensitivity for the last case of table 1. In each case the a priori $\sigma$ of the estimated accelerations is kept fixed at a value of $10^{-6}$ $m/s^2$, which is close to the expected value of the parameters, and the observation data $\sigma$ varies. It needs to be stressed that the optimal value of the observation data $\sigma$ depends on the arc length and the estimation interval.

| 25 September 2001 Case | rms (nm/s$^2$) radial | along | cross | correlation along | cross | 3D orbit error (cm) |
|---|---|---|---|---|---|---|
| 24 hr arc + 20 min interval | 84.23 | 13.92 | 18.10 | 0.9975 | 0.9893 | 3.03 |
| 24 hr arc + 10 min interval | 37.71 | 17.47 | 12.13 | 0.9966 | 0.9882 | 1.57 |
| 24 hr arc + 5 min interval | 34.96 | 13.93 | 12.40 | 0.9981 | 0.9844 | 1.28 |
| 5 hr arc + 5 min interval | 18.94 | 11.35 | 6.06 | 0.9987 | 0.9979 | 0.71 |
| 5 hr arc + 2 min interval | 6.71 | 4.59 | 2.32 | 0.9998 | 0.9996 | 0.12 |
| 5 hr arc + 1 min interval | 5.20 | 3.29 | 1.38 | 0.9999 | 0.9999 | 0.05 |
| 5 hr arc + 30 sec interval | 5.28 | 2.37 | 1.15 | 0.9999 | 0.9999 | 0.05 |

**Table 1.** Error-free simulation recovery results obtained with different orbit arcs and estimation intervals.

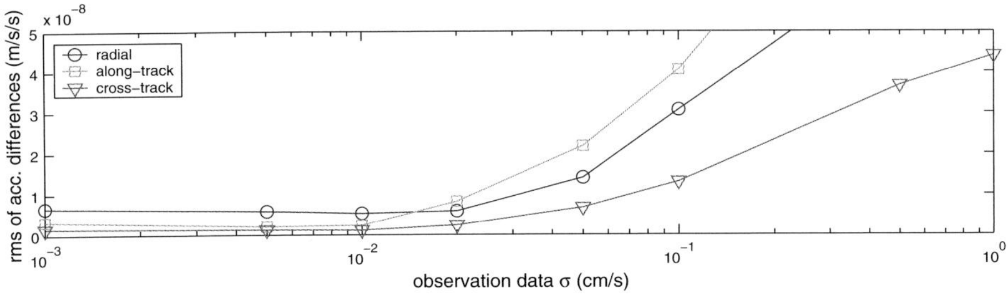


**Fig. 1.** Recovery errors in 3 directions for the error-free simulation using a 5-hour arc and a 30-seconds estimation interval.

The upper part of figure 2 shows for the last case of table 1 the estimated along-track accelerations as a function of time. The orbit arc is taken at the last 5 hours of the day, when a severe geomagnetic storm occurred. In the figure several distinctive peaks are visible, which occur when the satellite flies over the magnetic poles of the Earth. At these polar regions the geomagnetic storm causes sharp atmospheric density fluctuations. For comparison, the true accelerations averaged over the empirical acceleration estimation interval are also shown. Clearly there is a strong agreement between the true and the estimated accelerations. The differences between the true and the estimated accelerations are also shown, and from these differences it is clear that the recovery error is slightly larger during the sharp peaks caused by the storm. The recovery error also shows a small edge effect at the beginning and end of the orbit arc. This is caused by the fact that the orbit is less well constrained at the edge of the arc in the reduced-dynamic orbit determination. To avoid this effect in this study the first and last 40 minutes of each arc are eliminated.

### 3.2 Simulations using realistic error sources

Table 2 shows the effect of several realistic error sources on the recovery accuracy. It is clear that the largest recovery error is caused by the current gravity model error. The CHAMP clone has an accuracy that is expected of current available gravity models that include CHAMP data. When a GRACE clone is used, with an accuracy that is predicted for the GRACE mission [6],

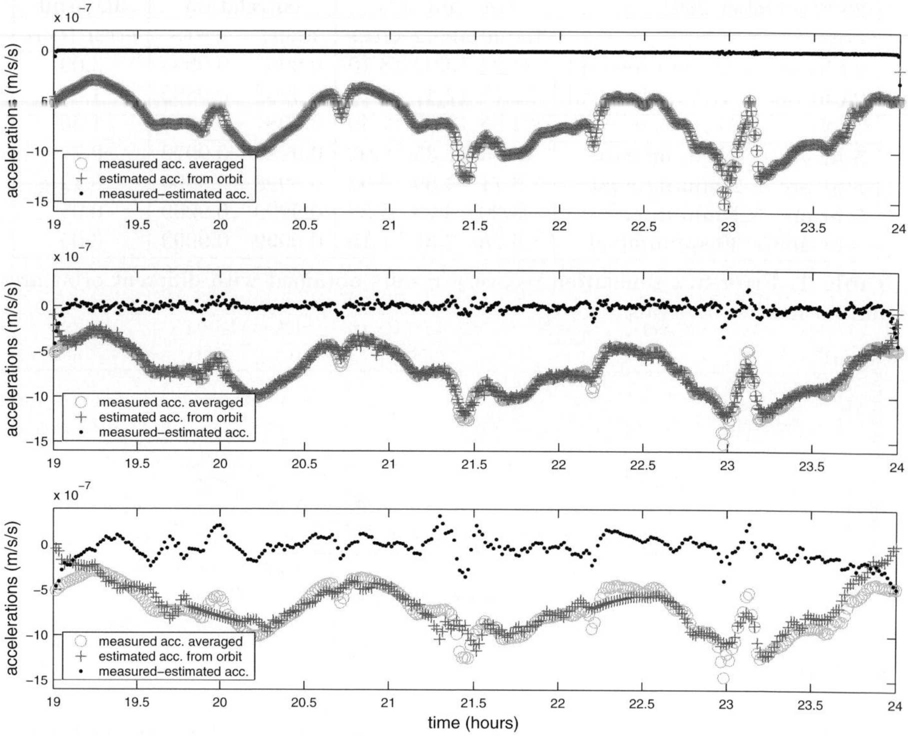


**Fig. 2.** Estimated along-track accelerations as a function of time. The accelerations in the upper and middle figure are obtained in respectively an error-free simulation and a simulation taking all error sources combined into account. In both simulations a 30-seconds estimation interval is used. The lower figure shows accelerations obtained with real CHAMP GPS data and a 1-minute estimation interval.

the recovery error reduces significantly. In that case the largest remaining recovery errors are due to observation noise and GPS ephemeris errors. The last lines of table 2 show that a recovery error of less than 50 $\mathrm{nm/s^2}$ seems possible when all error sources combined are taken into account, assuming a precise post-mission GRACE model is available. The middle part of figure 2 shows the estimated along-track accelerations for the last case of table 2. Compared to the error-free case the recovery error has become larger, with the largest errors again during the sharp peaks. Most of the peaks are quite well determined, however, the sharpest peaks are no longer properly estimated.

### 3.3 Real CHAMP data processing

In addition to the simulation study, several recovery experiments have been conducted using real CHAMP observation data with a 30-seconds time interval. Table 3 shows the results of these tests. The results for the radial direction are very poor, which is due to the well-known accelerometer prob-

| 25 September 2001 Case | rms ($nm/s^2$) radial | rms ($nm/s^2$) along | rms ($nm/s^2$) cross | correlation along | correlation cross | 3D orbit error (cm) |
|---|---|---|---|---|---|---|
| error-free | 5.20 | 3.29 | 1.38 | 0.9999 | 0.9999 | 0.05 |
| noise (0.6 mm iono-free) | 34.84 | 42.94 | 22.17 | 0.9868 | 0.9658 | 1.45 |
| gravity (GRACE clone) | 5.27 | 3.29 | 1.46 | 0.9999 | 0.9998 | 0.05 |
| gravity (CHAMP clone) | 141.83 | 102.19 | 157.40 | 0.9139 | 0.3964 | 0.06 |
| tides | 6.53 | 5.45 | 4.07 | 0.9997 | 0.9991 | 0.05 |
| troposphere | 11.55 | 7.80 | 5.80 | 0.9994 | 0.9976 | 0.18 |
| GPS ephem. (5 cm 1-cpr) | 45.12 | 29.28 | 22.37 | 0.9923 | 0.9657 | 1.93 |
| station (1 cm) | 7.40 | 4.17 | 2.62 | 0.9998 | 0.9996 | 0.21 |
| reference frame | 12.97 | 6.50 | 6.94 | 0.9996 | 0.9980 | 0.42 |
| total (GRACE clone) | 41.60 | 42.90 | 30.10 | 0.9852 | 0.9453 | 1.84 |
| total (GRACE clone)* | 44.75 | 47.91 | 37.26 | 0.9800 | 0.9154 | 1.84 |

**Table 2.** Recovery results for different simulated error sources using a 5-hour arc and a 1-minute estimation interval. The * indicates a 30-seconds interval is used.

| 25 September 2001 Case | rms ($nm/s^2$) radial | rms ($nm/s^2$) along | rms ($nm/s^2$) cross | correlation radial | correlation along | correlation cross |
|---|---|---|---|---|---|---|
| 1 min interval | 827.80 | 105.18 | 73.75 | 0.6071 | 0.8934 | 0.6371 |
| 5 min interval | 827.53 | 83.85 | 70.37 | 0.6169 | 0.9297 | 0.6648 |
| 10 min interval | 827.94 | 51.44 | 64.69 | 0.6346 | 0.9712 | 0.7046 |
| 20 min interval | 824.85 | 31.38 | 45.00 | 0.7264 | 0.9863 | 0.8103 |

**Table 3.** Recovery results obtained with real CHAMP data using a 5-hour arc and different estimation intervals.

lems in this direction. The results for the along-track and cross-track direction are much better. However, for small estimation intervals the recovery error is still quite large. In the simulation study it was already shown that for small estimation intervals the recovery error due to the current gravity model error is expected to be large. To assess the effect of the current gravity model error on the recovery accuracy, a covariance analysis has been conducted using the EIGEN-1S gravity model [4], which is the nominal model in the recovery experiments. The covariance analysis is based on the diagonals of this model only, but tests with the EGM96 model [2] have shown that using diagonals instead of the full matrix gives results that are of the same order of magnitude. The results of table 3 and table 4 show a reasonable agreement. Finally, the lower part of figure 2 shows the along-track accelerations for the first case of table 3. The agreement between the estimated and true accelerations is quite reasonable, although there are some significant differences, especially during the sharp peaks. However, several peaks are still well observed.

## 4 Conclusions and outlook

The concept of estimating non-conservative accelerations from precise GPS-based total accelerations has been evaluated by an extensive simulation study and an experiment using real CHAMP data. Both studies have shown the feasibility of the concept. It is shown that even high-frequency density pertur-

| Covariance model | averaging interval | rms ($nm/s^2$) radial | along | cross |
|---|---|---|---|---|
| EIGEN-1S diagonal | 1 min | 523 | 390 | 339 |
| EIGEN-1S diagonal | 5 min | 199 | 74 | 182 |
| EIGEN-1S diagonal | 10 min | 137 | 37 | 130 |
| EIGEN-1S diagonal | 20 min | 96 | 19 | 93 |
| GRACE clone diagonal | 1 min | 0.16 | 0.07 | 0.14 |

**Table 4.** Predicted gravity field induced satellite acceleration errors.

bations, as e.g. caused by magnetic storms, can be observed indirectly from orbit perturbations. The simulation study indicates that the current gravity model error is probably the dominant error source and that the impact of anticipated gravity model improvements will be significant: an accuracy level of 50 $nm/s^2$ seems feasible for 30-seconds averaged non-conservative accelerations after completion of the GRACE mission. The remaining dominant error sources are expected to be GPS carrier-phase noise and GPS ephemeris errors. The assessment study also showed that the non-conservative acceleration estimation is a complicated optimization problem. It requires optimization with respect to arc length and observation weighting in conjunction with the estimation interval of the accelerations and their constraint level. Finally, it is expected that the estimation problem can be further improved by properly taking into account the correlations between triple differenced observation errors, which are so far neglected.

*Acknowledgement.* The GeoForschungsZentrum is acknowledged for providing the CHAMP observations and the EIGEN-1S gravity model. The Goddard Space Flight Center kindly provided the GEODYN software. Part of the study was supported by ESA funding in the framework of the SWARM end-to-end simulator study.

## References

1. IJssel J van den, Visser P, Patiño Rodriguez E (2003) CHAMP precise orbit determination using GPS data. Adv Space Res *31(8)*: 1889–1895.
2. Lemoine F, Kenyon S, Factor J, et al. (1998) The development of the joint NASA GSFC and National Imagery and Mapping Agency (NIMA) geopotential model EGM96. NASA/TP-1998-206861.
3. Perosanz F, Biancale R, Loyer S, et al. (2003) On board evaluation of the STAR accelerometer. In: Reigber Ch, Luhr H and Schwintzer P (eds) First CHAMP Mission Results for Gravity, Magnetic & Atmospheric Studies, Springer: 11-18.
4. Reigber Ch, Balmino G, Schwintzer P, et al. (2002) A high quality global gravity field model from CHAMP GPS tracking data and accelerometry (EIGEN-1S). Geophys Res Lett *29(14)*: 10.1029/2002GL015064.
5. Rowlands D, Marshall JA, McCarthy J, et al. (1995) GEODYN II system description. *1-5*, Contractor Report, Hughes STX Corp., Greenbelt, MD.
6. Visser P, Rummel R, Balmino G, et al. (2002) The European Earth Explorer Mission GOCE: Impact for the Geosciences. Ice Sheets, Sea Level and the Dynamic Earth, Geodynamics Series *29*, American Geophysical Union, edited by Mitrovica J and Vermeersen L: 95-107.

# CHAMP and Resonances

Robert H. Gooding[1], Carl A. Wagner[2], Jaroslav Klokočník[3], Jan Kostelecký[4], Christoph Reigber[5]

[1] Univ. Surrey, Guildford GU9 8HY, England, *family.gooding@virgin.net*
[2] NOAA NESDIS, Lab. Sat. Altim., Silver Spring, MD, USA, *carl.wagner@noaa.gov*
[3] CEDR, Astronom Inst. Acad. Sci. Czech Rep., Ondřejov Obs., *jklokocn@asu.cas.cz*
[4] CEDR, Res. Inst. Geod., Zdiby, Czech Rep., *kost@fsv.cvut.cz*
[5] GFZ Potsdam, Dept. 1, Germany, *reigber@gfz-potsdam.de*

**Summary.** The technique of using the evolution of a satellite orbit through resonance to determine the values of appropriate *lumped* geopotential harmonic coefficients has recently been revived, and applied to the triple passage of the Champ orbit through 31:2 resonance. Preliminary results for four pairs of coefficients have been derived rapidly, without using the most precise data (which will be forthcoming). The values obtained are compared with those derivable from various global gravity models (to obtain which, vast amounts of data had to be analysed), and the comparison indicates that the resonance technique remains a competitive one.

**Key words:** gravity field, Champ, resonances, lumped coefficients

## 1 Introduction

About 30 years ago, a new orbit technique was developed at the (then) RAE (Royal Aerospace Establishment), at Farnborough in England, by means of which certain linear combinations of the geopotential (tesseral) harmonic coefficients, known as *lumped harmonics*, could be evaluated much more accurately than the values of the individual harmonics in the global models then available. The basis of the technique was the recognition that, due to orbital acceleration from the satellite's descent through the atmosphere (taking from months to years), significant resonances between the orbital motion and the Earth's rotation would in due course be encountered. The effective duration of such encounters would vary with the *order* of the resonance and the atmospheric density, but would normally be of a few months at most.

Rapid improvements in the accuracy and scope of the global models, towards the end of the century, had two effects: to confirm the accuracy of the early resonance results, in particular for 15$^{th}$ order resonance, but (in addition) to suggest that it was no longer possible for the resonance technique to generate superior results. More recently, however, it has appeared that, by use of more accurate orbital data and more sophisticated software, a revival of the technique would be justified, and efforts in this direction have been made in the UK, USA and Czech Republic. These efforts are concentrating on Champ, which, after passing through 46:3 and 77:5 resonances, has now passed through 31:2 three times (as a result of orbit manoeuvres). This triple passage has provided a unique opportunity for testing progress on the technique's revival, and a preliminary result is presented here.

## 2 Some detail

Considerable background material may be obtained from, in particular, the book of King-Hele (1992), which includes the historical development at the RAE, and a recent paper by three of the present authors (Klokočník et al., 2003), which relates to the technique's revival. The essence of the technique involves the concept of the resonant variable, $\Phi$, defined in terms of the usual orbit elements and the sidereal angle, $S$, by

$$\Phi = \alpha(\omega + M) + \beta(\Omega - S) \quad (1)$$

here $\beta$ and $\alpha$ are the pair of co-prime integers that define the particular resonance, written as either $\beta{:}\alpha$ or $\beta/\alpha$, whilst $\Omega$, $\omega$ and $M$ are the usual orbit elements specifying epochal positions for the ascending node, the perigee and the satellite itself.

We proceed in terms of the element $I$ (inclination), since the technique is most productively applied to this element. The *resonant* rate of change of *mean* $I$, for given $\beta{:}\alpha$, is expressible as a Fourier sum, the prototype of which is a term in $\gamma\Phi - q\omega$; in practice we are concerned with the *basic* term ($\gamma = 1$, $q = 0$), *overtone* terms ($\gamma > 1$, $q = 0$) and *sideband* terms ($\gamma = 1$, $|q| > 0$), usually at most one overtone ($\gamma = 2$) and two sidebands ($q = \pm 1$). The coefficient of a given term consists in the product of a particular lumped harmonic with functions (standardized) of $I$ and $e$ (eccentricity), the $e$-functions being of order $e^{|q|}$ (*cf* Gooding and King-Hele, 1989; Klokočník, 1983).

Each Fourier coefficient also involves a linear combination of the relevant tesseral harmonics, $C_{l,m}$ and $S_{l,m}$, for a fixed value of $m = \gamma\beta$; here $l$, in each combination, in principle takes (all) alternate values, from either $m$ or $m + 1$ as its minimum value. The concept of *lumping* now follows, since we can define $C_m$ (similarly $S_m$) via the sum of the effects of the relevant series; we can (as is usual) normalize these on the basis that $C_m$ would be exactly equal to the true $C_{l(min),m}$ if all subsequent $C_{l,m}$ were zero.

We cannot (without results from many satellites, at different orbital inclinations) separate the individual $C_{l,m}$ and $S_{l,m}$ from determinations of $C_m$ and $S_m$, but we *can* proceed in the opposite direction, by starting from a particular Earth model and comparing our values of $C_m$ and $S_m$ with the values implied by the model. Possible models include (pre-Champ) EGM96 and TEG4 (both US), Grim5-S1 and -C1 (European), the recent Champ-only models Eigen2 and 2ee, IAPG (Nice 2003) and PGS7772p24. This is the second main topic of this paper; but first we give the results (still to be regarded as preliminary) on which the comparison is based.

# 3 Data and results

We based our analysis on the so-called two-line element sets (*TLE*s) for Champ, which have become a universal and classic way of disseminating orbital data rapidly; we hope to analyse the potentially much more accurate 30-sec state vectors later. TLE accuracy we assess at about 0.00006 deg for the 'angular' elements, such as $I$ (equivalent to about 7 meters in position, when projected onto the orbit), which reflects extremely well on the improvements made in TLE generation over the years, bearing in mind that the width of field available for the angular elements allows only 4 decimal places!

The essence of resonance analysis (of Champ $I$'s, as we now assume) is the least-squares fitting of selected pairs ($C_m$ and $S_m$) of harmonics, together with a few other parameters as necessary, to the daily TLEs, over a period long enough to extract maximum information from resonance passage. Before fitting, the TLE values of $I$ are, as far as possible, cleared of known perturbations – in particular the direct lunisolar attraction, the long-period effects of the Earth's zonal harmonics (though uniquely very small for $I$), the effects of the upper atmosphere, and the rotation of the adopted reference axes themselves, due to precession and nutation. At the accuracy level now required, tidal effects (indirect lunisolar attraction) are also important, but suitable software for analytical modelling was not at our disposal, so the effects were removed empirically via additional fitted parameters.

Fig. 1 indicates the variation of the Champ inclination as it passed through the three significant resonances. It shows at once why we are currently presenting results for 31:2 (the change in $I$ was equivalent to more than 100 meters).

The independent approaches of the UK, US and Czech authors differed in nontrivial respects, of which details are not given here. In brief, the (original) UK approach (Gooding, 1971) at each stage uses the most recent TLEs (the complete

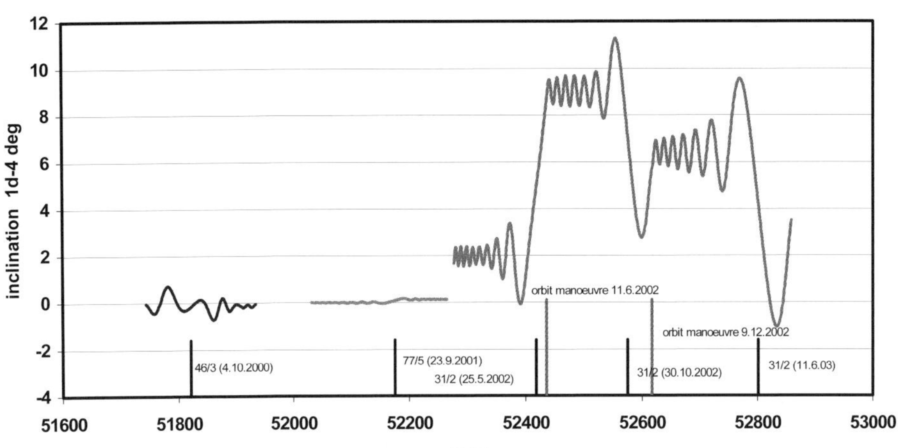


**Fig. 1.** Variation of inclination through three resonances (GRIM5 C1-based with arbitrary origin for each integration).

set) in computing the 'known' perturbations of $I$, but the US approach (Wagner, 1973) is a unified one in which the computation of these perturbations (and all the orbit arguments as well) is governed by a single orbit, assumed valid over perhaps several months; and the Czech approach (Kostelecký, 1984) applies a 'weighted numerical integration' technique after non-resonant perturbations have been removed. (When relevant, it is the first of the three approaches that should be assumed, since the results now to be presented were obtained by the first author's computer program.)

Our first analysis (Klokočník et al., 2003) was of the 46:3 resonance, where difficulties in extracting good values of lumped coefficients arose from the combination of a particularly small basic effect for the inclination of the Champ orbit (compare 46:3 with 31:2 in Fig. 1) with large sideband effects. The high order of the 77:5 resonance made any attempt to analyse this even more daunting, so (as already noted) we deal here only with the 31:2 resonance. At first it seemed that three separate analyses would be necessary, one for each of the three stages separated by the two manoeuvres. It was then realized that (thanks in part to using always the latest TLE set) a single fit should be possible, so long as two additional parameters were fitted (empirically), namely, values for the effective discontinuities in $I$ due to the manoeuvres.

In total, 558 TLE sets were used, starting from Jan 26, 2002 (MJD52300). There were a few gaps in the otherwise daily data, including (naturally enough) around the manoeuvres. For convenience, these gaps were dealt with by interpolation in the TLEs themselves; and empirical values of –0.000266 and –0.000278 deg were found for the 'effective discontinuities'. A total of 20 formal parameters were fitted, including 2 for an overall linear effect (normally essential

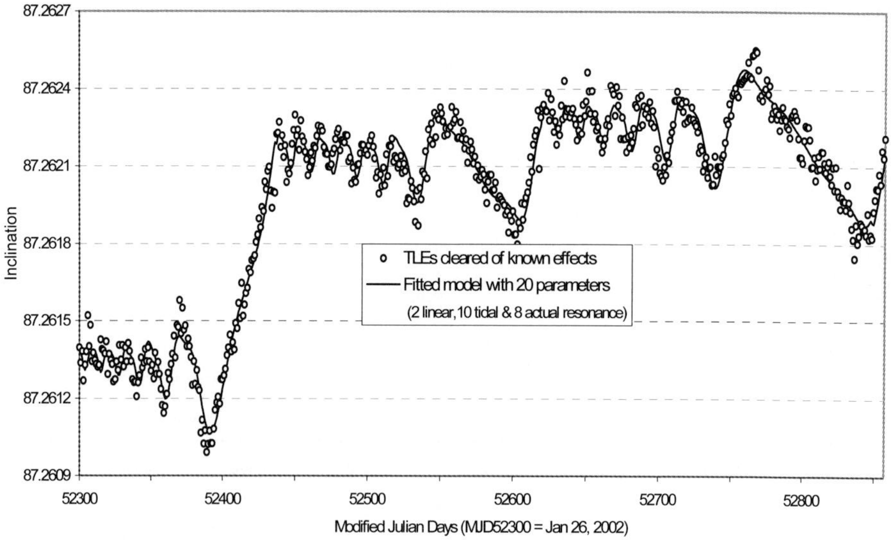


**Fig. 2.** Observed inclination (adjusted two-line elements), together with fitted curve.

in this approach) and 10 to cover five empirical periods for tidal effects; that left 8 parameters for the actual resonance.

Results for the basic (C, S) pair are (–15.05±0.58, –6.40±0.51), with the usual scaling factor of $10^{-9}$ implied; and for the 'first overtone' are (4.00±0.22, 2.20±0.33). For the only significant pair of sidebands, the results were: for $q = 1$, (–0.44±1.42, –8.61±1.14); and for q = –1, (0.68±3.07, 5.98±2.65). It is obvious that the sideband results, particularly for $q = -1$, are less accurate than the others, but in a way they are surprisingly good, since the $e^{|q|}$ factor degrades the sideband resonance analysis for $I$. For analysis of $e$, however (which we are not able to present yet), it is for $q = 0$, and NOT $|q| = 1$, that results are degraded, so analyses for $e$ and $I$ potentially complement each other. This is why it is normal to do both, making an appropriately weighted combination of the two sets of results.

Finally, the usual *a posteriori* estimate of *rms* was made, based on the 558 residuals and the number of *degrees of freedom*; the result was 0.00006 deg, whence our assessed accuracy at the beginning of this Section. Fig. 2 displays (as points) the *observed* values, as cleared of known perturbations (and the effective discontinuities as above), and (by the curve) the *fitted* evolution of $I$.

## 4 Comparisons

Are there external (independent) data of equivalent or better quality than the result just given, which could therefore validate it? Or alternatively, is this resonant result for Champ significantly better than those derived from general geopotential models, and could thus serve to calibrate *them*? The lumped harmonics ($C_{31}$, $S_{31}$) from $I$ (Champ) are the following linear sums of geopotential harmonics ($C_{l,m}$ and $S_{l,m}$):

$$\begin{aligned} C_{31}, S_{31} = {} & 1.0000(C_{32,31}, S_{32,31}) + 0.9096(C_{34,31}, S_{34,31}) + 0.7405(C_{36,31}, S_{36,31}) \\ & + \ldots = -15.05\pm0.58, -6.40\pm0.41 \end{aligned} \quad (2)$$

the directly measured result given in the previous section.

In the 1970's and 1980's, most of the resonant results were derived for orbits not used in comprehensive satellite-geopotential solutions. As a result, the resonant lumped harmonics for these orbits were generally superior (had much lower sd-estimates) than those computed for them from the comprehensive models, so they served as calibrating markers for them (eg., Wagner and Lerch, 1978). For Champ, however, there are already a number of high-degree geopotential models that have been computed from its GPS data, used roughly every 30 seconds for up to 6 months. These models are all complete to 120x120, with terms as high as 140,140. What are the lumped harmonics for this (31,2) resonance computed from them?

Table 1 gives these values from the above series, with projections of the covariance matrix for two of these Champ-only models, together with the series for a recent high-degree field computed from Grace-intersatellite tracking on a nearby orbit (altitude ~480 km, $I = 89.02$ degrees), as well as for the pre-Champ combination model Grim5-C1 (120x120; Gruber et al., 2000).

| $C_{31}$ | $S_{31}$ | Models | Data |
|---|---|---|---|
| | | *Champ's own 31:2 resonance* | |
| -15.05±0.58 | −6.40±0.41 | Analysis here | Champ TLEs (2002-3) |
| | | *Comprehensive pre-Champ* | |
| −15.71 | −8.54 | EGM96 | satellites + surface gravity |
| −16.47±1.68 | −7.33±1.56 | Grim5-C1 | 29 satellites + surface gravity |
| | | *Comprehensive Champ only* | |
| −16.91±0.45 | −9.43±0.37 | Eigen 2 | GPS ~2 cm Phases |
| −16.41±0.40 | −8.73±0.40 | PGS7772p24 | GPS ~2 cm Phases |
| −16.61 | −10.75 | IAPG(Nice 2003) | Geopotential Anomalies |
| | | *Comprehensive with Champ data* | |
| −15.80 | −10.03 | Eigen 1S | Grim5-S1 + Champ + SLR |
| | | *Comprehensive Grace only* | |
| −16.53 | −9.36 | GGM01S | 1 μm/sec range rates (111 days) |

**Table 1.** Lumped harmonics for Champ-type orbit (in $10^{-9}$ units, with standard deviations when known) (altitude = 393 km, inclination = 87.27 deg, eccentricity = 0.003)

Note the generally good agreement of all these independent results. (Among the Champ-only models, the data spans were wholly independent.)

Attesting to the method's efficiency, we also note that the precision of the Champ 31:2 resonance is roughly equal to that for the complete high-degree Champ-only models, while employing only a few hundred observations of (mean-)$I$, compared with more than a million GPS phases for the latter.

Formally, the resonance and Champ-only values all calibrate the less accurate Grim5-C1 values to within about 1-sd of the latter. Comparing values from the other four independent high-degree models with the resonance coefficients, we note that both $C$ and $S$ from resonance are numerically smaller, the discrepancy in $S$ being the more serious in terms of the stated precisions. In the resonance solution, the extra empirical parameters, especially those in the longer period tides, may be absorbing part of the resonant signal in $I$.

## 5 Conclusions

The variation of Champ's orbital inclination has been analysed over the period of a year and a half that covers three passages through 31:2 resonance with the geopotential. This has resulted in values of certain lumped harmonics that are in excellent agreement with those that can be inferred from comprehensive geopotential models. The latter are based on vast amounts of very precise tracking, followed by highly elaborate analysis, whereas our results have been obtained just from the (mean) orbital elements of Champ that continue to be issued daily.

This work is a preliminary stage of a programme in which it is hoped that more accurate resonance results can be obtained from the more precise state vectors being derived for Champ. If possible, accurate lumped harmonics will also be obtained for the higher order resonances (46th and 77th), through which the orbit passed before reaching 31st order.

*Acknowledgment.* MSMT of the Czech Republic are thanked for the grant # LN00A005.

## References

Gooding RH (1971) Lumped fifteenth-order harmonics in the geopotential. Nature Phys Sci *231*: 168-169.

Gooding RH, King-Hele DG (1989) Explicit form of some functions arising in the analysis of resonant satellite orbits. Proc R Soc Lond *A 422*: 241-259.

Gruber T, Bode A, Reigber Ch, Schwintzer P, Balmino G, Biancale R, Lemoine JM (2000) GRIM5-C1: combination solution of the global gravity field to degree and order 120. Geophys Res Lett *27*: 4005-4008.

King-Hele DG (1992) A tapestry of orbits. Cambridge University Press, Cambridge.

Klokočník J (1983) Orbital rates of Earth satellites at resonances to test the accuracy of Earth gravity field models. Celest Mech *30*: 407-422.

Klokočník J (1988) GRM, A contribution to the assessment of orbit accuracy, orbit determination and gravity field modelling. Bull Astronom Inst Czechosl *39*: 45-67.

Klokočník J, Kostelecký J, Gooding RH (2003) On fine orbit selection for particular geodetic and oceanographic missions involving passage through resonances. J Geod *77*: 30-40.

Kostelecký J (1984) A contribution to the method of determining values of the lumped coefficients from orbital inclination of artificial satellites passing through the resonance. Proc Res Inst Geodesy, Topography and Cartography (VUGTK) *15*: 44 – 49.

Lemoine FG, and 14 others (1998) The Development of the Joint NASA GSFC and the National Imagery and Mapping Agency (NIMA) Geopotential Model EGM96. NASA GSFC Greenbelt, NASA/TP-1998-206861.

Wagner CA (1973) Zonal gravity harmonics from long satellite arcs by a semi-numeric method. J Geophys Res *78*: 3271-3280.

Wagner CA, Lerch FJ (1978) The Accuracy of Geopotential Models. Planet Space Sci *26*: 1081-1140.

# CHAMP Gravity Field Solutions and Geophysical Constraint Studies

Shin-Chan Han[1], C.K. Shum[1], Christopher Jekeli[1], Alexander Braun[2], Yiqun Chen[1], and Chung-Yen Kuo[1]

[1] Laboratory for Space Geodesy and Remote Sensing, The Ohio State University, 2070 Neil Avenue, Columbus, Ohio 43210-1275, USA. Tel: 614-292-5701; Fax: 614-292-2957; ckshum@osu.edu

[2] Byrd Polar Research Center, The Ohio State University, 1090 Carmack Rd, Scott Hall, Columbus, Ohio 43210-1002, USA. Tel: 614-247-6839; Fax: 614-292-4697

**Summary**. We report results on the use of CHAMP observations for Earth's mean and time-varying gravity field solutions and the associated geophysical interpretation. 1.5 years of CHAMP data from May 2001 through February 2003 are used for monthly gravity solutions as well as a mean solution based on 6 months of data, employing the energy conservation principle and an efficient conjugate gradient inversion technique. The mean CHAMP gravity field model, OSU03a, is evaluated using other gravity fields model and various data including GPS leveling and Arctic 5'x5' gravity anomalies. It is shown that OSU03a agrees well with EIGEN2 (CHAMP) model, and both models exhibit improvement over the polar regions. The estimated 3x3 CHAMP time-varying gravity field model, with the exception of $J_3$ and 2nd order tesseral coefficients, are compared with available solutions using satellite laser ranging (SLR) for the semi-annual and annual components. It is shown that CHAMP temporal gravity solution agrees well with the SLR solutions as well as various geophysical fluid models, including atmosphere, hydrology, and ocean. The correlation coefficients between CHAMP and SLR solutions are 0.6~0.8, with RMS at 0.7~0.8 mm.

**Key words:** spherical harmonics, geopotential, geoid, time-variable gravity

## 1 Introduction

Since its launch on 15 July 2000, data from the CHAMP gravity mission have demonstrated notable improvement in the modeling of Earth's mean gravity field and time-varying gravity at the seasonal scale for $C_{20}$ geopotential coefficient (Reigber et al., 2002; 2003). In this investigation, we report CHAMP mean and temporal gravity field model solutions and their evaluations. We used the concept of the energy conservation to construct the *in situ* disturbing potential observations using the satellite position and velocity vectors (Jekeli, 1999), assuming that the 3-axis CHAMP accelerometer data adequately observe and model the non-conservative forces acting on the CHAMP satellite. This approach has been used for various gravity modeling studies, e.g., Han et al. (2002), Howe and Tscherning (2003) and Sneeuw et al. (2003). Our gravity field solution approach is based on an efficient conjugate gradient method which also approximates a solution variance and covariance matrix (Han, 2003). The resulting mean gravity field model, solution complete to degree 70, is evaluated for its accuracy. The low degree (3x3) temporal gravity field solution is compared at seasonal scale with solutions obtained from satellite laser ranging (SLR) as well as with geophysical models.

## 2 CHAMP Mean gravity solution, OSU03a

6 months worth of CHAMP Rapid Science Orbit (RSO) and 3-axis accelerometer data have been used to compute the kinematic, rotational, and frictional energy at the satellite altitude (Jekeli, 1999) to create a data set of disturbing potential for gravity field solution. Data covering the consecutive 6 months in year 2002 except data during maneuvers and with incomplete global sampling are used for the solution. Additional force modeling includes N-body perturbation, solid Earth, and ocean tides (CSR3.0), and non-conservative forces as observed by CHAMP accelerometer. The *in situ* disturbing potential observations along the CHAMP orbit can thus be obtained. The spherical harmonic coefficients are then estimated complete to degree 90 along with the corresponding approximate covariance matrix using the efficient iterative method (Han, 2003). The gravity field solution is obtained without any *a priori* variance or covariance constraints. The solution complete to degree 70 is referred as OSU03a.

Figure 1(a) shows the RMS degree differences of various gravity models and their associated errors per degree. Fig. 1(a) shows the power loss of two CHAMP gravity models, OSU03a and EIGEN2, starts from degree 40, as compared to the combination models, EGM96, TEG4 and GRIM5C1. The comparison between the errors of OSU03a and EGM96 indicates the possible improvement of the geopotential below degree 40 by CHAMP data. Figure 1(b) represents the RMS geoid differences as a function of longitude bands between various gravity models. Larger differences are found especially over the polar regions from the geoid differences between OSU03a and EGM96 or GRIM5C1. These 2~3 times larger differences over polar region (latitudes below 60°S and beyond 60°N) indicate CHAMP data significantly contributes to the polar gravity field improvement at wavelengths longer than 800 km (50x50). In order to check the improvement over the polar region with CHAMP data, external data sets are used to test the pre-CHAMP and CHAMP gravity models. First, NIMA's 5'×5' Arctic gravity anomaly data (Kenyon and Forsberg, 2001) are smoothed with a Gaussian filter (radius of 400 km) to reduce the omission error in the gravity model (complete to degree 50). Then, the gravity anomalies derived from the gravity models (truncated at $N_{max}$=50) are compared with the anomaly data. Figure 2 shows the gravity anomaly differences between data and the models. NIMA's data are filled-in with the EGM96 model over the regions covering 64°N~76°N in latitude and 60°E~190°E in longitude. Therefore the data over these regions were not included to compute the RMS of the differences. The data below latitude 68°N were not included because of the potential edge effects induced by Gaussian smoothing. While the RMS difference of GRIM5C1 is 7.4 mgal, both CHAMP

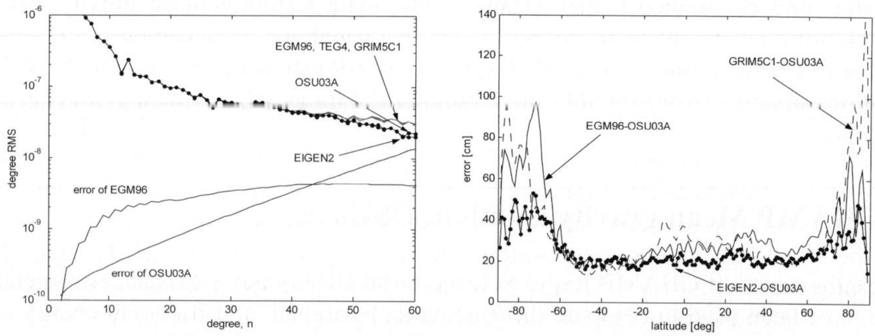


**Fig. 1**. (a) RMS degree differences of various gravity models and their errors (left), (b) RMS geoid differences between various gravity models as a function of longitude bands (right).

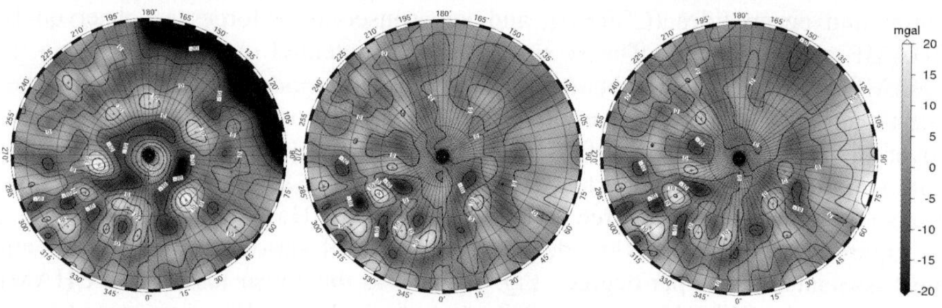


**Fig. 2**. Gravity anomaly differences (5° grids) between NIMA's 5'x5' Arctic gravity data and GRIM5C1 (left); OSU03a (middle); EIGEN2 (right).

gravity models show RMS of 4.9 mgal. The evident improvement is found over the high latitude regions (latitudes beyond 80ºN) and the regions between 90ºE~210ºE in longitude. Next, we use 6,169 points of GPS/leveling data over the U.S. to further evaluate the various gravity models ($N_{max}$=50). The gravity models which included CHAMP data, e.g., including TEG4, EIGEN1S, EIGEN2, and OSU03A, shows the smallest RMS differences at 101~102 cm, while the pre-CHAMP gravity models (e.g., GRIM5C1 and EGM96) show larger RMS at 104~105 cm.

## 3 Temporal variation of monthly CHAMP solutions

In order to investigate the temporal variation of the Earth gravity fields from CHAMP data, ~1.5 years of data spanning a 2 year time period have been processed at approximately monthly intervals. The gravitational signal whose period is less than two months would alias the monthly mean solution, however, this effect is neglected in this study assuming that the temporal aliasing effects are sufficiently small for the CHAMP solution. During the investigation period from May, 2001 through Feb., 2003, CHAMP data sometimes exhibits suboptimal coverage due to orbital decays. In this case, the data are edited and are not used for monthly solutions. The formal errors of the monthly CHAMP solutions are compared with the monthly mean variations of various geophysical fluid models such as ocean, continental surface water, and atmosphere. Figure 3 shows the differences of two consecutive monthly mean oceanic variations (based on T/P altimeter data), continental water variations (based on CDAS-1 soil moisture data and snow accumulation data), and atmospheric variations (based on NCEP reanalysis). In addition, the magnitudes of the time variable gravity field solutions for the annual component determined from satellite laser ranging (SLR) data are also shown in Figure 3. The SLR solution (4x4) shown is from Nerem et al. (2000). Figure 3 shows that the *individual* gravity signals from monthly variations of ocean, continental water, and atmosphere are in general less than the monthly CHAMP error. Therefore, they are hardly detectable from the current CHAMP monthly solutions. However, it is worth investigating the annual variations of the Earth's gravity field (longer than harmonic degree 3) from CHAMP monthly solutions, since the annual variation signal determined by SLR is slightly larger than the monthly CHAMP error. Further, the combined geophysical signals observed by CHAMP may be observable even the individual signals seem to be less than the current CHAMP monthly solution error.

Figures 4(a) and (b) show the time series of the 2$^{nd}$ and 3$^{rd}$ degree zonal coefficients of CHAMP monthly solutions and two independent solutions of the annual variations determined from SLR data analysis by Nerem et al. (2000) and Cheng et al. (2003). Using least-squares fits, components for the secular, annual, and semi-annual variations are estimated from the time series of CHAMP solutions. The annual estimates from the CHAMP solutions are shown in the same Figures. For the 2$^{nd}$ degree zonal coefficient, $C_{20}$, all three solutions (two SLRs and CHAMP) show similar magnitudes with a small phase difference. For the 3$^{rd}$ degree zonal coefficient, $C_{30}$, the two SLR solutions are anti-correlated but with similar magnitudes; the CHAMP solution seems to follow the phase of SLR1 solution (Nerem et al., 2000) well, but with a larger magnitude. For other coefficients, $C_{21}$, $S_{21}$, C31, $S_{31}$, $C_{32}$, $S_{32}$, $C_{33}$, and $S_{33}$, the CHAMP solution tends to follow the SLR1 solution with small phase shifts. The SLR2 solution (Cheng et

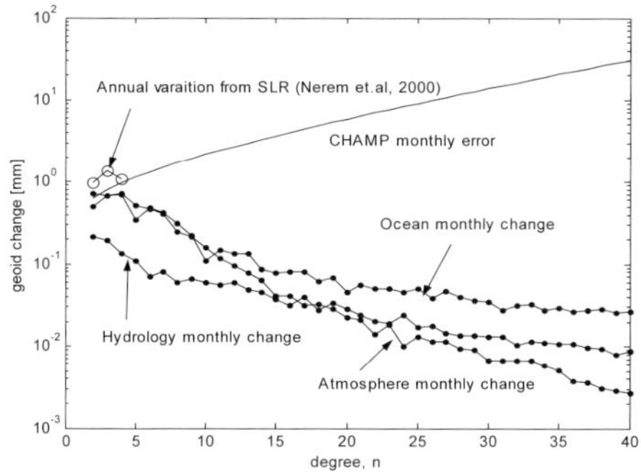


**Fig. 3**. Monthly mean variations of ocean, continental surface water, and atmosphere compared with the monthly CHAMP solution error and the annual variation determined from SLR.

al., 2003) in some cases shows larger magnitudes especially for $C_{32}$, $C_{33}$, and $S_{33}$. In addition to the coefficient-by-coefficient comparison, the temporal (annual) geopotential coefficients from these three solutions can be compared in terms of the annual geoid changes by lumping all coefficients. Both cosine and sine components of the annual geoid changes can be computed. However, in this geoid computation, the coefficients, $C_{22}$, $S_{22}$, and $C_{30}$ are excluded, because of CHAMP's unrealistically large solutions as compared to SLR. Figure 5 presents the cosine and sine components of the three annual geoid changes based on the SLR1, SLR2, and CHAMP solutions. For the cosine component, the SLR1 and SLR2 solutions show strong correlation of 0.96, while the CHAMP solution has a correlation of 0.6 with either of the solutions. For the sine component, the correlations between SLR1 and SLR2 and between CHAMP and SLR1 are 0.8, while the correlation between CHAMP and SLR2 is 0.4. The CHAMP temporal solution is more correlated with the SLR1 solution than with SLR2. The RMS variations from SLR1 and CHAMP are ~1 mm, however, the RMS variations with respect to SLR2, especially the cosine component, are large, approaching 2 mm. The SLR2 solution seems to have overestimated the annual geoid change. The difference between CHAMP and SLR1 is smaller at 0.7 mm, while the difference between CHAMP and SLR2 is larger at 1.4 mm.

## 4 Conclusion

The CHAMP mean gravity solution, OSU03a, shows the smallest difference with EIGEN2 in both spectral and spatial (geoid) domains. Its accuracy is commensurate to other CHAMP gravity models. The accuracy assessment using

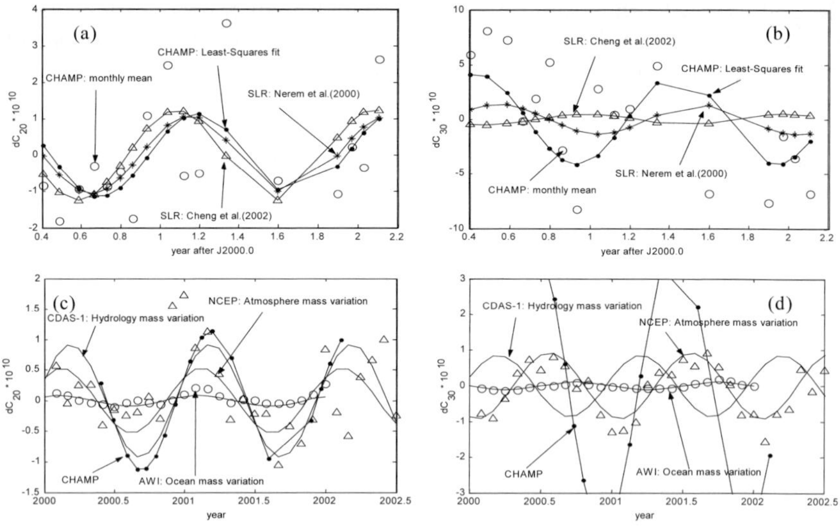


**Fig. 4**. The time series of temporal (annual) variations of the Earth's gravitational coefficients, $C_{20}$ (a) and $C_{30}$ (b) from SLR and CHAMP; $C_{20}$ (c) and $C_{30}$ (d) with some geophysical models.

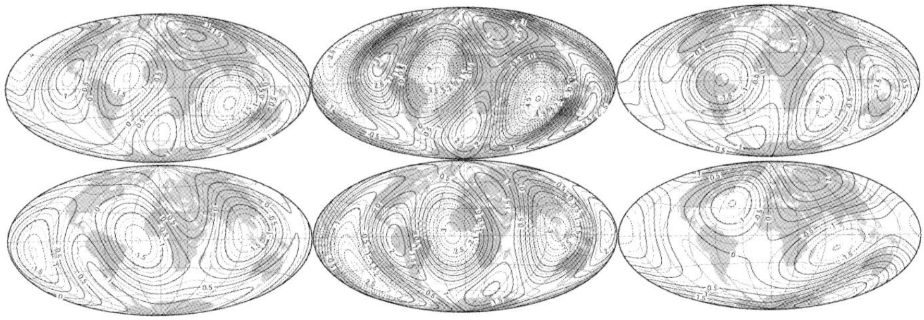

**Fig. 5**. The cosine (top three) and sine (bottom three) components of the annual variations of the Earth's gravitation in terms of the geoid change (unit: is mm, contour interval: 0.25mm); (Left) SLR, Nerem et al.(2000); (Middle) SLR, Cheng et al.(2003); (Right) CHAMP solutions

GPS/leveling and Arctic gravity anomaly data indicates that the gravity models including CHAMP data show significant improvements over polar regions than the pre-CHAMP gravity field models (e.g., GRIM5C1 and EGM96). The monthly Earth gravity change due to individual effects of oceanic, atmosphere and ground water is estimated to be less than the sensitivity of the monthly CHAMP solution, while the signal corresponding to annually averaged change is slightly larger. The time series of low degree coefficients are comparable with SLR solutions as well as geophysical models, except for $C_{22}$, $S_{22}$, $C_{30}$. The CHAMP annual geoid change (3x3) except $C_{22}$, $S_{22}$, $C_{30}$, has correlations of 0.6~0.8 and RMS differences of 0.7~0.8 mm when compared with SLR solutions (Nerem et al., 2000). The atmosphere and hydrological signals dominate in the annual Earth gravity variation. The ocean mass inferred from ocean bottom pressure seems to be one order of magnitude smaller than other signals. We conclude that CHAMP data are useful to the modeling of temporal gravity field solutions.

*Acknowledgements.* This research is supported by grants from the Center for Space Research, University of Texas under a prime contract from NASA, and from NASA's Solid Earth and Natural Hazards Program. The authors also acknowledge computational resources awarded by the Ohio Supercomputing Center.

## References

Cheng MK, Gunter B, Ries J, Chambers DP, Tapley BD (2003) Temporal variations in the earth's gravity field from SLR and CHAMP GPS data. in: Tziavos I N (Ed.), Gravity and Geoid 2002, 3rd meeting of the IGGC, Thessaloniki, Greece: 424–431.

Han SC (2003) The Efficient Determination of Global Gravity Field From Satellite-to-satellite Tracking. Celestial Mechanics and Dynamical Astronomy, in press.

Han SC, Jekeli C, Shum CK (2002) Efficient gravity field recovery using in situ disturbing potential observables from CHAMP. Geophys Res Lett *29(16)*: 10.1029/2002GL015180.

Howe E, Tscherning C (2003) Preliminary analysis of CHAMP state vector and accelerometer data for the recovery of the gravity potential. in: Reigber Ch, Lühr H and Schwintzer P (eds), First CHAMP Mission Results for Gravity, Magnetic and Atmospheric Studies, Springer-Verlag: 140-145.

Jekeli C (1999) The determination of gravitational potential differences from satellite-to-satellite tracking. Celestial Mechanics and Dynamical Astronomy *75*: 85-101.

Kenyon S, Forsberg R (2001) Arctic Gravity Project - a status. in: Sideris M (ed), Gravity, Geoid and Geodynamics 2000, International Association of Geodesy Symposia *123*, Springer-Verlag: 391-395.

Nerem RS, Eanes RJ, Thompson PF, Chen JL (2000) Observations of annual variations of the Earth's gravitational field using satellite laser ranging and geophysical models. Geophys Res Lett *27*: 1783-1786.

Sneeuw N, Gerlach C, Švehla D, Gruber C (2003) A first attempt at time-variable gravity recovery from CHAMP using the energy balance approach. in: Tziavos I N (ed), Gravity and Geoid 2002, 3rd meeting of the IGGC, Thessaloniki, Greece: 237–242.

Reigber Ch, Balmino G, Schwintzer P, Biancale R, Bode A, Lemoine JM, König R, Loyer S, Neumayer KH, Marty JC, Barthelmes F, Perosanz F, Zhu SY (2002) A high-quality global gravity field model from CHAMP GPS tracking data and accelerometry (EIGEN-1S). Geophys Res Lett *29(14)*: 10.1029/2002GL015064.

Reigber Ch, Jochmann H, Wünsch J, Neumayer KH, Schwintzer P (2003) First insight into temporal gravity variability from CHAMP. in: Reigber Ch, Lühr H and Schwintzer P (eds), First CHAMP Mission Results for Gravity, Magnetic and Atmospheric Studies, Springer-Verlag: 128-133.

# Application of Eigenvalue Decomposition in the Parallel Computation of a CHAMP 100x100 Gravity Field

Mark B. Hinga, Steve R. Poole, and Byron D. Tapley

Center for Space Research, The University of Texas at Austin, TX, 78712, USA
*hinga@csr.utexas.edu, poole@csr.utexas.edu, tapley@csr.utexas.edu*

**Summary.** To obtain an alternative gravity solution to that of EIGEN1S, the author's Singular Value Decomposition(SVD) tool, **P**arallel **LA**rge **S**vd **S**olver (PLASS), was applied to the CHAMP normal matrix **ngl-eigen-1s** [2] to perform an Eigenvalue Decomposition (EVD) analysis. The EIGEN1S solution is based on the Tikhonov regularization method of approximating the ill-conditioned system of equations in a subspace of lower rank. In the EVD solution, poorly determined linear combinations of parameter corrections are removed in the culpable eigenspace of the unconstrained least-squares normal equation. The selection of eigenvalues to be removed, is based upon a new method and four different common optimization (truncation) criteria. The new method, the Kaula Eigenvalue (KEV) relation, optimizes the removal of eigenvalues to best satisfy Kaula's Rule. The four other techniques are: inspection, relative error, norm-norm minimization, and finding the minimum trace of the mean square error (MSE) matrix. Analysis of the five different EVD gravity fields was performed. Two of them were shown to be comparable to the EIGEN1S CHAMP solution obtained by the GeoForschungsZentrum Potsdam (GFZ) [2]. The best of the five optimal solutions, that of the KEV, is presented. The number of estimated parameters is 11216.

**Key words:** eigenvalue disposal, Kaula's rule

## 1 Eigenvalue Disposal

To illustrate the effect of eigenvalue truncation on the inversion for solution, the inspection analysis, which is based on the graph of the eigenvalues versus number (where "number" is the ith eigenvalue), is discussed first. Figure 1 shows the spread of eigenvalues. The largest and smallest eigenvalues are $7.17x10^{25}$ and $8.08x10^{9}$, respectively, which yields a condition number of $8.87x10^{15}$; indicating an ill-conditioned system of equations. The eigenvalues are displayed from largest to smallest.

Because the smallest eigenvalue is much greater than zero and there is a smooth transition throughout most of the graph, it is difficult to determine which of the eigenvalues are responsible for the ill-conditioned nature of the normal matrix. To illustrate the stabilization effect of eigenvalue disposal, solutions were calculated in which the smallest 4000, 6000, and 10000 eigenvalues were set to zero. These are compared to a solution retaining all

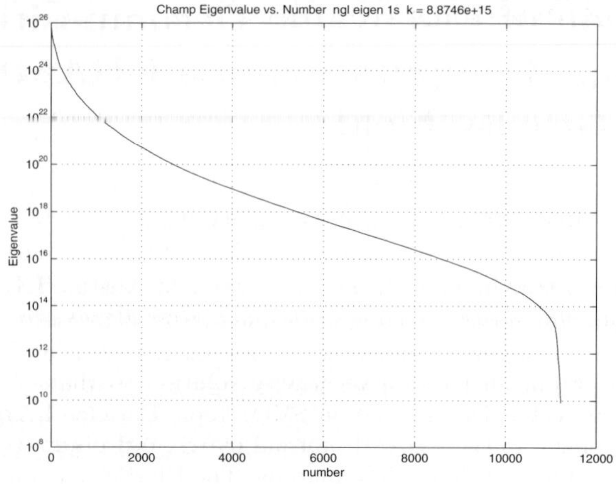


**Fig. 1.** CHAMP Eigenvalue vs. Number.

eigenvalues. Figure 2 is a graph of degree amplitude versus harmonic degree overlayed with Kaula's rule. This shows that much of the excess power above harmonic degree of about 35 is removed, when the culpable eigenvectors contributing to this inflation (through their linear combination in the eigenspace), are eliminated by setting their eigenvalues to zero. Because the inspection method is somewhat subjective, it is useful only for illustrative purposes.

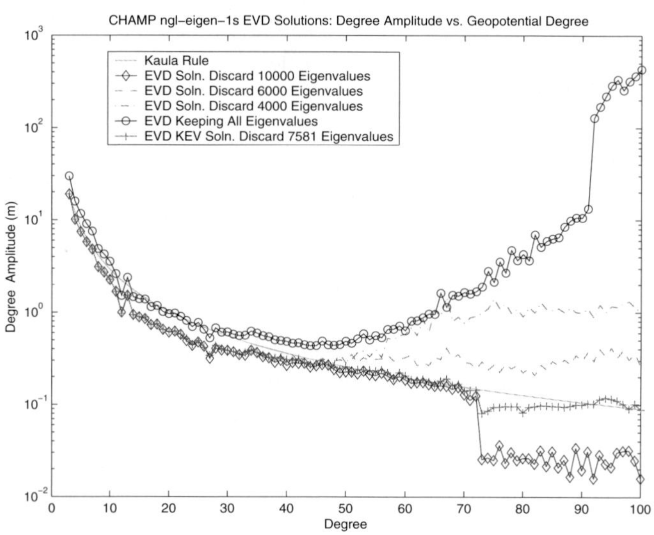


**Fig. 2.** CHAMP: All Eigenvalues Soln., Inspection Soln.'s, KEV EVD Soln.

## 2 Kaula Eigenvalue (KEV) Relation

The KEV method relates the disposal of eigenvalues, in the EVD stabilization of a gravity field solution, to Kaula's power rule of thumb. Since an EVD solution is affected by eigenvalue inclusion/exclusion, there must exist a relation between eigenvalue truncation and the equations of gravity field estimation. A gravity field solution is an estimated parameter vector, whose elements are the scaled dimensionless coefficients, $C_{l,m}$ and $S_{l,m}$. These parameters are the constants that are multiplied against the basis functions appearing in the spherical harmonic expansion, which is used in the equation to describe a three dimensional gravitational potential in the free (zero density) space above the Earth. The connection between eigenvalues and these spherical harmonic coefficients is revealed through the use of the degree variance equation, $\sigma_l^2 = \sum_{m=0}^{l}(C_{l,m}^2 + S_{l,m}^2)$. Since the power of these coefficients at a particular harmonic degree l are closely approximated by Kaula's rule, eigenvalue truncation/disposal can be manipulated to best satisfy this criterion. Thus, a series of these scalar power values can be monitored as eigenvalues are truncated (one at a time) for each inversion case and a minimum difference between the generated power curves of Kaula's rule and that of the truncated EVD solutions, can be found. The following illustrates this concept.

For ease of description, let us label the one dimensional storage array containing the Kaula power at all harmonic degrees and another which contains the degree variance as determined by an EVD solution, as the Kaula and EVD "vector", respectively. Thus the "Kaula vector" $\overline{v}_K$ can be constructed using Kaula's rule, and the estimated coefficients of the EVD solution defines the elements of the "EVD-vector", $\overline{v}_{EVD}$. The *ordering* of the elements for both vectors are identical and is based upon the sequence of the estimated coefficients in the EVD solution. A relation between Kaula's power rule and eigenvalue truncation/disposal is discovered by taking the two-norm of the difference of these two vectors, yielding the scalar, $\alpha = \|\overline{v}_{EVD} - \overline{v}_K\|_2$. This is equivalent to taking the square root of the sum of the squares of the differences between the vectors for every "jth" solution for a particular number of used eigenvalues "u". The following equations illustrate this.

$$\alpha(j;u) = \left\{ \sum_{l=1}^{lmax} \left[v_{EVD\,l} - v_{K\,l}\right]^2 \right\}^{\frac{1}{2}} , \tag{1}$$

where,

$$v_{EVD\,l} = \left[ \sum_{m=0}^{l} (C_{l,m}^2 + S_{l,m}^2) \right] \text{ and } v_{K\,l} = \left\{ \frac{10^{-10}(2l+1)}{l^4} \right\}. \tag{2}$$

By constructing this "vector" pair for each new EVD solution, according to each new combination of eigenvalues, the behavior of the dimensionless scalar

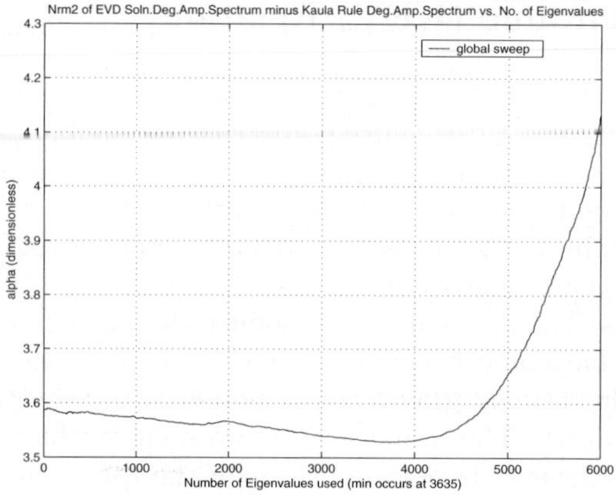


**Fig. 3.** CHAMP: Global Minimum of Used Eigenvalue Cases.

$\alpha(j; u)$ may be plotted against u, the number of used eigenvalues. Thus a function relating Kaula's rule to eigenvalue truncation may be plotted. It is the minimum of this function that corresponds to the optimal choice of eigenvalues, for the gravity solution that best satisfies Kaula's rule. By sweeping through many solutions, the global optimum (minimum) is very quickly found. Figure 3 displays the magnified view of the area where the global minimum occurs for "ngl-eigen-1s", which is at 7581 discarded eigenvalues (3635 included). Figure 2 also includes this KEV EVD gravity solution expressed as degree amplitude versus harmonic degree overlayed with Kaula's rule. It is the EVD solution which closely follows the entire length of Kaula's power curve. About 68 percent of all eigenvalues were discarded for this optimum solution.

## 3 Evaluation of Gravity Field

**Orbital Arc Fit Computations.** The satellites selected to fly through the estimated EVD gravity fields are shown in Table 1. All arc fits were computed using UTOPIA [1] and compared with the actual observation data for a chosen satellite. Table 1 shows the SLR orbit fits in centimeters RMS of all five candidate gravity fields on all selected satellites. Notice, that for the Inspection case (1216 eigenvalues used), all satellites fall out of orbit, leading to its rejection. The case EIGEN1S is the gravity field produced by the GFZ from the same CHAMP normal matrix "ngl-eigen-1s" used in this investigation. It is this EIGEN1S gravity field to which all EVD fields of this investigation were compared. Other than the GFZ1 satellite, the RMS fits are fairly similar for all cases of the EVD gravity fields and the EIGEN1S. However, the EVD deflation effects are best seen in the orbit fit residual of the low altitude satellite GFZ1.

| Case | GFZ1 | Lageos 1 | Lageos 2 | Starlette | Stella | Topex |
|---|---|---|---|---|---|---|
| Inspection | crash | crash | crash | crash | crash | crash |
| KEV | 11.52 | 8.12 | 10.77 | 3.08 | 3.64 | 2.32 |
| MSE | 15.65 | 10.77 | 10.77 | 2.95 | 3.64 | 2.32 |
| Norm-Norm | 11.34 | 8.13 | 10.77 | 3.08 | 3.64 | 2.39 |
| Relative Error | 11.22 | 8.12 | 10.77 | 2.97 | 3.64 | 2.34 |
| EIGEN1S | 74.03 | 8.11 | 10.76 | 3.07 | 3.31 | 2.37 |

**Table 1.** Orbital Arc Fits of Candidate Gravity Fields (cm. radial RMS).

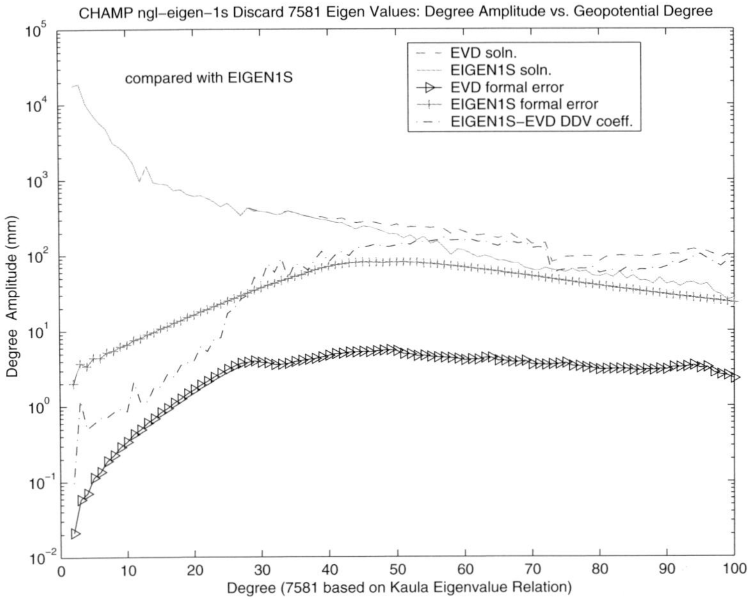


**Fig. 4.** CHAMP: KEV Degree Error Var. and Var. Geopotential Difference to EIGEN1S.

**EVD Degree Error Variance and Geopotential Variance Difference vs. EIGEN1S.** The covariance matrix corresponding to a truncated Eigenvalue Decomposition (EVD) estimated solution, is not an adequate measure of error for an estimate. Because not all eigenvectors were included into the estimation process, the estimate is biased. The calculated gravity field is "shifted" by some amount away from the true gravity field and the confidence in the estimated coefficients may be too optimistic, i.e. their variances are not an accurate indication of the difference between the estimated gravity field and the true gravity field. However, if the bias introduced by an EVD estimate is "small", its covariance may be considered unbiased in an approximate sense. Although the KEV EVD estimate is biased, its variances and those of an unbiased gravity solution (reference field) may be compared to evaluate the difference between the two fields with respect to the error variance of the reference field. Figure 4 shows the degree error variance of the KEV EVD solution in comparison to that of the EIGEN1S reference field. The differences between their geopotential coefficient variance spectra along with the EIGEN1S solution, is shown (Degree Difference Variance (DDV)). The formal error of the biased KEV solution are all within the error variances of EIGEN1S, implying that this candidate EVD field is within the uncertainty (in a random sense) of the EIGEN1S gravity field. However the differences in the coefficients between the two solutions, as seen in the DDV curve, become larger than the EIGEN1S formal error above a harmonic degree of about 30. This indicates that the KEV EVD solution may be too optimistic in this region.

**Conclusion. PLASS** demonstrates a new feasibility in the application of the EVD in the solution for large gravity fields. Employing the KEV technique, the removal of 7581 eigenvalues, was deemed optimum. The bias in this solution caused no deleterious effects detected by the analyzes of this investigation.

*Acknowledgement.* Mark Hinga thanks the Deutscher Akademischer Austauschdienst (DAAD) for a research fellowship and thanks the GFZ for the use of their facilities.

## References

1. Tapley BD, Schutz BE, Ries JC, and Rosborough GW (1984) The UT/Center for Space Research Orbit Determination Systems. The XXVth Cospar Plenary Meeting, Graz, Austria.
2. Reigber Ch, Balmino G, Schwintzer P *et al.* (2002) A high quality global gravity field model from CHAMP GPS tracking data and Accelerometry (EIGEN-1S). Geophys Res Lett *29(14)*: 10.1029/2002GL015064.

# Time-Variable Gravity Seen by Satellite Missions: On its Sampling and its Parametrization

Martin Wiehl and Reinhard Dietrich

TU Dresden, Institut für Planetare Geodäsie, 01062 Dresden, Germany
*wiehl@ipg.geo.tu-dresden.de*

**Summary.** Temporal variations of the gravity field may act either as a signal or as a source of noise for the current satellite gravity missions. This depends, to some extent, on the parametrization of the gravity field solution. We discuss qualitatively how temporal variations affect satellite gravity products and how their effects may be controlled by an adequate parametrization. We describe a mechanism how unparametrized temporal variations may alias into orbit-parallel spatial patterns of a gravity field solution. While the effect is too small to corrupt static gravity field models like EIGEN-2 or EIGEN-3p it may be a concern for studies on time-variable gravity from consecutive GRACE period solutions. Moreover, time-varying errors in non-gravity parameters such as CHAMP accelerometer corrections may, due to correlations with gravity parameters, cause similar effects as geophysical variations. These issues suggest that an adequate parametrization of the gravity field as a function of space and time needs further study. Eigenvalue analyses of solution normal matrices may be a useful tool for these studies.

**Key words:** time-variable gravity, aliasing, parametrization, accelerometer calibration parameters, EIGEN, GRACE

## 1 Aliasing of temporal variations into spatial variations

Consider the determination of a mean gravity signal (e.g., the total gravity field or its difference from another model) from CHAMP, GRACE or GOCE mission data over a certain solution period. During the solution period, there are temporal gravity variations which are not corrected or parametrized. Their possible effect on the mean solution is schematically demonstrated in Figure 1:

We start from a temporal variation (Fig. 1, top) which is large-scale in time and in space. It is observed along the gravity mission's free orbit (Fig. 1, center, where we assume for simplicity that the disturbing potential $T$ is a direct observable). The sampling along tracks close in space may be distant in time and may thus differ by the meantime evolution. Then, in the mean signal solution adjusted from all observations, orbit-parallel spatial patterns like those in Figure 1 (bottom) will appear. These patterns will have only a

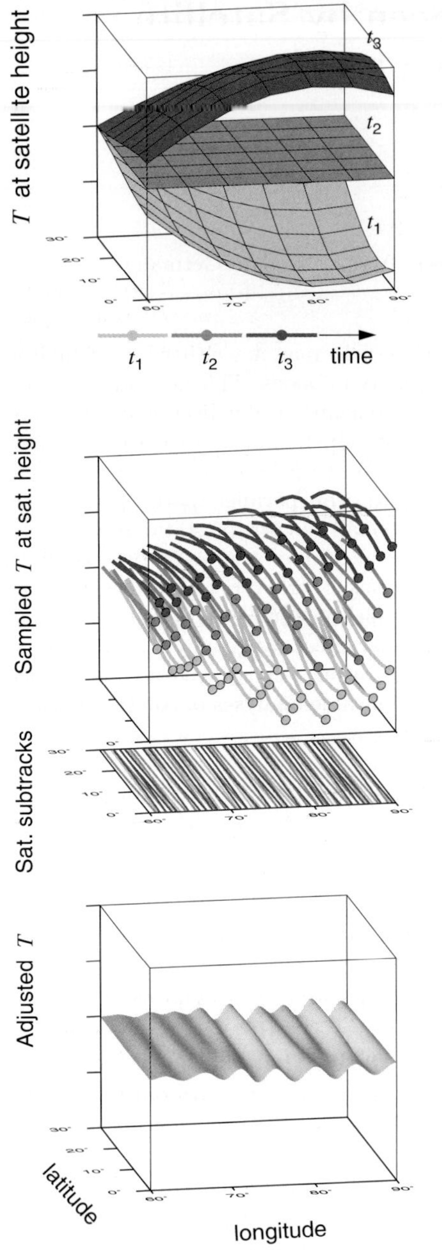


**Fig. 1.** Aliasing of temporal variations into spatial variations of the adjusted mean gravity field

part of the variance of the temporal variations since another part enters the adjustment residuals. But the features will have considerable variance at high spherical harmonic degrees even if the temporal variations are only at low degrees. Our theoretical consideration is supported by numerical simulations for GRACE [2] which obtained orbit-parallel features as an effect of temporal variations.

We call the described effect an aliasing effect: Due to, basically, an incomplete sampling one kind of signal (temporal variations) is misinterpreted as a different kind of signal (spatial variations).

A spectral view on the relative amplitudes is given in Figure 2. The temporal variations are small relative to the searched mean signal. (Fig. 2 uses the ratio $10^{-4}$ which may apply between intra-annual to inter-annual variations and the total gravity field.) Nevertheless, the low degree temporal variations may be larger than the high degree mean signal. The adjustment assigns a part of the temporal variance to the mean signal. This might dominate over the actual signal at high degrees.

Note that this dominance depends on the actual mission observable's spectral characteristics. The situations differ for the different gravity missions and even for the different CHAMP processing approaches. Moreover, describing the observable as a function of space provides only a rough scheme. In fact, the dependence is on the orbit, too,—especially for the "classical" CHAMP processing approach where, roughly speaking, orbit perturbations are observed.

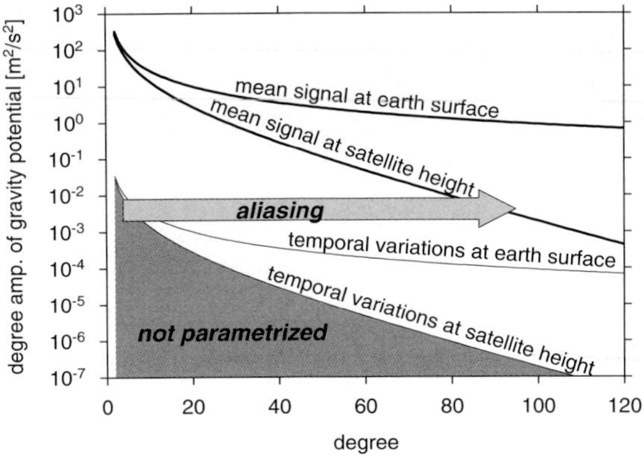


**Fig. 2.** Aliasing of temporal variations into spatial variations: spectral view

## 2 Relevance for current gravity field products

EIGEN-2 [6] and EIGEN-3p [5], the latest CHAMP models by the GeoForschungsZentrum Potsdam (GFZ), fully resolve the gravity field up to around degree $l = 40$ and $l = 65$, respectively. However, their computational resolution extends to $l = 120$ and partly to $l = 140$. The high degree parts of both models consist of distinct orbit-parallel patterns (Figure 3) which dominate above about $l = 60$ for EIGEN-2, and $l = 80$ for EIGEN-3p. They resemble the patterns predicted in the previous section. This suggests the hypothesis that these patterns are an aliasing effect of temporal gravity field variations. However, a quantitative assessment rejects this hypothesis.

For this assessment we simulate orbit perturbations which roughly represent the CHAMP observables in GFZ's processing. According to the hypothesis, the high-degree patterns entered the EIGEN models to explain a part of the orbit perturbations that was in fact induced by temporal variations. Consequently, the perturbations induced by temporal variations must be larger than those induced by the EIGEN patterns. The intra-annual to inter-annual variations reported in the literature (e.g., [7]) have magnitudes below $10^{-4}$ times the static field's magnitude. Hence, we generate random signals that obey $10^{-4}$ times Kaula's rule and simulate their effect on a CHAMP-like orbit. The perturbations are on the decimeter level. In contrast, the EIGEN-2 spectral part above $l = 70$ (and as well, the EGM96 part above $l = 70$) induces perturbations on the meter level. Hence, geophysical temporal variations are too small to induce the EIGEN model high-degree patterns. More general, spatial variations induced by the aliasing mechanism of Section 1 seem negligible compared to the total field's actual spatial variations in the respective spectral band. We expect that this result also holds for other CHAMP processing approaches and for the GRACE and GOCE missions, as their observables are more sensitive to high degrees which makes the curves in Figure 2 more flat.

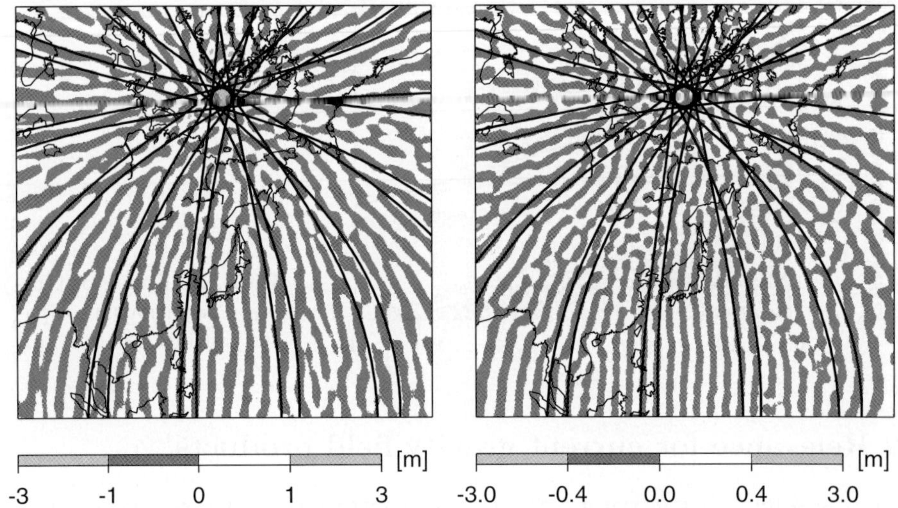


**Fig. 3.** Geoid contribution from (**left**) the EIGEN-2 spectral part above degree 70; (**right**) the EIGEN-3p part above degree 90. A 1-day CHAMP subtrack is added

The above "negative result" concerns the effect of geophysical temporal variations on static gravity field solutions. However, in two other contexts the aliasing effect of Section 1 may, indeed, play a role.

The GRACE mission is designed to determine temporal variations of the gravity field, in addition to its static part. A standard approach is to analyze a sequence of, e.g., monthly period solutions. Then, the relevant signal is the small difference between two period means which may be not larger than unparametrized variations within the solution periods. Hence, aliasing of unparametrized variations may be essential.

Together with the gravity field parameters, non-gravity parameters are estimated for the gravity field solution. In particular, calibration parameters for the CHAMP onboard accelerometer are estimated in certain time intervals [6, 5]. Errors in such parameters cause a wrong assignment of satellite accelerations to either gravitational or non-gravitational forces and may thus correlate with errors in the solved gravity field. By such correlations, time-varying errors of non-gravity parameters may appear like temporal gravity field variations. These apparent variations could, in turn, alias into spatial patterns as described in Section 1. Indeed, for EIGEN-3p the accelerometer corrections were parametrized in shorter time intervals than for EIGEN-2, and the orbit-parallel features in EIGEN-3p are reduced compared to EIGEN-2. This supports the suggestion that the parametrization of non-gravity parameters has an influence on the small-scale structure of the gravity models.

## 3 Discussion on gravity field parametrization

Owing to their dedicated design and unprecedented accuracy, the satellite gravity missions CHAMP, GRACE and GOCE are able to sense gravity field variations not only with space but also with time. Temporal variations may thus be monitored, in particular by GRACE. But they are also an additional error source. Irrespective of the correction for some temporal signals within the mission data processing [1], the missions solve for a basically new kind of signal. The discrete sampling of a mission can not fully resolve this signal, so that ambiguities between different kinds of variations are an inherent problem. But reducing the solution space to only spatial variations up to a certain spherical harmonic degree, as it is done in the "traditional" static gravity field modelling, may now mean an underparametrization which unnecessarily promotes the aliasing effect discussed before. The resulting task is to parametrize the gravity field as a function of both space and time. It is more complex and may involve a trade-off between temporal and spatial resolution.

A guideline is to retrieve as much geophysical variance as possible. In this context it is simple but not optimal to fix the temporal resolution at, e.g., one month over all spatial scales. Instead, different spatial components should be determined with different temporal resolutions according to their temporal variability on the one hand and their temporal resolvability by the mission on the other hand. For example, Perosanz's and others' approach [4] to choose a degree-dependent temporal resolution of the Stokes coefficients accounts for the fact that, roughly, for low-degree coefficients the geophysical variability and, as well, CHAMP's sensitivity are higher than for high degrees.

For a further refinement, with the goal of determining every spatial component with its highest possible temporal resolution, an eigenvalue analysis of gravity field solutions' normal matrices can be used to find an adapted parametrization: A solution's eigenvectors are linear combinations of Stokes coefficients and form a new basis. The factors for these eigenvectors are a new set of gravity field parameters. Their—uncorrelated—errors are obtained from the related eigenvalues. These errors, together with the expected geophysical variability of the parameter, may indicate its appropriate temporal resolution.

An eigenvalue analysis of the normal matrices (here, before the reduction of non-gravity parameters) can also reveal error correlations between gravity and non-gravity parameters. Note that the error covariance matrix contains correlations only between individual parameter pairs. Even if they are low, linear combinations of gravity parameters may still be highly correlated with linear combinations of non-gravity parameters ([3], ch. 12). The eigenvectors now contain both gravity and non-gravity parameters and the eigenvalues indicate how well these linear combinations are determined. Briefly, a badly determined eigenvector that contains significant proportions of both gravity and non-gravity parameters indicates a considerable error correlation between the involved combinations of gravity and non-gravity parameters.

## 4 Conclusions

We have described a mechanism how unparametrized temporal variations may alias into orbit-parallel spatial patterns in gravity field solutions. The effect is too small to corrupt solutions of the static gravity field, and in particular to explain high degree patterns of the EIGEN-2 and EIGEN-3p models. The mechanism might, however, affect analyses of temporal variations from series of GRACE period solutions. Furthermore, if errors in non-gravity (e.g., accelerometer) parameters correlate with gravity field parameters, time-varying errors of the non-gravity parameters can be seen as apparent gravity field variations and their effect may be similar as an aliasing of geophysical variations.

These insights raise the question of an adequate parametrization of the gravity field as a function of space and time. Certainly, different spatial components should be solved with different temporal resolutions according to their temporal variability and to their resolvability by the mission. For further studies, an eigenvalue decomposition of a mission's normal matrices may be a useful tool. It may give a decomposition of the spatial field to components with uncorrelated errors, and give an indication of their adequate temporal resolutions. It can also reveal error correlations between non-gravity and gravity parameters in order to find adapted parametrizations of both.

*Acknowledgement.* This research was funded by the German Research Foundation (DFG) under grant Di 473/13-1. The CHAMP gravity field models EIGEN-2 and EIGEN-3p were provided by GFZ Potsdam.

## References

1. Flechtner F (2003) Status of atmospheric and oceanic gravity field de-aliasing for GRACE. Geophys Res Abstr *5*: 10111.
2. Han S-C, Jekeli C, and Shum CK (2003) Time-variable gravity field solutions from GRACE: analysis and de-aliasing of tides, atmosphere, and hydrology signals and errors. Geophys Res Abstr 5: 13887.
3. Lawson CL and Hanson RJ (1995) Solving Least Squares Problems. Society for Industrial and Applied Mathematics, Philadelphia.
4. Perosanz F, Lemoine J-M, Loyer S, and Bourda G (2003) Long wavelength geoid variations from two years of CHAMP mission. Geophys Res Abstr 5: 07047
5. Reigber Ch, Jochmann H, Wünsch J, Petrovic S, Schwintzer P, Barthelmes F, Neumayer K-H, König R, Förste Ch, Balmino G, Biancale R, Lemoine J-M, Loyer S, and Perosanz F (2004) Earth Gravity Field and Seasonal Variability from CHAMP. *this volume.*
6. Reigber Ch, Schwintzer P, Neumayer K-H, Barthelmes F, König R, Förste Ch, Balmino G, Biancale R, Lemoine J-M, Loyer S, Bruinsma S, Perosanz F, and Fayard T (2003) The CHAMP-only Earth Gravity Field Model EIGEN-2. Adv Space Res *31(8)*: 1883–1888.
7. Wahr J, Molenaar M, and Bryan F (1998) Time variablity of the Earth's gravity field: Hydrological and oceanic effects and their possible detection using GRACE. J Geophys Res *103(B12)*: 30205–30229.

# Gravity Field Recovery by Analysis of Short Arcs of CHAMP

Karl Heinz Ilk, Torsten Mayer-Gürr, Martin Feuchtinger

Institute of Theoretical Geodesy, University Bonn, Germany, *ilk@theor.geod.uni-bonn.de*

**Summary:** The gravity field recovery strategy presented here enables the global recovery of the gravity field combined with a regional focus on geographical areas with rough gravity field features in a consistent way. The global gravity field is modeled by a series of spherical harmonics while the regional gravity field features are represented by space localizing base functions of harmonic spline type. The physical model of the orbit analysis technique is based on Newton's equation of motion, formulated as a boundary value problem in form of an integral equation of Fredholm type. The observation equations are established for short arcs of approximately 30 minutes length. The procedure can be applied either globally or regionally to selected geographical regions. For a regional application the coverage with short arcs should be slightly larger than the recovery region itself to prevent the solution from geographical truncation effects. A proper combination and weighting of the normal equations of every arc combined with a tailored regularization allows a stable solution for the field parameters. This procedure can be adapted to the roughness of the regional gravity field features, the discretization of the gravity field and the sampling rate of the observations. A global gravity field solution ITG-Champ01E has been derived based on kinematic orbits covering 360 days from March 2002 to March 2003. Regional gravity field solution have been determined for selected regions with rugged gravity field features.

**Key words:** Gravity field modeling – Regional gravity field recovery – CHAMP – Satellite-to-satellite-tracking – ITG-Champ01E

## 1 Introduction

The solution strategy presented here enables the global recovery of the gravity field combined with regional refinements in geographical areas with rough gravity field features in a consistent way. The physical model is based on Newton's equation of motion applied to short arcs of approximately 30 minutes and formulated as integral equations of Fredholm type. The integrands contain the reference and residual gravity fields and specific disturbing forces. The reference gravity field representing the low and medium frequency gravity field features are expressed by a series of spherical harmonics complete up to an appropriate degree, while the regional parts are represented by space localizing base functions of harmonic spline type. Especially for a regional refinement of the gravity field it is important to proof in an a-priori step whether there are residual gravity field signals in the kinematically determined orbits caused by rough gravity field features which are not modelled by spherical harmonics. In a post-processing step, the regionally refined gravity field is validated again by using the orbits. The validation procedure is based on the computation of an extended Jacobi integral along the satellite orbits and described in detail in Ilk and Löcher (2003).

## 2 Mathematical Model

The mathematical model of the gravity field analysis technique is based on Newton's equation of motion (Schneider, 1967),

$$\ddot{\mathbf{r}}(t) = \mathbf{f}(t;\mathbf{r},\dot{\mathbf{r}};\mathbf{x}), \tag{1}$$

formulated as a boundary value problem,

$$\mathbf{r}(t) - (1-\tau)\mathbf{r}_A - \tau\mathbf{r}_B = -T^2 \int_{\tau'=0}^{1} K(\tau,\tau')\,\mathbf{f}(t;\mathbf{r},\dot{\mathbf{r}},\mathbf{x})\,d\tau', \tag{2}$$

with the integral kernel

$$K(\tau,\tau') = \begin{cases} \tau(1-\tau'), & \tau \le \tau', \\ \tau'(1-\tau), & \tau' \le \tau, \end{cases} \tag{3}$$

satisfying the boundary values

$$\mathbf{r}_A := \mathbf{r}(t_A), \quad \mathbf{r}_B := \mathbf{r}(t_B), \quad t_A < t_B\,. \tag{4}$$

The specific force function,

$$\mathbf{f}(\tau';\mathbf{r},\dot{\mathbf{r}},\mathbf{x}) = \mathbf{f}_d(\tau';\mathbf{r},\dot{\mathbf{r}}) + \nabla V(\tau';\mathbf{r};\mathbf{x}) + \nabla T(\tau';\mathbf{r};\Delta\mathbf{x}). \tag{5}$$

can be separated in a disturbance part $\mathbf{f}_d$, which represents the non-conservative disturbing forces, in a reference part $\nabla V$, representing the global gravity field features and in an anomalous part $\nabla T$, modelling the corrections to the gravity field parameters. The mathematical model applied in this investigation is given by equation (2) together with the force function (5) (Ilk et al., 1995). The geocentric positions $\mathbf{r}(t)$ of the arcs over the analysis area represent the observations. The unknowns are the corrections $\Delta\mathbf{x}$ to the field parameters $\mathbf{x}$. These are in case of a global gravity field recovery corrections to the coefficients of a spherical harmonics expansion of the gravitational potential $V(\mathbf{r}_P)$ or in case of a regional recovery the parameters of space localizing base functions modelling the anomalous potential $T(\mathbf{r}_P)$:

$$T(\mathbf{r}_P) = \sum_{i=1}^{I} a_i\, \varphi(\mathbf{r}_P, \mathbf{r}_{Q_i}) \tag{6}$$

with the field parameters $a_i$ arranged in a column matrix $\Delta\mathbf{x} := (a_i, i = 1,...,I)^T$ and the base functions,

$$\varphi(\mathbf{r}_P, \mathbf{r}_{Q_i}) = \sum_{n=0}^{\infty} k_n \left(\frac{R_E}{r}\right)^{n+1} P_n(\mathbf{r}_P, \mathbf{r}_{Q_i})\,. \tag{7}$$

The coefficients $k_n$ are the degree variances of the gravity field spectrum to be determined,

$$k_n = \sum_{m=0}^{M} \left(\Delta\bar{c}_{nm}^2 + \Delta\bar{s}_{nm}^2\right). \tag{8}$$

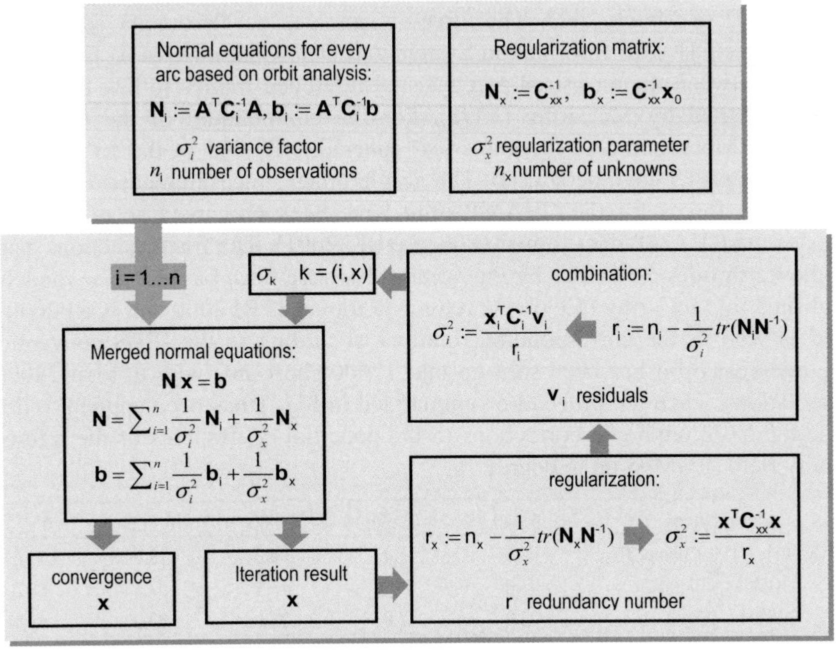


**Fig. 1.** Computation procedure including the determination of variance factors for every arc and the regularization parameter for the gravity field recovery.

$R_E$ is the mean equator radius of the Earth, $r$ the distance of a field point from the geo-centre and $P_n(\mathbf{r}_P, \mathbf{r}_{Q_i})$ are the Legendre's polynomials depending on the spherical distance between a field point $P$ and the nodal points $Q_i$ of the set of base functions. With this definition the base functions $\varphi(\mathbf{r}_P, \mathbf{r}_{Q_i})$ can be interpreted as isotropic and homogeneous harmonic spline functions (Freeden et al., 1998). The nodal points $Q_i$ are defined on a grid generated by a uniform subdivision of an icosahedron of twenty equal-area spherical triangles. In this way the global pattern of spline nodal points shows approximately uniform nodal point distances. Every short arc of approximately 30 minutes builds a normal equation. All normal equations are combined by estimating a variance factor for every arc as well as an accelerometer bias (Koch and Kusche, 2002). The regularization has been restricted to all potential coefficients from degree 40 upwards and the regularization parameter has been determined during the iteration as sketched in Fig. 1.

## 3 Global gravity field recovery based on kinematic orbits

The global gravity field recovery presented here is based on kinematic orbits of CHAMP with a sampling rate of 30 seconds provided by M. Rothacher and D. Svehla from the FESG of the Technical University Munich (Svehla and Rothacher, 2003). The orbits cover a time period of approximately 360 days from

March 2002 to March 2003. The three-dimensional accelerometer data are provided by the CHAMP Information System and Data Centre (ISDC). The transformations between the terrestrial and celestial reference frames follow the conventions published by McCarthy (1996). For the computation of the tidal forces caused by Moon and Sun the numerical ephemeris DE405 of the Jet Propulsion Laboratory (JPL) have been used. The accelerometer measurements to determine the surface forces for the CHAMP orbit have been processed according to the rules of the CHAMP data format (Förste et al., 2001). The force functions caused by the Earth tides as well as by the ocean tides have been based on the models as published by McCarthy (1996). As reference frames ITRF2000 and ICRF2000 are used as well as the corresponding rotations according to the IERS conventions. The one-year orbit has been split up into 17000 short arcs with in total 2400000 observations. Then the procedure summarized in Fig. 1 has been applied to determine the 5772 unknown corrections to the potential coefficients of the reference gravity field (EGM96 up to degree 75).

| spectral range | 0...40 | 0...50 | 0...60 | 0...65 | 0...70 |
|---|---|---|---|---|---|
| GGM01s – ITG-Champ01E | 3,5 | 7,4 | 14,3 | 18,8 | 21,8 |
| GGM01s – Eigen-2 | 7,6 | 20,7 | 47,3 | 57,5 | 68,0 |
| GGM01s – Eigen-3p | 3,9 | 7,6 | 23,1 | 32,0 | 42,1 |

**Table 1.** 1°x1°-grid comparisons of ITG-Champ01E, Eigen-2 and Eigen-3p with GGM01s: rms of undulation differences (cm).

| GPS data set | points | filled up with EGM96 from n=37 | | | filled up with EGM96 from n=73 | | |
|---|---|---|---|---|---|---|---|
| | | min | max | wrms | min | max | wrms |
| USA | 5168 | -122,2 | 179,4 | 43,4 | -129,4 | 183,5 | 44,0 |
| Canada | 1931 | -131,8 | 174,2 | 40,1 | -112,4 | 131,8 | 38,5 |
| EUVN (Europe) | 186 | -142,4 | 168,3 | 42,4 | -88,4 | 143,8 | 36,3 |
| BKG (Germany) | 575 | -114,8 | 73,9 | 29,6 | -95,8 | 63,4 | 20,5 |

**Table 2.** Global gravity field recovery: rms of geoid undulation differences (cm).

The result of the global recovery has been checked by determining the rms of the differences of 1°x1° grids of point geoid undulations between our solution ITG-Champ01E and the recent GRACE solution GGM01s (CSR, 2003) for various spectral bands (Table 1). For comparison the same tests have been performed for the gravity field models Eigen-2 (Reigber et al., 2003) and Eigen-3p (Reigber et al., 2004). Our model ITG-Champ01E is biased in the higher degrees by the reference gravity field EGM96 caused by a regularization from degree n=40 upwards. But another gravity field recovery solution with no a-priori information and no regularization at all confirmed the results shown in Tab. 1. While the results in the low-frequent spectral band are similar to Eigen-3p there are slight improvements in precision in the spectral band between degree 40 and 60 compared to alternative recovery solutions. Additional test computations have been performed by F.

Barthelmes from GFZ with the GFZ standard evaluation procedure. Tab. 2 shows selected results from this evaluation test, the comparisons with geoid undulations derived from GPS and levelling measurements.

# 4 Regional gravity field recovery

To demonstrate the regional recovery three regions with rough gravity field features have been selected (Fig. 2): South East Asia, South America and Europe. An additional strip of 10° around the recovery areas have been considered to prevent the solution from geographical truncation effects. The parameterisation of the residual field was based on harmonic spline functions as defined in equation (7); the corresponding recovery parameters are shown in Tab. 3. The mean distance between the base function nodal points amounts approximately 130 *km*. The same data set as in case of the global gravity field recovery has been used also for the regional recoveries. Instead of EGM96 Eigen-2 has been used as reference field up to degree 120. The regional recovery results are shown in Tab. 4.

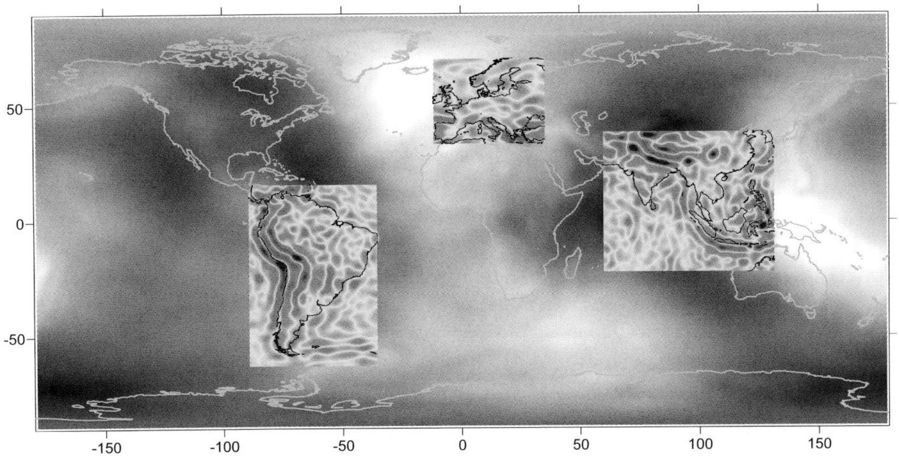


**Fig. 2.** Gravity field recovery for South America, Europe and South East Asia.

| region | South East Asia | South America | Europe |
|---|---|---|---|
| orbits | 3192 | 2340 | 1922 |
| observations | 317000 | 260000 | 135000 |
| unknowns | 4064 | 3046 | 1172 |

**Table 3.** Parameters for the regional gravity field recovery.

| region | South East Asia | South America | Europe |
|---|---|---|---|
| ITG – GGM01s | 92,2 | 85,5 | 65,4 |
| Eigen-2 – GGM01s | 141,2 | 114,7 | 106,0 |
| ITG – Eigen-2 | 104,7 | 101,3 | 88,9 |

**Table 4.** 1°x1°-grid comparisons of ITG solutions and Eigen-2 with GGM01s in the spectral band 36…120: rms of undulation differences (cm).

# 5 Conclusions

The use of short arcs for global gravity field recovery as well as for regional gravity field refinements is an adequate alternative recovery technique based on high-quality kinematic orbits as performed by the FESG Munich. Despite the fact that the CHAMP mission is designed to recover first of all the long and medium wavelength features it could be shown that also regional refinements are possible with surprising accuracy.

*Acknowledgements:* We are grateful for providing the data by the ISDC of the GeoForschungsZentrum (GFZ) Potsdam and for the kinematical orbits by D. Svehla and M. Rothacher from the FESG at the Technical University Munich. We also thank F. Barthelmes for his computations with the GFZ standard evaluation procedure. Thanks to T. Gruber and D. Svehla as well from the Technical University Munich for additional evaluation computations. The support by BMBF (Bundesministerium für Bildung und Forschung) and DFG (Deutsche Forschungsgemeinschaft) within the frame of the Geotechnologien-Programm is gratefully acknowledged.

# References

CSR (2003) GRACE Gravity Model 01 (GGM01). http://www.csr.utexas.edu/grace

Förste Ch, Schwintzer P, Reigber Ch (2001) The CHAMP Data Format. Internal publication, GFZ, http://op.gfz-potsdam.de/champ/docs_CHAMP.

Freeden W, Gervens T, Schreiner M (1998) Constructive Approximation on the Sphere. Oxford University Press, Oxford.

Ilk KH, Löcher A (2003) The Use of Energy Balance Relations for Validation of Gravity Field Models and Orbit Determination Results. presented at the Gen. Ass. of the IUGG 2003, Sapporo, Japan.

Ilk KH, Rummel R, Thalhammer M (1995) Refined Method for the Regional Recovery from GPS/SST and SGG. CIGAR III/2, ESA contract No. 10713/93/F/FL, European Space Agency.

McCarthy DD (ed) (1996) IERS Conventions 1996. Central Bureau of IERS – Observatoire de Paris, Paris.

Koch K-R, Kusche J (2002) Regularization of geopotential determination from satellite data by variance components. J Geodesy *76(5)*: 259-268.

McCarthy DD (ed) (1996) IERS Conventions 1996, Central Bureau of IERS – Observatoire de Paris, Paris

Reigber Ch, Schwintzer P, Neumayer K-H, Barthelmes F, König R, Förste Ch, Balmino G, Biancale R, Lemoine J-M, Loyer S, Bruinsma S, Perosanz F, Fayard T (2003) The CHAMP-only Earth Gravity Field Model Eigen-2, Adv Space Res *31(8)*: 1883-1888.

Reigber Ch, Jochmann, H, Wünsch, J, Petrovic S, Schwintzer P, Barthelmes F, Neumayer K-H, König R, Förste Ch, Balmino G, Biancale R, Lemoine J-M, Loyer S, Perosanz F (2004) Earth Gravity Field and Seasonal Variability from CHAMP. This issue.

Schneider M (1967) Lösungsvorschlag zum Bahnbestimmungsproblem. BWF Bericht W67-35.

Svehla D, Rothacher M (2003) Kinematic Precise Orbit Determination for Gravity Field Determination. presented at the Gen. Assembly of the IUGG 2003, Sapporo, Japan.

# Statistical Assessment of CHAMP Data and Models Using the Energy Balance Approach

Jürgen Kusche and Jasper van Loon

TU Delft/DEOS, 2600 GB Delft, P.O. Box 5058, The Netherlands
*j.kusche@lr.tudelft.nl, j.p.vanloon@lr.tudelft.nl*

**Summary.** The energy balance approach is used for a statistical assessment of CHAMP orbits, data and gravity models. It is known that the quality of GPS–derived orbits varies and that CHAMP accelerometer errors are difficult to model. This makes the selection of orbits for gravity recovery difficult. Here we identify the noise level present in in–situ potential values from the energy balance in an iterative variance–component estimation. This means we solve simultaneously for a spherical harmonic model, for polynomial coefficients absorbing accelerometer drift, and for sub–daily noise variance components. These should be understood in a sense of model consistency. Using dynamic GFZ orbits, results including 92 days in 2002 indicate that for most days the noise is at 0.25–0.3$m^2/s^2$, with notable exceptions. The corresponding gravity model is found close to EIGEN–2, after two iterations. With TUM kinematic orbits and Lagrange–interpolated velocities or TUM reduced-dynamic orbits, we found for preselected data the consistency at the 0.7-0.8$m^2/s^2$ (KIN), 0.3-0.35$m^2/s^2$ (RD) level; gravity models improve significantly on EGM96. Generally, iterative re–weighting improves the solutions significantly, and 'trackiness' is considerably reduced.

**Key words:** CHAMP, energy balance approach, statistical assessment, variance components

## 1 Energy Conservation Method

Various groups have demonstrated that the energy balance approach can be used to compute a gravity model from CHAMP reduced-dynamic or kinematic orbits, see Gerlach et al. (2003a,b) or Howe et al. (2003). The approach can also be used to verify the consistency of CHAMP accelerometer data, orbits, gravity field model and other (e.g. tidal) models, and to assess the magnitude of systematic and stochastic errors. We have analyzed CHAMP data for the in–situ potential following Jekeli's method. We have estimated simultaneously corrections to the spherical harmonic coefficients, sub–daily polynomial coefficients describing residual (after applying bias and scale factors from the ACC files) drift of the accelerometer, and sub–daily variance components of the in–situ potential values. A known obstacle for this type of analysis is the selection of 'good' orbits. In our approach, arcs showing spurious behaviour are effectively downweighted within an iterative maximum–likelihood estimation process, which improves our gravity field solution significantly. In turn, the estimation of the individual variance components is improved. As a by–product, we have investigated the correlations of estimated accelerometer drift parameters with the spherical harmonic coefficients.

The theory of the energy balance approach and its potential application to LEO satellite experiments goes back to the 60's, and has been considerably revived recently, see Jekeli (1999), Visser et al. (2003) Ilk and Löcher (in press). We use the following formulation, which is based on expressing all quantities of interest in an inertial coordinate system:

$$T(t) - V^{\text{ref}}(t) - \delta V(t) - R(t) = E_0 + \int_{t_0}^{t} \boldsymbol{f} \cdot \dot{\boldsymbol{x}}\, d\tau + \int_{t_0}^{t} \nabla V^{\text{tides}} \cdot \dot{\boldsymbol{x}}\, d\tau \quad (1)$$

Here $T = \frac{1}{2}|\dot{\boldsymbol{x}}|^2$ is the kinetic potential, $V^{\text{ref}}$ is a static reference potential appearing time–dependent in inertial coordinates, $\delta V$ is a residual geopotential that we parameterize by spherical harmonics whose coefficients $\delta c_{lm}, \delta s_{lm}$ are to be solved for, $R$ is the potential rotation term which approximates the potential contribution $\int_{t_0}^{t} \frac{\partial V}{\partial t}\, d\tau \approx -\omega_e(x_1\dot{x}_2 - x_2\dot{x}_1)$ (up to a constant, see Jekeli 1999) of the rotating earth in inertial space, and $E_0$ is a constant. Furthermore, $\boldsymbol{f}$ are corrected measurements from CHAMP's STAR accelerometer to account for non–conservative forces, and the last term on the right–hand side accommodates for tidal effects by evaluating the corresponding work integral. We found that neglecting the explicit time–variation of the tidal potentials by simply evaluating $V^{\text{tides}}(t)$ would be possible but introduces low–frequency drift effects that propagate into the low–order correction polynomial coefficients (see below). We model the direct attraction by sun and moon from JPL DE ephemeris, the solid earth tides following the IERS conventions, plus ocean tides (GOT 99.2).

## 2 Statistical Assessment and Estimation Procedure

The energy balance approach uses eq (1) for combining orbit, accelerometry data, reference geopotential model, tidal corrections and auxiliary information (e.g. earth rotation) into a preprocessed stream of pseudo–observations, $\delta V(t)$, which can be used without further linearization to estimate the $\delta c_{lm}, \delta s_{lm}$. Consequently, position errors $\boldsymbol{\epsilon_x}$, velocity errors $\boldsymbol{\epsilon_{\dot{x}}}$, accelerometry errors $\boldsymbol{\epsilon_f}$ and tide model errors affect the in–situ potential differences (see Visser et al., 2003) approximately as

$$\epsilon_V = \dot{\boldsymbol{x}}\cdot\boldsymbol{\epsilon_{\dot{x}}} - \nabla V^{\text{ref}}\cdot\boldsymbol{\epsilon_x} - \omega_e(\epsilon_{x_1}\dot{x}_2 - \epsilon_{x_2}\dot{x}_1) - \omega_e(x_1\epsilon_{\dot{x}_2} - x_2\epsilon_{\dot{x}_1}) + \int_{t_0}^{t} \boldsymbol{\epsilon_f}\cdot\dot{\boldsymbol{x}}\, d\tau + \delta V^{\text{tides}} \quad (2)$$

It is clear that accelerometer biases and scaling errors, predominantly in the in–flight axis, cause in first approximation a linear drift in the $\delta V$ measurement. It is also known that if an erroneous reference model went into the computation of (reduced–) dynamic orbits, resulting orbit errors will compensate to a certain extend for this in the energy balance (1), and an estimated gravity model will be biased. This is why we use the term noise variance in a sense of consistency. Knowing the variance of the potential difference error, $\sigma^2(\delta V) = E\{\epsilon_V^2\}$, we can set up a weighted least square adjustment which would suppress spurious arcs by downweighting. This noise variance, however, is nonstationary and difficult to assess a priori. The orbits, on the other hand, are given in batches of 0.5–1.5 day length dependent on the POD analysis strategy. Here we assign an unknown variance component $\sigma_k^2(\delta V)$ to each batch and estimate it jointly with the residual gravity field and with parameters that account for accelerometer drift. Written as a Gauss–Markov model, this is

$$\boldsymbol{A}_k \begin{pmatrix} \boldsymbol{x}_{\mathrm{SH}} \\ \boldsymbol{x}_{\mathrm{ACC}_1} \\ \cdots \\ \boldsymbol{x}_{\mathrm{ACC}_p} \end{pmatrix} - \boldsymbol{y}_k = \boldsymbol{\epsilon}_k \qquad \begin{matrix} D(\boldsymbol{\epsilon}_k) = \sigma^2_{(k)}\boldsymbol{I} \\ D(\boldsymbol{x}_{\mathrm{SH}}) = \sigma^2_{(0)}\boldsymbol{R} \end{matrix} \qquad k = 1 \ldots p \qquad (3)$$

where $\boldsymbol{x}_{\mathrm{SH}}$ contains SH coefficients, $\boldsymbol{x}_{\mathrm{ACC}_k}$ and $\sigma^2_{(k)}$ are drift parameters and variance components for the $k$–th data set, and $\sigma^2_{(0)}$ is a regularization parameter if needed. This requires an iterative strategy involving re–weightings of the contributions, synthesis of potential residuals, and repeated solutions of the overall LS problem. A fast Monte Carlo machinery for implementing maximum–likelihood estimation has been developed in Koch and Kusche (2002) and tested in Kusche (2003) on a simulated LEO gravity recovery problem. Weighting factors for combination solutions can be computed the same way. At convergence the results of ML estimation equal those of the iterated MINQUE technique. It should be emphasized that from the point of view of estimating drift parameters for each batch $k$ of data, it would be preferable to have short batches comprising an orbital revolution each. From the statistical analysis point, we must keep the number of solved–for variance components limited to maintain fast convergence of our algorithm.

## 3 Analysis of GFZ dynamic orbits

We used 92 days of GFZ's PSO and ACC data, of the first half 2002 and without preselection. These are broken up into daily/half–daily batches. We solve for a SH expansion to degree 75, for 4 polynomial parameters per day to account for accelerometer drift in the form $\epsilon(t) = \sum_{n=0}^{3} a_n^k (t - t_k)^n$, and for a daily variance $\sigma_k^2$. Arc–dependent $a_n^k$ parameters are eliminated from the normal system by the method of partitioning and backsubstitution. The gravity field model that went into GFZ's dynamic PSO orbits is not completely identical to EIGEN–2 (P. Schwintzer, pers. comm.), and we treat it as unknown here. In the first iteration, all batches are weighted equally. Fig. 1 shows the estimated $\sigma_k$'s after the second iteration where practically convergence is reached. Fig. 3 (left) shows the difference of the residual gravity field with respect to GFZ's published EIGEN–2 solution (Reigber et al., 2003) without re–weightings (0th iteration), in geoid heights. Fig. 3 (right) shows the difference, complete to degree 70, after arriving at convergence. No data has been removed, but all spurious orbits are downweighted. It should be noted that for EIGEN–2 data from a different time period was used than we used in this study.

## 4 TUM kinematic and reduced–dynamic orbits

We have further investigated about 100 days of kinematic CHAMP orbits, which were kindly provided by D. Svehla by IAPG, TU Munich. These orbits are processed following the zero–differencing strategy, see Svehla and Rothacher (2003), and were provided with full variance–covariance information per data point. In

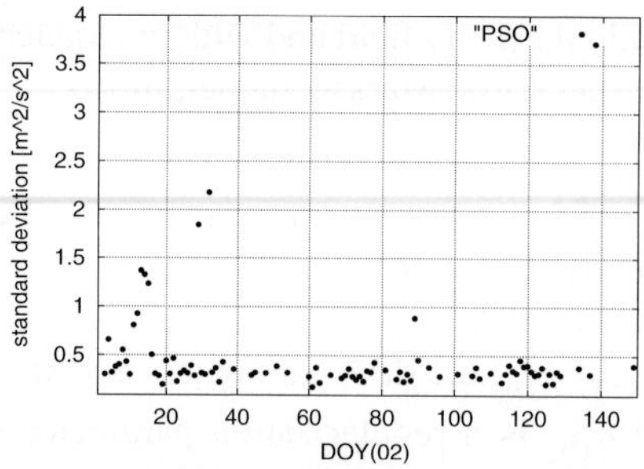


**Fig. 1.** $\sigma_k^2(\delta V)$ for DOY 002–148 (2002), PSO orbits

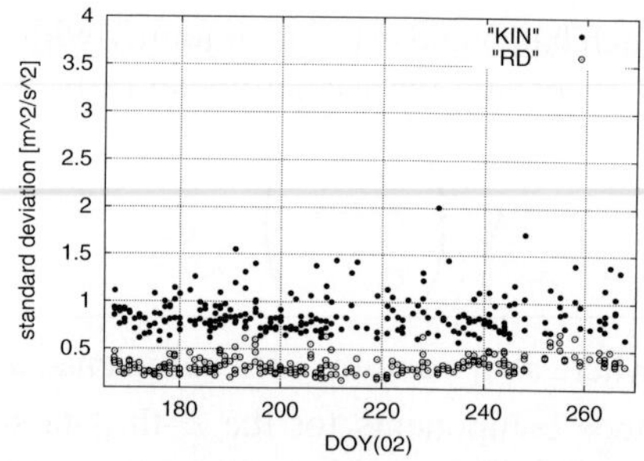


**Fig. 2.** $\sigma_k^2(\delta V)$ for DOY 167–268 (2002), KIN and RD orbits

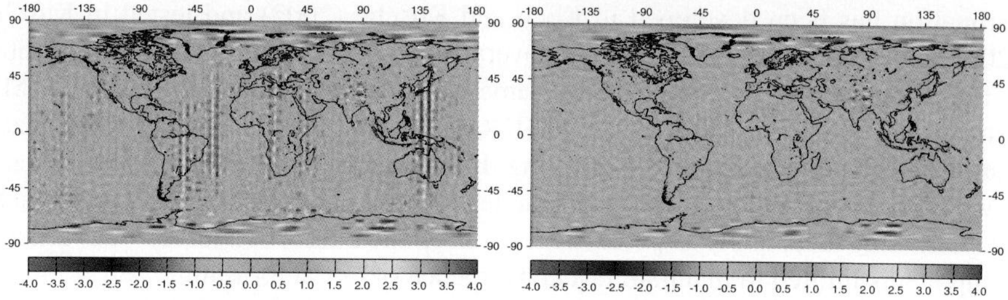


**Fig. 3.** Difference to EIGEN–2 [m]. Left: L=75, 0th iteration. Right: L=70, after 2nd iteration

a preprocessing step we have removed all kinematic positions to which a–priori sigma's of larger than 5cm in either $x$, $y$, or $z$ were assigned. Furthermore, only data segments of at least 2.5h have been selected. To allow a clean comparison afterwards, we have based our selection on purpose not on comparing KIN and RD orbits. After this, the used data corresponds to about 52 days. Kinematic orbit determination does not provide velocities, so we had to compute CHAMP velocities using a Lagrange 7–point interpolator. We avoid any smoothing at this step. In what follows, the same analysis has been performed as with the GFZ PSO data. One must, however, bear in mind that KIN orbits are completely independent from any prior gravity field, that they are spaced at 30s intervals, and that the interpolator used for deriving velocities is of influence. In addition, we have taken TUM RD orbits for the same time periods and repeated the procedure (apart from the Lagrange interpolation). Results for the noise levels are shown in Fig. 2.

In summary, one can state that the estimated variance components for the KIN orbits are roughly at the 0.7–0.8$m^2/s^2$ level, whereas those for the RD orbits are at the 0.3–0.35$m^2/s^2$ level, with exceptions and a slight increase at about DOY 250. According to eq. (2), assuming that the velocity error dominates, this indicates that velocities from Lagrange interpolation have an accuracy of about 1mm/s. We have compared these directly with RD velocities and found this error level confirmed. Note that we have based our orbit selection on purpose not on a–priori comparison of KIN and RD orbits; which would have given a more optimistic error level. Fig. 4 shows for the estimated gravity field solutions the signal de-

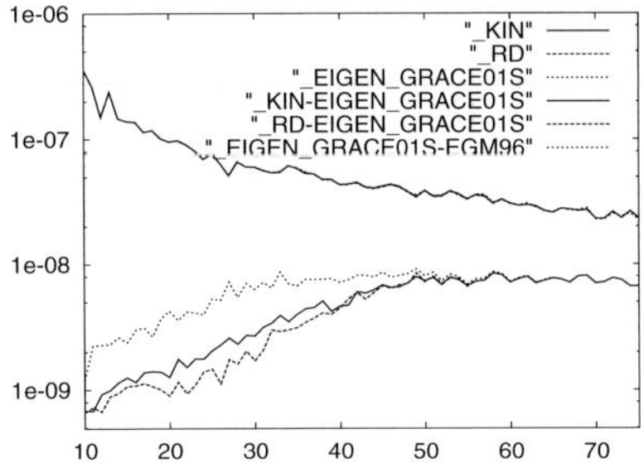


**Fig. 4.** Degree variances. Solutions with TUM KIN, RD orbits

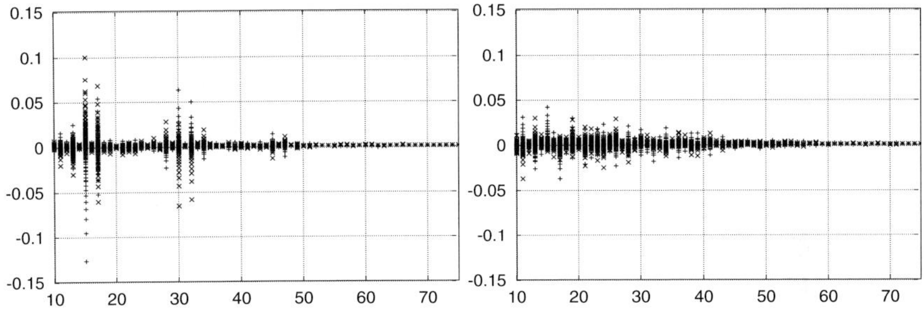


**Fig. 5.** Correlation of $a_0$ (+) and $a_1$ ($\times$) with the $\delta c_{lm}, \delta s_{lm}$. Left: arc length 16h. Right: 3h.

gree variances and the difference degree variance with respect to the recently published EIGEN–GRACE01S model (Reigber et al, in preparation). For comparison, EGM96 (Lemoine et al., 1998) is also shown, which has been used as a reference field in the computation. Both KIN and RD solution show clear improvements towards the EIGEN–GRACE01S, when compared with the EGM96 model. Due to regularization, there is almost no signal left beyond degree 50 and the reference solution EGM96 dominates. It is obvious that between degrees 15 and 40 the reduced-dynamic solution is somewhat closer to the GRACE model; probably because CHAMP data are already enclosed in the prior gravity model used for computing the RD orbits. We believe that the timespan we used is too short to draw further conclusions.

In our treatment the determination of arc–dependent accelerometer drift parameters $a_n^k$ is part of the estimation procedure and not a separate preprocessing step. It is therefore possible to study the correlation between the $a_n^k$ and the SH coefficients from the a–posteriori covariance matrix. In Fig. 5 we show as an example correlation coefficients of the estimated energy constant $a_0 \equiv E_0$ (+) and of the slope parameter $a_1$ ($\times$) with the $\delta c_{lm}, \delta s_{lm}$ per degree, for arc lengths of 16 hours and 3 hours. The correlation between estimates for $a_0$ and $a_1$ is much higher and increases for short arc length. However, as indicated in the figures, there appears to be no estimability problem so far. We plan to extend this type of correlation studies, in particular when incorporating longer data sets and time–variable low–degree SH coefficients.

## 5 Discussion and Outlook

We have discussed a statistical assessment of CHAMP orbits and data within the energy balance method. Non–stationary noise has been modelled with piecewise constant variance. We have proven that we can efficiently estimate individual noise levels for data batches, and that gravity solutions using an optimally weighted LS procedure are superior to heuristic weighting. Our results indicate that the consistency of GFZ PSO orbits is at 0.25–0.3$m^2/s^2$ for the time interval that we considered, with notable exceptions. TUM orbits were found at levels of 0.7–0.8$m^2/s^2$ (KIN) and 0.3–0.35$m^2/s^2$ (RD). This is due to the different POD processing strategies, and these figures should be interpreted with care. We have estimated gravity models which we believe clearly improve on pre–CHAMP models. More orbits have to be added to make final statements on the quality of these models. In particular, velocity derivation from KIN orbits needs to be investigated. Ongoing research includes using more data, and accounting for time–wise correlations. In the meantime, we added CHAMP data from KIN orbits to EGM96 by using its full variance–covariance matrix and determined weighting factors $\sigma_{(0)}$ by ML estimation.

*Acknowledgement.* We are grateful to GFZ Potsdam for providing CHAMP ACC and PSO data. Thanks go also to IAPG, TU Munich, for providing CHAMP orbits.

## References

1. Gerlach C, Sneeuw N, Visser P, Svehla D (2003) CHAMP gravity field recovery with the energy balance approach: first results. in: Reigber et al (Eds): First CHAMP Mission Results for Gravity, Magnetic and Atmospheric Studies, Springer, Berlin: 134–139.
2. Gerlach C and 11 authors (in press). A CHAMP–only gravity field model from kinematic orbits using the energy integral. Submitted to Geophys Res Lett.
3. Howe E, Stenseng L, Tscherning CC (2003) Analysis of one month of state vector and accelerometer data for the recovery of the gravity potential. Adv Geosciences *1*: 1–4.
4. Ilk KH, Löcher A (in press) The use of energy balance relations for validation of gravity field models and orbit determination. presented at IUGG Gen. Ass., 2003, Sapporo.
5. Jekeli C (1999) The determination of gravitational potential differences from satellite–to–satellite tracking. Cel Mech Dyn Astr *75*: 85–100.
6. Koch K–R, Kusche J (2002) Regularization of geopotential determination from satellite data by variance components. J Geodesy *76*: 259–268.
7. Kusche J (2003) A Monte–Carlo technique for weight estimation in satellite geodesy. J Geodesy *76*: 641–652.
8. Lemoine FG and 14 authors (1998) The Development of the Joint NASA GSFC and the National Imagery and Mapping Agency (NIMA) Geopotential Model EGM96, NASA/TP–1998–206861, NASA–GSFC, Greenbelt MD.
9. Reigber C and 12 authors (2003) The CHAMP–only Earth Gravity Model EIGEN–2. Adv Space Res *31(8)*: 1883–1888.
10. Reigber C, Schmidt R, Flechtner F, König R, Meyer U, Neumayer KH, Schwintzer P, Zhu S (in prep.) First EIGEN Gravity Field Model based on GRACE Mission Data Only
11. Svehla D, Rothacher M (2003) Kinematic and reduced–dynamic precise orbit determination of low earth orbiters. Adv Geosciences *1*: 47–56.
12. Visser P, Sneeuw N, Gerlach C (2003) Energy integral method for gravity field determination from satellite orbit coordinates. J Geodesy *77*: 207–216.

# Multiscale Geopotential Solutions from CHAMP Orbits and Accelerometry

Martin J. Fengler[1], Willi Freeden[1], and Jürgen Kusche[2]

[1] TU Kaiserslautern, Geomathematics Group, 67653 Kaiserslautern, P.O. Box 3049, Germany
[2] TU Delft/DEOS, 2600 GB Delft, P.O. Box 5058, The Netherlands

**Summary.** CHAMP orbits and accelerometer data are used to recover the long- to medium-wavelength features of the Earth's gravitational potential. In this study we are concerned with analyzing preprocessed data in a framework of multiscale recovery of the Earth's gravitational potential, allowing both global and regional solutions. The energy conservation approach has been used to convert orbits and accelerometer data into in-situ potential. Our modelling is spacewise, based on (1) non-bandlimited least square adjustment splines to take into account the true (non-spherical) shape of the trajectory (2) harmonic regularization wavelets for solving the underlying inverse problem of downward continuation. Furthermore we can show that by adapting regularization parameters to specific regions local solutions can improve considerably on global ones. We apply this concept to kinematic CHAMP orbits, and, for test purposes, to dynamic orbits. Finally we compare our recovered model to the EGM96 model, and the GFZ models EIGEN-2 and EIGEN-GRACE01s.

**Key words:** CHAMP, Kinematic Orbit, PSO Orbit, Regional Gravitational Field Recovery, Tikhonov-Wavelet Regularization

## 1 Introduction

In this paper high-low satellite-to-satellite tracking (hi-lo SST) of a low Earth orbiter (LEO) for gravity recovery purposes is discussed from an alternative point of view, as originally proposed by W. Freeden [4]. More specifically, we are concerned with the determination of the Earth's external gravitational field from a given set of potential values along the orbit of CHAMP. We have obtained these potential values by applying an energy conservation approach, basically following [8], to GFZ PSO as well as to TUM kinematic CHAMP orbits. In order to translate the hi-lo SST problem into a rigorous mathematical formulation we will make use of the geometrical situation outlined in Fig. 1.

We have the following problem (see [4]): *Let there be known the potential values $V(x), x \in \Gamma$, for a subset $\Gamma \subset \Sigma^{ext}$ of points at the orbit positions of the LEO. Find an approximation $U$ to the geopotential field $V$ on $\overline{\Sigma^{ext}}$, i.e. on and outside the Earth's surface, such that the difference of $V$ and $U$ is arbitrarily small on $\overline{\Sigma^{ext}}$ in terms of the underlying function spaces. In addition we require $V(x) = U(x)$ for all $x \in \Gamma$.* Existence, uniqueness, and well-posedness of the problem are discussed in [4] and the references therein.

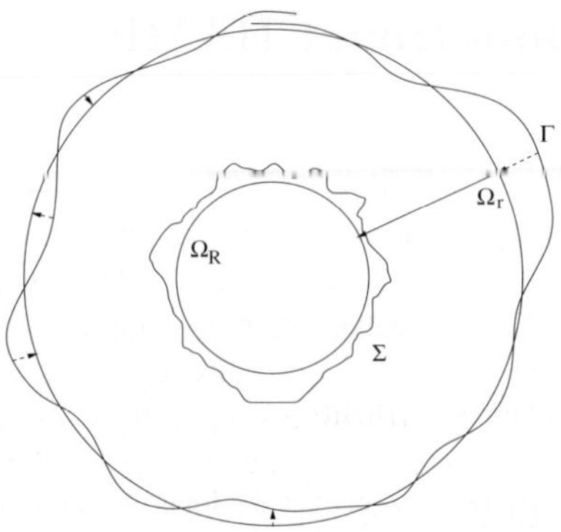


**Fig. 1.** Illustration of the geometrical configuration.

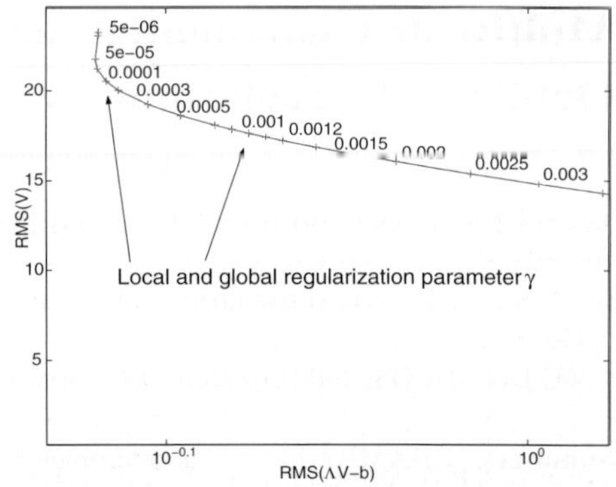


**Fig. 2.** Local L-curve for the region shown in Fig. 3, in $\mathrm{Harm}_{25,\dots,90}(\Omega_R)$.

In practice we are interested in computing global, and if possible, regionally improved gravity models from real CHAMP-data. In particular for the regional case our wavelet approach demonstrates advantages since it allows for the local choice of a regularization parameter. Thus, we apply locally adaptive regularization by virtue of multiresolution analysis using adequately constructed wavelets.

The three-dimensional coordinate system that we use throughout this study is the particular realization of the international terrestrial reference frame in which the CHAMP ephemeris are given. Let $r$ be CHAMP's mean orbital altitude. The sphere in $\mathbb{R}^3$ with radius $r$ around the origin is denoted by $\Omega_r$, i.e. $\Omega_r = \{x \in \mathbb{R}^3 | |x| = r\}$. For later use we reserve the name $\Omega_R$ to denote the Bjerhammar sphere (see Fig. 1). With $\Omega_r^{\mathrm{ext}}$ we denote the exterior of $\Omega_r$, while $\Omega_r^{\mathrm{int}}$ is the interior of $\Omega_r$. Further, in our modelling we will make use of two different spherical grids: The equiangular Driscoll-Healy grid [1] for numerical integration purposes, and the so–called Reuter grid [11] known from low discrepancy methods. The Reuter grid is important in our data selection strategy since it exhibits an almost equidistribution of the points on the sphere.

For extrapolation of the data $\{y_i, F_i\}, i = 1, \dots, M$, i.e. $M$ potential values along the orbit $\Gamma$ to an integration grid on the sphere $\Omega_r$ we use a least square adjustment spline $S$, see for more details [2, 3, 10]. For this purpose we employ a Reuter grid $\{x_j\}, j = 1, \dots, N \subset \Omega_R$ for locating the Abel-Poisson kernel $K$ which is defined by

$$K_{\mathcal{H}(A,\Omega_R)}(x,y) = \frac{1}{4\pi R^2} \frac{1-h^2}{(1+h^2-2h(x\cdot y))^{3/2}} . \tag{1}$$

The $N$ coefficients of the spline $S = \sum_{j=1}^{N} a_j K(x_j, \cdot)$ follow from solving the overdetermined linear system

$$\sum_{j=1}^{N} a_j^N K_{\mathcal{H}}(x_i, y_j) = F_i, \quad i = 1, \ldots, M.$$

**Sobolev Spaces:** We will work in Sobolev spaces, as introduced in [5, 4], and assume in particular that the measured potential can be characterized as an element of the space $\mathcal{H}(A, \Omega_R)$ with symbol $A^\wedge(n) = h^{-n/2}$, $n \in \mathbb{N}_0$, $0 < h < 1$. This again leads immediately to the Abel-Poisson kernel, representing the reproducing kernel in $\mathcal{H}(h^{-n/2}, \Omega_R)$. It should be remarked that this kernel can be directly identified with the *upward continuation* operator. This operator facilitates mapping between $\mathcal{H}(A, \overline{\Omega_R^{ext}})$ and $\mathcal{H}(\Lambda^{-1}A, \overline{\Omega_r^{ext}})$, and can be interpreted as a pseudodifferential operator (PDO) defined on $\mathcal{H}(A, \overline{\Omega_R^{ext}})$ with the symbol $\Lambda^\wedge(n) = \left(\frac{R}{r}\right)^{n+1}$.

**Tikhonov Regularization Scaling Functions:** Now we consider the solution of the inverse problem given by $\Lambda V = G$, $V \in \mathrm{L}^2(\Omega_R)$, and $G \in \mathrm{L}^2(\Omega_r)$. As is well-known, cf. [4], this equation possesses an exponentially ill-posed pseudodifferential equation with an unbounded inverse operator $\Lambda^{-1}$. The idea of regularization is to replace the inverse operator by a more 'friendly' operator which yields an approximate solution. By operating directly on the singular values of $\Lambda^{-1}$, wavelets appear as a very appropriate tool to solve this problem. We obtain the $J$-level regularization of the potential $V$ on $\Omega_R$ by evaluating the convolution $V_J = \Phi_J * G = \int_{\Omega_r} \Phi_J(\cdot, \eta) G(\eta) d\omega_r(\eta)$, where $\Phi_J$ denotes the $J$-level regularization and reconstruction Tikhonov scaling function:

**Definition 1 (Tikhonov Scaling Function).** *Let $P_n, n \in \mathbb{N}_0$ denote the Legendre polynomials as given by [5], and $\gamma_j, j \in \mathbb{N}_0$ a positive decreasing sequence with $\lim_{j\to\infty} \gamma_j = 0$. The Tikhonov regularization scaling function is defined by*

$$\Phi_j(x, y) = \sum_{n=0}^{\infty} \phi_j(n) \frac{2n+1}{4\pi R r} \left( \frac{Rr}{|x||y|} \right)^{n+1} P_n \left( \frac{x}{|x|} \cdot \frac{y}{|y|} \right),$$

*with $x \in \Omega_R, y \in \Omega_r$, and the symbol $\phi_j(n)$ is given by*

$$\phi_j(n) = \frac{\Lambda^\wedge(n)}{(\Lambda^\wedge(n))^2 + \gamma_j}, \quad n = 0, 1, \ldots; \; j \in \mathbb{N}_0.$$

Since $\phi_j(n)$ decreases for increasing $n$, we may regard these functions just as low-pass filters - which is similar to Tikhonov regularization for ill-conditioned linear systems. The Tikhonov regularization wavelets are analogously obtained as bandpass filters, i.e. by the difference of two subsequent low-pass filters (see [4]).

## 2 Multiscale Geopotential Modelling from CHAMP Data

In the present study we consider two different CHAMP orbit data sets, covering several months in the year 2002. We work with dynamic PSO orbits provided kindly by

GFZ Potsdam, and with kinematic orbits computed by Drazen Svehla (IAPG, TU Munich) following the zero-differencing strategy. Whereas dynamic orbits basically represent a best-fit of GPS observations within an a–priori gravity field and provide a kind of test for our method, kinematic orbits are free of a-priori gravity information and the recovered gravity model should be free of any biases. These orbits have been converted to in-situ potential values following the energy conservation approach (cf. [8]), corrected for nonconservative forces using GFZ's accelerometer data products, and for astronomical and solid Earth tides according to IERS conventions, and ocean tides using GOT 99.2. That the energy balance approach can be successfully applied to gravity recovery from CHAMP has been proven meanwhile by several groups, cf. [7], [6], or [9]. By subtracting a reference potential up to degree 24 (from EGM96) we obtain residual potential values along the orbit. Finally, by using cubic approximating splines we try to suppress trends in the data which we believe are induced by remaining accelerometer drift effects.

**Data Gridding with Splines:** First, using the Abel-Poisson kernel with $h = 0.95$, we fit a least square adjustment spline to the residual potential along the satellite orbit. All data has been weighted equally, after applying a selection procedure. It is then possible to extrapolate these values to a Driscoll-Healy integration grid defined on the mean altitude sphere $\Omega_r$ (see Fig. 1). This essential step preserves the harmonicity in the data.

**Solving the Inverse Problem and Reconstruction:** From the previous step, we have now a set of gridded, *predicted measurements* on $\Omega_r$. This allows us to compute a $j$-level regularization of the potential on $\Omega_R$. However, we need an appropriate criterion for stopping the regularization, i.e. we have to balance the regularization error decreasing with higher scales, and the the reconstruction/prediction error increasing with higher scales. For this purpose we use the L-curve method, which in turn requires that we predict residual potential values in the spherical integration grid on $\Omega_r$, by use of the recovered potential. Since $\|\cdot\|_{\mathcal{H}(A,\Omega_R)}$ is defined in the spectral domain, we prefer for regional applications the $\mathrm{L}^2(\Omega_R)-$norm which can easily be evaluated in the space domain. We plot the norm of the reconstructed potential (within $\mathrm{Harm}_{25,\ldots,90}(\Omega_R)$, see below) on the y-axis against the prediction residual from this potential to the orbit data on the x-axis (cf. Fig. 2). Locally, we calculate the RMS of the reconstructed potential values respectively the data residuals on a grid, which approximates the $\mathrm{L}^2$-norm. It should be remarked that the time-consuming numerical integration in the wavelet transform is intrinsically data parallel. We exploit this by using an efficient parallel implementation. For an easier comparison of our solution with spherical harmonic models, we project the solution globally on $\mathrm{Harm}_{25,\ldots,50}(\Omega_R)$ and locally on $\mathrm{Harm}_{25,\ldots,90}(\Omega_R)$; that is we extract those spectral bands from our solution which can then directly compared to spherical harmonic expansions. The difference in the globally recovered potential compared to several models is given in Tab. 1 and 2.

**Local Reconstruction Process:** The Tikhonov scaling functions are strongly localizing on $\Omega_R$. It is clear that only a small cap contributes in the reconstructing convolution, which resembles numerically a *local compact support*. It is the rea-

**Table 1.** Global potential differences from CHAMP-PSO data in $\mathrm{Harm}_{25,\ldots,50}$

| $[\mathrm{m}^2/\mathrm{s}^2]$ | EGM96 | EIGEN-1s | EIGEN-2 | EIGEN-GRACE01s |
|---|---|---|---|---|
| Median Abs. Diff. | 2.918 | 2.900 | 2.407 | 2.467 |
| Mean Abs. Diff. | 3.825 | 3.727 | 3.283 | 3.274 |

**Table 2.** Global potential differences from CHAMPTUM kinematic data in $\mathrm{Harm}_{25,\ldots,50}$

| $[\mathrm{m}^2/\mathrm{s}^2]$ | EGM96 | EIGEN-1s | EIGEN-2 | EIGEN-GRACE01s |
|---|---|---|---|---|
| Median Abs. Diff. | 2.888 | 3.257 | 2.620 | 2.552 |
| Mean Abs. Diff. | 3.785 | 4.027 | 3.416 | 3.258 |

son why we are able to compute a reconstruction of the gravitational potential in a desired area only. For a detailed description for the choices of the windows see [2].

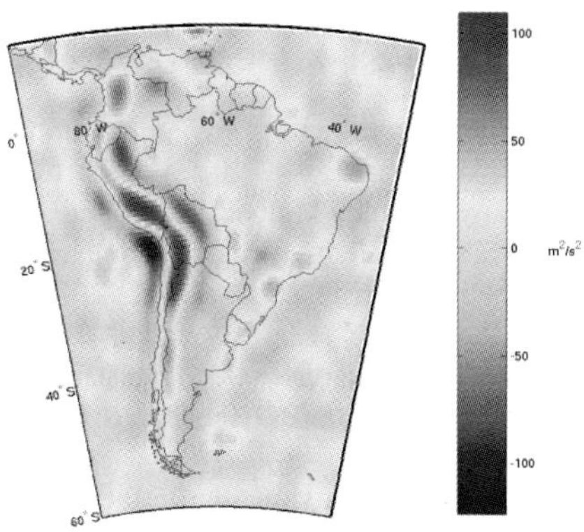


**Fig. 3.** Locally improved potential in $\mathrm{Harm}_{25,\ldots,90}(\Omega_R)$.

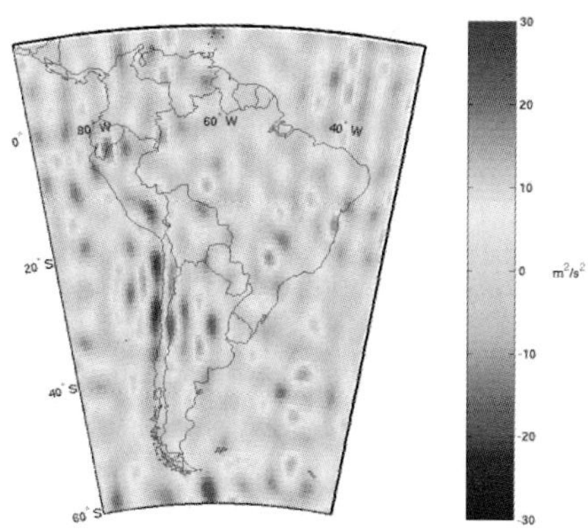


**Fig. 4.** Difference Fig. 3 and EIGEN-2 in $\mathrm{Harm}_{25,\ldots,90}(\Omega_R)$.

**Local CHAMP Data Analysis:** Global reconstructions show that differences to EIGEN-2 are located mainly in the high frequency parts, owing probably to regularization effects. Thus, we analyze a region of strong signal (see Fig. 3) and try to improve our global results locally. The locally obtained L-curve, Fig. 2, reveals that the regularization parameter obtained from a global L-curve is too large for this specific region, and high-frequent phenomena are over-smoothed. For more detail see [2]. Beyond this, the local L-curve indicates that $\gamma = 0.0001$ is an appropriate choice for this region. It turns out, that we are in fact able to improve our model locally. The maximum differences to EIGEN-1s and EIGEN-2 could be significantly decreased from ca. $60\frac{\mathrm{m}^2}{\mathrm{s}^2}$ to ca. $20\frac{\mathrm{m}^2}{\mathrm{s}^2}$ or even less, see Fig. 4. Remaining differences can be assigned to regions of high signal variability in the Andes. It will be subject of future work to investigate 'trackiness' in the solutions.

## 3 Conclusion

We believe our results show that a spline-based wavelet method can be applied successfully to real CHAMP data. Beyond this, it should be outlined that our method can improve satellite-only models regionally by adapting the regularization procedure to the regional variability of the Earth's gravity field.

*Acknowledgement.* We are grateful to GFZ Potsdam for providing access to CHAMP data products. Thanks go also to IAPG, TU Munich, for providing kinematic orbits. Further we are indebted to Research Center Jülich and to ITWM-Fraunhofer Institute Kaiserslautern, for granting access to high–performance computing facilities.

## References

1. Driscoll JR, Healy DM (1994) Computing Fourier Transforms and Convolutions on the 2-Sphere. Adv Appl Math *15*: 202–250.
2. MJ Fengler, Freeden W, Michel V (2003) The Kaiserslautern Multiscale Geopotential Model SWITCH-03 from Orbit Perturbations of the Satellite CHAMP and Its Comparison to the Models EGM96, UCPH2002_02_0.5, EIGEN-1s, and EIGEN-2. Geophys J Inter (accepted).
3. Freeden W (1981) On Approximation by Harmonic Splines. Manuscr Geod *6*: 193–244.
4. Freeden W (1999) Multiscale Modelling of Spaceborne Geodata. BG Teubner, Stuttgart, Leipzig.
5. Freeden W, Gervens T, Schreiner M (1998) Constructive Approximation on the Sphere (With Applications to Geomathematics). Oxford Science Publications, Clarendon, Oxford.
6. Gerlach C, Sneeuw N, Visser P, Svehla D (2003) CHAMP gravity field recovery with the energy balance approach: first results. in: Reigber C, Lühr H, Schwintzer P (eds), First CHAMP Mission Results for Gravity, Magnetic and Atmospheric Studies, Springer, Berlin: 134–139.
7. Han S-C, Jekeli C, Shum CK (2002) Efficient Gravity Field Recovery using in situ Disturbing Potential Observables from CHAMP. Geophys Res Lett *29(16)*: doi10.1029/2002GL015180.
8. Jekeli C (1999) The determination of gravitational potential differences from satellite-to-satellite tracking. Cel Mech Dyn Astron *75*: 85–101.
9. Kusche J, van Loon J (2004) Statistical assessment of Champ data and models using the energy balance approach. This volume.
10. Kusche J (2002) Inverse Probleme bei der Gravitationsfeldbestimmung mittels SST- und SGG-Satellitenmissionen. DGK Series C, No. 548, Munich.
11. Reuter R (1982) Über Integralformeln der Einheitssphäre und harmonische Splinefunktionen. Veröff Geod Inst RWTH Aachen, No. 33.

# Multiscale Modeling from EIGEN-1S, EIGEN-2, EIGEN-GRACE01S, UCPH2002_0.5, EGM96

Martin J. Fengler, Willi Freeden, Martin Gutting

University of Kaiserslautern, Geomathematics Group, 67653 Kaiserslautern, P.O. Box 3049, Germany

**Summary.** Spherical wavelets have been developed by the Geomathematics Group Kaiserslautern for several years and have been successfully applied to georelevant problems. Wavelets can be considered as consecutive band-pass filters and allow local approximations. The wavelet transform can also be applied to spherical harmonic models of the Earth's gravitational field like the most up-to-date EIGEN-1S, EIGEN-2, EIGEN-GRACE01S, UCPH2002_0.5, and the well-known EGM96. Thereby, wavelet coefficients arise and these shall be made available to other interested groups. These wavelet coefficients allow the reconstruction of the wavelet approximations. Different types of wavelets are considered: bandlimited wavelets (here: Shannon and Cubic Polynomial (CuP)) as well as non-bandlimited ones (in our case: Abel-Poisson). For these types wavelet coefficients are computed and wavelet variances are given. The data format of the wavelet coefficients is also included.

**Key words:** Multiscale Modeling, Wavelets, Wavelet Variances, Wavelet coefficients, Gravitational Field Model Convertion

## 1 Introduction

During the last years spherical wavelets have been brought into existence. (cf. e.g. [3], [2], [4] and the references therein). It is time to apply them to well-known models in order to offer easy access to the multiscale methods. Therefore, the spherical harmonics models EIGEN-1S, EIGEN-2, UCPH2002_0.5, EGM96 and also EIGEN-GRACE01S are transformed into bilinear wavelet models (see [3] or [2]) and the coefficients of these models are available via the worldwide web.

## 2 Wavelet Models

Due to the structure of the gravitational field we leave the first degrees and orders (up to $n = 2$) as an approximation by spherical harmonics which are denoted by $Y_{n,k}$. For the remaining parts of the models, a scaling function and its corresponding wavelets are applied. Thus, we can write the $J$-level representation of the geopotential $V$ on the sphere $\Omega_R$ in terms of a spherical harmonics part $V_{0..2}$ (which we neglect from now on), a low-frequent band $V_{j_0}$ and wavelet bands $W_j$ for $x = R\xi \in \Omega_R$:

$$V_J(x) = V_{j_0}(x) + \sum_{j=j_0}^{J-1} W_j(x)$$

$$= (\Phi_{j_0} * (\Phi_{j_0} * V))(x) + \sum_{j=j_0}^{J-1} (\tilde{\Psi}_j * (\Psi_j * V))(x), \quad (1)$$

where $*$ denotes the $\mathcal{L}^2(\Omega_R)$-convolution and $\lim_{J\to\infty} V_J = V$ in the sense of $\mathcal{L}^2(\Omega_R)$.
The kernel can be expressed by a sum over Legendre polynomials (cf. [3] for $x, y \in \Omega_R$:

$$\Phi_j(x,y) = \sum_{n=0}^{\infty} \frac{2n+1}{4\pi R^2} \varphi_j(n) P_n\left(\frac{x}{|x|} \cdot \frac{y}{|y|}\right).$$

where $\varphi_j(n)$ denotes its symbol. The kernels $\Psi_j, \tilde{\Psi}_j$ of the corresponding primal and dual wavelets are constructed analogously with their symbols $\psi_j(n)$, $\tilde{\psi}_j(n)$ given by the refinement equation:

$$\psi_j(n)\tilde{\psi}_j(n) = (\varphi_{j+1}(n))^2 - (\varphi_j(n))^2. \quad (2)$$

Since the considered gravitational field models are provided in terms of spherical harmonics, we may regard them to be bandlimited of degree $M$. For including the maximal information of the models, we choose the highest scale $J$ such that that $2^J > M$.
By using an equiangular grid we integrate exactly spherical harmonics up to the degree of the integrand (if it is bandlimited, otherwise we obtain an approximation) in (1). Thus, we write:

$$V_J(x) \approx \sum_{i=1}^{(N_{j_0}+1)^2} w_i^{j_0} \underbrace{\int_{\Omega_R} \Phi_{j_0}(z, y_i^{j_0}) V(z) d\omega(z)}_{a_i^{j_0}} \Phi_{j_0}(x, y_i^{j_0})$$

$$+ \sum_{j=j_0}^{J-1} \sum_{i=1}^{(N_j+1)^2} w_i^j \underbrace{\int_{\Omega_R} \Psi_j(z, y_i^j) V(z) d\omega(z)}_{c_i^j} \tilde{\Psi}_j(x, y_i^j).$$

In the latter formulae the $(y_i^{j_0}, w_i^{j_0})$ and $(y_i^j, w_i^j)$ denote the locations on $\Omega_R$ and corresponding weights of the Driscoll-Healy integration scheme (cf. [1]).
The scaling function and wavelet coefficients $a_i^{j_0}$ and $c_i^j$ are also obtained by numerical integration. Therefore, the coefficients are

$$a_i^{j_0} \approx \sum_{k=1}^{(\tilde{N}_{j_0}+1)^2} \tilde{w}_k^{j_0} \Phi_{j_0}(z_k^{j_0}, y_i^{j_0}) V(z_k^{j_0}), \quad c_i^j \approx \sum_{k=1}^{(\tilde{N}_j+1)^2} \tilde{w}_k^j \Psi_j(z_k^j, y_i^j) V(z_k^j).$$

## 3 Selected Examples

We have chosen three different types of wavelets: the bandlimited Shannon and Cubic Polynomial and the non-bandlimited Abel-Poisson kernel.

**Shannon Wavelets:** In the case of Shannon scaling functions the symbol $\varphi_j(n)$ reads as follows

$$\varphi_j^{SH}(n) = \begin{cases} 1 & \text{for} \quad n \in [0, 2^j) \\ 0 & \text{for} \quad n \in [2^j, \infty), \end{cases}$$

and for the corresponding wavelets we choose the P-scale version to resolve the refinement equation (2), i.e.

$$\tilde{\psi}_j^{SH}(n) = \psi_j^{SH}(n) = \sqrt{\left(\varphi_{j+1}^{SH}(n)\right)^2 - \left(\varphi_j^{SH}(n)\right)^2}.$$

**Cubic Polynomial (CuP) Wavelets:** In the CuP case the symbol takes the following form:

$$\varphi_j^{CP}(n) = \begin{cases} (1 - 2^{-j}n)^2(1 + 2^{-j+1}n) & \text{for} \quad n \in [0, 2^j) \\ 0 & \text{for} \quad n \in [2^j, \infty) \end{cases}$$

and for the corresponding wavelets we apply again the P-scale version. The parameters for the discretization of the integrals in (1) are also taken as above.

**Abel-Poisson Wavelets:** For the Abel-Poisson scaling function the symbol takes the following form:

$$\varphi_j^{AP}(n) = e^{-2^{-j}\alpha n}, \quad n \in [0, \infty), \quad \text{with some constant } \alpha > 0.$$

We choose $\alpha = 1$. Since $\varphi_j^{AP}(n) \neq 0$ for all $n \in \mathbb{N}$ this symbol leads to a non-bandlimited kernel.

It should be noted that the Abel-Poisson scaling function has a closed form representation which allows the omission of a series evaluation and truncation, i.e. for $x, y \in \Omega_R$, i.e. $|x| = |y| = R$ we have

$$\begin{aligned} \Phi_j^{AP}(x, y) &= \sum_{n=0}^{\infty} \frac{2n+1}{4\pi R^2} \varphi_j^{AP}(n) P_n\left(\frac{x}{|x|} \cdot \frac{y}{|y|}\right) \\ &= \frac{1}{4\pi R^2} \frac{1 - e^{-2^{-j+1}}}{\left(1 + e^{-2^{-j+1}} - 2e^{-2^{-j}}t\right)^{\frac{3}{2}}}. \end{aligned}$$

When constructing bilinear Abel-Poisson wavelets we want to keep such a representation as an elementary function. Thus, we decide to use M-scale wavelets whose symbols are deduced from the refinement equation (2) by the third binomial formula:

$$\begin{aligned} \psi_j^{AP}(n) &= \left(\varphi_{j+1}^{AP}(n) - \varphi_j^{AP}(n)\right) \\ \tilde{\psi}_j^{AP}(n) &= \left(\varphi_{j+1}^{AP}(n) + \varphi_j^{AP}(n)\right). \end{aligned}$$

Since the Abel-Poisson scaling function and its corresponding wavelets are non-bandlimited we obtain just a good approximation by the method based on an equiangular grid (we choose the parameter of polynomial exactness sufficiently large enough).

# 4 Wavelet Variances

The wavelet coefficients $c_i^j$ can also be used to compute the scale and space variance of $V$ at the positions $y_i^j$ and scale $j$. These variances at the positions $y_i^j$ are given by

$$\mathrm{Var}_{j\,;\,y_i^j}(V) = \int\limits_{\Omega_R}\int\limits_{\Omega_R} V(x)V(z)\Psi_j(x,y_i^j)\Psi_j(z,y_i^j)d\omega(x)d\omega(z) = \left(c_i^j\right)^2 .$$

The scale variance of $V$ at scale $j$, $\mathrm{Var}_j(V)$, is then defined as the integral over the scale and space variances which can be evaluated using the coefficients $c_i^j$. For more about wavelet variances, see [4].

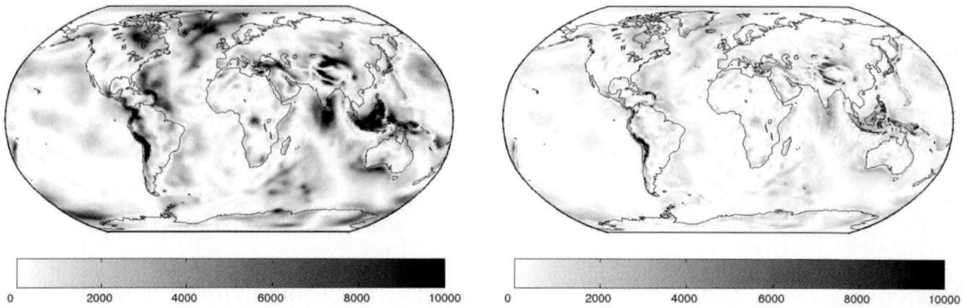


*CP-wavelet variances for scales 6 (left) and 7 (right) of EGM96, in* $[m^4/s^4]$

# 5 Reconstruction

We supply to the end-user the scaling function or wavelet coefficients, $a_i^{j_0}$ or $c_i^j$ corresponding to some locations on $\Omega_R$. One reconstructs the $J$-level representation of the potential $V$ by

$$V_J(x) \approx \sum_{i=1}^{(N_{j_0}+1)^2} w_i^{j_0} a_i^{j_0} \Phi_{j_0}(x, y_i^{j_0}) + \sum_{j=j_0}^{J-1} \sum_{i=1}^{(N_j+1)^2} w_i^j c_i^j \tilde{\Psi}_j(x, y_i^j). \qquad (3)$$

A full multiscale analysis of the EIGEN2 with CP-wavelets is exemplarily given below. (Note that all figures are in $[m^2/s^2]$.)

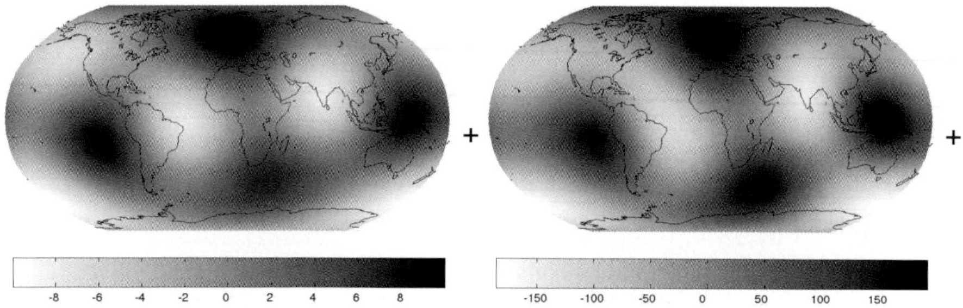


*The CP-scaling function with $j_0 = 2$ (left) and the CP-wavelet with $j = 2$ (right)*

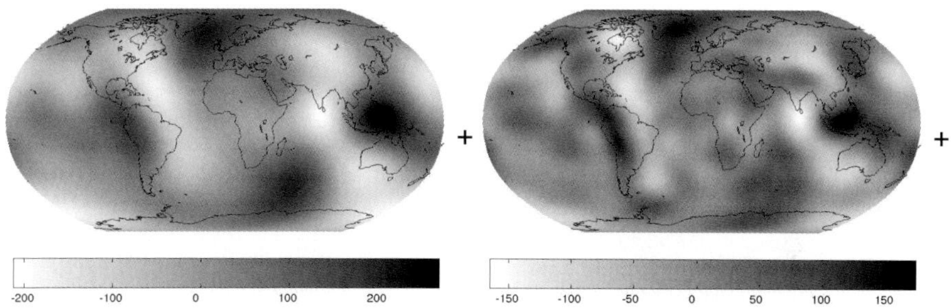


*The CP-wavelet with $j = 3$ (left) and $j = 4$ (right)*

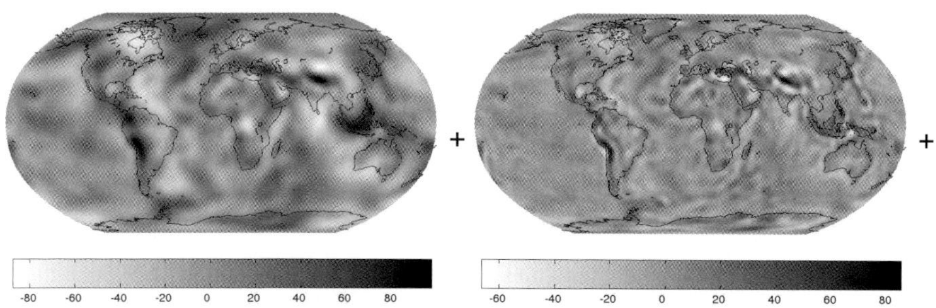


*The CP-wavelet with $j = 5$ (left) and $j = 6$ (right)*

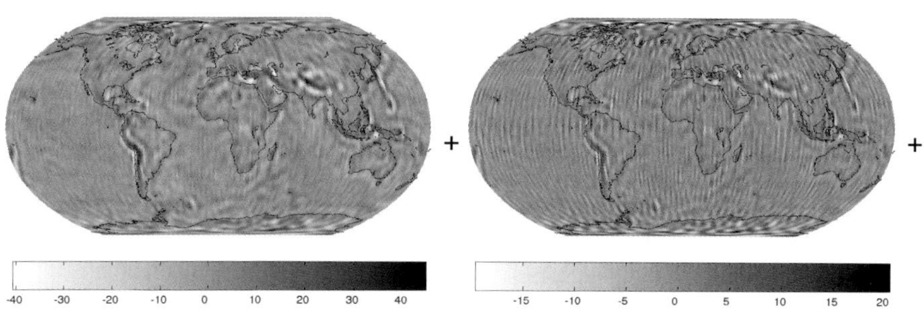


*The CP-wavelet with $j = 7$ (left) and $j = 8$ (right)*

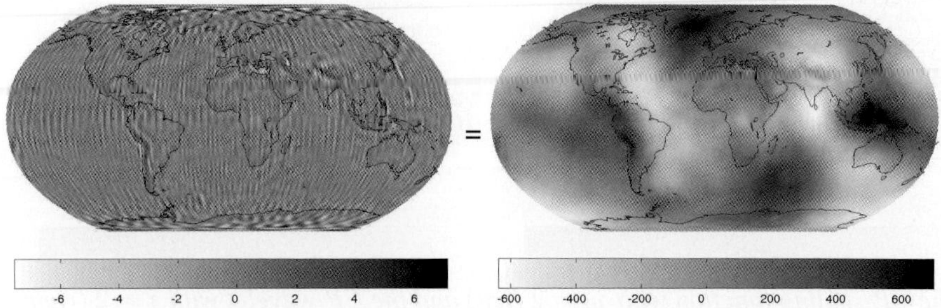


*The CP-wavelet with $j = 9$ (left) and the whole sum (3) (right)*

Moreover, we present some details of the Abel-Poisson multiresolution of EIGEN2. The sectorial parts of the model are resolved more and more by the higher scales.

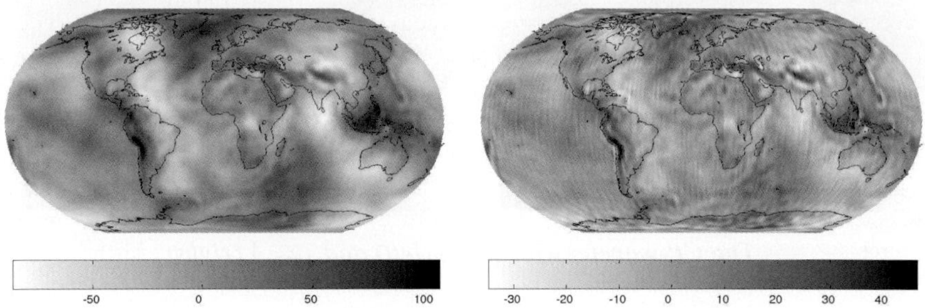


*The AP-wavelets of scale 6 (left) and 8 (right) in* $[m^2/s^2]$.

**Location in the Web**

The coefficients, a detailed model description as well as further figures can be found and downloaded at the following web page:

*http://www.mathematik.uni-kl.de/~wwwgeo/waveletmodels.html*

*Acknowledgement.* The authors want to thank the Fh-ITWM for granting them computing time on their cluster. Moreover, the DFG is gratefully acknowledged for its financial support.

## References

1. Driscoll JR, Healy DM (1994) Computing Fourier Transforms and Convolutions on the 2-Sphere. Adv Appl Math *15*: 202–250.
2. Freeden W (1999) Multiscale Modelling of Spaceborne Geodata. BG Teubner, Stuttgart, Leipzig.
3. Freeden W, Gervens T, Schreiner M (1998) Constructive Approximation on the Sphere (With Applications to Geomathematics). Oxford Science Publications, Clarendon, Oxford.
4. Freeden W, Maier T (2002) On Multiscale Denoising of Spherical Functions: Basic Theory and Numerical Aspects. Electron Trans Numer Anal *14*: 40-62.

# A Comparison of Various Procedures for Global Gravity Field Recovery from CHAMP Orbits

Torsten Mayer-Guerr[1], Martin Feuchtinger[1], Juergen Kusche[2]

[1] Institute of Theoretical Geodesy, University Bonn, Germany
*tmg@mail.geod.uni-bonn.de, feuchtinger@mail.geod.uni-bonn.de*
[2] DEOS, Delft University of Technology, Netherlands, *j.kusche@citg.tudelft.nl*

**Summary.** We compare selected techniques for recovering the global gravity field from precisely determined kinematic CHAMP orbits. The first method derives the second derivatives by use of an interpolation polynomial. The second procedure is based on Newton's equation of motion, formulated and solved as a boundary value problem in time equivalent to a corresponding integral equation of Fredholm type. It is applied to short arcs of the CHAMP orbits. The third method is based on the energy balance principle. We implement the analysis of in-situ potential differences following Jekeli's formulation. The normal equations from the three approaches are solved using Tikhonov–type regularization, where the regularization parameter is computed according to a variance component estimation procedure. The results are compared with the recent satellite-only model EIGEN2 and the first GRACE model GGM01s. All methods provide solutions of the gravity field which represent significant improvements with respect to the reference model EGM96 below degree 50. The quality of the solutions differs only slightly.

**Key words:** CHAMP, gravity field recovery, boundary value problem, polynomial interpolation, energy balance approach

## 1 Introduction

Various groups have introduced different approaches to determine the global gravity field from precisely determined kinematic CHAMP orbits. We compare three of those techniques. The first method derives the second derivatives by use of an interpolation polynomial. The second procedure is formulated as a boundary value problem in time. It is applied to short arcs of the CHAMP orbits. The third method is based on the energy balance principle. Apart from the different observation models the same procedure has been used for the calculation of all three methods. This includes the same data set and the same way of solving the normal equations using Tikhonov–type regularization, where the regularization parameter is computed according to a variance component estimation procedure.

## 2 Data settings

The global gravity field recovery presented here is based on kinematic orbits of CHAMP with a sampling rate of 30 seconds provided by M. Rothacher and

D. Svehla from the FESG of the Technical University Munich. The orbits cover a time period of approximately 100 days. These orbits are processed following the zero-differencing strategy, see (Svehla D, Rothacher M, 2003) and were provided with variance-covariance information per data point. In a preprocessing step we have removed all kinematic positions to which sigma's larger than 5cm in either x, y or z were assigned. Furthermore, only data segments of at leat 2.5h have been selected. After this, the used data corresponds to about 52 days.

EGM96 has been used as a reference field in the following denoted by $V$. The disturbing potential $T$ is modelled by a spherical harmonics expansion up to degree and order $L = 75$. The unknown coefficients $\Delta c_{nm}, \Delta s_{nm}$ can be estimated in an least squares adjustment.

The satellite's motion is also influenced by disturbing forces $\mathbf{f}$. We model the direct attraction by sun and moon from JPL DE ephemeris, the solid earth tides following the IERS conventions, and we implement an ocean tide model. CHAMP's STAR accelerometer measures the non-conservative forces like air drag and solar radiation pressure. Due to the spurious behavior of the accelerometer, bias parameters are estimated in all three models.

## 3 Observation equations

### 3.1 Polynomial differentiation

The functional model of the observations is based on Newton's equation of motion

$$\ddot{\mathbf{r}}(t) = \nabla V + \nabla T + \mathbf{f}. \tag{1}$$

The idea of this method is to approximate the orbit by an interpolation polynomial, in this case a Gregory-Newton $n$-point scheme (Austen et al., 2002)

$$\mathbf{r}(t) \approx \mathbf{r}_A + \sum_{i=1}^{n-1} \sum_{k=0}^{i} (-1)^{i+k} \binom{q}{i} \mathbf{r}_{k+1} \tag{2}$$

with

$$q = (t - t_1)/(t_2 - t_1).$$

To obtain the accelerations, we have to differentiate the interpolation scheme twice with respect to time $t$

$$\ddot{\mathbf{r}}(t) \approx \mathbf{r}_A + \sum_{i=1}^{n-1} \sum_{k=0}^{i} (-1)^{i+k} \binom{q}{i}'' \mathbf{r}_{k+1} \tag{3}$$

The accelerations are computed at the centre of the $n$-point scheme to get the smallest interpolation error. These approximated accelerations are used as pseudo observations for model (1). As they are linear combinations of positions, a full apriori variance-covariance matrix can be computed by linear error-propagation from given covariances per position.

### 3.2 Solving a boundary value problem in the time domain

This functional model is based on Newton's equation of motion as well, but formulated as a boundary value problem (Schneider, 1967),

$$\mathbf{r}(\tau) - (1-\tau)\mathbf{r}_A - \tau\mathbf{r}_B = -(t_B - t_A)^2 \int_0^1 K(\tau,\tau')\mathbf{g}(\tau',\mathbf{r},\dot{\mathbf{r}})\,d\tau', \tag{4}$$

with the integral kernel

$$K(\tau,\tau') = \begin{cases} \tau'(1-\tau) & \text{for } 0 \le \tau' \le \tau \\ \tau(1-\tau') & \text{for } \tau \le \tau' \le 1 \end{cases}$$

satisfying the boundary values

$$\mathbf{r}_A := \mathbf{r}(t_A), \quad \mathbf{r}_B := \mathbf{r}(t_B). \tag{5}$$

The function $\mathbf{g}$ contains all forces acting on the satellite's acceleration:

$$\mathbf{g} = \nabla V + \nabla T + \mathbf{f}. \tag{6}$$

Equation (4) is applied to short arcs after discretization in time (Ilk et al., 2003) and (P. Ditmer and A. van Eck van der Sluis, 2003). The linear combinations of three positions are used as pseudo observations. A full apriori variance-covariance matrix per arc is computed by linear error-propagation from given covariances per position.

### 3.3 Energy balance approach

The theory of the energy balance approach has been applied frequently, e.g. (Jekeli, 1999), (Gerlach et al., 2003), (Howe et al., 2003) or (Ilk and Loecher, 2003). We use the following formulation, which is based on expressing all quantities of interest in an inertial coordinate system:

$$E_{kin}(t) + V(t) + T(t) - R(t) - \int_{t_0}^{t} \mathbf{f}\cdot\dot{\mathbf{r}}\,d\tau = E_0 = const \tag{7}$$

with the kinetic energy

$$E_{kin} = \frac{1}{2}|\dot{\mathbf{r}}|^2, \tag{8}$$

and the potential rotation term

$$R(t) = \int_{t_0}^{t} \frac{\partial(V+T)}{\partial t}\,dt \approx -\omega_e(r_1\dot{r}_2 - r_2\dot{r}_1), \tag{9}$$

which approximates the potential contribution of the rotating earth in inertial space. The satellite's velocity $\dot{\mathbf{r}}(t)$ needed for the kinetic energy is computed similar to equation (3). As the energy does not depend linearly on velocities and positions, error propagation is more difficult and not implemented yet.

## 4 Robust parameter estimation

For all three techniques normal equations were computed per arc and accumulated. To make the solution robust against less accurate periods of the orbit, a variance component estimation (VCE) procedure is used. This can be done efficiently by re–weighting every orbital arc individually in an iterative Monte Carlo approach (MCVCE), see (Kusche 2003).

It is known that gravity field determination from satellite data poses an ill-posed problem. Downward continuation generally amplifies the measurement noise. To stabilize the systems of normal equations and overcome the ill-posedness of the problem a Kaula regularisation starting from degree $L = 40$ is applied, which requires a properly selected regularization parameter. Linking the regularization parameter to the variance of the gravity field parameters offers the possibility to determine it efficiently by means of MCVCE as well.

## 5 Results

All three methods were applied to kinematic CHAMP data of altogether about 51 days, and provide gravity solutions which we believe represent significant improvements with respect to the reference model EGM96 below degree 50. Above degree 50, these solutions are mainly determined by regularization and therefore biased towards the reference model. The time–wise BVP method and the polynomial differentiation appear as slightly superior to the energy balance method, which we believe is due to the neglected apriori variance information. Generally our solutions are almost free of spurious 'stripe patterns', due to the apparently high homogeneity of the data quality in combination with the MCVCE statistical technique.

## 6 Conclusions and Outlook

We have applied three non-conventional methods for recovering the Earth's gravity field from kinematic CHAMP orbits. All three methods can improve our knowledge of the gravity field from a very short time period, which is undoubtedly due to the high quality of the kinematic orbits. There are only slight differences between the three solutions. It remains to investigate whether these methods can supersede the traditional method of integrating the variational equations associated to the satellite's motion. A final statement cannot be given, as research in this matter has not been completed yet. There are possible improvements in methodology especially concerning the energy balance approach. First of all we plan to implement correct error propagation. In addition to this, other differentiation procedures like spline smoothing might improve the crucial velocity derivation step.

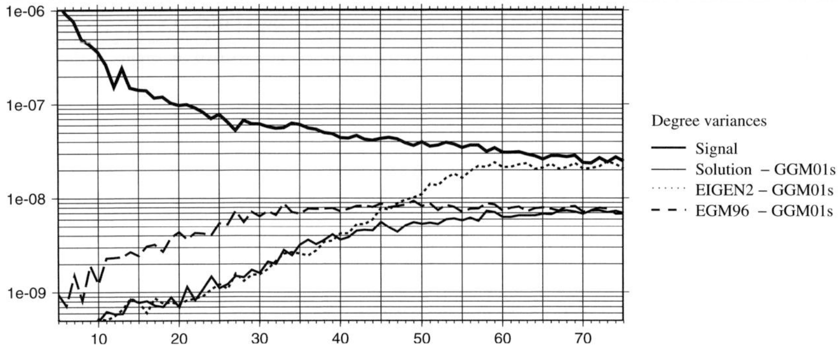


**Fig. 1.** Solution computed with BVP method.

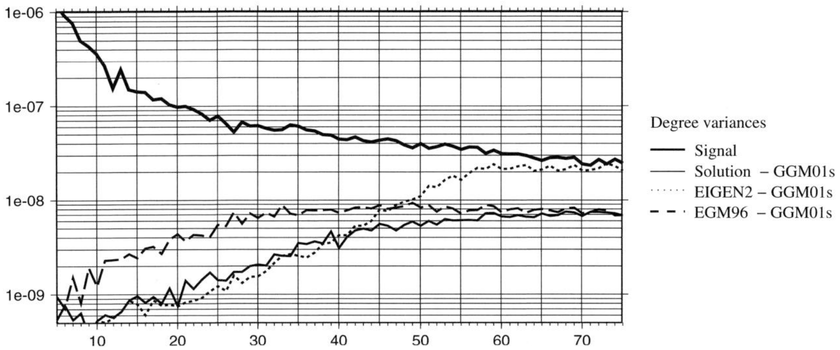


**Fig. 2.** Solution computed with polynom method.

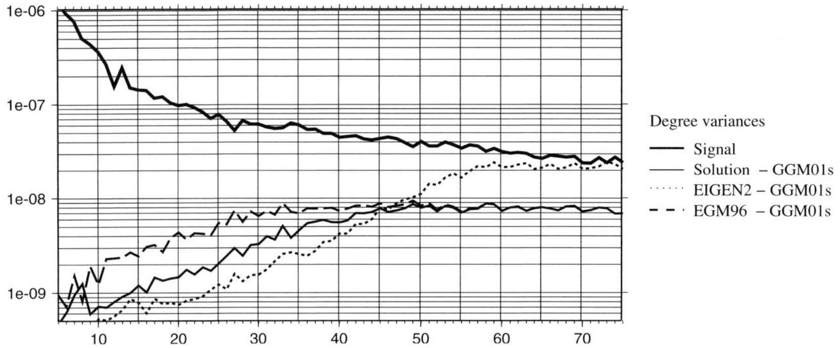


**Fig. 3.** Solution computed with energy method.

*Acknowledgement.* We are grateful to the ISDC of the GeoForschungsZentrum Potsdam (GFZ) and to the IAPG, TU Munich, for providing the data for this investigation. The support by BMBF (Bundesministerium für Bildung und Forschung) and DFG (Deutsche Forschungsgemeinschaft) within the frame of the Geotechnologien-Programm is gratefully acknowledged.

## References

1. Austen G, Grafarend E and Reubelt T (2002) Analysis of the Earth's Gravitational Field from semi-continuous Ephemerides of a low Earth Orbiting GPS-tracked Satellite of type CHAMP, GRACE or GOCE. In: Adam J and Schwarz KP (eds), Vistas for Geodesy in the New Millenium, Springer Verlag, Berlin-Heidelberg: 309–315.
2. Ditmer P and van Eckvan der Sluis A (2003) A Technique for Earth's Gravity Field Modeling on the Basis of Satellite Accelerations, accepted for J Geodesy.
3. Gerlach C, Sneeuw N, Visser P, Svehla D (2003) CHAMP gravity field recovery with the energy balance approach: first results. In: Reigber C, Lühr H, Schwintzer P (eds), First CHAMP Mission Results for Gravity, Magnetic and Atmospheric Studies, Springer, Berlin: 134–139.
4. Howe E, Stenseng L, Tscherning CC (2003) Analysis of one Month of State Vector and Accelerometer Data for the Recovery of the Gravity Potential. Adv Geosciences *1*: 1–4.
5. Ilk KH, Feuchtinger M, Mayer-Guerr T (2003) Gravity Field Recovery and Validation by Analysis of Short Arcs of a Satellite-to-Satellite Tracking Experiment as CHAMP and GRACE. Presented at the Gen. Ass. of the IUGG 2003, Sapporo, Japan.
6. Ilk KH, Loecher A (2003) The Use of Energy Balance Relations for Validation of Gravity Field Models and Orbit Determination Results. Presented at the Gen. Ass. of the IUGG 2003, Sapporo, Japan.
7. Kusche J (2003) A Monte–Carlo Technique for Weight Estimation in Satellie Geodesy. J Geodesy *76*: 641–652.
8. Schneider M (1967) Lösungsvorschlag zum Bahnbestimmungsproblem. BWF Bericht W67-35.
9. Svehla D, Rothacher M (2003) Kinematic Precise Orbit Determination for Gravity Field Determination. Presented at the General Assembly of the IUGG 2003, Sapporo, Japan.

# Precise Orbit Determination for CHAMP Using an Efficient Kinematic and Reduced-Dynamic Procedure

Heike Bock, Urs Hugentobler, Adrian Jäggi, Gerhard Beutler

Astronomical Institute, University of Berne, Sidlerstr. 5, CH-3012 Berne, Switzerland, *bock@aiub.unibe.ch*

**Summary.** Using an efficient and robust combination of a kinematic and reduced-dynamic orbit determination procedure CHAMP GPS data spanning about eleven months are processed and different aspects are addressed. Kinematic solutions are generated with and without elevation-dependent weighting of the observations in order to study the impact on the solution. GPS clock corrections with a sampling rate of 30 seconds and of 5 minutes, both interpolated to 10 seconds, are used. The orbit results are compared with the Post-processed Science Orbits from the GeoForschungsZentrum Potsdam, Germany, with orbit solutions from the Technical University of Munich, Germany, and they are validated with SLR measurements.

**Key words:** Efficient and precise orbit determination, Low Earth Orbiters, Elevation-dependent weighting, GPS clock interpolation

## 1 Introduction

An efficient algorithm is used for kinematic point positioning of a Low Earth Orbiter (LEO). Starting from epoch-wise code and epoch-by-epoch differenced phase observations code positions and phase position-differences are computed, which are then combined into precise positions in a subsequent step. An elaborate data screening algorithm is embedded in the procedure. These kinematic positions may be used as pseudo-observations to generate a reduced-dynamic orbit based on physical models. Pseudo-stochastic pulses may be set up in order to compensate for data problems and model insufficiencies. Comparisons with orbit results of the IGS Test Campaign 2001 showed that the procedure allows it to generate both, kinematic and reduced-dynamic orbits, for LEOs with an accuracy of the order of ten centimeters (see [1]). The procedure may be used for a quick data quality monitoring, for the robust generation of precise orbit solutions in cases where decimeter accuracy is sufficient, or to generate a priori orbits for boot-strapping other precise orbit determination (POD) procedures.

## 2 Orbit Determination Procedure and Solutions

The kinematic and the reduced-dynamic orbit determination procedures are used alternately in an iterative procedure composed of three steps. Figure 1 shows a flow diagram of this procedure for the determination of kinematic and reduced-dynamic orbits of LEOs. In a first step kinematic satellite positions are computed based on code observations only. An initial reduced-dynamic orbit is then adjusted to this relatively noisy trajectory. This orbit is used in the second step to allow for screening the code and the epoch-differenced phase observations. Based on code and phase-differences an improved kinematic trajectory is reconstructed which is then used in the third and final step as a priori orbit allowing for a refined data screening. The code positions and phase position-differences of this final kinematic solution are used as independent pseudo-observation types for the final reduced-dynamic orbit solution. The resulting procedure represents an efficient and robust method to determine precise orbits for a LEO. It can be set up in an automatic mode and could run in near-realtime. The algorithms of the kinematic and reduced-dynamic procedure are described in detail in [1].

The orbit solutions addressed in this paper emerge from the last step of the procedure and refer to a daily batch of data. For the kinematic solutions A and B an elevation-dependent weighting of the observations is applied, using a weight of $w = \cos^2 z$, where $z$ is the zenith distance. For solution A 30-second GPS clock corrections [2] and for solution B 5-minute GPS clock corrections are used and interpolated to the observation epochs spaced by 10 seconds. The reduced-dynamic solution RD can be characterized as follows:

- Code positions and phase position-differences of solution A are used as pseudo-observations and
- pseudo-stochastic pulses set up every ten minutes.

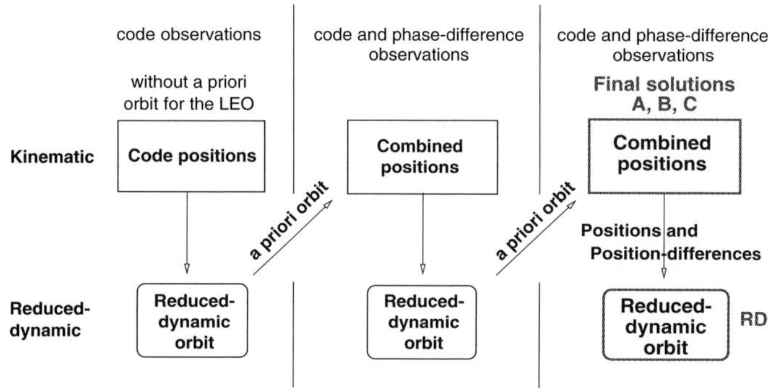


**Fig. 1.** Flow diagram for iterative procedure using the kinematic and reduced-dynamic procedure in a sequence.

Eleven months of CHAMP GPS data (day 075/2002 to 039/2003, 10-second sampling) were processed using the described procedure. CODE (Center for Orbit Determination in Europe) Final GPS orbits and Earth rotation parameters were used. The EIGEN2 gravity field model (120x120) [4] was adopted for the reduced-dynamic orbit determination.

## 3 Elevation-dependent Weighting

The GPS observations of terrestrial stations are usually weighted depending on the zenith distance of the observation w.r.t. the antenna pole to mitigate systematic effects at low elevations, such as troposphere anomalies and multipath. As LEO observations are not affected by tropospheric refraction it is not clear initially whether elevation-dependent weighting helps to improve the POD results. Detailed studies concerning this question were made in [1]. Whether or not elevation-dependent weighting is necessary depends on the satellite. For the CHAMP spacecraft weighting is appropriate for the code observations. Figure 2 shows the RMS errors per coordinate and per day in meters of the reduced-dynamic orbit adjustment using code positions either derived with elevation-dependent weighting of the observations (circles) or without (asteriks). Elevation-dependent weighting reduces the RMS of the code-derived positions by 30%. This indicates code data problems at low elevation that may be due to multipath or electronic interferences (see [3]). Figure 3 shows sky plots of the mean residuals of the observations. Data problems can be recognized for the code (Figure 3(a)) but not for the phase-difference observations (Figure 3(b)). As opposed to code observations the weighting of phase-difference observations has no impact on the results.

## 4 External Comparison

The final kinematic (A,B) and reduced-dynamic (RD) daily orbit solutions are compared with the pure dynamic Post-processed Science Orbits (PSO)

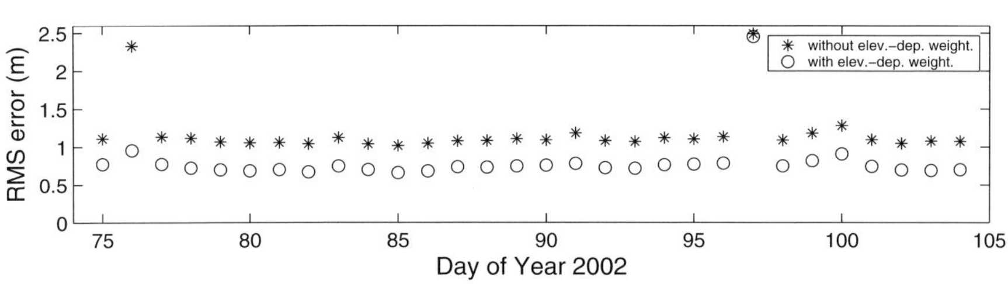


**Fig. 2.** RMS errors (m) per coordinate of the reduced-dynamic orbit determination using code positions derived either with or without elevation-dependent weighting of the observations.

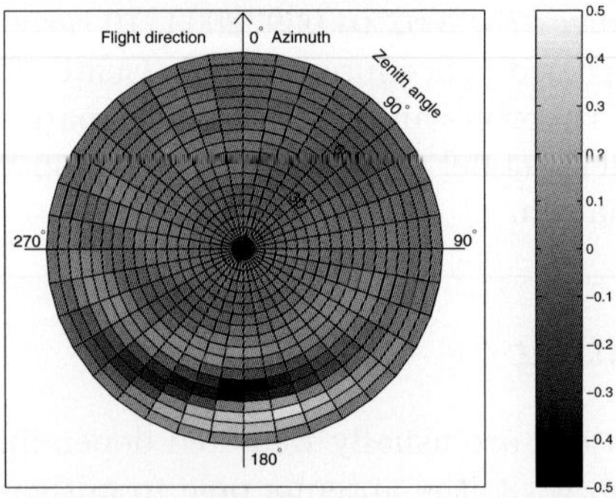


(a) Mean residuals of code observations (meter).

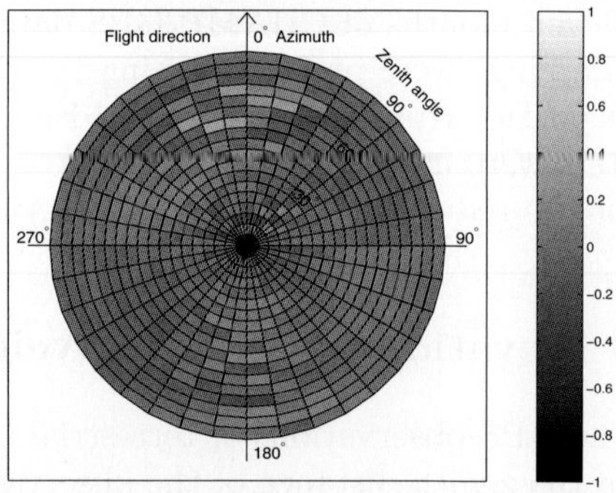


(b) Mean residuals of phase-difference observations (millimeter).

**Fig. 3.** Sky plots of mean residuals for kinematic solution A for CHAMP, average from day 075/2002 to 039/2003.

of CHAMP generated at GeoForschungsZentrum Potsdam (GFZ), Germany. Figure 4 shows the RMS errors (m) of the comparison with the PSO for the eleven months. The median values of these comparisons are listed in Table 1. There are some days with a large RMS error ($>$ 30 cm) due to several reasons:

– Jumps in the PSO in cases where arc boundaries of the 30-hour arcs fall into the middle of the processed 24-hour arcs (e.g., days 089, 172, 313/2002),
– modeling problems for our reduced-dynamic solution (e.g., days 147 and 350/2002), and
– problems to generate kinematic solutions due to few observations left by the data screening which causes additional problems by generating the reduced-dynamic orbit (e.g., days 163 and 164/2002).

For the great majority of the days the comparison is at a RMS level of 11-12 cm per satellite coordinate. For the kinematic solution B, based on 5-minute GPS clock corrections interpolated to 10 seconds, the RMS difference is increased by about 25% with respect to solution A (based on 30-second clock corrections). If an orbit quality of 15 cm is sufficient (e.g., for a quick data quality check or as a priori orbit for a double-difference procedure) precise 5-minute IGS GPS clock corrections may be interpolated to the data

**Table 1.** Median (m) of RMS errors of orbit comparisons between orbit solutions A, B, and RD and PSO-GFZ, TUM, and SLR.

| | A | B | RD |
|---|---|---|---|
| PSO-GFZ | 0.12 | 0.15 | 0.11 |
| TUM | 0.11 | 0.14 | 0.10 |
| SLR | – | – | 0.09 |

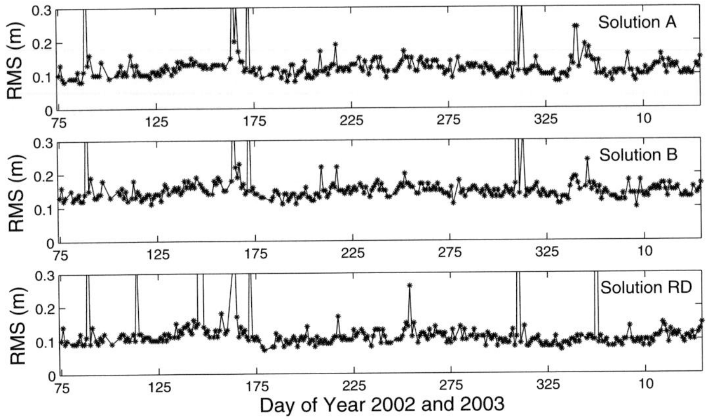


**Fig. 4.** RMS errors (m) per coordinate of the orbit differences between PSO-GFZ and orbit solutions A, B, and RD, CHAMP, day 075/2002 to 039/2003.

sampling and high-rate clock corrections are not needed, at least if SA is not active.

The reduced-dynamic orbit solution RD is marginally better than the kinematic solution A, because this solution is generated using code positions and phase position-differences as *independent* pseudo-observations. The pseudo-stochastic pulses which are set up every ten minutes can compensate for data problems in the kinematic positions resulting in a slightly better solution than the pure kinematic solution A.

Our orbits were also compared with zero-difference reduced-dynamic orbits from the Technical University of Munich, Germany (TUM) (see [5]). The median values of 1-dim RMS differences are listed in Table 1. The comparison with TUM shows slightly smaller median RMS errors than the comparison with PSO-GFZ.

Validation of the reduced-dynamic orbit RD with SLR observations from CHAMP show a median daily RMS error of nine centimeter for the considered time interval of eleven months (Table 1). Daily RMS values of the SLR residuals may be inspected in Figure 5.

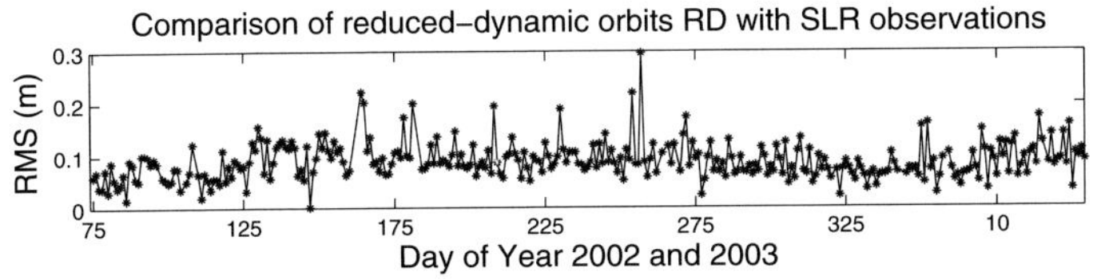


**Fig. 5.** RMS (m) of CHAMP SLR residuals w.r.t. the reduced-dynamic orbit solution RD, day 075/2002 to 039/2003.

## 5 Summary

Eleven months of CHAMP GPS data were processed and orbits were generated using an efficient and robust combination of a kinematic and reduced-dynamic orbit determination procedure. An orbit accuracy of a decimeter was achieved, if no serious data problems occurred and if enough GPS observations were available for the LEO satellite. The performance is confirmed by comparisons with independent orbits and by validation with SLR measurements.

The elevation-dependent weighting proportional to $\cos^2 z$ is appropriate for the code observations of CHAMP to cope with data problems at low elevations.

Three different solution types were compared with external orbit solutions. The reduced-dynamic orbit solutions show the best accuracy, the kinematic solutions A are, however, only slightly inferior. The quality of the kinematic solutions B (based on interpolated 5-minute GPS clock corrections) is degraded by 25% w.r.t. solution A (based on 30-second GPS clock corrections). In cases where a 15 cm-orbit quality is sufficient, 5-minute GPS clock corrections may be used for this kinematic point positioning approach. It is thus possible to generate LEO orbits of an accuracy of about 1-2 decimeters (per coordinate) using standard IGS products in a very efficient way.

*Acknowledgement.* We thank Drazen Švehla from TUM for making available one year of reduced-dynamic CHAMP orbits for comparison.

## References

[1] Bock H (2003) Efficient Methods for Determining Precise Orbits of Low Earth Orbiters Using the Global Positioning System, PhD thesis, Astronomical Institute, University of Berne.

[2] Bock H, Hugentobler U, Beutler G (2003) Kinematic and Dynamic Determination of trajectories for low Earth Satellites Using GPS. in: First CHAMP Mission Results for Gravity, Magnetic and Atmospheric Studies, edited by Reigber C et al., pp. 65–69, Springer, Berlin, ISBN 3-540-00206-5.

[3] Montenbruck O, Kroes R (2003) In-flight performance analysis of the CHAMP BlackJack GPS Receiver. GPS Solutions *7(2)*: 74–86.

[4] Reigber Ch, Schwintzer P, Neumayer K-H, Barthelmes F, König R, Förste Ch, Balmino G, Biancale R, Lemoine J-M, Loyer S, Bruinsma S, Perosanz F, Fayard T (2003) The CHAMP-only Earth Gravity Field Model EIGEN-2. Adv Space Res *31(8)*: 1883–1888 (doi: 10.1016/S0273–1177(03)00162-5).

[5] Švehla D, Rothacher M (2003) Kinematic and reduced-dynamic precise orbit determination of CHAMP satellite over one year using zero-differences. presented at EGS-AGU-EGU Joint Assembly, Nice, France, 06-11 April 2003.

# On Bias and Scale and Thrust Factors for CHAMP Accelerometry

Zhang Qiang and Philip Moore

School of Civil Engineering and Geosciences, Newcastle University, UK
*philip.moore@ncl.ac.uk*

**Summary.** Analysis of CHAMP GPS data through a reduced dynamic methodology has yielded precise Cartesian positioning for subsequent analyses for gravity field and geodynamic parameters. Results are accordingly presented of (i) a gravity field recovery, (ii) the bias and scale factors for the three axes accelerometer data and (iii) the apparent thrust forces estimated as parameters over a 13-month period. As a consequence we are able to consider the stability of the accelerometer parameters recovered daily or monthly. Analysis of the thruster data will quantify the magnitude of the additional forces involved and the stability, if any, of these forces over an extended period.

**Key words:** Gravity, Accelerometer, Bias and scale, Thrusters

## 1 Gravity Field Enhancement using CHAMP

As a prelude to gravity field studies precise orbits for CHAMP have been derived using the reduced-dynamic methodology (Moore et al, 2003) within the Jet Propulsion Laboratory's (JPL) software GIPSY-OASIS II. Table 1 summarises the reduced dynamic residuals relative to EGM-96 (Lemoine et al, 1998), EIGEN-1S (Reigber et al, 2002) and two inhouse gravity fields *NCL_champ*.2(3) (see below). The GPS tracking data processed in the reduced-dynamic mode provided Cartesian positioning in an Earth Centre Fixed (ECF) system at 30sec intervals. The ECF positioning was subsequently used as tracking data along with CHAMP Version 12 accelerometry (linear accelerations, quaternions, thrusts etc) within our inhouse orbit-determination software *Faust*. Table 2 summarises the CHAMP ECF positioning and SLR residuals using different gravity fields. Parameters estimated included the initial state vector and bias and scale factors for the linear accelerations for the daily arcs, residual 3D linear accelerations for each of the 6 pairs of thrusters and, for gravity field enhancement, geopotential harmonics constrained by the EGM96 normal equations complete to degree and order 70. Two strategies were employed that differed in the treatment of the thrust events (Strategy 1: 3*6 unknowns per 30 arcs; Strategy 2: 3*6 unknowns per arc) and the inclusion of once-per-revolution empirical radial accelerations (16 pairs per arc) in Strategy 2. The first strategy was applied to 515 days of CHAMP data and resulted in the new field gravity field NCL_champ.2. In

| Field | Pseudo range rms (cm) | # | phase rms (cm) | # |
|---|---|---|---|---|
| EGM96 | 346.6 | 11826 | 27.4 | 13629 |
| EIGEN-1S | 209.8 | 13517 | 9.9 | 13631 |
| *NCL_champ*.2 | 211.1 | 13513 | 10.3 | 13631 |
| *NCL_champ*.3 | 209.7 | 13518 | 11.0 | 13631 |

**Table 1.** GIPSY-OASIS CHAMP GPS residuals (cm) : day 151 2001

| Field | Xecf | Yecf | Zecf | SLR | strategy |
|---|---|---|---|---|---|
| EGM96 | 129.3 | 158.8 | 130.4 | 91.6 | 1 |
| EGM96 | 38.1 | 37.9 | 24.4 | 34.6 | 2 |
| EIGEN-1S | 37.9 | 42.3 | 53.2 | 35.3 | 1 |
| EIGEN-1S | 9.4 | 9.2 | 7.6 | 10.1 | 2 |
| *NCL_champ*.2 | 34.4 | 37.9 | 48.3 | 38.4 | 1 |
| *NCL_champ*.2 | 12.2 | 12.7 | 6.7 | 11.8 | 2 |
| *NCL_champ*.3 | 62.4 | 61.1 | 60.9 | 51.0 | 1 |
| *NCL_champ*.3 | 7.3 | 7.2 | 6.5 | 7.9 | 2 |

**Table 2.** RMS residuals (cm) of CHAMP ECF data (days 170-200, 2001)

contrast NCL_champ.3 derived using strategy 2 used just 30 days of tracking. Note that the inclusion of empirical radial accelerations introduced to compensate for known deficiency in this component leads a substantial improvement in fit to a level commensurate with the accuracy of the ECF Cartesian positioning of CHAMP which has been estimated to 5-10cm in each coordinate. Orbit computations with other satellites are presented in Table 3. $NCL - champ2(3)$ perform as well as EGM96 confirming that the additional CHAMP data has not affected global applicability.

| Satellite | Incl. (deg) | arcs | EGM96 (cm) | *NCL_champ*.2 (cm) | *NCL_champ*.3 (cm) |
|---|---|---|---|---|---|
| LAGEOS1 | 109.8 | 1*30d | 2.30 | 2.31 | 2.60 |
| LAGEOS11 | 52.6 | 1*30d | 1.94 | 1.95 | 2.16 |
| T/P | 6.5 | 1*10d | 4.87 | 4.90 | |
| STELLA | 98.7 | 3*5d | 6.18 | 5.93 | 5.29 |
| STARLETTE | 49.8 | 3*5d | 3.52 | 3.55 | 3.24 |
| ERS-2 | 96.5 | 7*5d | 7.44 | 6.91 | 9.33 |

**Table 3.** Test of the new gravity fields with other satellites; SLR data

## 2 Accelerometry: Bias and Scale

Bias (k0) and scale factors (k1) have been estimated using the EIGEN-1S gravity field model for a 13 month period using a number of different strategies that differed in the recovery periods of k0, k1 and the linear accelerations associated with thrust impulses. Using 24hr arcs, each strategy recovered the state vector at epoch. Strategies differed by considering k0, k1 and the thrust events as local parameters per 24hr arc or as global parameters estimated over 30day spans. A loose constraint implies that $k1 = 1.0 \pm 0.01$ was applied to all three scale factors while the tight constraint indicates that the along-track constraint was increased to $k1 = 1.0 \pm 0.001$. Table 4 shows that the radial

| Strategy | k0 | k1 | thrust | constraint | radial 1cy/rev |
|---|---|---|---|---|---|
| A | local | global | global | loose | no |
| B | local | local | global | loose | no |
| C | local | global | global | loose | yes |
| D | local | global | local | loose | yes |
| E | local | local | local | loose | yes |
| F | local | local | local | tight | yes |

**Table 4.** Strategies for bias, scale and thrust parameters

component of the accelerometer data has a clear linear trend over the period of this study. Furthermore, the scale factors k1 have significant variance for A-C, but stabilise when thrusters are estimated as local parameters. The estimation of radial 1 cy/rev empirical accelerations has little impact on the bias and scale, compare A and C, although their inclusion enhances the orbital fit in Table 2. Reducing the time period over which thrusters are estimated improves the radial and across-track stability, but adversely affects the along-track (see B and E; and C and D). The tighter constraint in F resolves this. Values of k0 and k1 for Strategy E are plotted as Figure 1. Note that k0 has a strong linear trend radially and a small trend cross-track. Scale factors are close to unity with the radial and cross-track components showing large deviations on occasions.

## 3 Accelerometry: Thrusters

An important question to ask is whether the estimated thruster parameters are meaningful at the current levels of accuracy. Our computational procedure for modelling such events is depicted in Figure 2. The resultant acceleration, c, over the time interval $\Delta t = t2 - t1$ is assumed to be represented by a constant acceleration d across the step length h= t3-t0. The total displacement at t3 due to accelerations c and d is easily shown to be given by

| | | Strategy | | | | | |
|---|---|---|---|---|---|---|---|
| | | A | B | C | D | E | F |
| Radial | $\alpha$ | -1.4e-06 | -1.2e-06 | -2.4e-06 | -1.1e-06 | -1.2e-06 | -1.2e-06 |
| | | ±7.1e-08 | ±7.8e-08 | ±9.0e-08 | ±1.7e-08 | ±9.1e-09 | ±1.0e-08 |
| | $\beta$ | 4.7e-09 | 3.7e-09 | 8.7e-09 | 3.6e-09 | 3.3e-09 | 3.3e-09 |
| | | ±3.1e-10 | ±3.5e-10 | ±4.0e-10 | ±7.7e-11 | ±4.1e-11 | ±4.5e-11 |
| | $k1$ | 0.744 | 0.869 | 0.952 | 0.806 | 0.991 | 0.992 |
| | | ±0.540 | ±0.677 | ±1.086 | ±0.247 | ±0.053 | ±0.054 |
| Along | $\alpha$ | 7.2e-09 | 2.9e-09 | -8.5e-09 | -2.6e-08 | -1.3e-08 | 1.5e-08 |
| | | ±2.0e-09 | ±4.5e-09 | ±2.2e-09 | ±3.6e-09 | ±5.7e-09 | ±3.3e-09 |
| | $\beta$ | 1.1e-11 | 6.2e-13 | 4.6e-11 | 1.6e-10 | 8.2e-11 | -6.6e-11 |
| | | ±8.8e-12 | ±2.0e-11 | ±9.7e-12 | ±1.6e-11 | ±2.5e-11 | ±1.5e-11 |
| | $k1$ | 0.998 | 0.989 | 0.986 | 0.984 | 0.986 | 1.000 |
| | | ±0.028 | ±0.083 | ±0.057 | ±0.070 | ±0.122 | ±0.004 |
| Cross | $\alpha$ | -2.9e-08 | -1.1e-08 | -1.5e-08 | -7.3e-09 | -9.7e-09 | -1.0e-08 |
| | | ±4.7e-08 | ±3.1e-08 | ±2.3e-08 | ±3.6e-09 | ±3.5e-09 | ±3.9e-09 |
| | $\beta$ | 4.2e-10 | 2.7e-10 | 3.1e-10 | 2.3e-10 | 2.7e-10 | 2.8e-10 |
| | | ±2.1e-10 | ±1.4e-10 | ±1.0e-10 | ±1.6e-11 | ±1.6e-11 | ±1.6e-11 |
| | $k1$ | 1.094 | 0.991 | 0.892 | 0.847 | 0.982 | 0.982 |
| | | ±0.383 | ±0.122 | ±0.156 | ±0.087 | ±0.038 | ±0.038 |

**Table 5.** Linear regression of Bias k0 (m/s**2) and constant k1 for 2001, May 20 2002, June 14: (Linear fit model: $k0 = \alpha + \beta * (T - T0), T0 = 52049$); $1\sigma$ standarderrors

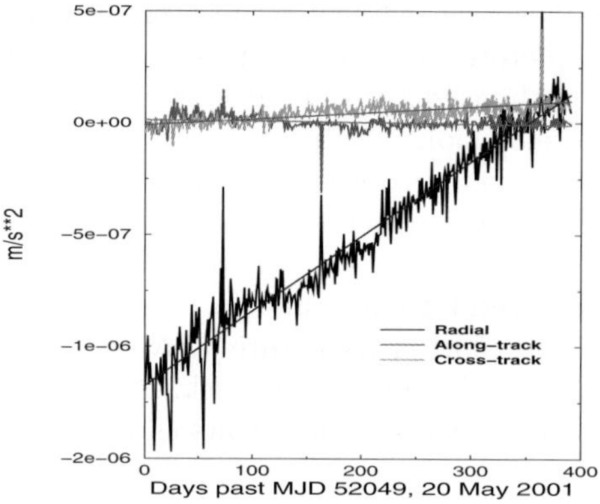


(a) k0 (m/s**2)

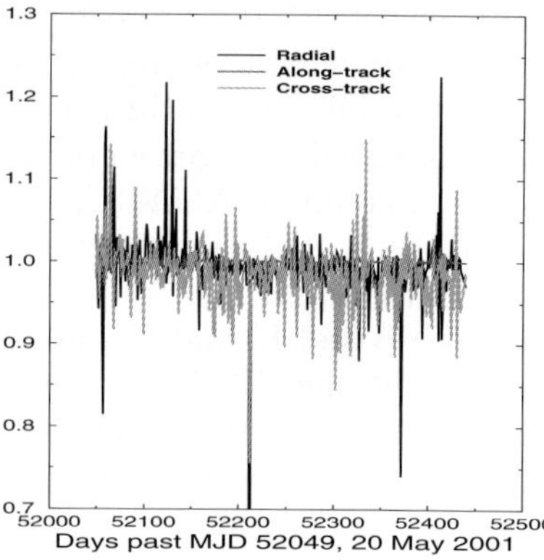


(b) k1

**Fig. 1.** Strategy E: linear accelerometer bias and scale

$$\delta = dh^2/2 = ct2/2 + t(t3 - t2) \tag{1}$$

For $t << h$ and taking average value t3-t2=h/2 gives $d \approx c\Delta t/h$. Utilising the this formulation of the thrust events we observed

– Thrusters +x, -x (roll) are a factor 2-3 larger than other four thrusters

- Radial component of resultant thruster accelerations larger than along and cross track components.
- Accelerations normally distributed with a mean near zero
- $c = 1.0e - 3m/s^2$ causes a displacement $\delta = 5\Delta t$mm with $\Delta t = 1s$ for thrusters +x and -x and $\Delta t = 10.1s$ for the other four.
- Thruster mismatch/misalignment typically 1-5 mm at the end of a 10s step length.

Figure 3 plots histograms of the estimated thruster accelerations for one pair, namely Thruster +x (roll) as computed over the 13month period. If the estimated thruster values are meaningfull then we would anticipate a skewed distribution centred on the required value. However, the figures show that the recovered thruster accelerations are normally distributed with a mean of zero. Thus, at the current level of accuracy, estimation of the thruster events applies small impulses that adjusts the orbital positioning within the computation rather than yielding a coherent solution.

## 4 Conclusions

The analyses undertaken in this study have allowed us to formulate the following conclusions:

- Addition of CHAMP to existing gravity models has produced fields that are superior for CHAMP and of comparable accuracy for other satellites.

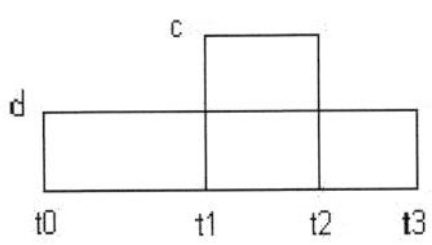


**Fig. 2.** Modelling of thrust events

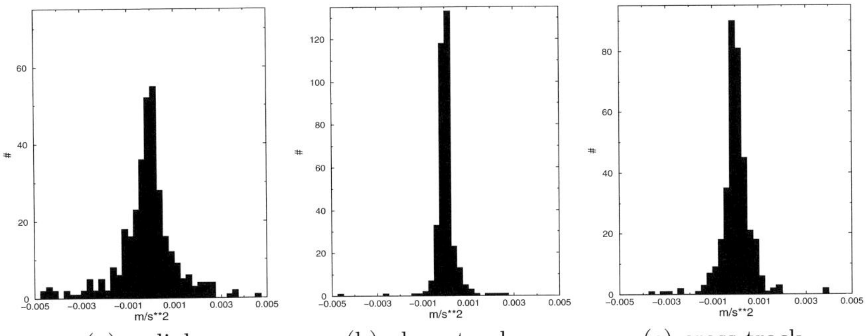


(a) radial (b) along-track (c) cross-track

**Fig. 3.** Histograms of estimated thruster accelerations for Th +x

- Agreement between the reduced dynamic orbits and dynamic orbits using the CHAMP based fields, accelerometry etc using the ECF positioning is 30-50cm. This discrepancy is probably due to known deficiency in the radial component of the accelerometry. Inclusion of empirical radial 1cy/rev accelerations enhances the orbital fit close to the accuracy of the ECF tracking data.
- The radial component of CHAMP version 12 accelerometry exhibits a clear linear trend over the 13month period of this study.
- Solution for daily thrust accelerations improves consistency of the bias and scale factors for the 3-axes accelerometer. However, the recovered thrust accelerations have a mean near zero with an approximate normal distribution indicating that the values are noise rather than a coherent signal
- Recovered thruster accelerations have the largest component radially. Another indication of the uncertainty in this component
- Over 10s time steps thruster accelerations may cause a displacement of up to 5mm in an ECF coordinate.

*Acknowledgement.* The authors would like to thank the U.K. Natural Environment Research Council (Grant No. NER/A/0000/00612) for financing this study and for GFZ for supplying the data and technical support.

## References

1. Lemoine FG et al (1998) The development of the joint NASA GSFC and the National Imagery and Mapping Agency (NIMA) geopotential model EGM96. NASA/TP-1998-206861.
2. Moore P, Turner JF, Qiang Z (2003) CHAMP Precise Orbit Determination and Gravity Field Recovery. In: First CHAMP Mission Results for Gravity, Magnetic and Atmospheric Studies, eds Reigber C, Lühr H, Schwintzer P, Springer-Verlag Berlin Heidelberg: 146–152.
3. Reigber Ch, Balmino G, Schwintzer P, Biancale R, Bode A, Lemoine J-M, Koenig R, Loyer S, Neumayer H, Marty J-C, Barthelmes F, Perosanz F, Zhu SY (2002) A high quality global gravity field model from CHAMP GPS tracking data and accelerometry (EIGEN-1S). Geophys Res Lett *28(17): 3263–3266.*

# CHAMP Accelerometer Preprocessing at GeoForschungsZentrum Potsdam

Christoph Förste and Sunchan Choi

GeoForschungsZentrum Potsdam (GFZ), Department 1, Geodesy and Remote Sensing, Telegrafenberg, D-14473 Potsdam, Germany, *foer@gfz-potsdam.de*

**Summary.** This paper presents an overview of the preprocessing of the CHAMP accelerometer measurements as carried out at GFZ Potsdam. The data are smoothed and the sample rate is changed from 1 to 10 seconds. Additionally, the accelerometer data are combined with attitude measurements and some housekeeping data. Two corrections for the linear acceleration are modelled: The so-called X3 correction and a model for the Lorentz force on the proof mass. The preprocessed data are available as a level-2 product at the CHAMP information system and data center at GFZ Potsdam.

**Key words:** accelerometer, accelerometer preprocessing, attitude preprocessing, X3 correction, Lorentz force, preprocessing

## 1 The purpose of the accelerometer preprocessing

The STAR accelerometer instrument onboard CHAMP [Grunwaldt et al. 2003, Bock et al. 2000] serves to measure the non-gravitational accelerations like air drag, earth albedo and solar radiation acting on the satellite. The accelerometer measurements are intended to improve the CHAMP orbit computation and gravity field determination. Before including into the orbit and gravity field computation, the raw accelerometer data (level-1) have to become preprocessed. First of all, this means the elimination of outliers and spikes and the compression of the sample rate from the originally 1 Hz to 0.1 Hz.

Furthermore two kinds of correction are modelled during the preprocessing and given together with the cleaned and sampled accelerometer data: **(1)** the so-called X3 correction to overcome the disturbance of the linear radial acceleration due to the anomalous behaviour of the X3 electrode [Perosanz 2003] and **(2)** the Lorentz force, which is assumed to act on the electrically charged proof mass.

The preprocessing also includes an evaluation of the attitude measurements from the CHAMP body mounted star cameras. This attitude measurements, the housekeeping information about the attitude control thrusters and the present cold gas mass are merged with the cleaned accelerometer data, the X3 corrections and the Lorentz force accelerations into a common data product (level-2 accelerometer data).

## 2 Cleaning and sampling algorithms for the accelerometer data

The acceleration measurements show various disturbances [Grunwaldt et al. 2003, Perosanz et al. 2003], mainly: **(1)** Spikes in the linear and angular accelerations at thruster pulses, due to thruster misalignments and thruster pulse induced boom oscillations, **(2)** Small spikes, induced by electric disturbances due to heater switches and **(3)** Several types of spikes of unknown sources, in some cases related to the orbit circulation frequency.

Due to the unknown sources or precise values most of the mentioned disturbances it was decided to clean the data regardless of the type of the individual spikes: **(1)** Cutting out the data during the thruster firing periods including the subsequent 3 seconds (as the thruster pulse duration is $< 3$ sec.), **(2)** Polynomial interpolation (incl. least square adjustment of the polynomial coefficients) as basis for outlier and spike detection, applied on overlapping 30 sec intervals (see Fig. 1) and **(3)** Data compression (1 Hz $\rightarrow$ 0.1 Hz) of the cleaned data, i.e. the computation of mean values over individual 10-sec. intervals, (see Fig. 1). This includes the filling of small data gaps up to 10 seconds. Fig. 2 shows a comparison between raw and preprocessed alongtrack acceleration data in different resolutions.

## 3 The attitude preprocessing

The attitude measurements for the CHAMP spacecraft are carried out by the Advanced Stellar Compass (ASC) unit mounted on the satellite body nearby the accelerometer sensor [Bock et al. 2000]. This ASC unit consists of two camera head units (CHU3 and CHU4) which produce individual raw orientation data expressed as quaternions. The main task of the attitude preprocessing is the combination of the single head measurements and the precise transformation of the combined data into the spacecraft reference system [Schwintzer et al. 2001].

The processing of the attitude data follows the scheme given in Fig. 3. The first processing step is a 3-1-3 rotation of the raw measurements back into the individual camera head systems by using the direction cosine matrix DC. The next step is

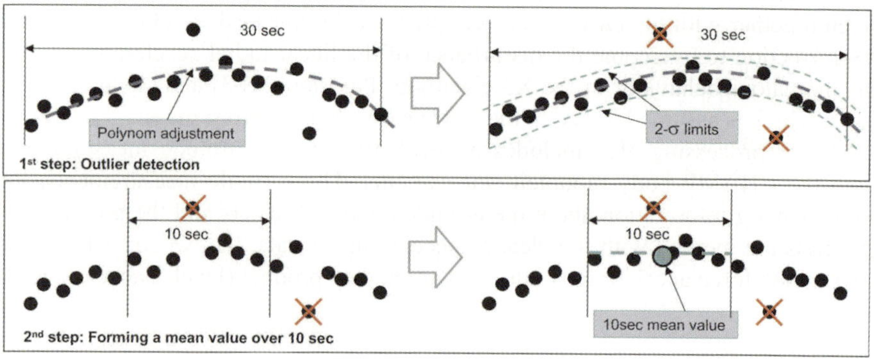


**Fig. 1.** The principle of cleaning and compression of the accelerometer data.

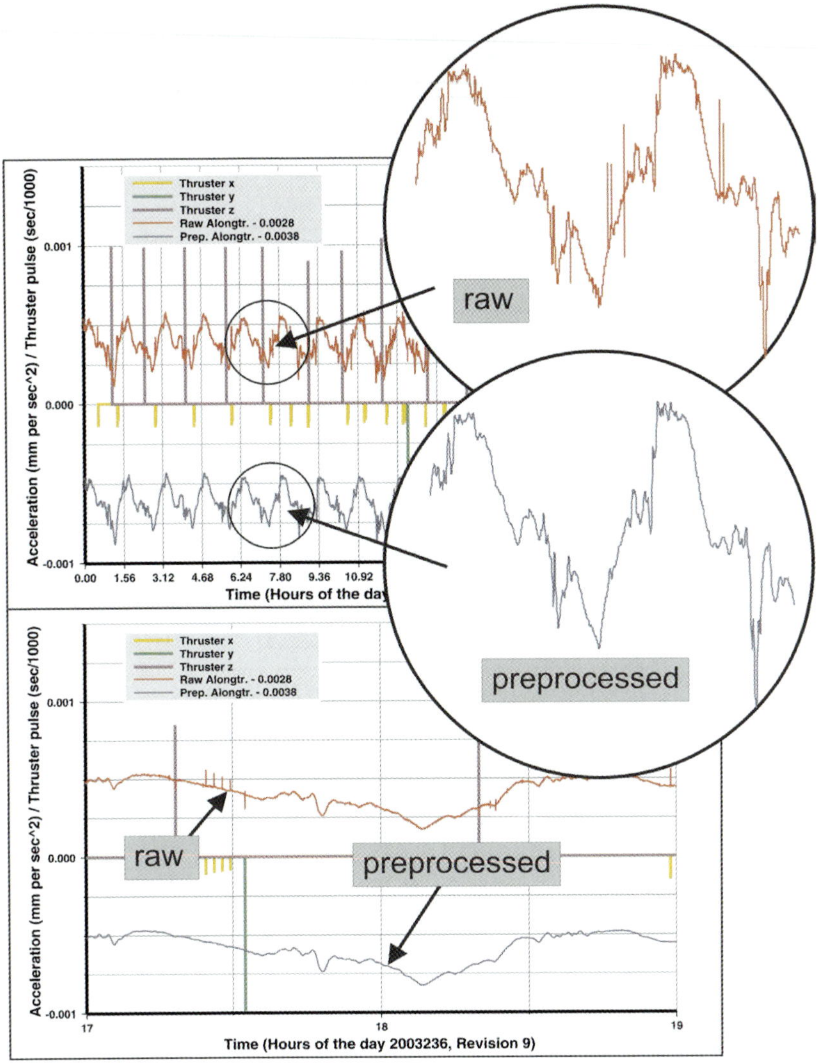


**Fig. 2.** Comparison of raw and preprocessed alongtrack acceleration data, including thruster pulses, presented in different resolutions and with different offsets.

the transformation of the both single head data into a common reference (CR) frame and the combination of the data using a so-called dual head vector operation. Finally, the combined attitude data are transformed from the common reference frame into the spacecraft system (SC) by using the rotation matrix given in Fig. 3. In the case of absence of one of the single head data, the transformation of the remaining data into the CR frame is done using the M3 or M4 matrices respectively. The combination of the dual head data yields the decrease of the signal-to-noise ratio of a factor of about 5 as shown in the example of Fig. 4.

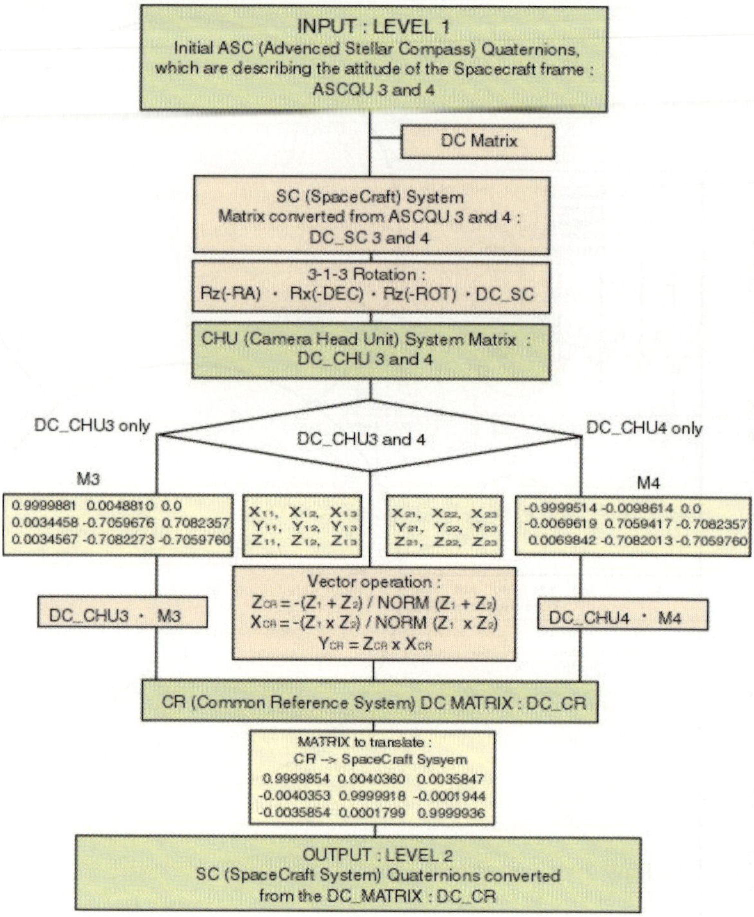


**Fig. 3.** The attitude processing scheme for the body ASC data.

## 4 The Lorentz force correction

The proof mass of the accelerometer is electrically charged. It can be assumed, that the movement of the accelerometer in the earth magnetic field causes a disturbing linear acceleration due to the Lorentz force. Therefore, this Lorentz force

$$\mathrm{a}_{\mathrm{L}} = \frac{q}{m}(v \times B)$$

acceleration $\mathbf{a}_{\mathrm{L}}$ is modelled during the preprocessing from the following formula: with $q$ = proof mass charge, $v$ = Velocity vector of the spacecraft, $m$ = mass of the proof mass and $B$ = Magnetic Field vector at the spacecraft position (from Fluxgate magnetometer data).

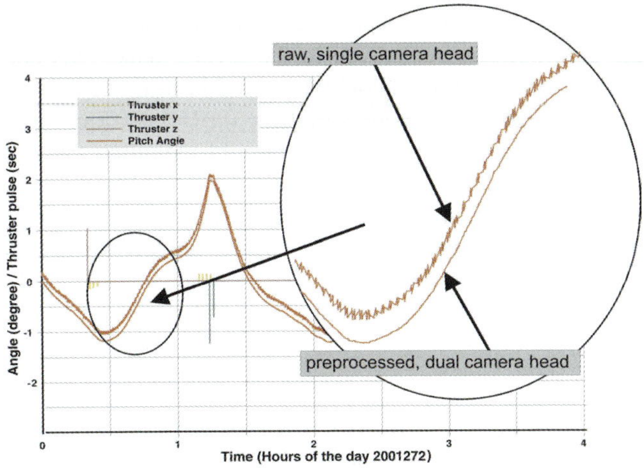


**Fig. 4.** Comparison of raw single head and preprocessed combined attitude data (from the body ASC, expressed in angles, example for the pitch angle).

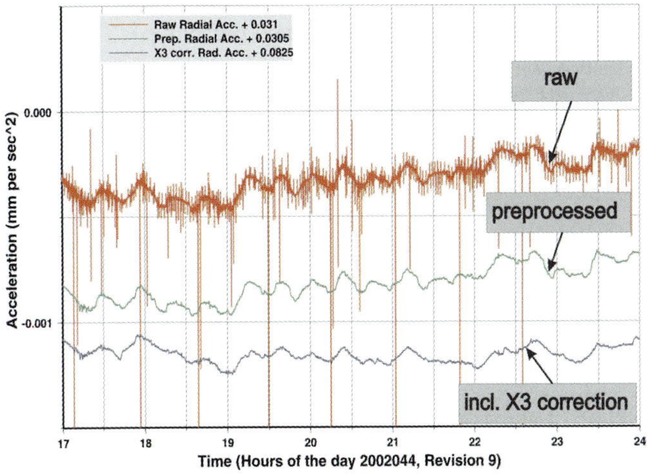


**Fig. 5** Comparison between raw, preprocessed and X3 corrected radial data.

The order of magnitude obtained for $a_L$ is of $5 \cdot 10^{-8}$ ms$^{-2}$ in the radial and cross track directions. But it must be noted, that possible screening effects of the conducting sensor cage are neglected - this topic is still under discussion. The $a_L$ vectors are modelled for every epoch of the 10sec-sampled accelerometer data.

## 5 The X3 correction

The measurements from the X3 electrode couple show an anomalous behaviour. This affects the radial linear ($a_x$) as well as the roll ($a_\Theta$) and pitch ($a_\Psi$) angular acceleration components [Perosanz 2003]. Following a report by the CNES/GRGS calibration team, a special correction to overcome this electrode failure is modelled during the preprocessing and given to the level-2 data. This correction $da_x$ regards the radial linear acceleration component and is computed from the roll angular acceleration channel $a_\Theta$. It amounts to $da_x = 0.015 \cdot a_\Theta$. Fig. 5 shows a comparison between the radial acceleration without and including the application of this so-called X3 correction.

A second correction term for $a_x$ can be obtained by differentiating the STAR sensor quaternions. This correction term is presently neglected (investigations are ongoing) and is considered of only secondary importance. The angular acceleration components $a_\Theta$ (roll) and $a_\Psi$ (pitch) are still fully affected by the electrode failure.

## 6 The GFZ level-2 accelerometer data product CH-OG-2-ACC

The output of the accelerometer preprocessing is the official CHAMP data product CH-OG-2-ACC, given in the so-called CHAMP Data Format [Förste et al. 2001]. This product is available at the CHAMP Information Center and Data System (ISDC) at GFZ Potsdam (http://op.gfz-potsdam.de/champ). The single data files consist of the following record types: **(1)** Linear and angular acceleration vectors (0.1 Hz), **(2)** Attitude quaternions from the body ASC (0.1 Hz), **(3)** Spacecraft mass data, **(4)** Lorentz Force correction vectors, **(5)** X3 correction records and **(6)** Accelerometer calibration parameters, i.e. scale factors and biases for the three linear acc. components, presently based on the period Sept. to Dec. 2001.

*Acknowledgement:* The support of the CHAMP exploitation phase within the German GEOTECHNOLOGIEN geoscientific R+D programme (grant 03F0333) is gratefully acknowledged.

## References

Bock R, Grunwaldt L, Lühr H, Reigber C (2000) CHAMP Satellite & Systems - Scientific Instruments., http://op.gfz-potsdam.de/champ/systems/index_SYSTEMS.html.

Förste C, Schwintzer P and Reigber C (2001) The CHAMP Data Format. GFZ Technical Note CH-GFZ-FD-001, Last Issue 2.0 2002-05-31, http://op.gfz-potsdam.de/champ/more/docs_CHAMP.html.

Grunwaldt G and Meehan T (2003) CHAMP Orbit and Instrument Status. in: Reigber C, Lühr H and Schwintzer P (Eds.): First CHAMP Mission Results for Gravity, Magnetic and Atmospheric Studies, Springer-Verlag, ISBN 3-540-00206-5, 3-10.

Perosanz F, Biancale R, Loyer S, Lemoine J-M, Perret A, Touboul P, Foulon B, Pradels G, Grunwaldt L, Fayard T, Vales N and Sarrailh M (2003) On Board Evaluation of the STAR Accelerometer. in: Reigber C, Lühr H and Schwintzer P (Eds.): First CHAMP Mission Results for Gravity, Magnetic and Atmospheric Studies, Springer-Verlag, ISBN 3-540-00206-5, 11-18.

Schwintzer P, Lühr H, Reigber C, Grunwaldt L and Förste C (2001) CHAMP Reference Systems, Transformations and Standards. GFZ Technical Note CH-GFZ-RS-002.PDF, Last Issue 2.3 2002-04-18, http://op.gfz-potsdam.de/champ/more/docs_CHAMP.html.

# CHAMP Clock Characterization Revisited

Rolf König, Grzegorz Michalak, Ludwig Grunwaldt and Christoph Reigber

GeoForschungsZentrum Potsdam (GFZ), Dept. 1 'Geodesy and Remote Sensing', Telegrafenberg, 14473 Potsdam, Germany, *koenigr@gfz-potsdam.de*

**Summary.** The behaviour of the space-borne CHAMP GPS receiver clock over nearly three years is inspected by the results of the Rapid Science Orbit (RSO) determination and the on-board navigation solution (NAV). The analysis completes and enhances the results of the clock characterization study carried out earlier on a limited number of data. It is shown, that CHAMP on-board events like software uploads and hardware reboots frequently lead to large clock offsets, which are reduced on-board within a few hours. Comparisons of RSO and NAV solutions show, that the RSO clock offset estimates are additionally affected by bias and drift of the reference clocks chosen in the processing. The two independent solutions, RSO and NAV, indicate, that the CHAMP clock is free of a drift in the long term. In the short term, the clock parameters show a standard deviation of 0.4 $\mu s$ and a small negative bias of 0.3 to 0.4 $\mu s$.

**Key words:** CHAMP, BlackJack GPS receiver, clock offset

## 1 Introduction and Data

The CHAMP on-board GPS receiver "BlackJack" is equipped with a steered quartz clock. The behaviour of the clock over the period 2001/04/17 - 2003/08/18 is analyzed based on two independent solutions for the clock offsets. One solution is generated by the GFZ Rapid Science Orbit (RSO) processing system [1].The second one is the real-time CHAMP on-board navigation solution (NAV). The CHAMP clock behaviour was already studied [2] based partially on Post-processed Science Orbits (PSO), RSO and NAV solutions. There however, the data covered a few small segments in the time frame 2000/07/30 to 2002/02/23. In this paper we present a study based on a continuous time series of RSO clock estimates since the beginning of the CHAMP RSO processing. Because the PSO solutions are rather incomplete over the chosen analysis period and because they are anyway rather similar to the RSO results, the PSO solutions are not considered here.

## 2 Clock Offset Time Series Analysis

A general view of the clock offset estimates is given in Fig. 1 (RSO solution) and Fig. 2 (NAV solution). Both figures are dominated by singular large clock

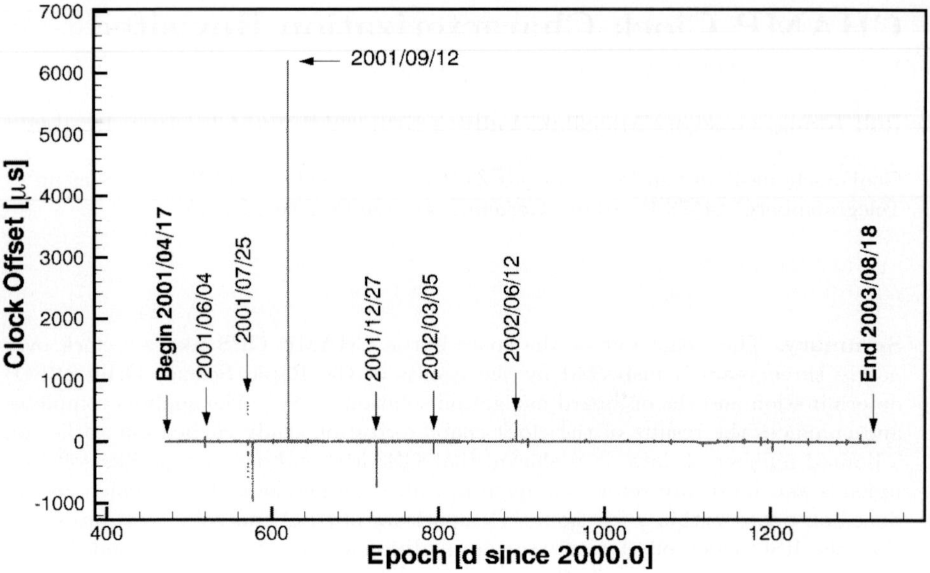


**Fig. 1.** CHAMP clock offset estimates of the RSO solution - overview

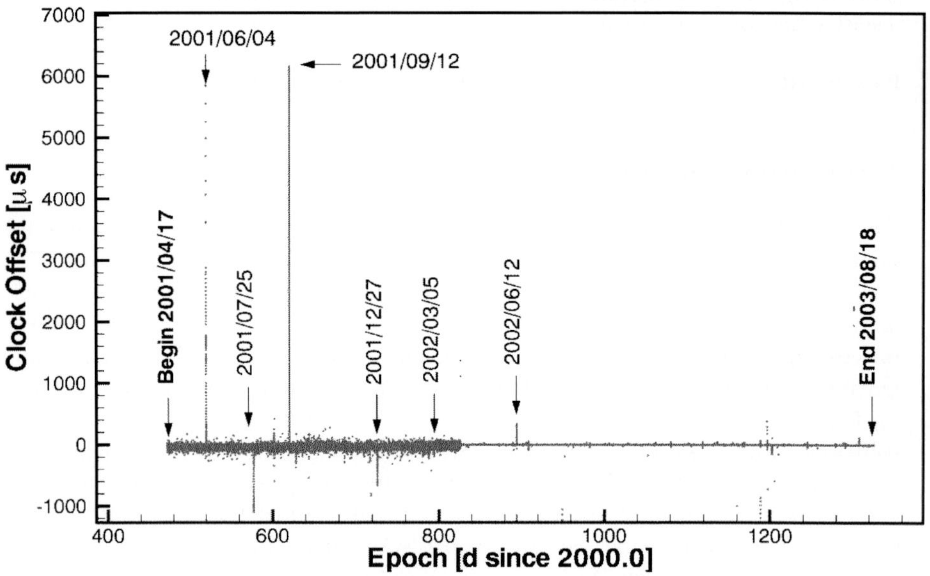


**Fig. 2.** CHAMP clock offset estimates of the Navigation solution - overview

offsets being congruently visible in both approaches. These clock offsets result from on-board events like software uploads and updates and receiver reboots.

In Fig. 3 and Fig. 4 a more detailed view of the clock offsets for both solution is given. A significant improvement of the clock behaviour after

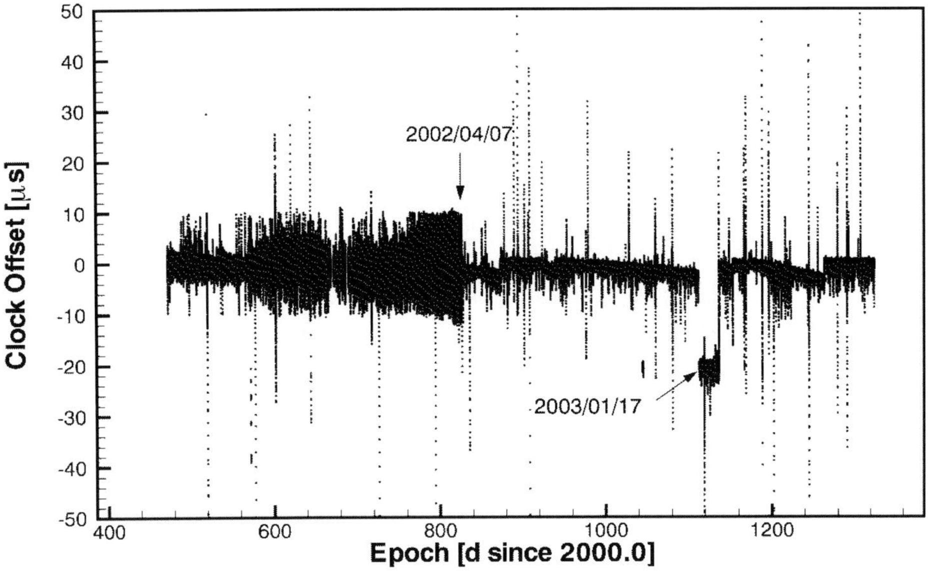


**Fig. 3.** CHAMP clock offset estimates of the RSO solution - detailed view

2002/04/07 can be noticed. This is due to an update of the receiver parameters (mainly causing a higher frequency of the NAV solution generation) that resulted in a more efficient correction of the clock drift. When comparing Fig. 3 to Fig. 4 it can also be seen, that the RSO solution is affected by biases and trends. These biases and trends are dependent on the characteristics of the ground reference clock being fixed in the generation of the GPS ephemerides in our two-step CHAMP RSO processing [1]. This effect is outstandingly visible on 2003/01/17, when the usual reference clock, station ALGO, had to be changed due to a station downtime. A list of the most important events, labelled by dates in Fig. 1 to 4, affecting significantly the CHAMP clock performance follows:

- 2001/06/04 - BlackJack GPS data flow discontinuities due to firmware software problems
- 2001/07/25 - Software upload, BlackJack receiver reboots
- 2001/09/12 - Reboot of the receiver, large clock offset reduced subsequently by on-board software
- 2001/12/27 - Reboot of the receiver, clock offset reduced subsequently by on-board software

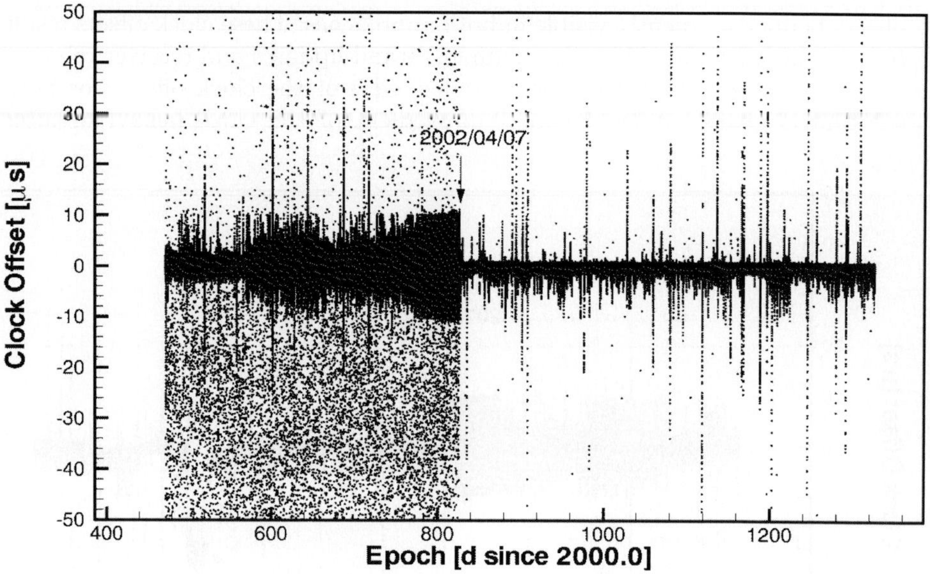


**Fig. 4.** CHAMP clock offset estimates of the NAV solution - detailed view

- 2002/03/05 - Receiver switch-off, major software upload
- 2002/04/07 - Reboot of the receiver, update of receiver parameters
- 2002/06/12 - Reboot of the receiver, switch of operation modes
- 2003/01/17 - Change of the GPS ground reference clock in the RSO solution due to down time of station ALGO

Looking at the most recent period of the CHAMP clock offsets time series, i.e. the last 30 days in Fig. 3 and Fig. 4, it can be seen, that the clock offsets obey values generally between -2 $\mu s$ and +1 $\mu s$, the bias in the NAV solution equals -0.3 $\mu s$, for the RSO solution the bias is larger at -0.4 $\mu s$. The standard deviation, however, is the same for both solutions with 0.4 $\mu s$.

A closer look into the time series indicates additionaly that the clock offset is not random around the mean value, but shows a periodic behaviour. The RSO solution is used to study these characteristics in more details. Since there is a significant improvement of the clock behaviour after 2002/04/07, the data were devided into 2 parts: before and after this date. All biases, trends and outliers are removed from these two data sets by means of iterative linear regression. The standard deviation of the clock offsets prior to 2002/04/07 is 1.7 $\mu s$, after 2002/04/07 it is 0.4 $\mu s$. Then the periodograms for the two data sets are calculated via a Fourier transformation and given in Fig. 5. Obviously strong periodicities show up in both periodograms, namely 1.37 h, 1.55 h and 1.77 h. They are close to the orbital period equal to 1.55 h or 93 minutes.

The amplitude of the periodic changes of the clock offset is approximately 1 $\mu s$ (equivalent to 300 m in position). This is too large to be attributed

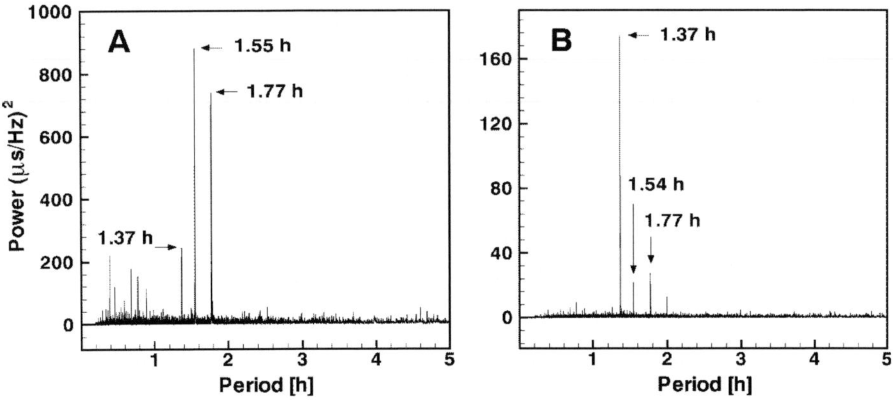


**Fig. 5.** Power spectra of the CHAMP clock estimates of the RSO solution. Plot A - prior to 2002/04/07, plot B - after 2002/04/07

to the orbit solution, neither in the RSO nor in the NAV case. The same periodicities are present in the RSO and in the NAV solution. The solutions are totally independent and follow distinct methodologies. The RSO is a dynamic batch solution, the NAV is a real-time filter approach. Therefore it could be suggested that the periodicities are not dependent on dynamics but rather on internal characteristics of the CHAMP clock parameter settings and steering algorithm. The BlackJack triggers its clock according to the GPS spacecraft clocks, after one revolution of CHAMP a similar GPS constellation is reached. Therefore the relation with the orbital frequency of CHAMP lies at hand. Exact explanations of the 1.37 h and 1.77 h periods however remain open.

## 3 Summary and Conclusions

The CHAMP clock characteristics can be summarized in the following way:

- The RSO clock parameter solution is less noisy than the real-time NAV solution.
- The RSO solution is affected by characteristics of the reference clock used in the processing.
- There is no long term drift of the CHAMP on-board clock. This is due to the clock steering loop, that updates the clock bias when it exceeds 1 $\mu s$.
- The clock offsets are in the range from -2 $\mu s$ to +1 $\mu s$. After removing biases and trends from the RSO solution, periodicities close to the orbital period can be detected. The periodicities are also present in the NAV solution. There is evidence that they are independent from dynamics, but may stemm from the clock itself.
- The recent bias for the NAV solution equals -0.3 $\mu s$, for the RSO solution the bias equals -0.4 $\mu s$.

– The recent standard deviation of the clock offsets equals 0.4 $\mu s$.

The time series of CHAMP clock solutions shows effects of software uploads and on-board receiver reboots. These events cause large jumps which are quickly recovered by the receiver. The RSO solution for the CHAMP clock offsets is affected by bias and trend of the GPS ground station clock chosen as reference. This does not affect the quality of the CHAMP orbits.

## References

1. Michalak G, Baustert G, König R, Reigber Ch (2003) CHAMP Rapid Science Orbit Determination. In: First CHAMP Mission Results for Gravity, Magnetic and Atmospheric Studies, eds. Reigber Ch, Lühr H Schwintzer P, p. 98-103.
2. König R, Neumayer KH, Michalak G, Grunwaldt L (2003) CHAMP Clock Error Characterization. In: First CHAMP Mission Results for Gravity, Magnetic and Atmospheric Studies, eds. Reigber Ch, Lühr H Schwintzer P, p. 32-37.

# How Baltic Sea Water Mass Variations Mask the Postglacial Rebound Signal in CHAMP and GRACE Gravity Field Solutions

Martin Wiehl[1], Reinhard Dietrich[1], and Andreas Lehmann[2]

[1] TU Dresden, Institut für Planetare Geodäsie, 01062 Dresden, Germany
*wiehl@ipg.geo.tu-dresden.de*

[2] Institut für Meereskunde an der Universität Kiel, Düsternbrooker Weg 20, 24105 Kiel, Germany

**Summary.** The postglacial rebound (PGR) in Fennoscandia is an attractive signal for the detection of time-variable gravity with CHAMP and GRACE. However, the separation from other geophysical signals is a challenge. We study the influence of oceanographic variations of the Baltic Sea. Using a precise high-resolution oceanographic model we analyze its water mass variations and the resulting geoid signal. The geoid signal is similar to the postglacial rebound signal in spatial pattern and in magnitude. Hence we show that the PGR detection from satellite gravity data will be corrupted by Baltic Sea oceanographic variations unless they are accounted for. Our study suggests to correct for the Baltic Sea signal using external data.

**Key words:** time-variable gravity, postglacial rebound, oceanographic variations, Fennoscandia, Baltic Sea

## 1 Introduction

The region of Fennoscandia (see Figure 1 for an overview) is attractive for the measurement of temporal gravity field variations with CHAMP, and in particular, GRACE: In this region the postglacial rebound (PGR), i.e., the solid Earth isostatic adjustment following the removal of the last glaciation's ice masses, provides a pronounced secular gravity change of regional scale. An observation of this signal by space gravity missions can be supported by extensive ground observations, such as relative sea level records and geodetic levelling [2], gravity measurements [3], and GPS observations [8].

However, the detection of the PGR signal from satellite gravity data includes the challenge of seperating this signal from other geophysical processes sensed by the mission. Gravity field variations due to, e.g., oceanographic, land hydrological and atmospheric processes may partly mask the PGR signal.

In this study we analyze water mass variations of the Baltic Sea and investigate their influence on the detection of PGR from satellite gravity data.

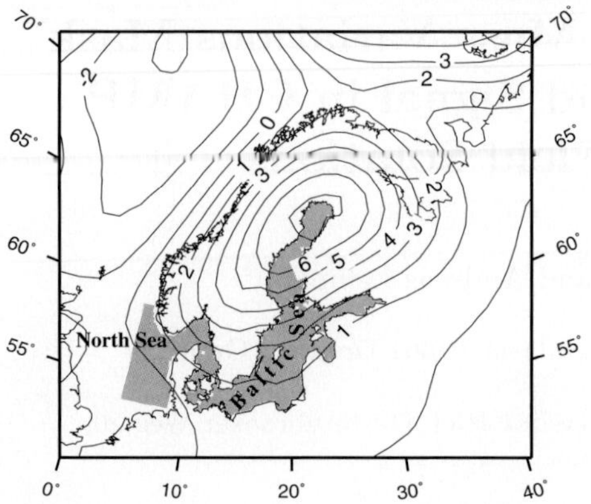


**Fig. 1.** The region of Fennoscandia with the PGR vertical uplift in mm/a after [11] (*contours*), and the Baltic Sea oceanographic model domain (*shading*)

## 2 Baltic Sea water mass variations

The Baltic Sea is a semi-enclosed basin which is connected with the North Sea by narrow and shallow channels. A precise high resolution oceanographic model of the Baltic Sea was developed by Lehmann [4, 5].

Comparisons with geodetic data [9, 10] confirmed that the model describes internal processes within the model domain (see Figure 1) well, including the control of water exchange with the North Sea by meteorological conditions in the transition region [6], and internal water oscillations. Due to its simple boundary conditions at the North Sea boundary the model does not include external effects like the propagation of sea level variations from the North Sea.

Prognostic model output are among others three-dimensional fields of temperature and salinity. From these data, density [7] and the corresponding water mass within 5×5 km$^2$ grid cells were calculated at 6 hourly resolution over 23 years from 1979 to 2001. The water mass data form the basis of our study.

After decreasing the spatial resolution to cells of approximately 50×50 km$^2$, a principal component analysis was performed on the water masses in these cells. Figure 2 shows the two first empirical orthogonal functions (EOFs), i.e., the two dominant spatial patterns of water mass variations, together with the corresponding principal components (PCs).

The first EOF is roughly a uniform change of the Baltic Sea fill level. This kind of variation (with peaks in the order of 50 cm water height deviation from the mean) already causes 70% of the total water mass variance. The second EOF corresponds to a north-south tilt of variable water masses causing another 9% of the total variance.

The first PC is a lower-frequency signal with a dominant annual contribution. The second PC includes larger variances in high frequencies, although it also shows a clear annual behaviour both in the monthly means and in the seasonally varying high-frequency signals.

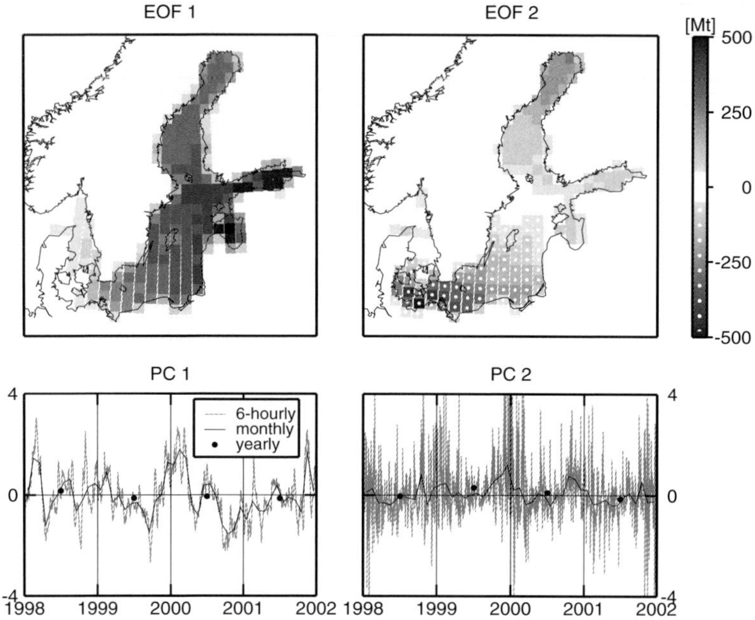


**Fig. 2.** Baltic Sea water mass variations: (**top**) the first two EOFs; (**bottom**) 4-year samples of the PCs. The product of EOF and PC gives the actual mass variations of the $\approx 50 \times 50$ km$^2$ cells in Megatons. 500 Mt correspond to $\approx 20$ cm water height.

Note that the variations do not average out over, e.g., one year: The rms of the annual means of PC1 is still 23% of the 6-hourly data rms.

## 3 Effect on the postglacial rebound detection from satellite gravity data

If we consider the resulting geoid variations instead of mass variations, then the first EOF—due to its large-scale spatial structure—is even more dominant. It causes about 95% of the total geoid variance induced by the modelled oceanography. Thus, it is the essential Baltic Sea geoid signal. In Figure 3, this Baltic Sea signal is compared to the PGR geoid signal. The two signals are similar in their spatial patterns and in their magnitudes. This similarity has an effect on the satellite detection of the PGR. In the following we will give a rough quantification of this statement.

For the estimation of the PGR signal we consider an approach where the signal's spatial pattern is given a priori and only its rate of change is estimated. (For the spatial pattern we use the ICE-4G viscoelastic earth model [11] regionally restricted by a Gaussian window of 1000 km 1-sigma-width.)

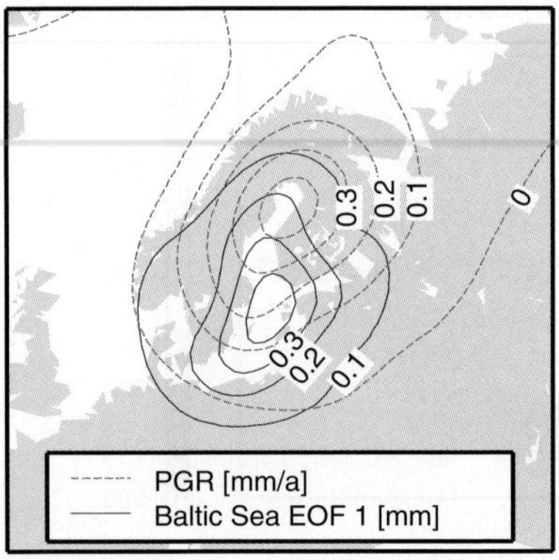


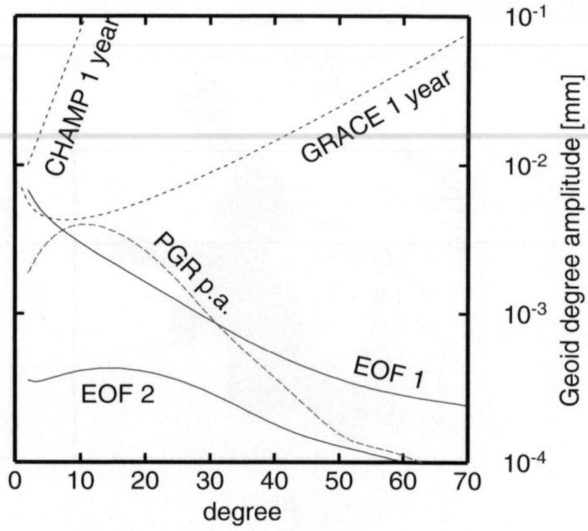


**Fig. 3.** Geoid signals of the PGR and the Baltic Sea (scaled to the rms of the annual means): spatial patterns (**left**) and geoid degree amplitudes (**right**)

For a sequence of years $\{t_n\}$ centered at zero, we express the difference between the yearly gravity models and their temporal mean by a vector $\Delta\boldsymbol{c}(t_n)$ of spherical harmonic coefficients. The spatial pattern of the PGR signal is expressed by a respective vector $\boldsymbol{a}$. We state $\Delta\boldsymbol{c}(t_n) = \alpha t_n \boldsymbol{a} + \Delta\boldsymbol{c}_{\text{other}}(t_n)$, where the PGR rate $\alpha$ has to be estimated, and $\Delta\boldsymbol{c}_{\text{other}}(t_n)$ contains all remaining effects. For a further simplification, let $\Delta\boldsymbol{c}_{\text{other}}(t_n)$ be the sum of solely two effects: the dominant EOF 1 contribution of the Baltic Sea signal with spatial pattern $\boldsymbol{b}$ and time-dependent coefficients $\beta(t_n)$, and the gravity model error $\boldsymbol{\varepsilon}(t_n)$ with covariance matrix $\sigma_\varepsilon^2 P^{-1}$. That is,

$$\Delta\boldsymbol{c}(t_n) = \alpha t_n \boldsymbol{a} + \beta(t_n)\boldsymbol{b} + \boldsymbol{\varepsilon}(t_n). \tag{1}$$

For the estimation of the PGR rate $\alpha$ by least squares adjustment we consider 4 cases A to D. The resulting statistical errors are shown in Table 1 for the GRACE mission where a diagonal error covariance matrix after [1] (see Figure 3) is assumed. For the CHAMP mission, the relative errors are always above 700% for a 5 year mission. They are not considered further.

Case A is the reference case without a Baltic Sea signal, i.e., $\beta \equiv 0$. The error variance of the estimate $\widehat{\alpha}$ can then be computed as

$$\sigma^2(\widehat{\alpha}) = \sigma_\varepsilon^2 \left(\boldsymbol{a}^T P \boldsymbol{a} \textstyle\sum t_n^2\right)^{-1}. \tag{2}$$

In case B we assume that the Baltic Sea signal is present but is ignored in the estimation model. Then, when modelling $\beta(t_n)$ as a random variable of variance $\sigma_\beta^2$, we find

$$\sigma^2(\widehat{\alpha}) = \left[\sigma_\varepsilon^2 + k^2\, \sigma_\beta^2(\boldsymbol{b}^T P \boldsymbol{b})\right] \left(\boldsymbol{a}^T P \boldsymbol{a} \textstyle\sum t_n^2\right)^{-1} \tag{3}$$

where, in the additional summand, $k = (\boldsymbol{a}^T P \boldsymbol{b})(\boldsymbol{a}^T P \boldsymbol{a})^{-\frac{1}{2}}(\boldsymbol{b}^T P \boldsymbol{b})^{-\frac{1}{2}}$ characterizes the similarity between the two signals' spatial patterns seen by the

**Table 1.** Errors of the PGR signal estimation from GRACE relative to the ICE-4G PGR signal [11] in %

| | 2 years | 5 years of mission |
|---|---|---|
| A no Baltic Sea signal | 43.2 | 9.7 |
| B Baltic Sea signal ignored | 87.3 | 19.5 |
| C Baltic Sea signal estimated | 49.9 | 11.1 |
| D Baltic Sea signal corrected | 45.8 | 10.2 |

mission, and $\sigma_\beta^2(\boldsymbol{b}^T P \boldsymbol{b})$ describes the Baltic Sea signal's magnitude. For our data, $k \approx 0.5$ and $\sigma_\beta^2(\boldsymbol{b}^T P \boldsymbol{b}) \approx 4\sigma_\varepsilon^2$.

In case C the Baltic Sea signal parameters $\beta(t_n)$ are estimated together with the PGR rate $\alpha$. In the resulting error variance

$$\sigma^2(\widehat{\alpha}) = (1-k^2)^{-1}\,\sigma_\varepsilon^2\left(\boldsymbol{a}^T P \boldsymbol{a} \sum t_n^2\right)^{-1}, \tag{4}$$

the factor $(1-k^2)^{-1} \approx 1.15^2$ depends only on the "similarity parameter" $k$.

In case D we assume that $\beta(t_n)$ is known a priori with a relative error of 20% and is corrected prior to the $\alpha$ estimation. The error variance is analogous to case B, with $\sigma_\beta$ reduced to $0.2\sigma_\beta$.

## 4 Discussion and conclusions

Based on a precise oceanographic model of the Baltic Sea we analyzed its temporal water mass variations. The dominant component is related to the Baltic Sea overall fill level. The induced geoid signal is similar to the PGR signal in spatial pattern and in magnitude. The results of the previous section show: Ignoring the effect of Baltic Sea water masses for an estimation of the PGR signal from 5 years of GRACE data (case B) would result in an additional error of 10% of the signal, independent from the mission error level $\sigma_\varepsilon$ (see Equ. 3). Including the Baltic Sea signal parameters in the estimation (case C) can reduce this error, but the weak separability between the two signals still results in an increase of the relative error from 9.7% to 11.1%, that is by a factor of 1.15. Using external data to correct for the Baltic Sea signal (case D) may best reduce its influence on the PGR estimation.

Of course, our results are affected by several simplifications in our treatment of the problem. First, the used oceanographic model does not include effects transmitted from the North Sea. These effects, again, mainly result in variations of the Baltic Sea's overall fill level with similar amplitudes as for the internal effects. (However, they might be satisfactorily represented in global ocean models.) Second, in view of the homogeneous spatial coverage of satellite missions it is desirable to infer the PGR effect's spatial pattern, too, instead of fixing an a priori known incorrect pattern. This approach will make the separation from other geophysical signals even more challenging.

Third, Baltic Sea oceanography is not the only signal corrupting the PGR estimation: Mass variations of the atmosphere and the ocean will be accounted for in the GRACE data processing relying on global models. Care should be taken for the inverse barometer effect which applies only partly to the semi-enclosed Baltic Sea. Variations in continental water are another signal to be inferred from the satellite data. For its distinction from the PGR signal our approach of identifying the different signals' dominant spatial patterns may be adapted. Last, an updating of the GRACE error model used here to the actual mission error covariance structure will modify the results.

The methods derived in this study can be applied to other signals of interest. One example is the Laurentide region where the Hudson Bay provides a large water body right in the center of the PGR effect. More generally, in presently ice-covered areas (Antarctica, Greenland) the same problem exists for the separation between ice mass changes and ice-induced crustal motions.

*Acknowledgement.* This research was funded by the German Research Foundation (DFG) under grant Di 473/13-1.

## References


1. National Research Council (1997) Satellite Gravity and the Geosphere. National Academic Press, Washington, DC.
2. Ekman M (1996) A consistent map of the postglacial uplift of Fennoscandia. Terra Nova *8*: 158–165.
3. Ekman M and Mäkinen J (1996) Recent postglacial rebound, gravity change and mantle flow in Fennoscandia. Geophys J Int *126*: 229–234.
4. Lehmann A (1995) A three-dimensional baroclinic eddy-resolving model of the Baltic Sea. Tellus *47A*: 1013–1031.
5. Lehmann A and Hinrichsen H-H (2000) On the thermohaline variability of the Baltic Sea. J Mar Sys *25*: 333–357.
6. Lehmann A and Hinrichsen H-H (2001) The Importance of Water Storage Variations for Water Balance Studies of the Baltic Sea. Phys Echem Earth (B) *26(5-6)*: 383–389.
7. Millero FJ and Kremling K (1976) The densities of Baltic Sea waters. Deep Sea Res *23*: 1129–1138.
8. Milne GA, Davis JL, Mitrovica JX, Scherneck H-G, Johansson JM, Vermeer M, and Koivula H (2001) Space-Geodetic Constraints on Glacial Isostatic Adjustment in Fennoscandia. Science *291*: 2381–2385.
9. Novotny K, Liebsch G, Dietrich R, and Lehmann A (2001) Sea-level variations in the Baltic Sea: Consistency of geodetic observations and oceanographic models. In: Ádám J and Schwarz K-P, eds, Vistas for Geodesy in the New Millenium, IAG Scientific Assembly, Budapest, September 2001, Heidelberg, Springer.
10. Novotny K, Liebsch G, Dietrich R, and Lehmann A (2004) Combination of Sea-Level Observations and an Oceanographic Model for Geodetic Applications in the Baltic Sea. In: Proc. IUGG 2003 General Assembly, Sapporo, in print.
11. Peltier WR (1998) Postglacial variations in the level of the sea: Implications for climate dynamics and solid-earth geophysics. Rev Geophys *36*: 603–689.

# The Impact of the New CHAMP and GRACE Gravity Models on the Measurement of the General Relativistic Lense–Thirring Effect

Lorenzo Iorio

Dipartimento di Fisica dell'Università di Bari, via Amendola 173, 70126, Bari, Italy

**Summary.** Here we wish to discuss the improvements obtainable in the measurement of the general relativistic Lense–Thirring effect with the LAGEOS and LAGEOS II satellites, in terms of reliability of the evaluation of the systematic error and reduction of its magnitude, due to the new CHAMP and GRACE Earth gravity models.

**Key words:** General Theory of Relativity, Lense-Thirring effect, LAGEOS-LAGEOS II orbits, Earth gravity models

## 1 Introduction

The linearized weak–field and slow–motion approximation of the General Theory of Relativity (GTR) [1] is characterized by the condition $g_{\mu\nu} \sim \eta_{\mu\nu} + h_{\mu\nu}$ where $g_{\mu\nu}$ is the curved spacetime metric tensor, $\eta_{\mu\nu}$ is the Minkowski metric tensor of the flat spacetime of Special Relativity and the $h_{\mu\nu}$ are small corrections such that $|h_{\mu\nu}| \ll 1$. Until now, many of its predictions, for the motion of light rays and test masses have been tested, in the Solar System, with a variety of techniques to an accuracy level of the order of 0.1% [2]. It is not so for the gravitomagnetic Lense–Thirring effect due to its extreme smallness. It can be thought of as a consequence of a gravitational spin–spin coupling.

If we consider the motion of a spinning particle in the gravitational field of a central body of mass $M$ and proper angular momentum $\boldsymbol{J}$, it turns out that the spin $\boldsymbol{s}$ of the orbiting particle undergoes a tiny precessional motion [3]. The most famous experiment devoted to the measurement, among other things, of such gravitomagnetic effect in the gravitational field of Earth is the Stanford University GP–B mission [4] which should fly in 2004.

If the whole orbit of a test particle in its geodesic motion around $M$ is considered as a sort of giant gyroscope, its orbital angular momentum $\boldsymbol{\ell}$ undergoes the Lense–Thirring precession, so that the longitude of the ascending node $\Omega$ and the argument of pericentre $\omega$ of the orbit of the test particle are affected by tiny secular precessions $\dot{\Omega}_{\rm LT}$, $\dot{\omega}_{\rm LT}$ [5, 1]

$$\dot{\Omega}_{\rm LT} = \frac{2GJ}{c^2 a^3 (1-e^2)^{\frac{3}{2}}},\ \dot{\omega}_{\rm LT} = -\frac{6GJ\cos i}{c^2 a^3 (1-e^2)^{\frac{3}{2}}}, \tag{1}$$

where $a$, $e$ and $i$ are the semimajor axis, the eccentricity and the inclination, respectively, of the orbit, $c$ is the speed of light and $G$ is the Newtonian gravitational constant.

Up to now, the only attempts to detect the Lense–Thirring effect on the orbit of test particles in the gravitational field of Earth are due to Ciufolini and coworkers [6] who analysed the laser data of the existing LAGEOS and LAGEOS II satellites over time spans of some years. The observable is a suitable combination of the orbital residuals of the nodes of LAGEOS and LAGEOS II and the perigee of LAGEOS II according to an idea exposed in [7]

$$\delta\dot{\Omega}^{\mathrm{L}} + c_1\delta\dot{\Omega}^{\mathrm{L\ II}} + c_2\delta\dot{\omega}^{\mathrm{L\ II}} \sim 60.2\mu_{\mathrm{LT}},\ c_1 \sim 0.295,\ c_2 \sim -0.35, \tag{2}$$

where the superscripts L and L II refer to LAGEOS and LAGEOS II, respectively. The quantity $\mu_{\mathrm{LT}}$ is the solved–for least square parameter which is 0 in Newtonian mechanics and 1 in GTR. The Lense-Thirring signature, entirely adsorbed in the residuals of $\dot{\Omega}$ and $\dot{\omega}$ because the gravitomagnetic force has been purposely set equal to zero in the force models, is a linear trend with a slope of 60.2 milliarcseconds per year (mas yr$^{-1}$ in the following). The standard, statistical error is evaluated as 2%. The claimed total accuracy, including various sources of systematic errors, is of the order of $20 - 30\%$.

The main sources of systematic errors in this experiment are

- the unavoidable aliasing effect due to the mismodelling in the classical secular precessions induced on $\Omega$ and $\omega$ by the even zonal coefficients $J_l$ of the multipolar expansion of geopotential
- the non–gravitational perturbations affecting especially the perigee of LAGEOS II [8, 9]. Their impact on the proposed measurement is difficult to be reliably assessed [10]

It turns out that the mismodelled classical precessions due to the first two even zonal harmonics of geopotential $J_2$ and $J_4$ are the most insidious source of error for the Lense–Thirring measurement with LAGEOS and LAGEOS II. The combination (2) is insensitive just to $J_2$ and $J_4$. According to the full covariance matrix of the EGM96 gravity model [11], the error due to the remaining uncancelled even zonal harmonics amounts to almost 13% [12]. However, if the correlations among the even zonal harmonic coefficients are neglected and the variance matrix is used in a Root–Sum–Square fashion[1], the error due to the even zonal harmonics of geopotential amounts to 46.6% [12]. With this estimate and the evaluations of [8, 9] for the impact of the non–gravitational perturbations the total error in the LAGEOS–LAGEOS II Lense–Thirring experiment would be of the order of 50%. If the sum of the

[1] Such approach is considered more realistic by some authors [10] because nothing assures that the correlations among the even zonal harmonics of the covariance matrix of the EGM96 model, which has been obtained during a multidecadal time span, would be the same during an arbitrary past or future time span of a few years as that used in the LAGEOS–LAGEOS II Lense–Thirring experiment.

absolute values of the individual errors is assumed, an upper bound of 83% for the systematic error due to the even zonal harmonics of geopotential is obtained; then, the total error in the LAGEOS–LAGEOS II Lense–Thirring experiment would become of the order of 100%. This evaluations agree with those released in [10].

The originally proposed LAGEOS III/LARES mission [13] consists of the launch of a LAGEOS–type satellite–the LARES–with the same orbit of LAGEOS except for the inclination $i$ of its orbit, which should be supplementary to that of LAGEOS, and the eccentricity $e$, which should be one order of magnitude larger in order to perform other tests of post–Newtonian gravity [14, 15]. The choice of the particular value of the inclination for LARES is motivated by the fact that in this way, by using as observable the sum of the nodes of LAGEOS and LARES, it should be possible to cancel out to a very high level all the contributions of the even zonal harmonics of geopotential, which depends on $\cos i$, and add up the Lense–Thirring precessions which, instead, are independent of $i$. The use of the nodes would allow to reduce greatly the impact of the non–gravitational perturbations to which such Keplerian orbital elements are rather insensitive [8, 9].

In [16] an alternative observable based on the combination of the residuals of the nodes of LAGEOS, LAGEOS II and LARES and the perigee of LAGEOS II and LARES has been proposed. It would allow to cancel out the first four even zonal harmonics so that the error due to the remaining even zonal harmonics of geopotential would be rather insensitive both to the unavoidable orbital injection errors in the LARES inclination and to the correlations among the even zonal harmonic coefficients. It would amount to 0.02%–0.1% only [16, 17] (EGM96 full covariance and variance RSS calculations).

In regard to the present status of the LARES project, unfortunately, up to now, although its very low cost with respect to other much more complex and expensive space–based missions, it has not yet been approved by any national space agency or scientific institution.

## 2 The impact of the CHAMP and GRACE Earth gravity models

From the previous considerations it could be argued that, in order to have a rather precise and reliable estimate of the total systematic error in the measurement of the Lense–Thirring effect with the existing LAGEOS satellites it would be better to reduce the impact of geopotential in the error budget and/or discard the perigee of LAGEOS II which is very difficult to handle and is a relevant source of uncertainty due to its great sensitivity to many non–gravitational perturbations.

The forthcoming more accurate Earth gravity models from CHAMP [18] and, especially, GRACE [19] will yield an opportunity to realize both these goals, at least to a certain extent.

In order to evaluate quantitatively the opportunities offered by the new terrestrial gravity models we have preliminarily used the recently released EIGEN2 gravity model [20].

With regard to the combination (2), it turns out that the systematic error due to the even zonal harmonics of geopotential, according to the full covariance matrix of EIGEN2 up to degree $l = 70$, amounts to 7%, while if the diagonal part only is adopted it becomes 9% (RSS calculation). Of course, even if the LAGEOS and LAGEOS II data had been reprocessed with the EIGEN2 model, the problems posed by the correct evaluation of the impact of the non–gravitational perturbations on the perigee of LAGEOS II would still persist.

A different approach could be followed by taking the drastic decision of canceling out only the first even zonal harmonic of geopotential by discarding at all the perigee of LAGEOS II. The hope is that the resulting gravitational error is reasonably small so to get a net gain in the error budget thanks to the fact that the nodes of LAGEOS and LAGEOS II exhibit a very good behavior with respect to the non–gravitational perturbations. Indeed, they are far less sensitive to their tricky features than the perigee of LAGEOS II. Moreover, they can be easily and accurately measured, so that also the formal, statistical error should be reduced. A possible observable is[2]

$$\delta\dot{\Omega}^{\mathrm{L}} + c_1 \delta\dot{\Omega}^{\mathrm{L\ II}} \sim 48.2\mu_{\mathrm{LT}},\ c_1 \sim 0.546. \tag{3}$$

According to the full covariance matrix of EIGEN2 up to degree $l = 70$, the systematic error due to the even zonal harmonics from $l = 4$ to $l = 70$ amounts to 8.5 mas $\mathrm{yr}^{-1}$ yielding a 17.8% percent error, while if the diagonal part only is adopted it becomes[3] 22% (RSS calculation). EGM96 would not allow to adopt (3) because its full covariance matrix up to degree $l = 70$ yields an error of 47.8% while the error according to its diagonal part only amounts even to 104% (RSS calculation). Note also that the combination (3) preserves one of the most important features of the combination (2): indeed, it allows to cancel out the very insidious 18.6-year tidal perturbation which is a $l = 2$, $m = 0$ constituent with a period of 18.6 years due to the Moon's node and nominal amplitudes of the order of $10^3$ mas on the nodes of LAGEOS and LAGEOS II [21]. On the other hand, the impact of the non–gravitational perturbations on the combination (3) over a time span of, say, 7 years can be quantified in just 0.1 mas $\mathrm{yr}^{-1}$, yielding a 0.3% percent error. The results of Table 2 and Table 3 of [16] have been used. It is also important to notice that, thanks to the fact that the periods of many gravitational and non–gravitational time–dependent perturbations acting on the nodes of the LAGEOS satellites are rather short, a reanalysis of the LAGEOS and LAGEOS II data over just a few

[2] A similar approach seems to be suggested in [19], although without quantitative details.

[3] The even zonal harmonics are much more mutually uncorrelated in EIGEN2 than in EGM96.

years could be performed. This is not so for the combination (2) because some of the gravitational [21] and non–gravitational [8] perturbations affecting the perigee of LAGEOS II have periods of many years. Then, with a little time–consuming reanalysis of the nodes only of the existing LAGEOS and LAGEOS II satellites with the EIGEN2 data it would at once be possible to obtain a more accurate and reliable measurement of the Lense–Thirring effect, avoiding the problem of the uncertainties related to the use of the perigee of LAGEOS II.

Very recently the first preliminary Earth gravity models including some data from GRACE have been released; among them the GGM01C model[4], which combines the Center for Space Research (CSR) TEG4 model with data from GRACE, seems to be very promising. Indeed, the error due to geopotential in the combination (2), evaluated by using the variance matrix only (RSS calculation), amounts to 2.2% (with an upper bound of 3.1% obtained from the sum of the absolute values of the individual terms). Instead, the combination (3) would be affected at almost 14% level (RSS calculation), with an upper bound of almost 18% from the sum of the absolute values of the individual errors. However, it should be pointed out that extensive calibration tests have still to be performed with the GGM01C model.

## 3 Conclusions

When more robust and complete terrestrial gravity models from CHAMP and GRACE will be available in the near future the combination (3) could hopefully allow for a measurement of the Lense–Thirring effect with a total systematic error, mainly due to geopotential, of some percent over a time span of a few years without the uncertainties related to the evaluation of the impact of the non–gravitational perturbations acting upon the perigee of LAGEOS II. On the other hand, the obtainable accuracy with the combination (2), whose error due to geopotential is smaller than that of (3), is strongly related to improvements in the evaluation of the non–gravitational part of the error budget and to the use of time spans of many years.

## References

1. Ciufolini I, Wheeler JA (1995) Gravitation and Inertia. Princeton University Press, New York.
2. Will CM (2001) The Confrontation between General Relativity and Experiment. Living Rev Relativity *4*, online article: http://www.livingreviews.org/lrr-2001-4.
3. Schiff L (1960) On Experimental Tests of the General Theory of Relativity. Am J Phys *28(4)*: 340–343.

[4] It can be retrieved on the WEB at http://www.csr.utexas.edu/grace/gravity/

4. Everitt CWF and other members of the Gravity Probe B team (2001) Gravity Probe B: Countdown to Launch. In: Gyros, Clocks, Interferometers ... : Testing Relativistic Gravity in Space, ed. by Lämmerzahl C, Everitt CWF, Hehl FW, Springer, Berlin: 52–82.
5. Lense J, Thirring H (1918) Über den Einfluss der Eigenrotation der Zentralkörper auf die Bewegung der Planeten und Monde nach der Einsteinschen Gravitationstheorie. Phys Z *19*: 156–163, translated by Mashhoon B, Hehl FW, Theiss DS (1984) On the Gravitational Effects of Rotating Masses: The Thirring-Lense Papers, Gen Rel and Gravit *16*: 711–750.
6. Ciufolini I (2000) The 1995-99 measurements of the Lense-Thirring effect using laser-ranged satellites. Class and Quantum Grav *17(12)*: 2369–2380.
7. Ciufolini I (1996) On a new method to measure the gravitomagnetic field using two orbiting satellites. Il Nuovo Cimento A *109(12)*: 1709–1720.
8. Lucchesi D (2001) Reassessment of the error modelling of non–gravitational perturbations on LAGEOS II and their impact in the Lense–Thirring determination - Part I. Plan and Space Sci *49(5)*: 447–463.
9. Lucchesi D (2002) Reassessment of the error modelling of non–gravitational perturbations on LAGEOS II and their impact in the Lense–Thirring determination - Part II. Plan and Space Sci *50(10-11)*: 1067–1100.
10. J Ries, RJ Eanes, BD Tapley (2003) Lense-Thirring Precession Determination from Laser Ranging to Artificial Satellites. In: Nonlinear Gravitodynamics. The Lense–Thirring Effect, ed by Ruffini R, Sigismondi C, World Scientific, Singapore: 201–211.
11. Lemoine FG et al. (1998) The Development of the Joint NASA GSFC and the National Imagery Mapping Agency (NIMA) Geopotential Model EGM96. NASA/TP-1998-206861.
12. Iorio L (2003) The impact of the static part of the Earth's gravity field on some tests of General Relativity with satellite laser ranging. Celest Mech and Dyn Astron *86(3)*: 277–294.
13. Ciufolini I (1986) Measurement of Lense–Thirring drag on high-altitude laser ranged artificial satellite. Phys Rev Lett *56(4)*: 278–281.
14. Iorio L, Ciufolini I, Pavlis E (2002) Measuring the Relativistic Perigee Advance with Satellite Laser Ranging. Class and Quantum Grav *19(16)*: 4301–4309.
15. Iorio L (2002) Constraints on a Yukawa gravitational potential from laser data of LAGEOS satellites. Phys Lett A *298(5-6)*: 315–318.
16. Iorio L, Lucchesi D, Ciufolini I (2002) The LARES Mission Revisited: An Alternative Scenario. Class and Quantum Grav *19(16)*: 4311–4325.
17. Iorio L (2003): A reassessment of the systematic gravitational error in the LARES mission. Gen Rel and Gravit *35(7)*: 1263–1272.
18. Pavlis E (2000) Geodetic Contributions to Gravitational Experiments in Space. In: Recent Developments in General Relativity, ed. by Cianci R, Collina R, Francaviglia M, Fré P, Springer, Milan: 217–233.
19. Ries JC, Eanes RJ, Tapley BD, Peterson GE (2002) Prospects for an Improved Lense-Thirring Test with SLR and the GRACE Gravity Mission. In: Proc 13th Intl Laser Ranging Workshop, Washington DC, October 7-11, 2002, Preprint *http://cddisa.gsfc.nasa.gov/lw13/lw_proceedings.html#science*
20. Reigber C et al. (2003): The CHAMP-only EIGEN-2 earth gravity field model. Adv Space Res *31(8)*: 1883–1888.
21. Iorio L (2001): Earth tides and Lense–Thirring effect. Celest Mech and Dyn Astron *79(3)*: 201–230.

# Recovery of Isostatic Topography over North America from Topographic and CHAMP Gravity Correlations

Laramie V. Potts[1] C. K. Shum[1], Ralph von Frese[2], Shin-Chan Han[1], and Rainer Mautz[1,3]

[1] Laboratory for Space Geodesy and Remote Sensing, Ohio State University, Columbus, USA *(potts.3@osu.edu)*
[2] Dept. of Geological Sciences, Ohio State University, Columbus, USA.
[3] Institute of Astronomical, Physical and Mathematical Geodesy, Technical University of Berlin, Germany.

**Summary.** We investigate North American crustal structure and mass loads from spectral correlation analysis of topographic, CHAMP and terrestrial gravity data. We use free-air and terrain gravity correlations to isolate tectonically driven vertical motions and mass imbalances of the crust and lithosphere. Specifically, we apply correlation filters to decompose the free-air gravity anomalies into terrain-correlated and terrain-decorrelated components to yield compensated terrain gravity effects that we evaluate for crustal thickness variations. Our results compare quite favourably with the seismically inferred global crustal thickness model Crust5.1 and a 3.4 km rms difference with LITH5.0 over North America. Terrain-correlated anomalies reveal mass excesses and deficits that are interpreted as uncompensated elements of the crust. For Hudson Bay, the average terrain-correlated free-air anomaly suggests that the crustal topography is depressed by about 400 m. Because glacial isostatic adjustment considerations can only marginally account for the depression, we speculate that it may reflect other effects such as a preglacial impact.

**Key words:** crustal thickness, correlations, gravity, topography, Hudson Bay

## 1 Introduction

Crustal modeling plays an important role in the prediction of isostatic surface topography, which is defined as the large-scale background topography that is in equilibrium at the surface of the Earth [1]. Non-zero free-air gravity anomalies over North America indicate lateral variations of uncompensated masses in the subsurface. In the mantle, these lateral density variations can reflect convective flows that generate vertical stresses at the base of the lithosphere, which by traction can sustain the observed topography as uncompensated crustal mass (e.g., [6]). Hence, the observed topography can involve superposed isostatic and dynamic topography contributions.

Seismic refraction and geodynamic modeling commonly provide regional characterizations of the isostatic and dynamic components of topography [1].

Specifically, analyses of the global Crust5.1 model of $5^0 \times 5^0$ crustal thickness variations [8] and the regional Canadian LITH5.0 crustal thickness model [10] suggest that most of the cratonic regions of North America lie at lower elevations than predicted by crustal isostasy. By subtracting the isostatic topography from the observed topography, dynamic topography is revealed that may constrain mantle flow. However, these complex seismic-geodynamic long wavelength flow models (e.g., [1]; [9]; [11]) overpredict by 25% the dynamic topography amplitudes [6]. These dynamic topography estimates may be limited by subjective choices of boundary conditions and depth-dependent profiles of mantle viscosity (e.g., [5]), sparseness of seismic stations, and uncertainties associated with densities of crustal layers (e.g., [1]; [8]; [10]).

Gravity data comprise the intergrated effects of uncompensated mass loads from the core to the crust. For example, the observed gravity anomaly minimum over Hudson Bay has been attributed to glacial isostatic adjustment (GIA) and mantle convection (e.g., [1]; [7]; [9]; [11]). However, topography data provide useful constraints to account for the gravity effects of crustal loads. Spectral correlation analysis between free-air and topographic gravity effects provide additional constraints on the lithosphere by extracting patterns of uncompensated mass variations [12]. The focus of this paper is to study constraints on dynamic and isostatic topography over North America from the joint analysis of gravity and topography data.

## 2 Gravity crustal modeling

Figure 1.a shows free-air gravity anomalies of North America evaluated at 20 km altitude from the TEG4 gravity model complete to degree 200 [3]. This model combines CHAMP terrestrial gravity data for the best estimate currently available of the free-air gravity anomalies at 20 km altitude (e.g., [15] this volume).

Figure 1.b shows the topography model from ETOPO5. The gravity effect of the terrain in Figure 1.c sums the contributions from the ice, water, and rock components assuming constant density values in each layer of 0.917 $g/cm^3$, 1.028 $g/cm^3$, and 2.8 $g/cm^3$, respectively (e.g., [1]). We conduct our analysis at 20-km altitude to minimize the effects from local terrain density and elevation errors. Dominated by rock topography, the terrain gravity effect is about five times larger than the poorly correlated (CC=0.2) free-air anomalies.

We separate the terrain-correlated and -decorrelated components of the free-air gravity anomalies on the basis of the correlation spectrum that we obtained from the Fourier transforms of the free-air (Figure 1.a) and terrain gravity effects (Figure 1.b) [14]. Inversely transforming the wavenumber components of the free-air anomalies that were strongly correlated both positively ($CC(k) \geq 0.6$) and negatively ($CC(k) \leq -0.6$) with the wavenumber components of the terrain gravity effects, determines the terrain-correlated

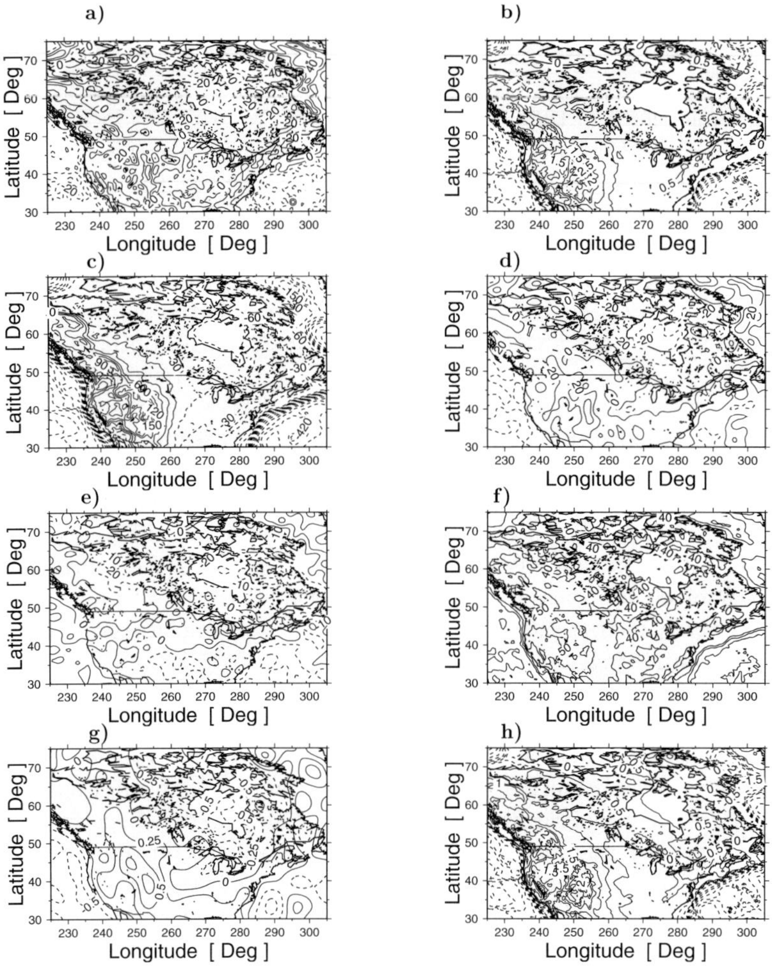


**Fig. 1.** Crustal modeling using TEG-4 free-air gravity anomalies (a) and topography (b). Terrain gravity effects (c) are the integerated effects of ice, water and rock. Free-air gravity components are divided in terrain-correlated (d) and terrain-decorrelated (e) anomalies. The modeling yields estimates of crustal thickness variations (f), dynamic topography (g), and isostatic topography (h).

free-air anomalies shown in Figure 1.d. The correlation coefficient cutoff values (CC(k)) for passing the k-th wavenumber component were chosen to nullify the correlation between the compensated terrain gravity effects (not shown) and the terrain-decorrelated free-air anomalies in Figure 1.e.

Subtracting the terrain-correlated components from the free-air anomalies yields the terrain-decorrelated free-air components. The longer wavelength components may best reveal the effects deep subcrustal mass variations related to deep mantle flow because of the abscence of the relatively strong interfering long wavelength free-air anomalies from uncompensated crustal terrain [12]. The higher frequency components can also reflect uncompensated density variations within the crust [12], but these signals are typically close to the noise level of the gravity data.

For estimating Moho and related crustal thickness variations of North America, we subtract the terrain-correlated free-air anomalies (Figure 1.d) from the terrain gravity effects (Figure 1.b) for the compensated terrain gravity effects. The lack of compensated terrain effects in the free-air anomalies reflects annihilating signals that may be analyzed for Moho variations in the context of an appropriate compensation model. Reversing the polarities of the compensated terrain effects yields the annihilating signals for obtaining Moho estimates by spherical coordinate inversions using Gauss-Legendre quadrature integration. The inversions estimated the thicknesses of spherical prisms about a reference depth of 30 km below mean sea level assuming a 0.5 g/cm$^3$ density contrast for the mantle relative to the crust (e.g., [1]). Our crustal thickness model, referenced to the EGM96 geoid, compares quite favourably with Crust5.1. Relative to LITH5.0, our results have a rms difference of 3.4 km over mid-latitudes which is well within the acceptable error limits of the seismically-inferred thicknesses [10]. At high latitudes where seismic observations are relatively sparse, our model over-estimates seismic crustal thickness estimates by about 15%.

Assuming that the terrain-correlated free-air anomalies reflect mostly uncompensated crustal topography, the inversion of these anomalies about the surface topography with a density contrast of 2.8 g/cm$^3$ furnishes the dynamic topography in Figure 1.g. Subtracting Figure 1.g from Figure 1.b yields the isostatic topography in Figure 1.h. The gravity effect from the dynamic topography matches in a least-squares sense the terrain-correlated free-air gravity anomalies of Figure 1.d. The crust presumably is attempting to achieve isostatic equilibrium and a zero-mean free-air gravity anomaly. Hence, where the dynamic topography is above (positive) or below (negative) the observed topography, the crust may be under pressure to subside or rise, respectively.

## 3 Hudson Bay Results

For Hudson Bay, the mean isostatic topography estimated in Figure 1.h is about 420 m higher than the observed topography so that the crust is in isostatic uplift. Absolute gravity measurements over a twelve year period near Hudson Bay show an average decrease in gravity of about $\pm 0.2$ $\mu$Gals/yr which corresponds to $\pm 1.3$ mm/yr uplift [4]. Vertical crustal motion over Hudson Bay has been attributed to the visco-elastic response of the lithosphere to the load of the Laurentide ice sheet that covered northern North America for thousands of years until its retreat some 10,000 years ago [7]. Present models on Laurentide deglaciation that satisfy uplift data predict only 15% to 30% of the observed -50 mGal free-air gravity anomaly [2] while geodynamic models attribute up to 50% of the total long wavelength geoid to GIA (e.g., [9]; [11]; [13]), depending on the choice of model parameters [5]. Assuming the crust was in isostatic equilibrium prior to ice loading, the -50 mGal gravity anomaly accomodates a surface depression of about 1.7 km due to ice loading (assuming the infinite-slab approximation) that promotes in turn problematic mantle viscosity and lithosphere thickness estimates (e.g., [2]). However, the terrain-correlated anomalies (Figure 1.d) represent about 70% of the original free-air anomaly signal (Figure 1.a) suggesting that the topographic deficit explains a significant percentage of the negative free-air gravity anomaly. Hence, the gravity effect of uncompensated crustal loads accounts for more than half of the observed gravity. The terrain-correlated free-air anomalies thus may mostly reflect other isostatic contributions than those due to GIA.

The terrain-decorrelated free-air gravity components (Figure 1.e) were assumed to not contribute to dynamic surface topography, but rather to arise from lateral subcrustal density variations. Seismic velocity data indicate that North America may lie above an anomalously cold, dense region that acts to depress the surface (e.g., [5]; [6]). Deep mantle convective flows may dynamically support the localized surface depression, although flow models that reconcile uplift data and gravity anomalies grossly overestimate uplift rates near Hudson Bay [7]. Hence, the terrain-decorrelated anomalies may help constrain the deep mantle flows in the context of the subcrustal density stratigraphy of Hudson Bay (e.g., [12]).

The broad dynamic topography depression centered on Hudson Bay in Figure 1.g includes a localized small central peak flanked by annuli of depressed and raised topography. The dynamic topography signature for Hudson Bay provokes the impression of a crustal impact basin. The possible location of the impact is indicated by the red triangle in Figure 1.g. The isostatic effects of the exogeneously disrupted crust dominate the terrain-correlated free-air anomalies that can only marginally be accounted for by GIA considerations. Hence, we speculate that an impact may have excavated crustal material prior to ice loading which subsequently eroded down much of the original topography of the impact basin.

## 4 Conclusion

We investigated topographic and gravity correlations using ETOPO5 and the TEG4 gravity model (which includes CHAMP data) at 20 km altitude to characterize possible crustal thickness variations (Figure 1.f) and related isostatic adjustments of the topography (Figure 1.g). Spectral correlation analysis decomposed the free-air anomalies (Figure 1.a) into terrain-correlated (Figure 1.d) and terrain-decorrelated (Figure 1.e) components. Terrain-correlated anomalies were interpreted mainly for dynamic surface topography (Figure 1.g). Our analysis indicates that the terrain-correlated components represent about 70% of the total free-air gravity anomalies of North America.

The gravity minimum over Hudson Bay reflects topography undercompensated by about 400 m due possibly more to a preglacial impact or some other effect than to the retreat of the Laurentide ice sheet. The residual terrain-decorrelated free-air gravity anomalies (Figure 1.e) may constrain uncompensated mass anomalies within the crust as well as in the deeper interior where density-driven mass flows operate.

## References


1. Forte A, Peltier W, Dziewonski A, Woodward R (1993) Dynamic surface topography: A new interpretation based upon mantle flow models derived from seismic tomography. Geophys Res Lett *20*: 225–228.
2. James T (1992) The Hudson Bay free-air gravity anomaly and glacial rebound. Geophys Res Lett *19(9)*: 861–864.
3. Kang Z, Bettadpur S, Tapley B, Cheng K, Ries J (2003) Determination of CHAMP accelerometer calibration parameters. In: Reigber Ch, Luehr J, Schwintzer P (eds.), First CHAMP Mission Results for Gravity, Magnetics, and Atmospheric Studies. Springer-Verlag: 19–25.
4. Lambert A, Courtier N, Sasagawa G, Klopping F, Winester D, James T, and Liard J (2001) New constraints on Laurentide postglacial rebound from absolute gravity measurements. Geophy Res Lett *28(10)*: 2109–2112.
5. Li A, Fischer K, Wysession M, Clarke T (1998) Mantle discontinuities and temperature under the North American continental keel. Nature *395*: 160–163.
6. Lithgow-Bertelloni C, Silver P (1998) Dynamic topography, plate driving forces and the African superswell. Nature *395*: 269–272.
7. Mitrovica J (1997) Going halves over Hudson Bay. Nature *390*: 444–446.
8. Mooney W, Laske G, Masters T (1998) CRUST5.1: A global crustal model at $5^0 \times 5^0$. J Geophys Res *103*: 727–747.
9. Pari G, Peltier W (2000) Subcontinental mantle dynamics: A further analysis based on the joint constraints of dynamic surface topography and free-air gravity. J Geophys Res *105*: 5635–5662.
10. Perry H, Eaton D, Forte A (2002) LITH5.0: A revised crustal model for Canada based on Lithoprobe results. Geophys J Int *150*: 285–294.
11. Perry H, Forte A, Eaton D (2003) Upper-mantle thermochemical structure below North America from seismic-geodynamic flow models. Geophys J Int *154*: 279–299.
12. Potts L, von Frese R (2003) Lunar subcrustal mass differentiation from spectral free-air and terrain gravity correlations. J Geophys Res *108(E4)*: 10.1029/2000JE001440.
13. Simons M, Hager B (1997) Localization of the gravity field and the signature of glacial rebound. Nature *390*: 500–504.
14. von Frese R, Jones M, Kim J, Li W (1997) Spectral correlation of magnetic and gravity anomalies of Ohio. Geophys *62(1)*: 365–380.
15. von Frese R, Kim H, Taylor P, Asgharzadeh M (2003) Reliability of CHAMP anomaly contributions. this volume.

# Dynamic Topography as Reflected in the Global Gravity Field

Mikhail K. Kaban, Peter Schwintzer, Christoph Reigber

GeoForschungsZentrum Potsdam (GFZ), Department 1 'Geodesy and Remote Sensing', Telegrafenberg A17, 14473 Potsdam, Germany
*kaban@gfz-potsdam.de, psch@gfz-potsdam.de, reigber@gfz-potsdam.de*

**Summary.** The long-wavelength part of non-isostatic topography is supposed to be generated by mantle dynamics and is up to now not well studied. In order to separate the dynamic part from the residual topography (that part of the Earth's topography which is not explained by the crustal model providing isostatic compensation) a correlation analysis is performed between the residual topography and long-wavelength isostatic gravity anomalies. It is found that the correlation between these quantities is positive for spatial wavelengths larger than 4000 km. The resulting correlated part of residual topography is regarded to represent the dynamic topography. Its amplitude was estimated to range from -0.4 to 0.5 km and is on the lower limit of what was estimated from a direct modelling of mantle convection. The calculated dynamic topography may be used as a strong constraint in further numerical simulations of mantle dynamics.

**Key words:** dynamic topography, isostatic gravity anomalies, global gravity field model

## 1 Introduction

Most of the Earth's topographic masses are isostatically balanced by lithosphere density inhomogeneities or variations of the crustal/lithosphere thickness (e.g. Pratt 1858). The remaining part of the surface topography, not isostatically compensated, contains information about various geodynamic processes. However, the non-isostatic topography may not be directly deduced from the Earth's relief because the lithosphere structure is not sufficiently known.

One of the main problems of present-day geodynamics is to estimate the dynamic effect of deep inner density, viscosity and thermal inhomogeneities on the Earth's surface (dynamic topography), i.e. the long-wavelength non-isostatic component of the surface topography which is supposed to be generated and supported by mantle flow. The wavelengths which are of interest here are larger than 2000 km, or in terms of spherical harmonics up to degree 20. The forward-computation method implies the construction of a complete global dynamic model of the Earth, including the density and viscosity distribution. Due to the large uncertainties in the knowledge of these parameters this kind of studies shows largely varying results. Different authors give amplitudes of the dynamic topography ranging from 0.5 km (e.g. Hager and Richards 1989, Čadek and Fleitout 2003) to 1 km (e.g.

Steinberger et al. 2001) and 2 km (e.g. Pari and Peltier 2000). Here an empirical approach based on topography, lithospheric density and gravity data is applied to estimate the dynamic topography.

## 2 Long-wavelength residual topography and isostatic gravity anomalies

To characterize the isostatic state of the lithosphere, the residual topography, which is equivalent to the total sum of anomalous masses with respect to a reference column above a level of compensation, is evaluated (Kaban et al. 2003b). The adopted level of compensation is the bottom of the Moho discontinuity under continents and the bottom of the lithosphere at 130 km depth under ocean areas according to the cooling plate model (Hager 1983).

The calculated residual topography is shown in Figure 1a. Its variations are as large as ± 2 km. The residual topography arises from two sources. First, residual topography is due to unmodelled compensating mass anomalies in the uppermost mantle that is part of the lithosphere: highs are supported by low-density lithosphere roots while lows are balanced by high density anchors. Second, residual topography arises from isostatic disturbances at different length scales of the lithosphere. The long-wavelength component of these disturbances is controlled by normal stress at the base of the lithosphere due to mantle flow. This part of the residual topography is defined as 'dynamic' topography. Dynamic topography is important to know for constraining geodynamic models.

The actual geographic distribution of dynamic topography is practically unknown. Here, isostatic and dynamic topography are separated using long-wavelength isostatic gravity anomalies, which also reflect the mantle density structure and, by this, mantle dynamics including dynamic topography. The isostatic gravity anomalies are obtained after subtracting from the observed gravity anomalies the field that is produced by the isostatically compensated litho-

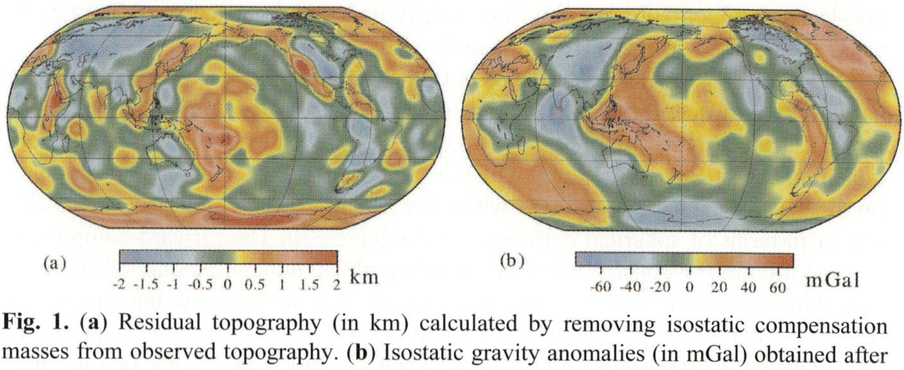


**Fig. 1.** (**a**) Residual topography (in km) calculated by removing isostatic compensation masses from observed topography. (**b**) Isostatic gravity anomalies (in mGal) obtained after subtracting the gravity effect of the isostatically compensated lithospheric model from the observed field EIGEN-1S. Both fields are truncated after degree/order 20.

sphere (Kaban et al. 2003b). Actually the computations are done in terms of gravity disturbances (Heiskanen and Moritz 1967) to account for the masses between the reference ellipsoid and geoid. But, for convenience the term 'gravity anomaly' is used throughout the text. The calculated isostatic gravity anomalies, based on the CHAMP global gravity field model EIGEN-1S (Reigber et al. 2002), are shown in Figure 1b. The isostatic gravity anomalies hardly correlate with topographic and bathymetric features, as the unreduced gravity anomalies do. For example, the Indian and Siberian minima previously separated by the Tibet 'bridge' are joined now into one global minimum.

The dynamic topography generated by mantle flow considerably contributes to the isostatic gravity anomalies. However, these depend not only on the dynamic topography but also on deep density variations in the Earth's mantle. To extract the purely dynamic effect we are going to analyse both quantities: residual topography and isostatic gravity anomalies.

## 3 Global dynamic topography

In order to discriminate in the computed residual topography (Figure 1a) the dynamic part and the isostatic one, a correlation analysis between the residual topography and the isostatic gravity anomalies (Figure 1b) has been performed, both fields developed in spherical harmonics and truncated after degree/order 20. The idea is, if the isostatic condition, applied to compute the isostatically compensated lithosphere, is justified, then the residual topography is completely compensated within the crust and upper mantle and contrary to dynamic topography no gravitational signal should be left in the isostatic gravity anomalies.

The residual topography reflects a mixture of dynamic and isostatic residual topography and the isostatic gravity anomalies also reflect other factors. Therefore a correlation analysis between both fields shall reveal the broad features of the dynamic topography. The applied approach not necessarily picks up the overall dynamic topography, as, depending on the viscosity distribution in the Earth interior, the joint gravity contribution of dynamic and isostatic residual topography may cancel each other or the deep density inhomogeneities may dominate the signal (e.g. Hager and Richards 1989). However, it is supposed that these cases are only deviations from the regular case.

The relationship between both fields (admittance) as a function of the harmonic degree is shown in Figure 2. It was found that the relationship between the residual topography and isostatic gravity anomalies is slightly positive up to degree 10 and varies within determination errors around an average value of $0.65 \cdot 10^{-2}$ km/mGal, while for higher degrees the correlation is practically lost. Thus, this single admittance coefficient is used to scale the isostatic gravity anomalies computed from the spherical harmonic coefficients up to degree 10. The result, the correlated part of residual topography is regarded to represent the dynamic topography, and the remaining (uncorrelated) part is considered to reflect the isostatic

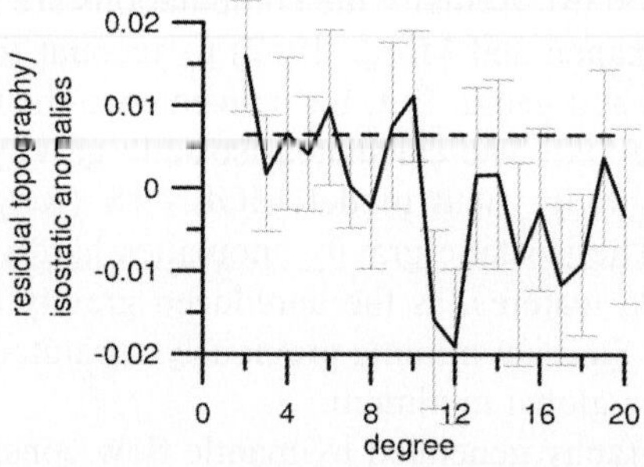


**Fig. 2.** The relationship (admittance) between residual topography and isostatic gravity anomalies (Figure 1) per spherical harmonic degree. The error estimates are based on 95% confidence level. Dashed line shows the regression coefficient ($0.65\cdot10^{-2}$ km/mGal) obtained for all coefficients within the degree interval from 2 through 10.

part of the residual topography which is due to an initially incomplete knowledge of the crustal/upper mantle structure.

Both correlated and non-correlated parts of the residual topography up to degree 10 are shown in Figure 3. The correlated part (Figure 3a) is supposed to be generated by mantle flow and represents most of the dynamic topography. It was found that the dynamic topography varies between –0.5 km (down-welling) and +0.4 km (up-welling). This order of magnitude is in accordance with Hager and Richards (1989), and Čadek and Fleitout (2003) and is at the lower limit obtained from forward computations of mantle convection models. Some portion of the dynamic topography that, as was mentioned before, is hidden in the global isostatic gravity anomalies, will not be present in the resulting field. Further analysis with an advanced inversion technique should clarify this question.

The larger part of the residual topography with amplitudes of about 2 km could be attributed to the initially unresolved isostatic topography (Figure 3b). The most pronounced lows are associated with the Andes, the East European Platform, the Alpine-Mediterranean fold belt, and the central and south-eastern parts of North America. These areas are characterised by dense mantle material, which could be attributed to a large extent to deep lithospheric roots as discussed in Kaban et al. (2003a). Pronounced highs are associated with vast Cenozoic regions of rifting and crustal extension: the East-African Rift and the Basin and Range Province of western USA. These topography heights are balanced by hot mantle material just below the lithosphere. In addition, a strong low is found over the South-Western Pacific. This area represents a large-scale disturbance (Kaban and Schwintzer 2001) with respect to the 'normal' ocean model (Hager 1983), where lithospheric density and ocean floor depth depend on the lithosphere age.

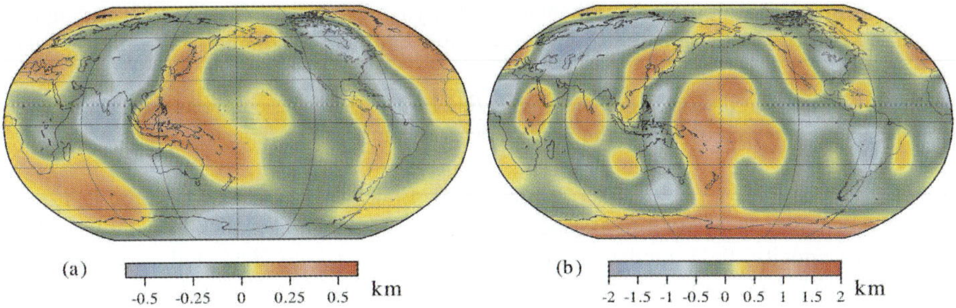


**Fig. 3.** Residual topography (km) with a resolution corresponding to a spherical harmonic expansion up to degree/order 10: (**a**) Portion of the residual topography correlated with the long-wavelength isostatic gravity anomalies (dynamic part); (**b**) Portion of the residual topography not correlated with the long-wavelength isostatic gravity anomalies, which is presumably balanced by upper mantle density variations.

## 4 Conclusions

Dynamic undulations of the Earth's surface are produced by mantle convection, subducting slabs and uprising mantle plumes. These phenomena are responsible for the overall Earth's dynamics, thus the identification of the dynamic topography is crucial when modelling deep Earth processes. However, forward computations direct estimates based on mantle convection modelling give amplitudes that differ by a factor of four (e.g. Čadek and Fleitout 2003; Pari and Peltier 2000). Also the geographical distribution of dynamic topography largely varies depending on the model assumptions in the forward computations as discussed recently in Čadek and Fleitout (2003). Thus, the present empirically derived results can be used to constrain further modeling of mantle dynamics.

## References

Čadek O, Fleitout L (2003) Effect of lateral viscosity variations in the top 300 km on the geoid and dynamic topography. Geophys J Int *152*: 566–580.

Hager BH (1983) Global isostatic geoid anomalies for plate and boundary layer models of the lithosphere. Earth Planet Sci Lett *63*: 97-109.

Hager BH, Richards MA (1989) Long-wavelength variations in Earth's geoid; physical models and dynamical implications. Philos Trans R Soc London *A 328*: 309-327.

Heiskanen W, Moritz H (1967) Physical geodesy, W.H. Freeman, San Francisco.

Kaban MK, Schwintzer P (2001) Oceanic upper mantle structure from experimental scaling of Vs and density at different depths. Geophys J Int *147*: 199-214.

Kaban MK, Schwintzer P, Artemieva IM, Mooney WD (2003a) Density of the continental roots: compositional and thermal contributions. Earth Planet Sci Lett *209*: 53-69.

Kaban MK, Schwintzer P, Reigber Ch (2003b) A new isostatic model of the lithosphere and gravity field. Journal of Geodesy, submitted.

Pari G, Peltier WR (2000) Subcontinental mantle dynamics: A further analysis based on the joint constraints of dynamic surface topography and free-air gravity. J Geoph Res *105*: 5635-5662.

Pratt JH (1858) The deflection of the plumb-line in India and the compensatory effect of a deficiency of matter below the Himalaya mountains. Philos Trans R So London, *Vol. CXLIX*: 745-778.

Reigber Ch, Balmino G, Schwintzer P, Biancale R, Bode A, Lemoine J-M, König R, Loyer S, Neumayer H, Marty J-C, Barthelmes F, Perosanz F, Zhu SY (2002) A high-quality global gravity field model from CHAMP GPS tracking data and accelerometry (EIGEN-IS). Geophys Res Lett *29(14)*: 10.1029, GL015064.

Steinberger B, Schmeling H, Marquart G (2001) Large-scale lithospheric stress field and topography induced by global mantle circulation. Earth Planet Sci Lett *186*: 75-91.

# Impact of the CHAMP Mission on Estimating the Mean Sea Surface

Verena Seufer[1,2], Jens Schröter[1], Manfred Wenzel[1], and Wolfgang Keller[3]

[1] Alfred Wegener Institute for Polar and Marine Research, Bremerhaven
[2] now at GeoForschungsZentrum Potsdam (GFZ), Dept. 1 'Geodesy and Remote Sensing', *vseufer@gfz-potsdam.de*
[3] Geodetic Institute, Stuttgart University

**Summary.** In physical oceanography the slope of the sea surface can be used to calculate a surface geostrophic velocity as a reference for the general circulation and its associated transports. Here we calculate a new mean sea surface (MSS) by combining oceanography, altimetry and gravity data. A first guess MSS is calculated from a dynamical ocean model into which measurements of temperature, salinity, property fluxes and tide gauges have been assimilated. The large scales of this surface are subsequently improved by adding information from satellite altimetry and a 'satellite only' geoid model like the CHAMP derived EIGEN-1S. The combination takes into account the different error structures of the three sources of data. Simultaneous we estimate large scale corrections for the respectively used geoid models. Comparing the corrections of different geoid models makes it possible to verify the improvement of 'satellite only' gravity models by the CHAMP mission.

**Key words:** Ocean model, satellite altimetry, geoid, mean sea surface

## 1 Introduction

The mean sea surface (MSS) is estimated by combining ocean modelling, altimetry and a state of the art geoid. A first guess MSS is calculated from a dynamical ocean model into which measurements of temperature, salinity, property fluxes and tide gauges have been assimilated. This surface is subsequently improved by adding information from satellite altimetry and a geoid model. The combination takes into account the different error structures of the three sources of data: altimetry, oceanography and geodesy. It seems to be reasonable to employ a geoid model that is independent of altimeter data, i.e. we must use a so called 'satellite only' geoid which is estimated without utilizing altimeter data. As the 'satellite only' geoid is very accurate on the largest scales, it should determine these scales of the combined solution.
The assimilation of geoid heights into ocean models has been done only in few cases because of the low accuracy of the available 'satellite only' gravity models. With the CHAMP satellite we got the highly improved geoid EIGEN-1S and later on the EIGEN2.

## 2 Filtering the data

Since altimeter data have a higher spatial resolution than the geoid data and our global ocean model they have to be smoothed. While the geoid data are usually given as a set of spherical harmonic coefficients the altimeter data are available as absolute heights. An ideal filter for smoothing the altimeter data would cut off spherical harmonics with degrees higher than the given geoid resolution. Such a filter is unlimited in physical space and is not suitable to smooth oceanographic data which do not cover the whole sphere. Therefore we use as a compromise the Hanning filter which is both in the spatial and the spherical domain similar to a Gaussian bell [2].

## 3 Estimating an optimized mean sea surface

For estimating an optimized mean sea surface we begin with the EIGEN-1S geoid which was derived from the CHAMP mission and the altimeter data set CLS_SHOM98.2. The global ocean model we use is the Hamburg LSG model [3] [4] with a spatial resolution of 3.5°. Due to the coverage of TOPEX/Poseidon and the resolution of the ocean model the observed region is limited to ±60°.
To get an optimized sea surface the misfit between the observed sea surface (e.g. the difference between the altimeter and geoid height) and the MSS from the ocean model has to be distributed in the sense of a least squares fit. We achieve this by minimizing the cost function $J$:

$$J = \frac{1}{2}(TP - N - OPT)^T * P_{TP-N} * (TP - N - OPT) + \\ + \frac{1}{2}(\bar{\zeta}_{LSG} - OPT)^T * P_{LSG} * (\bar{\zeta}_{LSG} - OPT) \stackrel{!}{=} min. \tag{1}$$

with
$TP$: CLS_SHOM.98.2
$N$: geoid undulation
$\bar{\zeta}_{LSG}$ : MSS from ocean model LSG
$OPT$: optimized MSS
$P_{LSG}$: weighting matrix ocean model
$P_{TP-N}$: weighting matrix for the difference CLS_Shom98.2 - geoid height

The weighting matrices $P$ are the inverted covariance matrices. The covariance matrix of the geoid $C_N$ can be easily calculated from the given errors of the spherical harmonic coefficients and the related correlation matrix. The errors have to be filtered in the same way as the data.
The covariance matrix of the altimeter data $C_{TP}$ is calculated from the annual means of seven years (1993-1999) TOPEX/Poseidon altimeter data. To this covariance matrix we have added the local error of the CLS_SHOM98.2

MSS.
Summing up the covariance matrices of the altimeter and the geoid data we get the covariance matrix for the difference between altimeter measurements and geoid height

$$C_{TP-N} = C_{TP} + C_N \tag{2}$$

and the weighting matrix

$$P_{TP-N} = C_{TP-N}^{-1}$$

The most subtle point is estimating the error of the MSS derived from the ocean model. The dynamical model is highly complex, so it is nearly impossible to estimate a formal error from the model run. A reasonable way to estimate this error is to compare the model output with another independent model. As an approximation of the model error we take half of the pointwise difference between the LSG model and the ECCO2 model. To avoid the error becoming zero at points where both models show the same sea surface height we choose the median of the halves of the differences as minimal error. The correlations of the model points are calculated from the annual means from 1993 to 1999 of the model output.
In our case it is possible to invert the covariance matrices directly. In cases where they become ill conditioned we have to use a singular value decomposition as we have described in [5].
The optimized mean sea surface can be calculated as

$$OPT = C_{OPT} * (P_{TP-N} * (TP - N) + P_{LSG} * \bar{\zeta}_{LSG}) \tag{3}$$

with the related error covariance matrix

$$C_{OPT} = (P_{TP-N} + P_{LSG})^{-1}.$$

The formal variance of the optimized sea surface corresponds to the diagonal of $C_{OPT}$.

## 4 Improvement of the geoid

The optimized MSS has been estimated by correcting the output of the ocean model with the measured altimeter and geoid data. In the same way we calculate a correction for the geoid model from altimeter and ocean model data. For this experiment we use not only the EIGEN-1S but also the GRIM5S1 geoid. It is known that the accuracy of the GRIM5S1 is far below the accuracy we need for oceanographic purpose. This makes it very suitable to prove our method of combining geodetic and oceanographic data. The corrections we get for the respectively used geoid are shown in figure 2.

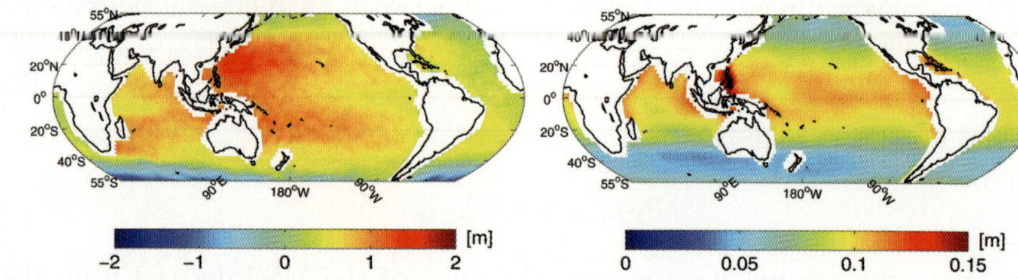


**Fig. 1.** Optimized mean sea surface height (left) and error of the optimized mean sea surface height (right)

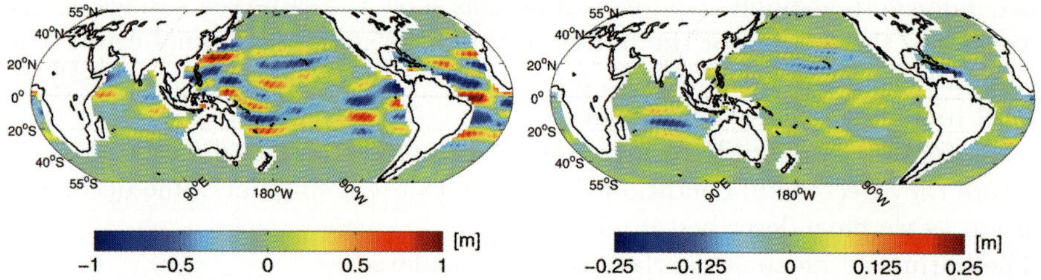


**Fig. 2.** Correction for the GRIM5S1 (left) and the EIGEN1S (right)

## 5 Verification of results

The only possibility to verify the results is to compare the corrected geoids with independent gravity models. We use the gravity model EIGEN-GRACE01S (herein after GRACE01S) from the GFZ and the combined model EGM96. The rms values for the differences between the various corrected and uncorrected geoid models are shown in table 1. They show a clear correction of the GRIM5S1 towards the EGM96 and towards the GRACE01S. A similar improvement although in a smaller range is achieved for the EIGEN-1S. Both corrected gravity models are closer to the GRACE01S geoid than to the EGM96.

## 6 Conclusions

The results show that it is possible to correct the large scales of the GRIM5S1 geoid both towards the EIGEN-1S and the GRACE01S geoid by using altimeter data and the mean sea surface height from a global ocean model.

**Table 1.** rms [m]

| | GRACE01S | EGM96 | GRIM5S1 | EIGEN-1S | corrected GRIM5S1 | corrected EIGEN-1S |
|---|---|---|---|---|---|---|
| GRACE01S | 0 | | | | | |
| EGM96 | 0.0564 | 0 | | | | |
| GRIM5S1 | 0.3776 | 0.3781 | 0 | | | |
| EIGEN-1S | 0.0849 | 0.1026 | 0.3690 | 0 | | |
| corr. GRIM5S1 | 0.1361 | 0.1354 | 0.3301 | 0.1358 | 0 | |
| corr. EIGEN-1S | 0.0688 | 0.0837 | 0.3698 | 0.0444 | 0.2300 | 0 |

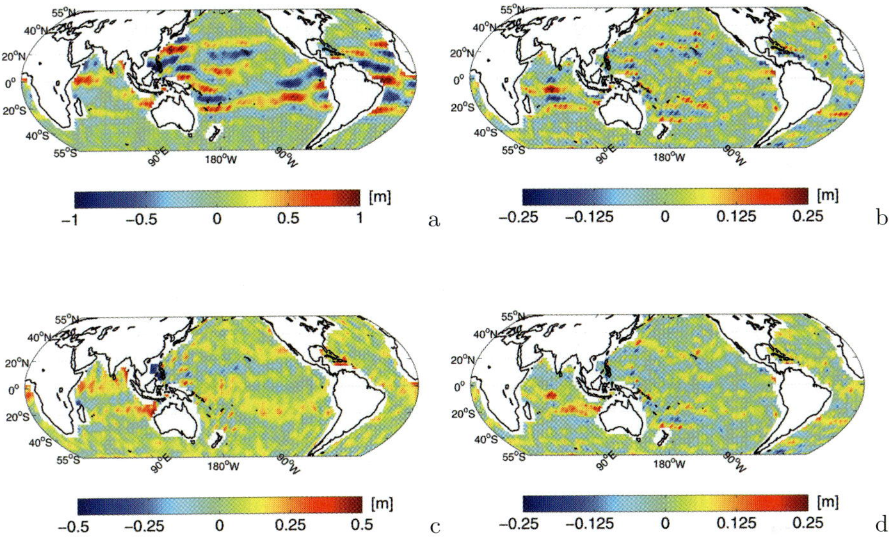


**Fig. 3.** Differences between geoid models and improved geoid models: GRIM5S1 - GRACE01S (a), EIGEN1S - GRACE01S (b), improved GRIM5S1 - GRACE01S (c) and improved EIGEN1S - GRACE01S (d). Note the different colorbars.

The improvement towards the GRACE01S is also shown for the EIGEN1S. The correction of the GRIM5S1 is clearly significant while the change of the EIGEN-1S is quite small. The results indicate that the accuracy of the ocean model is at the large scales comparable to the accuracy of the newest 'satellite only' geoid models derived from CHAMP and GRACE. We conclude that in future we will be able to use the difference between altimeter data and satellite only geoid models to constrain large scale ocean models by assimilation. Another effort will be the simultaneous estimation of ocean circulation and gravity field.

## References


1. Hernandez F, Schaeffer P (2000) Altimetric Mean Sea Surface and Gravity Anomaly maps Inter-Comaprisons. Technical Report no. AVI-NT-011-5242-CLS, preliminary version, CLS.
2. Jekeli C (1989) Alternative methods to smooth the earth's gravity field. Report no. 327, Ohio State University, Department of Geodetic Science and Surveying, 1958 Neil Avenue, Columbus, Ohio 43210.
3. Maier-Reimer E, Mikolajewicz U, Hasselman K (1993) Mean circulation of the Hamburg LSG OGCM and its sensitivity to the thermohaline surface forcing. J Phys Oceanogr *23(4)*: 731–754.
4. Wenzel M, Schröter J, Olbers D (2001) The annual cycle of the global ocean circulation as determined by 4D VAR data assimilation. Progress in Oceanogr *48*: 73–119.
5. Seufer V, Schröter J, Wenzel M, Keller W (2003) Assimilation of Altimeter and Geoid Data into a Global Ocean Model. In: Reigber Ch, Lühr H, Schwintzer P (eds) First CHAMP Mission results for Gravity, Magnetic and Atmospheric Studies. Springer, Berlin Heidelberg New York: 187–192.

# Improved Estimates of the Oceanic Circulation Using the CHAMP Geoid

Gennady Kivman, Sergey Danilov, Bernadette Fritzsch, Sven Harig, Christian Reick, Jens Schröter, Verena Seufer, Dmitry Sidorenko, Joanna Staneva, and Manfred Wenzel

Alfred Wegener Institute for Polar and Marine Research, Postfach 120161, 27515 Bremerhaven, Germany

**Summary.** Ocean general circulation models which are constrained by altimetry data usually assimilate only temporal sea-surface height anomalies. It is known that this is not enough to correct the mean ocean state. Here we present first results of assimilating the full (absolute) dynamical topography into a steady state version of a finite element ocean model (FEOM) for the North Atlantic. This makes it possible to notably reduce the model-data misfit especially in the western part of the basin and in the southern Labrador Sea.

**Key words:** satellite geodesy, satellite altimetry, ocean circulation, Kalman Filtering

## 1 Introduction

Ocean data assimilation is a synthesis of a physical model and observations that improves our knowledge of the ocean state and understanding the ocean dynamics. In most studies dealing with assimilation of altimetry data into ocean models only the time varying part of the signal (eg. anomalies) were used to constrain the ocean circulation. The reason for that was insufficient accuracy of geoid estimates in the past. However, Killworth et al. (2001) point out that assimilation of sea surface height (SSH) anomalies cannot correct the mean ocean state that is misrepresented in numerical ocean circulation models due to erroneous water masses induced by some poorly resolved processes.

With improvement of geoid estimation having been gained recently in the CHAMP mission (Reigber, 2004, this volume) we may now try to start assimilating the full ocean sea surface height. This paper presents our first results. It is worth noting that running modern ocean circulation models (let alone assimilating data into them) is very costly. Therefore, to get insight into the impact of the mean SSH assimilation onto the mean ocean state we begin with a steady state model and a simple data assimilation scheme before proceeding to more complicated applications.

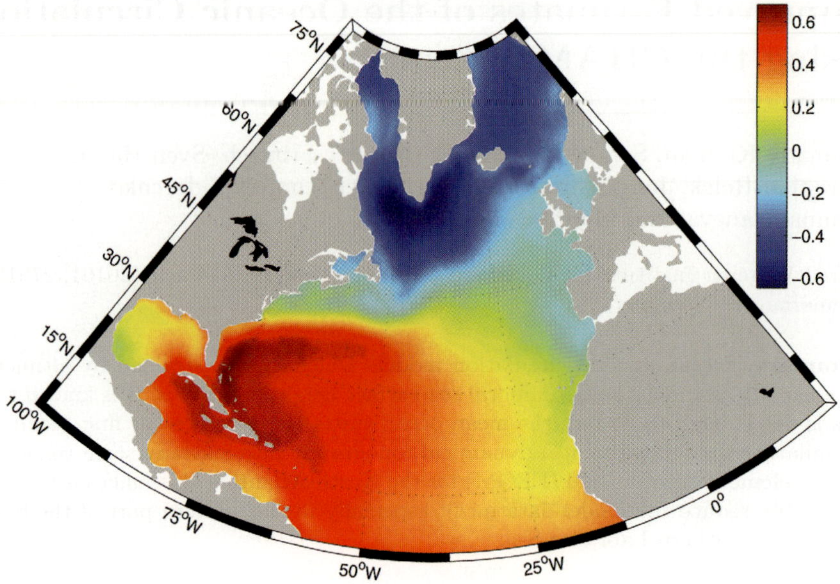


**Fig. 1.** Model dynamical topography in m.

## 2 Ocean model

We use the finite element ocean model (FEOM) described in Danilov et al. (2004). The model is based on the primitive equations discretized on an unstructured surface mesh covering the North Atlantic from $7^0$N to $80^0$N. The average horizontal resolution is $0.5^0$ and is refined in regions of steep bathymetry and in the Gulf Stream area. Due to the use of implicit time stepping, the model can be easily applied to diagnose the velocity field and the dynamical topography (DT) from the density field by solving the steady state primitive equations with the momentum advection neglected.

The model DT diagnosed from annual mean temperature and salinity (Levitus and Boyer, 1994) and wind stress (Trenberth et al., 1989) is shown in Fig. 1. Comparison of the model and CHAMP DT's reveals (Fig. 2) that the major differences are located near the coast where the altimetry is definitely not very precise. Apart from that, one can see much stronger recirculation of the Gulf Stream in the observed mean DT as compared with the model DT and some indication of the Azores Current that is absent in the model. Also, the model underestimate the overall slope of the mean DT which may be caused by inadequate no-flow boundary conditions imposed at open boundaries.

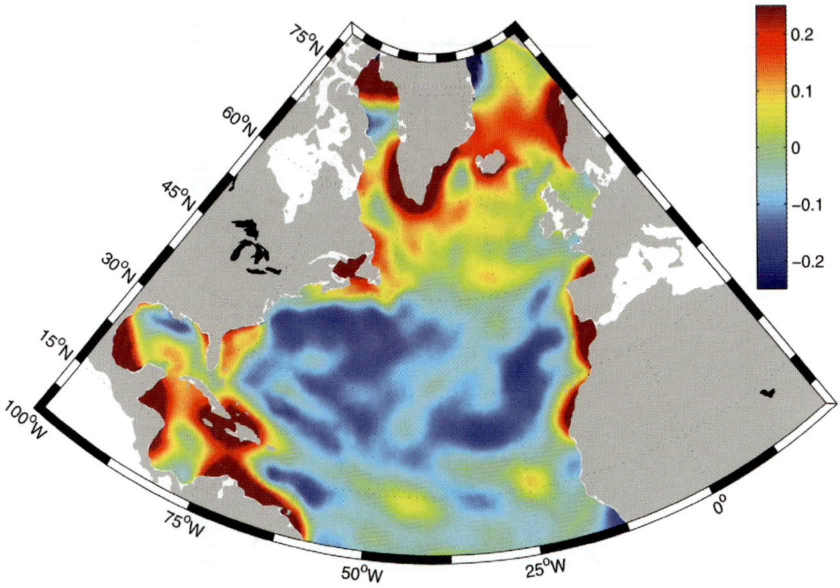


**Fig. 2.** Differences between the model and CHAMP dynamical topographies in m.

## 3 Assimilation methodology

In the present study to assimilate the mean DT we utilized a method similar to that of Mellor and Ezer (1991). The scheme consists of three basic steps. First, having the mean DT of the model $\zeta_m$ and observations $\zeta_o$ together with their variances $\delta\zeta_m$ and $\delta\zeta_o$ we compute the best linear unbiased estimate as

$$\zeta = \frac{(\delta\zeta_o)^2\zeta_m + (\delta\zeta_m)^2\zeta_o}{(\delta\zeta_m)^2 + (\delta\zeta_o)^2} . \tag{1}$$

The update (1) is performed at each surface model grid point where the data $\zeta_o$ are available. Then pointwise estimates of the DT increment $\zeta - \zeta_m$ are smoothed and interpolated over the whole computational domain.

Second, we solve a one-dimensional problem and project this new smooth increment onto ocean temperature $T$ and salinity $S$

$$T = T_m + F_{\zeta,T}(\zeta - \zeta_m) , \qquad S = S_m + F_{\zeta,S}(\zeta - \zeta_m) , \tag{2}$$

with use of predefined correlation factors

$$F_{\zeta,T} = \frac{Cov(\zeta, T)}{Var(\zeta)^2} , \qquad F_{\zeta,S} = \frac{Cov(\zeta, T)}{Var(\zeta)^2} , \tag{3}$$

which relate DT anomalies to anomalies of subsurface temperature and salinity $T_m$ and $S_m$.

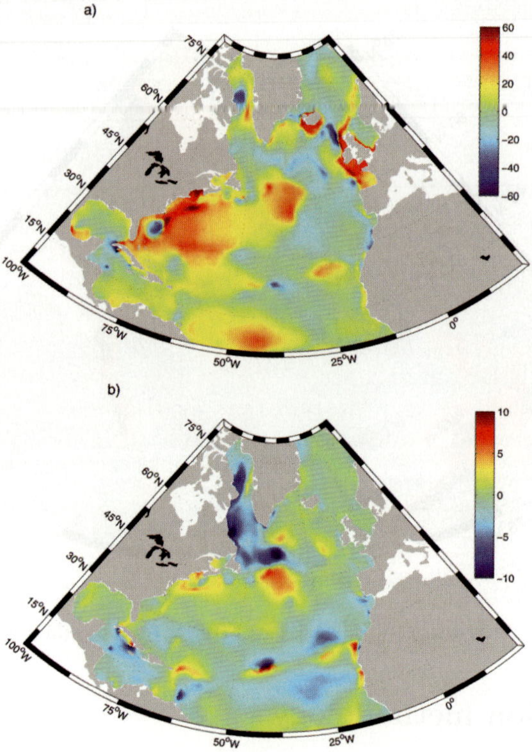


**Fig. 3.** Correlation factors at the depth of 100m: a) $F_{\zeta,T}$ in $^0$C/m; b) $F_{\zeta,S}$ in psu/m.

In Mellor and Ezer (1991), the correlation factors $F_{\zeta,T}$ and $F_{\zeta,S}$ were estimated from a segment of the model trajectory as temporal regression coefficients between $\zeta$ and $(T, S)$. However, as it was pointed out in Oschlies and Willebrand (1998) present ocean circulation models cannot provide us with reliable correlations. Instead, here we adopted another scheme and diagnosed the DT for each month of Levitus (1998) climatology. Using such an ensemble of ocean climatological states we computed $F_{\zeta,T}$ and $F_{\zeta,S}$ and the variance $\delta\zeta_m$.

The correlation factors $F_{\zeta,T}$ and $F_{\zeta,S}$ for 100 m depth are presented in Fig. 3. One can see very distinct positive correlations between the DT and ocean temperature in the area of the Gulf Stream and the North Atlantic Current and anti-correlations between the DT and salinity in the Labrador Sea. Thus, we can expect that assimilation of the mean DT will have a strong impact in these regions.

We augmented the scheme of Mellor and Ezer (1991) with a third step that is necessary when dealing with a steady-state model. After $T$ and $S$ having been updated according to (2), we diagnose the velocity field and DT which correspond to the updated $T$ and $S$. This is done by solving once more the steady-state primitive equations forced by the climatological mean wind stress (Trenberth et al., 1989) and the density field calculated from the updated $T$ and $S$.

## 4 Results and conclusions

The differences between the DT after assimilation and the original data are presented in Fig. 4. It is seen that assimilation notably reduces the model-data misfit. The major improvement is achieved in the western North Atlantic and in the southern Labrador Sea where the correlation factor $F_{\zeta,T}$ and $F_{\zeta,S}$ respectively reach extreme values and thus the differences between the model and data result in larger values of the density increment.

Improvement in the ocean SSH results in a slightly higher heat transport (Fig. 5) which is still notably below other estimates (MacDonald, 1998). These discrepancies indicate weakness in using a steady state model for assimilating the mean SSH. Stepping the model in time produces values of the meridional heat transport (Danilov et al., 2004) which are in agreement with other estimates. Our next step will be assimilating the mean SSH into the prognostic version of FEOM.

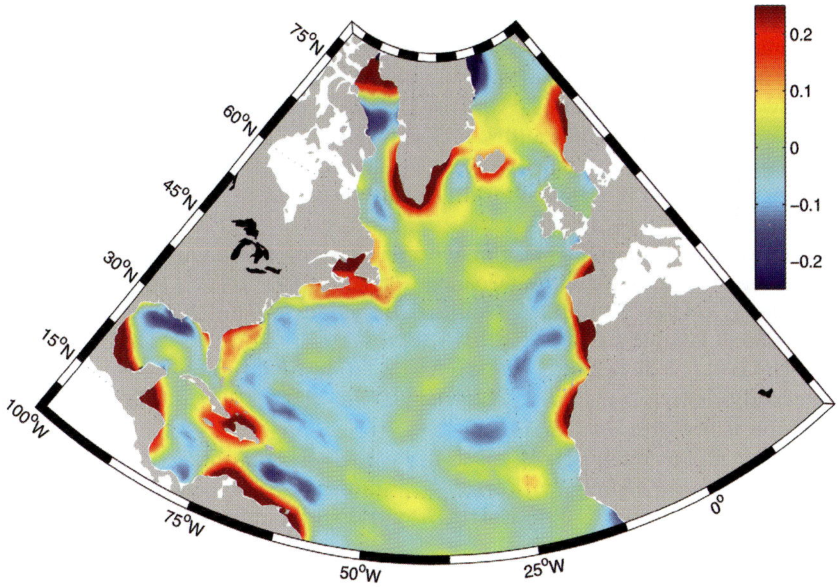


**Fig. 4.** Differences between the optimised and CHAMP dynamical topographies in m.

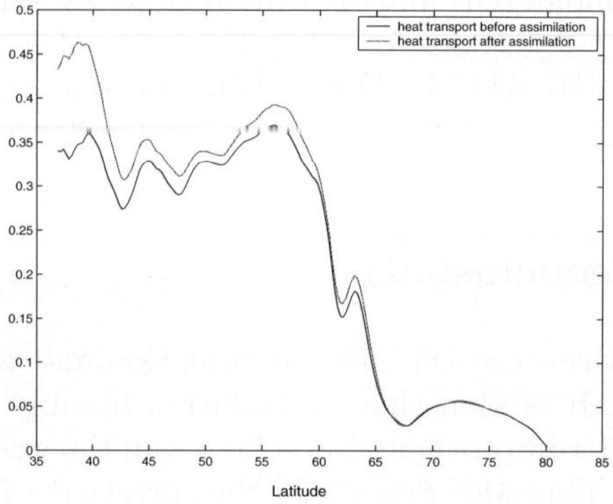


**Fig. 5.** Meridional heat transport before (blue) and after (green) assimilating the CHAMP DT in $10^{15}$ W

Another point where assimilating the mean SSH could improve the estimate of the ocean circulation is adjusting the open boundary conditions. However, this calls for a much more sophisticated variational data assimilation approach which is on the way.

## References


1. Danilov S, Kivman G, Schröter J (2004) A finite element ocean model: principles and evaluation. Ocean Modelling *6*: 125–150.
2. Killworth PD, Dietrich C, Le Provost C, Oschlies A, Willebrand J (2001) Assimilation of altimetric data and mean sea surface height into an eddy-permitting model of the North Atlantic. Prog Oceanogr *48*: 313–335.
3. Levitus S, Boyer T (1994) World Ocean Atlas 1994. U.S. Department of Commerce, Washington, D.C.
4. MacDonald, AM (1998) The global ocean circulation. A hydrographyic estimate and regional analysis. Prog Oceanogr *41*: 281–382.
5. Mellor GL, Ezer T (1991) A Gulf Stream model and an altimetry assimilation scheme. J Geophys Res *96*: 8779–8795.
6. Oschlies A, Willebrand J (1996) Assimilation of Geosat altimeter data into an eddy-resolving primitive equation model of the North Atlantic Ocean. J Geophys Res *101*: 14,175–14,190.
7. Reigber C, Jochmann H, Wünsch J, Petrovic S, Schwintzer P, Barthelmes F, K-H Neumayer, König R, Förste C, Balmino G, Biancale R, Lemoine J-M, Loyer S, Perosanz F (2004) Earth Gravity Field and Seasonal Variability from CHAMP. (this issue).
8. Trenberth K, Olson J, Large W (1989) A Global Ocean Wind Stress Climatology based on ECMWF Analyses. Tech. Rep. NCAR/TN-338+STR, National Center for Atmospheric Research, Boulder, Colorado.

# Contemporary Changes in the Geoid About Greenland: Predictions Relevant to Gravity Space Missions

Kevin Fleming, Zdeněk Martinec, Jan Hagedoorn, and Detlef Wolf

GeoForschungsZentrum Potsdam, Dept. Geodesy and Remote Sensing, Potsdam, Germany, *kevin@gfz-potsdam.de*

**Summary.** We have examined contemporary changes in the geoid about Greenland that result from glacial-isostatic adjustment. These may be divided into contributions from ice-load changes that occurred outside of Greenland following the Last Glacial Maximum and changes in the Greenland Ice Sheet (GIS). The GIS's contribution may itself be divided into past and current parts. For past ice-load changes, the resulting geoid displacement is more dependent upon the recent history of the GIS than on the earth model used. Considering an estimated accuracy for the GRACE temporal geoid signal, regional variability in the present-day mass balance of the GIS may be resolved. This variability significantly affects the geoid power spectrum, giving a signal that may be detected by measurements from gravity space missions more easily than has been proposed by other authors.

**Key words:** Geoid change, CHAMP and GRACE satellite missions, glacial-isostatic adjustment, Greenland Ice Sheet.

## 1 Introduction

The mass balance of the global continental ice cover, particularly the ice sheets of Antarctica and Greenland, is a crucial element when determining present-day sea-level change. However, because of the size and inaccessibility of these ice masses, such knowledge is currently lacking in detail.

Alternatively, information about the mass balance of these ice sheets may be gained by resolving temporal and spatial changes in the *geoid*, *i.e.* the equipotential surface of the Earth's gravity field that approximates mean sea level. These changes arise from the redistribution of mass within the Earth and upon its surface, in particular from the waxing and waning of ice sheets during glacial cycles.

We have assessed the contemporary geoid change about Greenland resulting from past changes in the major ice sheets and the current behaviour of the Greenland Ice Sheet (GIS). This is examined within the context of the gravity space missions now underway (CHAMP and GRACE).

## 2 Contribution of ongoing GIA

We first assess geoid change arising from ongoing glacial-isostatic adjustment (GIA) due to changes in ice loading following the Last Glacial Maximum (LGM, *ca.* 21 ka BP). We apply a gravitationally self-consistent GIA model based on the spectral finite-element method (1). The sea-level equation is also implemented and accommodates changes in the Earth's rotation, moving shorelines and the transition between grounded and floating ice.

A four-layer earth model is used, described by the elastic-lithosphere thickness, $h_L$, the upper-mantle viscosity, $\eta_{UM}$, the lower mantle-viscosity, $\eta_{LM}$, and an inviscid core. The mantle is an incompressible, Maxwell-viscoelastic fluid and the mass-density and elastic shear-modulus values are from PREM. Two viscosity models are used: EARTH1, similar to that preferred by (2), and EARTH2, similar to that used by (3).

The ice models are the global ice model ICE-3G (4) without its Greenland component (termed ICE-3G–GR), the Greenland ice model GREEN1 (5), which accommodates changes in the GIS since the LGM, but excludes a neoglacial component (when the GIS had retreated behind its present-day margin and readvanced over the past *ca.* 4 ka), and GREEN1+NEO, which is GREEN1 including neoglaciation in the southwest.

Fig. 1 presents the resulting spectra (where spectrum refers to the square root of the degree-power spectrum) from the individual ice and earth model combinations. We find there is a relatively small dependence on the viscosity model used. In contrast, the inclusion or exclusion of the neoglacial component is more important to Greenland's predicted contribution. This is further shown in Fig. 2, where the rates of uplift and geoid change are presented. Again, the neoglaciation is more significant, especially for the EARTH1 model, owing to the more-rapid reaction of its lower upper-mantle viscosity.

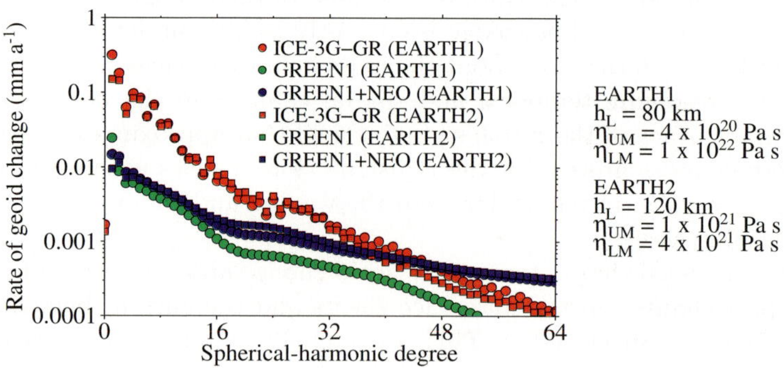


**Fig. 1.** The geoid-change spectra resulting from the deglaciation following the LGM.

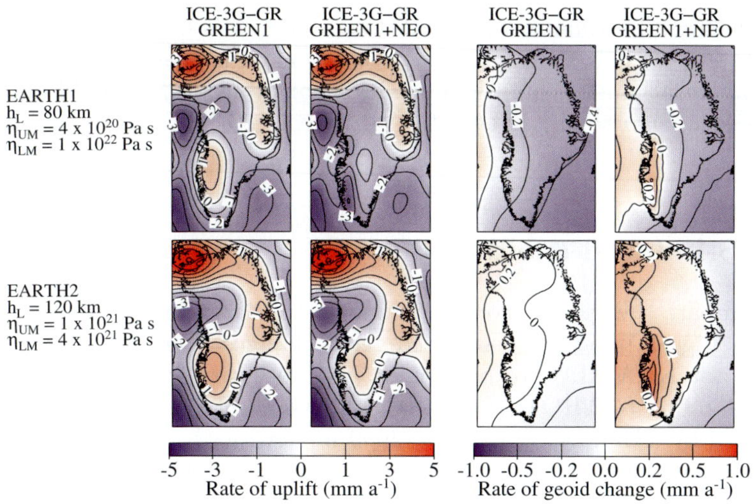


**Fig. 2.** The present-day rates of uplift and geoid change resulting from the deglaciation following the LGM.

## 3 Present-day changes in the GIS

Two estimates of the current mass balance of the GIS are examined (Fig. 3). The first, KRABILL (Fig. 3a) (6), is based on airborne laser-altimeter measurements of changes in ice-surface elevation and covers the entire ice sheet. We have corrected the changes in ice-surface elevation for ongoing GIA using results for EARTH1 and ICE-3G–GR+GREEN1+NEO. These corrections are usually relatively small, especially when one considers the measured changes along the ice margin.

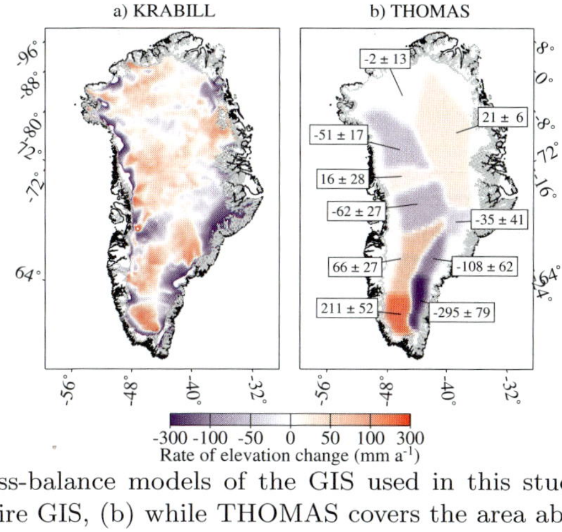


**Fig. 3.** The mass-balance models of the GIS used in this study. (a) KRABILL considers the entire GIS, (b) while THOMAS covers the area above *ca.* 2000 m.

The second estimate, THOMAS (Fig. 3b) (7), is based on GPS measurements and ice-flow calculations. It represents the mass balance above *ca.* 2000 m elevation. Its equivalent sea-level contribution (*ca.* 0.015 mm $a^{-1}$) is opposite in sign to the KRABILL value (*ca.* –0.033 mm $a^{-1}$) for the same area.

## 4 GIS-induced geoid change

We compare the spectra corresponding to the Greenland mass-balance scenarios (Fig. 4a) and the associated geoid changes (Fig. 4b). Three series of results for KRABILL are presented: (a) the entire ice sheet, (b) the area corresponding to THOMAS (above *ca.* 2000 m elevation) and (c) the area excluded by THOMAS (below *ca.* 2000 m elevation). Spectra for THOMAS are calculated for all combinations of the uncertainties about the nominal values.

Changes below 2000 m elevation dominate the KRABILL results. We also note that the upper limit of the range of the THOMAS results is comparable to the total KRABIL response. Some spatial variability in the rate of geoid change can be seen when the spectra are summed up to degree and order 32 (Fig. 5). The opposite signs of the equivalent sea-level contributions for THOMAS and KRABILL (>2000 m) are also apparent (Fig. 5c and d).

## 5 Discussion and summary

Some investigators have assumed uniform ice-mass changes when calculating geoid signals (8). We have examined this assumption using versions of KRABILL and THOMAS where the net ice-volume changes are uniformly distributed over the respective areas. The resulting geoid-change spectra (Fig. 6) fall off more quickly with increasing degree than those for more realistic spatial distributions.

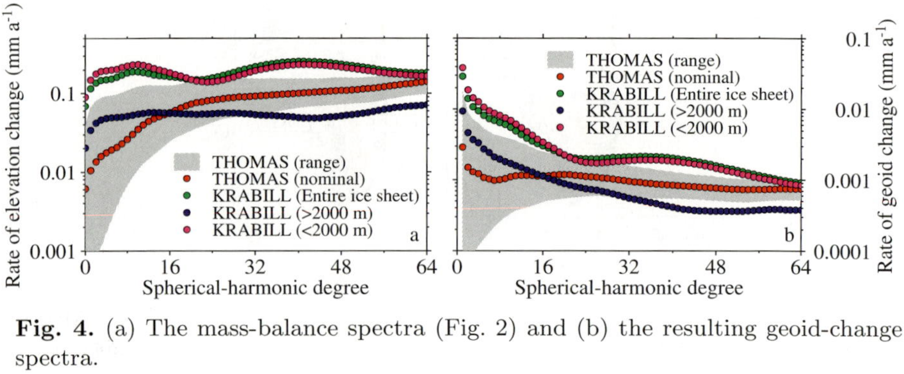


**Fig. 4.** (a) The mass-balance spectra (Fig. 2) and (b) the resulting geoid-change spectra.

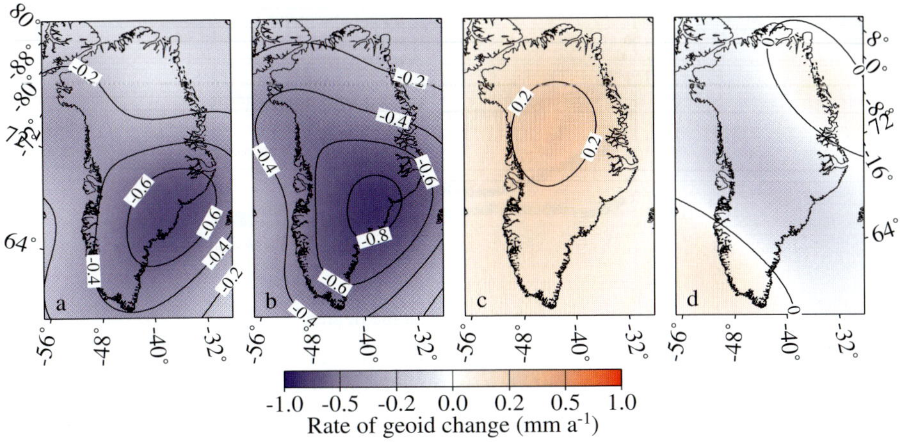


**Fig. 5.** Rates of geoid change for the various mass-balance scenarios: (a) KRABILL (entire GIS), (b) KRABILL (below *ca.* 2000 m elevation), (c) KRABILL (above *ca.* 2000 m elevation) and (d) THOMAS.

Finally, we compare the predicted spectra with an accuracy estimate for GRACE (Fig. 7). The total ongoing-GIA response is largest at lower degrees, but falls off quickly. The GIS signal, specifically for KRABILL, remains above the uncertainty up to degree *ca.* 64, but the signal for THOMAS is of similar magnitude. We also find that, while the geoid-change signal for Antarctica's present-day ice-mass balance (9) is much greater than for Greenland's, it has a similar shape.

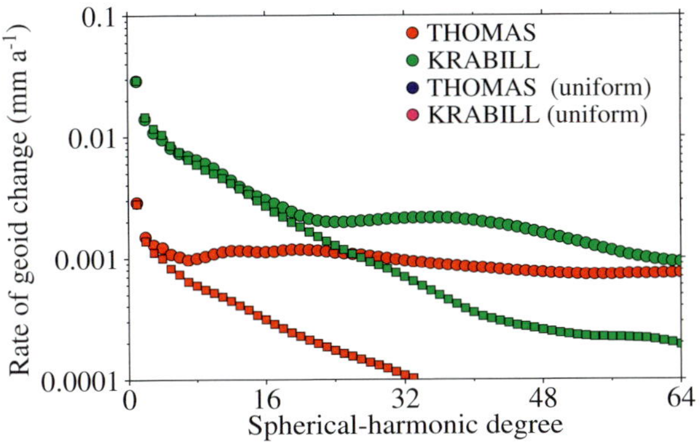


**Fig. 6.** The geoid-change spectra for the KRABILL and THOMAS mass-balance scenarios for realistic and uniform spatial descriptions.

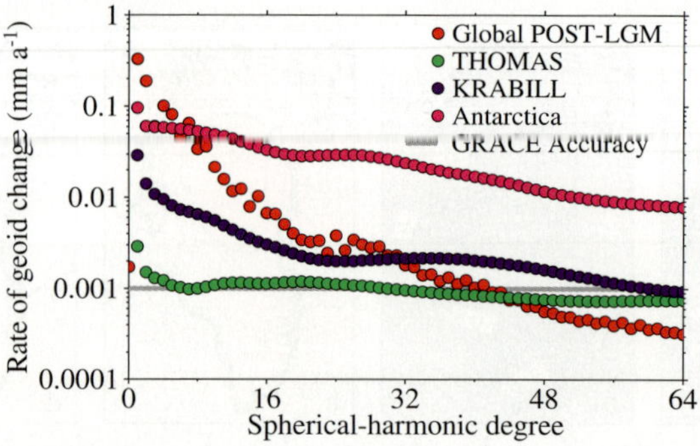


**Fig. 7.** The geoid-change spectra for the various models discussed in this study, and for an estimate of the present-day ice-mass changes in Antarctica. An estimate of the accuracy expected from the GRACE satellite mission is included.

*Acknowledgement.* This study was supported by the SEAL (Sea Level Change) project of the Hermann von Helmholtz Association of German Research Centres.

## References

[1] Martinec Z (2004) Time-domain, spectral-finite element method to viscoelastic relaxation of a self-gravitating, incompressible, Maxwell-viscoelastic, radially symmetric sphere. Computers and Geosciences, submitted.

[2] Lambeck K, Smither C and Ekman M (1998) Tests of glacial rebound models for Fennoscandinavia based on instrumented sea- and lake-level records. Geophys J Int *135*: 375–387.

[3] Mitrovica JX and Peltier WR (1989) Pleistocene deglaciation and the global gravity field. J Geophys Res *94*: 13,651–13,671.

[4] Tushingham AM and Peltier WR (1998) Ice 3G: a new global model of late Pleistocene deglaciation based on geophysical predictions of postglacial relative sea level change. J Geophys Res *96*: 4,497–4,523.

[5] Fleming K and Lambeck K (2004) Constraints on the Greenland Ice Sheet since the Last Glacial Maximum from sea-level observations and glacial rebound models. Quat Sci Rev, in press.

[6] Krabill W, Abdalati W, Frederick E, Manizade S, Martin C, Sonntag J, Swift R, Thomas R, Wright W and Yungel J (2000) Greenland Ice Sheet: high-elevation balance and peripheral thinning. Science *289*: 428–430.

[7] Thomas R, Akins T, Csatho B, Fahnestock M, Gogineni P, Kim C and Sonntag J (2000) Mass balance of the Greenland Ice Sheet at high elevations. Science *289*: 426–428.

[8] Velicogna I and Wahr J (2002) Postglacial rebound and Earth's viscosity structure from GRACE. J Geophys Res *107*: 2,376–2,387.

[9] Rignot E and Thomas R (2002) Mass balance of polar ice sheets. Science *297*: 1,502–1,506.

# Mantle Viscosity and S-Wave to Density Conversion Profiles using CHAMP Geoid Data

Gabriele Marquart[1] and Radboud Koop[2]

[1] SRON and Institute of Earth Science, University of Utrecht, The Netherlands
*marquart@geo.uu.nl*
[2] SRON National Institute for Space Research, Utrecht, The Netherlands
*R.Koop@sron.nl*

**Summary.** The long wavelength geoid height undulations are the result of density variations inside the Earth and the dynamic response of the viscous mantle due to the buoyancy forces resulting in dynamic topography, plus the contribution of isostatic topography due to crustal and lithospheric structure. The dynamic topography is a function of the effective stress transmission inside the earth and is linked to the viscosity of the mantle. We solve the equation of motion for a viscous Earth's mantle assuming an incompressible 6-layer shell model and determine the dynamic response function for geoid, dynamic topography, and (poloidal) surface velocity. The internal density distribution can be estimated from seismic tomography, but since density variations might be of thermal and chemical nature, the s-wave velocity to density conversion factor, $R_{\rho\backslash v_s}$, varies throughout the mantle. Based on CHAMP gravity field coefficients and four new seismic s-wave tomography models we search for ranges of radial profiles of viscosity and $R_{\rho\backslash v_s}$ resulting in a correlation to better than 0.85 to the long wavelength hydrostatic geoid, and to a fit better than 0.6 for gravity, dynamic topography and (poloidal) plate velocity. For purely thermal origin $R_{\rho\backslash v_s}$ should be between 0.2 and 0.4. Successful models however, show a small or even negative value for $R_{\rho\backslash v_s}$ between 100 and 300 km depth and a low value of $\sim .1$ between 700 and 1200 km, but is otherwise roughly constant with values of $\sim .28$. The viscosity is slightly reduced in the asthenosphere and even stronger decreased in the mantle transition zone between 410 and 670 km. Resolution for both, viscosity and conversion factor, is poor below the transition zone down to $\sim$1500 km, but well confined in deeper parts of the mantle, where a viscosity between 30 to 40 $10^{21}$ Pa s and a conversion factor of 0.28 to 0.32 is found.

**Key words:** mantle viscosity, mantle scaling factor, geoid anomalies

## 1 Introduction

Solid-state flow in the Earth is a direct consequence of density inhomogeneities, which are of thermal or chemical origin. These density anomalies drive a creeping flow, causing dynamic topography at the surface and internal boundary layers and is visible by plate motion. Gravity potential observations reflect the combined density effects due to internal variation in density and deflected boundaries with a density jump, e.g. surface topography. Topography itself results from the superposition of isostatic topography due to thickness variations of the crust and lithosphere maintained by the high strength of the lithosphere and dynamic topography due to pressure forces of the bouyant mantle rocks. If the gravity potential and the dynamic topography are known and independently the internal density distribution, the viscous flow model, i.e. the viscosity profile in the Earth can be determined. Additionally plate velocities directly represent a measure for the flow velocity in the uppermost part of the convection system.

Density heterogeneities in the mantle can be deduced from variations in seismic wave velocities, tomography, but the relation between seismic velocity and density is different for thermal and chemical origin. A thermal origin must be accompagnied by a reduction in density due to thermal expansivity, but a chemical origin might not reflect any density changes or even with opposite effects. For example, if seismic velocity anomalies are caused by variations in iron content, slow seismic velocity regions imply high iron content and thus higher density.

Dynamic topography strongly depends on the ability of the Earth's mantle to deform under stress and is related to the viscosity $\mu(r)$. Thus, the geoid is a consequence of the viscosity structure and the conversion factor $R_{\rho\backslash v_s}$ between seismic wave velocity variations and density anomalies. Here we model the Earth as a radially layered linear viscous sphere, with the internal load based on 4 different recent tomography models (sb4l18, [1]; s362d1, [2]; s20rts, [3]; saw24b16, [4]). For each tomography model we vary the seismic wave velocity-density conversion factor and the viscosity with depth within reasonable ranges until the best correlation between modeled geoid and the CHAMP geoid is found (and additionally for gravity, dynamic topography and surface plate velocity). Finally we discuss the consequences for the chemical layering of the Earth.

## 2 Observation Data

For the gravity potential we used the CHAMP spherical harmonic coefficients related to the hydrostatic reference figure. We use both the long wavelength sensitive geoid, which mainly reflects the large scale circulation of the mantle and the short wavelength sensitive gravity field, which essentially probes the shallow effects. Dynamic topography is not directly observable but has to be deduced from measured topography and a model of isostatic compensated crust and lithosphere. For the isostatic topography we use the spherical harmonic coefficients of [5], which are based on the assumption of isostatic balance of crustal and lithosphere thickness on continents and of thermal isostatic balance of the topography of oceanic ridges. Dynamic topography is estimated by subtracting the isostatic part from ETOPO05. However, since crustal thickness and densities are not well constrained in many continental areas and the estimate of lithospheric thickness even more relies on model assumptions, the error budget for dynamic topography is large.

For plate velocities we use the values given by [6]. However, the observed vector field of plate velocity can be separated in a poloidal and a toroidal part, whereby the toroidal part is about a third of the poloidal one. Toroidal motion is mainly created by strike-slip faults and cannot be produced by an Earth model with constant parameter shells, thus, only the poloidal part of plate velocity is used to constrain our models.

The density variations inside the Earth are deduced from four new s-wave tomography models, i.e. seismic velocity variations relative to a radial standard model (see Section 1). Since the tomography models are given in different non-equidistant grids, we gridded the data sets on 20 radially equidistant layers (145 km apart) down to the core-mantle boundary. Finally all data sets are expanded in spherical harmonics up to degree and order 30 (if not given to a higher degree) and filtered for a maximum degree 15 by applying a cosine taper. The maximum degree of 15 is motivated by the resolution of the tomography data on a global scale and also by the approach of radial (radial global average) profiles for viscosity and conversion factor.

# 3 The Method

To determine the linear response functions for a viscous Earth under loading we follow the classic approach were the equations of motions for a viscous incompressible spherical shell are solved by separation of the variables in the spectral domain (e.g. [8], [7]). The resulting ODEs are solved separately for each degree for response functions (kernels) which only depend on the viscosity $\mu(r)$. The kernels are derived for geoid $G_l(r, \mu(r))$, dynamic topography at the surface and the core-mantle boundary, and poloidal surface velocities. The (synthetic) geoid $N_{lm}$ (and in a similar way topography and surface velocity) is found by convolving the kernel with the density contrast $\delta\rho_{lm}$ derived from tomography and s-wave velocity to density conversion factor $(R_{\rho\backslash v_s}(r))$

$$N_{lm} = \frac{4\pi\gamma a}{2l+1} \int\limits_{CMB}^{a} G_l(r, \mu(r))\, \delta\rho_{lm}(r)\, dr \qquad (1)$$

$\gamma$ is the gravitational constant, $a$ the Earth radius, $CMB$ the core radius. The calculated (synthetic) fields can then be compared to observations.

The Earth model itself consists of six viscous layers for which viscosity and $R_{\rho\backslash v_s}$ are specified. For each layer we prescribe an upper and lower bound, and vary viscosity and $R_{\rho\backslash v_s}$ within these bounds in 5 steps to establish an equidistant model parameter space grid. Additionally $10^6$ random parameter combinations are tested within the prescribed bounds. The different layers of the Earth's mantle model and the parameter bounds are given in Table 1.

**Table 1.** Radial model for viscosity and $R_{\rho\backslash v_s}$ (multiply viscosity by $10^{21} Pas$)

| layer | depth [km] | $(R_{\rho\backslash v_s})_{min}$ | $(R_{\rho\backslash v_s})_{max}$ | $\mu_{min}$ | $\mu_{max}$ |
|---|---|---|---|---|---|
| lithosphere | 0 – 100 | −.20 | 0.20 | 10. | 40. |
| asthenosphere | 100 – 280 | −.20 | 0.20 | 0.2 | 1.8 |
| upper mantle | 280 – 410 | 0.15 | 0.35 | 0.5 | 3. |
| transition zone | 410 – 670 | 0.15 | 0.35 | 0.5 | 3. |
| lower mantle 1 | 670 – 1150 | 0.15 | 0.40 | 5. | 40. |
| lower mantle 2 | 1150 – 2900 | 0.2 | 0.40 | 20. | 80. |

As an example Fig. 1 shows the kernels as a function of depth and degree for geoid, dynamic topography and surface velocity for a particular viscosity profile. It is clearly indicated that density variations of higher degrees in the upper mantle strongly effect geoid, topography and plate velocity, but from the lower mantle only the very long wavelength density variations are important, mainly for the geoid. The uppermost lower mantle around 1000 km depth is particularly bad resolved, due to kernel values around zero and also weak tomographic signals.

For each viscosity parameter set kernels are derived and convoluted with the density functions obtained from a tomography model multiplied by the various $R_{\rho\backslash v_s}(r)$ profiles. For each parameter set the total correlation $c$

$$c = \frac{\sum_{l=0}^{l_{max}} \sum_{m=0}^{l} ((C_{l,m} \cdot V_{l,m}) + (S_{l,m} \cdot W_{l,m}))}{(\sum_{l=0}^{l_{max}} \sum_{m=0}^{l} (C_{l,m}^2 + S_{l,m}^2) \cdot (V_{l,m}^2 + W_{l,m}^2))^{1/2}} \quad (2)$$

between observed $(C_{lm}, S_{lm})$ and synthetic field $(V_{lm}, W_{lm})$ is estimated.

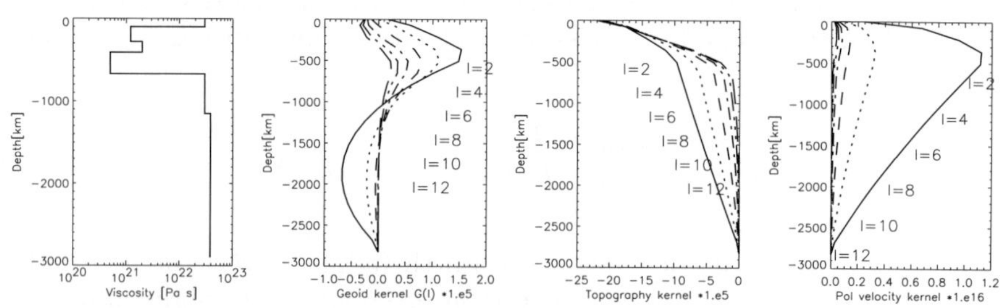


**Fig. 1.** Viscosity profile and geoid, topography and surface velocity kernels as used for the parameter combination which gives the highest correlation to the CHAMP geoid for the tomography data s20rts; kernels are given for even degrees l=2-12

## 4 The Fitting Models and Consequences for the Earth's Mantle

For the four tomography models a maximum total correlation between CHAMP geoid and synthetic geoid of 0.86 to 0.91 can be achieved (but considerable less correlation ($< 0.75$) for the other observational data sets). In Fig. 2 the viscosity and $R_{\rho\backslash v_s}$ profiles are shown for the 4 tomography data sets.

If only the fit to the geoid is considered, the "best" parameter combination (thick lines in Fig. 2), demands a negative $R_{\rho\backslash s}$ value between 100 and 280 km depth of $\sim$ -.1 (values between -.12 and -.06) independent of the tomography model. Three of the four tomography models also indicate a reduced value for $R_{\rho\backslash s}$ below the mantle transition zone down to $\sim$ 1100 km. The results for the viscosity show a weak decrease in the asthenosphere and a stronger decrease to values around .4 to .8 $10^{21}$ Pa s in the mantle transition zone.

The tomography model with the best fit to the observations is s20rts (upper left graphs in Fig. 2) for which we show here in Fig. 3 the synthetic geoid, gravity and dynamic topography in comparison the the observations.

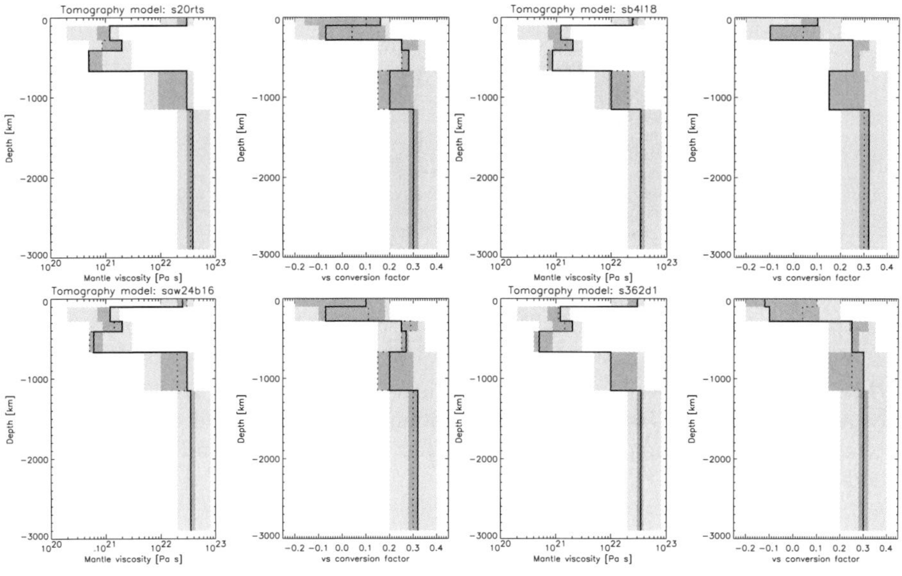


**Fig. 2.** Viscosity and $R_{\rho\backslash s}$ profiles for the 4 tomography models. The search range for viscosity and $R_{\rho\backslash s}$ at different depth levels is shown in light gray, the ranges for parameter combinations with a correlation to better than 0.85 to the hydrostatic CHAMP geoid are shown in dark gray. The thick lines indicate the parameter combinations which give the highest correlation to the geoid only, the thin dotted lines the parameter combinations with the best correlation to all 'observation' data (geoid, gravity, poloidal plate velocity and dynamic topography)

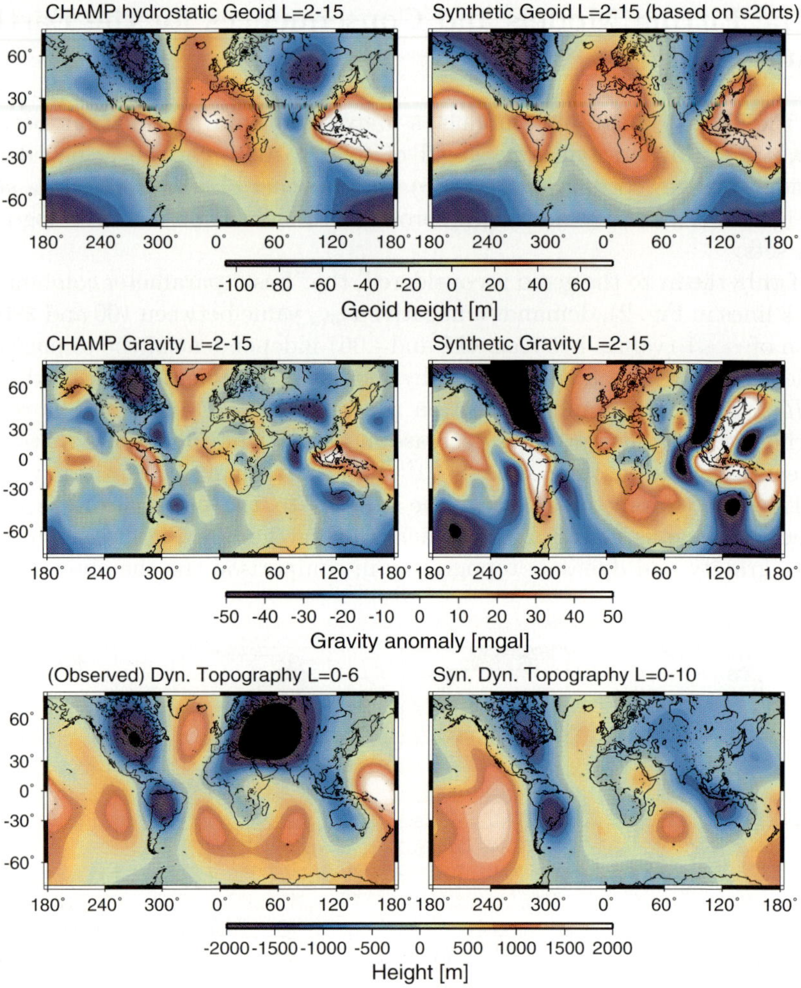


**Fig. 3.** Observed (left) and synthetic (right) geoid (top), gravity (center) and dynamic topography (bottom). The synthetic fields are derived using the s20rts tomography [3] and the parameters according to the thick line on Fig. 2 (upper left model).

The synthetic geoid matches well the observed data with the exception of the North Atlantic and South Indian Ocean regions where wide spread hotspot activity of the Iceland and Kerguelen plume might impair the fit. The pattern of the synthetic and the observed gravity field (Fig. 3 center) generally agree, but modeled amplitudes are considerably too strong. This cannot be attributed to an unresolved thin low viscosity layer at the base of the lithosphere, since then also synthetic dynamic topography could be expected to be too high, which is not the case. More likely are lateral chemical variations in the upper 300 km, attributed to cratons or different degree of melt extraction from oceanic regions.

Negative $R_{\rho\backslash s}$ values for the asthenosphere had been reported earlier and are explained by depletion in iron contents, causing high seismic velocity, but low density. Depletion of iron can be linked to the degree of melt extractions from mantle rocks, which is considered to be high for old continental lithosphere formed in a hotter Earth or (to a less degree) for oceanic plateaus formed by a hot mantle plume. A low viscosity transition zone might be explained by dehydration of water-saturated rocks. If water is brought into the lower mantle by slab subduction, dehydration should occur at mid mantle temperatures of $\sim$1400°C at the spinel-perovskite transition [10]. While the transition zone itself should be highly water-saturated, rising mantle material might undergo dehydration-induced partial melting, containing considerable amounts of incompatible elements and free water. While these rocks are denser than the depleted source mantle they might be trapped close to the upper (410 km) phase transition ([9]). Both effects, partial melt and water contents will lead to a reduction in viscosity.

## References


1. Masters G, Laske G, Bolton H, Dziewonski AM (2000) The relative behavior of shear velocity, bulk sound speed, and compressional velocity in the mantle: implications for chemical and thermal structure. In: Karato SI, Forte AM, Liebermann RC, Masters G, Stixrude L (eds) Mineral physics and tomography from the atomic to the global scale. AGU, Washington DC.
2. Gu YJ, Dziewonski AM, Su WI, Ekstroem G (2001) Models of the mantle shear velocity and discontinuities in the pattern of lateral heterogeneities. J Geophys Res *106*: 11,169–11,199.
3. Ritsema J, van Heijst HJ (2000) Seismic imaging of structural heterogeneity in Earth's mantle: Evidence for large-scale mantle flow. Sci Progr *83*: 243–259.
4. Mègnin C, Romanowicz B (2000) The shear velocity structure of the mantle from the inversion of body, surface, and higher modes waveforms. Geophys J Int *143*: 709-728.
5. Panasyuk SV, Hager BH (2000) Models of isostatic and dynamic topography, geoid anomalies, and their uncertainties. J Geophys Res *105*: 28,199–28,209.
6. Mueller RD, Royer JY, Lawver LA (1993) Revised plate motions relative to the hotspots from combined Atlantic and Indian ocean hotspot tracks. Geology *21*: 275-278.
7. Richards MA, Hager BH (1984) Geoid anomalies in a dynamic earth. J Geophys Res *89*: 5,987–6,002.
8. Ricard Y, Fleitout L, Froidevaux C (1984) Geoid height and dynamic stresses for a dynamic Earth. Ann Geophys *2*: 267-287.
9. Bercovici D, Karato S (2002) Whole-mantle convection and the transition-zone water filter. Submitted to Nature.
10. Litasov K, Ohtani E (2002) Water Solubility in Lower Mantle Minerals and Role of Peridotite and Basalt in Water Storage in the Lower Mantle. Abstract AGU fall meeting.

# Regional Geoid Undulations from CHAMP, Represented by Locally Supported Basis Functions

Rainer Mautz[1,2], Burkhard Schaffrin[1], C. K. Shum[1], Shin-Chan Han[1]

[1] Laboratory for Space Geodesy and Remote Sensing, The Ohio State University, Columbus, Ohio 43210, U.S.A. *schaffrin.1@osu.edu*
[2] Institute of Astronomical, Physical and Mathematical Geodesy, Technical University of Berlin, Germany. *rainer@mca.bv.tu-berlin.de*

**Summary.** Regional geoid undulations are determined from CHAMP data using various locally supported basis functions to assess their respective efficiency, accuracy and multi-resolution representation properties. These functions include (biharmonic) B-spline tensor wavelets (with or without compression), multiquadrics (with or without flexible centering and predetermined smoothing) and radially symmetric truncated polynomials.
It is concluded that the B-spline wavelet model is the computationally most efficient approach. The non-periodic variation of the B-spline wavelets allows one to handle data on a bounded domain with small edge effects, and the piecewise linear version allows one to model the geoid using a patch-wise approach. The use of multiquadrics without centering in the data points and predetermined smoothing constant allows handling of heterogeneously distributed data using global optimization. The linear multiquadrics model fits the data best when comparing the residuals of different models with a fixed number of unknowns. For an efficient data synthesis the nonlinear models are best suited due to their far smaller number of basis functions. The smoothest surface was obtained using the nonlinear polynomial approach, whereas the multiquadrics show peaks and the wavelet models show horizontal and vertical edges in their representations. The linear B-spline wavelets are biharmonic, and the approach is capable of an efficient multi-resolution representation of regional gravity field models combining satellite (CHAMP, GRACE, GOCE) and in-situ data.

**Key words:** geoid undulations, B-spline wavelets, multi-quadrics, nonlinear models

## 1 Introduction

Recent global gravity models are based on spherical harmonic functions which are excellent for representing the geopotential up to a certain degree of detail or resolution. Even though the global spherical harmonic representation is adequate for low degree global gravity modeling using satellite data (e.g. CHAMP), but for a detailed regional representation of the earth gravity field determined by satellite as well as terrestrial observations, spherical harmonics may not be the best basis functions. In this context, see Schmidt et al. (2002) who have been using spherical wavelets to represent the finer details of the gravity field.

Our goal is to use harmonic or biharmonic locally supported basis functions in order to construct patch-wise models to enhance the high frequency parts of the signals on the sphere leading to a multi-resolution representation of the gravity field. Towards selecting the most suitable functional model, some approaches are compared numerically using regional undulations derived from CHAMP disturbing potential data. For more details regarding CHAMP data processing, we refer to Han et al. (2002; 2003).

The first functional model in this investigation is the B-spline wavelet; see Chui and Quak (1992). These wavelets have become well known due to their useful properties such as compact support, semi-orthogonality and simplicity. Algorithms and applications for computer graphics can be found in Stollnitz et al. (1996).

The second model is based on the multiquadric method which fits a set of quadric (e.g., hyperbolic or conical) functions to the observations. It was introduced by Hardy (1971) and further developed by Hardy and Göpfert (1975) in order to interpolate gravity anomalies.

The comparison also includes polynomial radial symmetric basis functions with local support, which have been used successfully in Mautz et al. (2003).

Furthermore, the linear models described above are compared with their nonlinear counterparts having flexible positioning. Geodetic global models have been studied by Mautz (2001; 2002) while nonlinear models for surface data were discussed by Kaschenz (2002; 2003).

For a comparison of different surface representations on a large scale, see Franke (1982). Here, we focus on the data-fit, the computational effort, the number of basis functions and the smoothness/roughness of the surface as criteria.

## 2 Functional Models

The models discussed here can be classified as linear or nonlinear. The systems for solving the unknowns in capital letters (e.g., the amplitudes $A$, $B$) are linear if these are the only parameters. In this case the estimated parameters, the residuals, and the estimated variance component can be obtained using BLUUE (Best Linear Uniformly Unbiased Estimate) and BIQUUE (Best Invariant Quadratic Uniformly Unbiased Estimate). A detailed discussion of linear models is provided by Grafarend and Schaffrin (1993).

Less common are models where the positions or scaling coefficients are considered as unknown. With the models becoming nonlinear due to the flexible centering and scaling, the solving techniques require global optimization methods. Gradient methods like the Gauss-Newton iteration are not applicable as reliable starting values cannot usually be provided. Thus, we stick to global optimization techniques such as heuristic methods, interval strategies or genetic algorithms. The idea of optimized centering dates back to Barthelmes (1986).

(a) *2-dimensional B-spline wavelets:* If the 2-D signal is given by $f(x, y)$ the model function reads

$$f(x,y)=\sum_{k_x=1}^{2^{j_{\min}}+g}\sum_{k_y=1}^{2^{j_{\min}}+g} A_{j_{\min},k_x,k_y}\tilde{\phi}_{j_{\min},k_x,k_y}(x,y)+\sum_{j=j_{\min}}^{J}\sum_{\eta=1}^{3}\sum_{k_x=1}^{2^j+g}\sum_{k_y=1}^{2^j+g} B_{j,\eta,k_x,k_y}\tilde{\psi}_{j,\eta,k_x,k_y}(x,y), \quad (1)$$

where $\phi(x, y)$ is the 2-D scaling function, $\tilde{\phi}(x, y)$ its dual; $\psi(x, y)$ are the 2-D wavelet functions, and $\tilde{\psi}(x, y)$ their duals. Their polynomial degree is expressed by $g$. The different levels of detail are denoted by the index $j$. Wavelet coefficients with a larger $j$ indicate higher detail levels, essentially representing the high-frequency part. The index $\eta$ denotes the three directional components (horizontal, vertical and diagonal), and the indices $k_x, k_y \in {}_0$ denote the shift of the wavelets to different locations on the $(x, y)$-patch. The variables $A$ and $B$ denote the unknown coefficients for the scaling function and the wavelet functions. The problem of estimating $A$ and $B$ is linear; due to orthogonal subspaces, it is not necessary to solve one big system with linear equations of problem-size, but a sequence of smaller systems. This model has been discussed in more detail by Mautz et al. (2002) and Schaffrin et al. (2003).

(b) *Compressed 2-D B-spline wavelets:* In contrast to model (a), the hierarchical structure is now developed to the maximum level $J_{\max}$, fulfilling the condition

$$J_{\max} \le \ln\left(\sqrt{n} - g\right) / \ln 2, \tag{2}$$

where $n$ is the number of data points. The number of coefficients is then reduced by neglecting terms with coefficient values smaller than a predetermined bound. Figure 1 shows the shape of a linear B-wavelet according to model (a) and (b).

(c) *Multiquadric basis functions:* The multiquadric functional model, resp. its inverse, reads

$$f(x,y) = \sum_{k=1}^{n} A_k K(r_k), \tag{3}$$

where the kernel functions $K(r_k)$ are given by

$$K(r_k) = \left(r_k^2 + c^2\right)^t, \tag{4}$$

with typically $t = ½$, resp. $t = -½$ for the inverse case. The radial distances $r_k$ between the evaluation point $(x, y)$ and a fixed center position $(x_k, y_k)$, which could be chosen from the locations of the observations, are given by

$$r_k(x,y) = \sqrt{(x - x_k)^2 + (y - y_k)^2}. \tag{5}$$

The planar distance $r_k$ may be replaced with the spherical distance $\psi_k$ using the spherical coordinates $\phi$ and $\lambda$, along with the relation

$$\cos(\psi_k) = \sin(\phi)\sin(\phi_k) + \cos(\phi)\cos(\phi_k)\cos(\lambda - \lambda_k). \tag{6}$$

$A_k$ are the unknown parameters and $c \in$ is a predefined constant. The unknown parameters $A_k$ are estimated by solving a linear system. See Figure 1 for a graph of the radial multiquadric function according to model (c) and (d).

(d) *Multiquadric basis functions, with flexible positioning:* We now introduce $x_k \in$ and $y_k \in$ as unknowns for every $k \in \{1, 2, ..., n\}$. The resulting model becomes nonlinear, and the solving technique requires global optimization.

(e) *Locally supported radial functions, linear model:* Local support allows function values other than zero only within a certain distance from the center point location. Thus, the continuous model function needs to be truncated. The model

$$f(x,y) = \begin{cases} \sum_{k=1}^{n} A_k \left(\frac{r^2}{c_k^2} - 1\right)^2, & r < c_k, \\ 0, & r \ge c_k. \end{cases} \tag{7}$$

avoids a discontinuity in the function and in its first derivative at the cut-off location. With fixed center positions $(x_k, y_k)$ and parameters $c_k$, the model is linear.

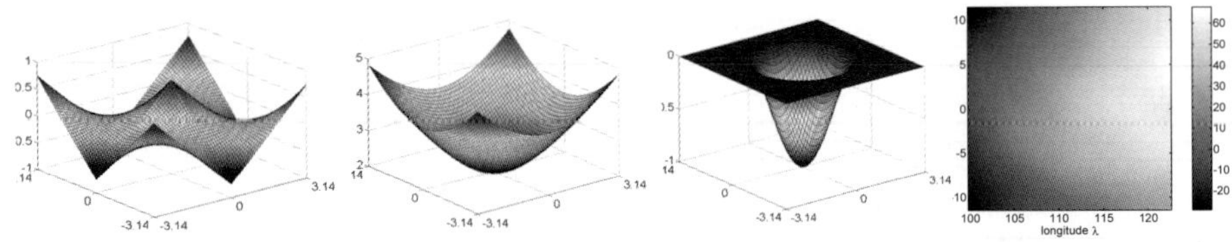


**Fig. 1.** Left panel: Diagonal linear B-spline wavelet of level 0. Mid left: Multiquadric function with $c = 2$. Mid right: polynomial function with $c = 2$. Right panel: Geoid undulations in [m] from CHAMP only solution, between [100°; 122.5°] longitude and [-11.25°; +11.25°] latitude; sampled onto a 65 × 65 grid. The range of the data values is [–29.0 m; +67.8 m].

(f) *Locally supported radial functions, nonlinear model:* Introducing the center positions $(x_k, y_k)$ as unknowns for every $k \in \{1, 2, \ldots, n\}$ the model becomes nonlinear. In addition to flexible centering, the parameters $c_k$, serving as scaling parameters, are also considered as unknown. Figure 1 shows a graph for $c = 2$.

## 3 Model Comparison Based on CHAMP Geoid

In order to make proper comparisons the models' special requirements have to be taken into account. The multiresolution representations (a) and (b) necessitate observations in form of a $2^j$ by $2^j$, ($j \in$   ) grid for efficient handling. Their application requires a prior adjustment to the grid. The compressed wavelet model (b) needs extra memory for storing the locations of the remaining terms. All comparisons are based on the specific dataset shown in Figure 1.

As shown in Table 1, the residual information is used to rate the models numerically. The criteria are the standard variation, the maximum deviation, the squared sum of the residuals and the average deviation. The number of unknowns is kept fixed.

Generally, linear models involve a normal equations system of problem size. Setting up the system requires a complexity of  $(n\, m^2)$ and its inversion  $(m^3)$, where $n$ is the number of observations and $m$ the number of parameters. However, properties like semi-orthogonality and local support causing banded matrix struc-

| Model | No. of un-- knowns | No. of basis func- tions | $m_0 = \sqrt{\hat{\sigma}_0^2} = \sqrt{\tilde{\mathbf{e}}^T\tilde{\mathbf{e}}/(n-m)}$ [m] | $\max\{\lvert\tilde{\mathbf{e}}_j\rvert\}$ $j = 1, 2 \ldots n$ [m] | $\tilde{\mathbf{e}}^T\tilde{\mathbf{e}}$ [m$^2$] | $\frac{\sum_{j=1}^{n}\lvert\tilde{\mathbf{e}}_j\rvert}{n}$ [m] | CPU on a 200 Mhz PC |
|---|---|---|---|---|---|---|---|
| (a) Wavelet, 4 levels | 289 | 289 | 0.08 | 0.41 | 31 | 0.06 | **1s** |
| (b) Wavelet, compression | 289 | 289 | 0.06 | 0.26 | 12 | 0.04 | **1s** |
| (c) Multiquadric (linear) | 289 | 289 | **0.05** | 0.36 | **9** | **0.03** | 3 min |
| (d) Multiquadric (nonlin.) | 288 | 96 | 0.09 | 0.58 | 34 | 0.07 | 10 h |
| (e) Local radial fct. (linear) | 289 | 289 | 0.12 | 0.56 | 61 | 0.12 | 3 min |
| (f) Local radial fct. (nonlin.) | 288 | **72** | 0.07 | **0.25** | 19 | 0.05 | 10 h |

**Table 1.** Comparison of various surface representation models with a constant number of unknowns resp. 93% redundancy; units: [m] or [m$^2$]. The models fulfilling a criterion best are highlighted in bold.

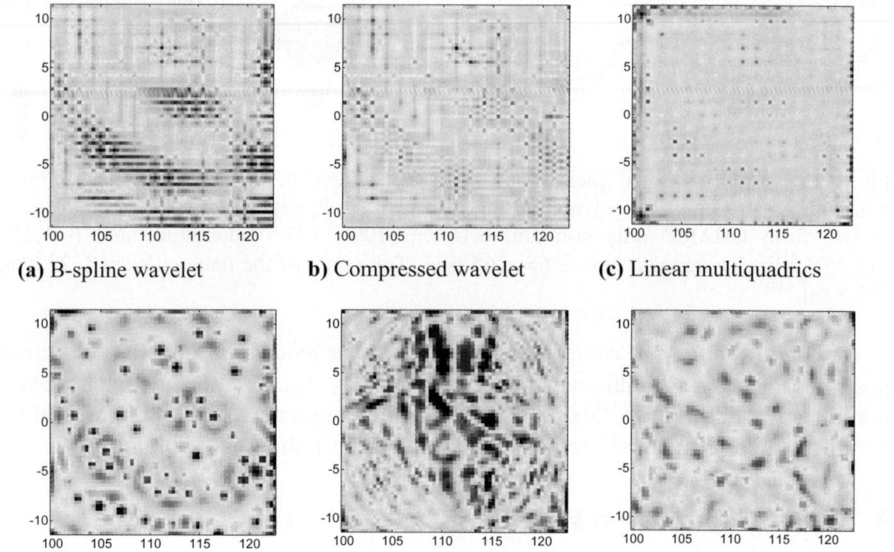


**(a)** B-spline wavelet **b)** Compressed wavelet **(c)** Linear multiquadrics

**(d)** Nonlinear multiquadrics **(e)** Linear polynomials **(f)** Nonlinear polynomials

**Fig. 2.** Residuals [m] of 6 different models. All models have 288 or 289 parameters. The number of basis functions is 289 for the linear and 72 for the nonlinear models. The unit for the colorbar is [m] and for the axes it is [°] latitude and longitude.

tures as it is the case for the B-spline wavelet model. The astonishing result in Table 2 for the wavelet model is due to a reduction of a factor $m^{1/2}$ by usage of the tensor product. The multiquadrics, having global support are of complexity $(n\,m^2)$. If a hierarchical data thinning algorithm is used according to Hales and Levesley (2000), the algorithm may be of linear complexity, but only if the data are spaced equidistantly. However, the computational burden decreases drastically for locally supported functions when handling very large $m$ and $n$, assuming that an optimized algorithm is implemented.

The computing time for nonlinear models may be higher in general, since a sequential series of inversions has to be performed. Nevertheless, making use of heuristic strategies and adaptation of the algorithm to the problem at hand, the computational effort can be reduced drastically.

The behavior of the six different models (a)-(f) is visualized in Figure 2, for a patch between 100° and 122.5° longitude and -11.25° to +11.25° latitude respectively. All models have 288 (or 289) parameters. The wavelet model shows

| complexity | B-splines | multiquadrics | locally supported fcts. | nonlinear models |
|---|---|---|---|---|
| linear system: | $(n\,m^{1/2})$ | $(n\,m^2)$ | $<(n\,m^2)$ | involve solutions of |
| inversion: | $<(m^{3/2})$ | $(m^3)$ | $<(m^3)$ | many linear systems |

**Table 2.** Computational effort of surface models with the number of observations $n$ and the number of parameters $m$.

distinct horizontal features in the residual plot. At the center point locations of the multiquadric models, some peaks can be seen. The locally supported radial basis functions show a smooth surface throughout, particularly in the nonlinear case.

## 4 Conclusions

Due to unequal premises and different rating at various criteria, a strict ranking of the models is not feasible. Nevertheless, it has been verified, that the B-spline wavelet model is computationally the most efficient approach. The linear multiquadrics model fits the data best when comparing the residuals of different models with a fixed number of unknowns. For an efficient data synthesis the nonlinear models are best suited due to their far smaller number of basis functions. The smoothest surface was obtained using the nonlinear polynomial approach, whereas the multiquadrics show peaks and the wavelet models show horizontal and vertical edges in their representations.

Towards modeling the geopotential it can be outlined that the B-spline wavelets are biharmonic for the linear case and the multiquadric functions fulfill conditions for a harmonic upwards continuation in the case $c = 0$ and exponent $t = -0.5$.

The CHAMP only data do not have significant detail information for the multiresolution analysis discussed here. Nevertheless, this approach could be quite useful for the determination of a regional high-resolution gravity field model by combining CHAMP, GRACE, GOCE ($N_{max}$= 300) and in-situ terrestrial gravity data.

*Acknowledgements.* R. Mautz acknowledges the partial support of a Feodor-Lynen scholarship of the Alexander-von-Humboldt Foundation (Germany). R. Mautz, C. Shum, and S. Han acknowledge the support of a grant from NIMA's University Research Initiative program.

## References

Barthelmes F (1986) On the approximation of the outer earth gravity field by point masses with optimised positions (in German). Veröffentlichungen des Zentralinstitut für Physik der Erde, *92*, Potsdam, Germany.

Chui CK, Quak E (1992) Wavelets on a bounded interval. in: Braess D, Schumaker L, eds, Numerical Methods in Approximation Theory, *Vol 9*: 53-77.

Franke R (1982) Scattered data interpolation: Tests of some methods. Math Comps *38 (157)*: 181-199.

Grafarend E, Schaffrin B (1993) Adjustment Computations in Linear Models (in German), Bibliograph. Inst., Mannheim, Germany.

Hales S, Levesley J (2000) Multi-level approximation to scattered data using inverse multiquadrics. in: Cohen A, Rabut C, Schumaker L, eds, Curve and Surface Fitting: Saint-Malo 1999, Vanderbilt University Press, Nashville/TN: 247-254.

Han S-C, Jekeli C, Shum CK (2002) Efficient gravity field recovery using in-situ disturbing potential observables from CHAMP. Geophys Res Letters *29 (16)*: 36-41.

Han S, Shum CK, Jekeli C, Braun A, Chen Y, Kuo C (2003) CHAMP mean and temporal gravity field solutions and geophysical constraint studies. 2nd CHAMP Sci Meeting, GeoForschungsZentrum (GFZ), Potsdam, Sept 2003.

Hardy R (1971) Multiquadric equations of topography and other irregular surfaces. J Geophys Res *76*: 1905-1915.

Hardy R, Göpfert W (1975) Least-squares prediction of gravity anomalies, geoidal undulations, and deflections of the vertical by multiquadric harmonic functions. Geophys Res Letters *2 (10)*: 423-426.

Kaschenz J (2002) Multidimensional spectral analysis with problem-oriented frequencies (in German). Diploma Thesis; Inst of Astronomical, Physical and Math Geodesy, Technical University of Berlin, Germany.

Kaschenz J (2003) Data modelling using various frequency series (in German). Z für Vermessungswesen (ZfV) *128 (4)*: 260-265.

Mautz R (2001) On the determination of frequencies in time series solving nonlinear adjustment problems (in German). German Geodetic Comm, C-532, Munich, Germany.

Mautz R (2002) Solving nonlinear adjustment problems by global optimization. Bollettino di Geodesia e Scienze Affini *61 (2)*: 123-134.

Mautz R, Schaffrin B, Schmidt M, Shum CK (2002) Representation of spatial functions in geodesy using B-spline wavelets with compact support. Proc of the Heiskanen Symposium in Geodesy (Jekeli C, Shum CK, eds), Columbus/Ohio, Oct 2002.

Mautz R, Schaffrin B, Kaschenz J (2003) Biharmonic spline wavelets versus generalized multi-quadrics for continuous surface representations. IUGG General Assembly, IAG-Symp G4, Sapporo, Japan, July 2003.

Schaffrin B, Mautz R, Shum CK, Tseng HZ (2003) Towards a spherical pseudo-wavelet basis for geodetic applications. Computer-Aided Civil and Infrastructure Engineering *18 (5)*: 369-378.

Schmidt M, Fabert O, Shum CK (2002) Multi-resolution representation of the gravity field using spherical wavelets. Proc of the Heiskanen Symposium in Geodesy (Jekeli C, Shum CK, eds.), Columbus/Ohio, Oct 2002.

Stollnitz E, DeRose T, Salesin D (1996) Wavelets for Computer Graphics. M Kaufmann, San Francisco/CA.

# Earth Magnetic Field

Main Field Studies, Lithospheric Magnetisation, Induction Studies and Ionospheric Current Systems, Instruments

# Ionospheric Plasma Effects for Geomagnetic LEO Missions at Mid- and Low-Latitudes

Matthias Förster[1], Martin Rother[2], and Hermann Lühr[2]

[1] Max–Planck–Institut für extraterrestrische Physik (MPE), P.O. Box 1312, 85471 Garching, Germany *mfo@mpe.mpg.de*
[2] GeoForschungsZentrum Potsdam, Telegrafenberg, 14473 Potsdam, Germany

**Summary.** Some of the main plasma characteristics are reviewed that a LEO satellite with high orbital inclination encounters during its travel across the terrestrial ionosphere of mid- and low-latitudes. It is the region of highest plasma density in the near-Earth environment. Its properties are predominantly ruled by the geomagnetic field. It will be shown how different ionospheric layers - first of all the E- and F-layer - contribute in different ways to the electrodynamic and thermodynamic behaviour of the highly interacting, complex system comprising the ionosphere, thermosphere, and plasmasphere. The physical description of its phenomena and data interpretation have nowadays to rely to a substantial part on numerical methods and models. New observational methods and space missions have essentially contributed to the recent progress in this field. The CHAMP mission takes part in this progress just as much as the IMAGE, TIMED, and other satellite projects as well as ground-based observation programs. The paper summarises recent developments in ionospheric studies as, e.g., the plasma transport at mid- and low-latitudes, the regular Sq-dynamo and the contribution of the F–region dynamo, the interhemispheric coupling by current systems and plasma flows, pulsations, the equatorial electrojet and the plasma fountain effect, the Appleton anomaly, the near-equatorial plasma bubbles, and further open issues.

**Key words:** mid- and low-latitude ionosphere, thermosphere, plasmasphere, electric fields, thermospheric winds and composition, equatorial effects, modelling

## Introduction

The low-Earth orbiting (LEO) satellite CHAMP in its circular, near-polar orbit is a suitable platform for studying ionospheric plasma and electrodynamic phenomena on a global scale. With a cruising altitude from about 450 km at the begin of the mission to around 250 km at the end, it is just the ionospheric F-layer, i.e. the region of the highest plasma densities in the near-Earth environment, where its diagnostic instruments are operating. The high-precision measurements onboard CHAMP are therefore affected by this environment. For the correct data interpretation of, e.g., the vector magnetic field data with respect to the diamagnetic effect of the plasma [13] one needs to know the in-situ plasma parameters of the surrounding media where the measurements are performed. These high precision (and adjusted) magnetic field measurements, on the other hand, contribute together with the other diagnostic instruments to highly valuable insights into ionospheric electrodynamics and plasma physical processes which were inaccessible previously.

The ionosphere at mid- and low-latitudes has been explored for more than 80 years - first by ground-based ionosondes and later, beginning in the 1960-ies, by more improved radar techniques and early satellite missions. The general ideas about ionospheric physics, its global frame of the acting processes are supposed to be well-known now and the interest in ionospheric physics at mid- and low-latitudes declined. But during recent years, space based technologies have been developed that are sensibly dependent on more precise predictability of near-Earth conditions. Radio wave transmissions are disturbed, in particular, by small and medium scale phenomena in that region which need much more detailed studies. New experimental techniques appeared which allow more accurate and higher resolution measurements as well as global coverage. These are for instance remote sounding such as GPS occultation experiments as described, e.g., by [5], imaging techniques as onboard the Imager for Magnetopause-to-Aurora Global Exploration (IMAGE) satellite [1] and various new in-situ explorations as, e.g., CHAMP and the Thermosphere Ionosphere Mesosphere Energetics Dynamics (TIMED) mission [8].

## Ionospheric/plasmaspheric characteristics

The ionosphere is a multiconstituent plasma embedded in the high-altitude atmosphere [21, 7]. Traditionally, it is divided into several layers or regions (see Fig. 1, right panel) which are distinguished according to different main ionic constituents and different physical conditions. At $F_2$ region heights, the layer with the maximum plasma densities in near-Earth space, the portion of charged to neutral particles is of the order of 0.1% or less, i.e. the neutral gas and its motion and waves play an essential role in the plasma dynamic. The ionosphere is created by photoionization of solar extreme ultraviolet (EUV) radiations with wavelengths $<$102.7 nm and X-rays (see Fig. 1, left side). Ionization due to collisions, caused by precipitating high energetic particles, as the second important source, acts mainly at auroral and polar regions. Due to the absorption in the high-altitude atmosphere, solar EUV radiation is not observable by ground-based observatories. The large variability of radiation fluxes in the course of varying solar activity became obvious by extraterrestrial observations as, e.g., by the EIT instrument of the SOHO mission (see, e.g., http://sohowww.nascom.nasa.gov/). During solar flares, the X-radiation can increase by more than a factor of 200 while the EUV increase is more modest ($\approx$ 50%) which was shown, e.g. for the Bastille Storm Event in July 2000 [15], to resulting in an F–region plasma density enhancement of the order of 40%.

The mid- and low-latitude ionosphere is closely connected, via the Earth's magnetic field, to the plasmasphere above. During magnetically quiet condi-

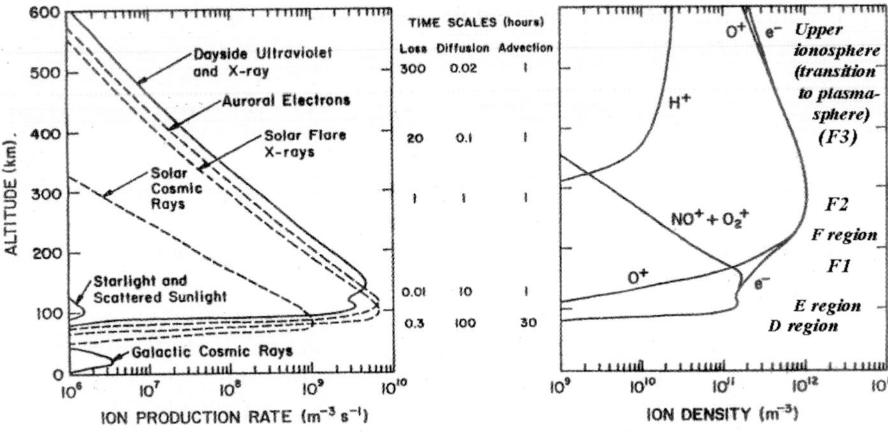


**Fig. 1.** Ionisation rates (left) and typical ion densities of the n(h) profile (right). Adapted from: A. Richmond and Gang Lu, CEDAR Workshop 1999, tutorial lecture.

tions, the plasmaspheric flux tubes are filled up from the ionospheric source region as they corotate together with the upper atmospheric layers. The outer boundary of the plasmasphere, the plasmapause, is established as the equilibrium between corotation and global magnetospheric convection electric fields. Its projection into the ionosphere marks the transition between mid-latitudes and subauroral (or ionospheric trough) latitudes which corresponds to an average McIlwain L-shell position of L=4 during undisturbed conditions. During geomagnetically disturbed days it can come as close as L=2 or less for very strong storms [6, 10, 3].

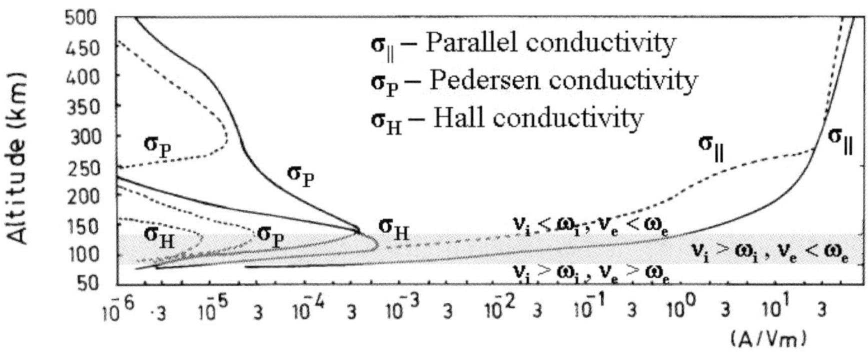


**Fig. 2.** Typical daytime conductivity profiles for the ionosphere with the parallel conductivity $\sigma_{\|}$, the Pedersen $\sigma_P$ and the Hall conductivity $\sigma_H$; dashed lines indicate nighttime conditions. After: [22, Figure 1].

The most fundamental equation in any ionospheric theory is the continuity equation (1):

$$\frac{\partial n_i}{\partial t} + B\nabla\left(\frac{n_i\,\mathbf{v}_{i\|}}{B}\right) + \nabla\,(n_i\mathbf{v}_{i\perp}) = P_i \; - L_i\,n_i \qquad (1)$$

with $P_i$ as the source (or production) terms of the individual species $n_i$, $L_i$ their loss function, $B$ is the ambient magnetic field, and $\mathbf{v}_i = \mathbf{v}_{i\perp} + \mathbf{v}_{i\|}$ the particle's bulk flow velocity. All these terms contribute differently to the plasma density distribution at different altitudes; Fig. 1 gives in its middle part some estimation of the characteristic times for the different terms. At altitudes up to the $F_1$–layer the right-hand terms of (1) dominate; during daytime they balance each other and during sunset the plasma density at this altitude range falls quickly by several orders of magnitude.

## Dynamics: neutral winds, electric fields, and currents

In the $F_2$–layer, the upper ionosphere and plasmasphere the transport terms (diffusion and advection, left side of equation 1) are the most important. Any physical description of the plasma behaviour there depends on nonlocal physical processes as the diffusion (fluxes) along the magnetic flux tubes, the plasma drift induced by ion drag with the neutral air, and the electromagnetic drift perpendicular to the magnetic field. These processes are described by the equations of motion and energy or by kinetic equations for the suprathermal particles [3, and references therein].

The collisions between neutrals and various charged particles play an important role in the partially ionized ionospheric plasma. They determine the electrodynamical behaviour of this medium (see Fig. 2). At E–region altitudes, there exists a region (indicated in grey in Fig. 2: from about 80 km to 130 km), where the electrons are more tied to the geomagnetic field than the ions. Electrons and ions are subject to different forces and, consequently, charge separations occur and electric polarization fields are built up. This process is well-known for a long time as the atmospheric dynamo resulting in the Sq (i.e. solar quiet) electric fields (see, e.g., [14]) and the corresponding current system (Fig. 1 in [14]). This generator was originally described by a 2–D dynamo model that could explain at that time nearly all observational facts of the mean Sq currents found from ground-based magnetometer observations. It is interesting to note that these electric fields, which are due to neutral wind modes at a restricted height range in the E–region and which reflect planetary, tidal, and gravity waves propagating upward from the lower atmosphere, map along the geomagnetic field lines up to the upper ionosphere and plasmasphere because of the high parallel conductivity (cf. Fig. 2).

With new observations in the late 60-ies and 70-ies, summarized in the first global mid-to-low latitude electric field model by [18], it became clear

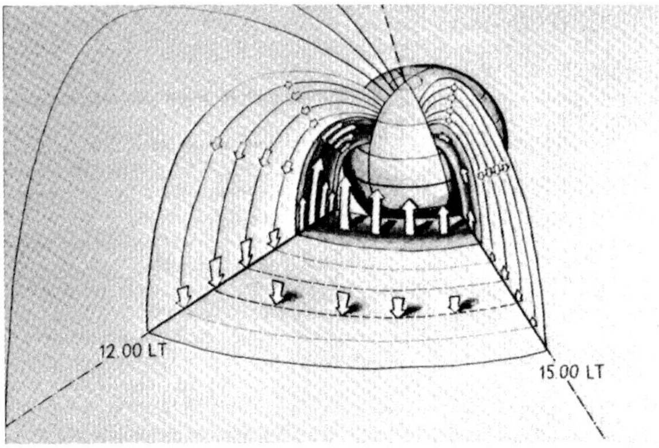


**Fig. 3.** Scheme of the 3-D interhemispheric current system for Northern summer conditions, looking into the afternoon plasmasphere region with cuts at the geomagnetic equator, the 15 LT and the noon meridian. (After: [22, Figure 20]).

that a 3–D model which includes the interhemispheric coupling for asymmetric (e.g. seasonal) conditions is necessary (see, e.g., [22] and references therein). One of those models is schematically shown in Fig. 3 showing a concentrated region of FACs near local noon at L≈1.5 flowing steadily into the summer hemisphere and more diffuse FAC current regions toward the morning and evening side and at other latitudes with the opposite flow direction.

While the dynamo described above is mainly confined to daytime hours (because of E–region conductivity), during nighttime hours further generator mechanisms at F–region heights contribute the major part to mid-latitude currents. One is associated with the gravitational equilibrium of the plasma, in which the ions drift eastward and electrons westward (i.e. an east-west current), and the other with the dynamo action of the neutral wind circulation at F–region heights [19, and references therein].

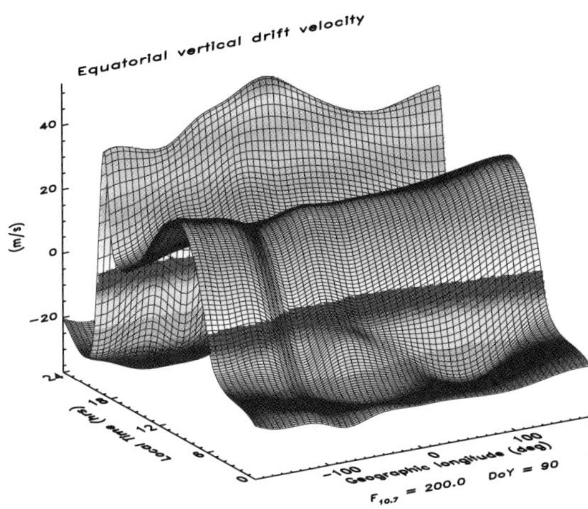


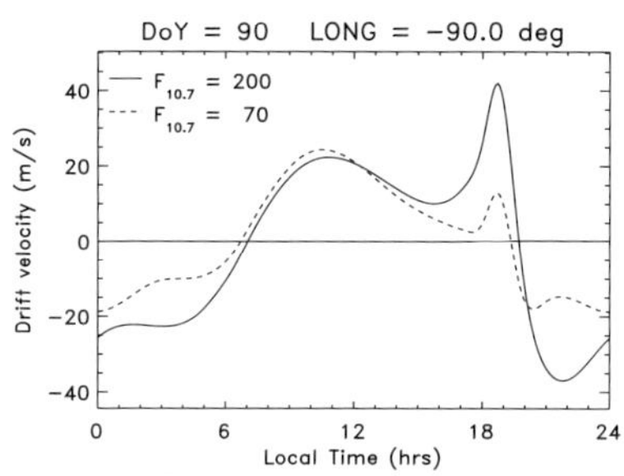


**Fig. 4.** Scherliess' model of equatorial upward drift [20]: perspective view for high solar activity (left) and comparison of low and high solar activity (right).

Considering the equatorial ionosphere, it is quite a modelling challenge up to now, to describe the complicated current circuit situation adequately owing to the many coupled processes operating in a complicated geometrical situation [19]. The drift peculiarities near the equator like the fountain effect which results in the Appleton anomaly and the post-sunset upward drift maximum [21, 7] are well-known empirically for long time. The most advanced empirical model of the equatorial zonal electric field component (upward drift) was published by [20] and Fig. 4 illustrates this vertical drift behaviour.

## Disturbed conditions and geospheric storms

Large perturbations of the Earth's space environment result from solar events and perturbations in the solar wind with the subsequent coupling of their energy and mass to the magnetospheric-ionospheric-thermospheric domain [6]. The ionospheric storm response manifests itself in dramatic variations of plasma densities at the $F_2$ region electron density peak [17, and references therein]. Those variations can exceed 100% from averages and involve enhancements (positive storm phases) and depletions (negative storm phases). A part of these variations result from disturbances of the high atmosphere like composition changes, generation of planetary waves, and additional neutral gas heating mainly in the auroral regions. Other important factors are the disturbance dynamo, transient prompt penetration electric fields [2], and the transport of plasma with enhanced electron density caused by particle precipitation. New techniques and recent satellite missions like IMAGE and TIMED brought about much progress in their study, as was demonstrated with event analyses like, e.g., for the large storm of 15 July 2000 [9, 15, 24].

Global first principle numerical modelling offers the possibility to study theoretically the complex behaviour of the whole system: the thermosphere, ionosphere, plasmasphere, and its electrodynamics (as performed by, e.g., [4, 16, and references therein]. Fig. 5 shows an exemplary model result of a

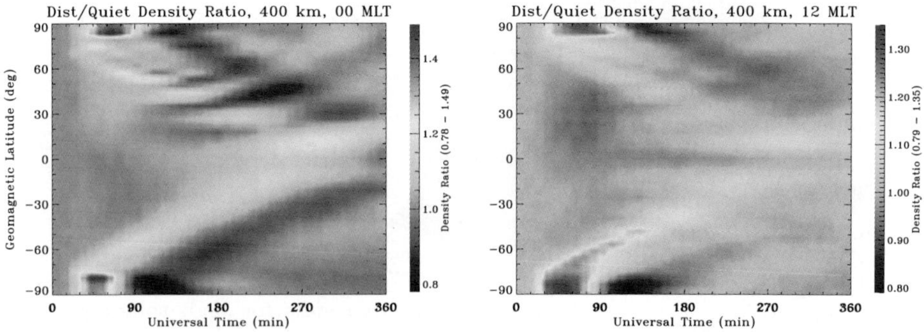


**Fig. 5.** Numerical simulation study of global TAD propagation (see text).

storm event: the propagation of a Travelling Atmospheric Disturbance (TAD) wave due to a moderate auroral energy input which corresponds to a 20-min increase of the cross-polar potential drop up to 155 kV or an AE index of about 1000 nT followed by the subsequent relaxation to the quiet (Namgaladze 1998, private communication). A numerical storm study during the preparation phase of the CHAMP mission [11, 16] pointed to the unique possibility to combine high-resolution accelerometer and magnetometer measurements to investigate density variations and neutral wind responses to geomagnetic forcing. The simulations showed that it even should be possible to resolve TADs (gravity waves) and to draw conclusions about the validity of thermospheric models.

## Equatorial anomaly, plasma bubbles and instabilities

The post-sunset near-equatorial ionosphere undergoes dramatic changes due to eastward electric field variations (see Fig. 4) and is therefore the realm of several interesting plasma processes [23]:

- the Appleton anomaly which is characterized by upward plasma motion followed by diffuse downward transport at both sides of the equator,
- spread F echoes as a plasma instability process causing irregularities at the bottomside F region, and
- plasma bubble generation due to rapid upward plasma motion causing turbulent conditions in the F–region and upper ionosphere.

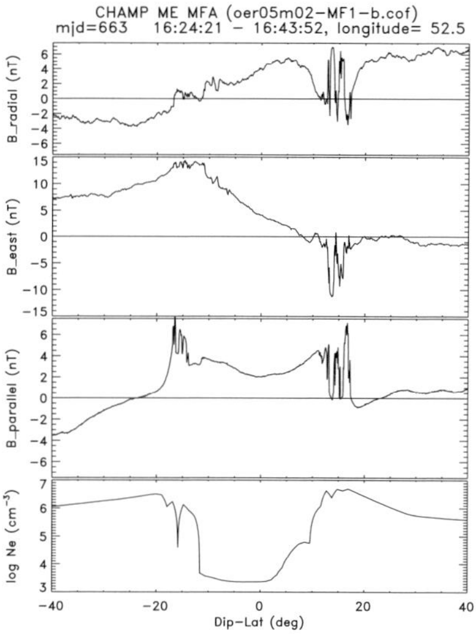


**Fig. 6.** Measurements of magnetic variations (deviations from a model in MFA coordinates, upper three panels) and DIDM plasma density (bottom panel, log. scale) onboard CHAMP for an equator crossing on the geomagnetically quiet day 25 Oct 2001.

CHAMP can contribute a lot for further elucidating these processes and first results have already been published [12, 13]. In combination with plasma measurements, the high precision magnetic field sensors can be used to detect ionospheric F–region currents. The global occurrence of F–region currents and their spatial confinement to the near-equatorial region bounded by the Appleton anomaly was found with a predominant appearance in the premidnight sector [12]. Fig. 6 shows an example of geomagnetically quiet time records at about 20 LT together with the DIDM plasma density measurements. The deep density trough at the equator is due to the large post-sunset

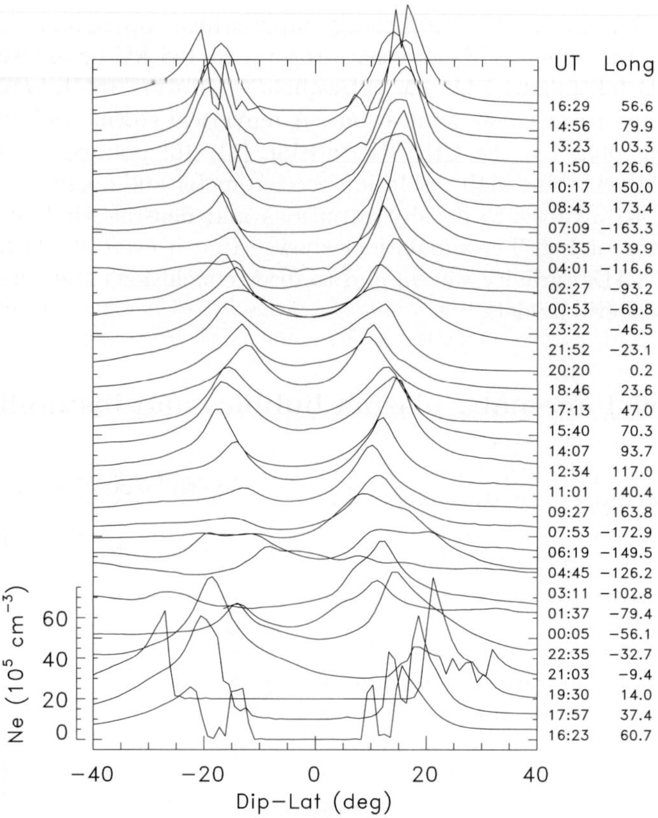


**Fig. 7.** Stack plot of DIDM plasma density measurements (from bottom up, offset $10^6$ cm$^{-3}$) during successive orbits during the geomagnetic storm of 21-23 Oct 2001. Equatorial crossings (UT and geogr. long.) of each orbit are indicated on the right.

upward plasma drift (Fig. 4) with the subsequent diffusive transport to the crest regions (Appleton anomaly). The small-scale magnetic fluctuations observed there are interpreted as the appearance of F region currents, coupled with the presence of plasma bubbles (see density fluctuations in the bottom panel, Fig. 6).

Fig. 7 displays the near-equatorial density changes during a two-day storm interval on 21-23 Oct 2001 with Kp values up to 8- and a minimum Dst of -187 nT on Oct 21, 22 UT. The shift of the crests toward higher latitudes in the first few hours of the storm is probably due to penetration electric fields while they move back and forth during the later storm development. The depth of the equatorial trough flattens which is due to the combined effect of disturbance dynamo and neutral wind action.

## Summary

During recent years, new LEO missions like the CHAMP and TIMED satellites have provided new and much more refined data about the near-Earth space concerning the electromagnetic, ionospheric and high-atmospheric environment. They brought about a wealth of new data for interesting studies which were not possible previously. These studies serve a dual purpose: they allow on the one side new insights into the complex physical processes of the ionospheric plasma and current systems at all latitudes during quiet and disturbed conditions. On the other side, they are the prerequisite for precise modelling of the Earth's magnetic field as well as the high atmosphere and the ionosphere itself from space. They allow novel science with important new and interesting studies. Among these are:

- investigations on the role of the diamagnetic effect and its relation to measured or modelled ionospheric plasma parameters;
- detailed analysis of the influence of the ionosphere on magnetic field modelling from space;
- systematic survey of plasma bubble signatures and of other plasma instabilities using global search algorithms and elucidation of their generation and development;
- fine-tuned ionospheric current studies at mid- and low-latitudes including the equatorial electrojet and the 3-D dynamo current system;
- case studies of storm events and the accompanying change of high-atmospheric, ionospheric and electrodynamic properties including travelling atmospheric disturbances, planetary waves as well as disturbance dynamo and penetration electric fields;
- large-scale pulsation studies from space simultaneously with ground observations of magnetometer networks to test theoretic ideas and models and aiming at remote control of the near-Earth plasma environment.

## References


1. Benson RF et al. (1998) Magnetospheric radio sounding on the IMAGE mission. Radio Sci Bull *285*: 9–20.
2. Fejer BG (1997) The electrodynamics of the low–latitude ionosphere: Recent results and future challenges. J Atmos Solar–Terr Phys *59*: 1465–1482.
3. Förster M and Jakowski N (2000) Geomagnetic storm effects on the topside ionosphere and plasmasphere: A compact tutorial and new results. Surveys in Geophys *21*: 47–87.
4. Fuller-Rowell TJ, Codrescu MV, Fejer BG, Borer W, Marcos FA, and Anderson DN (1997) Dynamics of the low-latitude thermosphere: quiet and disturbed conditions. J Atmos Solar–Terr Phys *59*: 1533–1540.
5. Jakowski N (1996) TEC monitoring by using satellite positioning systems. In: Kohl H, Rüster R, and Schlegel K, eds, Modern Ionospheric Science, European Geophysical Society, Katlenburg–Lindau, ProduServ GmbH Verlagsservice, Berlin: 371–390.

6. Kamide Y et al. (1997) Magnetic storms: Current understanding and outstandig questions. In: Magnetic storms, Geophysical Monograph *98*, American Geophysical Union: 1–19.
7. Kelley MC (1989) The Earth's Ionosphere: Plasma Physics and Electrodynamics, International Geophysics Series *43* Academic Press, Inc, with contributions from Heelis RA.
8. Kelley MC et al. (2003) The first coordinated ground- and space-based optical observations of equatorial plasma bubbles. Geophys Res Lett *30(14)*: 1766, doi:10.1029/2003GL017301.
9. Kil H et al. (2003) Case study of the 15 July 2000 magnetic storm effects on the ionosphere-driver of the positive ionospheric storm in the winter hemisphere. J Geophys Res *108(A11)*: 1391, doi:10.1029/2002JA009782.
10. Lemaire JF and Gringauz KI (1998) The Earth's Plasmasphere. Cambridge University Press, with contributions from Carpenter DL and Bassolo V.
11. Lühr H, Förster M, Reigber C, König R, Namgaladze AA, and Yurik RY (1998) Monitoring thermospheric density variations with the CHAMP satellite. In: Proc 'Space Weather' Workshop, ESTEC: 249–252.
12. Lühr H, Maus S, Rother M, and Cooke D (2002) First in-situ observation of night-time F region currents with the CHAMP satellite. Geophys Res Lett *29(10)*: 1489, doi: 10.1029/2001GL013845.
13. Lühr H, Rother M, Maus S, Mai W, and Cooke D (2003) The diamagnetic effect of the equatorial Appleton anomaly: Its characteristics and impact on geomagnetic field modeling. Geophys Res Lett *30(17)*: 1906, doi: 10.1029/2003GL017407.
14. Matsushita S (1969) Dynamo currents, winds, and electric fields. Radio Sci *4*: 771–780.
15. Meier RR et al. (2002) Ionospheric and dayglow responses to the radiative phase of the Bastille Day flare. Geophys Res Lett *29(10)*: 1461, doi: 10.1029/2001GL013956.
16. Namgaladze AA, Förster M, and Yurik RY (2000) Analysis of the positive ionospheric response to a moderate geomagnetic storm using a global numerical model. Ann Geophys *18*: 461–477.
17. Prölss GW (1997) Magnetic storm associated perturbations of the upper atmosphere. In: Magnetic storms, Geophysical Monograph *98*, American Geophysical Union: 227–241.
18. Richmond AD (1976) Electric field in the ionosphere and plasmasphere on quiet days. J Geophys Res *81*: 1447–1450.
19. Rishbeth H (1981) The F–region dynamo. J Atmos Terr Phys *43*: 387–392.
20. Scherliess L and Fejer BG (1999) Radar and satellite global equatorial F region vertical drift model. J Geophys Res *104*: 6829–6842.
21. Schunk RW (1983) The Terrestrial Ionosphere. D Reidel Publishing Company: 609–676.
22. Wagner CU, Möhlmann D, Schäfer K, Mishin VM, and Matveev MI (1980) Large–scale electric fields and currents and related geomagnetic variations in the quiet plasmasphere. Space Science Reviews *26*: 391–446.
23. Whalen JA (2000) An equatorial bubble: Its evolution observed in relation to bottomside spread F and to the Appleton anomaly. J Geophys Res *105(A3)*: 5303–5315.
24. Zhang Y et al. (2003) Negative ionospheric storms seen by the IMAGE FUV instrument. J Geophys Res *108(A9)*: 1343, doi:10.1029/2002JA009797.

# Interpretation of CHAMP Crustal Field Anomaly Maps Using Geographical Information System (GIS) Technique

Kumar Hemant, Stefan Maus and Volker Haak

GeoForschungsZentrum Potsdam, Section 2.3, Telegrafenberg, D-14473 Potsdam

**Summary.** Crustal field models from CHAMP magnetic measurements are increasingly stable and reliable. In particular, they now allow for quantitative geological studies of crustal structure and composition. Here, we use a forward modeling technique to infer deep crustal structure of continental regions overlain by younger sediments. For this, a Geographical Information System (GIS) based technique has been developed to model the various geological units of the continental crust. Starting from geologic and tectonic maps of the world and considering the known rock types of each region, an average magnetic susceptibility value is assigned to every geological unit. Next, a vertically integrated susceptibility (VIS) is computed for each unit, taking into account the seismic crustal thickness, as given by models 3SMAC and CRUST2.1. From this preliminary VIS model, an initial vertical field anomaly map is computed at a satellite altitude of 400 km and compared with the corresponding CHAMP vertical field anomaly map. We demonstrate that significant geological inferences can be made from the agreement and the discrepancies between the predicted and observed anomaly maps. In particular, the lateral extent of Precambrian provinces under Phanerozoic cover is revealed.

**Key words:** Global, magnetic, crustal, GIS

## Introduction

Scalar magnetometers aboard satellites have been into orbit for more than three decades now. POGO (1965-1971) measured only scalar data, while Magsat (1979-1980) and Ørsted (since February, 1999) missions measured the vector components as well. Though these missions led to the derivation of many crustal field models, the accuracy was limited due either to the higher orbital altitude of satellites (POGO and Ørsted) or to the inaccuracies in the star cameras that corrupted the vector magnetic field components (Magsat). In July-2000, the CHAMP satellite (http://op.gfz-potsdam.de/champ/main_CHAMP.shtml) was launched into a low Earth orbit of 450 km. The present altitude of its almost circular orbit is 400 km and is particularly suited to map the crustal field anomalies much more accurately than any of its predecessors. Maus et al. (2002) derived a new global crustal magnetic anomaly map using the CHAMP satellite scalar and vector data. We use the first revision MF2 of this map (http://www.gfz-potsdam.de/pb2/pb23/SatMag/model.html) for all further analysis.

The aim of deriving these crustal field anomaly maps is to reveal new geological and tectonic information of the subsurface. The interpretation strategy followed here is not that of direct inversion, which is non-unique due to inherent ambiguities of the magnetic inverse problem.
The present method is a GIS based forward modelling technique, which takes geologic and tectonic maps of the world and considering the known rock types of each region, an average susceptibility value is assigned to every geological unit. Next, a vertically integrated susceptibility (VIS) is computed by taking product of average susceptibility value and the seismic crustal thickness, as given by global seismic models like 3SMAC (Nataf and Ricard, 1996) and CRUST2.1, the latest model by Mooney et al. (1998). From this VIS model, the vertical field anomaly map is predicted at a satellite altitude of 400 km and compared with the corresponding CHAMP vertical field anomaly map. Geological inferences are drawn on the basis of discrepancy between the extent of anomalies and their strength in the two maps.

## GIS modelling

The sources of the continental magnetic anomalies primarily consist of rock types formed early in the geological history of the earth. These rocks are Precambrian in age. Our primary interest is to generate a global crustal magnetisation model based on the detailed information of the rock types exposed in the Precambrian provinces, their magnetic susceptibility values and a known stratigraphy for that region.

For this all the known rock types for a particular geological region are compiled and using their maximum volume susceptibility value (Clark and Emerson, 1991; Hunt et al., 1995) an average maximum susceptibility value is computed. The assigned susceptibility of the region is some percentage of this maximum value. This factor (0.55) is kept as a global constant for all the rock types. It was derived on the basis of minimum rms difference between the Gauss coefficients of predicted and the observed field. Next, the crustal thickness known from the stratigraphical information for each geological layer within a vertical column is multiplied with the average susceptibility of that layer. For regions where stratigraphical information are not available, the average maximum susceptibilities are multiplied with the crustal thickness of the upper crust known from the global seismic models like 3SMAC and CRUST2.1. The susceptibility value for the lower crust in any geologic region is computed by multiplying the average susceptibility of the upper crust with a factor accounting for the difference in the iron oxide content in the average composition of the upper and the lower crust (Taylor and McLennan, 1985). Integrating the product of susceptibility and thickness for various layers in a vertical column provides the vertically integrated susceptibility value for that region. Each geological province is modelled following the above steps and all of the VIS values are used to generate an initial VIS model (Fig. 1). The above modeling steps were done on the GIS ArcInfo 8.1 platform. One of the important

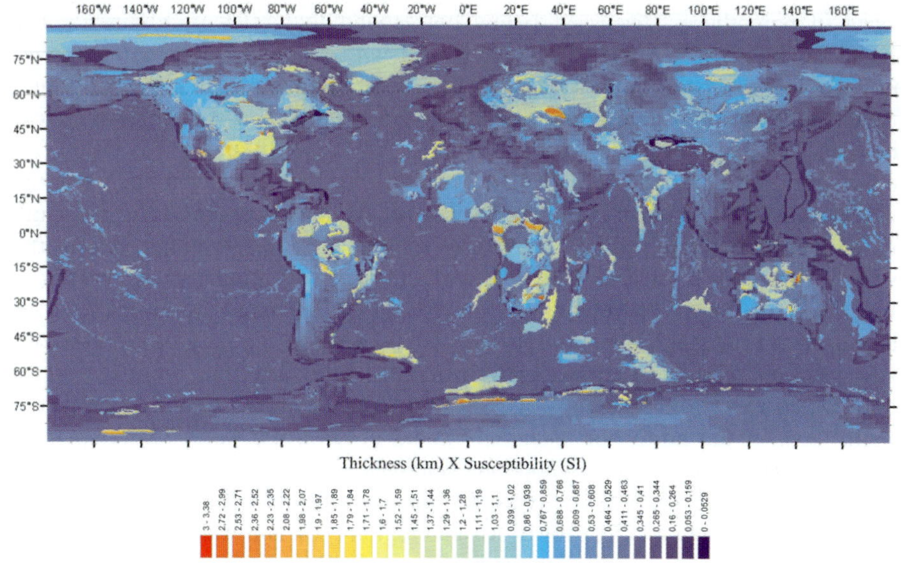


**Fig. 1.** The vertically integrated susceptibility (VIS) model.

assumptions in the present work is to consider the seismic Moho as a magnetic boundary (Wasilewski and Mayhew, 1992). Thus, the upper mantle is considered to be non-magnetic (Meyer et al., 1983; Purucker et al., 1998).

This VIS model is replaced with a distribution of equivalent dipoles on the Earth induced in the direction of the main field of the Earth. No remanent magnetization is considered here and the sources for the anomalies are considered to be induced magnetization. From the spherical grid of dipoles the magnetic field is computed at a height of 400 km and expanded into spherical harmonic to derive the Gauss coefficients of the predicted field. The vertical field anomaly map is computed from these Gauss coefficients and spherical harmonic degrees 1-15 are set to zero because this long wavelength crustal field is masked by the main field and therefore is not observable. Finally, only spherical harmonic degrees 16-80 of the predicted crustal magnetic field are compared with the corresponding degrees 16-80 of the observed crustal field.

## Results

The CHAMP observed vertical field anomaly map for spherical harmonic degrees 16-80 at an altitude of 400 km is shown in Figure 2 and the corresponding predicted anomaly map or the initial model computed using the initial VIS model is shown in Figure 3. The two maps are compared visually. On comparison it is apparent that there are anomalies over the regions where the predicted and observed maps match well. Over some regions of the Precambrian provinces, the predicted anomalies display weaker amplitudes than the observation. The numbers referenced in parentheses discussed below correspond to circles in Figure 2. The shapes of the predicted anomalies agree well with those of observed anomalies over the Turkmenistan shield (1), partly over the Songliao massif (2), and eastern (3) and southern (4) shield regions of the Indian craton. Over the Siberian craton, the strong anomaly over the Anabar shield (5), in the north and Aldan shield (6), in the southeast, correlate well with the observations. Predicted anomalies over the Kiruna mines (7), Sweden, in the north and over the Kursk region (8), in the south of the European craton show a good agreement with observations. Along the western region of East European platform, a partial agreement in anomalies is seen along the Tornquist-Teisseyre Zone (TTZ) (9). The magnetic anomalies over the Superior (10) and Slave (11) province show a weak anomaly both in the observed and the predicted map but a strong anomaly is evident over the Ungava craton (12), in the northern region of North American craton. Regions of Guyana shield (13), in the north and Brazilian shield (14), in the north-central part of South American craton show moderate anomaly features in the observed map, which are in agreement with the predicted anomaly. Major anomaly features over the west African craton (15), central African region (16), and southern Africa (17), are reproduced well in the predicted map. Precambrian provinces of Australia, the Pilbara (18), and the Yilgarn craton (19) show weak anomalies, which are only partially reproduced, in the predicted map. However, strong anomalies over the largely buried Nullarbor block (20), in the south and Mt Isa inlier (21), in the north of Australian craton agree well with predictions. Observed anomaly over the Archean block in the southern Greenland (22), is largely in agreement with predictions. Antarctica is mostly covered with ice except at the periphery, hence, the observed anomalies agree only over some regions of the predicted map especially over the Rayner complex (23) and its surrounding Precambrian provinces. Much of results discussed above are consistent with the conclusions of various researchers and are summarized in Langel and Hinze (1998).

The comparisons made above between predicted and the observed magnetic anomalies for the vertical component show agreement over many Precambrian provinces of the world but the discrepancies between the two maps are also evident in some regions. For instance, most of the negative anomaly pattern and its extent over the Himalayan fold belt and the Tibetan plateau are in agreement with observations, however, the observed anomaly shows a more intense negative anomaly. Some of the regions like southwest USA show a large stretch of predicted anomalies but it does not completely match with the extension of the observed anomalies. Similar disagreement in extent and shapes of anomalies are also seen over the Kolyma-Omolon regions in the eastern Siberia, Tarim craton in China and north-central Greenland. Over the west African craton and central African region, the trend of the predicted anomalies are largely in agreement with observations but the predicted anomaly is comparatively much weaker in strength than the observed anomaly. Another significant disagreement between the anomalies in the predicted and observed map lies in the northwestward region of Anabar shield. The observed map shows the anomaly over the known geological boundary of Anabar shield to extend northwestward over the Phanerozoic sediments filled Khatanga trough. This extended anomaly is absent in the predicted map as this strong anomaly cannot have a source in the upper crust of Khatanga trough filled with 15 km of sediments (Goodwin, 1991). The regions shown in ovals in Figure 3 show some of the regions mentioned above.

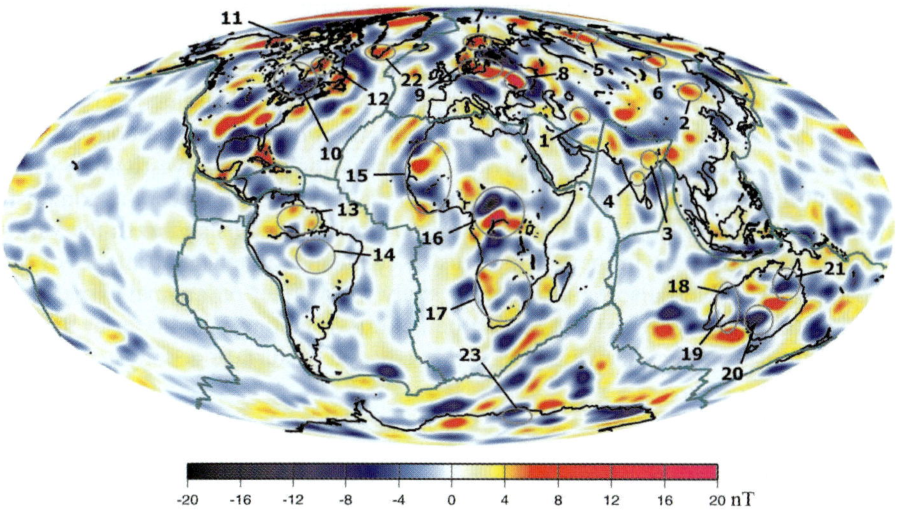


**Fig. 2.** Lithospheric field model (MF2) for vertical component derived from CHAMP scalar and vector data at 400 km for degrees 16-80 (http://www.gfz-potsdam.de/pb2/pb23/SatMag /model.html). Numbers marked represent some of the major Precambrian provinces which are described in the text.

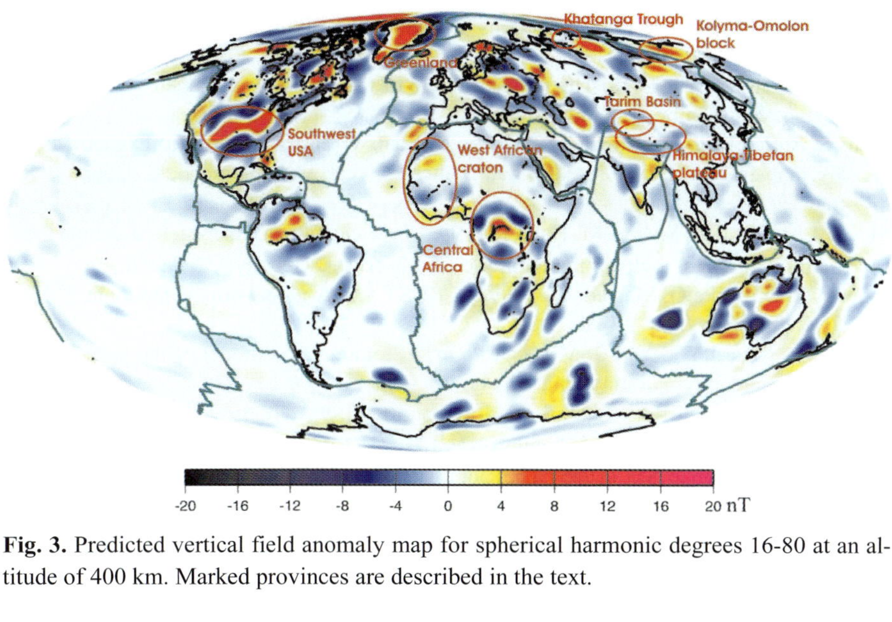


**Fig. 3.** Predicted vertical field anomaly map for spherical harmonic degrees 16-80 at an altitude of 400 km. Marked provinces are described in the text.

## Conclusion

Our GIS based magnetic modelling technique demonstrates the possibilities of this technique of inferring geological and tectonic information from the differences between the predicted and the observed magnetic anomaly maps. The overall agreement in the predicted and the observed maps show that sources to magnetic anomalies are indeed geological in origin. This also supports our assumptions employed during the derivation of vertically integrated susceptibility map. The discrepancies between the magnetic anomalies of the predicted and the observed map over certain regions have shown that the assumptions used in the present VIS map may not be true globally. This provides the basis for further investigation especially in the context of extent of subsurface Precambrian provinces, the composition of the lower crust, accuracy of the seismic crustal models and even the variation of Curie-isotherm in the Earth's crust. Remanent magnetisation model of the oceanic crust should also be included to study their effect on anomalies near the continental edges. These studies could be helpful in getting insights in to the nature of the crust in regions which are poorly covered by surface geophysics.

## References

CGMW (2000) Geological Map of the World, Commission for the Geological Map of the World (CGMW) / © UNESCO Publishing.

Clark DA, Emerson DW (1991) Notes on rock magnetization characteristics in applied geophysical studies, Exploration Geophysics *22*: 547-555.

Goodwin, AM (1991) Precambrian Geology, Academic Press, London.

Hunt CP, Moskowitz BM, Banerjee SK (1995) Magnetic properties of Rocks and Minerals, In: Rock Physics and Phase Relations -A handbook of Physical constants, AGU Reference shelf 3, ed. Ahrens TJ.

Langel RA, Hinze WJ (1998) The magnetic field of the Earth's Lithosphere – The satellite perspective, Cambridge University Press.

Maus S, Rother M, Holme R, Lühr H, Olsen N, Haak V (2002) First scalar magnetic anomaly map from CHAMP satellite data indicates weak lithospheric field. Geophys Res Lett *29(14)*.

Meyer J, Hufen J–H, Siebert M, Hahn A (1983) Investigations of the internal geomagnetic field by means of a global model of the earth's crust. J Geophys *52*: 71-84.

Mooney WD, Laske G, Masters TG (1998) CRUST 5.1: A global crustal model at $5^0$ X $5^0$, J Geophys Res *103*: 727-747.

Nataf H-C, Ricard Y (1996) 3SMAC: An a priori tomographic model of the upper mantle based on geophysical modeling. Phys Earth Planet Int *95*: 101-122.

Purucker ME, Langel RA, Rajaram M, Raymond C (1998) Global magnetization models with a priori information. J Geophys Res *103*: 2563-2584.

Taylor SR, McLennan SM (1985) The Continental Crust: Its composition and Evolution. Oxford: Blackwell Scientific Publ. 312pp.

Wasilewski PJ, Mayhew MA (1992) The Moho as a magnetic boundary revisited, Geophys Res Lett *19*: 2259-2262.

# Magnetic Crustal Thickness in Greenland from CHAMP and Ørsted Data

Cathrine Fox Maule[1], Michael E. Purucker[2], and Nils Olsen[1]

[1] Center for Planetary Science, Juliane Maries Vej 30, DK-2100 Copenhagen Ø, Denmark, *foxmaule@gfy.ku.dk*

[2] Raytheon ITSS at Geodynamics Branch, Goddard Space Flight Center, Greenbelt, Maryland, USA

**Summary.** The magnetic crustal thickness of Greenland and the surrounding area is determined by inversion of gridded values of the magnetic radial component as given by the IDEMM model, which is based on CHAMP and Ørsted data alone, and by the Comprehensive Model (CM4), which is based on satellite and observatory data.

After correcting for the remanent magnetization, we determine the vertically integrated magnetization of the crust. Making some simplifying assumptions about the susceptibility, the thickness of the magnetic crust is determined by iteratively improving an initial crustal thickness model using the equivalent source magnetic dipole method.

**Key words:** Magnetic crustal thickness, Greenland

## 1 Introduction

The new high-precision data from the CHAMP satellite has improved the accuracy of the short wavelength part of the magnetic field models. These short wavelengths reflect the magnetic field of crustal origin, and it has thus become possible to extract more information about the crustal field from the satellite data than previously. We have taken advantage of this significant improvement and attempted to extract information about the magnetic structure of the crust in Greenland from the improved field models.

## 2 Estimation of the magnetic crustal thickness

A scheme of the iterative inversion method that we have used is shown in Figure 1.

*Data* We use magnetic field models as the data for our inversion in order to minimize the effects of the large time-variable external field in the Greenland region. To determine the magnetic crustal thickness we have chosen only to use the radial component of the magnetic field, which make the inversion linear. We use $B_r$ from the International Decade Earth Magnetic

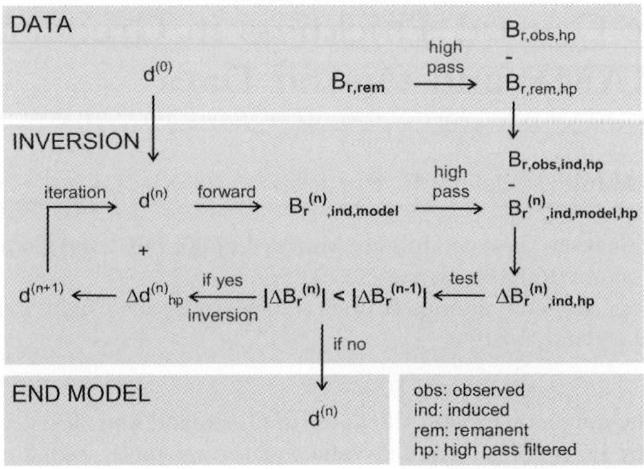


**Fig. 1.** The inversion scheme. The details of the inversion scheme is explained in the text.

Model (IDEMM) [http://www.dsri.dk/Oersted/Field_models/IDEMM/] and the Comprehensive Model (CM4) [9] at an altitude of 400 km. IDEMM is derived using Ørsted and CHAMP data for the longer wavelengths (spherical harmonic degree ≤25) and CHAMP data alone for the shorter wavelengths (degree >25). As IDEMM is determined until degree and order 80 the resolution scale is 4.5° or about 500 km. CM4 is obtained from observatory data and satellite data (POGO, Magsat, Ørsted and CHAMP), and is determined to degree and order 65, corresponding to about 615 km.

The inversion is done globally but interpretation of the results is only done at high latitudes. At low latitudes the results of our inversion are poor as we only invert from the radial component, which makes it difficult to resolve the here dominant horizontal component of the crustal magnetization.

*Data processing* As the Earth's main field dominates the long wavelength part of the magnetic field, the data (from the field models) are high-pass filtered, thus contributions from degrees less than 15 are discarded. This means that the very long wavelength part (>2650 km) of the crustal field also is discarded, but this is unavoidable as it cannot be separated from the core field.

To account for remanent magnetism in the crust, a model of remanent magnetism for the oceanic crust $B_{r,rem}$ [3, 6] is subtracted. It is assumed that remanent magnetism in the continental crust is negligible, recent work [10] does however support this for Greenland. The remanent field is high-pass filtered and subtracted from the observed field to give the observed induced field $B_{r,obs,ind,hp}$, which is used as base for the inversion.

*Initial model* Since the long wavelength component of the crustal magnetic field is unknown the long wavelength part of the magnetic crustal thickness cannot be resolved from the inversion. Even if the long wavelength magnetic field from the crust was available, it still would be difficult to resolve the corresponding magnetic crustal thickness because it lies close to the null-space (the long undulations of the magnetic crustal thickness only give a very small contribution to the magnetic field at satellite altitude). The extreme case is a shell of constant thickness and susceptibility and a magnetic field induced by interior sources, which has no magnetic field outside the shell [8]. Therefore we introduce an initial model for the magnetic crustal thickness (labelled $d^{(0)}$ in Figure 1). Since the final solution of the inversion depends on the initial model it is important to choose an initial model where the long wavelength variations is close to the true structure of the crust. We use the 3SMAC model [4], a crustal thickness model based on seismic measurements and a thermal model, as the initial model.

*Equivalent source magnetic dipole method* From the initial model of the magnetic crustal thickness we calculate the induced field $B_{r,ind,model}$ that this crust would give by applying the equivalent source magnetic dipole method (ESMD). Assuming that the inducing field is the Earth's main field, the dipole moments are aligned in the direction of the main field.

In the ESMD method it is assumed that the crustal field can be represented by a finite number of dipoles evenly distributed over the surface of the Earth. The idea is to determine the dipole moments of these dipoles such that the magnetic field due to the dipoles matches the observed field at some given altitude above the surface. Thus the first step is to calculate the moments of these dipoles from the magnetic crustal thickness model. We assume that the susceptibility is constant with 0.035 SI in the continental areas and 0.040 SI in the oceanic areas [7]. The moment $\overline{m}_j$ of a dipole representing the magnetism of the j'th crustal block is

$$\overline{m}_j = \frac{\kappa_j}{\mu_0}\overline{B}_{MF}Ah_j \qquad (1)$$

where $\kappa_j$ is the susceptibility, $\mu_0$ is the vacuum permeability, $\overline{B}_{MF}$ is the inducing field, $A$ is the surface area per dipole, and $h_j$ is the magnetic crustal thickness. Under the given assumptions the magnitude of the dipole moments thus become proportional to the magnetic crustal thickness.

We use 11562 dipoles distributed on a spherical icosahedron grid to represent the crustal field. The dipole spacing is 1.89° corresponding to about 210 km at the surface. The dipole locations in the Greenland area are shown in Figure 2. To avoid a geographically uneven weighting of the data points, we have chosen the data points to be the same as the dipole locations, but at 400 km altitude. Once the dipole moments of the dipoles are determined it is straightforward to calculate the magnetic field due to the dipoles, the relevant equations can e.g. be found in [2]. Because the dipoles are aligned in the main field direction the only variable is the magnitude of the dipole moment $m_j$. Let $\overline{B}_r$ be the vector of $B_r$ at location 1 to 11562, and similarly let $\overline{m}$ be the magnitude of the dipole moment at the 11562 locations, the problem can then be written in matrix form as

$$\overline{B}_r = G\,\overline{m} \qquad (2)$$

where $G$ is a 11562×11562 matrix. $G$ is ill-conditioned as the 11562 data are not linearly independent. They are created from 80×82 (IDEMM) or 65×67 (CM4) spherical harmonics coefficients, thus there are only 6560 or 4355 independent parameters, respectively. So the problem is both over- and underdetermined.

*Inversion* The induced magnetic field from the crustal thickness model is calculated ($B_{r,ind,model}$ in Figure 1) and high-pass filtered, so it can be compared to the observed induced field. The high-pass filtering is done by first regridding the field from the dipole grid to an equal grid, then a spherical harmonic analysis is made and the coefficients up to degree 15 are removed, thereafter a spherical harmonic synthesis is made to calculate the high-pass filtered field in the dipole grid. The high-pass filtered modelled field ($B_{r,ind,model,hp}$) is subtracted from the observed field to find the difference ($\Delta B_{r,ind,hp}$ in Figure 1). If the root mean square (rms) error of $\Delta B_{r,ind,hp}$ is less than in the previous iteration, then $\Delta B_{r,ind,hp}$ is inverted to obtain a correction to the dipole moment $\Delta d_{hp}$. The inversion is done with the LSQR-algorithm, which is an iterative conjugate gradient method for solving large, sparse, linear, ill-conditioned systems. For the inversion $G$ is made sparse in order to speed up the calculation by only letting dipoles within a distance of 2500 km contribute to the magnetic field at a given location (for the forward calculation of determining $\overline{B}_r$ from the crustal model the full matrix is used). The correction to the dipole moment $\Delta d_{hp}$ is then added to the previous model for the dipole moment $d$, giving a new improved estimate of the dipole moments $d^{(n+1)} = d^{(n)} + \Delta d^{(n)}_{hp}$. The iteration is stopped when the rms error of $\Delta B_{r,ind,hp}$ no longer decreases, this happens after about 15-20 iterations for CM4 and about 37-40 iterations for IDEMM. The end model for the dipole moments is then converted into crustal thickness through equation 1.

## 3 Results and discussion

Figure 2 shows the radial component of the magnetic field at 400 km altitude from IDEMM and CM4. The magnetic crustal thickness that we have obtained are shown in Figure 3. The differences between the two magnetic models are reflected in the results for the magnetic crustal thickness, but there are some general features. A large region in the northwestern part of Greenland has relatively large magnetic crustal thickness locally reaching values above 50 km. Our results predicts a relatively low magnetic crustal thickness along the east coast and across the middle of Greenland, and intermediate thickness around the southern tip of Greenland.

The thicknesses of the magnetic crust should not exceed the Moho thickness as the mantle rocks in general are believed to be non-magnetic. Thus

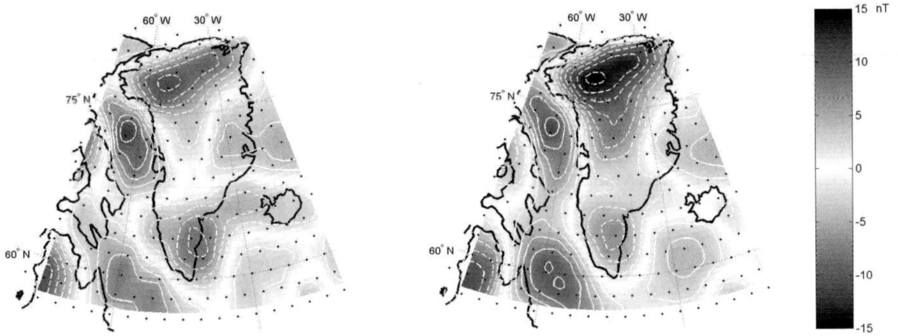


**Fig. 2.** The radial component of observed induced field from IDEMM (left), degree 15-80, and CM4 (right), degree 15-65 in nT. Contour lines are drawn at every 2 nT, solid (dashed) lines present positive (negative) values. The black dots are the dipole locations.

we have compared our results with the results of a recent seismic study in Greenland where the Moho thickness were determined at 16 locations [1]. At all locations the magnetic crustal thicknesses that we obtained are less than or equal to the Moho thicknesses. However, there are no Moho measurements in the northwestern part of Greenland where our study suggests a very thick magnetic crust.

Although we have no evidence suggesting that the crust is not thick, there remains the possibility that one of the assumptions that went into our model-building is faulty. It could be that this region contains significant amounts of unrecognized large-scale remanent magnetization, perhaps associated with the banded ironstones of the Thule region. Our assumption of constant magnetic susceptibility might also be suspect for similar reasons, although based on the aeromagnetics the banded ironstones do not appear to be extensive.

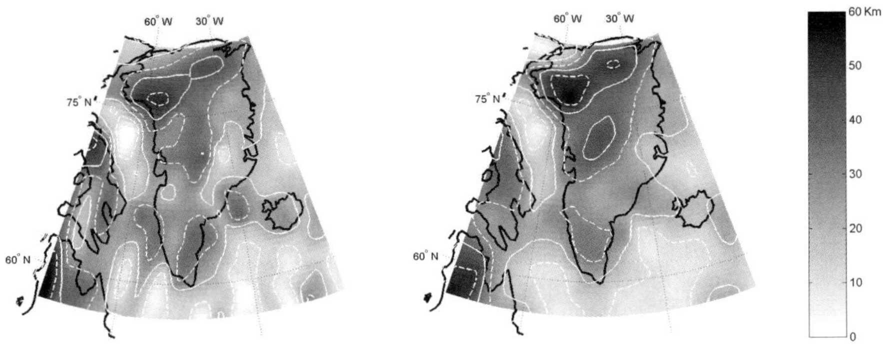


**Fig. 3.** The magnetic crustal thickness in km from IDEMM (left), and CM4 (right). Contour lines at 10 (dashed), 20 (solid), 30 (dashed), 40 (solid), and 50 (dashed) km.

It might also be argued that our initial model significantly overestimated the magnetic crustal thickness here, leading to a bias in the final solution. However, the initial model in NW Greenland has thicknesses of 40 km or less, a reasonable figure for an old cratonic region such as this. More than 13% of the magnetic crustal thicknesses in the initial model are in excess of 35 km, and the thickest is in excess of 58 km.

*Acknowledgement.* We would like to thank Trine Dahl-Jensen for making the Moho data from Greenland available to us.

## References

1. Dahl-Jensen T, Larsen TB, Woelbern I, Bach T, Hanka W, Kind R, Gregersen S, Mosegaard K, Voss P and Gudmundsson O (2003) Depth to Moho in Greenland: Receiver-Function Analysis Suggests Two Proterozoic Blocks in Greenland. Earth Planet Sci Letters *205*: 379–393.
2. Dyment J and Arkani-Hamed J (1998) Equivalent Source Magnetic Dipoles Revisited. Geoph Res Letters *25(11)*: 2003–2006.
3. Dyment J and Arkani-Hamed J (1998) Contribution of Lithospheric Remanent Magnetization to Satellite Magnetic Anomalies over the World's Oceans. J Geophys Res *103(B7)*: 15,423–15,441.
4. Nataf H-C and Ricard Y (1996) 3SMAC: An a Priori Tomographic Model of the Upper Mantle Based on Geophysical Modeling. Phys Earth and Planet Int *95*: 101–122.
5. Purucker M, Langlais B, Olsen N, Hulot G and Mandea M (2002) The Southern Edge of Cratonic North America: Evidence from New Satellite Magnetometer Observations. Geoph Res Letters *29(9)*: Art. No. 1342, 10.1029/2001GL013645.
6. Purucker ME and Dyment J (2000) Satellite Magnetic Anomalies Related to Seafloor Spreading in the South Atlantic Ocean. Geoph Res Letters *27(17)*: 2765–2768.
7. Purucker M, Langlais B, Olsen N, Hulot G and Mandea M (2002) The Southern Edge of Cratonic North America: Evidence from New Satellite Magnetometer Observations. Geoph Res Letters *29(9)*: Art. No. 1342, 10.1029/2001GL013645.
8. Runcorn SK (1975) An Ancient Lunar Magnetic Dipole Field. Nature *253*: 701–703.
9. Sabaka T, Olsen N and Purucker M (2003) Extending Comprehensive Models of the Earth's Magnetic Field with Ørsted and CHAMP Data. Submitted to Geophys J Int.
10. Whaler K and ME Purucker (2003) Global Models of Lithospheric Magnetic Field from Satellite Data. Poster at IUGG, Sapporo, Japan.

# CHAMP Magnetic Anomalies of the Antarctic Crust

Hyung Rae Kim[1], Luis R. Gaya-Piqué[2,3], Ralph R. B. von Frese[4], Patrick T. Taylor[5], and Jeong Woo Kim[6]

[1] GEST, Univ. of Maryland, Baltimore County at GSFC/NASA, Greenbelt, MD, USA,
[2] Istituto Nazionale di Geofisica e Vulcanologia, Roma, Rome, Italy,
[3] Observatori de l'Ebre, 43520 Roquetes, Spain,
[4] Dept. of Geol. Sci., The Ohio State Univ., Columbus, OH 43210, USA
[5] Geodynamics Branch, NASA/GSFC, Greenbelt, MD 20771, USA,
[6] Dept. Geoinfo. Eng., Sejong Univ., Seoul, Republic of Korea.

**Summary.** Regional magnetic signals of the crust are strongly masked by the core field and its secular variation components and hence difficult to isolate in the satellite measurements. In particular, the un-modeled effects of the strong auroral external fields and the complicated behavior of the core field near the geomagnetic poles conspire to greatly reduce the crustal magnetic signal-to-noise ratio in the polar regions relative to the rest of the Earth. We can, however, use spectral correlation theory to filter the static lithospheric anomalies and core field components from the dynamic external field effects. To help isolate regional lithospheric from core field components, the correlations between CHAMP magnetic anomalies and the pseudo-magnetic effects inferred from gravity-derived crustal thickness variations can also be exploited. Employing these procedures, we processed the CHAMP magnetic observations for an improved magnetic anomaly map of the Antarctic crust. Relative to the much higher altitude Ørsted and noisier Magsat observations, the CHAMP magnetic anomalies at 400 km altitude reveal new details on the effects of intracrustal magnetic features and crustal thickness variations of the Antarctic.

**Keywords:** CHAMP, magnetic anomalies, Antarctica, crustal thickness.

## 1 Introduction

Satellite magnetic data from the POGO, Magsat, Ørsted and CHAMP missions play a critical role in understanding regional or global magnetic anomaly variations due to petrological variations of the crust and upper mantle, and crustal thickness and thermal perturbations (e.g., [1], [2], [3], [4]). Data from satellite observations provide uniform magnetic coverage to within about 3 degrees of the poles. However, the relatively enhanced density of satellite data at the poles is definitely useful for polar geologic studies that generally lack coverage by conventional near-surface anomaly surveys due to the remoteness and harsh environmental conditions of the polar regions.

CHAMP magnetic observations offer significant advantages for Antarctic crustal studies over the data from NASA's Magsat mission. For example, the Magsat data collected only over a 6-month period during austral summer when external field activity is strongest are highly contaminated by non-lithospheric noise in the Antarctic. The CHAMP satellite, by contrast, has been operating since July 2000 and collected with considerably greater accuracy a much larger quantity of relatively low-noise data over several austral winters. The improved CHAMP magnetic data offer significant opportunities for developing new insight on the features and development of the poorly understood south polar crust.

Satellite magnetic crustal sources, in general, reflect both inductive and remanent components of magnetization. As crustal depth increases, remanent and thermal overprints are diminished while viscous magnetization is enhanced as temperature increases to within about 100-150°C of the Curie point of magnetite. The effects of viscous magnetization are in-phase with the induced component and may have a strong effect at satellite altitudes.

Thermo-remanently magnetized sources, on the other hand, are located predominantly in the upper crust. These sources produce relatively high frequency signals that are particularly strong at the near-surface, but considerably attenuated at satellite altitudes. Exceptions are the regional signals from extensive areas of oceanic crust that were produced during a long normal polarity interval in the Cretaceous. The effects of the oceanic Cretaceous Quiet Zones are commonly observed at satellite altitudes (e.g., [5]).

Satellite anomalies from typical global spherical harmonic models, however, can be problematic for polar lithospheric studies because the strong and complex effects from the auroral external fields are only marginally accounted for in the anomaly estimates. Hence, in this study, we use advanced spectral correlation theory to process CHAMP magnetic observations south of 60°S latitude for anomalies with maximum crustal signal-to-noise ratios, where the noise components are composed of external field and other dynamic non-crustal effects.

## 2 Antarctic Crustal Magnetic Anomalies From CHAMP Observations

Our processing of the south polar CHAMP magnetic observations follows the flow chart in Figure 1 ([6], [7]). These procedures were initially developed to extract crustal anomaly components from Antarctic Magsat [8] and Ørsted [2] magnetic data. Figure 2.A shows the crustal thickness model of the Antarctic produced by [9] using the spectral correlation analysis of free-air and computed terrain gravity effects at satellite altitude. This model agrees well with available seismic Moho estimates for the Antarctic. Figure 2.C gives the related total magnetic anomalies at 400 km that we modeled from crustal thickness variations by Gauss-Legendre quadrature integration using the core field attributes shown in Figure 2.B. Here, the continental and oceanic crustal components were modeled using 0.01 and 0.03 SI-susceptibilities, respectively [11].

Figure 2.D shows the total field anomalies (Fig. 2.C) reduced differentially to the radial pole to enhance the relationships between the satellite altitude magnetic anomalies and crustal thickness variations such as along the Antarctic continent-ocean margins. Comparing the differentially reduced to pole (DRTP) anomalies (Fig. 2.D) with the total field anomalies (Fig. 2.C) reveals the strong and complicated influence of the Antarctic core field on crustal anomalies even in the vicinity of the geomagnetic pole that is located just off the coast of the Wikes Subglacial Basin (WSB in Fig. 2.A).

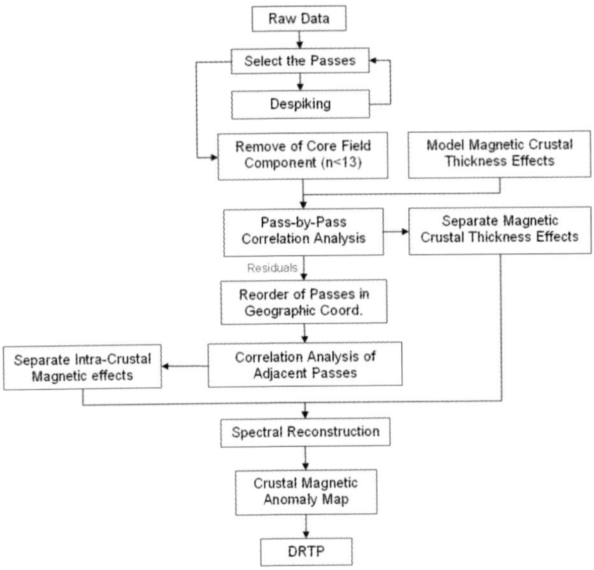


**Fig. 1.** Flowchart for processing the CHAMP Antarctic magnetic data for crustal anomalies.

These results show that Antarctic crustal thickness variations may involve substantial regional magnetic effects in the core field component estimates (e.g., [2] [11]). To extract these possible effects from the CHAMP data on a track-by-track basis, we represented the magnetic effects in Figure 2.C by a spherical cap harmonic (SCH) model (e.g., [12]). The maximum spatial index of 30 was chosen for the SCH model defined by a total of 250 coefficients on a half-angle cap of 30 degrees.

Following the processing flowchart in Figure 1, we first removed estimates of the core field from the CHAMP observations using CM3 model from GSFC/NASA [13]. The residuals were then spectrally correlated with the predictions from the SCH model of the crustal thickness effects along each satellite track. Signals in the CHAMP anomaly residuals that were positively correlated with the SCH predictions were extracted and mapped at 450 km altitude as shown in Figure 3.A.

The remaining relatively higher frequency satellite measurements that were not correlated with the crustal thickness magnetic effects, were next sorted into orbital pairs of nearest-neighbor tracks to extract the static, correlated features that presumably reflect the effects of additional crustal sources ([8], [14]). Negative or null correlated signals between tracks separated by distances that are small clearly cannot be related to features of the crust, but rather must be related to the dynamic effects of external fields and other non-crustal noise. Figure 3.C gives the estimates resulting from the track-by-track spectral correlation analysis that mostly reflect the intracrustal magnetization variations. The magnetic effects from crustal thickness variations and the intracrustal magnetization variations were then added

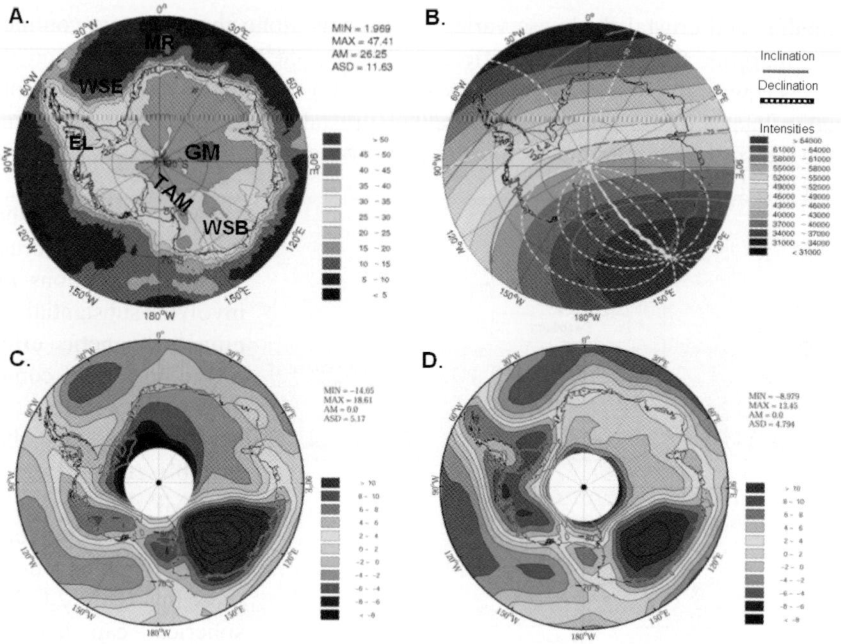


**Fig. 2.** A) Antarctic crustal thickness variations from gravity and topography data (von Frese et al., 1999). Annotations include the maximum (MAX) and minimum (MIN) amplitude values, the amplitude mean (AM) and standard deviation (ASD). B) Core field attributes from the CHAMP spherical harmonic model [10]. The magnetic effects (nT) of the crustal thickness variations are evaluated as total field and differentially reduced-to-pole (DRTP) anomalies at 400 km in maps (C) and (D), respectively.

together and plotted in Figure 3.E. The DRTP anomalies for the respective maps are also presented in Figure 3.B, 3.D and 3.F.

## 3 Discussion and Conclusions

Figures 3.A and 3.B show the resulting CHAMP magnetic anomaly map that can reflect the Antarctic continent-ocean edge effects. The dominant portion of the large anomalies over WSB appear to indicate the integrated edge effect of the underlying thinned crust (Fig. 2.A) that involves the continent-ocean boundary and thicker crust of the Transantarctic Mountains (TAM) on the west and the Gamburtsev Mountains (GM) to the east. The intracrustal anomalies over this area are relatively weakly expressed in Figures 3.C and 3.D.

A prominent elongated positive intracrustal anomaly overlies Maude Rise (MR) that reflects the remanent magnetization effects of this Cretaceous Quiet Zone (e.g., [11], [15]). This long linear positive anomaly extends from Maud Rise along the eastern margin of the Weddell Sea Embayment (WSE) to Ellsworth Land (EL) where a series of microplates are assembled (e.g., [16]). This linear

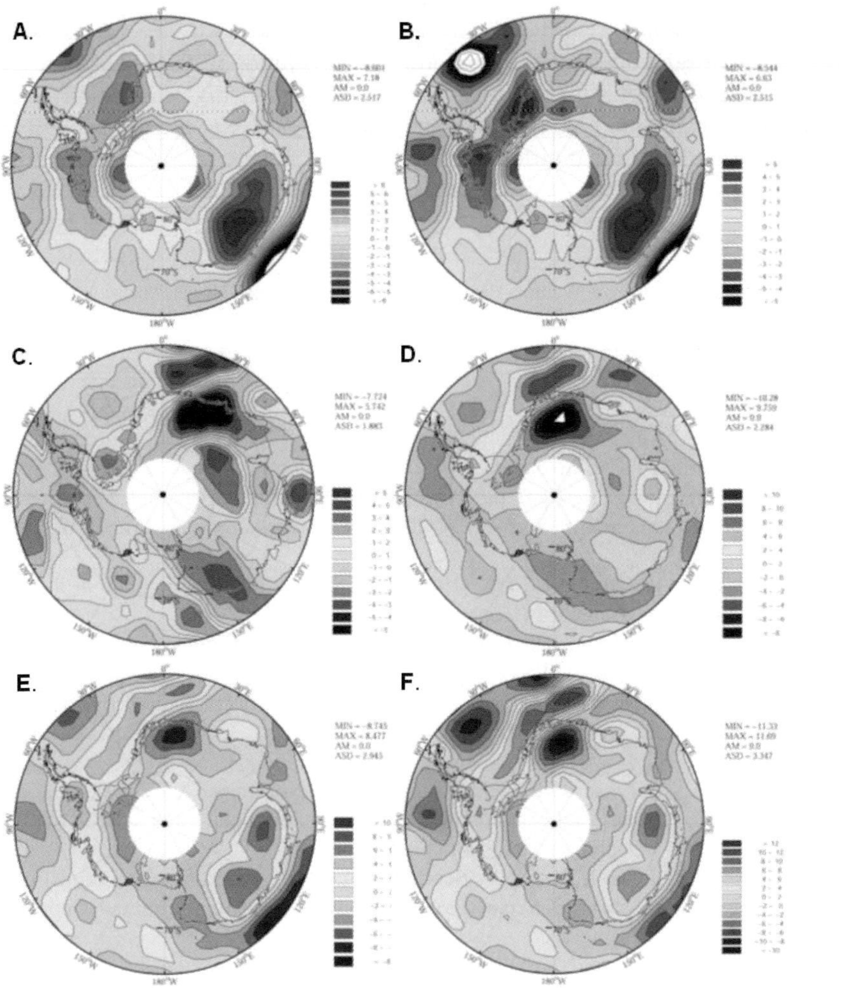


**Fig. 3.** CHAMP magnetic anomalies at 450 km altitude include A) total field and B) DRTP crustal thickness magnetic effects at contour interval CI = 1nT; C) total field and D) DRTP intracrustal magnetic anomalies at CI = 1 nT and 2 nT, respectively; and E) total field and F) DRTP comprehensive crustal magnetic anomalies at CI = 2 nT.

anomaly is a relatively robust feature because it is also seen in degree 15-90 components of the global spherical harmonic map of the CHAMP magnetic data produced by [17]. Further insight on the crustal significance of this feature can result from jointly modeling both the satellite and near-surface magnetic observations (e.g., [2], [11]).

We are currently extending this study to include the regional remanent magnetization effects from the Cretaceous Quiet Zones about Antarctica (e.g., [4]). We are also investigating the joint use of satellite and near-surface magnetic anomalies to constrain Curie isotherm estimates and related heat flow variations of the

Antarctic crust that may affect the dynamics of the overlying ice sheets. Our efforts to understand the magnetic properties and evolution of the lithosphere will be significantly enhanced by the new high-precision and high-resolution measurements of the magnetic field that the proposed SWARM satellite project will obtain from four satellites in two different polar orbits.

## References

1. von Frese RRB, Hinze WJ and Braile LW (1982) Regional North American gravity and magnetic anomaly correlations. Geophys J R Astron Soc *69*: 745-761.
2. von Frese RRB, Kim HR, Tan L, Kim JW, Taylor PT, Purucker ME, Alsdorf DE and Anderson AJ (1999) Satellite magnetic anomalies of the Antarctic crust. Annali di Geofisica *42*: 293-307.
3. Mayhew MA, Johnson BD and Wasilewski PJ (1985) A review of problems and progress in studies of satellite magnetic anomalies. J Geophys Res *90*: 2511-2542.
4. Purucker ME, von Frese RRB and Taylor PT (1999) Mapping and interpretation of satellite magnetic anomalies from POGO data over Antarctic region. Annali di Geofisica *42*: 215-228.
5. LeBrecque JL and Raymond CA (1982) Seafloor spreading anomalies in the Magsat field of the North Atlantic. J Geophys Res *9*: 250-253.
6. von Frese RRB and Kim HR (2003) Satellite magnetic anomalies for lithospheric exploration. Proceedings of OIST-4: 115-118.
7. Kim HR, von Frese RRB, Kim JW, Taylor PT, and Neubert T (2002) Ørsted verifies regional magnetic anomalies of the Antarctic lithosphere. Geophys Res Lett *29*: 8002-8005.
8. Alsdorf DE, von Frese RRB, Arkani-hamed J and Noltimier HC (1994) Separation of lithospheric, external, and core components of the south polar geomagnetic field at satellite altitudes. J Geophys Res *99*: 4655-4667.
9. von Frese RRB, Tan L, Kim JW, Bentley CR (1999) Antarctic crustal modeling from the spectral correlation of free-air gravity anomalies with the terrain. J Geophys Res *104*: 25275-25297.
10. Olsen N, Sabaka T and Tøffner-Clausen L (2000) Determination of the IGRF 2000. Earth, Planets and Space *52*: 1175-1182.
11. Kim HR (2002) Antarctic lithospheric anomalies from Orsted satellite and near-surface magnetic observations (unpubl.). PhD Thesis, The OSU, Columbus, Ohio.
12. Haines GV (1985) Spherical cap harmonic analysis. J Geophys Res *90*: 2583-2591.
13. Sabaka TJ, Olsen N, and Langel RA (2002) A comprehensive model of the quiet-time, near-Earth magnetic field: phase 3. Geophys J Int *151*: 32-68.
14. von Frese RRB, Jones MB, Kim JW and Kim J-H (1997) Analysis of anomaly correlations. Geophysics *62*: 342-351.
15. Fullerton LG, Frey HV, Roark JH and Thomas HH (1994) Contributions of Cretaceous Quiet Zone natural remanent magnetization to Magsat anomalies in the Southwest Indian Ocean. J Geophys Res *99*: 11923-11936.
16. Danziel IWD and Elliot DH (1982) West Antarctica: Problem child of Gondwanaland. Tectonics *1*: 3-19.
17. Maus S. Rother M, Lühr H, Balasis G, Olsen N & Kuvshinov A (2003) (preparing for a manuscript presently) http://www.gfz-potsdam.de/pb2/pb23/SatMag/litmod3.html

# Magnetic Petrology Database for Interpretation Satellite Magnetic Anomalies

Katherine A. Nazarova

Raytheon ITSS at Geodynamics Branch, Code 921, NASA/GSFC, Greenbelt, MD 20771

**Summary.** A Global Magnetic Petrology Database (MPDB) is now being compiled at NASA/Goddard Space Flight Center and consists of many thousands of records. The prototype database is located at
http://core2.gsfc.nasa.gov/research/terr_mag/php/MPDB/frames.html. The purpose of this database is to provide the geomagnetic community with a comprehensive and user-friendly method of accessing magnetic petrology data via Internet for a more realistic interpretation of satellite (as well as aeromagnetic and ground) magnetic anomalies. MPDB is focused on lower crustal and upper mantle rocks and includes data on mantle xenoliths, serpentinized ultramafic rocks, granulites, iron quartzites and rocks from Archean-Proterozoic metamorphic sequences from all around the world. The definition of crustal magnetic anomalies is improving due to the mini-constellation of three satellites - Oersted, Champ, and SAC-C. Recent lithospheric field models (CM4, MF1, MF2) reveal magnetic anomalies with better resolution, for example in the areas of Iceland, Polar Urals Mountains, and Anabar Shield where we have an excellent magnetic petrology records.

**Key words:** Database, Magnetic Petrology, Lithospheric

## 1 Introduction

One of the major outstanding problems in geophysics is the origin of magnetic anomalies observed by satellites in near Earth orbit. The wavelength of these anomalies suggests that they are crustal in origin. To construct sensible geophysical models of the likely sources of these anomalies comprehensive magnetic petrology information is required. Until now there has been no easily accessible central repository of such information. A Magnetic Petrology Database (MPDB) is now being compiled at NASA/Goddard Space Flight Center. The purpose of this database is to provide the geomagnetic community with a comprehensive and user-friendly method of accessing magnetic petrology data via Internet for more realistic interpretation of satellite magnetic anomalies. Magnetic Petrology data has been acquired from the NASA/Goddard Space Flight Center (Codes 691 and 921), United Institute of Physics of the Earth (Russia) and Institute of Geophysics (Ukraine) and from other worldwide sources. This data was accumulated over several decades and now consists of many thousands of records of data in our archives at GSFC. However most investigators do not have access to these data sets, which consisted, in part, of unpublished data. In addition, it is now recognized that magnetic petrology data is useless unless it is supported by the necessary metadata: exact sample names and locations, tectonics settings, reference data, etc. There is a real need for a database that contains all the magnetic petrology data and supporting metadata in a form in which it can be accessed by the scientific community.

To understand the nature of long wavelength satellite magnetic anomalies it is necessary to have information about the magnetization in the lower crust and upper mantle. Therefore MPDB is focused on lower crustal and upper mantle rocks and includes data on mantle xenoliths, serpentinized ultramafic rocks, granulites, iron quartzites and rocks from Archean-Proterozoic metamorphic sequences from all around the world. The locations of available magnetic petrology datasets do not always coincide with the major satellite magnetic anomalies (except in the areas of Kursk and Kiruna Magnetic Anomalies). But such data for lower crustal and upper mantle rocks around the world is necessary to determine if magnetic rocks can or cannot be the sources of satellite and high altitude aeromagnetic and stratospheric magnetic data and for developing realistic global magnetic models. Moreover the data on crustal magnetic anomalies is improving due to mini-constellation of three satellites - Oersted, Champ, and SAC-C. Recent lithospheric magnetic models –CM3e (Sabaka et al., 2002), MF1 (Maus et al., 2001) reveal magnetic anomalies with better resolution (about 1 nT) for example in the areas of Iceland, Polar Urals Mountains, Anabar Shield where we have an excellent magnetic petrology records. A substantial amount of data is coming from deep boreholes located in the area of the unique Kursk Magnetic Anomaly (Krivoy Rog Borehole) and not far from Kiruna Magnetic Anomaly (Kola SuperDeep Borehole which recovered 12 km of continental crust).

## 2 Database Description

The development of Magnetic Petrology Database was initiated at Geodynamics Branch of NASA/GSFC in the end of 2001. A comprehensive magnetic petrology database includes much more than just the results of magnetic and petrological measurements. Supplementary data, usually called metadata (data about data) include sampling description, tectonic setting, geographical position of the sampling site, data of satellite magnetic measurements and reference metadata that give bibliographical information for the reference which reports the measured values. All of the information listed above represents multidimensional, related data that cannot be handled reasonably or efficiently in a two-dimensional flat file format database consisting of a number of separate spreadsheets. Searches for specific data would be complicated and inefficient because data need to be retrieved from each individual file separately. The optimal method for organization and delivery of such complex but related data is a relational database. Development of a relational database structure requires that all data that will be stored in the database are broken down into small logical units corresponding to the tables of the database. Primary keys need to be defined for each table, and relations among the tables need to be established through foreign keys. The logical structure of a relational database is represented in the schema that illustrates tables, fields, and their relations. The schema of the Magnetic Petrology Database we are developing is shown in Figure 1.

The database consists of 6 interrelated tables. Logically, the central unit of the schema is the table SAMPLE_DESCRIPTION. Each sample for which magnetic petrology data are stored in the database is identified by a unique sample number, which serves as a foreign key throughout the database and links sample metadata in tables such as TECTONICS and REFERENCE to the measured data. All measured data are stored in two tables PETROPHYSICAL_DATA and PETROCHEMICAL_DATA. Metadata on satellite magnetic anomalies are planned to be included into database and stored in the tables Satellite_Magnetic_Anomaly.

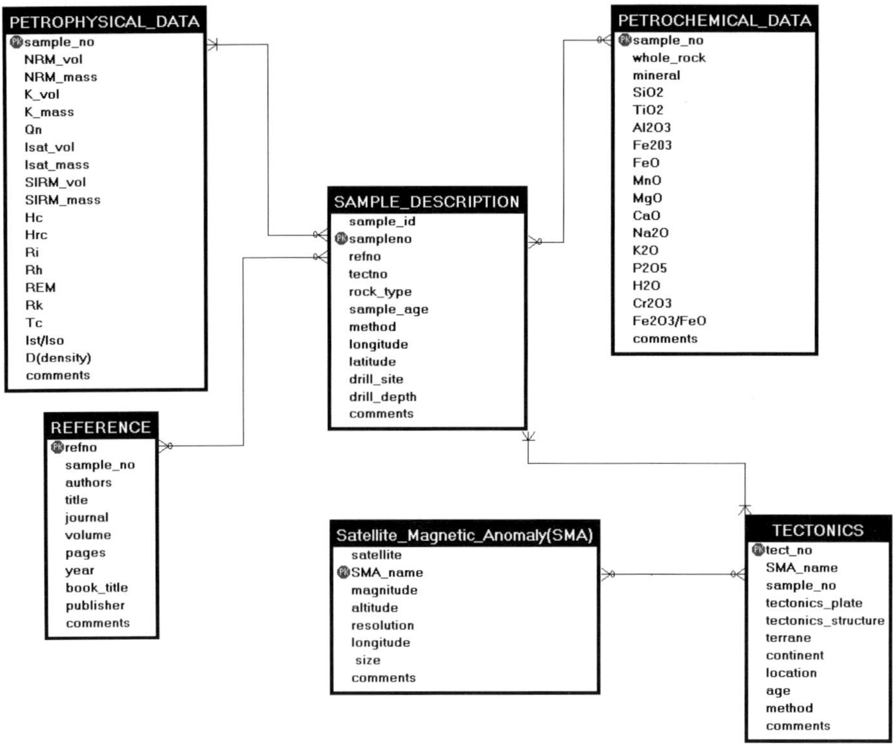


**Fig. 1.** The Schema of Magnetic Petrology Database.

## 3 Database management and WEB presentation

The schema is the conceptual structure of a database. In order to build a working database version, the physical structure of the database needs to be created with a Relational Database Management System (RDBMS). The database then needs to be populated with data, and finally interfaces have to be developed to allow users to work with the data. We have successfully implemented the database in MySQL database management system under LINUX operation system. MPDB main page includes clickable map for 89 magnetic petrology data locations.

Data entry forms and applications are essential to facilitate the complex process of loading data into the tables while ensuring the uniqueness of primary keys, data integrity, and correct referencing of primary keys by foreign keys.

Submission data entry forms have been developed and can be found at: *http://core2.gsfc.nasa.gov/research/terr_mag/php/MPDB/doc.html* Access to the database for searching, viewing, and downloading data is best provided over the Internet in order to make it independent of computer platforms and locations. For the current applications, Web interfaces have been developed using PHP and CGI scripts. The interfaces allow the user to select samples and data by means of graphical tools such as frames. menus, scrollable lists, or buttons that generate and submit dynamic SQL statements to the database (see *http://core2.gsfc.nasa.gov/research/terr_mag/php/MPDB/frames.html* query database section).

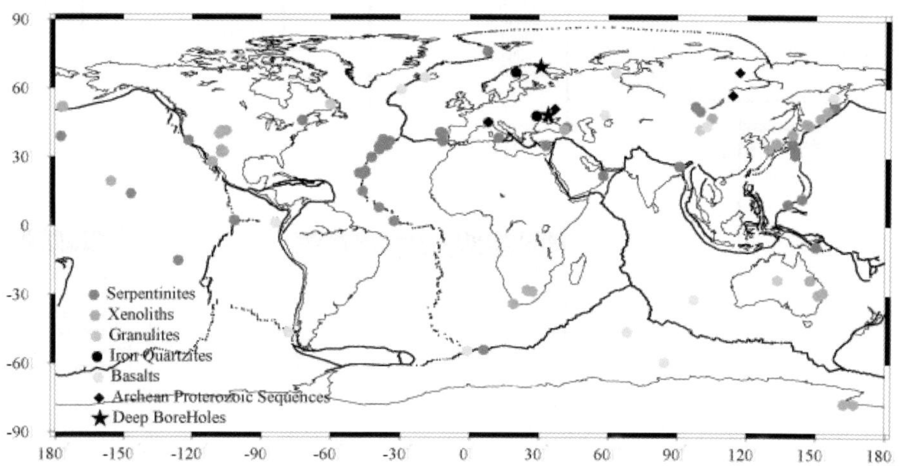


**Fig. 2.** Clickable Map for Magnetic Petrology Data Locations.

## 4 Statistics

Submitted and included into database data are shown in Table 1.

| ROCKS | TECTONIC STRUCTURE | # OF SAMPLES |
|---|---|---|
| Iron Quartzites (BIF) | Ukrainian Shield | 10000 |
| Sedimentary and BIF | Krivoy Rog BoreHole | 303 |
| Serpentinites, gabbro, sedimentary, volcanic | Baltic Shield Kola Deep BoreHole | 400 |
| Continental Serpentinites | worldwide | 210 |
| Oceanic Serpentinites | worldwide | 460 |
| Xenoliths | worldwide | 2000 |
| Basalts | Kamchatka | 800 |
| Ferrobasalts | Iceland | 2000 |
| Granulites | Urals Mountains | 900 |
| Archean Crustal Sections | worldwide | 480 |

**Table 1.** Data included into Magnetic Petrology DataBase.

## 5 Database applications

MPDB is a research oriented database and several database applications have been developed. (1) Magnetic data on Kola and Krivoy Rog BoreHoles were compared. Extremely high values of remanent magnetization (NRM), magnetic susceptibility (K) and Konigsberger ratio (Qn) are found at approximately the same depths of about 2000 m for both boreholes. Highly magnetic serpentinized peridotites and sedimentary rocks affected by sulfide mineralization were recovered at the Kola borehole for a depth interval of 1540-1940 m. The Krivoy Rog borehole recovered highly magnetic iron quartzites of Band Iron Formations at depths of 1853-2040 m. There is no obvious reason why high magnetizations should occur in two boreholes with different lithologies at the same depth. Magnetic surveys and surface sampling in the nearby Krivoy Rog and Kursk Magnetic Anomaly areas have revealed iron quartzites with high magnetization, similar to values given here. AF demagnetization tests suggest that hard and stable NRM component which is caused by hematite occurred in iron quartzites in different forms and grain sizes ranges. (2) Recent magnetic model (Maus et al., 2002) revealed quite intense positive lithospheric magnetic anomaly (> 5 nT) over Iceland. Magnetic data for about 2000 Icelandic rocks were analyzed and we concluded that ferrobasalts (NRM aver =8.8 A/M and K aver= 0.054 SI) are most likely the source of satellite magnetic anomaly over Iceland. (3) Petrological data for basalts from Kamchatka (content of Fe and Ti oxides) were used to assign the level of basalts magnetization. These values were utilized for interpretation the results of balloon geomagnetic survey in this area.

## 6 Status of development

- Magnetic Petrology Database (MPDB) contains data about rocks from all around the world (about 17,000 samples)
- MPDB is successfully designed, managed and presented on WEB.
- MPDB is a research oriented database and several database applications were developed.
- MPDB is expected to be the part of the system of existing geomagnetic and paleomagnetic databases under National Geophysical Data Center umbrella and will be used for more reasonable interpretation of high altitude magnetic survey products including satellite, stratospheric and aeromagnetic surveys and constructing more realistic magnetic models.

*Acknowledgements.* This project was supported by John LaBrecque, manager of SENH program, NASAHQ.

## References

Maus S, Rother M, Holme R, Lühr H, Olsen N and Haak V (2002) First Scalar Magnetic Anomaly Map from Champ Satellite Data Indicates Weak Lithospheric Field. Geophys Res Lett *29(14)*: 10.1029/2001GL013685.

Sabaka TJ, Olsen N & Langel RA (2002) A Comprehensive Model of the Quiet-Time near-Earth Magnetic Field: Phase 3. Geophys J Int *151*: 32-68.

# Balloon Geomagnetic Survey at Stratospheric Altitudes

Katherine A. Nazarova[1], Yuri Tsvetkov[2], James Heirtzler[3], Terence Sabaka[1]

[1] Raytheon ITSS at Geodynamics Branch NASA/GSFC, Greenbelt, MD 20771, USA
[2] Izmiran, 142190 Troitsk Moscow Region, RUSSIA
[3] NASA/GSFC, Greeenbelt, MD 20771, USA

**Summary.** A stratospheric balloon flight at 30 km altitude measured the geomagnetic field intensity along a 6000 km track extending from Kamchatka to near the Ural Mountains. When the CM model was used to remove the main and external fields from the observed data, magnetic anomalies of several 100 nT amplitude and 250 to 750 km wavelength are observed. In the eastern part of the track these anomalies appear to be due to the bodies of up to 5 km depth and magnetizations of 0.12 SI (0.01 cgs).

**Key words:** Geomagnetism, balloon, lithosphere, stratosphere

## 1 Introduction

Magnetic measurements by stratospheric balloons provide wavelengths intermediate between those registered by aeromagnetic and satellite magnetic surveys. Aeromagnetic profiles provide information about anomalies whose shortest wavelength is comparable to the distance from the source, namely from few hundred meters to a few kilometers. Although aeromagnetic profiles also provide information about long wavelengths these long wavelengths are usually discarded in making aeromagnetic maps. Large geologic structures, with dimensions of a few hundred kilometers, cannot always be inferred from their shorter wavelength surface expression.

On the other hand present day lithospheric models of the geomagnetic field derived from satellite magnetic data at the altitude about 400 km do not have resolution to show magnetic features with wavelength shorter than about 1000 kilometers. These long wavelength anomalies are considered to be caused by the sources located in the deep crust and upper mantle both in continental and oceanic areas.

Stratospheric balloons which fly at the altitude of about 30 km: (a) register magnetic signal from the whole thickness of the earth's crust, (b) fill the gap between aeromagnetic and satellite magnetic data, (c) using vertical gradient measurements allow reliable separation of the external and internal components of the Earth's magnetic field, (d) provide long term coverage of hard to access areas, (e) allow identification of large and significant tectonic structures.

There have been relatively few geomagnetic field measurements at stratospheric altitudes. However, in recent years, there have been several stratospheric balloon flights by France (Cohen et al., 1986; Achache et al., 1991) and Japan

(Tohyama et al., 1992). One of the longest and most successful of these, made by Russian scientists is reported here.

## 2 Balloon magnetic survey

In July 1996 the total intensity of the geomagnetic field was measured at an altitude of approximately 30 km by a balloon flight from Kamchatka (56.29N, 159.75E) on the east to the Caspian Sea covering the Sea of Okhotsk, the Central Siberian Platform, the West Siberian Plains and Urals Mountains (54.00N, 50.5E). This 6000 km traverse was made at latitude of about $55^0$N in 6 days. The measurements were made with a scalar proton precession magnetometer suspended 1 km below the gondola with accuracy about 0.2 nT. All readings were recorded each 8 minutes, providing a complete record of 990 readings. The altitude fluctuations were similar day by day for the whole flight. At 5 am local time the altitude rose from 27 km to a maximum average altitude about 33 km. It leveled off at a maximum altitude a few hours later. In the late afternoon it fell to the lower altitude again. Therefore for present purposes we will consider the altitude to be a constant 30 km. No attempt was made to adjust the magnetic readings for time variations of the geomagnetic field strength. A typical amplitude for the diurnal variation in the total field in 1996 in the area surveyed by the balloon was about 50 nT as observed at Novosibirsk (Klyuchi) geomagnetic observatory located at $55^o$N, $82.9^o$E. Over a limited portion of the track a second proton precession magnetometer was suspended 1 km below the first magnetometer to provide measurements of the vertical geomagnetic field gradient. Results of gradient study are not discussed here.

## 3 Aeromagnetic survey

In 1979 the Ministry of Geology of the USSR issued aeromagnetic anomaly maps of geomagnetic field intensity (Z.A. Makarova, editor). These 18 maps were digitized by the US Naval Oceanographic Office and edited by Conoco. The digital data is deposited at the National Geophysical Data Center (NGDC). We analyzed digital aeromagnetic anomaly maps and came to conclusion that long wavelength anomalies were lost during compilation when local magnetic maps were consolidated into larger regional maps. The balloon track covers the easternmost of these aeromagnetic maps and over the central Sea of Okhotsk aeromagnetic data is missing (Fig. 1).

**Fig. 1.** Orthographic map showing balloon track (black line) and aeromagnetic map coverage.

## 4 Isolation of the Anomaly Field

Because of its altitude this survey is different from other geomagnetic field measurements and provides unique data. It does not show the short wavelength anomalies of near surface surveys but it does show wavelengths too short to be included in global models of the geomagnetic field. The track is the first over land which shows these intermediate wavelength anomalies which are common in oceanic areas. The track of the balloon passed near one of the highs on global maps of geomagnetic field intensity, but south of the region dominated by time varying field aligned currents. Geomagnetic field models show long wavelength anomalies which are thought due to the Earth's main with sources deep within the earth. The models also show shorter wavelength anomalies ($n>14$ in spherical harmonic expansion series) which are thought to be due to crustal sources. Figure 2 shows regional crustal anomalies of the CM3e model at the altitude of 30 km. (Sabaka et al., 2002). CM3e is a magnetic reference model able to represent not only main and crustal magnetic fields but also the ionospheric and magnetospheric (primary and induced) fields. A distinct magnetic low near the eastern end of the track over the Sea of Okhotsk was observed.

Ravat et al., (2003) clearly showed the advantage of CM magnetic model (Sabaka et al., 2002) relative to IGRF in the processing of aeromagnetic data in the central US (Kansas area). In the area of balloon track there are few ground magnetic observatories and, since ground magnetic observatory data is integrated into the models, there have been some difficulties with models here in the past. Total

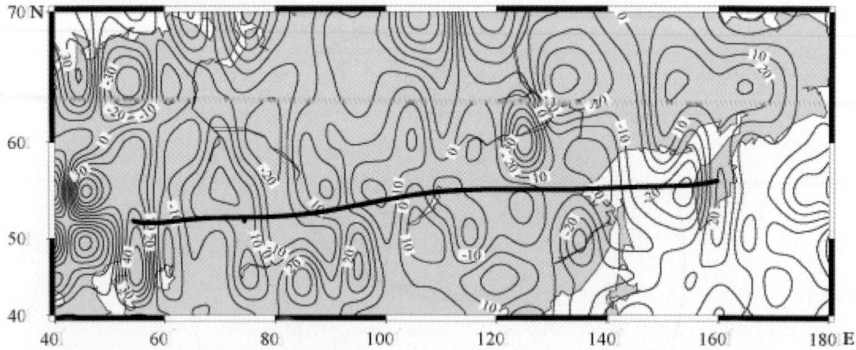


**Fig. 2.** Contours of crustal anomalies of CM3e (Sabaka et al., 2002) at the altitude of 30 km. Black line is the track of balloon.

field maps made from the IGRF show a maximum for this region in the northern hemisphere. Only the CM model compensate for magnetic disturbances, which are part of quiet daily variations. If one subtracts the main field components of CM3e (n<15) from the observed data one obtains the profile shown in Figure 3. For illustrative purposes the data is also shown with a best fit straight line removed. Anomalies of length 5 to 15 degrees of longitude (250 to 750 km) stand out.

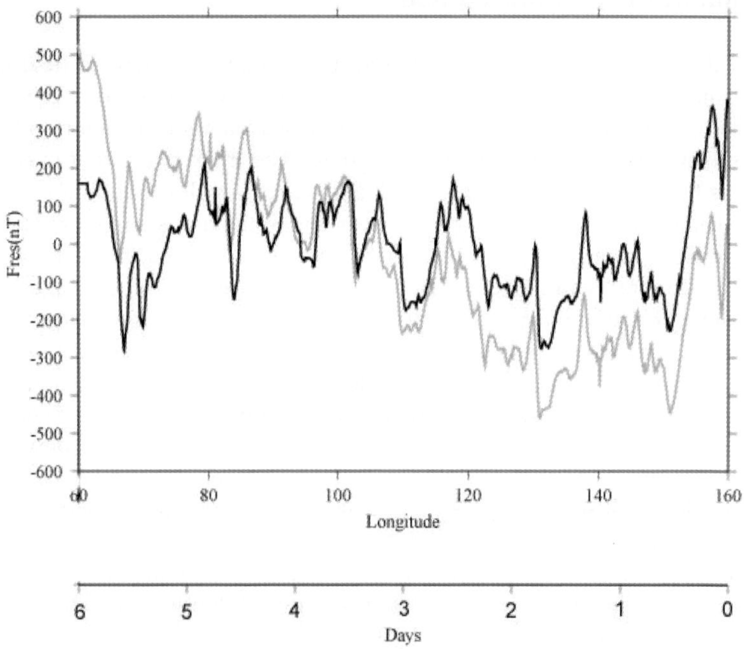


**Fig. 3.** Magnetic profile with CM3e (n<15) removed from observed data (grey line) and with CM3e (n<15) and straight line fit removed from observed data (black line) versus longitude and time.

## 5 Magnetic modeling of balloon lithospheric anomalies

To understand what crustal bodies might cause the anomalies observed by the balloon, we used a forward modeling technique. Since we have only a one-dimensional survey, rather than an area survey we used a 2D modeling technique (Heirtzler et al., 1964). This assumes that the anomalies are linear for a length equal to about the height of the survey (30 km). The Russian aeromagnetic survey in other parts of Siberia suggests that this may be approximately true. For this model study we chose the large amplitude anomaly on the eastern end of the track covering western Kamchatka and the Sea of Okhotsk. Figure 4 shows the observed anomaly (measured anomaly minus main field from CM3e) and the calculated anomaly for the body shown. This body has a deep root of about 5 km and a strong magnetization of 0.12 SI (0.01 cgs). The topographic data show the deep root is under the western side of Kamchatka and the root to be less deep under the eastern side of the Sea of Okhotsk. In this calculation we assumed the body to strike north-south, the magnetization to be induced and the inducing field direction to be like that of the present geomagnetic field.

Preliminary checks on 2D models for other anomalies along the profile show that there are other deep rooted bodies of high magnetization. A correlation with major tectonic or structural features in Siberia is planned.

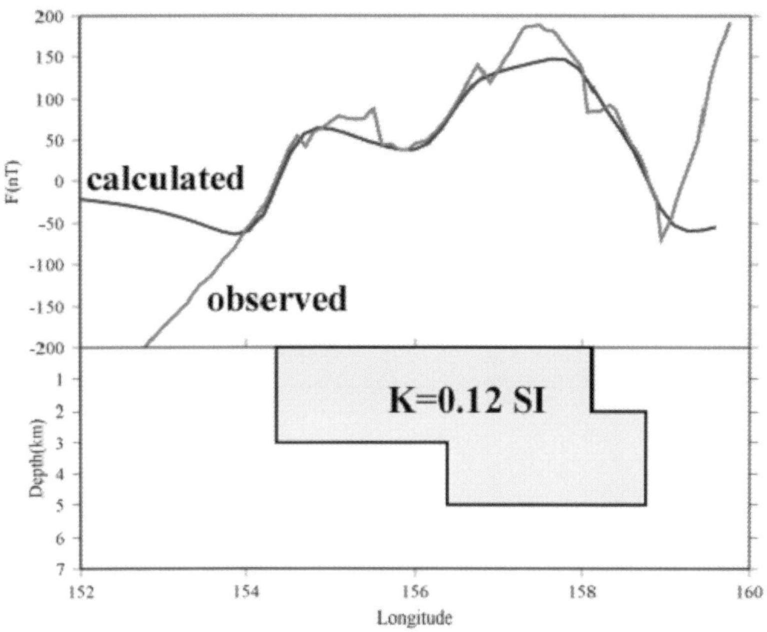


**Fig. 4.** Observed and calculated magnetic anomalies for bodies whose cross sections are shown. The bodies are assumed to be two dimensional, striking north-south with a susceptibility 0.12 SI (~0.01 CGS).

# 6 Conclusions

Magnetic measurements on a stratospheric balloon flight shows that there are many magnetic bodies with dimensions of 250 to 750 km in Siberia. Bodies of this size cannot be identified with present satellite geomagnetic field models and have not been identified from surface surveys.

*Acknowledgements.* This project was supported by NASA Headquarters code Y, and Russian Academy of Sciences (IZMIRAN)

# References

Achache J, Cohen Y and Unal G (1991) The French Program of Circumterrestrial Magnetic Surveys Using Stratospheric Balloons. EOS Transactions of the American Geophysical Union *72*: 97-101.

Cohen YM, Menvielle M and LeMouel JL (1986) Magnetic Measurements aboard a Stratospheric Balloon. Phys Earth Planet Int *44*: 348-357.

Heirtzler JR, Peter G, Talwani M, and Zurflueh EG (1964) Magnetic Anomalies Caused by Two-Dimensional Structures: Their Computation by Digital Computers and Their Interpretation. Technical Report No. 6, CU-6-62, Nonr-Geology, Columbia University.

Ravat D, Hildenbrand TG, Roest W (2003) New Way of Processing Near-surface Magnetic Data: The Utility of the Comprehensive Model of the Magnetic Field. The Leading Edge *22(8)*: 784-785.

Sabaka TJ, Olsen N & Langel RA (2002) A Comprehensive Model of the Quiet-Time Near-Earth Magnetic Field: Phase 3. Geophys J Int *151*: 32-68.

Tohyama FY, T Takahashi (1992) Observation of the Geomagnetic Field by Polar Patrol Balloon. J Solar Terr Environm Res Japan *16*: 60.

# Effect of Varying Crustal Thickness on CHAMP Geopotential Data

Patrick T. Taylor[1], Károly I. Kis[2], Ralph R. B. von Frese[3], Juha V. Korhonen[4], Géza Wittmann[5], Hyung Rae Kim[6], Laramie V. Potts[7]

[1] Geodynamics Branch NASA/GSFC, Greenbelt, MD 20771 USA
[2] Geophysical and Environmental Physics Research Group of the Hungarian Academy of Sciences, Eötvös Loránd University, 1117 Budapest, Hungary
[3] Dept. of Geological Science, The Ohio State University, Columbus, OH 43210 USA
[4] Geological Survey of Finland, 02151Espoo, Finland
[5] MOL Hungarian Oil and Gas Co., Budapest H-1311, Hungary
[6] UMBC/GEST and Geodynamics Branch NASA/GSFC, Greenbelt, MD 20771 USA
[7] Laboratory for Space Geodesy and Remote Sensing Research, The Ohio State University, Columbus, OH 43210 USA

**Summary.** To determine the effect of crustal thickness variation on satellite-altitude geopotential anomalies we compared two regions of Europe with vastly different values, South and Central Finland and the Pannonian Basin. Crustal thickness exceeds 44 km in Finland and is less than 26 km in the Pannonian Basin. Heat-flow data indicate that the crust of the Pannonian Basin has a value nearly three times that of the Finnish Svecofennian Province. A positive CHAMP gravity anomaly (~4 mGal) is quasi-coincidental with the CHAMP magnetic anomaly across the Pannonian Basin. CHAMP gravity data indicates a minimum of 3 mGal in southwest Finland. CHAMP magnetic data reveal elongated semi-circular negative anomalies for both regions with South-Central Finland having larger amplitude (<-6 nT) than that over the Pannonian Basin, Hungary (<-5 nT). In this latter region subducted oceanic lithosphere has been proposed as the anomalous body. In the former the central part of the negative gravity anomaly covers the northern part of the Baltic Sea basin and Gulf of Finland and underlying two rapakivi provinces plus it coincides with an area of lower crustal thickness. The magnetic anomaly directly correlates with the crustal thickness and inversely with the heat flow and, hence, may be caused either by variation of concentration of magnetite or by the elevated Curie-isotherm of magnetite in the lower crust - upper mantle region.

**Key words:** Comparing satellite magnetic anomalies, Pannonian Basin and South and Central Finland, crustal magnetization, crustal thickness.

## 1 Introduction

Crustal magnetic anomalies are proposed to be the result of both horizontal and vertical contrasts in magnetization. To evaluate this concept we have chosen two continental regions of Europe, South-Central Finland (Fig 1a) and the Pannonian Basin (Fig 1b), with very different crustal thickness. For the former the crust is

42-46 km thick and extends down to 62 km in its deepest parts (Bock et al., 2001; Fig 1c) while the latter has a thickness of <25 km (Fig 1d). Crustal gravity (Fig. 1e and 1f) and magnetics (Fig 2a and 2b) anomalies from the CHAMP mission, computed at 400 km, were used together with regional geologic/tectonic and heat flow data (Fig 2e and f). Geologically South and Central Finland belong to the exposed Precambrian area of the NW Europe (Fennoscandian Shield). Together with the Ukrainian Shield and East European platform this basement establishes the Fennosarmatian Shield, bordered to the southwest by Törnquist-Teisseyre fault zone, on the east by Ural Mountains and by North Atlantic and Barents Sea on the west and north. The eldest dated rocks in the Fennoscandian Shield are 3500 Ma old (Mutanen and Huhma 2003). The core of the Shield was established by plate collisions of both Archaean and juvenile material during the Svekofennian orogeny 1900-1800 Ma ago. New material was subsequently added to the western part of the Shield. Vertical neotectonic processes are active in the Fennoscandian Shield due to postglacial rebound. The Pannonian basin is bordered on the north, east, and southeast by the Carpathian Mountains, west and southwest by the Eastern and Dinaric Alps. Pre–Neogene Pannonian basement consists of sedimentary, igneous and metamorphic complexes and is represented by two megatectonic units, Pelsonia Composite and Tisia Terranes (Kovács et al. 2000, Fig 1b). Present geological structures developed mainly in the Neogene and are represented by the Carpathian fold and trust belt formed during the Cretaceous–Miocene collision of the African and Eurasian plates (Horváth 1993; Horváth and Cloetingh 1966). These geodynamic processes have produced a thin crust and lithosphere. Crustal (Fig 1d) and lithospheric thickness maps of the central part Pannonian basin (Horváth 1993; Lenkey 1999) show a below average 22–26 km thick crust. The lithosphere thins below the Pannonian basin with present crustal and lithospheric structures the probable result of Neogene evolution.

## 2 CHAMP Geopotential Data Processing and Presentation

CHAMP gravity field anomaly data were obtained from the global gravity model, degree and order 75. Negative CHAMP gravity anomalies (<-2 mGals) in South and Central Finland are semi-circular, with a radius of some 700 km and centered near 60° N latitude and 23° E longitude in the Svecofennian geological province (Fig 1e). However, the gravity field in this region may be dominated by the effect of the last de-glaciation (Olesen and Skilbrei 2003). CHAMP positive gravity anomalies cross the Pannonian Basin are in the shape of a northwest-southeast trending trough (~900 km long and 660 km wide) and open to the southeast with the maximum value (4-5 mGals) at the southeastern end (Fig 1f).

CHAMP magnetic data were processed using the method of Alsdorf et al. (1998). A total of 96 orbits were selected for the region of South and Central Finland while 90 were used over the Pannonian Basin. Magnetic anomalies in South and Central Finland are dominated by an enclosed oblong shaped low of

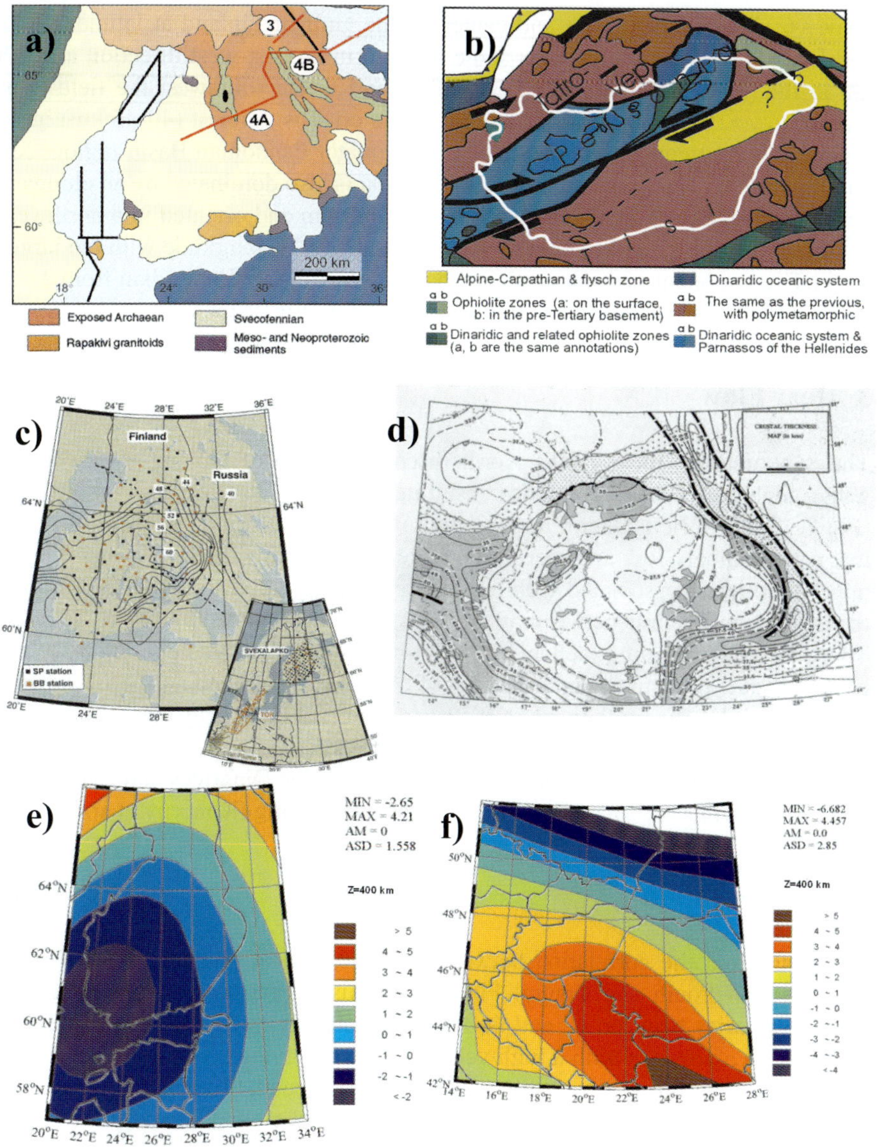


**Fig. 1.** a) Tectonic Map of Eastern Scandinavia (Hjelt and Daly 1996)
b) Tectonic/terrane sketch of the Pannonian Basin and its surroundings (Kovács et al. 2000)
c) Crustal thickness map of South and Central Finland (Bock et al. 2001, Reproduced by permission of American Geophysical Union)
d) Crustal thickness map of the Pannonian Basin region (Reprinted from Horváth, F (1993) page 338 with permission from Elsevier)
e) CHAMP gravity anomaly map of South and Central Finland.
f) CHAMP gravity anomaly map of the region of the Pannonian Basin

some –6 nT coincident with and center on the Gulf of Finland at latitude 60° N and longitude 25° E and extends some 660 km in an east-west direction and 200 km in the North-South (Fig 2a). At 63° N and to the southwest the field is in creases to >+10 nT north of 68° latitude. The crust is thinnest (40-46 km) in the region of these geopotential anomalies. The greater Pannonian Basin region, centered mainly on Hungary and northern Romania is also dominated by an elongated negative magnetic anomaly some 670 km by 400 km and oriented in a northwest-southeast direction and centered at 46° N latitude, 20° E longitude with the largest (<-5 nT) component running north south along the 20°-21° E meridian from 45° to 47° N latitude where the crust is thinnest (<22.5 km, Fig 2b).

## 3 Heat Flow

The Svekofennian rocks of the Northern coast of Gulf of Finland are associated with a high crustal heat production resulting in high crustal heat flow >60mW/m$^2$ (Fig 2e). The Curie isotherm may be high in the Gulf of Finland where there is a magnetic anomaly low and there may be relative negative density contrast in the upper lithosphere associated with the negative gravity anomaly. High heat flow in the Pannonian Basin is between 50 and 130 mW/m$^2$, average 90 mW/m$^2$ (Fig 2f). There is, however, an excellent correlation between the magnetic anomaly and heat flow data particularly in the region of central Hungary and northern Romania. These anomalies lie on the axis of the basins thinnest crust (Fig 1d). The correlation between high-heat flow, thin crust and CHAMP magnetic anomalies no doubt reflects the dynamics of the plate collision between the Dinarides and Carpathians Mountains in this region and/or mantle diapiric upwelling (Konecny et at. 2002).

## 4 Interpretation

We considered the correlation between geology, crustal thickness and CHAMP anomalies in South and Central Finland. In this region the background crustal thickness varies between 42 and 46 km. Anomalously thick crust (48 to 62 km), however, is associated with a high velocity zone overlying the upper mantle. The upper surface of this high velocity zone is sub-horizontal, dividing the lower crust in two parts. The high velocity zone is absent in background crustal thickness areas (Korja et al. 1993; Heikkinen et al. 2003). This thickest crust is located 500 km to the NE of the negative CHAMP gravity anomaly and 400 km to north of the negative magnetic anomaly associated with the Shield. This thick crust is everywhere overlain by partly allocthonous Svekofennian crust. In deeper levels it corresponds to both Proterozoic and Archaean crustal blocks, however, and the thickest part is located under the Achaean craton. Both negative gravity and magnetic anomalies are associated with relatively thinner crust (40-46 km). The gravity and magnetic anomalies are centered with the younger (1550 Ma) and older (1650 Ma) Rapaķivi granite provinces respectively. This gravity anomaly is on the deeper

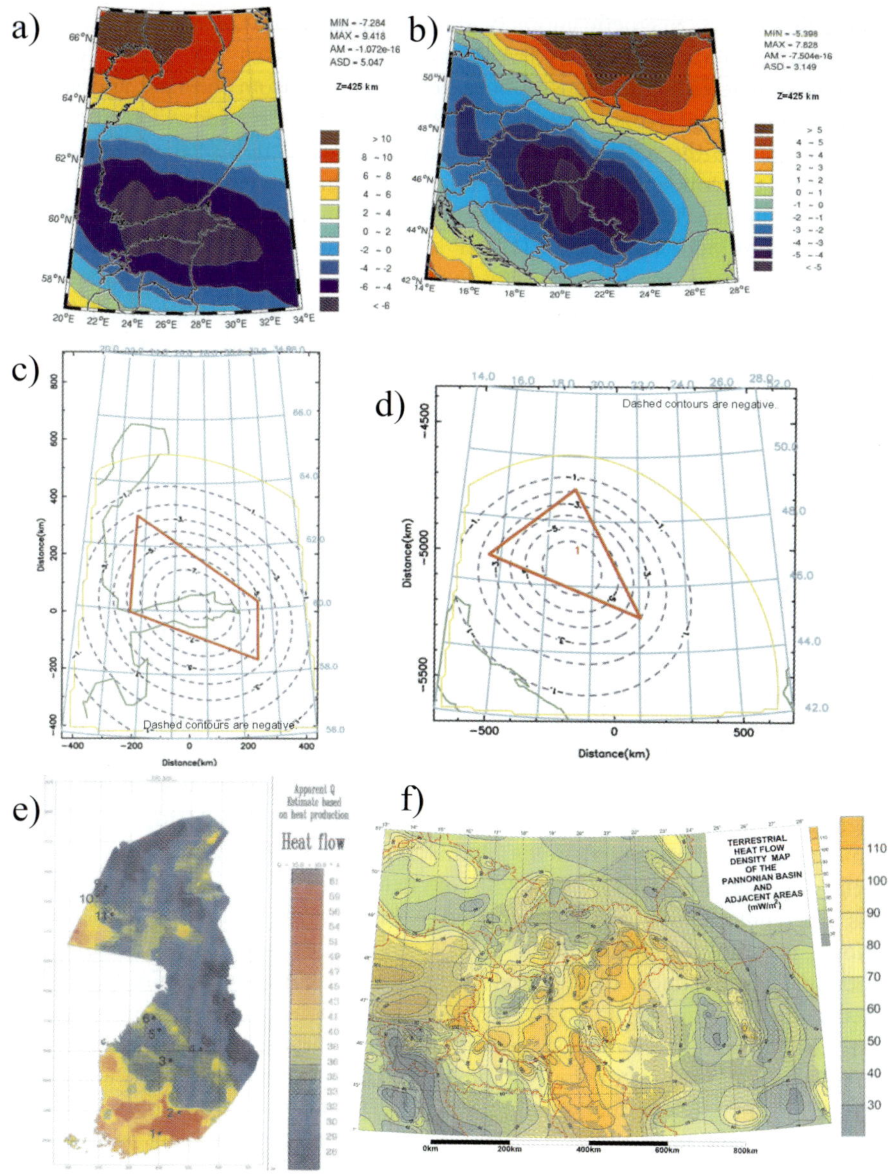


**Fig. 2.** a) CHAMP magnetic anomaly map of South and Central Finland
b) CHAMP magnetic anomaly map of the region of the Pannonian Basin
c) Model of the anomaly in figure 2 a.), details given in text.
d) Model of the anomaly in figure 2 b.), details given in text.
e) Apparent heat-flow map of Finland based on rock heat production (Reprinted from Kukkonen (1993) page 7, with permission from Elsevier)
f) Heat flow map of the Pannonian Basin (Dövényi 2003 and Dövényi et al. 2003).

Baltic Sea and the magnetic anomaly on the shallower Gulf of Finland.

In the Pannonian basin the CHAMP gravity high strikes NW–SE and probably reflects subducted higher density oceanic lithosphere from the East and asthenospheric upwelling (Konecny et al. 2002) below the Dinarides. The low in the northern part of the region most likely indicates subduction of the Eurasian plate caused by the collision with the Adriatic plate. Due to crustal thinning and high heat flow in the Pannonian basin the crust is thin relative to the adjacent areas. However, our modeling of the CHAMP magnetic anomalies show that contrasting crustal thickness alone is insufficient to reproduce this low magnetic anomaly and that the anomalous body is required to have a remanent and reverse magnetization.

Models of these CHAMP satellite altitude magnetic data, from both regions were computed using vertically sided, uniformly magnetized, irregular shaped prisms at an altitude of 400 km (Fig 2c and 2d). The surrounding area is considered to have zero magnetization. The method of Plouff (1976) was used. Body parameters for the South and Central Finland-Northern Europe block: Depth 0 to 50 km; Magnetization, Dip 90 deg, Dec 0 deg, -1.0 A/m (reversed, Fig 2c). Our observed anomalies may be explained by variations in Moho or Curie-isotherm relief. Considering a magnetization contrast of 3 A/m at the relief, we obtain a height of ca. 17 km for it. This is comparable to the known Moho relief between the anomaly area and surroundings. Pannonian Basin: Depth 0 to 25 km; Magnetization, Dip 90 deg, Dec 50 deg., -1.5 A/m (reversed, Fig 2d). The assumptions used in our modeling method are, obviously, a greatly over simplified representation of the magnetization of the crust. Unsuccessful attempts were made to model these data with positively magnetized bodies surrounded by regions of zero magnetization. Reversed magnetization, however; was the easiest way of reproducing significant negative anomalies at relatively high latitudes, due to the non-uniqueness of geopotential fields alternative models not requiring reverse magnetization may be possible. You will note that, at least to a first order, these models are representative of the anomalously thin and thick crusts of our study regions. In both regions the heat flow is anomalous high for the crustal ages, therefore the thickness of our models may be too large and require higher magnetizations. There is a strong inverse correlation between largest negative CHAMP magnetic anomalies (<-6 nT) with the largest heat flow values (>110 $mW/m^2$) in the region of the Pannonian Basin-Makó trough (46° N and 20° E). These results suggest that satellite magnetic anomaly data not only reflect the thickness of the magnetized crust but the geothermal environment as well, especially where this crust is anomalously thick or thin and the two parameters are related.

## 5 Conclusions

While the object of this study was to determine if regions having significantly different crustal thickness would display different magnetic anomaly values at satellite altitude, our results were unable to completely establish this result. We did, however, find that both regions displayed significantly higher than normal (world average) heat flow values. The CHAMP gravity anomalies for South and Central Finland and northern Baltic Sea were associated with a relatively thinner part of the crust but the positive gravity anomaly over the Pannonian Basin was correlated with the negative magnetic anomaly. Simple model studies suggest that both negative magnetic anomalies result from negative magnetization contrast with respect to the background and may be the result of reversely magnetized lower crust. In the Pannonian Basin Tertiary subduction of the Eurasian plate produced the crustal thinning with greatest amount in the central part of this basin. The anomalously high-density body resulted from east-west compression with subsequent lithospheric mantle upwelling.

*Acknowledgements.* The Hungarian part of this research was supported by the OTKA (Hungarian Scientific Research Fund) Project No. T 038008. The award of the project is highly acknowledged.

## References

Alsdorf D, Taylor P, von Frese R., Langel R., and Frawley J. (1998) Arctic and Asia lithospheric satellite magnetic anomalies. Physics of the Earth and Planetary Interiors *108*: 81-99.

Bock B. et al. (2001) Seismic Probing of Fennoscandian Lithosphere. Transactions of the AGU EOS *82*: 621-628-9.

Dövényi P (2003) Private communication.

Dövényi P, Horváth F, Drahos D (2002) Heat–flow density of Hungary. In: Hurter S, .Haenel R (eds) Atlas of Geothermal Resources in Europe. Leibniz Institute for Applied Geosciences Hanover, Germany

Heikkinen P, Kukkonen I T, Ekdahl E, Korja A. Lahtinen R, Hjelt S-E, Yliniemi J, Berzin R and FIRE Working Group (2003) Alakuoren rakenne FIRE-1 linjalla heijastavuuden ja nopeusjakautuman perusteella. In (Hyvönen, E and Sandgren, E. eds) Abstracts of Sovelletun Geofysiikan XIV Neuvottelupäivät, Rovaniemi 4-5-11 Vuorimiesyhdistys, Rovaniemi

Hjelt S-E and Daly S (1996) SVEKALAPKO. Europrobe Newsletter 9

Horváth F (1993) Towards a mechanical model for the evolution of the Pannonian basin. Tectonophysics *226*: 333–357

Horváth F, Cloetingh SAPI (1996) Stress-induced late stage subsidence anomalies in the Pannonian basin. Tectonophysics *266*: 287–300

Konecny V, Kovac M, Lexa J, Sefara J (2002) Neogene evolution of the Carpatho-Pannonian region: and interplay of subjuction and back-arc diapiric uprise in the mantle. European Geosciences Union Stephan Mueller Special Publication Series *1*: 105-123.

Korja A, Korja T, Luosto U, Heikkinen P (1993) Seismic and geoelectirc evidence for collisional and extensional events in the Fennoscandian Shield-implications fro Precambrian crustal evolution. Tectonophyscis *219*: 129-152

Kovács S, Haas J, Császar G, Szederkényi T, Buda Gy, Nagymarosy A (2000) Tectonostratigraphic terranes in the pre–Neogene basement of the Hungarian part of the Pannonian area. Acta Geologica Hungarica 43: 225–328

Kukkonen I T (1993) Heat flow map of northern and central parts of the Fennoscandian Shield based on geochemical surveys of heat producing elements. In: V.Cermák (ed) Heat flow and the strucuture of the lithosphere, Tectonophysics 225; 3-13

Lenkey L (1999) Geotermics of the Pannonian basin and its bearing on the tectonics of basin evolution. Ph.D. thesis, Vrije Universiteit

Mutanen T and Huhma (2003) The 3.5 Ga Siurua trondhjemite gneiss in the Archaean Pudasjärvi granulite belt in Northern Finland. Bulletin of the Geological Society of Finland 75: 51-68

Olesen O and Skilbrei JR (2003) Isostatic Compensation of the Scandinavian Mountains. EGS Geophysical Research Abstracts 5: 12711

Plouff D (1976) Gravity and magnetic fields of polygonal prisms and application to magnetic terrain corrections, Geophysics 41: 727-741.

# Reliability of CHAMP Anomaly Continuations

Ralph R.B. von Frese[1], Hyung Rae Kim[2], Patrick T. Taylor[3], and Mohammad F. Asgharzadeh[1]

[1] Dept. of Geological Sciences, The Ohio State University, Columbus, OH, USA
*vonfrese@osu.edu*
[2] UMBC/GEST and Geodynamics Branch, NASA/GSFC, Greenbelt, MD, USA
[3] Geodynamics Branch, NASA/GSFC, Greenbelt, MD, USA

**Summary:** CHAMP is recording state-of-the-art magnetic and gravity field observations at altitudes ranging over roughly 300-550 km. However, the non-uniqueness of the process and satellite anomaly errors severely limit anomaly continuation. Indeed, our numerical anomaly simulations from satellite to airborne altitudes show that effective downward continuations of the CHAMP data are restricted to within approximately 50 km of the observation altitudes while upward continuations can be effective over a somewhat larger altitude range. The great unreliability of downward continuation requires that the satellite geopotential observations must be analyzed at satellite altitudes if the anomaly details are to be exploited most fully. Given current anomaly error levels, multi-field inversion of satellite and near-surface anomalies is the best approach for implementing satellite geopotential observations for subsurface studies. We demonstrate the power of this approach using a crustal model obtained by the inversion of combined near-surface and satellite magnetic anomalies for Maud Rise, Antarctica, in the southwestern Indian Ocean. Our modeling, which also includes regional gravity constraints, suggests that crustal thickness variations and remanent magnetization of the normal polarity Cretaceous Quiet Zone produce the dominant satellite altitude magnetic anomalies.

**Key words:** anomaly continuation, Maude Rise, Antarctica, ADMAP, CHAMP, Ørsted.

## 1 Introduction

Satellite magnetic and gravity observations are commonly represented by geopotential models that are interrogated for anomaly estimates at or near the Earth's surface. For example, downward continued satellite magnetic anomalies have been compared with aeromagnetic anomalies (e.g., [1]), recommended for leveling aeromagnetic datasets with disparate survey parameters into regional anomaly compilations (e.g., [2]), and used to infer the magnetic field at the core-mantle boundary (e.g., [3]). Geological applications of satellite gravity observations also typically evaluate the anomalies at or near the Earth's surface (e.g., [4]).

However, as shown in the sections below, anomaly errors severely limit these applications. For example, just the satellite measurement errors alone restrict viable anomaly continuations to within roughly +/- 50 km of mission altitude. The non-uniqueness of anomaly continuation also confines the analysis of satellite geopotential anomalies to a very restricted range of altitudes about the mission orbits. Hence, for improved understanding of the geopotential fields below mission altitude, lower altitude surveying is necessary from shipborne, airborne, balloon, space shuttle, and space station platforms. As shown below, inversion of the combined low-altitude and satellite observations yields greatly improved, but not unique estimates of the intervening anomaly fields.

## 2 Spherical prism anomaly simulations

We illustrate the results described above using the spherical prism magnetic anomaly simulations in Figure 1, which we can directly extend to comparable gravity anomaly simulations by Poisson's relation (e.g., [5]). The left column in Figure 1 gives the total magnetic field anomalies for five arbitrary crustal prisms modeled to 16-decimal place accuracy directly in spherical coordinates by Gauss-Legendre quadrature integration [6]. The spherical prisms are outlined and labeled in black with cgs-volume magnetic susceptibility contrasts of 0.0061468 for prism A, - 0.0028831 for B, 0.0009452 for C, 0.0029529 for D, and – 0.0004456 for F. All prisms are 25 km thick with tops at 5 km below sea level. For each map, the observation grid is at Z km above sea level and has 80 nodes at an interval of about 77 km over the Balkan test region (35 - 42)°N, (22 - 28)°W. The Magsat 12/83 model [7] updated to 1980 was used to evaluate the core field attributes at the source and observation points.

The middle column of Figure 1 shows the utility of the inversion of combined

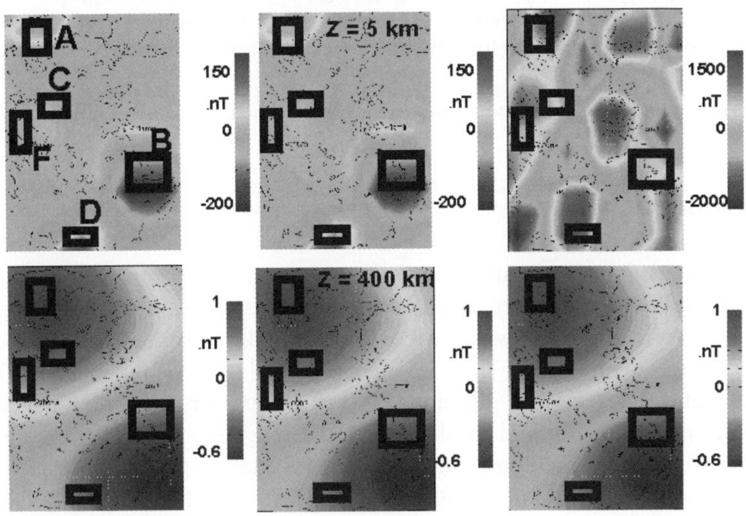


**Fig. 1.** Comparisons of magnetic anomalies for 5 prisms modeled directly in spherical coordinates (left column), the inversion of both the 5 km and 400 km altitude fields truncated to 0.0001 nT accuracy (middle column), and the inversion of only the 400 km altitude field truncated to the nearest 0.1 nT (right column).

near-surface and satellite altitude data of limited accuracy for effective anomaly continuation. Here, we used the 5 and 400 km altitude prism anomalies truncated to 0.0001 nT, which is three orders of magnitude better than CHAMP's measurement accuracy. Following [8], we modeled the truncated anomalies by equivalent point source (EPS) inversion [5] [9] using an EPS grid of 1079 dipoles located 50 km below sea level at an interval of about 30 km. Comparing the maps of the left and middle columns of Figure 1, the left performance diagram of Figure 2 shows how remarkably the EPS anomaly predictions are matched with the prism anomalies at all altitudes. This performance, however, requires anomaly accuracy that is not possible to achieve in practice.

The right column of Figure 1 gives the marginal results from EPS inversion where only the 400 km anomalies are considered truncated to the CHAMP measurement accuracy of 0.1 nT. The right performance diagram of Figure 2 summarizes these results, which show that viable anomaly continuations are essentially restricted to +/- 50 km of the truncated anomalies. Improving the accuracy of the satellite observations can only marginally enhance effective anomaly continuation.

Furthermore, the veracity of satellite anomalies continued great distances from the observations is very dubious due to the non-uniqueness of anomaly continuation. To see this, we note that the fitting of any model (e.g., spherical harmonics, Fourier transform, equivalent source, etc.) to a set of geopotential observations can be generically expressed by the matrix equation $\mathbf{AX} = \mathbf{B}$. It has the least squares solution $\mathbf{X} = (\mathbf{A}^t\mathbf{A})^{-1}\mathbf{A}^t\mathbf{B}$, where $\mathbf{A}^t$ is the transposed design matrix $\mathbf{A}$, $\mathbf{X}$ the column matrix of solution coefficients, and $\mathbf{B}$ the column matrix of n observations.

For inversion of the high- and low-altitude sets of observations, the observation vector can be expressed by $\mathbf{B} = \mathbf{B_h} + \mathbf{B_l}$, where $\mathbf{B_h} = (b_{hi}\, b_{h2} \ldots b_{hm}\, 0_{l1}\, 0_{l2} \ldots 0_{lk})^t$ is the high-altitude observation vector, $\mathbf{B_l} = (0_{h1}\, 0_{h2} \ldots 0_{hm}\, b_{l1}\, b_{l2} \ldots b_{lk})^t$ the low-altitude vector, and $n = m + k$. The solution $\mathbf{X} = (\mathbf{A}^t\mathbf{A})^{-1}\mathbf{A}^t(\mathbf{B_h} + \mathbf{B_l}) = \mathbf{X_h} + \mathbf{X_l}$ is clearly made up of partial solutions $\mathbf{X_h}$ and $\mathbf{X_l}$ that are basically independent of the actually observed anomaly values in $\mathbf{B_l}$ and $\mathbf{B_h}$, respectively. Thus, the anomaly prediction from inversion at an unmapped location is not unique because superposition can extend the solution to accommodate any anomaly value there.

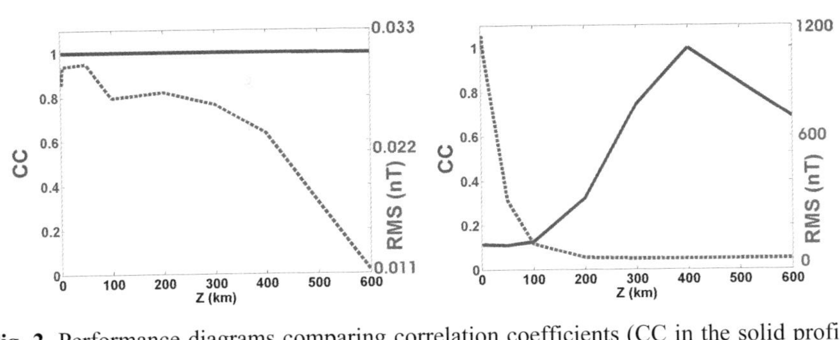


**Fig. 2.** Performance diagrams comparing correlation coefficients (CC in the solid profiles) and root-mean-squared differences (RMS in the dotted profiles) for the anomaly predictions from inversion of the 5 and 400 km prism anomalies at 0.0001 nT accuracy (left), and only the 400 km altitude anomalies truncated to the nearest 0.1 nT (right).

In practice, the computational limitations of our inversion codes [9] readily admit anomaly prediction ambiguities at distances of about 50 km and greater from the satellite observations. Thus, for the fullest understanding of the anomaly fields between the Earth's surface and satellite altitudes, we really must exploit every opportunity to directly map the lower altitude fields. In the next section, we demonstrate the exciting new dimension for interpreting satellite geopotential anomalies that is offered by this approach.

## 3 Surface-to-satellite altitude magnetic anomaly variations

For enhanced geological studies of Maude Rise, Antarctica, [10] produced a crustal magnetization model from the joint inversion of magnetic anomalies at 5 and 700 km altitudes. The 5 km altitude data were low-pass filtered anomalies ($\geq$ 400 km) from the grid of airborne and shipborne magnetic anomalies maintained by the Antarctic Digital Magnetic Anomaly Project [11]. The 700 km altitude anomalies were derived directly from the Ørsted magnetic observations [12].

Joint inversion of the two anomaly fields obtained a crustal magnetization model with the effects shown in Figure 3 [10]. This magnetization model accounted for thickness variations of the crust as inferred from seismic data supplemented by satellite altitude gravity observations [2], remanent magnetization variations of the oceanic crust, and the variable core field attributes over the study region. Maps A and H of Figure 3 show the modeled effects that match with negligible error the original 5 and 700 km input anomaly fields, respectively.

Consideration of the model predictions over the intervening altitudes suggests that the dominant magnetic effects at satellite altitudes are from crustal thickness variations and remanence involving the normal polarity Cretaceous Quiet Zone (KQZ). Modeling of the regional components of the residual Ørsted and near-surface magnetic anomalies supports extending the KQZ eastwards from Maude Rise (MR in Figure 3.A) towards the Astrid Ridge (AR), while the higher frequency residual near-surface anomalies can be related to crustal features with effects that are strongly attenuated at satellite altitudes.

The 8 slices of the geomagnetic field in Figure 3 show how the 3 or 4 satellite altitude anomalies break down with decreasing altitude into a complex multitude of anomalies at the near-surface. Alternatively, they show how the near-surface anomalies coalesce with increasing altitude into the roughly handful of anomalies that are observed at satellite altitude.

For example, at the near-surface altitudes (maps A-B) linear maxima over and along the Maude Rise (MR) with relatively well-defined interior minima characterize the KQZ. It is only at altitudes of about 100 km and greater (maps E-H) that the strong, regionally positive magnetic character of the KQZ becomes apparent.

Similarly, the near-surface magnetic minima along the coast of East Antarctica coalesce at altitudes of 100 km and higher with the Riiser-Larsen Sea (RLS) minimum. The near-surface continental minima are broken up by a maximum

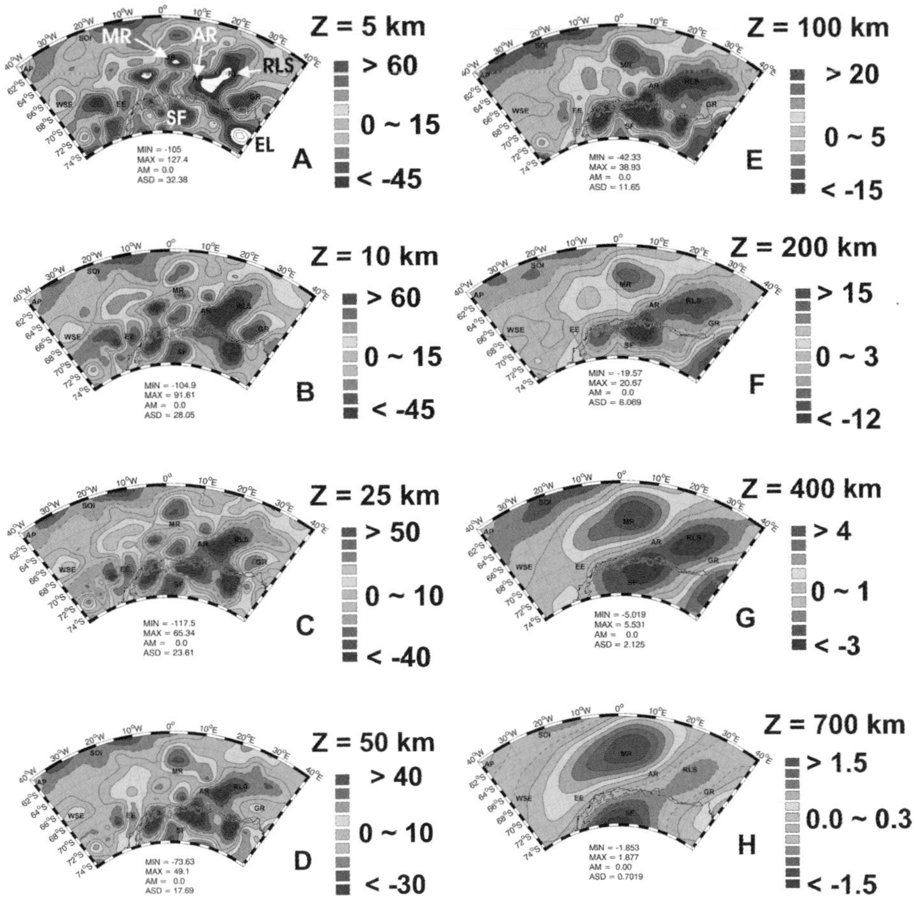


**Fig. 3.** Magnetic anomaly predictions in nT for the Maude Rise region of Antarctica from the inversion of regional ADMAP and Ørsted data at respective altitudes of 5 and 700 km.

over Sveshjfella (SF) that dies out at altitudes of nearly 200 km and higher. By then, however, the SF anomaly appears to connect with a positive anomaly over western Enderby Land (EL) that may reflect an Archean shield or platform [13]. The EL anomaly is weakly expressed in the near-surface data, but becomes increasingly prominent with altitude.

The anomaly behavior in Figure 3 suggested by the joint inversion clearly would not be revealed in the simple downward continuation of the satellite altitude data nor in the upward continuation of the near-surface magnetic data. Unfortunately, like for any inversion, these results are not unique. Thus, they do not obviate the need for supplemental magnetic measurements, especially at the intervening altitudes that may be accessed by high-altitude aircraft, balloons, and space shuttle tethers, to better define the geologic relationships in the near-surface and satellite altitude magnetic fields.

## References


1. von Frese R R B, Kim H R, Tan L, Kim J W, Taylor P T, Purucker M E, Alsdorf D E and Raymond C A (1999) Satellite magnetic anomalies of the Antarctic crust. Annali di Geofisica *42*: 309-326.
2. Gubbins D, and J Bloxham (1985) Geomagnetic field analysis, III, Magnetic fields on the core-mantle boundary. Geophys J R Astron Soc *80*: 695-713.
3. Llubes M, Florsch N, Legresy B, Lemoine J-M, Loyer S, Crossley D and Rémy F (2003) Crustal thickness in Antarctica from CHAMP gravimetery. Earth Planet Sci Lett *212*: 103-117.
4. von Frese R R B, Hinze W J and Braile L W (1981) Spherical earth gravity and magnetic anomaly analysis by equivalent point source inversion. Earth Planet Sci Lett *53*: 69-83.
5. von Frese R R B, Hinze W J, Braile L W and Luca A J (1981) Spherical earth gravity and magnetic anomaly modeling by Gauss-Legendre quadrature integration. J Geophys *49*: 234-242.
6. Langel R A, and Estes R H (1985) The near-earth geomagnetic field at 1980 determined from Magsat data. J Geophys Res *90*: 2,495-2,510.
7. Ravat, D, Whaler KA, Pilkington M, Sabaka T, and Purucker M (2002) Compatibility of high-altitude aeromagnetic and satellite-altitude magnetic anomalies over Canada. Geophysics *67:* 546-554.
8. Alsdorf D E, and von Frese R R B (1994) Fortran Programs for Processing Magsat Data for Lithospheric, External Field and Residual Core Components. NASA-GSFC TM104612, 196 pp.
9. Kim H R (2002) Antarctic Lithospheric Anomalies from Ørsted Satellite and Near-Surface Magnetic Observations. Ph.D. Dissertation (unpubl.), The Ohio State University, Columbus, 160 pp.
10. Golynsky A, Chiappini M, Damaske D, Ferraccioli F, Ferris J, Finn C, Ghidella M, Ishara T, Johnson A, Kim H R, Kovacs L, LaBrecque J, Masolov V, Nogi Y, Purucker M, Taylor P and Torta M (2001) ADMAP – Magnetic Anomaly Map of the Antarctic, 1:10,000,000 scale map. in: Morris P and von Frese R, eds, BAS (Misc.) 10, Cambridge, British Antarctic Survey.
11. Kim H R, von Frese R R B, Kim J W, Taylor P T, Purucker M E and Neubert T (2002) Ørsted verifies regional magnetic anomalies of the Antarctic lithosphere. Geophys Res Lett *29(15)*: ORS 3-1 to 3-3.
12. von Frese R R B, Tan L, Kim J W and Bentley C R (1999) Antarctic crustal modeling from the spectral correlation of free-air gravity anomalies with the terrain. J Geophys Res *104*: 25275-25296.
13. Bormann P, Bankwitz P, Bankwitz E, Damm V, Hurtig E, Kompf H, Menning M, Paech H-J, Schofer U and Stackegrandt W (1986) Structure and development of the passive continental margin across the Princess Astrid coast, East Antarctica. J Geodyn *6*: 347-373.

# Introducing POMME, the POtsdam Magnetic Model of the Earth

Stefan Maus[1], Hermann Lühr[1], Georgios Balasis[1], Martin Rother[1], and Mioara Mandea[2]

[1] GFZ, Telegrafenberg, 14473 Potsdam, *smaus@gfz-potsdam.de*
[2] Institut de Physique du Globe de Paris, 4 Place Jussieu

**Summary.** Models of the main and external field play a key role in the analysis and interpretation of satellite, airborne, marine and ground magnetic data. Here, we introduce a series of main field models with several new features: (1) External fields are parametrised in SM and GSM coordinate systems, accounting for the geometry of the ring current, the magnetosphere and the solar wind. (2) We use vector data globally, instead of the usual approach of using vector data at low latitudes and scalar data at high latitudes. (3) The angles between CHAMP's star camera and its vector magnetometer are co-estimated in a joint inversion with Ørsted data. (4) The model includes 2nd time derivatives of the magnetic field to account for the non-negligible secular acceleration in the current period of new magnetic satellite data. As inferred from the degree spectrum of the current version POMME-1.4, the secular variation is stable to degree 11 and the secular acceleration to degree 6.

**Key words:** geomagnetic field, field models, main field

## 1 Introduction

Analysis of geomagnetic data usually starts with subtracting a main field model in order to uncover weaker contributions to the magnetic field, such as from mantle induction [2], crustal magnetisation [8], ocean dynamo [11], ionospheric [5] and magnetospheric currents [3]. Main field models thus play a key role in the analysis and interpretation of geomagnetic data. Popular recent models include the comprehensive model CM3 [10], the Ørsted field models [9] and the first official CHAMP main field model CO2 [4]. In view of the general importance of accurate main field models and motivated by the call for candidate models for the 9th generation IGRF, we have developed a new series of main field models which are available at http://www.gfz-potsdam.de/pb2/pb23/SatMag/pomme.html and shall be updated regularly.

## 2 Model details

The POMME model includes a parametrisation of the internal field, external static magnetospheric field, ring current field (including induced part) and co-estimated corrections for the angles between the CHAMP star camera and vector magnetometer reference systems.

### 2.1 Internal field

The internal field is parametrised in the usual way as

$$V(r,\vartheta,\varphi) = a\sum_{\ell=1}^{\infty}\left(\frac{a}{r}\right)^{\ell+1}\sum_{m=-\ell}^{\ell} g_\ell^m Y_\ell^m(\vartheta,\varphi), \tag{1}$$

with the scalar potential $V$ of the magnetic field, $\mathbf{B} = -\nabla V$, colatitude $\vartheta$, longitude $\varphi$, degree $\ell$, order $m$, reference radius $a = 6371.2$ km, and Gauss coefficients $g_\ell^m$, where coefficients with negative order are often called $h$ in geomagnetism. Finally, $Y_\ell^m$ are Schmidt quasi-normalised spherical harmonics [1].

Accounting for the rapid change of the core field, each Gauss coefficient is given as a truncated Taylor expansion

$$g(t) = g + tg' + 0.5t^2 g''. \tag{2}$$

For the current version (POMME-1.4), $g$ and $g'$ are given to degree and order 15, while the secular acceleration $g''$ extends to degree and order 10 (Figure 1).

### 2.2 Static magnetospheric field

In all existing geomagnetic field models the magnetospheric fields are given in Earth fixed coordinates. Hence, they are assumed to co-rotate with the Earth. This means, in particular, that daily and seasonal variations have to be introduced in order to be able to describe static magnetospheric fields. This reminds of the pre-Kepler practice of describing the movement of planets in an Earth fixed coordinate system. In order to describe magnetospheric fields in the natural coordinate systems of their source currents, POMME forsees

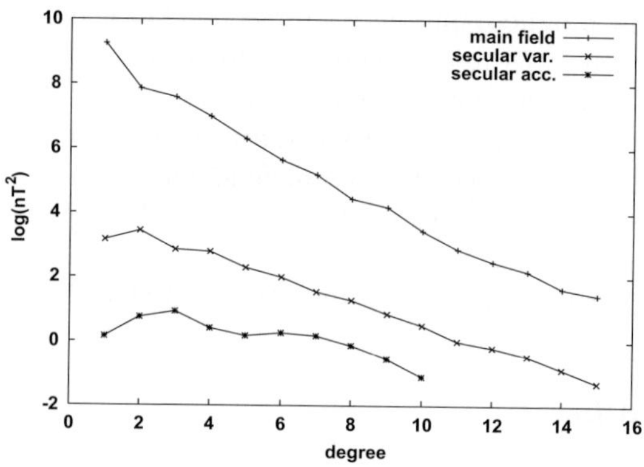


**Fig. 1.** Power spectra of the POMME-1.4 core field, its secular variation and the secular acceleration. The secular variation was damped for degrees 12-15, and the secular acceleration was damped for degrees 7-10.

contributions in Geocentric Solar Magnetospheric (GSM) and Solar Magnetic (SM) coordinates, which are the natural coordinate systems of the magnetosphere and the ring current, respectively. The current version (POMME-1.4) includes a static degree-2 field in GSM. This GSM field appears as an annually, semi-annually and daily varying external field in Earth fixed coordinates (Figure 2). Furthermore, an axisymmetric ring current field is given in SM.

## 2.3 Ring current field

In main field modelling it is common practice to co-estimate the coupling coefficients between the Dst index and the ring current field. However, since low Dst values are a data selection criterion, this co-estimation does not work very well. We therefore estimate the coupling coefficients from a data set including high Dst and then regard these coefficients as fixed. We find the optimum values for the external ($q_1^0$) and internal ($g_1^0$) dipoles in SM to be

$$q_1^0(t) = 0.76 \cdot Dst(t) \tag{3}$$

$$g_1^0(t) = 0.76 \cdot 0.32 \cdot Dst(t) \tag{4}$$

Figure 3 shows these values in comparison with typical values given by previous field models. Future versions of POMME will include non-symmetric ring current contributions in SM, consistent with the local time dependence found by Schwarte (personal communication).

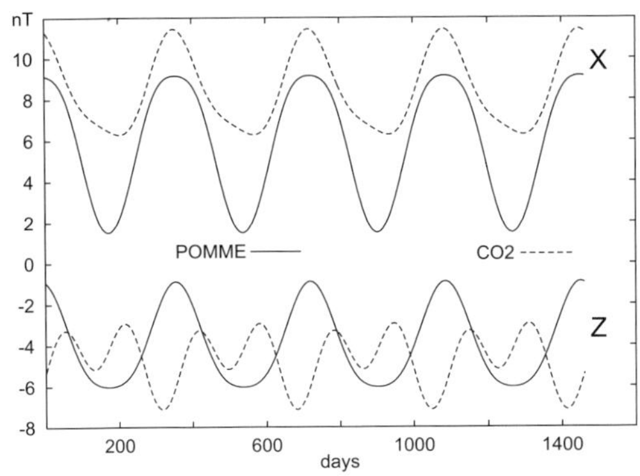


**Fig. 2.** Annually varying magnetospheric field (without ring current field) predicted by two field models for an observer at 0° longitude and 30° latitude. While the CO2 field model[4] explicitly contains annual and semi-annual coefficients in an Earth fixed reference frame, the model POMME has a static external field in magnetospheric (GSM) coordinates and the annual variations are merely a consequence of the seasonal orientation of the Earth axis in the GSM reference frame.

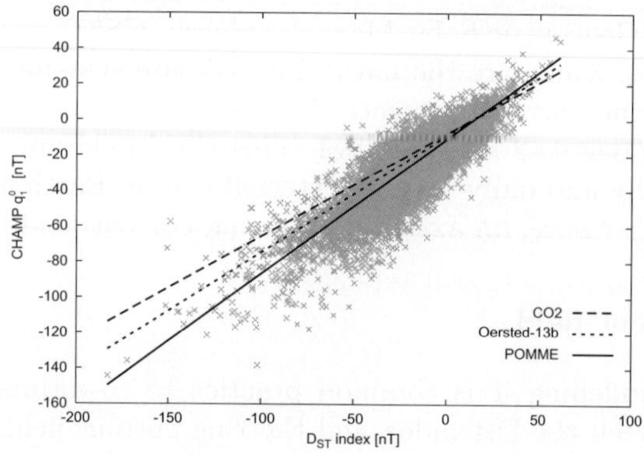


**Fig. 3.** Usually, the coupling coefficients of the Dst index to the external ($q_1^0$) and induced ($g_1^0$) ring current field are co-estimated with the field model coefficients. However, since data are selected for low Dst, the coupling coefficients tend to be biased to small values. For POMME, the coupling coefficients were determined by a visual best fit to a large data set of $q_1^0$ and $g_1^0$ estimates. Shown here is only $q_1^0$.

## 2.4 Star camera re-calibration for CHAMP

Accurate vector measurements require very accurate knowledge of the satellite orientation (attitude). Attitude uncertainty is the largest source of error in Magsat, Ørsted and CHAMP data. For CHAMP, with its stable attitude, an additional difficulty arises in the calibration of the angles between the star camera and magnetometer reference systems. Due to thermo-mechanical instabilities, these angles vary with time and have to be co-estimated in the inversion for the model parameters. Luckily, the angles appear to be rather stable over time spans of months. ¿From the visual inspection of field residuals, we identified 6 stable time intervals for POMME-1.4. For each interval, three rotation angles are given as a correction of the field vector orientation with respect to the CHAMP satellite. These correction angles should be used in the analysis of CHAMP data. Furthermore it is foreseen that these angles will be used as corrections in producing the final level-3 CHAMP vector data product.

## 2.5 Data selection for POMME-1.4

In geomagnetic field modelling it is general practice to use vector data at mid latitude and scalar data at high latitudes in order to avoid contamination by field aligned currents. However, the work on the crustal field model MF2 showed that the inclusion of vector data at high latitudes is very well possible if only the quietest tracks are selected by their RMS against an initial field model. Our primary motivation for using vector data at high latitudes for estimating the POMME coefficients is the following: If vector data are available

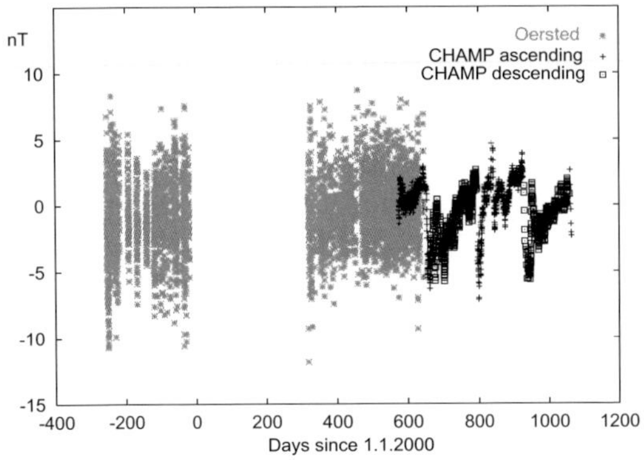


**Fig. 4.** Data residuals, averaged over 2000 seconds, for the vertical component of the magnetic field against POMME-1.4. During the long empty interval in the year 2000, Ørsted was outside of the 0-6 LT window, which was used here as a data selection criterion. The same LT requirement was applied to the CHAMP data. Hence, not all of the CHAMP data displayed here were actually used. Trends in the residuals may either be local time effects or indicate gradual changes in the angle between the star camera and magnetometer reference systems.

over the entire sphere, internal and external potential fields can be unambiguously separated from toroidal fields [1]. While we are not co-estimating toroidal fields, attitude errors are known to look like toroidal fields and vice versa. Thus, it may be difficult to co-estimate attitude angle corrections if there are no vector data at the high latitudes. For POMME-1.4 we therefore used only vector data, with CHAMP data from 15-May-2001 to 30-Sep-2002 and Ørsted vector data from 21-Apr-1999 to 30-Sep-2002. Due to the occurrence of F-Region currents in the pre-midnight hours [6] which are associated with the diamagnetic effect of ionospheric plasma [7], we used only data in the 0-6 local time (LT) window. Only for the 1999 Ørsted measurements we also included data at 22-24 LT. An earlier POMME version, which was the basis for our DGRF-1995 candidate model, also included observatory annual means. Observatory data help to stabilise secular variation and will again be included in the upcoming versions. Data residuals (Figure 4) are centered around zero. This indicates that the Taylor series to degree-2 for the temporal variation of the Gauss coefficients is adequate for the currently considered time span of 4 years.

## 3 Conclusion

With POMME we introduce a series of field models for the accurate description of the main and external field. New features include the parametrisation of magnetospheric fields in magnetospheric (instead of Earth fixed) coordinates, the imposing of realistic, fixed coupling coefficients between the Dst index and the ring current field, and the co-estimation of attitude corrections.

## References


1. Backus G, Parker RL, and Constable C (1996) Foundations of Geomagnetism. Cambridge Univ. Press.
2. Constable S and Constable C (2004) Observing geomagnetic induction in magnetic satellite measurements and associated implications for mantle conductivity. Geochem Geophys Geosyst *5(1)*: Q01006, DOI 10.1029/2003GC000634.
3. Daglis IA, Thorne RM, Baumjohann W, and Orsini S (1999) The terrestrial ring current: Origin, formation and decay. Reviews of Geophysics *37(4)*: 407–438.
4. Holme R, Olsen N, Rother M, and Lühr H (2003) CO2 - A CHAMP magnetic field model. In: Reigber C, Lühr H, Schwintzer P (Eds), First CHAMP Mission Results for Gravity, Magnetic and Atmospheric Studies, Springer, Berlin Heidelberg: 220–225.
5. Lühr H, Maus S, and Rother M (2003) The noon-time electrojet, its spatial features as determined by the CHAMP satellite. J Geophys Res *109(A1)*: A01306, 10.1029/2002JA009656.
6. Lühr H, Maus S, Rother M and Cooke D (2002) First in situ observation of night time F region currents with the CHAMP satellite. Geoph Res Lett *29*: 10.1029/2001GL013845.
7. Lühr H, Rother M, Maus S, Mai W, and Cooke D (2003) The diamagnetic effect of the equatorial Appleton anomaly: Its characteristics and impact on geomagnetic field modeling. Geoph Res Lett *30(17)*: 10.1029/2003GL017407.
8. Maus S, Rother M, Holme R, Lühr H, Olsen N, and Haak V (2002) First scalar magnetic anomaly map from champ satellite data indicates weak lithospheric field. Geoph Res Lett *29(14)*: 10.1029/2001GL013685.
9. Olsen N (2002) A model of the geomagnetic main field and its secular variation for epoch 2000 estimated from Ørsted data. Geophys J Int *149*: 454–462.
10. Sabaka TJ, Olsen N, and Langel RA (2002) A comprehensive model of the quiet-time, near-Earth magnetic field: phase 3. Geophys J Int *151*: 32–68.
11. Tyler R, Maus S, and Lühr H (2003) Satellite observations of magnetic fields due to ocean tidal flow. Science *299*: 239–241.

# Alternative Parameterisations of the External Magnetic Field and its Induced Counterpart for 2001 and 2002 Using Ørsted, Champ and Observatory Data

Vincent Lesur, Susan Macmillan and Alan Thomson

British Geological Survey, Murchison House, West Mains Road, Edinburgh, EH9 3LA, Scotland, *vbfl@bgs.ac.uk, smac@bgs.ac.uk, awpt@bgs.ac.uk*

**Summary.** A better parameterisation of the field generated by the large-scale magnetospheric current systems is required to avoid the need for further filtering of the data when modelling the crustal magnetic field. Using Ørsted and observatory data for year 2001, we investigate new parameterisations where internal and external degree one Gauss coefficients are computed daily. These models are compared with models where the external field is parameterised with a linear dependence on the Dst index. For both types of model the overall fits to the data are similar but the rapid variations in the magnetospheric currents and their induced counterparts are better captured using the new parameterisations.

**Key words:** Geomagnetic field modelling, magnetospheric currents

## Introduction

The large-scale magnetospheric current systems such as the ring current, magnetopause current and tail current, are always present and their magnetic signals are a constituent part of the magnetic field observed on and near the Earth's surface. When deriving models of the quiet night-time Earth's magnetic field it is therefore necessary to include parameters which characterise the sum total of these current systems. The usual approach is to include parameters which depend on the hourly Dst index [e.g. Sabaka *et al,* 2002]. However, there are some uncertainties with the Dst index. In particular the "preliminary" values (which are the only values available for about one year after the date of interest) can be noisy and contain baseline shifts. In this paper the usual approach is compared with three other parameterisations, namely (1) modelling the daily average magnetospheric field and its induced counterpart using degree one spherical harmonics, (2) same as (1) but including Dst dependence for the faster variations during each day, (3) same as (1) but imposing continuity in time through the use of cubic spline functions with the knot interval set at one day.

## Data selection

In order to have as even a coverage in time and space of suitable satellite and observatory data as possible, the year 2001 was chosen. During this year Ørsted provided almost continuous data on the night-side of the Earth and more than 100 observatories had submitted their final data to INTERMAGNET or the World Data Centre in Copenhagen. Night-side is defined as local time 23:00-06:00 and where the ionosphere is in darkness up to a height of 250 km. Data collected during quiet to moderately disturbed conditions are used, defined by the final hourly Dst index –60 nT to +20 nT. Every 20$^{th}$ Ørsted vector and scalar data point and all observatory hourly mean values were selected using these criteria. Only scalar satellite data were used at high latitudes, defined as greater than geomagnetic latitude 60°, in order to reduce the effects of field-aligned currents. Vector observatory data at high latitudes were only used to determine the crustal biases at these locations. Observatory data collected during days when there were less than 60 observatories providing valid data, were not included.

## Model parameterisations and determinations

Four models were derived. The parameters in each model are summarised in Table 1. The basis functions in space are internal or external spherical harmonics from minimum to maximum degree m to n (SHm/n) and in time they are polynomials of degree k (POk), cubic B-splines with knots at daily intervals (S), or periodic functions with period t years (PEt). If the dataset is split up into days in order to determine these basis functions this is tabulated as SHm/n-day, otherwise the whole dataset is used.

For the external spherical harmonics there may also be a linear dependence on the final hourly Dst index (SHm/n-Dst). This rapidly varying external field is

| | classic with Dst dependence | daily dipole | daily dipole with Dst dependence | spline |
|---|---|---|---|---|
| internal | SH1/10-PO2 | SH1/1-day | SH1/1-Dst-day | SH1/1-S |
| | SH11/20 | SH2/10-PO2 | SH2/10-PO2 | SH2/10-PO2 |
| | SH1/1-Dst-I | SH11/20 | SH11/20 | SH11/20 |
| | SH1/2-PE0.5-I | SH2/2-PE0.5-I | SH2/2-PE0.5-I | SH2/2-PE0.5-I |
| | SH1/2-PE1.0-I | SH2/2-PE1.0-I | SH2/2-PE1.0-I | SH2/2-PE1.0-I |
| | O306 | O306 | O306 | O306 |
| external | SH1/1-Dst | SH1/1-day | SH1/1-Dst-day | SH1/1-S |
| | SH1/2-PE0.5 | SH2/2-PE0.5 | SH2/2-PE0.5 | SH2/2-PE0.5 |
| | SH1/2-PE1.0 | SH2/2-PE1.0 | SH2/2-PE1.0 | SH2/2-PE1.0 |

**Table 1.** Summary of models computed. SHm/n-spherical harmonic degrees m to n, POk-polynomial in time of degree k, PEt-periodic function in time with period t, I-induced terms, Op-the p observatory biases.

assumed to induce an internal field with a linear dependence factor of 0.27 [Olsen *et al*, 2002] and this is tabulated as SHm/n-Dst-I. For the more slowly varying external field (PEt) there is also a corresponding induced internal field but a pre-set value for the dependence factor is not assumed. This is tabulated as SHn-PEt-I. The remaining parameters are the p biases for the observatories to account for the local crustal fields (Op).

The model parameters are determined by iterative least squares. At each iteration the parameter set is changed – this is necessary because some of the data are non-linear functions of the parameters and therefore must be linearised using an initial parameter set. The assigned uncertainty for low-latitude data is 2 nT and for high-latitude data is 5 nT. Satellite and observatory data are weighted (separately) according to the number of data in the equal-area tessera the data point lies within. The equal-area tesserae are equivalent to 5° latitude by 5° longitude boxes at the equator. The Ørsted star camera pointing uncertainty is assumed to be 10" and the rotation uncertainty, 40". These anisotropic uncertainties lead to off-diagonal terms in the data covariance matrix used in the least squares analysis.

## Analysis of results

The overall weighted misfits are shown in Table 2. Daily average coefficients and corresponding uncertainties for spherical harmonic degree 1 terms can be computed for the "classic with Dst dependence" model using the relevant coefficients, Dst and standard deviations. These are compared with those output directly from the "daily dipole" model solution in Figure 1. The same patterns are seen in both sets of uncertainties, e.g. the increase towards the end of the year when there is a poorer spatial distribution of data and the increases at the equinoxes resulting from increased magnetic activity at these times.

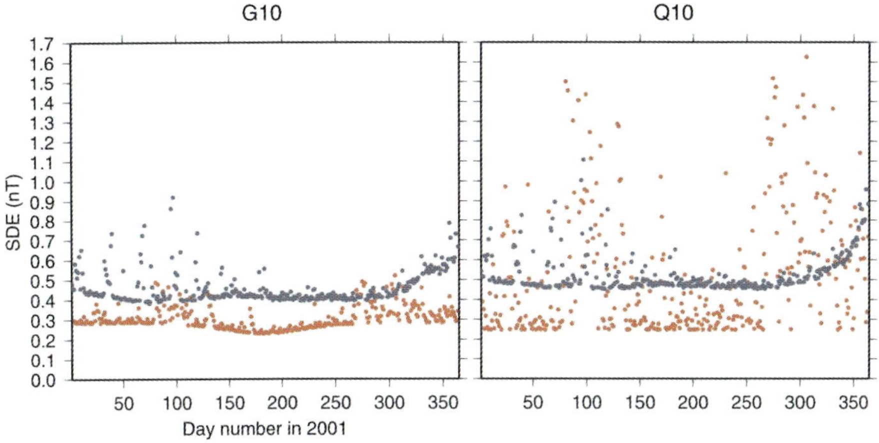


**Fig. 1.** Daily standard deviation estimates (SDE) for "classic with Dst dependence" model (red) and "daily dipole" model (blue) for first internal (G10) and external (Q10) spherical harmonic coefficients.

| | classic with Dst dependence | daily dipole | daily dipole with Dst dependence | spline |
|---|---|---|---|---|
| misfit (nT) | 2.43 | 2.59 | 2.22 | 2.48 |

**Table 2.** Overall weighted misfits for models in Table 1.

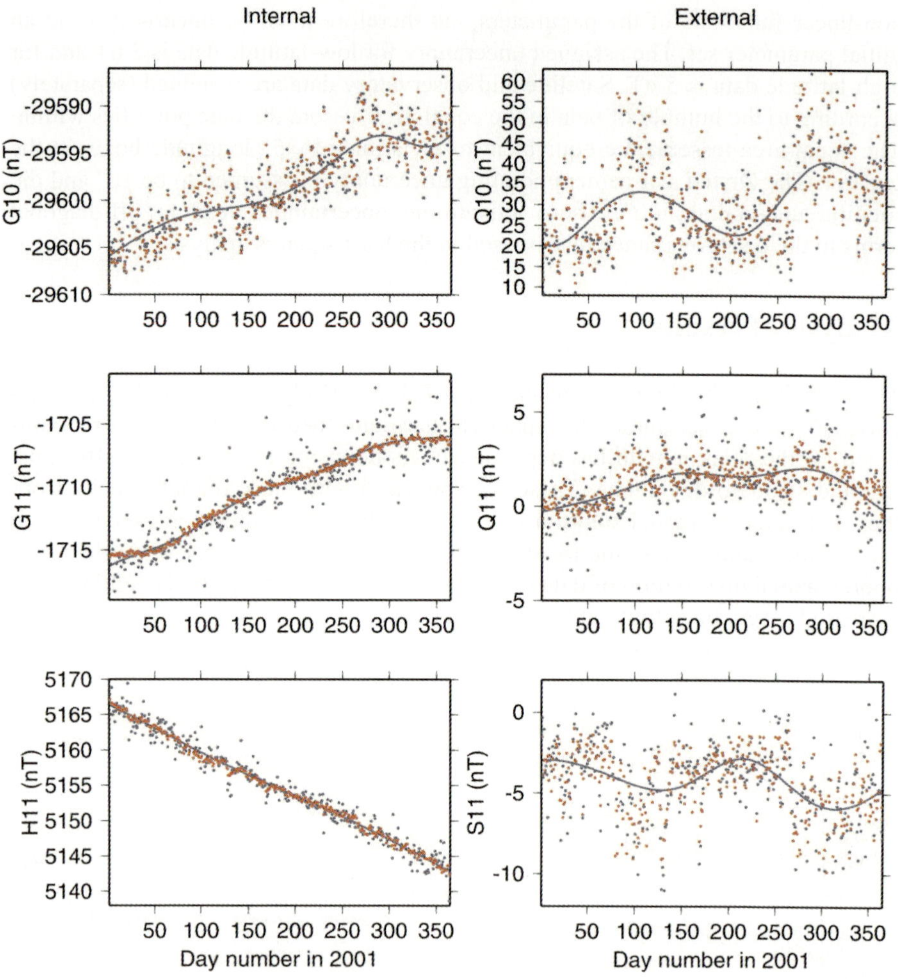


**Fig. 2.** The first three internal and external coefficients from the "classic with Dst dependence" model (red) and the "daily dipole" model (blue dots and spline line fitted to blue dots).

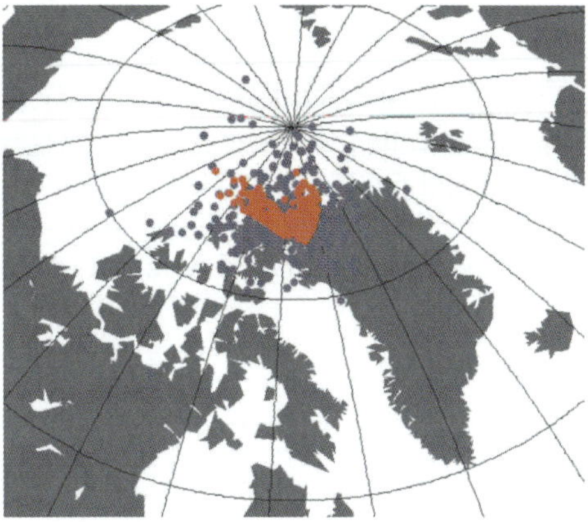

**Fig. 3.** Mean daily field directions from degree 1 spherical harmonic external coefficients from "classic with Dst dependence" model (red) and "daily dipole" model (blue).

The magnitude and direction of the daily external dipole from the "daily dipole" model can be analysed further. The magnitude, as determined from the first three external spherical harmonic coefficients, not surprisingly shows a high correlation with the Dst index but there is an offset of approximately 20 nT. This provides confirmation that the quiet-time level of the Dst index is non-zero. The direction, however, is highly variable and in Figures 2 and 3 we show the daily internal and external dipole coefficients and directions from both the "classic with Dst dependence" and "daily dipole" models.

For the internal dipole one can see drift, presumably associated with fluid flows in the core, and induced annual and semi-annual signals. For the external dipole one can see annual and semi-annual signals. There is considerably more scatter in the directions determined from the "daily dipole" model, particularly in longitude. Even when the more rapid variations are parameterised, as in the "daily dipole with Dst dependence" model, this scatter is not significantly reduced.

This scatter is very probably real signal. The "classic with Dst dependence" model cannot generate the rapidly varying signal in the Y component seen at observatories located on the magnetic meridian. This is because there are constraints imposed on the short-term magnetic field directions in this model. However, the other models all fit the rapid variations in the Y component at observatories, particularly those on the magnetic meridian, better than the "classic with Dst dependence" model. We note that the large deviations of the daily dipole directions do not coincide with large Dst values.

Finally, CHAMP vector and scalar data collected in 2001 and 2002 were selected in a similar way as the Ørsted data, and used to determine a "daily dipole" model, without observatory biases. However, this model could not be fitted because of calibration difficulties with the CHAMP vector data.

## Conclusions

- Models with rapidly varying spherical harmonic degree 1 terms included, fit Ørsted and observatory data well.
- The magnitude of the external field is correlated with the Dst index but there is an offset of ~20 nT.
- The external field is not necessarily aligned with the internal field, especially at quiet times.

## References

Olsen N (2002). A model of the geomagnetic field and its secular variation for epoch 2000 estimated from Ørsted data. Geophys J Int *149*: 454-462.

Sabaka T J, Olsen N, Langel R A (2002). A comprehensive model of the quiet-time, near-Earth magnetic field: phase 3. Geophys J Int *151*: 32-68.

# New Insight into Secular Variation Between MAGSAT and CHAMP/ØRSTED

Ingo Wardinski[1] and Richard Holme[2]

[1] GeoForschungsZentrum Potsdam, Sect. 2.3, Potsdam, Germany *ingo@gfz-potsdam.de*
[2] Department of Earth's and Ocean Sciences, University of Liverpool, United Kingdom *holme@liv.ac.uk*

**Key words:** secular variation, geomagnetic jerk, core-surface flows, CHAMP, ØRSTED, MAGSAT

## 1 Introduction

Over the two decades between 1980 to 2000, the secular variation of the Earth's magnetic field demonstrates rich behavior. Three geomagnetic jerks (sharp changes in the slope of the first time derivative of the geomagnetic field) were detected within this period: 1983 only seen at the southern hemisphere [Dowson *et al.*, 1988; Kotzé *et al.*, 1991], 1991 [Macmillan, 1996] and around 2000 [Mandea *et al.*, 2000] most clearly seen in the first time derivative of the East component, dY/dt, at European observatories. The non–global extent of the jerks raises questions about the processes which cause it. In this study we developed models for the secular variation and core–surface flows over this period. These models were derived from magnetic observatory quietest monthly means and satellite data. Our resulting models reveal small scale secular variation structure.

## 2 Data

For secular variation modelling we use observatory quietest monthly means; if not available, annual means are used instead. This gives us $\approx$ 100 monthly measurement locations and $\approx$ 175 annual measurement locations. Also repeat survey data are included.
These data are used as modelling input, and are processed in the following ways. Monthly means are treated by n-step difference filter

$$\text{e.g. } dX/dt = X(t + n/2) - X(t - n/2), \;\; n = 12.$$

Annual means are treated using,

$$\text{e.g. } dX/dt = X(t) - X(t - 1).$$

We refer these as secular variation estimates. Using secular variation estimates the contaminations due to static crustal and external field sources are generally reduced [Gubbins and Tomlinson, 1986]. However, the influence of the solar cycle variations could remain as the leading source of noise (work on this topic is going to be published by Wardinski & Mandea).

## 3 Secular Variation Modelling

Our method is based on an approach proposed by [Bloxham and Jackson, 1992]: a decomposition of the magnetic potential $V(r, \theta, \phi, t)$ into spherical harmonics up to degree and order 13

$$V = a \sum_{l,m} \left(\frac{a}{r}\right)^{l+1} \left(g_l^m \cos m\phi + h_l^m \sin m\phi\right) \cdot \mathrm{P}_l^m(\cos\theta) \tag{1}$$

and an expantion of each harmonic coefficient $g_l^{mn}$ on a basis of cubic B–Splines $M_n(t)$ in time.

$$g_l^{mn}(t) = \sum_{n=1}^{N} g_l^m \, M_n(t) \,, \quad h_l^{mn}(t) = \sum_{n=1}^{N} h_l^m \, M_n(t) \tag{2}$$

where the knot separation is 1 year.

Like Bloxham and Jackson [1992] we seek the smoothest models for a given fit to the data by seeking the model vector $\mathbf{m} = [g_l^{mn}, h_l^{mn}]$ that minimizes the misfit to the data, two model norms (measuring the spatial and temporal roughness). The essential new feature of our method is the use of main field models as constraints for the field at the endpoints, 1980 and 2000 [Wardinski and Holme, 2002]. The a priori main field models for 1980.0 and 2000.0 are derived from high quality satellite vector data (MAGSAT, CHAMP and ØRSTED) rather than using satellite data directly in the inversion.
The final model is archived in an iteratively re-weighting scheme. At the first step data are weighted equally to gain an initial model, than the data of each observation site are re-weighted by its misfit to the first model to get a second model. Finally we discard data which are outside 2 standard deviation from second model to earn a final model.

## 4 Core–surface flow modelling

Models of core-surface flow are usually computed by using some approximation of the radial component of the diffusion-less magnetic induction equation

$$\partial_t \mathrm{B_r} = -\nabla_h \cdot (\mathbf{v}\,\mathrm{B_r}) \,. \tag{3}$$

$\mathrm{B_r}$ is the radial component of the magnetic field at the core-mantle boundary (CMB), $\nabla_h$ the horizontal divergence operator. The apparent simplicity of eq. 3 is misleading, Backus [1968] showed that solutions for the flow velocity $\mathbf{v}$ are non–unique. The ambiguity is resulting from the use of one equation to determine two orthogonal components of the flow, namely $\mathrm{v}_\phi$ and $\mathrm{v}_\theta$. To reduce the non–uniqueness one has to make assumption about the nature of the flow [Bloxham and Jackson, 1991, and ref. therein].
If the flow is assumed to be steady, or the same at three timestamps of a given interval, than eq. 3 is fully determined.
The geostrophic assumption bases on the hypothesis that the Coriolis force dominates the Lorentz and the viscous force in the core fluid near the CMB. This leads to the geostrophic constraint

$$\nabla_h \cdot (\mathbf{v_h} \cos(\theta)) = 0. \tag{4}$$

In this study we derive flow models, which are not time–dependent, for single epochs (e.g. for each year) assuming the flow to be either steady or geostrophic. The flows are computed using the time–dependent secular variation model and are not constrained to fit decadal variation of the length of day.

## 5 Results and Discussion

Our time–dependent model shows good conformity with the globally observed secular variation. In figure 1 we show fits of the model to the secular variation estimates at five permanent magnetic observatories. The jerks occurred in 1991 and around 2000 are well reshaped by our model, as one can see in the graph for dY/dt for Chambon la Forêt. The latter is also visible in dY/dt for Resolute Bay (more clearly in fig. 2). The jerk of 1983, observed only in the southern hemisphere, is visible in dY/dt for Hermanus and to some extent in dY/dt for Scott Base observatory (but with different sign than in Hermanus, more clearly in fig. 2). Furthermore, the model reveals rapid secular variation on a sub-decadal time scale, easiest seen in figure 2, where a comparison of the time–dependent modeled secular variation and flow–predicted secular variation is given for the same sites as in fig. 1. Secular variation predicted from geostrophic motion agrees well with the secular variation model, whereas steady flow is incapable of fitting the temporal complexity of observed secular variation. This finding is not new [Jackson, 1997].
We believe that the short term variations observed in figure 2 are of internal origin as: 1,) by using quietest monthly means and how we determine the secular variation estimates, external contributions (e. g., semi–annual and annual variations) are filtered out, 2,) applying a further weighting scheme to minimize the anisotropic error of the residuals [Holme and Bloxham, 1996].

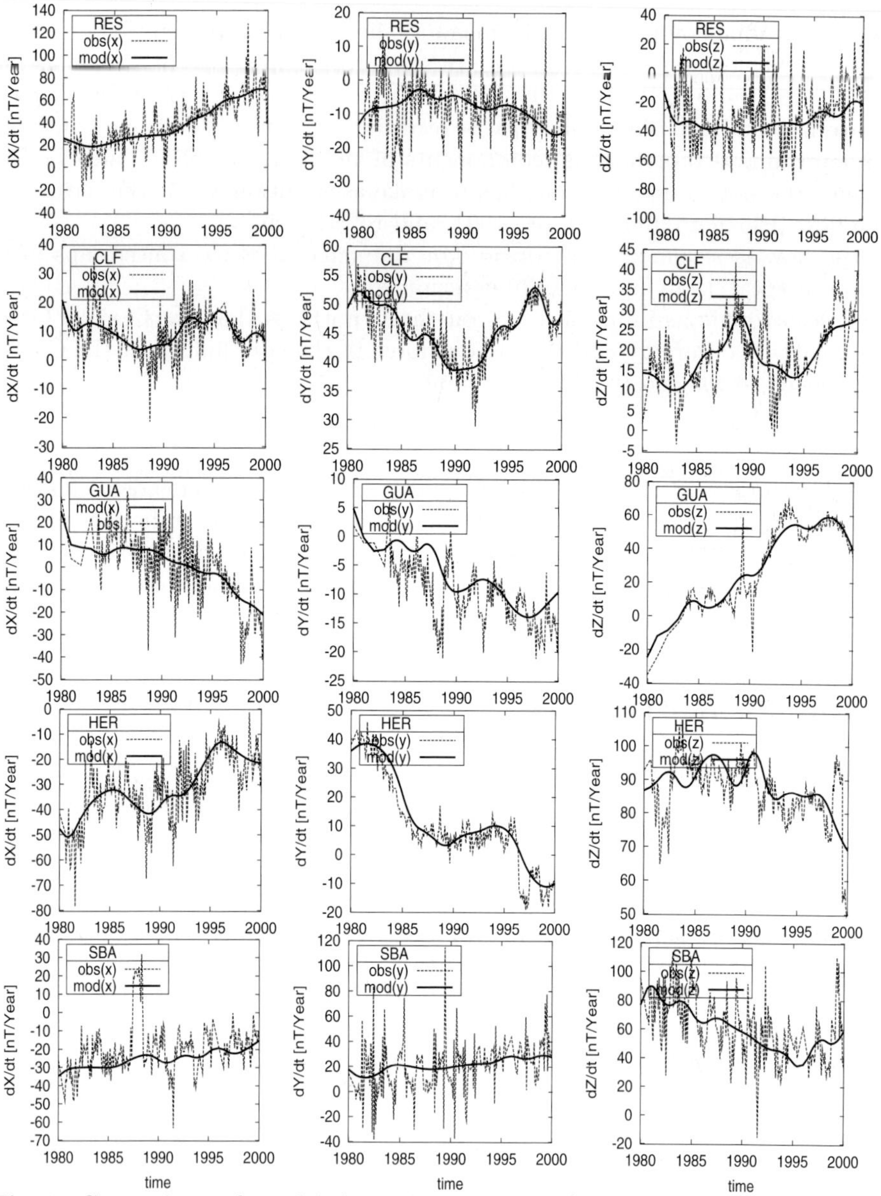


**Fig. 1.** Comparison of modeled secular variation (solid line) to the secular variation estimates at selected permanent observatories. From left to right: dX/dt, dY/dt, dZ/dt. From top to bottom: Resolute Bay (Canada), Chambon la Forêt (France), Guam (Pacific region), Hermanus (South Africa), Scott Base (Antarctica).

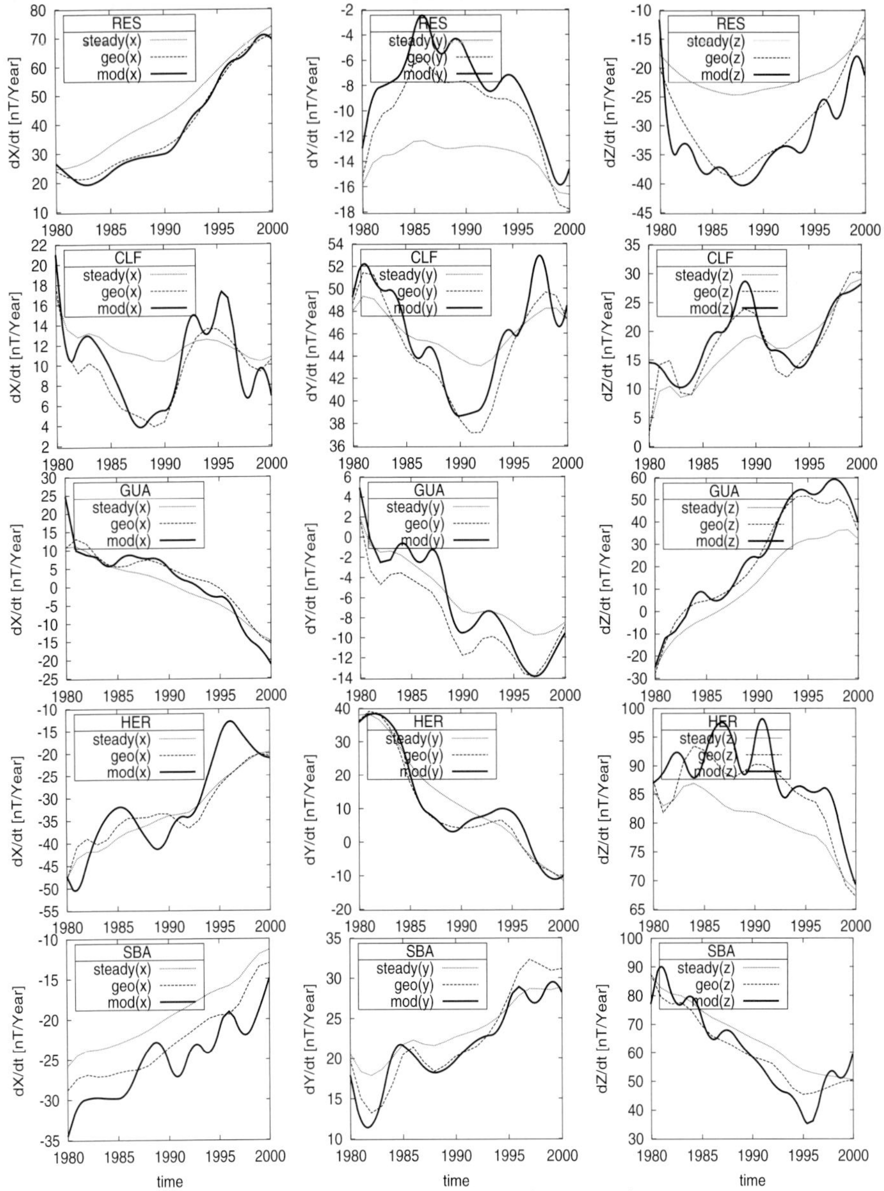


**Fig. 2.** Comparison of modeled secular variation (solid line) to flow–predicted secular variation, from a geostrophic tangentially flow (dashed line) and from a steady flow (dotted line) same components and observatories as in Figure 1.

*Acknowledgement.* We would like to record our gratitude to S. Macmillan, M. Mandea and N. Olsen for providing us with observatory means and ØRSTED field model. Funded by DFG special research program SPP 1097.

## References

Backus GE (1968) Kinematics of geomagnetic secular variation in a perfectly conducting core. Philos Trans R Soc London A *263*: 239-266.

Bloxham J and Jackson A (1991) Fluid flow near the surface of the Earth's outer core. Rev Geophys *29*: 97-120.

Bloxham J and Jackson A (1992) Time-dependent mapping of the magnetic field at the core-mantle boundary. J Geophys Res *97*: 19,537–19,563.

Dowson MJ, Buckingham JP, Simmons DA (1988) Notes on geomagnetic repeat measurements at Grytviken, April 1987. Br Antarc S Bull *71*: 49–52.

Gubbins D and Tomlinson L (1986) Secular Variation from monthly means from Apia and Amberly magnetic Observatory. Geophys J R astr Soc *86*: 603–616.

Holme R and Bloxham J (1996) The treatment of attitude errors in satellite geomagnetic data. Phys Earth Planet Int *98*: 221–233.

Kotzé PB, Kühn GJ, Scheepers GLM (1991) IUGG, Vienna, session GAM 1.1.

Jackson A (1997) Time–dependency of tangentially geostrophic core surface motions. Phys Earth Planet Int *103*: 293–311.

Macmillan S (1996). A geomagnetic jerk for the early 1990's. Earth Planet Sci Lett *137*: 189–192.

Mandea M, Bellanger E, and Le–Mouël J-L (2000) A geomagnetic jerk for the end of the 20th century? Earth Planet Sci Lett *183*: 369–373.

Wardinski I and Holme R (2002) Decadal and Subdecadal Secular Variation of Main Geomagnetic Field. In: Reigber C, Lühr H, Schwintzer P (eds) $1^{st}$ CHAMP Mission Results for Gravity, Magnetic and Atmospheric Studies, Springer: 226–232.

# Time Structure of the 1991 Magnetic Jerk in the Core-Mantle Boundary Zone by Inverting Global Magnetic Data Supported by Satellite Measurements

Ludwig Ballani, Ingo Wardinski, Dietrich Stromeyer, and Hans Greiner-Mai

GeoForschungsZentrum Potsdam, Telegrafenberg, D-14473 Potsdam, Germany, *bal@gfz-potsdam.de*

**Summary.** New global magnetic data - Gauss coefficients up to degree and order 5, monthly values from 1980 to 2000, fitted to global data and partly based on high-quality satellite vector data of Magsat and CHAMP/ØRSTED - are processed with a recent non-harmonic downward continuation method (Ballani et al. 2002). Using a weakly conducting mantle and the highly conducting fluid in the outer core we investigate the temporal structure of the 1991 jerk below some geomagnetic stations calculating the component dY/dt at the core-mantle boundary and underneath in different depths of the fluid outer core assuming fluid velocity there. The jerk structure dissolves and differs considerably in magnitude and in phase from the harmonically downward continued component.

**Key words:** CHAMP, Magsat, ØRSTED, geomagnetic jerk, Gauss coefficients, downward continuation, core-mantle boundary, fluid outer core

## 1 Introduction

Geomagnetic jerks are rapid changes in the secular variation of the Earth's magnetic field and are most easily seen in the first time derivative of the East component Y (Courtillot et al. 1978, Macmillan 1996). New global model data (Gauss coefficients) are the prerequisite for applying a recently developed rigorous inverse approach for downward continuation of the poloidal field to the core-mantle boundary (CMB). This approach facilitates to investigate the jerk morphology at the CMB and the region beneath and provides a new view of the features of the field near to the sources, if a passive layer or a simple-structured velocity distribution is assumed there.

## 2 Global model data

The input quantities for the non-harmonic downward continuation method (Ballani et al. 2002) are the Gauss coefficients $g_{nm}(t), h_{nm}(t)$. Their calculation is done in a separate modelling process (Wardinski & Holme 2003). As geomagnetic data we use observatory quiet monthly means, if not available

**Table 1.** Conductivity ($\sigma$) values for different shells and velocity $\omega$ in the upper core layer ($R_E$ = 6370 km, $R_\sigma$ = 5480 km, $R_c$ = 3485 km (CMB))

| shell | $\sigma$ [$\Omega^{-1}m^{-1}$] | $\omega$ [$10^{-3}$ rad/year] |
|---|---|---|
| $R_E \geq r > R_\sigma$ | 0 | 0 |
| $R_\sigma \geq r > R_c + 2$ km | $10 * (R_c/r)^5$ | 0 |
| $R_c + 2$ km $\geq r \geq R_c$ | $1 * 10^5$ | 0 |
| $R_c > r \geq R_c - 20$ km | $2 * 10^5$ | -1.745 |

annual means are used instead. With these data first–differences ('secular variations') are formed : Monthly means are treated by the n-step difference filter $dY/dt = Y(t + n/2) - Y(t - n/2)$, n = 12. Annual means are treated using $dY/dt = Y(t) - Y(t - 1)$. The geomagnetic potential is decomposed to spherical harmonics up to degree and order 5 and expanded in time by a basis of cubic B–splines with a knot separation of 1 year (Bloxham & Jackson 1992). The model is fitted to the data in the linear least squares sense, where we minimize the misfit between data and model, the spatial and temporal roughness and the deviation from *a priori* models at 1980 (Magsat) and 2000 (CHAMP and ØRSTED) using special norms and damping parameters. The final model is achieved by an iteratively re-weighting scheme (for more details see Wardinski & Holme in this volume).
Further input quantities (see table 1) are the core radius $R_c$, $R_\sigma$ as assumed upper bound for the conducting mantle shell and conductivity values, which are based on laboratory experiments for $R_\sigma \geq r > R_c + 2$ km (Shankland et al. 1993) while the value in $R_c + 2$ km $\geq r \geq R_c$ is suggested by D" theories and electromagnetic core-mantle coupling (Buffett 1992, Holme 1998). The angular velocity $\omega$ in the layer beneath the CMB was selected corresponding to the mean westward drift as major part of the secular variation.

## 3 Downward continuation theory

We apply the theory, presented in detail in (Ballani et al. 2002) and in extended form for a rotating fluid upper core layer with prescribed velocity in (Greiner-Mai et al. 2004), for the downward continuation of the quantity dY/dt. The vectorial induction equation (**B** magnetic flux, $\mu_0$ permeability of vacuum, $\sigma = \sigma(r)$ electrical conductivity, **v** velocity field)

$$- \nabla \times ( 1/(\mu_0 \, \sigma) \, \nabla \times \mathbf{B} ) + \nabla \times ( \mathbf{v} \times \mathbf{B} ) = \dot{\mathbf{B}}, \qquad \nabla \cdot \mathbf{B} = 0 \quad (1)$$

is decomposed to poloidal and toroidal parts by $\mathbf{B} = \mathbf{B_p} + \mathbf{B_t} = \nabla \times (\nabla \times \mathbf{r} S) + \nabla \times \mathbf{r} T$ with scalar functions $S$ and $T$. Only the poloidal part represented by the function $S$ expanded into spherical harmonics $S(r, \vartheta, \varphi, t) = \sum_{n,m} \left( S^c_{nm}(r,t) \cos m\varphi + S^s_{nm}(r,t) \sin m\varphi \right) P_{nm}(\cos\vartheta)$ is considered here. Strictly, the downward continuation problem is an initial-boundary value problem with two boundary conditions, both at the same (outer) radius, here $R_\sigma$ (see table 1). Prescribing $\mathbf{v} = (0, 0, \omega(r)) \times \mathbf{r}$ and $\sigma(r)$ we search the solution at $R_{oc}$ in the complex form $U(r,t) := S^c_{nm}(r,t) + i\, S^s_{nm}(r,t), \quad R_{oc} \le r \le R_\sigma, \quad 0 \le t \le T,$ from

$$U_{rr} + (2/r)\, U_r + (-n(n+1)/r^2 + im\mu_0\sigma(r)\omega(r))\, U = \mu_0\sigma(r)\, U_t$$
$$U(R_\sigma, t) = \phi(t)\,, \quad U_r(R_\sigma, t) + ((n+1)/R_\sigma)\, U(R_\sigma, t) = 0, \quad U(r, 0) = \psi(r)\,,$$

where $U_r := \partial U/\partial r$ etc. The input function $\phi(t)$ is derived from Gauss coefficients. By the output functions $S^{c,s}_{nm}(R_{oc}, t),\ R_{oc} \le R_c$, all field components of $\mathbf{B_p}$ at the CMB (taking $\omega(r) \equiv 0$) and beneath, in the fluid outer core upper layer, can be calculated. For the jerk studies we first compute $Y = B_\varphi$ with the spherical coordinates $(\vartheta, \varphi)$ for the corresponding observatory

$$Y = \frac{1}{r} \sum_{n=1}^{N} \sum_{m=0}^{n} \left( \left(\frac{\partial}{\partial r}(r\, S^s_{nm}) \cos m\varphi - \frac{\partial}{\partial r}(r\, S^c_{nm}) \sin m\varphi\right) m\, \frac{P_{nm}(\cos\vartheta)}{\sin\vartheta} \right)$$

Finally, the derivative dY/dt is realized by the n-step difference filter.

## 4 Time structure of dY/dt at the CMB – the 1991 jerk

Results for dY/dt at four geomagnetic stations can be seen in the figs. 1 and 2. In (a), the secular variation at the Earth's surface predicted by the model are compared with the observatory data to see the ability to reproduce the jerk or in general the low-frequency course of the model data between 1985 and 1995. Their limited degree 5 guarantees that the visible effect mainly originates from the deep earth interior. Figs. 1 and 2 (b) compare the non-harmonic downward continuation to the CMB and the application of the often used harmonic downward continuation which ignores any mantle conductivity. The high sensibility of the conductivity on dY/dt can be seen in the clearly differing order of magnitudes and the phase shifts which is in addition based on a different weighting of the single degree and order parts. As an application of the extended non-harmonic downward continuation theory to the outer core upper layer figs. 1 and 2 (c) give the possible jerk component for three depths. High conductivity combined with an assumed velocity (for the first 20 km $\omega$ = const is sufficient) changes the amplitude to higher scales of magnitude. These results can be considered as an illustrative numerical experiment possibly near to the "sources" of jerks and therefore useful for further modelling. It can be observed that the difference in the jerk data of relatively near stations, as Niemegk and Chambon-la-Foret, disappears nearer to the supposed source region where the calculated curves get a greater similarity.

By fig. 3 we show some synthetic single oscillation results to capture the "spectral characteristic" of a jerk: value of amplitude and phase shift. This feature is two-faced: the jerk can be understood as a "longperiodic" trend changing process and, on the other hand, concerning the decadal variations as a very short periodic event most visible in the dY/dt component. Fig. 3 compares related oscillations with periods between 3 years and 100 years.

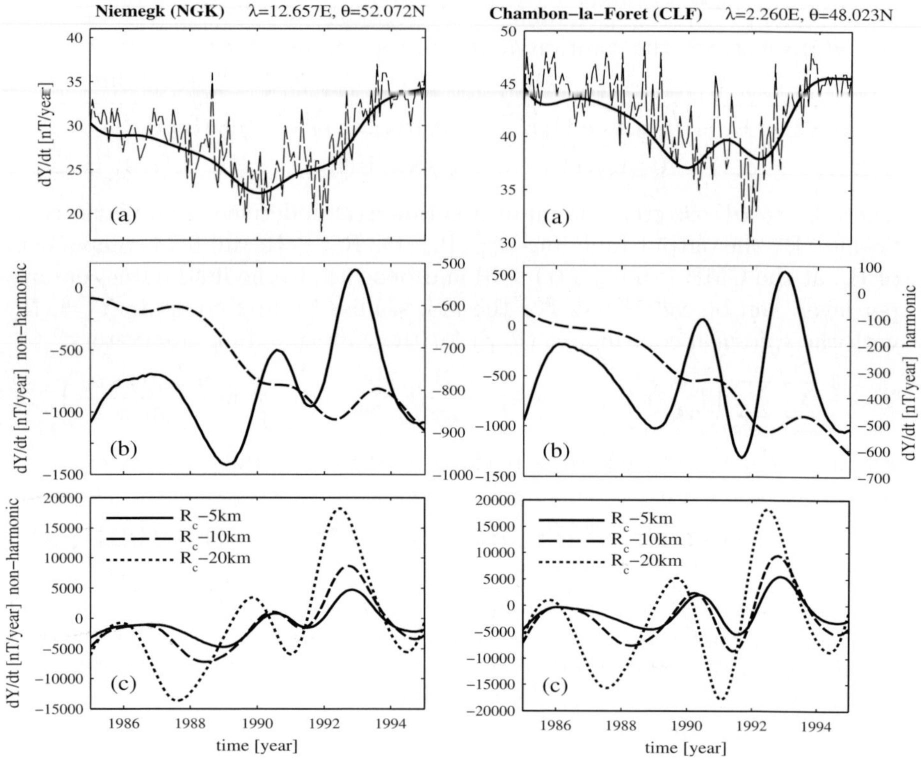


**Fig. 1.** Component dY/dt at the stations Niemegk and Chambon-la-Foret: (a) Derived from observatory data (dashed line) and global model data derived from Gauss coefficients $(n, m \leq 5)$ at the earth surface (solid line) (b) Harmonic (dashed line, right scale) and non-harmonic (solid line, left scale) downward continuation of the global model data to the CMB (c) Non-harmonic downward continuation of global model data to three depths below the CMB.
(note the different vertical scales)

# 5 Conclusions

The dY/dt component $(\dot{B}_\varphi)$ derived from the Gauss coefficients has the typical jerk structure at geomagnetic observatories. We obtained the jerk structure by a spherical harmonic expansion limited to degree and order n=m=5 from globally distributed observatory data. This indicates that the origin of this signal is coming with the utmost probability from the deep Earth's interior.
Secondly, we apply the method of non-harmonic downward continuation to calculate the poloidal part of $\dot{B}_\varphi$ at the CMB in a rigorous inverse way. This creates a basis for a geophysically founded study of the magnetic jerk of 1991 and can contribute to finding the causing mechanism. Compared with the

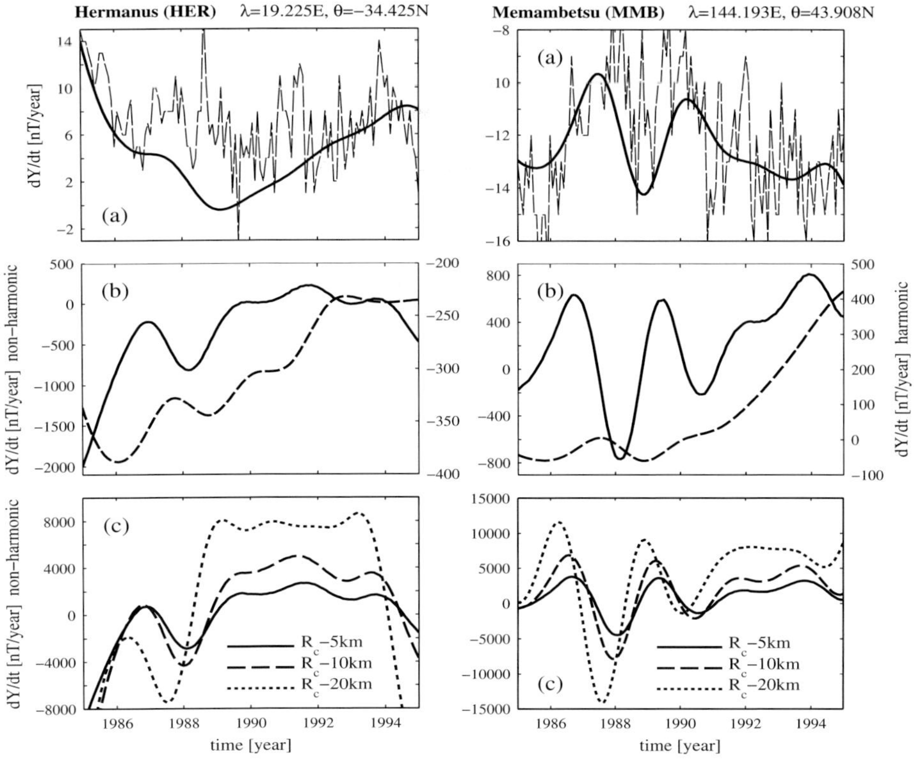


**Fig. 2.** Component dY/dt at the stations Hermanus and Memambetsu: (a) Derived from observatory data (dashed line) and global model data derived from Gauss coefficients ($n, m \leq 5$) at the earth surface (solid line) (b) Harmonic (dashed line, right scale) and non-harmonic (solid line, left scale) downward continuation of the global model data to the CMB (c) Non-harmonic downward continuation of global model data to three depths below the CMB. (note the different vertical scales)

*harmonically* downward continued jerk there is a completely other weighting of the degree portions by the *non-harmonic* procedure which, in addition, allows more realistic geophysical assumptions to be included. Thus, the results for the sensitive $\dot{B}_\varphi$ component are much larger changes in the order of magnitude combined with significant phase shifts. Moreover, the initial jerk structure dissolves and vanishes completely, so that no simple function type can be assigned to it, and some earlier speculations about its morphology in the CMB region and/or physical causes should be reconceived.

Although some basic assumptions (about $\sigma(r)$ and $\omega$) are not proven yet, our study of the jerk in different depths of a fluid outer core upper layer can be considered as a first numerical experiment giving indications for a better understanding of the jerk phenomenon.

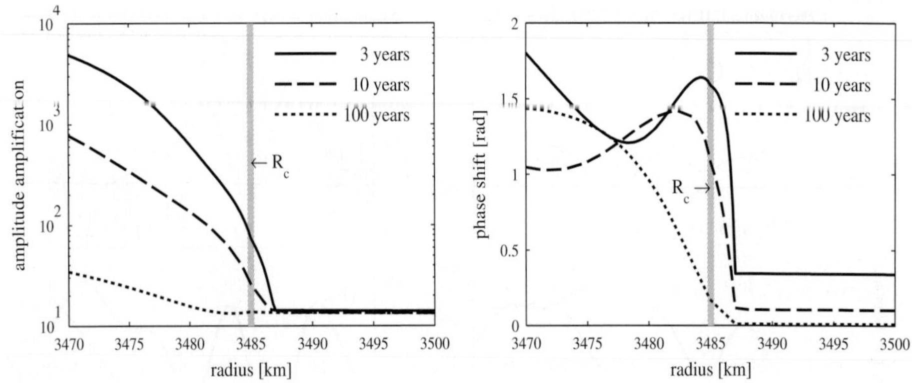


**Fig. 3.** dY/dt set up as a single oscillation: Amplitude and phase for periods of 3, 10 and 100 years in depths around the CMB, normalized at $R_E$ by 1 and 0, respectively.

*Acknowledgement.* We thank our colleagues, Veronika Söllner and Hans-Joachim Kühn, for their kind technical help and an unknown reviewer for some helpful suggestions of improvement.

## References

1. Ballani L, Greiner-Mai H & Stromeyer D (2002) Determining the magnetic field in the core-mantle-boundary zone by non-harmonic downward continuation. Geophys J Int *149*: 374–389.
2. Bloxham J & Jackson A (1992) Time-dependent mapping of the magnetic field at the core-mantle boundary. J Geophys Res *97*: 19,537–19,563.
3. Buffett BA (1992) Constraints on magnetic energy and mantle conductivity from the forced nutations of the Earth. J Geophys Res *97*: 19,581–19,597.
4. Courtillot V, Ducruix J & Le Mouël J-L (1978) Sur une accélération récente de la variation séculaire du champ magnétique terrestre. C R Acad Sci D *287*: 1095–1098.
5. Greiner-Mai H, Ballani L, & Stromeyer D (2004) The poloidal geomagnetic field in a differentially rotating upper core layer. Geophys J Int, accepted.
6. Holme R (1998) Electromagnetic core-mantle coupling - II. Probing deep mantle conductance. The Core-Mantle Boundary Region. In: Gurnis M, Wysession ME, Knittle E, Buffett BA (eds), The Core-Mantle Boundary Region, American Geophysical Union, Washington, pp. 139–151.
7. Macmillan S (1996) A geomagnetic jerk for the early 1990's. Earth Planet Sci Lett *137*: 189–192.
8. Shankland TJ, Peyronneau J, & Poirier JP (1993) Electrical conductivity of the Earth's lower mantle. Nature *366*: 453–455.
9. Wardinski I & Holme R (2003) Decadal and Subdecadal Secular Variation of Main Geomagnetic Field. In: Reigber C, Lühr H, Schwintzer P (eds), First CHAMP Mission Results for Gravity, Magnetic and Atmospheric Studies, Springer-Verlag, Berlin et al, pp. 226–232.

# Use of Champ Magnetic Data to Improve the Antarctic Geomagnetic Reference Model

Luis R. Gaya-Piqué[1,2], Angelo De Santis[1] and Joan Miquel Torta[2]

[1] Istituto Nazionale di Geofisica e Vulcanologia, Via di Vigna Murata 605, 00143 Roma, Italy, *pique@ingv.it*
[2] Observatori de l'Ebre, 43520 Roquetes, Spain

**Summary.** Champ total field measurements have been used to develop the new version of the Antarctic geomagnetic Reference Model (ARM). The model was conceived as a tool to evaluate the main field in Antarctica, facilitating the merging of different magnetic surveys carried on in the region from 1960 onwards. Spherical cap harmonic analysis was used to produce the model. Together with data coming from POGO, Magsat, and Ørsted satellite missions, a suitable selection of Champ data based on different criteria was performed to minimise the effect of external fields. The comparison of ARM and other global models with regard to real data demonstrates the validity of our regional model, specially for the representation of the secular variation of the geomagnetic field. Since Champ satellite tracks cover the Geographical South Pole better than other satellite missions, this fact contributed to improve the model in the central region of the cap.

**Key words:** regional modeling, main geomagnetic field, spherical cap harmonic analysis

## 1 Introduction

The beginning of the new millennium coincided with the launch of two satellite missions that opened new sights for the development of global magnetic reference models, the Danish Ørsted and the German Champ. The most clear example is the Comprehensive Model by Sabaka et al. (2002), which gives account of the main and the crustal fields up to degree 65, but also of the ionospheric and magnetospheric (primary and induced) fields. Nevertheless regional models can still be considered a good approach when we try to reproduce the geomagnetic field for those epochs in which no satellite missions were in course (between Ørsted and the previous Magsat magnetic mission there exists a 20 years gap), and especially in those regions devoid of magnetic observatories, like Antarctica and the surrounding seas.

In this paper we present the updating of the Antarctic geomagnetic Reference Model (ARM). The first version of this regional model (De Santis et al. 2002) was developed as an answer to the request of the Antarctic Digital Magnetic Anomaly Project (Chiappini et al. 1998) for a more accurate model than the International Geomagnetic Reference Model (IGRF) to reduce the magnetic surveys carried on in Antarctica during the last forty years odd. Satellite data have played an important role in this new version of the model.

## 2 Data used

The amount of observatory data used has been highly increased with respect to the previous secular variation model (Torta et al. 2002), and even with respect to the first complete version (main field plus secular variation) of the ARM (De Santis et al. 2002), which already included satellite data as well.. A total of 500 annual means for X, Y, and Z magnetic components from 25 Antarctic observatories have been used, with a temporal coverage from 1960 to nowadays (we refer the reader to figure 1 in Torta et al. (2002) for the details about the location of the observatories). The new ground data used correspond to the latest annual means sent to the World Data Center for Geomagnetism in Edinburgh by each observatory, whereas some old data have been also revisited and disagreements corrected.

The novelty with regard to the satellite data has been the use of total intensity values F from six different missions, in contrast to the few measurements from Ørsted satellite used in the previous version of ARM. Data from OGO-2, OGO-4, OGO-6 (in flight from 1965 to 1971), Magsat (1980), Ørsted, and Champ (from 2000 to present) missions have been considered. A detailed description of the first four missions can be found in Langel and Hinze (1998). The criteria used for the satellite data selection was based on the indices of external magnetic activity (see for example Mayaud 1980). The Kp index was chosen to be lower or equal to 1- for the time of observation and to 2o for the previous interval, as proposed by Coles (1985) for polar areas, a more restrictive criteria than the one ($Kp \leq 1+$) used by Olsen et al. (2000). The maximum absolute value of Dst index was fixed in 10 nT with a maximum temporal variation of 3 nT/h. The AE index value, a more accurate parameter to describe the activity in auroral regions, was chosen less than 50 nT (Ravat et al. 1995). Since AE values were not available from 2002 onwards, a final selection was applied based on the PC index (Troshichev et al. 1979), given that there exits a good correlation between both PC and AE indices (Vennerstrøm et al., 1991). The threshold value chosen by us was equal to 0.3 for the Ørsted and Champ data. Finally, data were limited to a maximum altitude of 900 km. Table 1 shows the number of data coming from each mission after the progressive application of these criteria, and figure 1 represents the spatial distribution of the data.

| Satellite | Years | Altitude | Data | Satellite | Years | Altitude | Data |
|---|---|---|---|---|---|---|---|
| OGO-2 | 1966-67 | 430-900 | 72 | Magsat | 1980 | 330-460 | 341 |
| OGO-4 | 1967-70 | 420-900 | 1116 | Ørsted | 2000- | 650-870 | 2076 |
| OGO-6 | 1969-71 | 400-900 | 3414 | Champ | 2000- | 400-500 | 2274 |

**Table 1.** Number of data used to develop the ARM together with information for each satellite mission (altitude in km above Earth's surface).

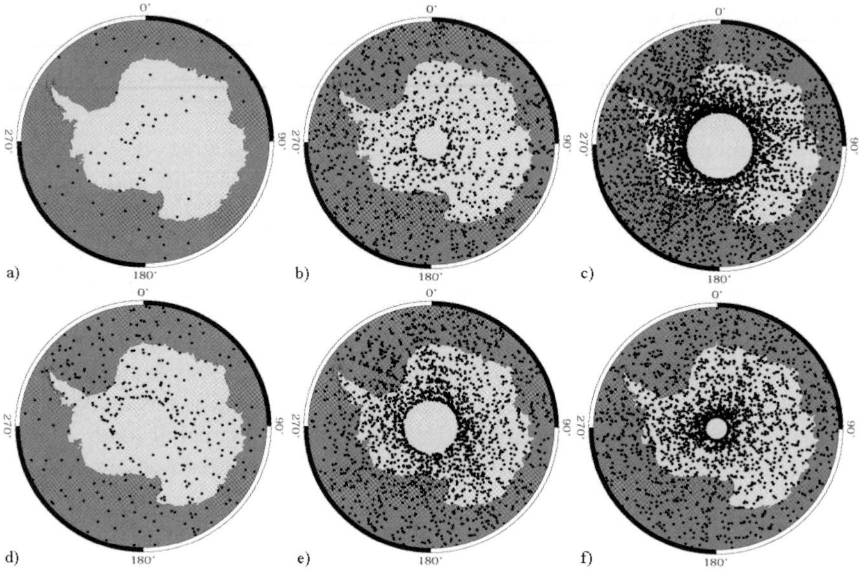


**Figure 1.** Spatial distribution of the satellite data used in the model. The figures correspond to: a) OGO-2; b) OGO-4; c) OGO-6; d) Magsat; e) Ørsted; and f) Champ missions.

# 3 Model

Spherical cap harmonic analysis (SCHA, Haines 1985) was applied to obtain the reference model. The magnetic potential due to internal sources over a spherical cap in spherical coordinates is given as solution of Laplace's equation in terms of a harmonic expansion similar to that one of the global case, but with the main difference that the associated Legendre functions present an integer order $m$ but a generally non-integer degree $n_k$, being $k$ the index used to sort the different degrees for a given order. The expansion is developed for a maximum spatial $K$ and temporal $L$ indices. The coefficients were obtained using the main field differences method proposed by Haines (1993). This method seems to provide better results for the modelling of the secular variation when the input data series present many discontinuities, like in the Antarctic observatory database case. The reader can refer for example to De Santis et al. (2002) for details.

The non-linearity of the total field in the coefficients introduced by the inclusion of the scalar satellite data was surpassed applying an iterative linearization (Cain et al. 1965); in our case, the value of the IGRF2000 at the epoch of each measurement provided the starting value for the iteration, which continued until the change in the model coefficients was negligible.

Each data set was weighted differently according to the reciprocal of the expected crustal field and the error contained in the measurements (Haines and Newitt 1997). For the observatory data set, the crustal anomaly was set to

$\sigma_c$=200 nT and the measurement error to $\sigma_m$=10 nT. For the satellite missions, the values chosen were 0 nT for $\sigma_c$ (except for Magsat and Champ, where $\sigma_c$=5 nT was chosen because of their lower altitude), whereas $\sigma_m$ fluctuated between 10 nT for the newest missions and 25 nT for the oldest.

The maximum spatial index was fixed in $K$=8. For a 30 degrees half angle cap, this choice represents a maximum degree $n_k$ for the expansion approximately equal to global spherical harmonic degree 25 (IGRF model goes up to degree 10). It means that the minimum wavelengths that the ARM model is theoretically able to represent are about 1600 km. Cosine functions were chosen to represent the temporal variation of the model, since these functions have a derivative equal to zero at the borders of the temporal interval of the model, reducing the problems related to extrapolation. A variable maximum temporal order was decided depending on the value of the spatial index. In this way, $L$=4 was assigned for $K$=0 to 4, $L$=1 for $K$=5 to 7, and $L$=0 was given to the functions with spatial index $K$=8. In this way, the temporal variation of the higher degree harmonics was avoided. After applying a statistical stepwise regression procedure (Haines and Torta 1994), the number of coefficients that define the model was 123.

## 4 Results and discussion

The utility of the ARM can be demonstrated comparing the fit to the real data with that given by other global models. Table 2 shows the fit to the observatory annual mean values for ARM, IGRF, and for CM3, the latest version of the Comprehensive Model by Sabaka et al. (2002). For this last model only the main field and crustal contributions (up to degree 65) have been considered. All three models present more or less the same behavior. ARM fits better than the other models X, Z, and F magnetic elements, whereas it is slightly worst with regard to the Y component. The big residuals found can be explained due to the high biases presented by the Antarctic observatories. The biases modeled by the CM3 model for the observatories used in this study are similar to the residuals between the real annual means and the values predicted by the ARM (result not shown in this paper). Major discrepancies are found for those observatories which only present data for few years or that are located in poorly covered regions.

| MODEL | X | Y | Z | F |
|---|---|---|---|---|
| IGRF | 691 | 599 | 1363 | 1401 |
| CM3 | 686 | 611 | 1350 | 1389 |
| ARM | 680 | 616 | 1345 | 1384 |

**Table 2.** RMS fit (in nT) of IGRF, CM3, and ARM models to observatory annual means.

A look at table 3 gives an idea about the importance of a model like ARM. The root mean square fits for IGRF and ARM to the observatory differences relative to the means over all data, shows an improvement of the regional model with regard to the global one that is about 45% for X and Y components, reaching 70% for the Z component. Taking into account that IGRF has been the model traditionally used to reduce magnetic surveys to a same epoch for the isolation of the crustal field, it can be said that ARM represents an improved tool for these studies.

| MODEL | X | Y | Z |
|---|---|---|---|
| IGRF | 42.6 | 55.3 | 121.8 |
| ARM | 23.9 | 28.5 | 35.2 |

**Table 3.** RMS fit (in nT/year) of IGRF and ARM to observatory differences with respect to the respective mean values.

On the other hand, ARM fits also better the satellite data than IGRF and CM3 when only the internal contributions for the latter one are considered, as can be seen in table 4. CM3 is superior when it is allowed to take into account the external field through the (previously known) values of Dst and f10.7 indices. The inclusion of external field contributions has not been considered in our model. In any case, the comparison of the magnetic field values predicted by all three models for a given satellite track (figure 2) shows that differences between ARM and CM3 (including internal and external contributions) are much less than between IGRF and the Comprehensive Model.

| Satellite | OGO-2 | OGO-4 | OGO-6 | Magsat | Ørsted | Champ |
|---|---|---|---|---|---|---|
| IGRF | 21.9 | 21.7 | 22.5 | 14.7 | 27.2 | 27.2 |
| $CM3_{internal}$ | 16.6 | 22.3 | 22.2 | 21.4 | 24.5 | 24.8 |
| $CM3_{internal+external}$ | 6.3 | 5.8 | 5.3 | 3.7 | 3.3 | 4.6 |
| ARM | 10.1 | 8.7 | 7.7 | 9.5 | 6.9 | 6.8 |

**Table 4.** RMS fit (in nT) of IGRF, CM3, and ARM to satellite data.

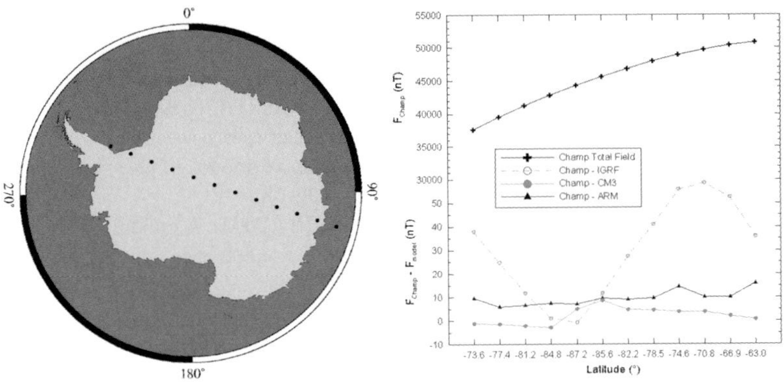


**Figure 2.** Residuals (right) of IGRF, CM3, and ARM models with respect to a Champ satellite track for May 19th, 2002 (left).

The quality of the ARM fits to the real data allows us to affirm that the Antarctic geomagnetic Regional Model can be considered a useful tool to develop magnetic studies in Antarctica. The choice of a low maximum spatial index assures avoiding overfitting, so the behavior of the model is expected to be good for epochs (from 1960 onwards) or regions (below 60 degrees South latitude) in which magnetic data were not available. The future work will be focused in demonstrating the validity of the ARM to reduce magnetic surveys carried on in Antarctica for the last forty years. Once the fit to the real ground data used in the model has shown that ARM presents a better behavior than IGRF, it is expected that the isolation of the Antarctic anomaly field will be improved through the use of the regional reference model.

## References


Cain JC, Daniels WE, Hendricks SJ, Jensen DC (1965) An evaluation of the main geomagnetic field, 1940-1962. J Geophys Res *70*: 3647-3674.

Chiappini M, von Frese RRB, Ferris J (1998) Effort to develop magnetic anomaly database aids Antarctic research. EOS Trans Am Geophys Union *79*: 290.

Coles RL (1985) Magsat scalar magnetic anomalies at Northern high latitudes. J Geophys Res *90 (B3):* 2576-2582.

De Santis A, Torta JM, Gaya-Piqué LR (2002) The first Antarctic geomagnetic Reference Model (ARM). Geophys Res Lett *29 (8):* 33.1-33.4.

Haines GV (1985) Spherical Cap Harmonic Analysis. J Geophys Res *90 (B3):* 2583-2591.

Haines GV (1993) Modellig geomagnetic secular variation by main-field differences. Geophys J Int *114*: 490-500.

Haines GV, Torta JM (1994) Determination of Equivalent Currents Sources from Spherical Cap Harmonic Models of Geomagnetic Field Variations. Geophys J Int *118*: 499-514.

Haines GV, Newitt LR (1997) The Canadian geomagnetic reference field 1995. J Geomag Geoelectr *49*: 317-336.

Langel RA, Hinze WJ (1998) The magnetic field of the Earth's lithosphere. The satellite perspective. Cambridge University Press, Cambridge.

Mayaud PN (1980) Derivation, meaning and use of geomagnetic indices. Geophys Monograph Am Geophys Union 22, Washington DC.

Olsen N et al. (2000) Ørsted Initial Field Model. Geophys Res Lett *27*: 3607-3610.

Ravat D, Langel RA, Purucker M, Arkani-Hamed J, Alsdorf DE (1995) Global vector and scalar Magsat magnetic anomaly maps. J Geophys Res *100 (B10):* 20111-20136.

Sabaka TJ, Olsen N, Langel RA (2002) A comprehensive model of the quiet-time, near Earth magnetic field: phase 3. Geophys J Int *151*: 32-68.

Torta JM, De Santis A, Chiappini M, von Frese RRB (2002) A model of the Secular Change of the Geomagnetic Field for Antarctica. Tectonophysics *347*: 179-187.

Troshichev OA, Dmitrieva NP, Kuznetsov BM (1979) Polar cap magnetic activity as a signature of substorm development. Planet Space Sci *27*: 217-221.

Vennerstrøm S, Friis-Christensen E, Troshichev OA, Andersen VG (1991) Comparison between the Polar Cap index, PC, and the Auroral Electrojet Indices AE, AL, and AU. J Geophys Res *96 (A1):* 101-113.

# Secular Variation of the Geomagnetic Field from Satellite Data

Vadim P. Golovkov, Tatiana I. Zvereva and Tatiana A. Chernova

Institute of Terrestrial Magnetism, Ionosphere and Radio Wave Propagation, IZMIRAN, Troitsk, Moscow Reg., 142190, Russia, *golovkov@izmiran.rssi.ru*

**Summary.** Two methods of utilizing data from magnetic satellites for the secular variation modelling are proposed. Both methods are based on the parametric expansion in time, where natural orthogonal components are used as parameters. As data for such expansion are used time series of coefficients of spherical harmonic models, developed from satellite data for each day.

**Key words:** secular variation, satellite data, natural orthogonal components.

## Introduction

Modern harmonic models of the geomagnetic field include the coefficients of the main field, the secular variation, Dst-variation and sometimes the annual variation (for example, [*Olsen et al., 2002*]). The misfit of such approximation is 3-5 nT and can be considered as sufficient for the main field description. But errors of the satellite models of SV are relative high.

That is the reason to search an alternative approach to SV determination from satellite data, which is based upon an idea of representation of the measured field with a sum of the fields of different origins. These fields are to be uncorrelated both in space and in time. Earlier we have investigated daily data from magnetic observatory network and succeeded in a similar separation of the measured field by means of expanding in a series of numerical natural orthogonal functions [*Golovkov and Zvereva, 1998*]. The first term of such the expansion describes the secular variation, the second term the Dst-variation and the third one the annual variation. The next components describe a white noise. The study of different 2-year intervals corresponding to different levels of solar activity showed that only these three field components of expansion are significant and approximate the field with enough good accuracy.

The main aim of this paper is to apply similar approach for the satellite survey data analysis, and to estimate the accuracy of models, obtained in such way.

## Data and modelling

We have used daily data X and Z from 01-January-2002 to 31-December-2002. From all observatories available for this interval we have taken 16 more or less uniformly distributed on the globe. Our former research has shown that this number is enough for precise separation of the field [*Golovkov and Zvereva, 1998*]. The global CHAMP satellite 1 second vector data (without averaging) were used (all 365 days). Every day consisted of about 85000 (±1000) measurements.

Combinations of two methods of the determination of the secular variation were used. The first was the method of expansion of the data on series of natural orthogonal components. The second was the well-known method of spherical harmonic analysis. These two methods are briefly described below.

The natural orthogonal component analysis [*Faynberg, 1975; Langel, 1987*]. If we have a number of temporal series of data that can be described by a rectangular matrix, in which each line $j \in [1,J]$ contains elements $H_{ij} (i \in [1, I])$, they can be represented as

$$H_{ij} = \sum_{k} C_{kj} \cdot T_{ki} \tag{1}$$

where numerical functions $C_{kj}$ do not depend on time and numerical functions $T_{ki}$ do not depend on the point position. Main quality of obtained $T_{ki}$ and $C_{kj}$ is their full orthogonality on the data set.

Hence, as it follows from (1), a variation of the field in some point $j$ is represented by means of linear combination of $k$ temporal functions $T_{ki}$ with coefficients $C_{kj}$. These temporal functions are common for all observation points and do not correlate one with others.

Taking into account that the condition of the orthogonality is satisfied we can say that the method NOC permits to separatee the observed variations of the geomagnetic field into parts, each of which has particular structure and varies independently in time.

The commonly used spherical harmonic expansion of the potential can be written in the form:

$$U(r,\theta,\lambda,t_i) = a \cdot \sum_{k=1}^{K} T_{ki} \sum_{n=1}^{N} \sum_{m=0}^{n} \left(\frac{a}{r}\right)^{n+1} (g_{nk}^{m} \cos m\lambda + h_{nk}^{m} \sin m\lambda) \times P_n^m(\cos\theta) \tag{2}$$

where *a* is the mean radius of Earth; $P_n^m(\cos\theta)$ is the associated Legendre function of degree n and order m, normalized according to the convention of Schmidt. The $g_n^m$ and $h_n^m$ in this approach are functions depending only on time. They can be expanded in time as that follows from (1):

$$g_n^m = \sum_{k=1}^{K} g_{nk}^m T_{ki} \,. \tag{3}$$

$K \ll \min [J, I]$ and $N(N+2) <$ full amount of data. It is similar for $h_{nk}^m$.

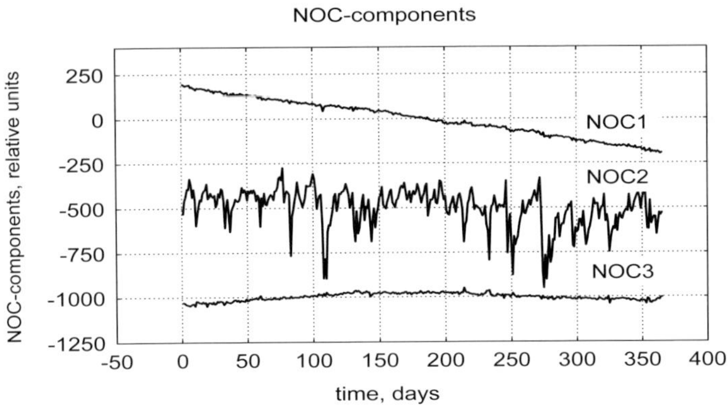


**Fig. 1.** Three temporal orthogonal components NOC1, NOC2, NOC3, received from NOC-analysis of the daily mean values of X- and Z-components of the magnetic field in observatories in 2002. The NOC2 and NOC3 curves are set off by -500 and -1000 units respectively.

Therefore we can use some numerical functions describing $T_{ki}$ as well as analytical ones, traditionally used, to describe the temporal changes of $g_n^m, h_n^m$. Requirements of these numerical functions are the same as to analytical ones. They must (i) be orthogonal to each other on the whole time interval, (ii) approximate the observed series with the required accuracy, (iii) be the rapidly converging series of expansion terms.

## Models

**Model SV_1**. The model was obtained as follows. At the first stage we used the daily data X, Z of 16 observatories. The data of these observatories were processed by the method NOC. The result is demonstrated in Fig. 1. The first term of the NOC-expansion describes the secular variation, the second term the Dst-variation and the third one the annual variation.
At the second stage the data of the satellite CHAMP for each day of the year 2002 were used. Daily spherical harmonic models up to n=m=8 were constructed. For all coefficients $g_n^m$, $h_n^m$ of these 365 models the average values for 2002 $\bar{g}_n^m$ and $\bar{h}_n^m$ were found. According to results of a stage 1 difference $\Delta g_n^m(i) = g_n^m(i) - \bar{g}_n^m$ can be presented as:

$$\Delta g_n^m(i) = g_{n1}^m * NOC1(i) + g_{n2}^m * NOC2(i) + g_{n3}^m * NOC3(i)$$

and the same for $\Delta h_n^m(i)$.
The solution of this over-determined system of linear equations by the method of least squares gives three sets of coefficients $g_{n1}^m(i), h_{n1}^m(i)$, $g_{n2}^m(i)$, $h_{n2}^m(i)$ and

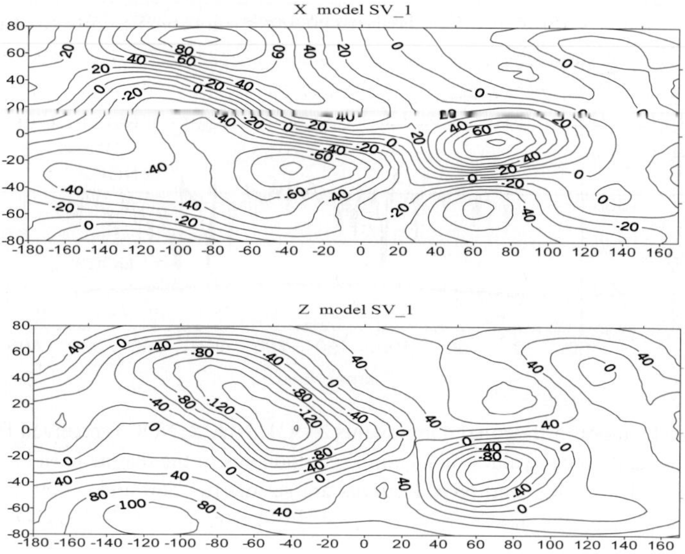


**Fig. 2.** Maps of X and Z of the secular variation for model SV_1. Contour intervals are 10 T/yr(X) and 20 nT/yr(Z).

$g_{n3}^m(i), h_{n3}^m(i)$ for each n, m. The first set of coefficients $g_{n1}^m(i), h_{n1}^m(i)$ is the spherical harmonic model of the secular variation for the year 2002.5 (model SV_1). The maps of X and Z of this model are presented in Fig. 2.

**Model SV_2**. The one-year interval from January 2002 to December 2002 was selected. In each month the most geomagnetically quiet day was selected on $k_p \leq 2$. Therefore 12 independent collections of the SH-coefficients were obtained. They were used for processing by NOC method. The matrix for this method was composed as follows: the columns of the matrix were the sets of obtained SH-coefficients. In equation (1), $C_{kj}$ corresponds in the given NOC-expansion to one $g_n^m$ or $h_n^m$, so that each j reflects a unique combination (n, m, g, h). The result of NOC-analysis, namely the secular variation received as the first orthogonal component, is presented in Fig. 3.

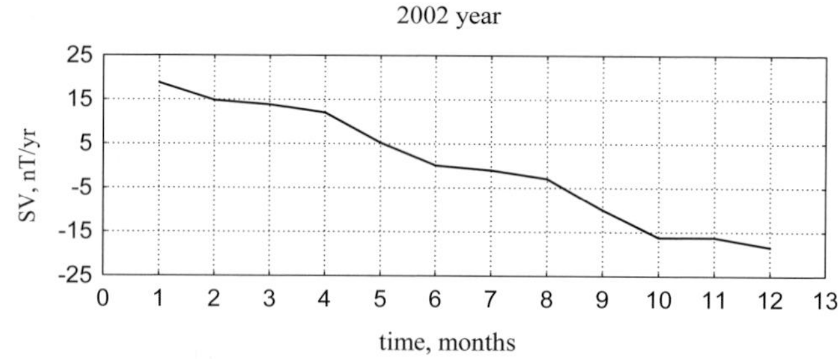


**Fig. 3.** The secular variation is received as the first orthogonal component in the expansion by NOC-analysis on CHAMP data.

Thus we received the sets of the spherical harmonic coefficients for the secular variation. The pseudo-energy spectrum from spherical coefficients of secular variation of the model SV_2 shows, that significant part of this spectrum is limited by n=7. Higher degrees reflect rather errors of obtained coefficients than any physical phenomenon.

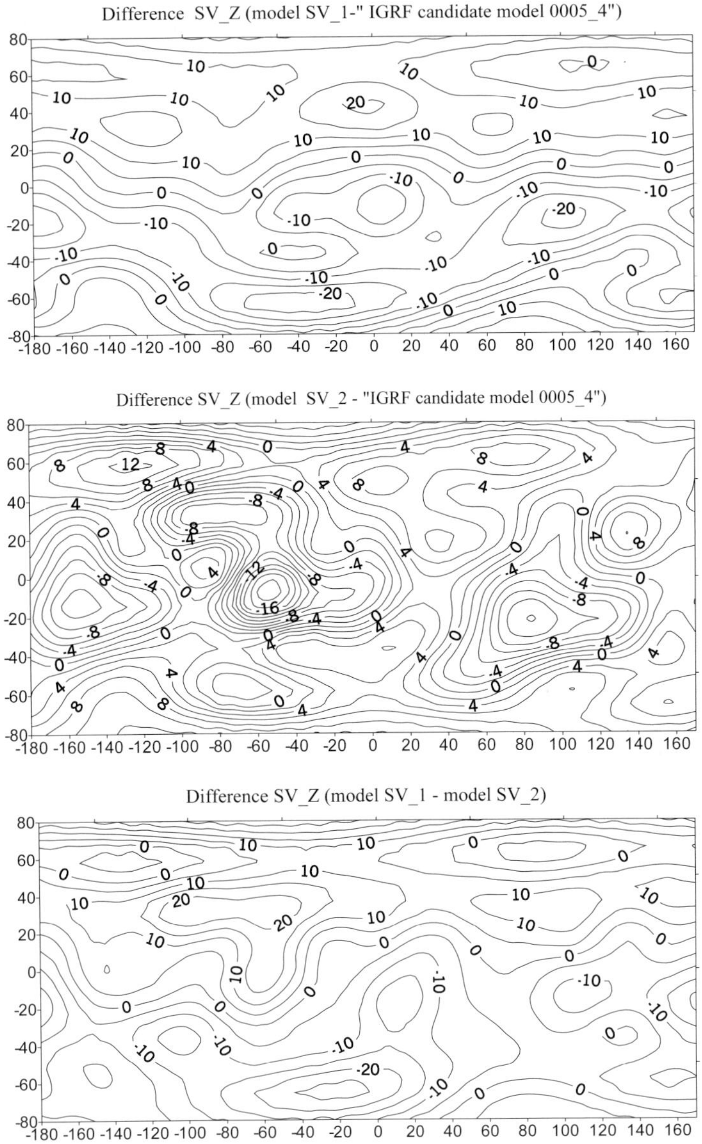


**Fig. 4.** Differences of the secular variation Z for models SV_1, SV_2 and "IGRF candidate model 0005_4" are presented. Contour intervals are 5 nT/yr (the first and third maps) and 2 nT/yr (the second map).

In Fig. 4 are presented differences Z of the secular variation for models SV_1, SV_2 and "IGRF candidate model 0005_4" *[Maus, Lühr et al.]*.

## Conclusion

Two presented methods applied natural orthogonal components as basic functions for an expansion on time series of the SH coefficients. Comparing results of these methods as well as both of them with the secular variation from IGRF candidate model 005_4, obtained by *[Maus, Lühr et al.]* we have found, that the differences between the models are of the order of 10% of the mean global SV. Consequently the suggested methods of SV determination are equally accurate. Their accuracy is comparable with the accuracy of SV models offered as prognostic ones for epoch 2002.5 in spite of method used. Model SV_1 is based on data of all local times, geographical location and geomagnetic activity. Model SV_2 was created using only limited set of quiet day data. The method 1 does not demand a selection of the initial data. Because of the averaging of data within a day we use in SHA all satellite data: both the day and night sides of the orbit. Only use observatory annual means could fulfill a function of the choice.

*Acknowledgment.* The work was supported by grants of INTAS 01-0142, ISTC KR-214 and RFBR 03-05-64518.

## References

Bondar TN, Golovkov VP, Yakovleva SV (2002) Secular variations around 2000, obtained from satellite and observatory data. OIST-4 Proceedings, Copenhagen.

Faynberg EB (1975) Separation of the geomagnetic field into a normal and anomalous part. Geomagn Aeron Engl Trans. *15*: 117.

Golovkov VP, Zvereva TI (1998) Decomposition of the geomagnetic variations inside a year on natural orthogonal components. Geomagnetism and Aeronomia *38*: 140-145.

Langel RA (1987) Main field in Geomagnetizm edited by Jacobs JA, Academic Press, London.

Maus S, Lühr H, Rother M, Mai W: Information on GFZ candidate model for SV-2002.5. "Candidate model IGRF_0005_4". http://www.ipgp.jussieu.fr/~mioara/dgrf/dgrf_page.html.

Olsen N (2002) A model of the geomagnetic field and its secular variation for epoch 2000 estimated from Oersted data. Geophys J Int *149*: 454-462.

Zvereva TI, Kasjanenko LG (1991) Influence of spatial distribution and data errors on secular change at sea stations on making the global models of secular variations. Geomagnetism and Aeronomia *31*: 1084-1093.

# The Spectrum of the Magnetic Secular Variation

Richard Holme[1] and Nils Olsen[2]

[1] University of Liverpool, UK *holme@liv.ac.uk*
[2] Danish Space Research Institute, Copenhagen, Denmark

**Summary.** New satellite data have led to increased resolution models of the geomagnetic secular variation (SV). Simple plots of the spectra of these models suggest an unphysical source depth above the core-mantle boundary (CMB). By taking a ratio of the main field and SV spectra, we argue that this result comes from the chosen spectral definition. The models are consistent with a CMB source for SV.

**Key words:** Geomagnetic field modelling, Geomagnetic spectrum, Secular variation

## 1 Introduction

Magnetic spectra provide a useful diagnostic tool for examining the geomagnetic field. The most commonly used such spectrum is motivated by the calculation of the mean square field over a spherical surface, radius $r$:

$$< \mathbf{B}^2 >= \sum_{l=1}^{\infty} \left(\frac{a}{r}\right)^{2l+4} (l+1) \sum_{m=0}^{l} \left(g_l^{m\,2} + h_l^{m\,2}\right) \tag{1}$$

where $a$ is the radius of the Earth, and $\{g_l^m, h_l^m\}$ are the Gauss coefficients of spherical harmonic degree $l$, order $m$ which parameterise the magnetic field model [5]. A spectrum is defined by plotting individual contributions to $< \mathbf{B}^2 >$ from each spherical harmonic degree $l$. As an example, Figure 1 gives such a spectrum plotted at $r = a$ for a main field model obtained from Magsat data [2]. The "depth to source" is interpreted as the depth $r$ at which spectrum is "white" (power approximately independent of harmonic degree). The broad interpretation is that degrees 1-13 arise from a source near the CMB (the core field), degrees 15-40 have a source depth near the Earth's surface (the magnetised lithosphere), and degrees 41+ from above the Earth, consistent with random errors in satellite data.

In this paper, we use power spectra to investigate the SV derived from high-quality data obtained by the Ørsted and CHAMP satellites.

## 2 New Field Models

We name the model considered here CO2003. The modelling methodology follows that of [8]. We use Ørsted scalar and vector data spanning almost 4.5

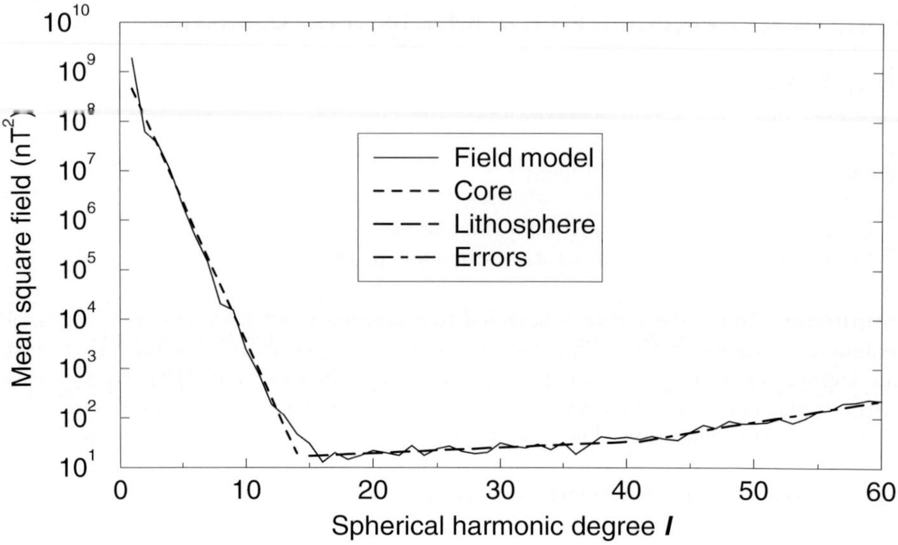


**Fig. 1.** The power spectrum of a field model derived from Magsat data, with simple division into parts from the core, lithosphere and data errors.

years (March 1999 - July 2003), and CHAMP scalar data spanning 3 years (August 2000 - July 2003). We restrict attention to night-side data, with local times between 19:00 and 07:00, to reduce contributions from ionospheric currents at middle and low latitudes. We select data by magnetic activity indices, requiring $\mathrm{Kp} \leq 1+$ (as well as $\mathrm{Kp} \leq 2$ for previous 3 hour interval), $|\mathrm{Dst}| < 10\mathrm{nT}$, $|d(\mathrm{Dst})/dt| < 3\mathrm{nT/hr}$, and IMF $|By| < 3\mathrm{nT}$ for data within the polar caps ($|\mathrm{lat}_{\mathrm{dip}}| > 75°$). We use Ørsted vector data at low latitudes, for $|\mathrm{lat}_{\mathrm{dip}}| < 60°$, with scalar data at high latitudes, or if no attitude information is available. Data are sampled every 120 s (to limit serial correlation) and weighted $\propto$ sin(colatitude) (to simulate an equal area distribution). Unlike in [8], we do not use observatory data. The model is parameterised as a truncated spherical harmonic expansion of the static field $\{g_l^m, h_l^m\}$ up to $l = 40$, SV $\{\dot{g}_l^m, \dot{h}_l^m\}$ up to $l = 16$, and secular acceleration $\{\ddot{g}_l^m, \ddot{h}_l^m\}$ up to $l = 8$, giving time dependent Gauss coefficients

$$g_l^m = g_l^m|_{t=t_0} + \dot{g}_l^m(t - t_0) + \frac{1}{2}\ddot{g}_l^m(t - t_0)^2 \quad (2)$$

We include external field contributions up to $l_{\max} = 2$, allowing annual and semi-annual periodicity of coefficients $q_1^0, q_2^0$, and a Dst-dependence for $l = 1$ terms with an a priori ratio of induced/external coefficients of $Q_1 = 0.28$. Data covariance estimates allow for attitude errors [4] and unmodelled sources in sunlit areas. Iteratively Reweighted Least Squares (with Huber weights) is used to account for non-Gaussian data errors.

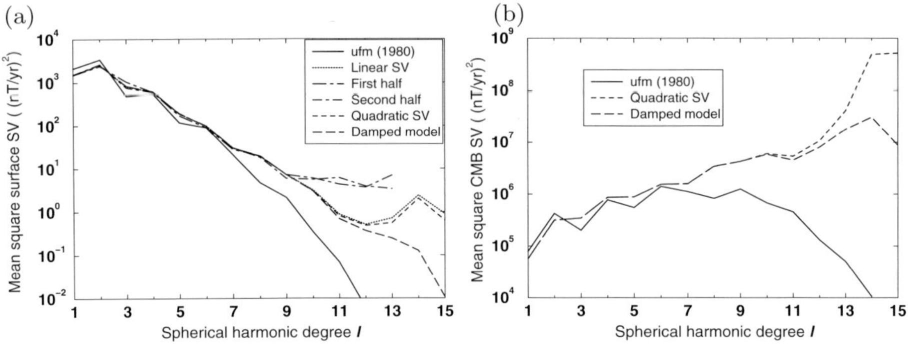


**Fig. 2.** Secular variation spectra at (a) Earth's surface and (b) CMB for various field models

# 3 Secular Variation Spectrum

We define a SV power spectrum analagous to Eq. (1), plotting

$$\left(\frac{a}{r}\right)^{2l+4}(l+1)\sum_{m=0}^{l}\left((\dot{g}_l^m)^2+(\dot{h}_l^m)^2\right) \quad \text{against} \quad l \tag{3}$$

We consider six field models:

1. ufm (1980) [1], a good quality pre-CHAMP/Ørsted model.
2. A new model from CHAMP/Ørsted data, calculated as described above, but without solving for secular acceleration (no $\ddot{g}_l^m, \ddot{h}_l^m$ terms).

3,4 As model 2, but using only the first and second halves of the data set respectively.

5 As model 2, but with secular acceleration terms.

6 A regularised version of model 5: the SV is damped at the CMB $r = c$.

In Figure 2(a), we plot the spectra at the Earth's surface $r = a$. ufm (1980) follows power law behaviour to $l \approx 7$, above which the spectrum falls off. This is due to model regularisation, or more specifically, shows that the data used to derive the model were of insufficient quality to resolve SV above $l = 7$. The new model shows power law behaviour to $l \approx 12$, then levels off, consistent with data errors dominating the spectrum at $l = 13$ and above. The peak at degree 14 is due to $\dot{g}_{14}^{14}, \dot{h}_{14}^{14}$, and so may be related to the Backus effect (perpendicular error). The spectra of the two half-period models level off at only $l = 10$: as they are calculated with fewer data, the effect of errors is seen at longer wavelength. Including the secular acceleration terms reduces the SV power only very slightly. Regularisation controls noisy high-degree SV, supressing the levelling off of the spectrum.

In Figure 2(b), we plot the SV spectra at the CMB ($r = c$) for models 1,5,6. Again ufm (1980) falls off above degree 7, while the spectrum of the new model rises after degree 12, again consistent with sources of error dominating the spectrum at these wavelengths. Plotting at $r = c$ emphasises additional interesting details, for example "steps" in the spectra: they show a sharp increase from $l$ = odd to $l$ = even, and are flat or decrease from $l$ = even to $l$ = odd. Most obviously, however, the spectra are "blue" – their power increases with degree. A simple interpretation of this based on "white noise" sources would suggest a SV source depth above CMB – the prediction from the CO2003 model is a depth of $0.66a$ (in middle of lower mantle). There is no obvious physical explanation for this.

The $< \dot{\mathbf{B}}^2 >$ spectrum (Eqn. 3) we have considered is convenient, but not strongly physically motivated. Many other spectra are possible, for example

$$\left(\frac{a}{r}\right)^{2l+4} \frac{l+1}{2l+1} \frac{1}{l^n} \sum_{m=0}^{l} \left( (\dot{g}_l^m)^2 + (\dot{h}_l^m)^2 \right) \tag{4}$$

Division by $(2l + 1)$ is motivated to give power per degree of freedom (spherical harmonics order $m$ per degree $l$), while $n$ is arbitrary. More physically motivated spectra are provided in [6, 7, 9]. Choosing $n = 2$ results in a source depth below the CMB, but such a choice is highly arbitrary.

How may we avoid this arbitrary choice of spectrum? In common with [6, 7, 9, 3], we consider the ratio of the main field and SV spectra. We take the ratio of $< \dot{\mathbf{B}}^2 >$ to $< \mathbf{B}^2 >$ term by term, plotting

$$R(l) = \frac{\sum_{m=0}^{l} \left( (\dot{g}_l^m)^2 + (\dot{h}_l^m)^2 \right)}{\sum_{m=0}^{l} \left( (g_l^m)^2 + (h_l^m)^2 \right)} \quad \text{against} \quad l \tag{5}$$

This function is independent of definition of spectrum, and has the further advantage of having no free depth parameter. Degrees 1 and 2 are known to be anomalous in both the main field power spectrum and the SV spectrum [9], whilst degree 13-14 are affected by errors, and controlled by regularisation. We fit the remaining degrees ($l$ = 3-12) with two different functions. First, we fit an exponential curve $R = Ae^{Bl}$, which would be consistent with fitting a different source depth for the main field and SV; the source depth ratio would be $\exp[B/2]$. Secondly, we fit a power law $R = Al^B$, which would be consistent with a different spectral definition for the main field and SV. The results of this fit are presented in Figure 3. The power law curve fits the model better than the exponential curve, and in addition predicts the dipole term ($l = 1$) much better. The exponent of this curve is $B = 2.9$. From this we conclude that there is no reason to assume a different source depth for the SV and main field, but instead their spectra take different forms.

Why should this be? A different spectral form for field and SV has been proposed before: in particular a difference $\sim l^4$ has been derived from theoretical considerations [7, 9], and the difference has also been discussed in terms of time scales [3]. We choose to take a simpler approach. Secular variation is

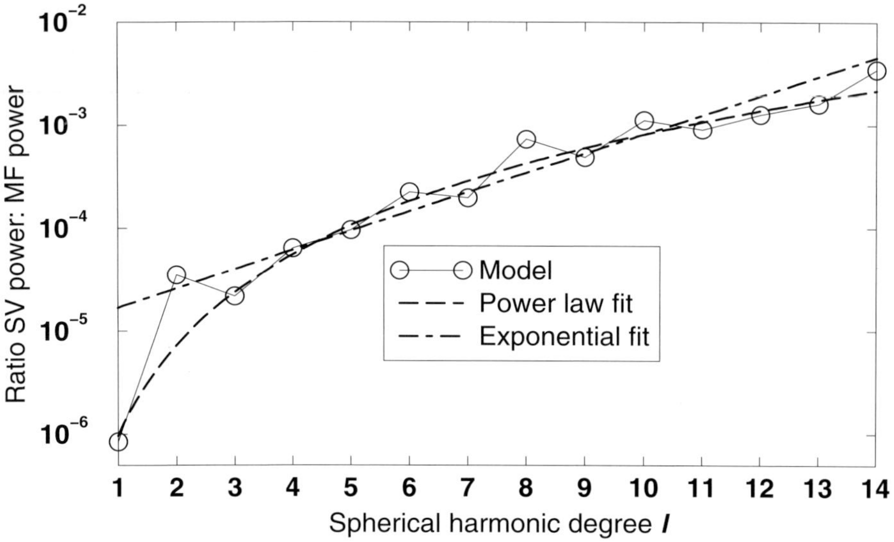


**Fig. 3.** Fit to ratio of spectral power in SV and main field. Exponent for power law fit is $B = 2.9$

generated by the advection and diffusion of the field at the CMB. The radial component of the induction equation gives

$$\dot{B}_r + \nabla_H(\mathbf{u}B_r) = \frac{\eta}{r}\nabla^2(rB_r) \tag{6}$$

where $B_r$ is the radial field component, **u** the flow at the top of the core, and $\eta$ the magnetic diffusivity. If horizontal and radial length scales are similar then scaling analysis would suggest that for diffusion $\nabla^2 \sim l(l+1)$ leading to $R \sim l^4$, whilst for advection $\nabla_H \sim l$, suggesting $R \sim l^2 f(l)$ where $f(l)$ would be a complicated function depending on field/flow interactions, and in particular the evaluation of Gaunt and Elsasser integrals. Despite this uncertainty, a power law relation for $R$ between $l^2$ and $l^4$ seems plausible.

## 4 Conclusion:

We conclude that the SV spectrum is consistent with a source depth of close to the CMB, as might be expected. However, we caution against overinterpretation of the value of the exponent that we obtained. A preliminary field model, from only slightly fewer data than the model used here, predicted an exponent of $B = 3.4$. Thus, the value of this coefficient appears uncertain. More detailed analysis should await detailed field models from longer data series. Even with better models, it is unclear whether small differences in the power law (of order $l^{0.5}$) will be robust.

## References

1. Bloxham J and Jackson A (1992) Time-dependent mapping of the magnetic field at the core-mantle boundary. J Geophys Res *97*: 19,537–19,563.
2. Cain JC, Wang Z, Schmitz DR, and Meyer J (1989) The geomagnetic spectrum for 1980 and core-crustal separation. Geophys J *97*: 443–447.
3. De Santis A, Barraclough DR, and Tozzi R (2003) Spatial and temporal spectra of the geomagnetic field and their scaling properties. Phys Earth Planet Int *135*: 125–134.
4. Holme R and Bloxham J (1996) The treatment of attitude errors in satellite geomagnetic data. Phys Earth Planet Int *98*: 221–233.
5. Lowes FJ (1974) Spatial power spectrum of the main geomagnetic field, and extrapolation to the core. Geophys J R Astron Soc *36*: 717–730.
6. McLeod MG (1985) Statistical theory of the geomagnetic field and its secular variation. EOS Trans. AGU *66(46)*: 878.
7. McLeod MG (1996) Spatial and temporal power spectra of the geomagnetic field. J Geophys Res *101*: 2745–2763.
8. Olsen N (2002) A model of the geomagnetic field and its secular variation for epoch 2000 estimated from Ørsted data. Geophys J Int *149*: 454–462.
9. Voorhies CV (2004) Narrow scale flow and a weak field by the top of Earth's core: evidence from Ørsted, Magsat and secular variation. J Geophys Res, *in press*.

# Geomagnetic Induction Modeling Based on CHAMP Magnetic Vector Data

Heather McCreadie[1] and Zdeněk Martinec[2]

[1] Graduate School of Science, Kyoto University, Kyoto, 606-8502, Japan, *bilby@kugi.kyoto-u.ac.jp*
[2] GeoForschungsZentrum-Potsdam, Telegrafenberg, D-14473 Germany

**Summary.** Using an induction method developed by Martinec et al., (2003) which was refined by Martinec and McCreadie (submitted) to include satellite data, calculations of the magnetic effect of induced currents at CHAMP satellite heights for a region over the Pacific Ocean are shown. The method generally reproduces the data for the northern hemisphere but underestimates the inductive effect in the southern hemisphere. This suggests that other known inductive currents should be incorporated and/or a more complex conductivity structure than the 5-layer 1-D model used here should be implemented for this region.

**Key words:** electromagnetic induction, electrical conductivity of mantle, Dst variation

## Introduction

The CHAMP satellite orbits the globe every 90 minutes. It shifts longitude by 11.75 degrees west every orbit. Therefore almost full coverage of the globe occurs every 24 hours. A 3-component vector magnetometer on board the satellite measures the Earth's magnetic field. Assuming that the Z-component of the magnetic induction field is used to measure the signatures from inductive currents, a method of computing the Z-component using the X-component at ground level was established by Martinec et al., (2003). Modifying this method to satellite height is shown in Martinec and McCreadie (submitted). Here, the first results of this calculation are shown for a region over the Pacific Ocean.

## The Induction Method

The method, TISFEM, which is shown in Martinec et al., (2003), was modified to include a response of magnetic induction of the Earth from the Ring Current excitation at satellite height in Martinec and McCreadie (submitted). The theory is described in detail in the two papers. The method of computing the induced field requires an input that is the expansion coefficients of the X component of the magnetic field from one track of CHAMP in terms of the series of Legendre Derivatives, a conductivity structure for the Earth (from Pecova and Pec 1982) and the time from the onset of the excitation function. The output is the Z-component of the magnetic field at satellite height.

## Method of extracting the usable part of a track

The method of computing the induced field requires an input of an excitation function which is, the amplitudes of the derivatives of the long wavelength harmonics in the X-component of a satellite track, and the time since the initial inducing current began. It is well known that the magnitude of the Dst index is an excellent indicator of a magnetic storm and that the major component of this index is the contribution of the equatorial ring current. It is assumed that the inducing current flows in the vicinity and/or in the equatorial plane. Thus, the choice of tracks was dependent upon the magnitude of the Dst index for each track over a consecutive time period. A small magnetic storm with a highly variable Dst index occurred in the time September 25 to October 7, 2001.

Fig. 1 is a plot of a night-time satellite pass. The model oer05m02-MF1 up to degree 65 was used to remove the main and lithospheric fields (Olsen, 2002, and Maus, et al., 2002). Large oscillations in the X component at high latitudes indicate the presence of polar currents. The signatures of the polar currents mask any contributions from near-equatorial currents making the signatures of the near-equatorial currents indistinguishable in this region. If it is assumed that the field due to the polar currents doesn't encroach upon the field produced by the near-equatorial currents, then the part of a track from [40,140] colatitude should only

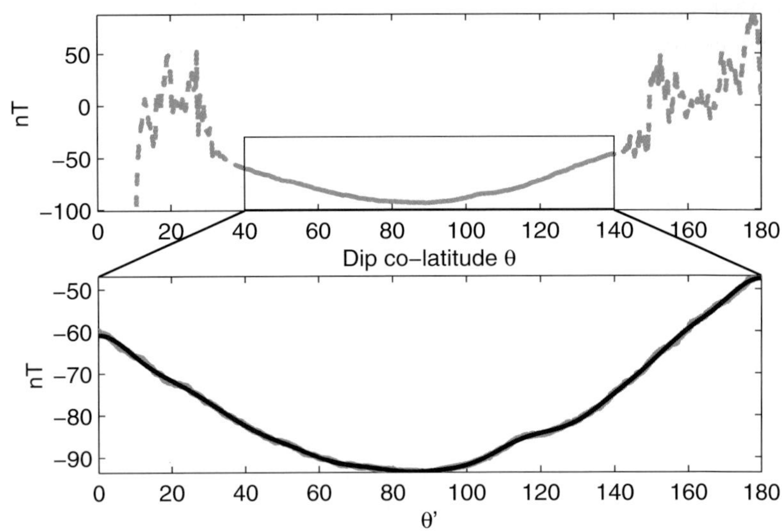


**Fig. 1.** Top graph shows a full night-time track (dashed grey) and part of track (grey line). Bottom graph shows the linearly extrapolated base of part track (grey line) and track computed from Associated Legendre functions with cut-off degree $N'$=25 (black line).

have signatures contributed by the near-equatorial currents. However, data from 0° to 180° colatitude is required in order to perform a spherical harmonic analysis correctly. If the colatitude base is linearly extended from [40,140] to [0,180], the Legendre coefficients up to degree $N'$ of this new track can be computed. Choosing $N'=25$ at the cut-off degree was arbitrary. It filters very high frequency signals that cannot penetrate the Earth's crust. These coefficients are transformed back into the original base [40,140] and in the same step converted to Associated Legendre coefficients (The input required by the induction program). The following equations show this transformation.

$$\sum_{j=1}^{N} K_j^{(X)} \frac{\partial Y_j(\theta)}{\partial \theta}\Big|_{\theta=(\theta'-\beta)/\alpha} \overset{!}{=} \sum_{k=0}^{N'} A_k' Y_k(\theta') \tag{1}$$

$$2\pi \sum_{j=1}^{N} K_j^{(X)} \int_0^{\pi} \frac{\partial Y_j(\theta)}{\partial \theta}\Big|_{\theta=(\theta'-\beta)/\alpha} Y_k(\theta') \sin\theta' \, d\theta' = A'_k \tag{2}$$

where $\theta' \in \langle 0,180 \rangle$, $\theta \in \langle 40,140 \rangle$, and $\alpha$ and $\beta$ are the linear transformation coefficients. $N'$ is the number of coefficients needed to map the X track using the Legendre Polynomials $Y_k(\theta')$.

The left hand side of equation (1) contains the derivatives of the Legendre polynomials mapped for the interval $\theta$. Using the orthonormal property of the spherical harmonics yields equation (2), where the only unknowns are the amplitudes of the coefficients $K_j^{(X)}$.

Using this method, the cut-off degree (N) can be arbitrarily chosen. The problem is that when the energy spread between the coefficients is unbalanced spurious amplitudes develop in the polar regions. To reduce this effect a method was developed, called the Minimum Energy Method. Fig. 2 displays the power spectrum in the right hand column for increasing cut-off degrees N. The power for each angular degree is given by

$$A_j = \log_{10}\left(abs\left(K_j^{(X)}\right)\right) \tag{3}$$

When the power spectrum of the coefficients of a particular cut-off degree is not monotonically decreasing then the previous cut-off degree is chosen. In the example shown here a cut-off degree of N=5 is chosen.

## Results for a track which is mostly over land

Fig. 3 shows the results of the TISFEM method shown in Martinec and McCreadie (submitted) using a 5-layer 1-D conductivity profile displayed in table 1 for the night side (local time near 2200 hours) CHAMP track with number 6863. The Dst index for this pass is -174. The filtered X-track (solid grey line) has a cut-off degree of 3. The derived Z-component (solid black line) and the CHAMP Z-component (dashed black line) are in good agreement. Differences occur for the short wavelengths seen in the southern hemisphere. These wavelengths were not input into the model and so cannot be resolved using this method. For CHAMP tracks that cover land-based regions the derived Z-data and the CHAMP Z-data are in good agreement.

## Results for a region in the Pacific Ocean

Plots of Z CHAMP tracks and the computed Z-tracks using the 5-layer conductivity model for tracks centered on 190° longitude (±10°) are shown in figure 4. These were chosen because the upper mantle conductivity over this area is believed to be more complex than most areas of the globe. The tracks range in Dst so the input excitation function for the model (i.e. the long wavelength X components) varies considerably, thereby, testing the sensitivity of the model. The dashed line is the Z CHAMP track and the solid line is the Z calculated from the model. Note the change in range of the axis as the Dst (shown in the upper right hand corner of each track) increases.

The model generally recreates the original data in the northern hemisphere. However, in the southern hemisphere the track is always more positive than the model. A possible reason for the differences between the Z-data and the model over the Pacific region may include the exclusion of ionospheric currents from the

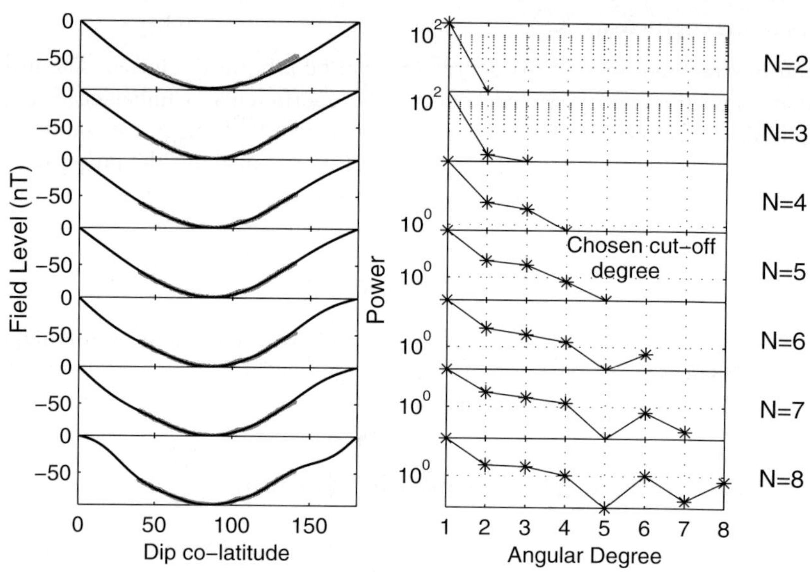


**Fig. 2.** Left graph contains the usable part of a track (grey) and the full track recreated from the Associated Legendre polynomials with cut-off degree shown on the far right. Right graph is the corresponding power spectrum.

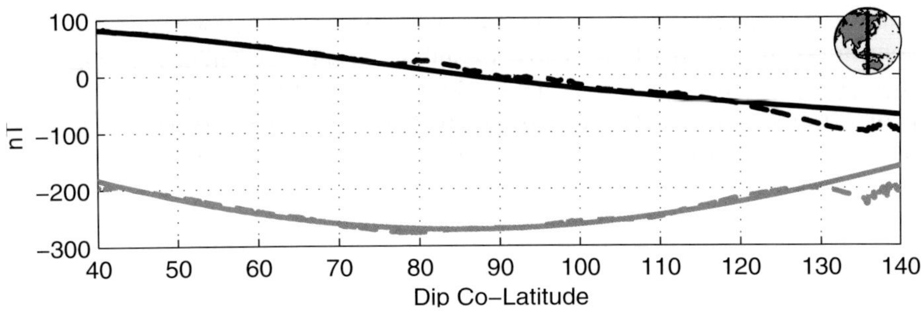


**Fig. 3.** Tracks (dashed lines), Computed tracks (solid lines), X-component (grey), Z-component (black). Line across the picture of the Earth shows longitude of track.

| Depth (km) | 0 | 70 | 420 | 670 | 2891 till centre |
|---|---|---|---|---|---|
| Conductivity (S/m) | 0.001 | 0.01 | 0.1 | 1.0 | 10 |

**Table 1.** 5-layer 1-D conductivity profile. Depth shown is the initial depth where the conductivity changes. Hence, a conductivity of 0.001 is from 0 km until 70 km (from Pecova and Pec, 1982).

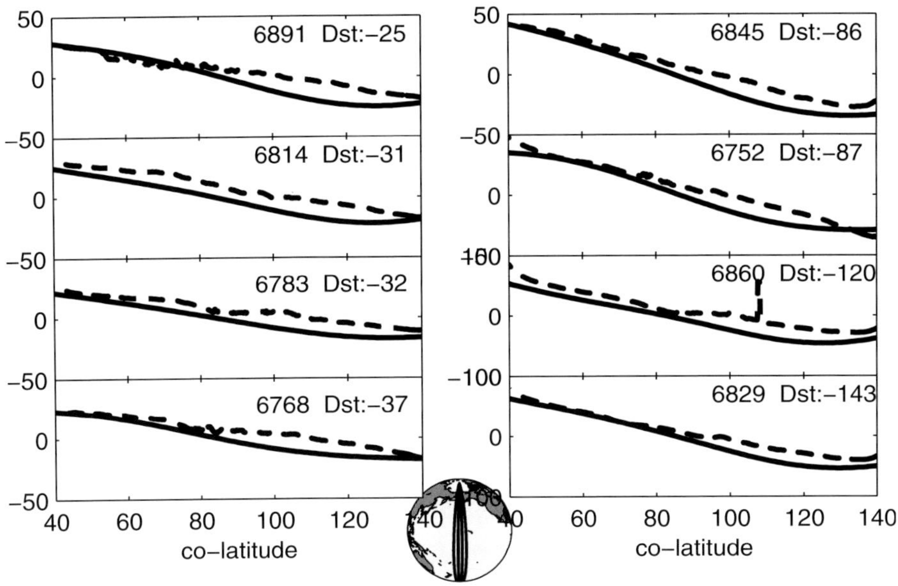


**Fig. 4.** Pacific Region Z-tracks computed from the model. Z-CHAMP track (dashed line), computed track (solid line).

model. The ionospheric currents in the day are well known and have wavelengths comparable to the Dst effect currents. Little is understood about the night-time ionospheric currents. However, the contributions from these currents are considered insignificant at the ground level. Perhaps this isn't so at satellite heights. Also not accounted for in the model are oceanic currents. However, the induced effects from the currents are considered insignificant at satellite altitudes (Weiss and Everett, 1998). Recently, Everett et al. 2003 suggested that near surface conductance may contribute significantly to induced currents at satellite altitudes when a resistive upper mantle is present. Hence, a more complex upper mantle conductivity structure below the South Pacific is required.

## Conclusion

The model reproduces the long wavelengths in the Z CHAMP vector data for regions over land. It generally reproduces the Z CHAMP vector data for the region chosen over the Pacific Ocean. However, other known inductive currents must be incorporated. Also, a more complex conductivity structure should be implemented. Clearly, this is only the initial stages of the analysis.

*Acknowledgments.* We are grateful for the support of the CHAMP project at GFZ-Potsdam who supplied the data and funding for this study.

## References

Everett M, Constable S and Constable CG (2003) Effects of near surface conductance on global satellite induction responses. Geophys J Int *153*: 277-286.

Martinec Z, Everett ME and Velímský J (2003) Time-domain, spectral-finite element approach to transient two-dimensional geomagnetic induction in a spherical heterogeneous earth. Geophys J Int *155*: 33-43.

Martinec Z and McCreadie H (2004) Magnetic induction modelling based on satellite magnetic vector data. Geophys J Int (submitted).

Maus S, Rother M, Holme R, Lühr H, Olsen N, and Haak V (2002). First scalar magnetic anomaly map from CHAMP satellite data indicates weak lithospheric field. Geophys Res Lett *29*: 10.1029/2001GL013685.

Olsen N (2002) A model of the geomagnetic field and its secular variation for Epoch 2002 estimated from Oersted data. Geophys J Int *149*: 453-461.

Pecova J and Pec K (1982) Models of electrical conductivity in the upper mantle. Studia Geoph et Geod *26*: 196-209.

Weiss CJ and Everett M (1998) Geomagnetic induction on a heterogeneous sphere: Three-dimensional test computation and the response of a realistic distribution of oceans and continents. Geophys J Int *135*: 650-662.

# Electromagnetic Induction by *Sq* Ionospheric Currents in a Heterogeneous Earth: Modeling Using Ground-based and Satellite Measurements

Jakub Velímský and Mark E. Everett

Department of Geology and Geophysics, Texas A&M University, College Station, TX 77843, USA. *velimsky@geo.tamu.edu*

**Summary.** We have created a database consisting of hourly means of the geomagnetic field components observed on quiet days in years 2001–2002 on ground observatories, and Ørsted and CHAMP satellite measurements covering the same time intervals. In the first part of our study, we use the potential method to estimate the model of external inducing field. Following 3-D simulations are used to evaluate the effect of heterogeneous surface conductance map on Ørsted and CHAMP satellite measurements and to compare the results with observations. Improvement of up to 15% with respect to the best 1-D model was observed in surface observatory data as well as in the Ørsted and CHAMP measurements.

**Key words:** electromagnetic induction, solar-quiet variations

## 1 Introduction

Effect of distribution of resistive continents and conductive oceans on electromagnetic induction in the Earth induced by daily variations of ionospheric currents (*Sq*) has been targeted by recent studies. Detailed study of the coastline effect on surface observatories was done in [3]. Authors of [1, 2] estimated the influence of near-surface heterogeneities at satellite altitudes. In the first part of our study, we use the potential method to estimate the model of external inducing field on selected quiet days in years 2001–2002. Following 3-D simulations are used to evaluate the EM induction signal on Ørsted and CHAMP satellites and compare the results with observations.

## 2 Data selection

Our study is based on data from three sources. The first dataset consists of the hourly mean values of magnetic induction vector from 126 permanent observatories provided by the World Data Center in Kyoto. The two other sources are the magnetic vector measurements by the CHAMP and Ørsted satellites sampled approximately at 1 s. Following the discussion in [5], we limit ourselves to the $Q^*$ quiet days. A day is denoted by $Q^*$ when the sum of eight $ap$ indices on this day, four $ap$ indices on previous half-day, and four $ap$ indices on following half-day does not exceed 60. There are 52 $Q^*$ days in the period of interest 2001–2002.

Data from polar and equatorial observatories whose geomagnetic latitude is outside of the interval $5° \le |\lambda_m| \le 60°$ are excluded, since they are possibly affected by polar and equatorial jets. Data from the CHAMP and Ørsted satellites are resampled at 30 s, polar and equatorial data points are excluded using the same criterion as in the case of surface observatories. We use the Ørsted model of geomagnetic field and its secular variation (OSVM, [4]) to remove the main field, the crustal field, the contribution of annual and semi-annual variations, and that of the magnetospheric ring current from the observatory and satellite data.

## 3 Harmonic analysis and 1-D inversion

The potential method is used to determine the inductive responses. We use the parameterization of magnetic potential by partial sums of products of spherical and time harmonics [5]. Namely,

$$U_1(r,\vartheta,\varphi;t) = 2\,a\,\mathcal{R}\left\{\sum_{p=1}^{P}\sum_{m=p-M}^{p+M}\sum_{j=m}^{m+J-1}\left[G^{(e)}_{jm,p}\left(\frac{r}{a}\right)^{j} + G^{(i)}_{jm,p}\left(\frac{a}{r}\right)^{j+1}\right] P_{jm}(\cos\vartheta)\,e^{i\,m\,\varphi}\,e^{i\,\omega_p\,t}\right\}, \quad \text{for } a \le r \le a+h, \tag{1}$$

$$U_2(r,\vartheta,\varphi;t) = 2\,a\,\mathcal{R}\left\{\sum_{p,j,m}\left[-G^{(e)}_{jm,p}\,\frac{j}{j+1}\left(\frac{a+h}{a}\right)^{2j+1} + G^{(i)}_{jm,p}\right]\left(\frac{a}{r}\right)^{j+1} P_{jm}(\cos\vartheta)\,e^{i\,m\,\varphi}\,e^{i\,\omega_p\,t}\right\}, \quad \text{for } r > a+h, \tag{2}$$

where $\mathcal{R}$ denotes the real part, $r$, $\vartheta$, and $\varphi$ are respectively the geocentric radius, colatitude, and longitude, $a = 6371.2\,\mathrm{km}$ is the Earth's radius, $P_{jm}$ are the associated Legendre polynomials, and $\omega_p = 2\,\pi\,p\,\mathrm{rad/day}$ are the angular frequencies corresponding to the 1 day period and its harmonics. The limits of summation in (2) are the same as in (1). The ionosphere is assumed to be a thin layer of horizontal currents at altitude $h = 110\,\mathrm{km}$ above the surface. Therefore the magnetic field is expressed as $\boldsymbol{B}_\mathrm{S} = -\operatorname{grad} U_1|_{r=a}$ at the surface observatories and $\boldsymbol{B}_{\text{Ø},\mathrm{C}} = -\operatorname{grad} U_2$ at the Ørsted and CHAMP satellites. Note that potential $U$ and the radial component of magnetic induction vector $B_r$ change continuously across $r = a+h$. The series are truncated at $P = 6$,

$M = 1$, and $J = 4$, yielding 72 complex coefficients of the external and internal field $G^{(e)}_{jm,p}$ and $G^{(i)}_{jm,p}$, respectively. The most stable terms are the principal terms $\left\{ G^{(e,i)}_{jm,p} \middle| j = p + 1,\ m = p \right\}$ which represent the largest part of local-time daily variations.

The initial intent was to find the harmonic coefficients $G^{(e,i)}_{jm,p}$ by fitting into the surface *and* the satellite data. However, preliminary results showed that the OSVM model failed to remove the non-$Sq$ components of the field satisfactorily, leaving constant, or long-periodic baselines of amplitudes up to several hundreds nT on surface observatories. Similar results were obtained with the IGRF 2000 and CHAMP (CO2) models. Following [5], we removed the baselines from the surface data using the local midnight value for each $Q^*$ day and each observatory. However, similar procedure is not possible with data observed by a moving satellite. Therefore we decided to use only the surface observatories in the spatio-temporal harmonic analysis. The satellite data are used in the next section for comparison of 3-D forward modeling results.

For each quiet day $k = 1, \ldots, 52$ in our database, we determined $G^{(e,i)}_{jm,p,k}$ by minimizing the misfit,

$$\chi^2_k \left( G^{(e,i)}_{jm,p,k} \right) = \sum_{l \in \mathcal{M}_k} \sum_{n=1}^{24} \left| \boldsymbol{B}_{\mathrm{S},ln} + \operatorname{grad} U_1 (G^{(e,i)}_{jm,p,k}; a, \vartheta_l, \varphi_l; t_n) \right|^2 , \qquad (3)$$

where index $l$ runs through the set $\mathcal{M}_k$ of observatories available on the $k$-th day, $\boldsymbol{B}_{\mathrm{S},ln}$ is the $n$-th hourly value observed at the $l$-th observatory, and $t_n = (n - 1/2)\,\mathrm{h}$ UTC. The problem of minimization is solved using the singular-value decomposition technique.

Using the univariate linear regression analysis,

$$G^{(i)}_{jm,p,k} = q_{jm,p}\, G^{(e)}_{jm,p,k} + \delta G^{(i)}_{jm,p,k}, \qquad k = 1, \ldots, 52, \qquad (4)$$

we find the response functions $q_{jm,p}$ [6]. Assuming spherically symmetric conductivity, and taking into account only the principal terms of the harmonic expansion, we invert the observatory-based responses $q_{jm,p}$ in terms of a layer of resistivity $\rho^*$ and thickness $z^*$ above a perfect conductor, see Fig. 1. The same responses are also inverted using a three-layer model, which creates the basis for following 3-D modeling.

## 4 3-D forward modeling

In order to investigate the sensitivity of satellite data to the lateral conductivity heterogeneities in $Sq$-driven EM induction, we overlay the three-layer model derived in the previous section by the surface conductance map by [1]. Then, using the time-domain spherical harmonic-finite element approach

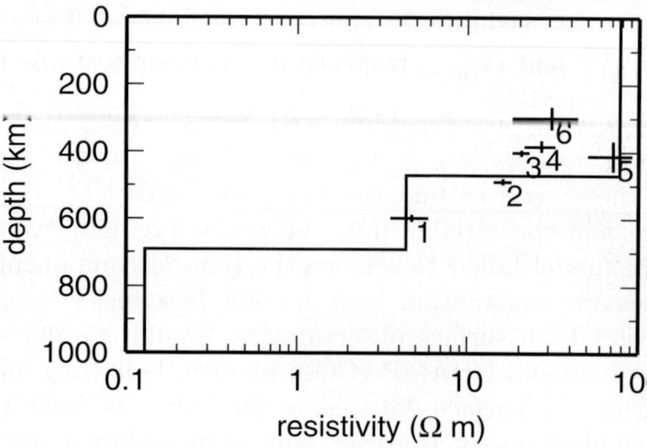


**Fig. 1.** Results of the 1-D inversion of observatory-based response functions. Crosses show the results of the inversion of principal spherical harmonic terms for time harmonics $p = 1, \ldots, 6$ in terms of a layer of resistivity $\rho^*$ and thickness $z^*$ above a perfect conductor. Error bars correspond to 68% level of confidence. The best-fitting three-layer model is shown by solid line.

[7], we excite the model by the external field coefficients $G^{(e)}_{jm,p,k}$ introduced above, and for each day we compute the time series of $\boldsymbol{B}$ on surface observatories and along the Ørsted and CHAMP trajectories. This is done separately for conductivity model without and with the conductance map, referenced respectively as 1-D and 3-D from now on.

Figure 2 shows the effect of the heterogeneous surface conductance map on observatories. We evaluate the total $\chi^2$ misfit between the observed and

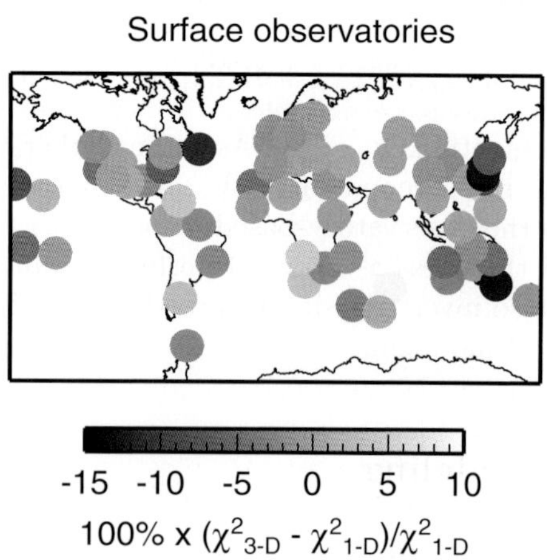


**Fig. 2.** Relative improvement of the total misfit of data on surface observatories due to the surface conductance map.

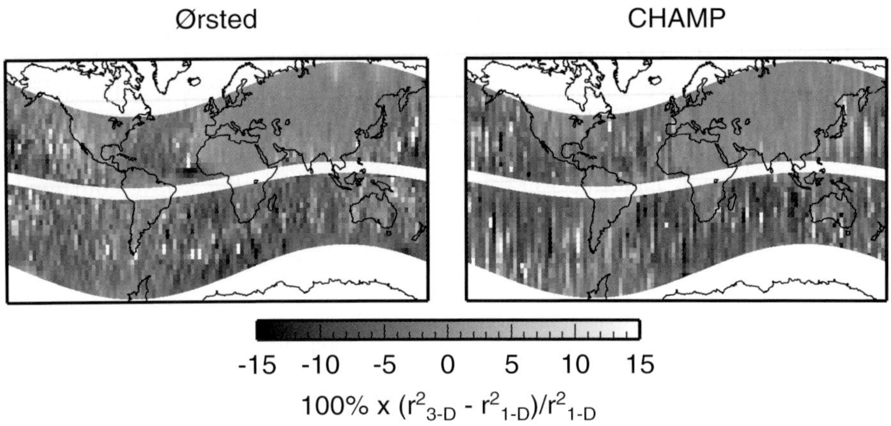


**Fig. 3.** Relative improvement of the residua due to the conductance map observed by the Ørsted and CHAMP satellites. The relative changes in residua are evaluated along the satellites paths and binned into 3° × 3° bins.

modeled field over all 52 $Q^*$ days at each station, separately for 1-D and 3-D models. The conductance map yields significantly better, up to 15% lower misfit on some coastal observatories, e.g. St. Johns (Newfoundland), Canberra (Australia), and most of the Japanese stations. On the other hand, some of the stations on the western African coast, San Juan in the Caribbean, and Trelew in South America, yield better results without the conductance map. As expected, most of the continental stations in Eurasia are insensitive to the distant conductivity jumps across the coastlines.

Similar comparison is done using the satellite data. Local $r^2$ residua between observed and modeled 3-D and 1-D data are evaluated point-by-point along the satellite paths. Relative differences in residua are binned into $3° \times 3°$ bins and shown in Fig. 3 separately for Ørsted and CHAMP satellites. The results are similar to the surface observatories. Conductance map improves the fit to the data above the oceans by up to 15%. Smaller differences are observed above the continents far from the coastlines.

## 5 Conclusions

In the first part of our study we used the potential method [5, 6] with observatory data from period 2001–2002 and obtained similar results in terms of $\rho^*(z^*)$ (see Fig. 1 in [6]). The problem of removal of non-$Sq$ signals from the satellite data prevented their use in the derivation of external $Sq$ currents model. Further research to avoid this obstacle is highly desirable. The surface observatory-based, day-by-day model of ionospheric currents was used as input for 3-D modeling studying the influence of large conductance contrasts between oceans, continents and coastal shelves on surface observatory and satellite data. Our results confirm the importance of heterogeneous surface conductance in the EM induction modeling. Improvement of up to 15% with respect to the best 1-D model was observed in surface observatory data as well as in the Ørsted and CHAMP measurements. Satellite data acquired above oceans and coastal areas are highly sensitive to the conductance map. The differences between 1-D and 3-D solutions are much smaller above continents.

## References

1. Everett ME, Constable S, and Constable C (2003) Effects of near-surface conductance on global satellite induction responses. Geophys J Int *153*: 277–286.
2. Grammatica N and Tarits P (2002) Contribution at satellite altitude of electromagnetically induced anomalies arising from a three-dimensional heterogeneously conducting Earth, using Sq as an inducing source field. Geophys J Int *151*: 913–923.
3. Kuvshinov AV, Avdeev DB and Pankratov OV (1999) Global induction by Sq and Dst sources in the presence of oceans: bimodal solutions for non-uniform spherical surface shells above radially symmetric earth models in comparison to observations. Geophys J Int *137*: 630–650.
4. Olsen N (2002) A model of the geomagnetic field and its secular variation for epoch 2000 estimated from Ørsted data. Geophys J Int *149*: 454–462.
5. Schmucker U (1999) A spherical harmonic analysis of solar daily variations in the years 1964–1965: response estimates and source fields for global induction — I. Methods. Geophys J Int *136*: 439–454.
6. Schmucker U (1999) A spherical harmonic analysis of solar daily variations in the years 1964–1965: response estimates and source fields for global induction — II. Results. Geophys J Int *136*: 455–476.
7. Velímský J (2004) Time-domain, spherical harmonic-finite element approach to transient three-dimensional geomagnetic induction in a spherical heterogeneous Earth. Submitted to Geophys J Int.

# Wavelet Analysis of CHAMP Flux Gate Magnetometer Data

Georgios Balasis, Stefan Maus, Hermann Lühr and Martin Rother

GeoForschungsZentrum Potsdam, Section 2.3, Potsdam, Germany
*gbalasis@gfz-potsdam.de*

**Summary.** Wavelet spectral analysis permits quantitative monitoring of the signal evolution by decomposing a time-series into a linear superposition of predefined mathematical waveforms, each with finite duration and narrow frequency content. Thus, the frequency range of the analyzing wavelets corresponds to the spectral content of time-series components. We present a wavelet analysis of 3 years of vector magnetic data from the CHAMP satellite mission. We have detected, identified and classified not only artificial noise sources (e.g. instrument problems and preprocessing errors) but also high frequency natural signals of external fields (e.g. F-region instabilities and pulsations). The results of this analysis will be used for: (a) consequent correction and flagging of the data, (b) derivation of a clean (undisturbed) dataset suitable for the purposes of crustal and main field modeling, and, (c) study of natural signals (e.g. F-region instabilities, pulsations) contained in the data.

**Key words:** Wavelets, CHAMP FGM Data, Noise, F-region instabilities, Pulsations

## 1 Introduction

Wavelet transforms began to be used in geophysics in the early 1980s for the analysis of seismic signals. In geophysics the power of wavelets for analysis of nonstationary processes that contain multiscale features, detection of singularities, analysis of transient phenomena, fractal and multifractal processes and signal compression is nowadays being exploited for the study of several mechanisms [1]. Wavelet analysis is becoming a common tool for analyzing localized variations of power within a time-series. By decomposing a time-series into time-frequency space, one is able to determine both the dominant modes of variability and how those modes vary in time [4]. Unfortunately, many studies using time-frequency analysis have suffered from an apparent lack of quantitative results. The wavelet transform has been regarded by many as an interesting diversion that produces colorful pictures, yet purely qualitative results. This misconception is in some sense the fault of wavelet analysis itself, as it involves a transform from a one-dimensional time-series to a diffuse two-dimensional time-frequency image.

The advantage of analyzing a signal with wavelets as the analyzing kernels, is that it enables one to study features of the signal locally with a detail

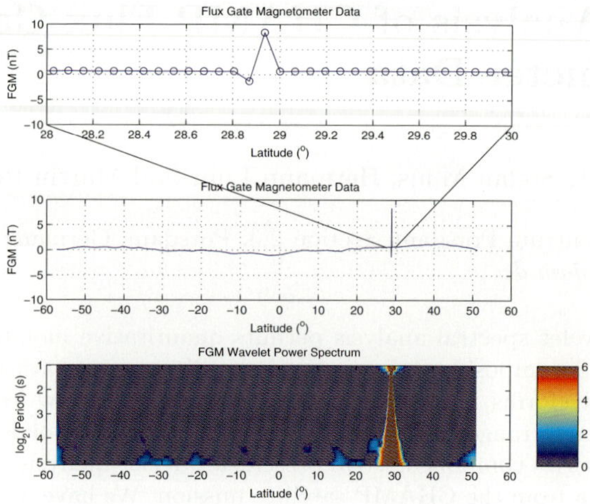


**Fig. 1.** Orbit with pre-processing error (2 s pulse): variation of the total vector magnetic field with latitude and the associated fingerprint in the wavelet power spectrum. The top diagram zooms in on the 2 s feature. (In all figures the power of the spectrum is given in a $log_2$ scale.)

matched to their scale. Owing to its unique time-frequency localization [1], wavelet analysis is especially useful for signals that are non-stationary, have short-lived transient components, have features at different scales, or have singularities. The lack of this property makes Fourier transforms inapplicable to the characterization of time-varying signals. Wavelet transforms allow us to identify time-varying frequency content, while Fourier transforms imply a constant frequency content of a time-series.

## 2 Data and appropriate wavelet basis selection

We performed a time-frequency analysis of the magnetic field magnitude data derived from the CHAMP 1 Hz Flux Gate Magnetometer (FGM) measurements, which were collected from August 2000 to May 2003. From all the CHAMP 1 Hz FGM measurements we selected night time (22:00–06:00 local time, LT), and quiet (Kp<2) orbits. This gave a dataset of ∼5000 orbits from which the latitude range from -60° to +60° was considered. The total field was computed from the three vector components. Prior to the analysis, the GFZ main field model POMME-1.4 (*http://www.gfz-potsdam.de/pb2/pb23/index_e.html*) and crustal field model MF2 were subtracted from the data [3].

Wavelet transforms enable us to obtain orthonormal base, as well as non-orthogonal expansions of a signal using time-frequency kernels called "wavelets" that have good properties of localization in time and frequency

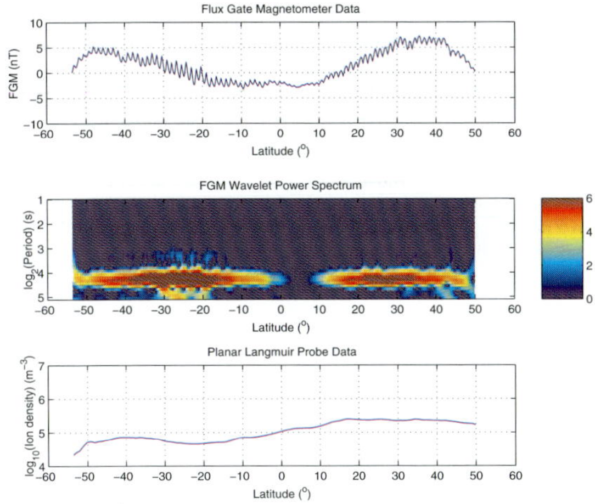


**Fig. 2.** Orbit with instrument problem (missing torquer correction): variation of the total vector magnetic field with latitude and the corresponding signature in the wavelet power spectrum. The bottom diagram shows that there are no significant variations in the ion density.

domains. The basic idea can be understood as a time-frequency plane that indicates the frequency content of a signal at every time. In any such decomposition the time-frequency plane is layered with cells, called Heisenberg cells, whose minimum area is determined by the uncertainty principle [1]. Heisenberg's uncertainty principle dictates that one cannot measure simultaneously with arbitrarily high resolution in both time and frequency space. The decomposition pattern of the time-frequency plane is predetermined by the choice of the basis function.

In our case, we have used the continuous wavelet transform [4] with the Morlet wavelet as the basis function. Our results were, however, checked for consistency using the Paul and DOG mother functions, as well. Next, we have classified the disturbed segments into orbits: (a) with instrument problems and pre-processing errors, (b) with F-region instabilities, and, (c) with pulsations.

## 3 Artificial source noise

The wavelet transform can be used to analyze time-series that contain nonstationary power at many different frequencies. For the purposes of our analysis we have focused on the period range between 2 and 32 s. In Fig. 1 we see the fingerpint of an abrupt 2 s pulse in the time-frequency domain. It is depicted as a narrow column of maximum power that dominates all shown frequencies and corresponds to a peak in the time-series at the part of the orbit where

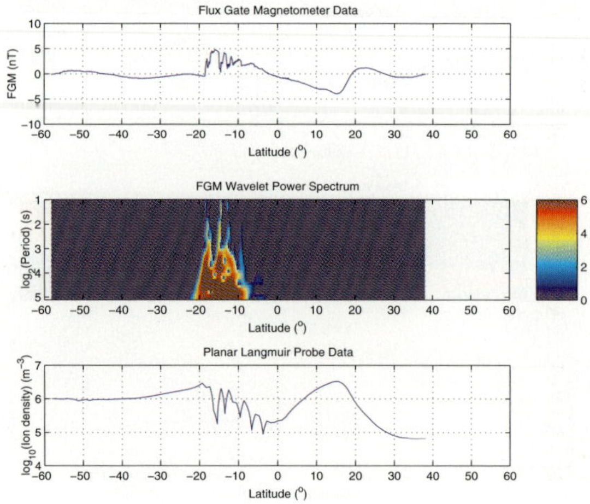


**Fig. 3.** Orbit with F-region instabilities: total vector magnetic field and its fingerprint in the time-frequency domain. Note that in this and in the following cases the instability is accompanied by a perturbation in the ion density.

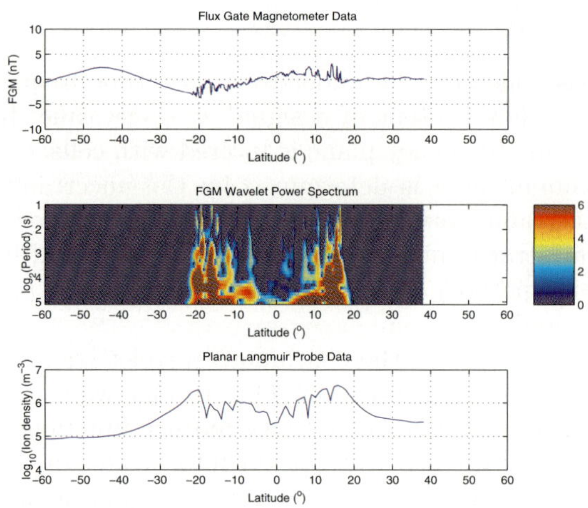


**Fig. 4.** Orbit with F-region instabilities: this instability region crosses the equator and has an extent of -20° to +20°.

the sudden change occurs in the wavelet power spectrum. The signature of a missing torquer correction looks quite different, from the first case, in the time-frequency space (Fig. 2): it is a zone of maximum power symmetrically located at a 20 s period, almost along the whole orbit. The forms of these two wavelet power spectra directly imply that they cannot be results of physical processes.

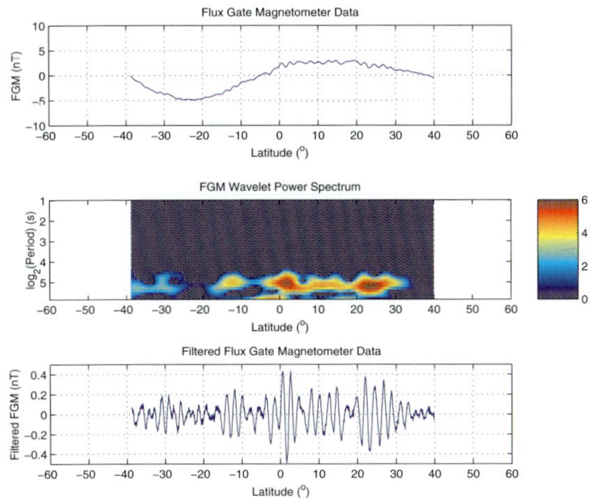


**Fig. 5.** Orbit with a Pc3 pulsation: total vector magnetic field and its signature in the wavelet power spectrum. The bottom diagram represents the same orbit after applying a 32 s running mean filter.

## 4 Natural source signals

The CHAMP satellite in its polar, low Earth orbit (below 450 km altitude) is a suitable platform for observing ionospheric instabilities in the F-region [2]. Based on half a year of CHAMP scalar magnetic data it has been suggested that F-region instabilities events are mainly confined in LT to before midnight. Our wavelet analysis of 3 years of vector magnetic data reveals that in principle there is an appreciable occurence rate of F-region instabilities in the LT sector from 22 to 06 (Table 1). These instabilities are generally accompanied by local depletions of the electron density.

In comparison with the artificial signals fingerprints in the time-frequency domain of Sec. 3, we observe physically plausible dispersion in the maximum power regions of the corresponding wavelet power spectra that are signatures of F-region instabilities (Figs. 3–4). These instabilities can be associated with the formation of plasma bubbles [5] that are visible in the Planar Langmuir Probe (PLP) density measurements.

The CHAMP satellite is also capable of monitoring ultra-low-frequency (ULF) magnetospheric waves, called geomagnetic pulsations. In Fig. 5 we observe the wavelet power spectrum of such an event. The form of the maximum power region of this pulsation is somehow similar to the case of the missing torquer correction. However, the frequency that is observed is different ($\sim$30 s instead of 20 s for the artificial signal) and the physically expected dispersion is evident in the shape of the highest energy part of the spectrum.

**Table 1.** Statistics of the examined orbits.

| Description | Number of orbits |
|---|---|
| Total | 5078 |
| F-region instabilities | 914 |
| Pulsations | 105 |
| Pre-processing errors | 91 |
| Instrument problems | 54 |

## 5 Conclusions

We have examined 5078 orbits of FGM (22:00–06:00 LT) data and found 1164 orbits ($\simeq$23%) contaminated by different kinds of artificial (instrument problems, pre-processing errors) and natural sources (F-region instabilities and pulsations) signal. We have managed to derive a clean dataset suitable for users and modellers of the CHAMP vector magnetic data. We have also established a large dataset of $\sim$900 orbits dominated by F-region instabilities. This promising sample will provide a basis for a better understanding of this external, ionospheric effect, in subsequent work.

*Acknowledgement.* This work was supported by DFG's research grant MA 2117/3, as part of the Priority Program: "Geomagnetic Variations: Spatio-temporal structure, processes and effects on system Earth", SPP 1097.

## References

1. Kumar P, Foufoula-Georgiou E (1997) Wavelet analysis for geophysical applications. Rev Geophys *35*: 385–412.
2. Lühr H, Maus S, Rother M, Cooke D (2002) First in-situ observation of night-time F region currents with the CHAMP satellite. Geophys Res Letters *29*: 127-1–4.
3. Maus S, Rother M, Holme R, Lühr H, Olsen N, Haak V (2002) First scalar magnetic anomaly map from CHAMP satellite data indicates weak lithospheric field. Geophys Res Letters *29*: 47-1–4.
4. Torrence C, Compo GP (1998) A Practical Guide to Wavelet Analysis. Bull Am Meteor Soc *79*: 61–78.
5. Whalen JA (2000) An equatorial bubble: Its evolution observed in relation to bottomside spread *F* and to the Appleton anomaly. J Geophys Res *105(A3)*: 5303–5315.

# Modelling the Ocean Effect of Geomagnetic Storms at Ground and Satellite Altitude

Alexei Kuvshinov and Nils Olsen

Danish Space Research Institute/Center for Planetary Science, Juliane Maries Vej 30, DK-2100 Copenhagen, Denmark, *alexei@dsri.dk; nio@dsri.dk*

**Summary.** Recently Olsen & Kuvshinov (2003) presented an approach for modelling the ocean effect of geomagnetic storms by solving the induction equation given the conductivity distribution of the Earth's interior and the time-space structure of the storm. The results for several major storms show much better agreement between the observed and the simulated magnetic vertical component at coastal sites if the oceans are considered and contributions from spherical harmonics other than $P_1^0$ are included. In that paper we report on the results for selected ground observatories. Here we investigate the time-space distribution of the anomalous induction (ocean) effect of geomagnetic storms at ground as well as at satellite altitudes in order to answer the following question: Where and during which phase of the storm can we expect to observe the ocean effect from ground and from satellite altitudes, and what is the expected amplitude? Our simulations show that during the main phase of major geomagnetic storms the ocean effect in the radial component, $B_r$, and in the scalar field, $dF$, should be clearly visible at CHAMP and Ørsted altitudes and might reach magnitudes of tens of nT. The effect is especially prominent near the Pacific coast of North America, and near the southern edges of Africa, Australia and South America.

**Key words:** ocean effect, magnetic field variations, electromagnetic induction

## 1 An approach for modelling the ocean effect of geomagnetic storms

We used the following procedure to model the ocean effect in time domain (Olsen & Kuvshinov, 2003):

(1) For a specific geomagnetic storm time series of hourly mean values of the horizontal components, $X^{obs}(t)$ and $Y^{obs}(t)$, at worldwide distributed observatories equatorward of $\pm 60^\circ$ geomagnetic latitude are used (distribution shown in Figure 1), and a time segment of 10 days length for each component and each observatory is Fourier transformed after subtracting the baseline.

(2) A robust spherical harmonic analysis (cf. Olsen, 2002) of $X^{obs}(\omega_i)$ and $Y^{obs}(\omega_i)$ is performed for each frequency $\omega_i$, using the 1-D conductivity model of Schmucker (1985) (but with $3\times10^{-4}$ S/m for the upper 100 km), to separate external and induced contributions. As a result, the expansion coefficients

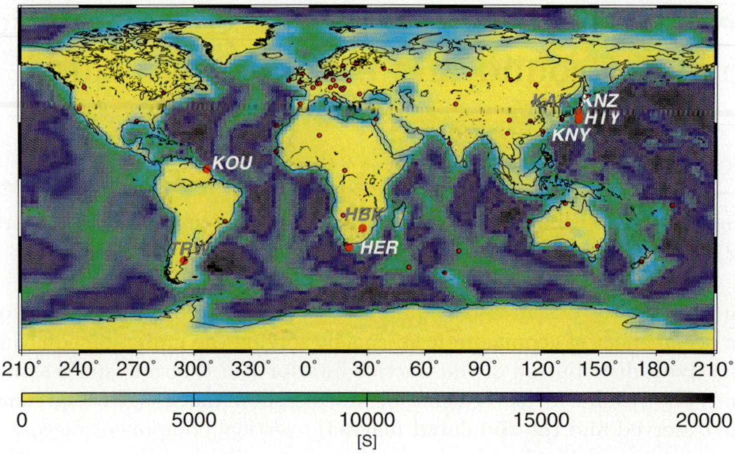


**Fig. 1.** Conductance of the surface shell and locations of the used observatories (redrawn from (Olsen & Kuvshinov, 2003)).

$\epsilon_n^m(\omega_i)$ of the external potential for $n, m \leq 3$ are estimated for 120 frequencies between 1/240 $\text{hrs}^{-1}$ and 1/2 $\text{hrs}^{-1}$.

(3) For each frequency $\omega_i$, electromagnetic (EM) induction simulations are performed, using a three-dimensional (3-D) conductivity model that consists of a thin shell of conductance $S(\vartheta, \varphi)$ (cf. Figure 1) and the 1-D conductivity $\sigma(r)$ (the same as in Step 2) underneath. The inducing external field is given by $\epsilon_n^m(\omega_i)$ found in Step 2. Maxwell's equations are solved numerically according to the integral equation scheme presented in Kuvshinov et al. (1999, 2002) on a mesh of $1^\circ \times 1^\circ$ in longitude and colatitude. For each of the 120 frequencies we thus obtain the components of the geomagnetic field $X^{mod}(\omega_i)$, $Y^{mod}(\omega_i)$ and $Z^{mod}(\omega_i)$ at ground for the adopted mesh.

(4) At each frequency these magnetic field components are upward continued to CHAMP and Ørsted altitudes.

(5) Finally, in each cell of the mesh and at ground as well as at satellite altitudes, the results are Fourier transformed to the time domain to obtain synthetic time series $X^{mod}(t)$, $Y^{mod}(t)$ and $Z^{mod}(t)$.

## 2 Results for observatory data

Figure 2 presents time series of observed (black) and modelled $X$(left) and $Z$(right) at selected observatories for the storm of 15-16 July, 2000. 3-D model results are shown in red, 1-D results in blue, and values based on the $Dst$-index

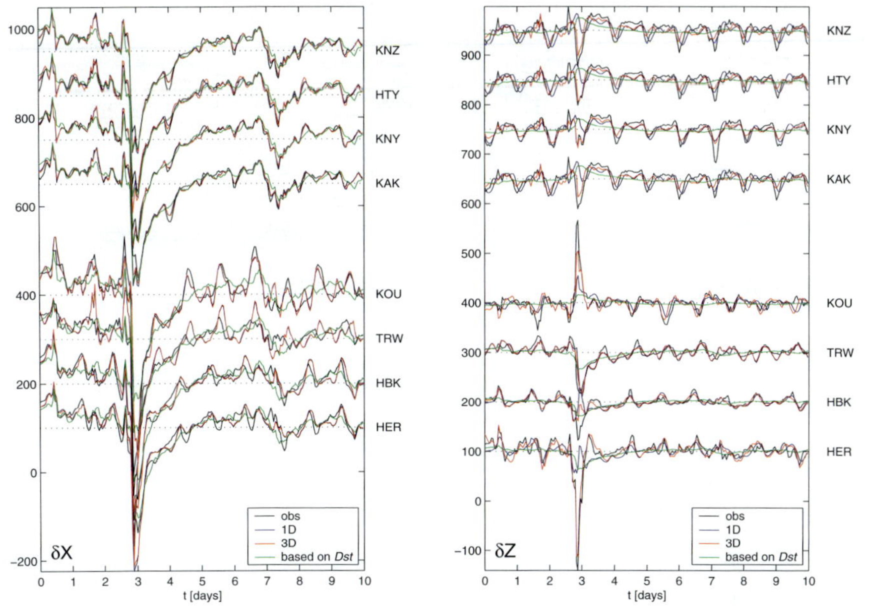


**Fig. 2.** Time series of observed and modelled $X$ and $Z$ at selected observatories for the storm of July 15-16, 2000. $t$=0 corresponds to July 13, 2000, 00:00 UT (redrawn from (Olsen & Kuvshinov, 2003)).

in green. Here the 1-D results stand for the case of a 1-D conductivity without oceans. 3-D and 1-D results are very similar in the horizontal component and agree with the observed values rather well, confirming the use of a 1-D model when obtaining $\epsilon_n^m$ in Step 2 of the previous section. However, there are considerable differences between the 1-D and 3-D results in $Z$ for most of the displayed observatories, and much closer agreement between the 3-D results and the observations.

## 3 Space-time structure of the ocean effect at ground and satellite altitudes

Fig. 3 shows global maps (hour by hour) of the anomalous vertical geomagnetic field at ground for the storm of 15-16, July, 2000. Here we determine as anomalous field at a specific location the difference between the field obtained using 3-D modelling and the local normal field defined as that we would get using a 1-D model with the conductivity profile of that location. It is clearly seen that the effect is especially prominent during the peak in *Dst* (21:30 and 22:30 UT). As expected, the effect is strongest near the coasts, and reaches magnitudes of 140 nT. The largest amplitudes are found near the coasts of South Africa and South Australia. Fig. 4 presents in a similar

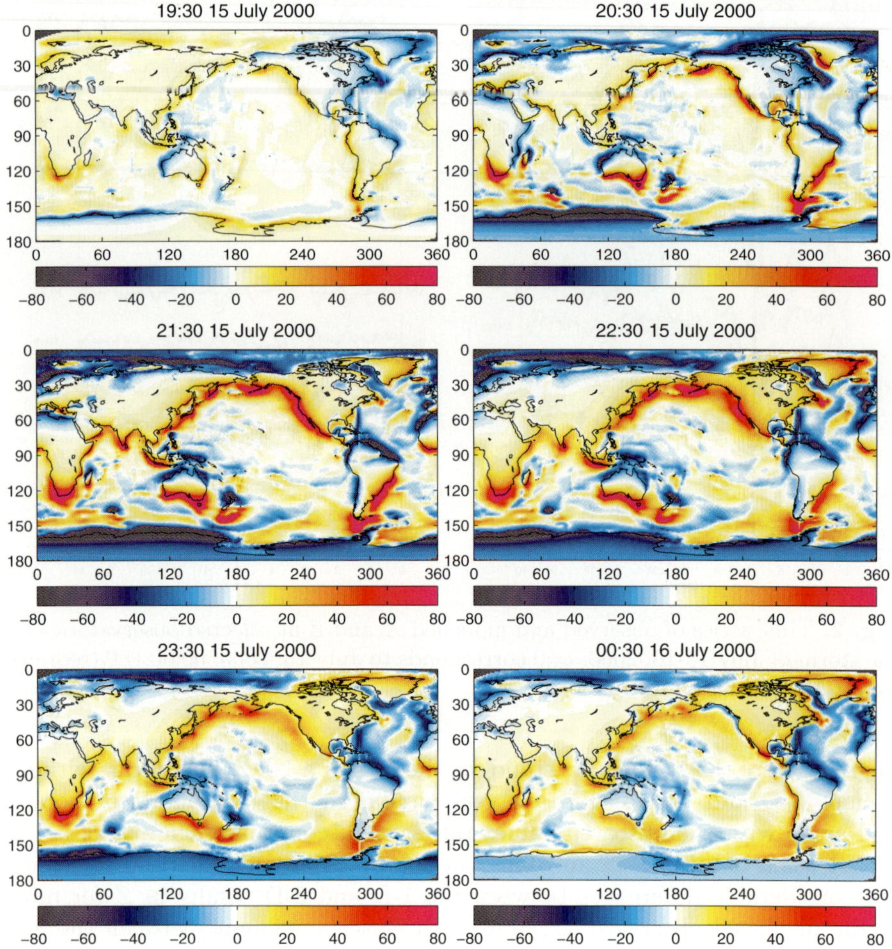


**Fig. 3.** The anomalous $B_r$ (in nT) at ground for the 6 hours between 19:30 UT on July 15 until 00:30 UT on July 16, 2000.

way the anomalous effect in $B_r$ at CHAMP ($h$=400 km; upper panel; 60 nT maximum amplitude) and Ørsted ($h$=800 km; lower panel; 30 nT maximum amplitude) altitudes. The effect in $dF$ at CHAMP altitude is of the same order as in $B_r$ (cf. Fig. 5). Fig. 6 shows the anomalous effect in $dF$ at CHAMP altitude for another strong storm of 6-7 April, 2000. The space structure of the effect remains rather similar, whereas amplitudes are smaller compared to the storm of 15-16 July, 2000.

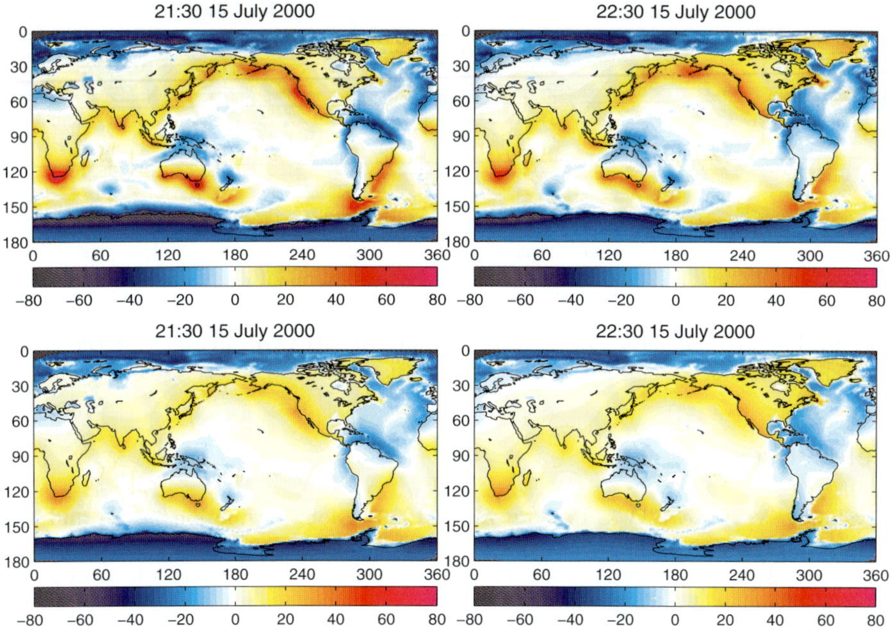


**Fig. 4.** The anomalous $B_r$ (in nT) at CHAMP altitude (upper panel) and at Ørsted altitude (lower panel) for the storm of 15-16 July, 2000.

## 4 Conclusion

During the main phase of major geomagnetic storms the ocean effect in the radial component, $B_r$, and in the scalar field, $dF$, should be clearly visible at CHAMP and Ørsted altitudes and might reach magnitudes of tens of nT. The effect is especially prominent near the Pacific coast of North America, and near the southern edges of Africa, Australia and South America.

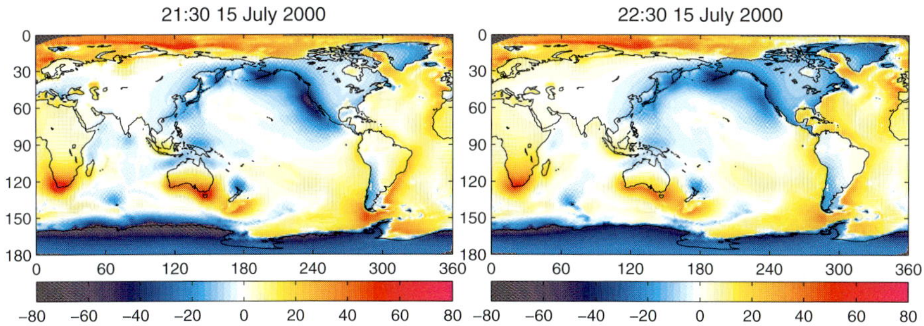


**Fig. 5.** The anomalous scalar field $dF$ (in nT) at CHAMP altitude for the storm of 15-16 July, 2000.

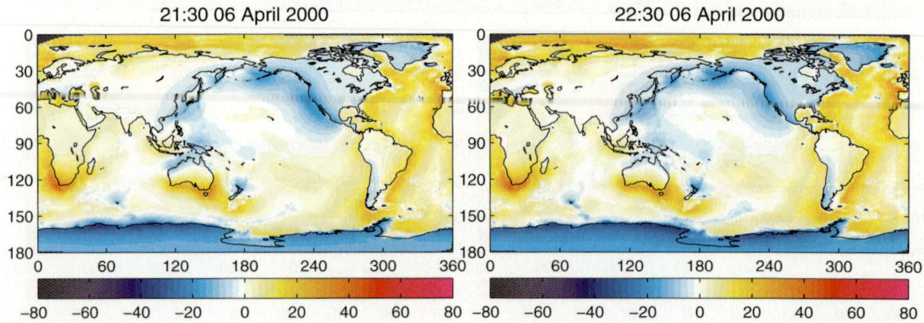


**Fig. 6.** The anomalous scalar field $dF$ (in nT) at CHAMP altitude for the storm of 6-7 April, 2000.

## References

Kuvshinov A, Avdeev D & Pankratov O (1999) Global induction by Sq and Dst sources in the presence of oceans: bimodal solutions for non-uniform spherical surface shells above radially symmetric earth models in comparison to observations. Geophys J Int *137*: 630–650.

Kuvshinov A, Avdeev D, Pankratov O, Golyshev S & N Olsen (2002) Modelling electromagnetic fields in a 3D spherical Earth using fast integral equation approach. In: 3D Electromagnetics, Elsevier, Holland, Chapter 3, 43–54.

Olsen N (2002) A Model of the Geomagnetic Field and its Secular Variation for Epoch 2000 Estimated from Ørsted Data. Geophys J Int *149(2)*: 454–462.

Olsen N & Kuvshinov A (2003) Modelling the ocean effect of geomagnetic storms. Earth, Planets, Space, submitted.

Schmucker U (1985) Magnetic and electric fields due to electromagnetic induction by external sources, electrical properties of the earth interior. In: Landolt-Boernstein New Series Group *5/2b*, Springer Verlag.

# 3-D Modelling of the Magnetic Fields Due to Ocean Tidal Flow

Alexei Kuvshinov and Nils Olsen

Danish Space Research Institute/Center for Planetary Science, Juliane Maries Vej 30, DK-2100 Copenhagen, Denmark, *alexei@dsri.dk; nio@dsri.dk*

**Summary.** Recently Tyler et al. (2003) demonstrated that the magnetic fields generated by the lunar semidiurnal ($M_2$) ocean flow can be clearly identified in magnetic satellite observations. They compared their numerical simulations of magnetic fields due to the $M_2$ tide with CHAMP observations and found close agreement between observations and predictions. Their three-dimensional (3-D) conductivity model consists of a surface thin shell of variable conductance and an insulating mantle underneath. Some discrepancies between observations and predictions have been addressed to the absence of a coupling between the surface shell and the mantle. Here we performed model studies of the magnetic signals due to ocean tidal flow in order to answer the following questions. (1) How does the inclusion of a conducting mantle affect the magnetic signals of the $M_2$ tide at CHAMP altitude? (2) Are magnetic signals from other tidal components detectable at CHAMP altitude? (3) What amplitude has the magnetic $M_2$ tide at Ørsted altitude? The 3-D conductivity model that we consider incorporates a thin shell and either a radially symmetric or a 3-D mantle underneath. Our model studies demonstrate that including a conducting mantle yields significant changes of the magnetic $M_2$ oceanic signals, with peak-to-peak values at CHAMP altitude of order 3 nT. The magnetic signals due to other prominent ocean tidal modes (like $K_1$ and $O_1$) are below 0.5 nT at CHAMP altitude. The $M_2$ peak-to-peak magnetic signal at Ørsted altitude is of order 1 nT.

**Key words:** ocean tides, magnetic fields variation, electromagnetic induction, three-dimensional modellings

## 1 Modelling approach

To calculate the magnetic fields due to global ocean tides we used the numerical solution of Kuvshinov et al. (2002a), but with the modifications introduced by Avdeev et al. (2002). The solution is based on a volume integral equation (IE) approach and allows simulating the electromagnetic (EM) fields, excited by arbitrary sources in three-dimensional (3D) spherical models of electric conductivity. These 3-D models consist of a number of anomalies of conductivity $\sigma_{3D}(\vartheta,\varphi,r)$, embedded in a host section of conductivity $\sigma_b(r)$. Here $\vartheta,\varphi$ and $r$ are co-latitude, longitude and the distance from the Earth's centre, respectively. In IE solution discussed, the frequency domain Maxwell's equations

$$\nabla \times \mathbf{H} = \sigma \mathbf{E} + \mathbf{j}^{\text{ext}}, \tag{1}$$

$$\nabla \times \mathbf{E} = i\omega\mu_{\text{o}} \mathbf{H}, \tag{2}$$

are reduced, in accordance with the modified iterative-dissipative method (Singer, 1995), to a scattering equation of specific type (cf. Pankratov et al., 1997)

$$\boldsymbol{\chi}(\mathbf{r}) - \int\limits_{\text{V}^{\text{mod}}} \text{K}(\mathbf{r},\mathbf{r}')\text{R}(\mathbf{r}')\boldsymbol{\chi}(\mathbf{r}')\text{dv}' = \boldsymbol{\chi}_{\text{o}}(\mathbf{r}), \tag{3}$$

which is solved by the generalized bi-conjugate gradient method (Zhang, 1997). Here $\mathbf{j}^{\text{ext}}$ is the exciting current density, the time-harmonic dependency is $e^{-i\omega t}$, $\mu_{\text{o}}$ is the magnetic permeability of free space, $i=\sqrt{-1}$, $\omega = 2\pi / T$ is the angular frequency, $T$ is the period of variations, $\sigma$ $(\mathbf{r})$ is the model conductivity distribution, $\mathbf{r} = (\vartheta, \varphi, r)$, $\mathbf{r}' = (\vartheta', \varphi', r')$, $\text{V}^{\text{mod}}$ is the modelling region and

$$R = \frac{\sigma - \sigma_{\text{o}}}{\sigma + \sigma_{\text{o}}}, \tag{4}$$

$$K(\mathbf{r},\mathbf{r}') = \delta(\mathbf{r} - \mathbf{r}')I + 2\sqrt{\sigma_{\text{o}}(r)}G^{e}_{\text{o}}(\mathbf{r},\mathbf{r}')\sqrt{\sigma_{\text{o}}(r')}, \tag{5}$$

$$\boldsymbol{\chi}_{\text{o}} = \int\limits_{\text{V}^{\text{mod}}} K(\mathbf{r},\mathbf{r}') \frac{\sqrt{\sigma_{\text{o}}(r')}}{\sigma(\mathbf{r}') + \sigma_{\text{o}}(r')} \mathbf{j}^{\text{s}}(\mathbf{r}')dv', \tag{6}$$

$$\boldsymbol{\chi} = \frac{1}{2\sqrt{\sigma_{\text{o}}}}((\sigma + \sigma_{\text{o}})\mathbf{E}^{\text{s}} + \mathbf{j}^{\text{s}}), \tag{7}$$

$$\mathbf{j}^{\text{s}} = (\sigma - \sigma_{\text{o}})\mathbf{E}^{\text{o}}, \tag{8}$$

$$\mathbf{E}^{\text{o}} = \int\limits_{\text{V}^{\text{ext}}} G^{e}_{\text{o}}(\mathbf{r},\mathbf{r}')\mathbf{j}^{\text{ext}}(\mathbf{r}')dv', \tag{9}$$

where $\delta(\mathbf{r} - \mathbf{r}')$ is Dirac's delta function, $I$ is the identity operator, $\text{V}^{\text{ext}}$ is the volume occupied by exciting current density, $\mathbf{j}^{\text{ext}}$, $\mathbf{E}^{\text{s}} = \mathbf{E} - \mathbf{E}^{\text{o}}$ is the scattered electric field, $G^{e}_{\text{o}}$ is the 3×3 Green's tensor of the 1-D reference conductivity $\sigma_{\text{o}}(r)$. Note that, in order to provide optimum efficiency of scattering equation solution, $\sigma_{\text{o}}(r)$ is chosen in a special way (cf. Singer, 1995) and generally differs from $\sigma_{b}(r)$ at depths where the anomalies are located.

Once $\boldsymbol{\chi}$ is determined from the solution of (3), the scattered electric field, $\mathbf{E}^{\text{s}}$, is obtained from (7) and the magnetic field, $\mathbf{H}$, at the observation points, $\mathbf{r} \in \text{V}^{\text{obs}}$, is calculated as

$$\mathbf{H} = \int\limits_{\text{V}^{ext}} G^{h}_{\text{o}}(\mathbf{r},\mathbf{r}')\mathbf{j}^{\text{ext}}(\mathbf{r}')dv' + \int\limits_{\text{V}^{\text{mod}}} G^{h}_{\text{o}}(\mathbf{r},\mathbf{r}')\mathbf{j}^{\text{q}}(\mathbf{r}')dv', \tag{10}$$

with

$$\mathbf{j}^{\text{q}} = (\sigma - \sigma_{\text{o}})(\mathbf{E}^{\text{o}} + \mathbf{E}^{\text{s}}). \tag{11}$$

The explicit expressions to calculate the elements of Green's tensors $G_o^e$ and $G_o^h$ are presented in Appendix of Kuvshinov et al. (2002a). For our problem the exciting current density, $\mathbf{j}^{\text{ext}}$, is calculated as

$$\mathbf{j}^{\text{ext}} = \sigma_w (\mathbf{V} \times \mathbf{B}^{\text{m}}), \tag{12}$$

where "×" denotes the vector product, $\sigma_w = 3.2$ S/m is the sea water conductivity, $\mathbf{V}$ is the water transport (depth integrated velocity) due to ocean tides, taken from the TPXO6.1 global tidal model of Egbert & Erofeeva (2002), and $\mathbf{B}^{\text{m}}$ is the main magnetic field taken from IGRF 2000.

## 2 Comparison of two solutions

To check our IE solution we compare it against the finite difference solution of Tyler et al. (2003). In both cases the conductivity model consists of a surface thin shell and an insulating mantle underneath. A realistic distribution of the shell conductance is obtained by considering contributions from sea water and from sediments as described in Kuvshinov et al. (2002b). Fig. 1 presents the vertical component of the magnetic field, $B_r$, due to the $M_2$ tide (period = 12.42 hours) at ground, as given by Tyler et al. (2003) (upper panel) and by IE solution (lower panel). Tyler's results are for a mesh of 2°×2° resolution, while our results are for a mesh of 1°×1° resolution. The good agreement between the two results verifies both approaches.

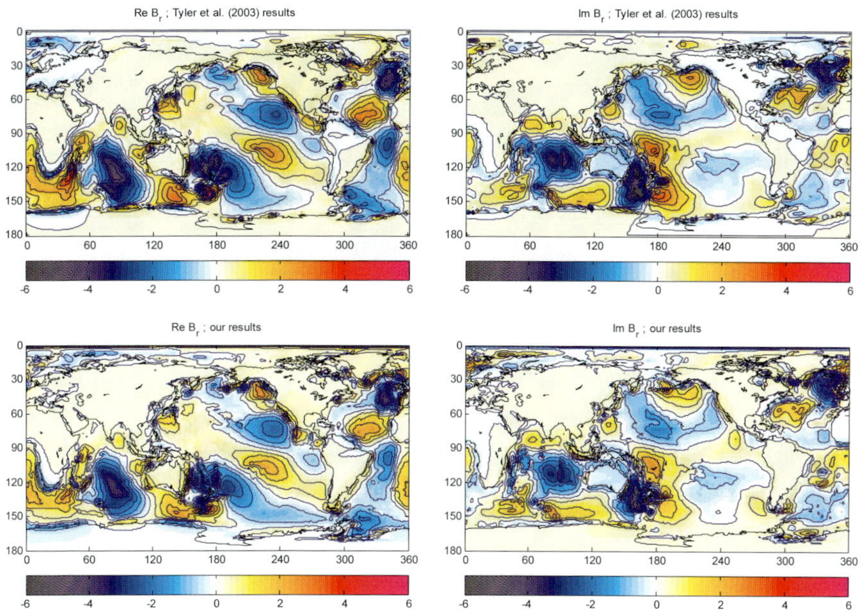


**Fig. 1.** $B_r$ (nT) at the ground from $M_2$ tide obtained by Tyler et al. (2003) (upper panels) and with our IE solution (lower panels). The results are for an insulating mantle.

## 3 Simulations with conducting mantle

In order to investigate how the inclusion of a conducting mantle affects the magnetic signal of the $M_2$-tide at CHAMP altitude we performed simulations using a conductivity model that consists of a laterally inhomogeneous surface shell (same as in previous section) and a 1-D mantle section underneath. Fig. 2 presents the $M_2$ tide scalar anomalies *dF* at CHAMP altitude (430 km); model resolution is 1°×1°. The upper panels show the results obtained for an insulating mantle, whereas the lower ones show those for a 1-D conducting mantle which is based on the 3-layer model of Schmucker (1985), but with a 100 km lithosphere of 3000 Ωm. Comparing the results it is seen that in general the pattern remains the same, but the magnitude of the signals decreases when a conducting mantle is incorporated in the model.

Next were carried out simulations for two other prominent tidal components. Fig. 3 shows the scalar anomalies *dF* at CHAMP altitude due to the diurnal tides $K_1$ (period = 23.9 hours; upper panel) and $O_1$ (period = 25.8 hours; lower panel) for a conducting mantle. One can see that the magnetic signals due to these tides are below 0.5 nT at CHAMP altitude.

Finally we calculated the $M_2$ tide scalar anomalies *dF* at Ørsted altitude (800 km) for the same conductivity model. As it is expected (see Fig. 4) the magnitude of the signal is decreased and becomes smoother compared with that at CHAMP altitude.

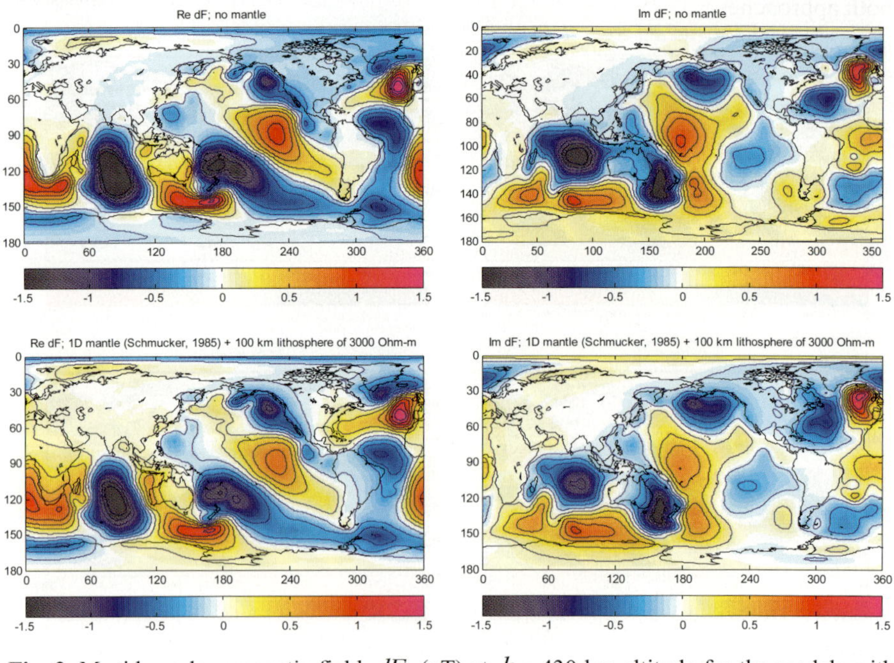


**Fig. 2.** $M_2$ tide scalar magnetic field $dF$ (nT) at $h$= 430 km altitude for the models with insulating (upper panels) and conducting (lower panels) mantle, respectively.

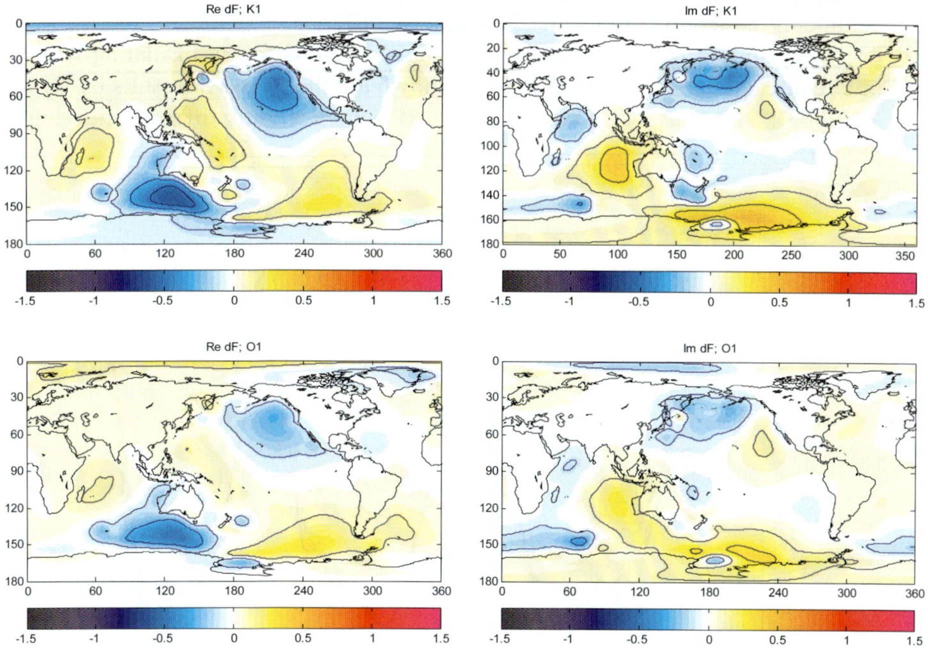


**Fig. 3.** Scalar anomaly $dF$ (nT) at $h = 430$ km altitude due to the $K_1$ (upper panels) and $O_1$ (lower panels) tidal modes.

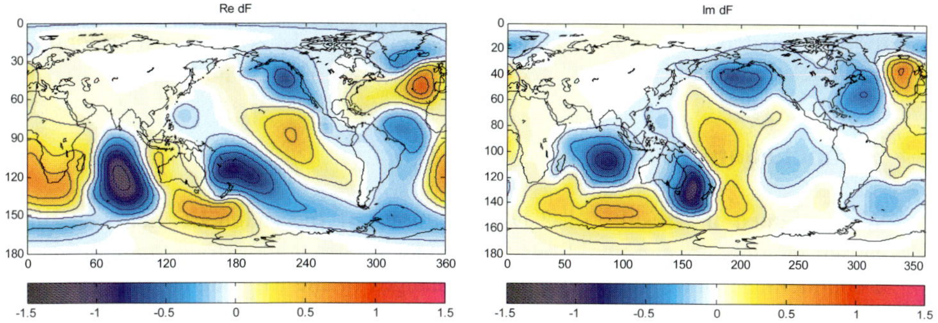


**Fig. 4.** $M_2$ tide scalar magnetic field $dF$ (nT) at $h = 800$ km altitude.

# 4 Comparison with CHAMP observations

Fig. 5 shows the amplitudes of high-pass filtered modelled magnetic $M_2$ scalar anomaly, averaged over all latitudes between –60° and +60°, in dependence on longitude. These results are for different mantle conductivity models beneath the surface shell. For the "nonuniform lithosphere" model, the resistivity of the uppermost 100 km is set to 300 Ωm beneath the oceans, whereas that beneath the continents is set to 3000 Ωm. The largest difference in the magnetic field at CHAMP altitude is found when comparing the results for an insulating with that of a realistic (conducting) mantle. Also shown are the CHAMP results by Tyler et al. (2003). Note that they have been derived using a slightly different filtering approach.

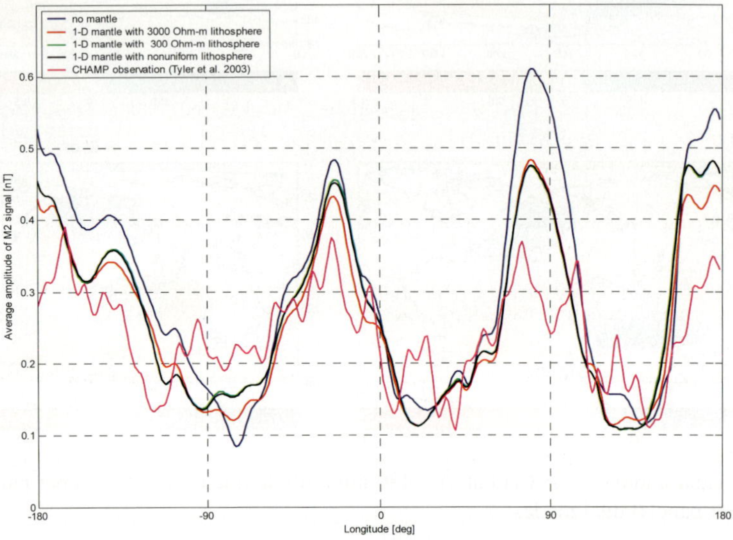


**Fig. 5.** Meridionally averaged magnetic signal amplitudes (nT) at CHAMP altitude, calculated for various mantle conductivity models. Also shown are observational results, obtained from CHAMP scalar data by Tyler et al. (2003).

## 5 Conclusions

We presented the results of 3-D simulations of the magnetic field variations caused by global ocean tidal flow. Our model studies demonstrate that considering a conducting mantle yields significant changes of the $M_2$ magnetic signals, with peak-to-peak values at CHAMP altitude of order 3 nT. The magnetic signals due to other prominent ocean tidal modes (like $K_1$ and $O_1$) are below 0.5 nT at CHAMP altitude. The $M_2$ peak-to-peak magnetic signal at Ørsted altitude is of order 1 nT.

*Acknowledgements.* We would like to thank Stefan Maus and Rob Tyler for providing us with data for the study comparisons.

## References

Avdeev D, Kuvshinov A, Pankratov O, Newman G (2002) Three-dimensional induction logging problems, Part I: An integral equation solution and model comparisons. Geophysics *67*: 413-426.

Egbert G, Erofeeva S (2002) Efficient inverse modeling of barotropic ocean tides. J of Oceanic and Atmosph. Technol. *19*: 183-204

Kuvshinov A, Avdeev D, Pankratov O, Golyshev S, Olsen N (2002a) Modelling electromagnetic fields in 3D spherical Earth using fast integral equation approach. 3D Electromagnetics, Elsevier, Chapter 3: 43-54.

Kuvshinov A, Olsen N, Avdeev D, Pankratov O (2002b) Electromagnetic induction in the oceans and the anomalous behaviour of coastal C-responses for periods up to 20 days. Geophys Res Lett *29*: 2001GL014409.

Pankratov O, Kuvshinov A, Avdeev D (1997) High-performance three-dimensional electromagnetic modeling using modified Neumann series. Anisotropic case. J Geomagn Geoelectr *49*: 1541-1547.

Schmucker U (1985) Magnetic and electric fields due to electromagnetic induction by external sources, electrical properties of the earth interior. Landolt-Boernstein New Series Group, 5/2b, Springer Verlag, Berlin.

Singer B (1995) Method for solution of Maxwell's equations in non-uniform media. Geophys J Int *120*: 590-598.

Tyler R, Maus S, Lühr H (2003) Satellite observations of magnetic field due to ocean tidal flow. Science *299*: 239-241.

Zhang S-L (1997) GPBi-CG: generalized product-type methods based on Bi-CG for solving non symmetric linear systems. SIAM J Sci Comput *18*: 537-551.

# The Enhancement of the Thermospheric Density During the Sept. 25–26, 2001 Magnetic Storm

Huixin Liu, Hermann Lühr and Wolfgang Köhler

GeoForschungZentrum Potsdam, Potsdam, Germany, *huixin@gfz-potsdam.de*

**Summary.** The CHAMP satellite observed large enhancement of the neutral atmospheric density in the southern auroral region during the Sept. 25–26, 2001 magnetic storm. This enhancement was accompanied by intense auroral electrojets and strong small-scale field-aligned currents. Joule heating rate and the heating rate by these small-scale field-aligned currents are compared. Results show that the heating rate by small-scale field-aligned currents is about 2∼3 orders of magnitude higher than that by the Joule heating, indicating its importance as a heating source.

**Key words:** Thermosphere, magnetospheric storm, Joule heating, small-scale field-aligned current

## 1 Introduction

Neutral atmospheric density increases during geomagnetic storms were first noted by Jacchia (1959) [1] from satellite drag studies. The density enhancement has been interpreted as being caused by global atmospheric heating. Joule heating produced by the Pedersen component of the auroral electrojet has been considered to be the main heating source. Field-aligned current (FAC) is normally regarded to be less important due to large field-aligned conductivity. Small-scale field-aligned currents, however, have so far not been considered. This is partly due to the lack of observation of these currents. The CHAMP satellite [3] provides continues simultaneous data of neutral density and ionospheric currents with high time and spatial resolution, thus make it possible to examine quantitatively the heating produced by different currents.

In this paper, we present the observations during the Sept. 25-26, 2001 magnetic storm. Large enhancements of the thermospheric density occurred shortly after the onset of the magnetic storm [2]. These enhancements were accompanied by increase of the auroral electrojets and the small-scale field-aligned currents. A comparison shows that the heating rate produced by small-scale field-aligned currents can be considerably larger than the Joule heating rate, indicating its importance as a heating source.

## 2 Observations

### 2.1 Geomagnetic Conditions

The Sept. 25-26, 2001 geomagnetic storm started at 2025 UT with a storm sudden commencement (SSC). The Dst index reached a minimum of -101 nT at 0200 UT on Sept. 26, 2001. The maximum value of AE was about 1,500 nT. Figure 1 shows the IMF and the solar wind data from WIND satellite between 1800 UT and 2400 UT on Sept. 25, 2001. A time delay between 10–15 minutes should be taken into account for comparison with ground observations.

### 2.2 The Thermospheric Density and Ionospheric Electric Currents

The satellite CHAMP with its sensitive accelerometer on board provides the opportunity to investigate the thermospheric dynamics in great detail. The neutral atmospheric density observed by CHAMP satellite on Sept. 25, 2001 is shown in Fig. 2 for three consecutive orbits 6742, 6743 and 6744. The data are normalized to an altitude of ~430 km. The vertical line indicates the magnetic storm onset, which occurred in orbit 6743 at 2025 UT on Sept. 25, 2001. The air density showed large enhancements in polar regions during subsequent orbits 6743 and 6744. Particularly in the southern hemisphere, the density is doubled to prestorm condition, reaching $1.8 \times 10^{-11}$ kgm$^{-3}$. Note that the density enhancement in orbit 6743 followed the storm onset very

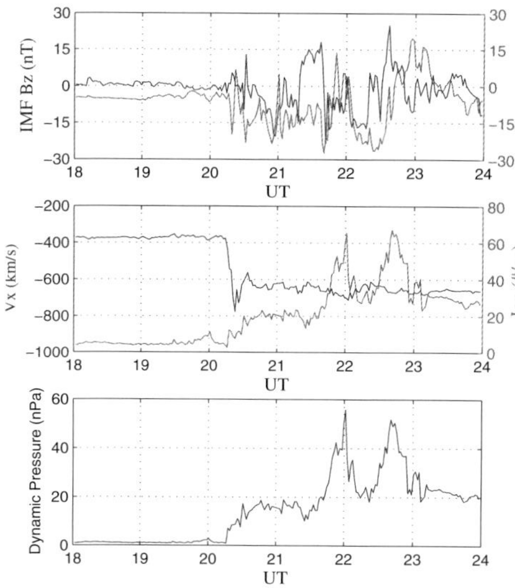


**Fig. 1.** The IMF and the solar wind data from WIND satellite during 1800–2400 UT on Sept. 25, 2001. A time delay about 10–15 min should be taken into account for comparison with ground observations.

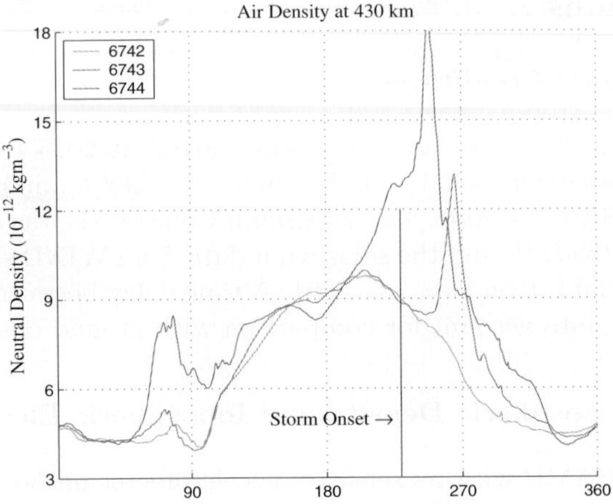


**Fig. 2.** The neutral atmospheric density observed by CHAMP on Sept. 25, 2001. The x-axis is geomagnetic latitude, where the equator is at 0, the north pole at 90, and the south pole at 270.

closely (a few minutes). A comparison of Orbit 6743 and 6744 clearly shows an equatorward propagation of the density enhancement. This propagation is an important feature of the thermospheric response to magnetic storms. However, this feature is not the focus of the present paper, and therefore will not be discussed further.

The ionospheric Hall and FAC currents were derived from the CHAMP magnetometer data [4]. Figure 3 presents the results during period of 2205–2225 UT (in orbit 6744), when the largest density enhancement occurred. The third and fourth panels show the current density of the small-scale (50 Hz) and large-scale FACs, respectively. It is clear that both the Hall and FAC currents were enhanced during periods of high neutral density. The Hall current density increased to more than 1.2 $\mathrm{Am}^{-1}$. The small-scale FAC density was particularly large, reaching 300 $\mu\mathrm{Am}^{-2}$ around 2044 UT and almost 1 $\mathrm{mAm}^{-2}$ around 2211 UT.

## 3 Heating

In this section, we briefly compare different heating rate produced by Pedersen currents, large and small-scale field-aligned currents.

The heating rate produced by the Pedersen and FAC currents can be calculated using Eq. (1). The 1st term of Eq. (1) is the Joule heating rate. The 2nd and 3rd terms are respectively the transverse and field-aligned heating rate produced by the small-scale FAC. Here $\sigma_p$ and $\sigma_\|$ are the Pedersen and the parallel ionospheric conductivity, respectively. And $\Sigma_H$ and $\Sigma_P$ are the height-integrated Hall and Pedersen conductivity. Here $J_H$ is the hight-integrated

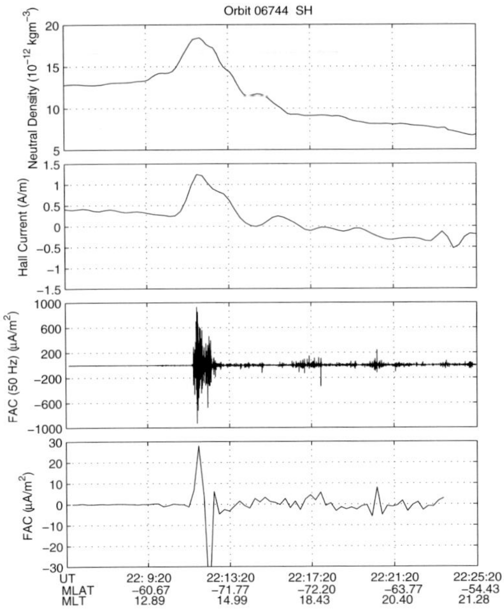


**Fig. 3.** Simultaneous measurements of the neutral density, Hall and FAC currents during 2205–2225 UT. The third and fourth panels show the small-scale (50 Hz) and large-scale field-aligned current, respectively.

Hall current density, which is about 1.24 $\mathrm{Am}^{-1}$ around 2212 UT on Sept. 25, 2001. $J_{\|}$ is the horizontally-integrated field-aligned current density, is about 12.98 $\mathrm{Am}^{-1}$. It is calculated from the small-scale field-aligned currents ($j_{\|}$) using Eq. (2), where $x_1$ and $x_2$ are the starting and ending point of the enhanced small-scale FACs, N is the number of these FAC filaments along the satellite track. As the ratio of $\Sigma_H$ and $\Sigma_P$ is normally between 1 and 10, the second term is therefore 2∼3 orders of magnitude higher than the first term. This means that the heating rate by the small-scale field-aligned currents dominated the Joule heating rate for our case.

$$
\begin{aligned}
\mathbf{j} \bullet \mathbf{E} &= \sigma_p\left(\frac{J_H}{\Sigma_H}\right)^2 + \sigma_p\left(\frac{J_{\|}}{\Sigma_P}\right)^2 + \sigma_{\|}\left(\frac{j_{\|}}{\sigma_{\|}}\right)^2 \\
&= \frac{\sigma_P}{\Sigma_H^2}\left(J_H^2 + \frac{\Sigma_H}{\Sigma_P} J_{\|}^2\right) + \sigma_{\|}\left(\frac{j_{\|}}{\sigma_{\|}}\right)^2 \qquad (1)
\end{aligned}
$$

$$
\begin{aligned}
J_{\|} &= \frac{1}{2}\int_{x_1}^{x_2} |j_{\|}| dx \\
&= \frac{1}{2}\sum_{i=1}^{N} |j_{\|i}| \Delta x
\end{aligned}
$$

$$= \frac{1}{2N} \sum_{i=1}^{N} |j_{\parallel i}| N \Delta x$$

$$= \frac{1}{2N} \sum_{i=1}^{N} |j_{\parallel i}| (x_2 - x_1), \quad (2)$$

## 4 Concluding Remarks

The neutral atmospheric density increases in response to excess heating in the auroral zone during disturbed periods. A comparison of different heating rate shows that the heating rate by small-scale field-aligned currents is 2 to 3 orders of magnitude higher than the Joule heating rate in the Sept. 25-26, 2001 magnetic storm. However, as the small-scale FAC is normally enhanced in a very narrow region, the total heating it produced is not necessarily higher. Nevertheless, heating produced by small-scale FACs should be taken into account when heating sources are concerned.

## References

1. Jacchia LG (1959) Corpuscular radiation and the acceleration of artificial satellites. Nature *183*: 1662.
2. König R and Neumayer KH (2003) Thermosperic events in CHAMP precise orbit determination. In: First CHAMP Mission Results for Gravity, Magnetic and Atmospheric Studies, ed. by Reigber C, Lühr H and Schwintzer P, Springer, Berlin-Heidelberg, 112–119.
3. Reigber C, Lühr H and Schwintzer P (2002) CHAMP mission status. Adv Space Res *30(2)*: 129.
4. Ritter P, Lühr H, Viljanen A, Amm O, Pulkkinen A and Sillanp I (2004) Ionospheric currents estimated simultaneously from CHAMP satellite and IMAGE ground-based magnetic field measurements: A statistical study at auroral latitudes. Ann Geophys *22*: 417–430.

# On the Modelling of Field-Aligned Currents from Magnetic Observations by Polar Orbiting Satellites

Peter Stauning, Freddy Christiansen, and Jürgen Watermann

Danish Meteorological Institute, Copenhagen, Denmark. *pst@dmi.dk, fch@dmi.dk, jfw@dmi.dk*

**Summary.** The comprehensive data from Ørsted and CHAMP adding to the data base from the earlier Magsat magnetometry mission have made it possible to develop sophisticated models for the high-latitude field-aligned currents (FAC) relating to solar wind and magnetospheric parameters. However, there are still large uncertainties involved in using the new models primarily due to the lack of precise knowledge of the temporal and spatial development of the currents in relation to the highly variable solar wind parameters and the still unpredictable effects of magnetospheric substorms. A further problem is the lack of proper account of fine-structure currents in the statistical models. By considering separately the upward and downward FAC intensities the paper presents some first attempts to examine the possible diminishing effects on modelled FAC magnitudes from averaging oppositely directed currents.

**Key words:** field-aligned currents, FAC modelling, polar ionosphere

## 1 Introduction

In many recent works the abundance of high-quality magnetic data from Ørsted and CHAMP satellites have been added to the data base available from the Magsat mission. Using appropriate binning of IMF and other parameters, the magnetic perturbation vectors have been derived for a number of polar passes to serve as a data base for FAC calculations at a variety of conditions. The methods used to calculate FAC distributions from observed geomagnetic data are derivatives of the Ampère-Maxwells law, $\mu_0 I = \nabla \times B$. Usually the stationary main field components given by a precise model are first subtracted from the measured values and only the magnetic field perturbations $B$ are considered. We distinguish:

- Gradient $B_T$ method for current sheets. (e.g., [3])
- Curl of bivariately interpolated 2-D $B$-field. (e.g., [2])
- Curl of Spherical Harmonics 2-D $B$-field. (e.g., [1])
- Surface Laplacian on Magnetic Euler Potential of 2-D $B$-field. ([4])

An example of bivariate interpolation to provide a continuous 2-D perturbation vector field over the polar cap is shown in Fig. 1 (left) for the *NBZ* case: $+5 < B_Z < +10$ nT. The FAC distribution (right) is then derived from taking the curl of the vector field.

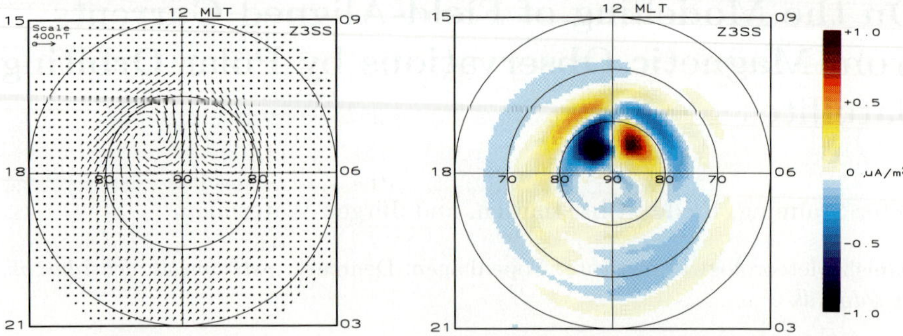


**Fig. 1.** Left: Example of using bivariate interpolation to derive the 2-D vector field of geomagnetic perturbations over the polar cap during NBZ conditions. Right: resulting FAC distribution.

In Fig. 1 the *NBZ* FAC regions are the two rounded regions at latitudes of around 84°located at local magnetic times of 08-11 MLT for the upward current (yellow-red) and 13-16 MLT for the downward current (blue) respectively. The *R1* currents in the morning sector (blue) and in the evening sector (yellow-red) are clearly visible. The oppositely directed *R2* currents further equatorward are just discernible.

Rather than using interpolation a spherical harmonics (SH) functional representation of the magnetic perturbation vector field can be derived for each specific bin of the data base. The FAC distribution is then found analytically from the curl of the magnetic vector functions (e.g., [1]). Recently, a new technique has been developed [4] for mapping the distribution of polar FAC and ionospheric currents using satellite magnetometer data to derive the magnetic Euler potential distribution from spherical cap harmonic analysis (SCHA). The FAC distribution is now derived from the surface Laplacian of the Euler potentials.

## 2 Reliability of statistically averaged FAC distributions

Statistical FAC models now derived on basis of Magsat, Ørsted and CHAMP satellite data use some kind of averaging over space and time and over many cases. Even in the large-scale structures the FAC's most often occur in the form of pairs of oppositely directed currents. Hence there is an obvious risk that the averaging diminishes the resulting current intensities if the spatial averaging is made over pairs of oppositely directed FAC sheets or if the selected cases alternately have oppositely directed currents at the same location.

These problems are amplified if one considers the fine-scale variations where oppositely directed currents of up to more than 1000 $\mu$A/m$^2$ are encountered at distances of only a few hundred m corresponding to a few ms

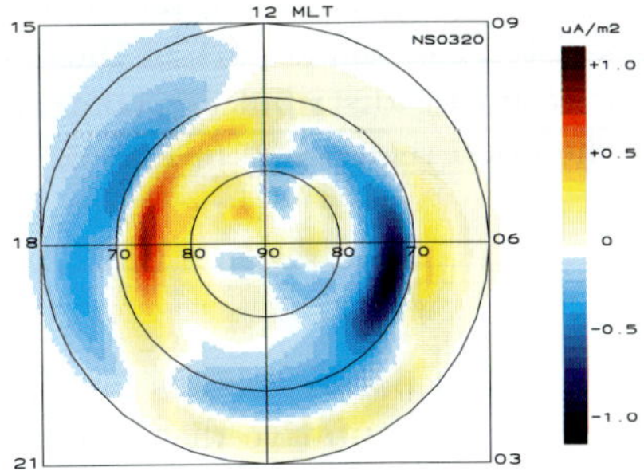


**Fig. 2.** FAC distribution over the northern polar cap for the case: -5<IMF $B_X$ <+5, +2< $B_Y$ <+5 nT, -5< $B_Z$ <-2 nT, 300< $V_{SW}$ <600 km/s, 3< $D_{SW}$ <10 p/ccm, summer.

flight time. It is clear that the alternating closely spaced upward and downward currents will not be represented at all in any of the available statistical models.

Fig. 2 presents a typical case of FAC mapped over the northern polar cap. IMF $B_Y$ is weakly positive and $B_Z$ is weakly negative. Solar wind density and velocity have moderate values. The FAC distribution has been derived by using the interpolated perturbation field method (e.g., [2]).

From largely the same data set but now from using the field gradients (2-sec averages) measured during near-polar passes we have derived the FAC distribution separately for upward and downward currents. From the results presented in Fig. 3 it is clear, that there are strong FAC not accounted for in the average distribution presented in Fig. 2.

This is particularly the case for the FAC currents near local magnetic noon. In this region and for IMF $B_Y$ >0 the FAC currents usually comprise a downward directed current sheet equatorward of the cusp region and an upward directed current sheet poleward of the cusp (e.g., [3]). These currents sheets are closely spaced and at the same time their positions depend on past

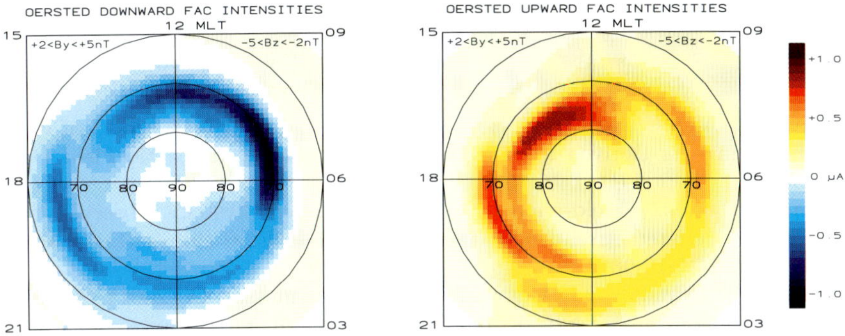


**Fig. 3.** Separate distributions for the downward (left diagram) and upward (right diagram) FAC intensities derived for the same case as that of Fig. 2.

history and present values of the IMF $B_Z$ component as well as solar wind dynamic pressure. Hence the averaging will tend to make part of these FAC currents cancel each other such that the resulting distribution, contrary to reality, indicates fairly weak currents in the noon region like in Fig. 2.

## 3 Discussions and conclusions

The results presented strongly argue for making separate models for the upward and downward FAC in order to derive representative distributions in statistical FAC models. This is a simple task when using the gradient-$B_T$ method. The 2-D methods will need strong modifications. The presently used FAC models are bound to produce diminished FAC intensities in regions where the upward and downward currents may shift positions within the binning constraints. This problem will, in particular, pose difficulties for the proper modelling of the IMF $B_Y$ -related FAC at the dayside and for the substorm-related FAC in the nightside high-latitude regions. A further problem area is the description of the fine-scale FAC where upward and downward currents at intensities up to several hundred $\mu A/m^2$ may alternate at distances of only few hundred m. Such fine-scale currents are particularly intense in the cusp regions. These currents should be described either in unipolar distributions like the example in Fig. 3 or in terms of their AC characteristics, for instance, through the distribution of their rms intensities. Among the consequences from underestimating the FAC intensities are corresponding underestimates of related ionospheric Pedersen currents causing heating of the atmosphere and Hall currents responsible for the major part of magnetic perturbations observed at ground level in polar regions.

*Acknowledgement.* We gratefully acknowledge the use of ACE data supplied from the ACE Science Center at Caltech and WIND satellite data from the National Space Science Data Center. The Ørsted satellite is operated by Terma A/S and the Danish Meteorological Institute (DMI).

## References

1. Papitashvili VO, Christiansen F, and Neubert T (2002) A New Model of Field-Aligned Currents Derived from High-Precision Satellite Magnetic Field Data. Geophys Res Lett *29(14)*: 10.1029.
2. Stauning P (2002) Field-aligned ionospheric current systems observed from Magsat and Ørsted satellites during northward IMF. Geophys Res Lett *29*: 10.1029/2001GL013961.
3. Stauning P, Primdahl F, Watermann J, and Rasmussen O (2001) IMF $B_Y$-related Cusp currents observed from the Ørsted satellite and from ground. Geophys Res Lett *28*: 99.
4. Weimer DR (2001) Maps of ionospheric field-aligned currents as a function of interplanetary magnetic field derived from Dynamics Explorer 2 data. J Geophys Res *106*: 12,889.

# The Low-Altitude Cusp: Multi-Point Observations During the February 2002 SIRCUS Campaign

Jurgen Watermann[1], Hermann Lühr[2], Kristian Schlegel[3], Peter Stauning[1], Jeffrey P. Thayer[4], Freddy Christiansen[1], and Patrick T. Newell[5]

[1] Danish Meteorological Institute, Copenhagen, Denmark, *jfw@dmi.dk, pst@dmi.dk,fch@dmi.dk*
[2] GeoForschungsZentrum Potsdam, Potsdam, Germany, *hluehr@gfz-potsdam.de*
[3] Max-Planck-Institut für Aeronomie, Katlenburg-Lindau, Germany, *schlegel@linmpi.mpg.de*
[4] SRI International, Menlo Park, CA, U.S.A., *thayer@sri.com*
[5] Applied Physics Laboratory, Johns Hopkins University, Laurel, MD, U.S.A., *patrick.newell@jhuapl.edu*

**Summary.** The first SIRCUS campaign covered the days February 16–22, 2002. It involved the CHAMP and Ørsted LEO satellites and the EISCAT (Tromsø, Svalbard) and Sondrestrom incoherent scatter radars. The primary data set was supplemented by DMSP-F13 and -F14 particle flux and ACE solar wind measurements. On February 21 all instruments were operated in campaign mode and delivered usable data. We identify the low-altitude cusp based on signatures in our multi-instrument data set and derive a consistent time-dependent mapping of the cusp in terms of magnetic local time and latitude. We demonstrate that small-scale variations of the magnetic field resp. field-aligned currents (spatial scales of several hundred meters) can be used to identify the cusp. However, their general ability as cusp signature has not yet been proven. An extended analysis of several SIRCUS campaigns is expected to give a qualified answer.

**Key words:** high-latitude ionosphere, low-altitude cusp, field-aligned currents

## 1 Introduction

The acronym SIRCUS (**S**atellite and **I**ncoherent scatter **R**adar **CU**sp **S**tudies) identifies a research program which is built upon a sequence of observational campaigns of a few days each. The campaigns were initiated by K. Schlegel and H. Lühr and are coordinated efforts to get the best data return from the CHAMP and Ørsted satellites and to benefit from temporal and spatial coincidence with operation of the EISCAT and Sondrestrom incoherent scatter radars (ISR). The object of investigation is the low-altitude magnetospheric cusp and its footpoint in the ionosphere. Ionospheric cusp signatures and dynamics have not yet been reliably and comprehensively established. More information on SIRCUS, its objectives, activities and future plans are found in [7]. One winter and one summer campaign have been conducted so far, and we focus here on February 21, 2002, the winter campaign day with the most complete observational coverage by participating facilities and instruments.

The Ørsted and CHAMP satellites offer the chance to record magnetic field variations at sampling rates of 50 Hz (CHAMP) and of 25 and 100 Hz (Ørsted). In consequence, investigations into the spatio-temporal distribution of small-scale field-aligned currents (FAC) have been initiated. Recent statistical results ([8] and subsequently [3]) indicate that small-scale FAC are particularly intense in the nominal cusp region but exist also in other regions.

## 2 Geophysical Conditions

Solar wind conditions were largely "normal". Velocity and density fluctuated around 450 km/s and 5 $H^{+}cm^{-3}$, respectively. The total interplanetary magnetic field (IMF) strength was around 10 nT, with a predominantly negative (eastward) $B_y$ component and a $B_z$ component hovering around zero (Fig. 1). Campaign hours (09–15 UT) are highlighted in dark orange and yellow. Because of the time needed for the solar wind to propagate from the ACE location (L1 point) to the bow shock, ACE measurements considered relevant to the campaign interval are those taken from about 0730 UT until 1400 UT (highlighted in light and dark orange).

Although the interplanetary medium parameters were in general relatively steady, some moderate variations of the IMF-$B_y$ and -$B_z$ components observed by ACE were expected to result in moderate spatial shifts of the cusp location in terms of magnetic local time and magnetic latitude in accordance with established statistical results [4] [5] [1]. Under the given conditions (slightly negative $B_y$ and approximately zero $B_z$ components of the IMF) we expect the low-altitude cusp most likely to be centered around 11 MLT (magnetic local time) with its equatorward boundary near 76° corrected magnetic latitude (we use AACGM for reference), see [5] and [1].

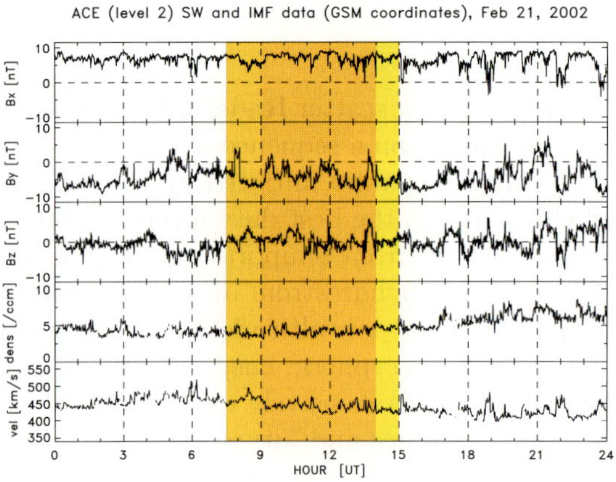


**Fig. 1.** IMF and solar wind variations recorded by the ACE spacecraft on 2002-02-21 (level 2 data kindly provided by the ACE Science Data Center). Hours relevant to the SIRCUS winter campaign are highlighted, starting with light orange 1.5 hrs prior to campaign start, turning dark orange at campaign start and yellow 1.0 hrs prior to campaign end.

## 3 Incoherent Scatter Radar Observations

Cusp particle precipitation is characterized by intense 0.2–2 keV ion and low-energy electron fluxes [5] [1]. Such fluxes, in particular the electrons, lead to local electron heating in the upper atmosphere at time scales of seconds (to a lesser degree also to ion heating) and sometimes but not always to a measurable increase of the plasma density (at time scales of minutes). The 250–450 km altitude range is typically most affected. This kind of cusp particle energy deposition has been observed and tracked with the Sondrestrom ISR [10] [6] [2]. Ion outflow is an additional cusp signature observed by the ISR [6]. However, ISR cusp signatures are sometimes ambiguous because electron flux intensities and spectral characteristics similar to cusp particles are occasionally found to occur in the open low latitude boundary layer [9] [2].

The EISCAT Svalbard Radar (ESR) 42-m dish – which is always pointing up along the local magnetic field – observed a cusp-typical signature (strongly elevated electron and slightly elevated ion temperatures and intense ion outflow) for a brief period of time around 09 UT (11.8 MLT), see Fig. 2.

The Sondrestrom ISR performed 60° azimuth scans at 30° antenna elevation and observed between 1320 and 1340 UT (≈11.3 MLT) in the northwest-most section of the scan area electron temperature elevation coincident with persistently high electron density and a large electron/ion temperature ratio (Fig. 3). This observation coincided spatially and temporally with a DMSP-F14 cusp identification. High plasma density was also observed in other scan sectors and at other times but not accompanied by elevated temperatures.

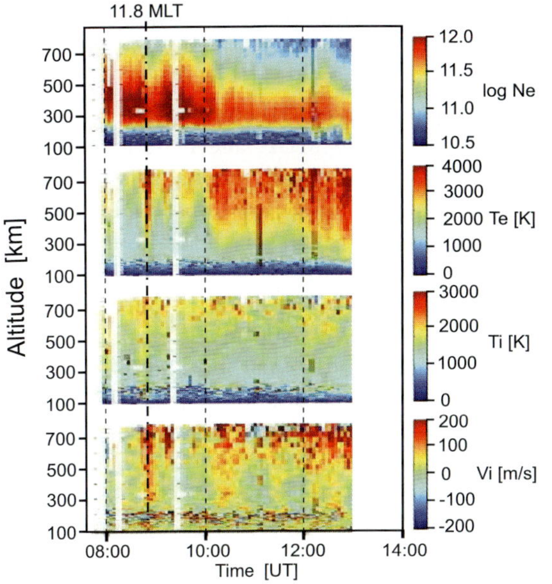


**Fig. 2.** ESR measurements of prime parameters made along the local magnetic field line on Feb 21, 2002, 08–13 UT. The thin vertical line segments between the panels (at 0850 UT ≃ 11.8 MLT) mark a short period of elevated electron temperature (second panel) and ion outflow (bottom panel). They are characteristic though ambiguous signatures of the cusp footpoint.

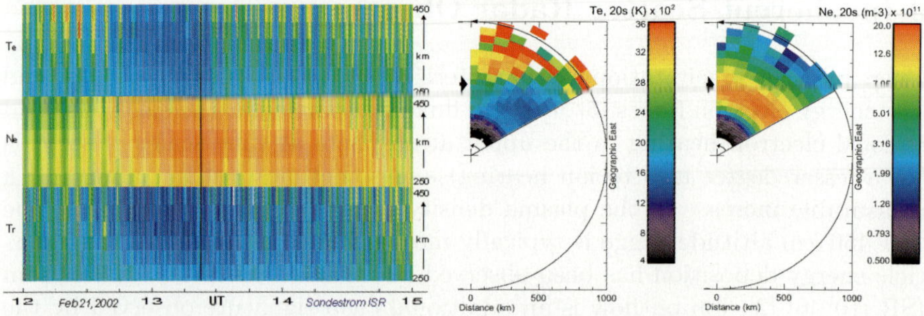


**Fig. 3.** Left: Electron temperature (top), electron density (middle) and electron/ion temperature ratio (bottom) measured by the Sondrestrom ISR. Right: Electron temperature and density from a single azimuth scan at 30° antenna elevation (at the time marked by a vertical line in the left panel). A typical cusp signature, namely elevated electron temperature on a background of high electron density, is seen in the leftmost (green) part of the scan area (see arrow marker) between 250 km and 500 km ground distance (200–400 km altitude).

## 4 Synoptic View from LEO Satellites and ISR

We have combined cusp signatures observed with the ESR-42m and the Sondrestrom ISR with small-scale (1–50 Hz) FAC densities inferred from high-resolution CHAMP magnetometer measurements and absolute values of small-scale (0.5–25 Hz) magnetic field variations recorded by the Ørsted magnetometer. The results are shown in a semi-quantitative way in Fig. 4. Observations ordered according to Universal Time progress, generally speaking, from east to west (right to left). MLT numbers refer to the equatorward boundary of the observed signatures.

The first set of coincident observations (marked ①) comprises the DMSP-F13 particle cusp detection (green bar) at 10.8 MLT, the ESR ionospheric cusp signature (cyan circle) at 11.8 MLT and the locally enhanced small-scale FAC density along the CHAMP trajectory (broadened red line) at 11.8 MLT, all approximately at the same UT. They are consistent with the nominal cusp location expected to be around 11 MLT and poleward of ≈76° CGML.

100 min later (marked ②) we find again the DMSP-F13 particle cusp, now at 11.0 MLT, and CHAMP enhanced small-scale FAC density at 12.3 MLT. Both are shifted to higher magnetic latitudes. At the same UT we have the first Ørsted pass of this sequence (rightmost blue line) which shows no signs of enhanced small-scale magnetic field perturbations in the cusp region. Ørsted crossed the 75° CGML contour at 14.0 MLT which is probably too far into the afternoon sector and outside the cusp.

Another 100 min later (marked ③) CHAMP observes locally enhanced FAC densities at 11.7 MLT and Ørsted records magnetic field perturbations near 13.5 MLT. However an expected DMSP particle cusp signature is absent.

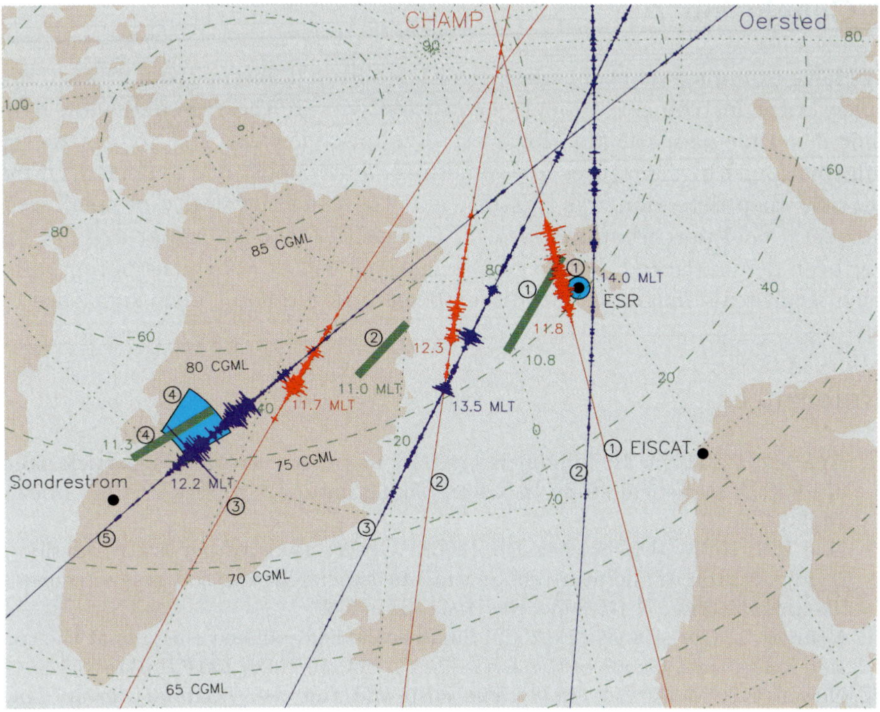


**Fig. 4.** Synoptic view of LEO satellite and ISR measurements suggested to mark the cusp: DMSP particle cusp (green bars), ISR ionospheric plasma parameters (cyan areas), CHAMP small-scale FAC (broadened red segments), Ørsted small-scale magnetic field variations (broadened blue segments)

One hour later (marked ④) DMSP-F14 finds the particle cusp (green bar) at 11.3 MLT and poleward of 75° CGML, coincident with Sondrestrom ISR observations of an F region cusp signature (cyan four-cornered fan-shaped segment). Shortly after, (marked ⑤) the third Ørsted pass reveals intense small-scale magnetic field perturbation at 12.2 MLT, poleward of ≈75° CGML.

We have thus followed a sequence of observations all of which show signatures of the low-altitude cusp: precipitating particle characteristics (DMSP), ionospheric plasma parameters (ISR), enhanced small-scale FAC densities (CHAMP) and small-scale magnetic field perturbations (Ørsted). The signatures extend over the 10.8–13.5 MLT interval, and the equatorward boundary, initially at 75° CGML, moves poleward up to 77° CGML and returns to 75° CGML. The motion occurs over a time span of ≈5 hrs, probably in response to IMF variations.

## 5 Conclusion

We have examined an event when signatures from different ground-based and space borne instruments provide a consistent view of the cusp and allow mapping it in time and latitude. However, a synoptic view was able to provide this information. ISR signatures are sometimes ambiguous, one expected DMSP particle cusp detection was missing, and the general validity of small-scale magnetic field perturbations resp. FAC densities as cusp signatures has not yet been demonstrated. Further investigations will establish the potential for such small-scale magnetic variations to serve as a reliable cusp signature.

## References

1. Aparicio B, Thelin B, Lundin R (1991) The polar cusp from a particle point of view: A statistical study based on Viking data. J Geophys Res *96*: 14,023–14,031.
2. Doe RA, Kelly JD, Sánchez ER (2001) Observations of persistent dayside F region electron temperature enhancements associated with soft magnetosheath-like precipitation. J Geophys Res *106*: 3615–3630.
3. Neubert T, Christiansen F (2003) Small-scale, field-aligned currents at the topside ionosphere. Geophys Res Lett *30(19)*: doi: 10.1029/2003GL017808.
4. Newell PT, Meng C-I (1988) The cusp and the cleft/boundary layer: Low-altitude identification and statistical local time variation. J Geophys Res *93*: 14,549–14,556.
5. Newell PT, Meng C-I, Sibeck DG, Lepping R (1989) Some low-altitude cusp dependencies on the interplanetary magnetic field. J Geophys Res *94*: 8921–8927
6. Nilsson H, Kirkwood S, Eliasson L, Norberg O (1994) The ionospheric signature of the cusp: A case study using Freja and the Sondrestrom radar. Geophys Res Lett *21*: 1923–1926.
7. The SIRCUS Science Team (2003) SIRCUS Workshop Report. In: Stauning P et al (eds) OIST-4 Proceedings, DMI Scientific Report 03-09, 327–334, Copenhagen, Denmark.
8. Stauning P, Primdahl F, Christiansen F, Watermann J (2003) Detection of high-latitude fine-scale field-aligned current structures from the Ørsted satellite. In: Stauning P et al (eds) OIST-4 Proceedings, DMI Scientific Report 03-09, 167–174, Copenhagen, Denmark.
9. Watermann J, Lummerzheim D, de la Beaujardiére O, Newell PT, Rich FJ (1994) The ionospheric footprint of magnetosheath-like particle precipitation observed by an incoherent-scatter radar. J Geophys Res *99*: 3855–3867.
10. Watermann J, de la Beaujardiére O, Lummerzheim D, Woch J, Newell PT, Potemra TA, Rich FJ, Shapshak M (1994) The dynamic cusp at low altitudes: a case study utilizing Viking, DMSP-F7, and Sondrestrom incoherent scatter radar observations. Ann Geophys *12*: 1144–1157.

# Detection of Intense Fine-Scale Field-Aligned Current Structures in the Cusp Region

Peter Stauning, Freddy Christiansen, Jürgen Watermann

Danish Meteorological Institute, Copenhagen, Denmark
*pst@dmi.dk, fch@dmi.dk, jfw@dmi.dk*

**Summary.** The Ørsted high-precision magnetic measurements made at high temporal resolution of 25 and occasionally 100 samples/sec corresponding to spatial resolutions down to less than 100 m have demonstrated the occurrences of highly structured field-aligned currents (FAC) in the low-altitude Cusp region. The observed magnetic perturbations indicate structures of very intense but thin sheets or narrow filaments of mixed up- and downward currents up to several hundreds of $\mu A/m^2$ embedded in large-scale FAC structures of only up to a few $\mu A/m^2$. The intensities and locations of fine-scale FAC structures are closely related to solar wind conditions.

**Key words:** field-aligned currents, fine-scale FAC, cusp region, polar ionosphere

## 1 Introduction

Satellite-based observations of magnetic perturbations in space have provided a comprehensive picture of the quasi-steady magnetosphere-ionosphere field-aligned (FAC) coupling currents systems (e.g., [3, 4]). These observations have revealed the existence of the so-called "Region 1" (*R1*) and "Region 2" (*R2*) oppositely directed FAC systems which appear in the morning and evening sectors of the auroral regions at the border of the polar caps. The *R1* and *R2* FAC intensities depend strongly on the direction and magnitude of the interplanetary magnetic field (IMF) and on the solar wind (SW) velocity.

During strong positive (northward) IMF $B_Z$ (NBZ) conditions strong FAC currents may be observed in the central dayside polar cap (e.g., [6, 13]). In the noon sector at high latitudes a further FAC system is often observed to form oppositely directed pairs of east-west oriented current sheets. The

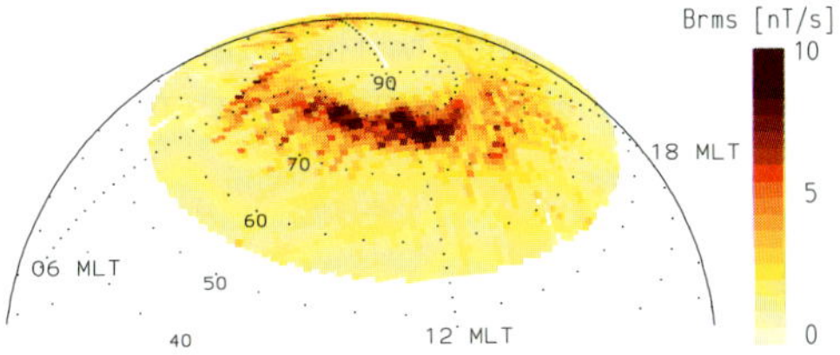


**Fig. 1.** Distribution of fine-scale FAC intensities over the polar cap mapped in magnetic latitude-local time coordinates.

intensity and the direction of these currents are related to the magnitude and sign of the IMF $B_Y$ component, while the latitude of the currents is strongly influenced by the sign and magnitude of the IMF $B_Z$ component (e.g., [4], [5], [12], [14]). The more equatorward FAC region appears to be the continuation across magnetic local noon of either dawn or dusk dayside *R1* FAC. The more poleward currents were termed "Region 0" (*R0*) or "cusp currents" in [4].

These large-scale quasi-static FAC systems have typical intensities of around one or a few $\mu A/m^2$. However, it is clear from many observations that the high-latitude FAC occurrences display highly complex and variable structures where the current densities may vary from a few to several hundred $\mu A/m^2$ [11]. An example of the spatial distribution of fine-structure FAC is displayed in Fig.1. The diagram is based on a selection of northern polar passes that represents all-year seasonal and average solar wind plasma and IMF conditions. The intensity parameter used here is the RMS value calculated at each second of the sample-to-sample magnetic vector variations $|\Delta B|$. The coordinates are magnetic latitude and local time.

## 2 Instrumentation, observations and data analysis

The Ørsted satellite was launched on 23 February 1999 into a near-polar orbit with an inclination of 96° , perigee at 640 km and apogee at 860 km. Its orbit drifts slowly in local time. The principal aim of the satellite is the precise mapping of the Earths magnetic field [9]. The most recent data included here are from October 2002. The Compact Spherical Coil (CSC) satellite vector magnetometer is normally sampled at 25 $s^{-1}$, occasionally at 100 $s^{-1}$. This temporal resolution corresponds to spatial resolution of around 300 and 70 m, respectively, along the orbit. In this work the position of the Ørsted satellite is defined in standard altitude adjusted corrected geomagnetic (aacgm) coordinates and magnetic local time.

We use a special version of Ampère-Maxwells law, $\mu_0 I = \nabla \times B$, to calculate FAC intensities from the varying magnetic perturbations observed along the orbit. The magnetic perturbations are found by subtracting Oersted (10c/99) reference field model values [9] from the measured data. In addition to the internal source terms, the contributions from the magnetospheric ring current have been included through Dst-dependent terms.

The FAC current density $J$ is now found from applying an Ampère's integral over a differential parallelogram with sides formed by the orbit increment $\Delta r$ and a homogeneity vector $h$ in the direction along assumed current sheets.

$$J = \frac{h \cdot \Delta B_1}{\mu_0 b \cdot \Delta r \times h}, \tag{1}$$

In eq.(1) $\Delta B_1$ is the increment in $B_1$ corresponding to $\Delta r$ while $b$ is a unit vector along the total field direction $B$. The direction of $h$ is determined for each segment of the orbit as the direction of maximum variance in the projected magnetic perturbations. It is usually (but not always) near magnetic east-west direction.

Data from the ACE satellite stationed upstream in the solar wind at the L1 point and from the orbiting WIND satellite have been used to define solar wind magnetic field and plasma parameters used for the analyses. The timing of IMF and SW data samples have been adjusted corresponding to the translation with solar wind velocity from the actual satellite positions to a reference position at $12R_E$ in front of the Earth. The delay of FAC intensities observed in the polar ionosphere relating to the IMF and SW parameters at the reference location is considered to be around 10 min (e.g., [12]).

The relevant IMF parameters to characterize the conditions are now defined to be the average values of the $12R_E$ reference data taken over an interval from $T_0$-25 min to $T_0$-10 min where $T_0$ is the time of the central north polar (N) or south polar (S) Ørsted pass. In order to specify limits for IMF variability we have also calculated the corresponding values for the preceding interval from $T_0$-40 to $T_0$-25 min and the RMS variations during the entire interval from $T_0$-40 to $T_0$-10 min.

## 3 Example case of fine-scale FAC

In this study we have focused on data obtained during Ørsted passes through the northern and southern low altitude cusp regions. These regions are located close to local noon at magnetic latitudes of around +75 and - 75° . In Figs.2 and 3 we show an example of the observed magnetic field variations along the orbit projection during a cusp pass. Fig.2 shows the orbit in a polar sector plot in aacgm latitude and magnetic local time. The orbit passes over the northern polar cusp at around 0415 UT on 30 March 2002. The horizontal magnetic perturbations derived from subtracting the reference field from the observed values are plotted as vectors extending from the track every 5 s. A signature for the fine-scale FAC intensities is plotted as bars extending perpendicular from the trace. Their length corresponds to the 1-sec RMS value of $|\Delta B|/\Delta t$. At the crossing of 75° latitude close to noon (cusp region) a marked increase in the RMS signature is seen. Fig.3 displays the magnetic observations plotted against UT in seconds after the start of the pass. The top panel of Fig. 3 presents the observed magnetic variations in the meridional (upper trace), zonal (middle) and parallel component (lower trace) referred to the local geomagnetic coordinate system.

The black squares in the upper field of Fig.3 represent the rotation from the zonal direction to a homogeneity direction derived from variance analysis. The second panel from the top presents the average FAC intensities. The FAC data have been smoothed using a Gaussian weighting function with a

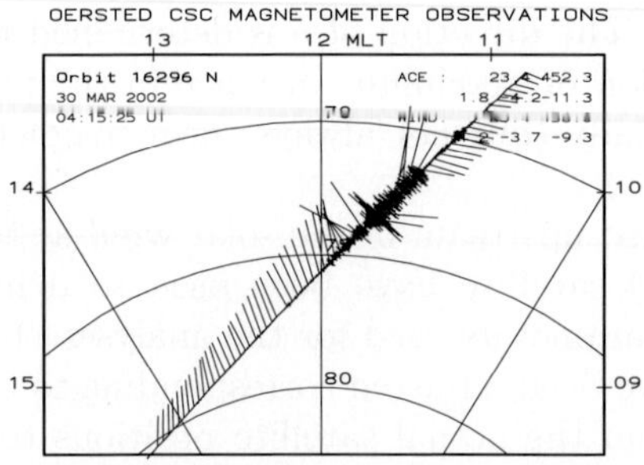


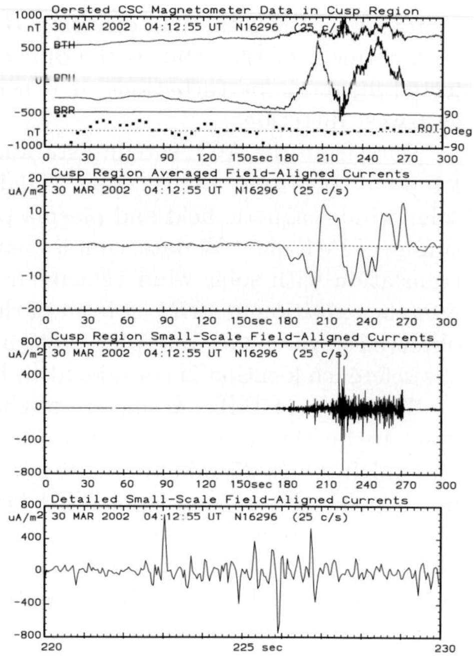


**Fig. 2.** (above) Magnetic 5-s perturbation vectors and signature for fine-scale FAC intensities plotted along the track during the northern cusp passage of orbit 16296 on 30 March 2002.

**Fig.3.** (right) Magnetic variations and derived FAC intensities during the cusp passage on 30 March 2002 starting 04:12:55 UT. IMF $B$: (1.8,-4.2,-11.3nT); SW: 23.4 p/cm$^3$, 452.3 km/s

window width of 5 s. The next lower panel displays the FAC values derived by applying eq. (1) to each step of the 25 c/s data. It is evident that there is a region with extent of around 70s (i.e. 500 km along the track) with strong and variable small-scale FAC structures. Outside the narrow region of disturbances the FAC trace is smooth and the values are very small.

In the bottom panel the variations have been plotted on a more detailed time scale. One may note several FAC peaks of 100-300 $\mu$A/m$^2$. The large peak to the left reaches up to over +600 $\mu$A/m$^2$ and the peak in the middle extends down below -700 $\mu$A/m$^2$. The width of the current-carrying region in the strong FAC peaks ranges from less than a tenth to a few tenths of a second, that is, from a few hundred m to a few km.

## 4 Statistical analysis of fine-scale FAC

In the present database the sample-to-sample magnetic vector variations $|\Delta B|$ have been calculated for around 5000 Ørsted orbits in data sets spanning from September 1999 to October 2002. For each polar pass the related IMF and SW parameters have been derived using the procedure outlined above. Now it is possible to study the effect of each parameter from appropriate selections of passes.

Fig.4 displays the north polar distributions of RMS fine-scale FAC intensities for cases selected to present the variations with IMF $B$, the $B_Y$ and

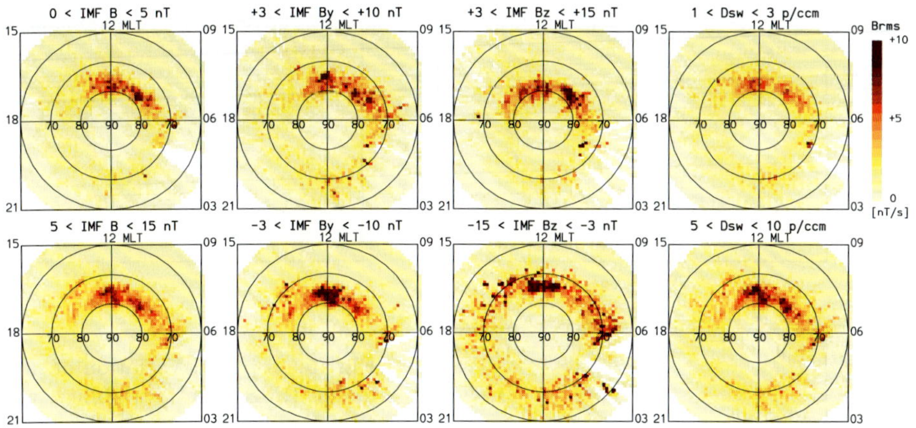


**Fig. 4.** Variation in the north polar distribution of RMS values of fine-scale FAC intensities from Ørsted data. Leftmost: variations with IMF $B$; middle left: variations with $B_Y$; middle right: variations with $B_Z$; rightmost: variations with SW density.

the $B_Z$ components and with the solar wind density. The color scale is the same as that used in Fig.1. The pixel size corresponds to 1 degree in latitude. ¿From the left column of Fig.4 it is clear that the fine-structure FAC intensities are stronger and more focused in the cusp region for stronger $B$ (lower panel) than for smaller values of IMF $B$ (upper panel). The next left column displays the variation from positive $B_Y$ (upper panel) to negative IMF $B_Y$ (lower panel). The main difference is the "tail" of events swinging dawnward for positive $B_Y$ and duskward for negative $B_Y$ values. The core of events remains at noon.

The FAC variations specifically with the sign of IMF $B_Z$ are displayed in the middle right column of Fig.4. It is clear that the intensities are stronger for negative (lower panel) than for positive (upper panel) values of IMF $B_Z$. The strongly negative IMF $B_Z$ cases enhance the fine-scale structures of *R1* and substorm FAC current systems seen at local MLT away from noon.

The middle right column, furthermore, displays the latitude variation in fine-scale FAC intensities with $B_Z$. The average latitude for the high-intensity structures in the upper panel (positive IMF $B_Z$) is around 78° while it is around 73° for the structures in the lower field (negative $B_Z$). The rightmost column of Fig.4 illustrates that the intensity of fine-scale FAC increases and is more concentrated within the assumed cusp region in response to increasing solar wind density going from the upper to the lower panel.

The variation of the intensity of fine-scale FAC structures against the transverse component $B_T$ of IMF is illustrated in the upper left panel of Fig.5. For each polar pass (north or south) the maximum intensity of fine-structure FAC variations is derived as the maximum running average value of 1-sec RMS samples taken over 1 min. This value is marked by a small dot against

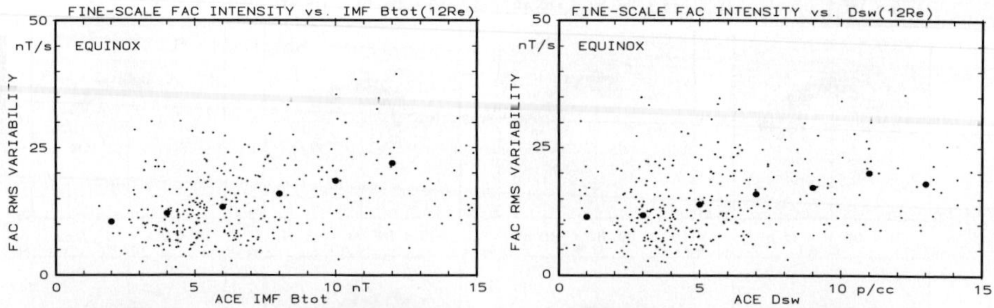


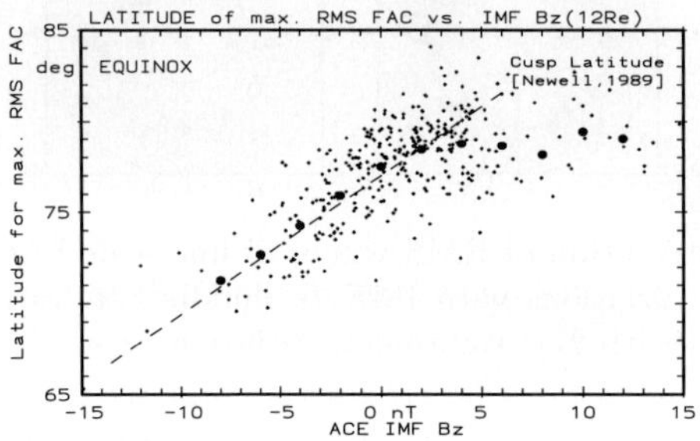


**Fig. 5.** Top left: Variation of maximum RMS FAC variability with IMF $B_T$ for cusp passes. Top right: Variation with solar wind density, $D_{SW}$. Left: Variation in average latitude of max. RMS fine-scale FAC intensities with IMF $B_Z$.

the actual IMF $B_T$ value. For groups of passes within intervals spanning 2 nT in $B_T$ value a large dot has been placed to mark the average value. The corresponding dependence on solar wind density, $D_{SW}$, is illustrated in the upper right panel of Fig. 5. The heavy dots, again, represent averages of values taken over intervals of $D_{SW}$.

The variations in latitude of the location of maximum 1-min running average intensity of fine-structure FAC with the IMF $B_Z$ component are clearly depicted in the bottom field of Fig.5. The dashed line in the diagram indicates the latitude of the equatorward boundary of the cusp region for negative or small positive values of IMF $B_Z$ according to [8]. It is seen that the fine-structure FAC intensities maximize just poleward of this boundary, that is, within the cusp. For the large positive values of IMF $B_Z$ the latitude remains almost constant between 78° and 80° magnetic latitude. This is also in agreement with the cusp boundary variations reported in [8].

## 5 Discussions

The FAC current structures observed in the cusp region are hardly ever in a stationary state. From single satellite observations is not possible to precisely define the persistence and the possible motions of the sheet-like or filamentary structures involved. In the example displayed in Fig.2 we cannot tell directly whether the observed structure relates to a sequence of quasi-steady current sheets or whether the observed variations arise from temporal intensity variations in larger current sheet.

The high-intensity fine-scale structures are most likely the consequences of patchy and transient processes in the dynamic interactions of the solar wind and the magnetosphere. Likely generators of fine-scale FAC are processes associated with compression, twisting or bending of geomagnetic field lines (e.g., [10]), with the braking of intruding solar wind plasma [2] or with merging processes between the IMF and the geomagnetic field (e.g., [7]). FAC patches generated by these processes may travel away from the interaction or reconnection region as Alfven wave packets propagating along cusp field lines toward the ionosphere. At the arrival to the ionosphere the waves are partly reflected and partly dissipated through the transfer of energy to local Joule heating processes and convective motions in the ionospheric plasma.

Possible effects of the strongly variable and filamentary FAC currents in the cusp region comprise the generation of patches of large ionospheric electric fields, strongly varying convection velocities and localized strong Joule heating of the atmosphere at altitudes between approximately 100 and 200 km. The large electric fields may create E-region two-stream instabilities whenever the horizontal field exceeds around 25 mV/m needed to generate ion drift in excess of the ion-acoustic velocity. Such plasma instabilities cause backscatter of radio waves. The occurrence of patchy backscatter and strong convection variability in the cusp region have been documented, among other, through analyses of SuperDARN coherent scatter radar observations (e.g., [1]).

## 6 Conclusions

The observations of fine-structure FAC systems have revealed thin sheets or narrow filaments of very strong field-aligned currents. Current intensities up to 1 mA/m$^2$ within structures of apparent width down to 200 m have been deduced from the high-precision magnetic observations made from the Ørsted satellite. These currents form multiple mixed up- and downward oriented FAC structures of localized intensities larger by one or two orders of magnitude than their average large-scale values.

- The high-intensity, fine-scale structures have highest occurrence probability at noon in magnetic local time. The magnetic latitude of the fine-scale FAC varies systematically with the IMF $B_Z$ component with a preferred location at or a little poleward of the cusp equatorward boundary depending on IMF $B_Z$ as defined by [8]. Hence they are considered part of cusp characteristics.

- The intensity of the fine-structure FAC was found to increase with increasing solar wind density and velocity and with increasing magnitude of the IMF. The fine-scale FAC intensities were found to be markedly stronger, more widespread in MLT and displaced toward lower latitudes for negative compared to positive values of the IMF $B_Z$ component.

- The possible consequences of the high and variable FAC intensities in terms of high ionospheric electric fields and strong localized currents have been experimentally verified through observations of cusp region features like patchy HF backscatter from ionospheric plasma instabilities and highly variable convection velocities

*Acknowledgement.* We gratefully acknowledge the use of ACE data supplied from the ACE Science Center at Caltech and WIND satellite data from the National Space Science Data Center. The Ørsted satellite is operated by TERMA Electronics and the Danish Meteorological Institute (DMI). The Science Data Center at DMI has processed the Ørsted data. The Ørsted vector magnetometer and star imager instruments were built at the Ørsted·DTU Institute at the Danish Technical University. The Ørsted satellite project is funded by the Danish Ministry of Transport, the Ministry of Research and Information Technology, the Ministry of Trade and Industry, and the Danish Research Councils.

## References

1. Baker KB, Greenwald RA, Pinnock M, Newell PT, Rodger AS, and Meng C-I (1995) HF-Radar signatures of the cusp and the low-latitude boundary layer. J Geophys Res *100*: 7671–7696.
2. Heikkila WJ (1982) Impulsive plasma transport through the magnetopause. Geophys Res Lett *9*: 877.
3. Iijima T and Potemra TA (1976a) The amplitude distribution of field-aligned currents at northern high latitudes observed by Triad. J Geophys Res *81*: 2165.
4. Iijima T and Potemra TA (1976b) Field-aligned currents in the dayside cusp observed by Triad. J Geophys Res *81*: 5971.
5. Iijima T and Potemra T (1982) The relationship between interplanetary quantities and Birkeland current densities. Geophys Res Lett *9*: 442.
6. Iijima T, Potemra TA, Zanetti LJ, and Bythrow PF (1984) Large-scale Birkeland currents in the dayside polar region during strongly northward IMF: A new Birkeland current system. J Geophys Res *89*: 7441.
7. Kan JR (1988) A theory of patchy and intermittent reconnections for magnetospheric flux transfer events. J Geophys Res *93*: 5613.
8. Newell PT, Meng C-I, Sibeck DG, Lepping R (1989) Some low-altitude cusp dependencies on the interplanetary magnetic field. J Geophys Res *94*: 8921.
9. Olsen N et al. (2000) Oersted Initial Field Model. Geophys Res Lett *7*: 3607.
10. Song Y and Lysak RL (1994) Alfvenon driven reconnection and the direct generation of the field-aligned current. Geophys Res Lett *21*: 1755.
11. Stasiewicz K and Potemra T (1998) Multiscale current structures observed by Freja. J Geophys Res *103*: 4315–4325.
12. Stauning P (1994) Coupling of IMF $B_Y$ variations into the polar ionospheres through interplanetary field-aligned currents. J Geophys Res *99*: 17.309.
13. Stauning P (2002) Field-aligned ionospheric current systems observed from Magsat and Oersted satellites during northward IMF. Geophys Res Lett *29*: 10.1029/2001GL013961.
14. Stauning P, Primdahl F, Watermann J, and Rasmussen O (2001) IMF $B_Y$-related Cusp currents observed from the Ørsted satellite and from ground. Geophys Res Lett *28*: 99.

# A Comparative Study of Geomagnetic Pi2 Pulsations Observed by CHAMP and on the Ground

Peter R. Sutcliffe[1] and Hermann Lühr[2]

[1] Hermanus Magnetic Observatory, Hermanus, South Africa, *psutcliffe@hmo.ac.za*
[2] GeoForschungsZentrum, Potsdam, Germany, *hluehr@gfz-potsdam.de*

**Summary.** We present initial results of a comparative study of Pi2 pulsations observed by the CHAMP satellite and at the Sutherland ground station [32° 24' S, 20° 40' E]. Times when a Pi2 pulsation was observed on the ground (predominantly during night-time) and when CHAMP was located within 30° of longitude of Sutherland and at latitudes less than 50° N and S were selected for study. Following pre-processing and inspection to exclude unsuitable events, the satellite vector magnetic field data were rotated into a field aligned coordinate system and band-pass filtered in the Pi2 frequency band (typically .005 - .05 Hz).

**Key words:** Pi2 pulsations, cavity modes, ULF waves, vector magnetometer data

## 1 Introduction

Using data from CHAMP, *Sutcliffe and Lühr [2003]* for the first time observed and reported clearly resolved Pi2 pulsations in vector magnetometer data at low Earth orbit (LEO). Previously, the difficulty has been to extract the relatively small pulsation perturbation from the relatively large background field due to the limited digitisation step size of the satellite magnetometer and to the environmental noise. The magnetic field measurements from CHAMP are of unprecedented accuracy and resolution, thus making it possible to extract and clearly resolve Pi2 pulsations. In this paper we describe the method used to extract the Pi2 pulsations from the satellite magnetometer data and then briefly describe the initial results of a comparison of Pi2 pulsations observed at LEO by CHAMP and on the ground in South Africa.

## 2 Data Set and Processing Techniques

The satellite data that were used for this study are the pre-processed (Level 2) fluxgate vector magnetometer data from CHAMP, transformed into the North-East-Center (NEC) coordinate system (product identifier: CH-ME-2-FGM-NEC). These are publicly available through the CHAMP Data Center (ISDC) [http://isdc.gfz-potsdam.de/champ/]. The ground based data used were the induction magnetometer data recorded at the Hermanus Magnetic Observatory's geomagnetic pulsation field station at Sutherland (32° 24’ S, 20° 40’ E) [*Sutcliffe et al.*, 2002]. Both data sets were sampled at 1 sec intervals.

Data selection commenced by determining times when CHAMP was located within 30° of longitude of Sutherland and at latitudes less than 50° N and S; the latter condition was specifically applied to exclude times when the satellite was crossing current systems associated with the auroral electrojets. The Sutherland data for these times were then scanned for suitable Pi2 pulsation events. The satellite data for the times spanning the ground station Pi2 events were then inspected. For reasons of quality, measurements were only utilised if both star cameras were operational; this ensured low attitude-related noise of the vector readings. The data were then further processed to determine whether Pi2 pulsations were present and observable.

The supplemental processing of the satellite data involved four stages. The first step was to initially subtract a main field model from the observed data in order to inspect the residual field. For this purpose the CO2 (CHAMP, Ørsted, Ørsted2) model [*Holme et al.*, 2003] was used, which is available through the ISDC. The reason for inspecting the residual field was to ensure that the interval did not contain disturbances (of tens of nT amplitude), which when filtered may have leaked into the Pi2 band and which could be erroneously interpreted as Pi2 pulsations. Examples of one type of disturbance that we wished to avoid are the night-time F-region currents, which *Lühr et al.* [2002] identified to occur in the near-equatorial region and which show up in the CHAMP data. Unfortunately, these currents occur with maximum probability during evening hours when the clearest Pi2 pulsations are typically observed at low latitudes.

The second step in processing was to subtract a lithospheric magnetic field anomaly model. Consideration of the scale size of lithospheric magnetic anomalies and the rate at which these are traversed by CHAMP, suggests that they will result in variations falling within the Pi2 band of frequencies. Tests using the spherical harmonic coefficients of the magnetic anomaly model MF1 determined from CHAMP observations [*Maus et al.*, 2002] confirmed this to be the case. A contour plot of the lithospheric magnetic field anomaly model for the southern Africa region determined from the MF1 coefficients is shown in Figure 1. The thick line along the 35° E meridian indicates the satellite track during the 10-minute interval when a Pi2 pulsation was observed at Sutherland on 17 May 2002 with onset shortly after 2248 UT. Figure 2 shows the components of the lithospheric magnetic field anomaly model along this satellite track, filtered in the Pi2 frequency band. Oscillations in the Pi2 frequency band and of similar amplitude are clearly observable. The top panel in Figure 3 shows the H component of the Pi2 pulsation observed at Sutherland. The centre and bottom panels in Figure 3 show the compressional component of the same pulsation observed by CHAMP for which no correction and a correction respectively have been applied for the lithospheric anomaly field. It is seen that the correction for the lithospheric field improves the correlation between H and the compressional component, a correlation that one would expect if the oscillations are due to a cavity mode resonance.

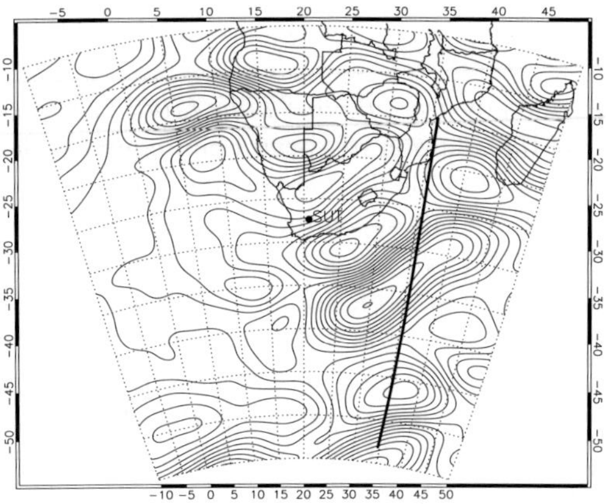


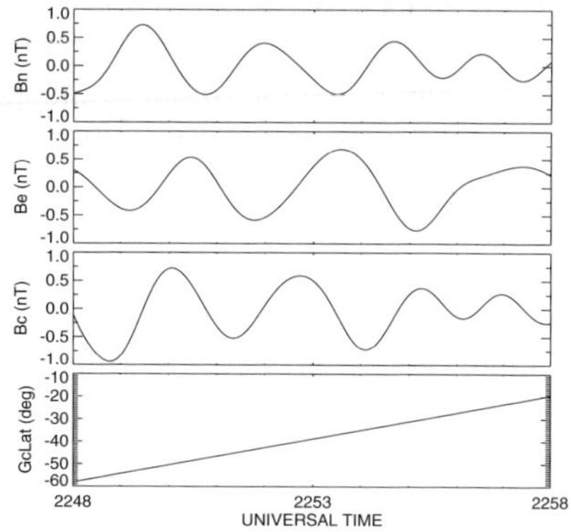


**Fig. 1.** Contour plot of lithospheric magnetic field anomaly model for southern Africa. The Sutherland ground station and the satellite track during the time of occurrence of a Pi2 pulsation are also shown.

**Fig. 2.** Components of the lithospheric magnetic field anomaly model filtered in the Pi2 frequency band.

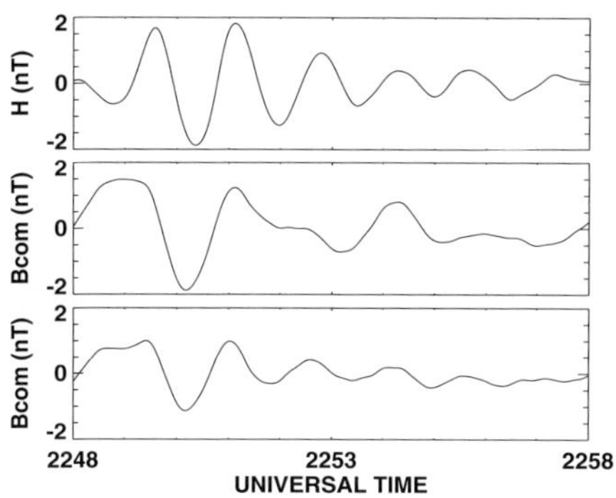


**Fig. 3.** The top panel shows the H component of a Pi2 pulsation observed at Sutherland on 17 May 2002. The centre and bottom panels show the compressional component of the same pulsation observed by CHAMP for which no correction and a correction respectively have been applied for the lithospheric anomaly field.

The third step of the supplemental processing was to rotate the data from the NEC frame into a field aligned coordinate system. We refer to the three orthogonal coordinates as the compressional, toroidal, and poloidal components, which are respectively parallel to the field line, perpendicular to the field line and directed azimuthally eastward, and perpendicular to the field line and directed inward. The coordinate transformation to execute this rotation was determined from the background field obtained by low-pass filtering (.0008 Hz) the observed data. Finally, the data were band-pass filtered in the Pi2 frequency band (typically .005 - .05 Hz). The ground station data were filtered with the same frequency bandpass.

## 3 Observations

The results for some of the events presented by *Sutcliffe and Lühr [2003]* are presented here in somewhat more detail by the addition of spectra and complex demodulates. The compressional, poloidal, and toroidal components of the 17 May 2002 Pi2 pulsation as observed onboard CHAMP are plotted in the upper three panels in Figure 4 as solid curves. The satellite's geocentric latitude is plotted in the bottom panel, which shows that it was located in the southern hemisphere during this event and travelled from approximately 60° to 20° S during the 10-minute interval. It is seen that the compressional and poloidal component oscillations are similar and in phase. The Sutherland H component is plotted as a dashed curve over the compressional and poloidal components and the D component over the toroidal component. It is seen that the H component oscillation on the ground is similar to the compressional and poloidal oscillations at CHAMP, that is, closely in phase but with larger and lesser amplitudes respectively. The amplitude spectra are plotted in Figure 5 in the same order as the component data in Figure 4. This figure confirms that the compressional, poloidal, and H components all oscillate at the same frequency and makes the relative amplitude relationships clear. The complex demodulates of the Pi2 are plotted in Figure 6. The poloidal component observed by CHAMP and H component on the ground are plotted in the upper panel, while the second and third panels show the amplitudes and relative phases of these oscillations as functions of time; the curves in the latter panel confirm that these signals are closely in phase.

The other event that we consider, shown in Figure 7, occurred on 2 February 2002 with onset shortly after 2310 UT. During this event CHAMP was located in the northern hemisphere and travelling southward. The top panel again shows

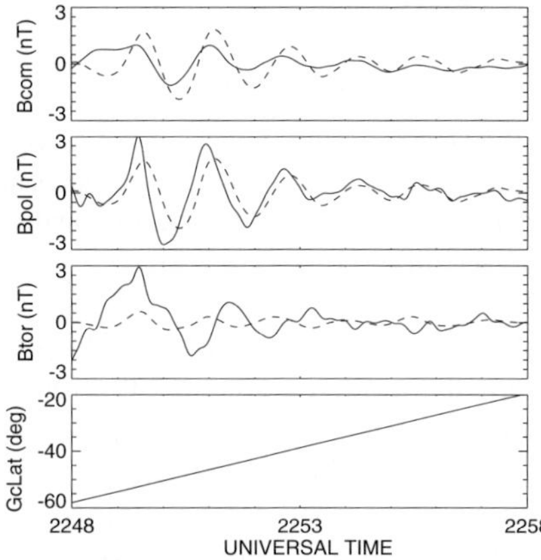


**Fig. 4.** The panels from top to bottom show the compressional, poloidal, and toroidal components of the Pi2 pulsation and the satellite latitude on 17 May 2002 (solid curves). The H (upper two panels) and D (third panel) components at Sutherland are over-plotted (dashed curves).

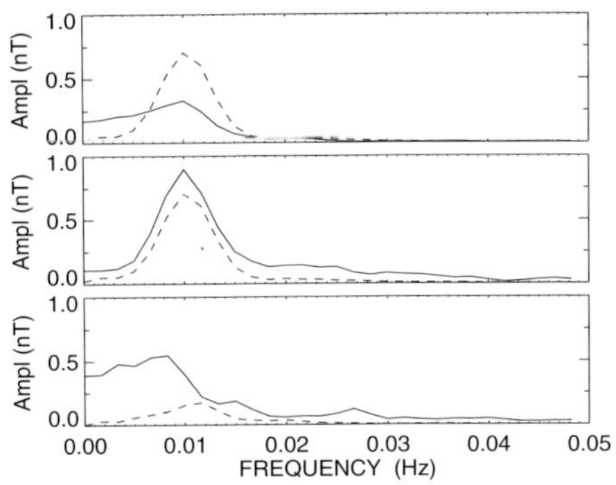


**Fig. 5.** Amplitude spectra of the Pi2 pulsation observed on 17 May 2002 onboard CHAMP (solid line) and at Sutherland (dashed line) plotted in the same ordes as the component data in Figure 4.

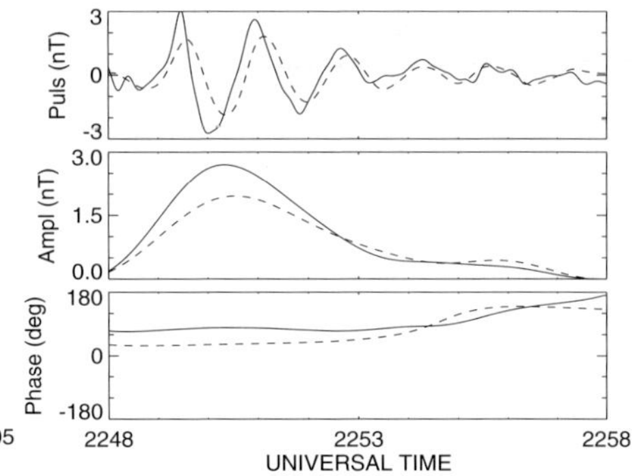


**Fig. 6.** Complex demodulates of the Pi2 pulsation poloidal component observed onboard CHAMP (solid line) and H component at Sutherland (dashed line) on 17 May 2002.

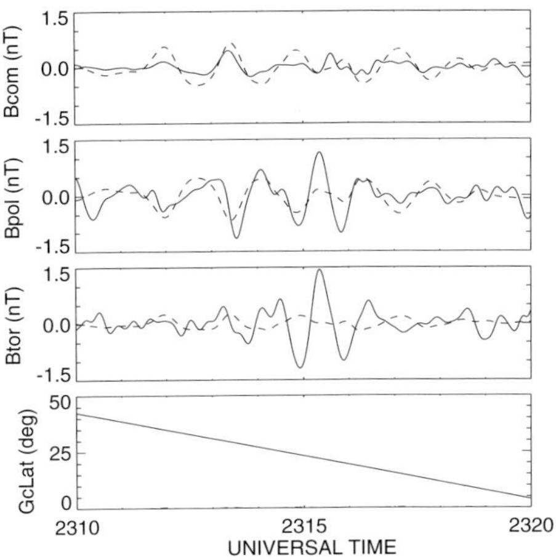


**Fig. 7.** Same as Figure 4, except that the inverse of the H component is plotted in the second panel, for a Pi2 pulsation observed on 2 February 2002.

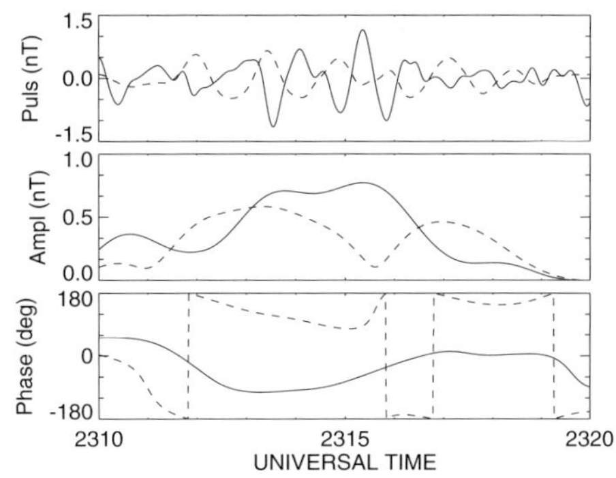


**Fig. 8.** Complex demodulates of the Pi2 pulsation poloidal component observed onboard CHAMP (solid line) and at Sutherland (dashed line) on 2 February 2002.

a reasonable in-phase correlation between the compressional and H components. In the second panel the dashed curve is the inverted H component, which appears correlated with the poloidal component, indicating that these two components are oscillating in anti-phase. There does not appear to be any correlation between the toroidal and D components. The complex demodulates of this Pi2 are plotted in Figure 8. The poloidal component observed by CHAMP and H component on the ground are plotted in the upper panel, while the second and third panels again show the amplitudes and relative phases of these oscillations as functions of time;

the curves in the latter panel confirm that in this case the poloidal and H components oscillate in anti-phase.

## 4 Conclusions

It has for the first time been possible to extract and clearly resolve Pi2 pulsations from low Earth orbit satellite data due to the unprecedented accuracy and resolution of the CHAMP magnetic field measurements. The data have been used to make an initial comparative study of Pi2 pulsations observed by CHAMP at LEO and at the Sutherland ground station in South Africa. Initial findings to date are the following:

- An important new finding is that the correlation between satellite and ground Pi2s is improved if a lithospheric magnetic field anomaly model is subtracted from the satellite data.
- The H-component of the Pi2 pulsation on the ground is well correlated with the compressional and poloidal components above the ionosphere.
- Typical H-component amplitudes on the ground are 0.5 to 2 nT, while at CHAMP the compressional and poloidal component amplitudes are roughly 0.7 and 1.4 times this respectively.
- In the southern hemisphere the compressional and poloidal components oscillate in phase with H; however, in the northern hemisphere the poloidal component appears to oscillate in anti-phase with the compressional and H components.

## References

Holme R, Olsen N, Rother M, and Lühr H (2003) CO2 – A CHAMP magnetic field model. in: Reigber C, Lühr H, Schwintzer P (eds), First CHAMP Mission Results from Gravity, Magnetic and Atmospheric Studies, Springer Verlag, Berlin: 220-225.

Lühr H, Maus S, and Rother M (2002) First in-situ observation of night-time F region currents with the CHAMP satellite. Geophys Res Lett *29(10):* 10.1029/2001GL013845.

Maus S, Rother M, Holme R, Lühr H, Olsen N, and Haak V (2002) First scalar magnetic anomaly map from CHAMP satellite data indicates weak lithospheric field. Geophys Res Lett *29(14):* 10.1029/2001GL013685.

Sutcliffe PR, duToit PJ, Julies EJJ, Pretorius BJ, Theron H, Wright DR, Holme R, and P Fourie (2002) Collaborative effort to record cleaner geomagnetic pulsation data in South Africa. in: Proceedings of the Xth IAGA Workshop on Geomagnetic Observatory Instruments, Data Acquisition, and Processing, Hermanus, South Africa, April 2002.

Sutcliffe PR and Lühr H (2003) A comparison of Pi2 pulsations observed by CHAMP in low Earth orbit and on the ground at low latitudes. Geophys Res Lett *30(21):* 2105, doi:10.1029/2003GL018270.

# ULF Wave Magnetic Measurements by CHAMP Satellite and SEGMA Ground Magnetometer Array: Case Study of July 6, 2002

Massimo Vellante[1], Hermann Lühr[2], Tie-Long Zhang[3], Viktor Wesztergom[4], Umberto Villante[1], Marcello De Lauretis[1], Andrea Piancatelli[1], Martin Rother[2], Konrad Schwingenschuh[3], Wolfgang Koren[3] and Werner Magnes[3]

[1] Dipartimento di Fisica, Università dell'Aquila, L'Aquila, Italy, *massimo.vellante@aquila.infn.it*
[2] GeoForschungsZentrum Potsdam, Germany
[3] Institut für Weltraumforschung, Graz, Austria
[4] Geodetic and Geophysical Research Institute of the Hungarian Academy of Sciences, Sopron, Hungary

**Summary.** We present the analysis of a Pc 3 geomagnetic pulsation event observed simultaneously by CHAMP and by the South European GeoMagnetic Array SEGMA (1.56 $< L <$ 1.88) during the conjunction of July 6, 2002. Both compressional and transverse oscillations were identified in CHAMP magnetic measurements. A close correspondence between the compressional component and the ground signals is observed. At the same time the joint analysis of space and ground observations clearly indicates the occurrence of a field line resonance at $L \cong 1.6$. A direct confirmation of the well known 90° rotation of the ULF wave polarization ellipse through the ionosphere is also provided.

**Key words:** Magnetosphere/ionosphere interaction, ULF waves, field line resonance

## 1 Introduction

ULF magnetohydrodynamic waves are an intrinsic signature of different processes taking place in the Earth's magnetosphere. Their observation and interpretation is therefore very important in the global context of the magnetospheric dynamics. For an unambiguous interpretation of the observations, simultaneous measurements in space and on a ground array are needed. So far, direct space-ground comparisons of ULF wave events in the frequency range 2-100 mHz (Pc 3-5 geomagnetic pulsations) have been conducted using data from near equatorial spacecraft (either with geosynchronous or eccentric orbit) and from magnetically linked ground sites, i.e. from very distant measurement points. This may cause low coherency between space and ground signals [1, 2]. Such combination of space and ground measurements has been also used to determine the effect of the transmission through the ionosphere with controversial results [3]. In this sense the use of a low-altitude satellite like CHAMP would be preferable although its polar orbit causes a rapid movement across different $L$-shells. Nevertheless, a ground-space comparison for Pc 3 pulsations in the plasmasphere can still be conducted. For instance, for a pulsation with a frequency of 50 mHz, CHAMP covers in one wave period a latitudinal range of ~1.5°, which is not a too dramatic variation. In addition, the use of an array of meridionally distributed ground stations may help to improve the situation. In the present study we used the South European GeoMagnetic Array (SEGMA, $1.56 < L < 1.88$) and found indeed that very interesting results can be obtained from the direct combination of magnetic measurements from a ground array and a low-altitude satellite for Pc 3 wave phenomena.

## 2 Instrumentation and data

SEGMA is a latitudinal magnetometer array recently established in southern Europe for continuous measurements of geomagnetic field variations in the ULF band. All stations are equipped with high sensitivity triaxial fluxgate magnetometers, recording the geomagnetic northward ($H$), eastward ($D$), and vertically downward ($Z$) components of the geomagnetic field variations. After low-pass data filtering, data are recorded at a sampling rate of 1 Hz. Precise timing is provided by a GPS receiver. For more details about instrumentation and data acquisition, see [4].

The CHAMP data used in this study are the pre-processed (Level 2) fluxgate vector magnetometer data. This instrument samples the magnetic field at a rate of 50 Hz at a resolution of 0.1 nT. In the standard processing the data are averaged to 1 Hz samples. The data are transformed into the Mean-Field-Aligned (MFA) coordinate system. This is well suited for studying wave phenomena. In this frame the azimuthal component, $B_\varphi$, points to magnetic east, the compressional component, $B_\mu$, is aligned with the ambient field, and the poloidal component, $B_\nu$, completes the triad, pointing outward. A model magnetic field consisting of the main field (Ørsted_05m02 [5]) and of the crustal part (MF1 [6]) is subtracted from the compressional component and subsequently all components are high-pass filtered with a cutoff frequency of 16 mHz to remove long-period variations.

## 3 Observations

During June-July 2002, CHAMP passed approximately every second day close to SEGMA during daytime hours. For some of these short time conjunctions clear Pc3 geomagnetic pulsations were simultaneously observed both in space and on ground. A representative event occurring on July 6 is presented here. Figure 1 shows the satellite track mapped down along the field lines to the E-layer and the locations of the SEGMA stations, both in corrected geomagnetic coordinates. As shown in the figure, CHAMP needs about 1.5 min to traverse the latitudinal range of the SEGMA array. At the closest approach, the longitudinal difference is less than 4° and the CHAMP altitude is 415 km.

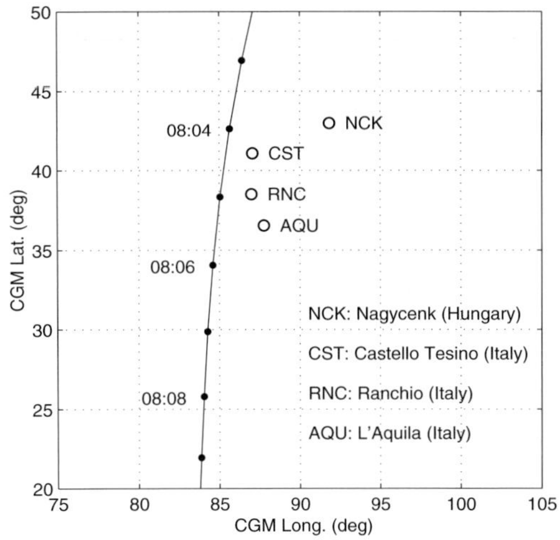


**Fig. 1.** CHAMP track mapped down along the field lines to the E-layer and the locations of the SEGMA stations for the event of July 6, 2002.

Figure 2 shows the time series of the different signals, filtered in the frequency band 20-100 mHz. As can be seen, the CHAMP signals contain a mixture of frequencies and show a low coherence between the different components. The azimuthal component exhibits the largest amplitude oscillations (as large as 5-6 nT peak-to-peak). The parallel component, although of much smaller amplitude (less than 0.5 nT peak-to-peak) shows, even outside the conjunction interval (shaded region), the best correspondence with the ground signals: note for example the almost in-phase relation with the NCK *H*-signal in the time interval 0805:30 – 0809:30 UT. The ground signals are in general more monochromatic and highly coherent between different stations. The maximum amplitude (about 3 nT peak-to-peak) is observed in the *H*-component at NCK in the time interval 0807 – 0808 UT. A dynamic spectral analysis of the different signals (figure not shown) confirms that several spectral features of the compressional component find a good correspondence on the ground while this is not true for the transverse components. This is a consequence of the rapid movement of CHAMP across field lines with different resonance responses (which affects mainly the transverse components).

An interesting behaviour of the azimuthal component can be noted when CHAMP is in between the RNC and AQU *L*-shells: the appearance of higher frequency oscillations with enhanced amplitude. At the same time, the *H*-ground signal maximizes just at RNC and AQU. This might be an indication of a field line resonance occurring at an *L*-shell between RNC and AQU. Note, however, that the frequency of the oscillations at both stations is lower than that of $B_\varphi$. In particular the dominant frequencies of the different signals (as provided by a spectral analysis) are: ~65 mHz for $B_\varphi$, ~55 mHz for $B_\mu$, and ~50-55 mHz for all ground stations (both *H* and *D* components). The higher frequency of $B_\varphi$ can be explained by the

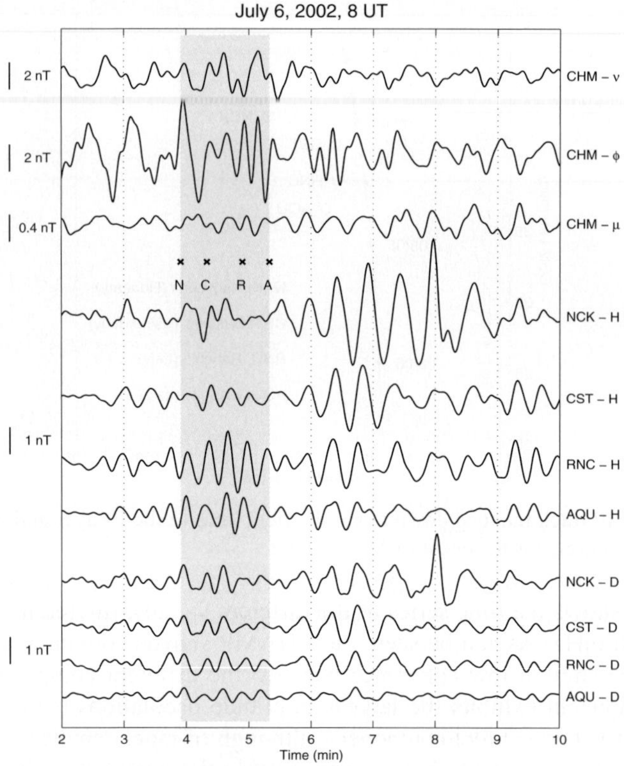


**Fig. 2.** Magnetic field data from CHAMP and SEGMA array. The data are filtered in the frequency band 20-100 mHz. The shaded region indicates the time interval of the CHAMP-SEGMA conjunction. The crosses indicate the conjunction to each ground station.

expected rapid phase increase of the azimuthal signal (by 180°) experienced when moving equatorward across the resonance region [7]. We can get an idea of the importance of this effect by comparing the CHAMP velocity ($V_{sat}$ = 7.6 km/s, directed equatorward) and the apparent meridional phase velocity ($V_{phase}$, directed poleward). From the theoretical expression of the latitudinal structure of the azimuthal component [8] we obtain, at the resonant location, the following expression $V_{phase} = 2\pi f \varepsilon$, where $f$ is the frequency and $\varepsilon$ is the resonance width at the ionospheric altitude. Using $f$ = 55 mHz and for $\varepsilon$ a reasonable value of 80 km, we get $V_{phase} \cong 30$ km/s which is only ~4 times larger than $V_{sat}$. Therefore a significant Doppler shift is expected which can well account for the observed difference in frequency.

It is also worth to mention that the interplanetary magnetic field magnitude $B_{IMF}$ prevailing during this event was ~10 nT (as obtained from the ACE satellite, taking into account the propagation delay from ACE to the magnetopause). According to the empirical relationship $f_{UW}$ (mHz) $\cong 6\ B_{IMF}$ (nT) [9], the expected frequency ($f_{UW}$) of upstream waves in the Earth's foreshock would be ~60 mHz,

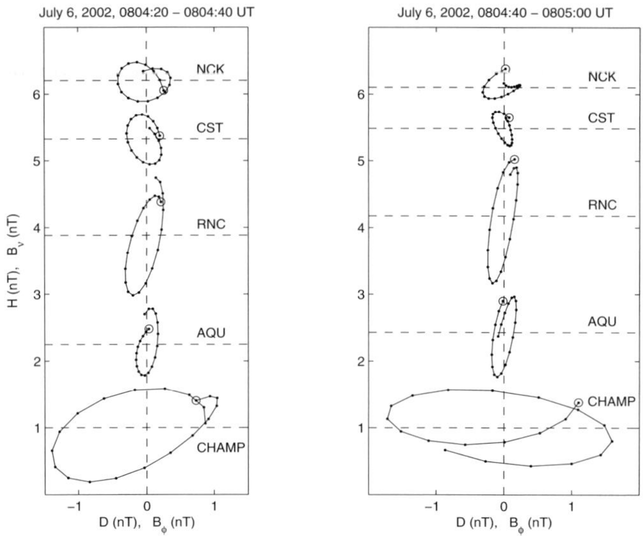


**Fig. 3.** Wave hodograms for CHAMP and SEGMA stations for two consecutive 20-s time intervals. The data are filtered in the frequency band 40-70 mHz. In each hodogram the first measurement is marked.

which is very close to the frequency of the compressional signal observed by CHAMP. It confirms previous observations that the upstream waves are the main candidate for the source of daytime Pc 3 geomagnetic pulsations [10].

We also performed an analysis of the polarization characteristics of the space and ground signals. Such a comparison can be very useful to test theoretical models on the ULF wave transmission through the ionosphere. In Figure 3 we show the wave hodograms in the ν-φ plane (for CHAMP measurements) and in the *H-D* plane (on the ground) for two consecutive 20-sec time intervals during the CHAMP–SEGMA conjunction. Data are filtered in the frequency band 40-70 mHz. It is worth to note the approximate 90° rotation of the polarization ellipse between CHAMP and the ground station at the closest conjunction (CST in the first interval, RNC in the second). This is a well known effect produced by the induced currents in the ionosphere as predicted by theoretical models [11]. Note also a reversal of the polarization sense, being clockwise (CW) at CHAMP position and counterclockwise (CCW) on the ground. From well established theoretical and experimental results we expect in the morning hours a CCW polarization, except in a more or less extended region poleward of the resonant line [12]. The numerical model by [13] predicts however that, because of a spatial integration effect, such polarization reversal can get lost on the ground. Our results seem to support, at least in this case, such a picture.

## 4 Conclusions

We have shown that the CHAMP mission provides an interesting opportunity for a close comparison between space and ground ULF wave magnetic measurements. This first analysis of a Pc 3 event observed simultaneously by CHAMP and the low-latitude SEGMA array has indeed provided an unprecedented direct test of theoretical predictions about the field line resonance phenomenon and the effects produced by the ionosphere on the transmission of ULF signals to the ground.

## References

1. Takahashi K, Anderson BJ, Newell PT, Yamamoto T, Sato N (1994) Propagation of compressional Pc 3 pulsations from space to the ground: A case study using multipoint measurements. In Solar Wind Sources of Magnetospheric ULF Waves, Geophys Monogr Ser vol. 81, edited by Engebretson MJ, Takahashi K, Scholer M, pp 355-363, AGU, Washington, D. C.
2. Ziesolleck CWS, Mc Diarmid DR, Feng Q (1997) A comparison between Pc 3-4 pulsations observed by GOES 7 and the CANOPUS magnetometer array. J Geophys Res *102*: 4893-4909.
3. Glassmeier K-H, Junginger H (1987) Concerning the ionospheric modification of magnetospheric hydromagnetic waves: case studies. J Geophys Res *92*: 12213-12219.
4. Vellante M, De Lauretis M, Villante U, Adorante N, Piancatelli A, Schwingenschuh K, Magnes W, Koren W, Zhang TL (2002) A new meridional magnetometer array in middle-south Europe for monitoring dynamic processes in the coupled system plasmasphere-ionosphere. In Proceedings of the Second Solar cycle and Space Weather Conference SOLSPA 20001, SP-477, pp 487-490, ESA, The Netherlands.
5. Olsen N (2002) A model of the geomagnetic main field and its secular variation for epoch 2000 estimated from Ørsted data. Geophys J Int *149*: 454-462.
6. Maus S, Rother M, Holme R, Lühr H, Olsen N, Haak V (2002) First scalar magnetic field anomaly map from CHAMP satellite data indicating weak lithospheric field. Geophys Res Lett *29* (14): 10.1029/2001GL013685.
7. Walker ADM, Greenwald RA, Stuart WF, Green CA (1979) STARE auroral radar observations of Pc 5 geomagnetic pulsations. J Geophys Res *84*: 3373-3388.
8. Green AW, Worthington EW, Baransky LN, Fedorov EN, Kurneva NA, Pilipenko VA, Shvetzov DN, Bektemirov AA, Philipov GV (1993) Alfven field line resonances at low latitudes (L=1.5). J Geophys Res *98*: 15693-15699.
9. Russell CT, Hoppe MM (1981) The dependence of upstream wave periods on the interplanetay magnetic field strength. Geophys Res Lett *8*: 615-617.
10. Yumoto K (1986) Generation and propagation mechanisms of low-latitude magnetic pulsations: A review. J Geophys *60*: 79-105.
11. Hughes WJ (1974) The effect of the atmosphere and ionosphere on long period magnetospheric micropulsations. Planet Space Sci *22*: 1157-1172.
12. Chen L, Hasegawa A (1974) A theory of long-period magnetic pulsations, 1. Steady state excitation of field line resonance. J Geophys Res *79*: 1024-1032.
13. Hughes WJ, Southwood DJ (1976) An illustration of modification of geomagnetic pulsation structure by the ionosphere. J Geophys Res *81*: 3241-3247.

# Classes of the Equatorial Electrojet

Heather McCreadie

Graduate School of Science, Kyoto University, Kyoto, 606-8502, Japan,
*bilby@kugi.kyoto-u.ac.jp*

**Summary.** The equatorial electrojet signal recorded on the Overhauser Magnetometer (OVM) on board CHAMP can be divided into four classes: the classical electrojet (EEJ), the counter electrojet (CEJ), no electrojet (NEJ) and a new term, formation of the electrojet (FEJ). A statistical study on the location, period and corresponding magnetic events shows that at longitudes near 60°, 165°, 255° and 345° the morning events are CEJ, which then slowly form into FEJ near noon, and finally, as local time increases, the full EEJ is realized. All other longitude zones have only EEJ events. The reason for the occurrence of these events at certain longitudes is as yet unclear.

**Key words:** equatorial electrojet, ionosphere, geomagnetism

## Introduction

Each day the CHAMP satellite passes over a current in the ionosphere called the equatorial electrojet. This current is in a narrow belt at approximately 108km above the surface of the Earth and generally flows in an eastward direction. The Overhauser Magnetometer (OVM) on board CHAMP records a signal from this current for each pass. The usual signal has an inverted bell shape with the peak in the height of the bell occurring near midday. However, the OVM also records other signals for some passes. A description of each type of event and a statistical analysis of location, period and corresponding magnetic events is described.

## Description of filter

The long wavelength daytime magnetic field variation at mid and equatorial latitudes is mainly comprised of effects from the equatorial ring current (RC), and the daily variation (Sq). It has been suggested that the equatorial electrojet (EJ) is part of the Sq system (Stening, 1995). Keeping this in mind, a method of extracting only the RC component was found. Generally, when using ground-based data to study ionospheric currents the effect of the RC is removed by using the Dst index (Hutton and Oyinloye, 1970). A similar technique can be employed using satellite data.

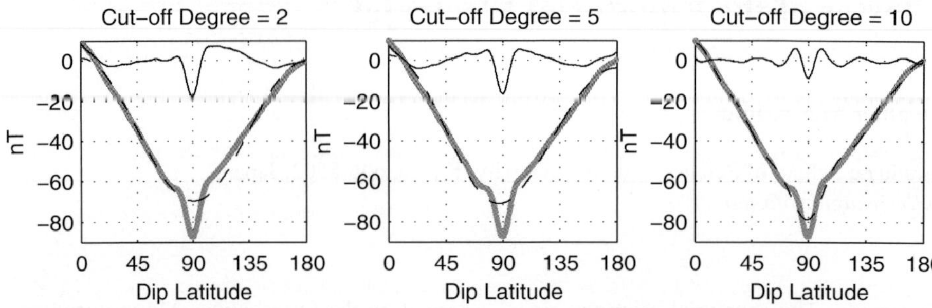


**Fig. 1.** A typical satellite pass showing a track with the main and lithospheric fields removed (grey) . Dashed lines are the computed track using Legendre Polynomials up to cut-off degree shown above each plot. The black lines are the filtered track..

Figure 1 is a plot of the day-time mid-latitude part of a satellite pass. The model oer05m02-MF1 up to degree 65 was used to remove the main and lithospheric fields (Olsen, 2002, and Maus, et.al., 2002). Shown in each window is the pass (grey line) with a Legendre polynomial of differing cut-off degree (dashed line). The residual between the two lines is the filtered data and is shown in black. The obvious problem is the amplitudes of the side lobes are affected by the type of filter. The plot on the right shows that when the wavelength of the filter is smaller than the width of electrojet signal the entire signal is affected. Therefore, a filter is required which will remove effects from currents that are not ionospheric and will not introduce spurious effects into the side lobes.

The choice of filter was dependent on the association of the amplitude of the angular degree and geomagnetic indices, and the cut-off degree. The correlation coefficient (R) between the amplitudes of the angular coefficients (n) for a satellite track and the associated Dst index when the cut-off degree is 2 are; (n=0, R=0.4), (n=1,R=-0.05), and (n=2, R=-0.6). The Tnull for each set was zero. The angular degree n=2 correlates very well with the Dst index. No other angular degree correlates as significantly with the Dst index and this correlation maximizes when the cut-off degree is 2. The n=0 angular degree corresponds to the field level and it is well known that an increase in the absolute Dst coincides with an increase in the ambient field level, as seen here. The amplitude of the n=1 angular degree shows no correlation with the aa index (R=0.05, Tnull=0.01) and Kp (R= 0.05, Tnull=0) which suggests no dependency on Sq effects. This angular degree does not correlate with Dst. However, it was included in the filter because it is clearly not linearly dependent on ionospheric effects either.

## Description of Events

As the CHAMP satellite passes over the area of the dip equator the OVM records four possible configurations described in Table 1 and shown in Figure 2. Each

| Name | Code | Simple description | Classification Criteria |
|---|---|---|---|
| Classical Electrojet | EEJ | Inverted bell shape | One minimum lies within two degrees of the quasi-dip equator and one maximum sits directly either side. These maximums must lie between ±10° of the quasi-dip equator. The resulting amplitude must be the maximum amplitude in the pass of this configuration. |
| Counter Electrojet | CEJ | Bell shape | One maximum lies within 4° of the quasi-dip equator and one minimum sits directly either side. |
| Formation Electrojet | FEJ | Combination of EEJ and CEJ | A minimum with a maximum on either side occurs near the quasi-dip equator but it may not have the largest amplitude in the pass. |
| No Electrojet | NEJ | None of the above | None of the above. |

**Table 1.** Classification criteria of equatorial electrojet events.

configuration has been labeled with events recorded in the review by Forbes, 1981. The classical electrojet (EEJ), counter electrojet (CEJ) and no electrojet (NEJ) events are well known. However, it is necessary to introduce a new event labeled the Formation Electrojet (FEJ). These events have similar characteristics to EEJ events except that the amplitude is much smaller and the local time of occurrence of the events is distinctive (see later section).

## Statistical Analysis of Events

Data were averaged into bins centered on each hour, both in local time (LT) and universal time (UT). This is plotted in figure 3a for each event type. Graphs of this sort show the Earth in a certain configuration with respect to the Sun. For example, a UT of 2400 means that the longitude of 180° is facing the Sun. The EEJ is always a maximum at local noon so these graphs allow the configuration of the EEJ to be shown for each UT bin. Plotting LT versus UT realizes both a spatial and a temporal aspect since longitude can easily be determined by {LT – UT}x15 (The most interesting longitudes are shown in the graphs by dashed grey lines) . Including a month versus UT graph (figure 3b) also realizes a seasonal aspect. Using two graphs we are able to determine spatial, daily-temporal and seasonal variations.

Figure 3a-4 shows that most NEJ events occur near 345° longitude at 1000, 1100, and 1500 LT. Figure 3b-4 shows that the occurrences of NEJ events in the early local times are independent of season, whilst, the number of afternoon events peak in September. The fact that most of these events occur in one longitude zone suggests a local phenomenon not accounted for. Closer inspection of individual

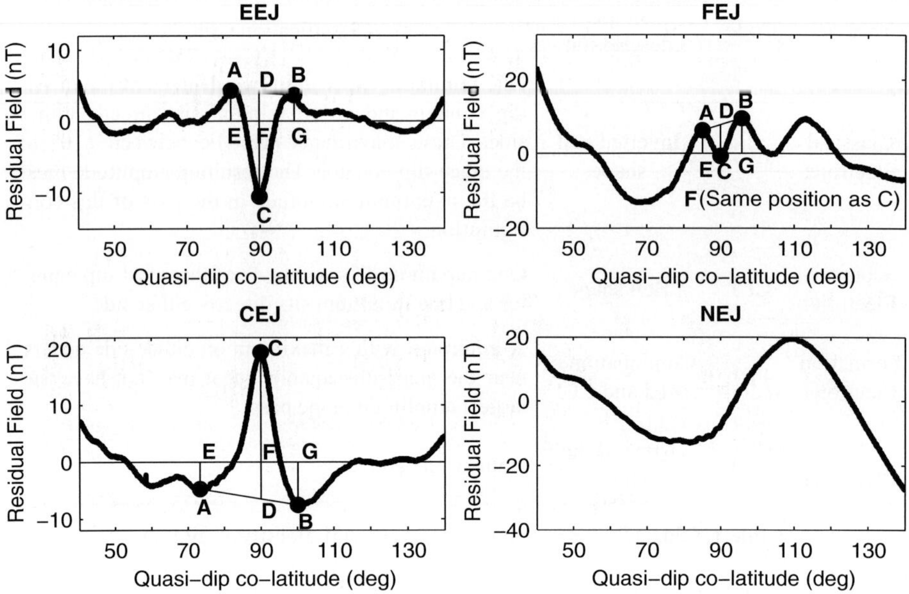


**Fig. 2.** Examples of classification of the equatorial electrojet observed using the Overhauser Magnetometer on board CHAMP. CD = Amplitude, AB = Width, AB′ = Gradient, |A| & |B| = Magnitude of side lobes, |C| = Magnitude of centre, E & G = positions of side lobes, F = position of centre.

tracks revealed that EEJ, CEJ and FEJ signatures are present in some NEJ tracks but the long wavelength sinusoidal feature seen in Fig. 2 (NEJ) obfuscates the signals. Clearly, for this longitude zone more analysis is required.

Figure 3a-2 shows that most CEJ events occur near longitudes 60°, 165° and 300° and the peak occurrences are in the morning and afternoon. Figure 3b-2 shows that the morning CEJ events occur only in July, whilst, the afternoon events are in December. Some events occur at midday in June near longitude 150°.

FEJ events (figure 3a-2 and 3b-2) and CEJ events have similar occurrence rates with longitude. The peak occurrence of FEJ events is midday in June and is centered on longitude 50°. FEJ events occur in all months where data is available with no propensity towards any one month.

Figure 3a-1 shows that all longitudes record EEJ events. The occurrence peaks at midday. Please note that the number of satellite passes between LT 0900 and 1500 are low in March so the lack events in this month is not an indication of no events. There just isn't enough available data to resolve this month. The occurrence of EEJ events peaks in May and September. Please also note that the total number of passes for each longitude bin is approximately equal, indicating a common total Earth coverage.

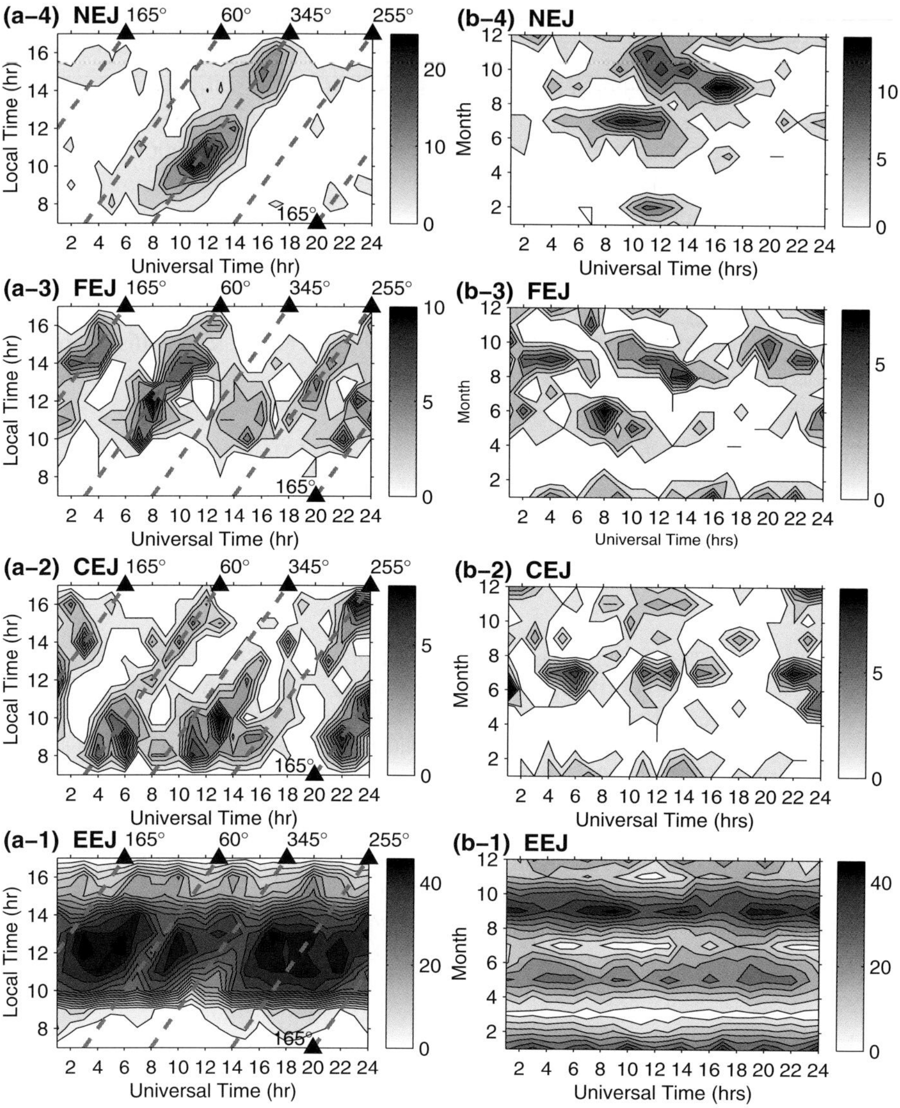


**Fig. 3.** The occurrence of each type of event using UT for the absicssae and LT for the ordinate (a) Season for the ordinate (b). Grey scale indicates number of events. Triangles point to the label of longitude, indicated by the dashed grey lines

## Conclusion

Each satellite pass gives us a snapshot of the equatorial electrojet at a longitude, local time and season. It is clear that at longitudes near 60°, 165° and 300° the morning events are CEJ, which then slowly form into FEJ near 1000LT and noon, and finally as local time increases full EEJ is realized. All other longitude zones have only EEJ events.

There is no correlation between the occurrence of any of these equatorial electrojet events and Dst index, aa-index, Kp or PCN. This suggests that the source of these events is not related to external current systems. The reason for the occurrence of these events at certain longitudes is as yet unclear. Clearly this is only the tip of the iceberg. More analysis is required.

*Acknowledgments.* I wish to thank Robert Stening, Richard Holme, and Zdeněk Martinec for many helpful discussions. I am grateful for the support of the CHAMP project at GFZ-Potsdam who supplied the data and funding for this study.

## References

Forbes JM (1981) The equatorial electrojet. Rev Geophys Space Phys *19*: 469-504.

Hutton R, and Oyinloye JO (1970) The counter electrojet in Nigeria. Ann Geophys *26*: 921.

Maus S, Rother M, Holme R, Lühr H, Olsen N, and Haak V (2002) First scalar magnetic anomaly map from CHAMP satellite data indicates weak lithospheric field. Geophys Res Lett *29*: 10.1029/2001GL013685.

Mayaud PN (1980) Derivation Meaning and Use of Geomagnetic Indices. AU Geophysical monograph *22*.

Olsen N (2002) A model of the geomagnetic field and its secular variation for Epoch 2002 estimated from Oersted data. Geophys J Int *149*: 453-461.

Stening RJ (1995) What drives the equatorial electrojet? J Atmos Terr Phys *57(10):* 1117-1128.

# The ESPERIA Project: a Mission to Investigate the near-Earth Space

Vittorio Sgrigna[1], Rodolfo Console[2], Livio Conti[1], Arkady Moiseev Galper[3], Valeria Malvezzi[1], Michel Parrot[4], Piergiorgio Picozza[5], Renato Scrimaglio[6], Piero Spillantini[7], and David Zilpimiani[8]

[1] Department of Physics and INFN Sect., University Roma Tre, Rome, Italy, *sgrigna@fis.uniroma3.it*
[2] Italian National Institute of Geophysics and Volcanology (INGV), Rome, Italy
[3] Institute of Cosmic Physics, MEPhI, Moscow, Russian Federation
[4] LPCE/CNRS, Orleans, France
[5] Department of Physics and INFN Sect., University Tor Vergata, Rome, Italy
[6] Department of Physics, University of L'Aquila, L'Aquila, Italy
[7] Department of Physics, University of Florence, Florence, Italy
[8] Institute of Geophysics, Georgian Academy of Sciences, Tbilisi, Republic of Georgia

**Summary.** ESPERIA is an equatorial space mission mainly concerned with detecting any tectonic and preseismic related signals. More in general, it has been proposed for defining the near-Earth electromagnetic, plasma, and particle environment, and for studying perturbations and instabilities in the ionosphere-magnetosphere transition region. The same multi-instrument payload also allows anthropogenic electromagnetic emissions to be investigated. To study earthquake preparation processes and anthropogenic impacts in the Earth's surface, a phase A study has been realized for the Italian Space Agency. Within this framework also the ARINA particle experiment is in progress.

**Key words:** geomagnetic field, earthquakes, magnetosphere, plasma, radiation belts.

## 1 Introduction

Earthquake prediction is a very fascinating tool in geophysics but the possibility to determine the space and time occurrence, and the magnitude of a forthcoming earthquake is still a good way from a solution. Also relevant is the possibility to investigate lithosphere-atmosphere-ionosphere-magnetosphere coupling mechanisms in the near-Earth space. In recent years it has been observed that both earthquakes and near-Earth space phenomena have a privileged and sensitive zone of investigation constituted by the ionosphere-magnetosphere transition region at altitudes ranging around 600÷1000 km. Sun and cosmic rays as well as Earth's interior processes, anthropogenic electromagnetic emissions and thunderstorm activity, influence the structure and dynamical behavior of the zone. These external and internal contributions play an important role in defining the particle and electromagnetic (EM) field character of the region, both in steady-state and perturbed-state conditions. A suitable monitoring of the ionosphere-magnetosphere transition zone may also give an help mapping the geomagnetic field and studying many important physical phenomena as earthquakes, anthropogenic EM emissions (EME), solar wind and flares, etc.

ESPERIA is an equatorial mission planned with a *Low-Earth-Orbit* (LEO) small-satellite and a multi-instrument payload ideally conceived to define the steady-state near-Earth EM, plasma, and particle environment, and to investigate perturbations and instabilities in the ionosphere-magnetosphere transition region. The ESPERIA *Phase A* study has been performed by an International Consortium lead by the University Roma Tre (Sgrigna, 2001), within the Earth observation program for small scientific missions of the Italian Space Agency (ASI). The ARINA particle experiment, has been realized within the PAMELA mission (Picozza, 2003). Specific ASI constrains restricted the above mentioned original scientific objectives of the ESPERIA project, but contacts with other missions and science teams give indications to reconcile the project to its original aims.

## 2 The ESPERIA mission

The scientific objectives of the mission are to detect tectonic, anthropogenic, and preseismic related EM signals. The scientific program is planned with strong emphasis on coordinated, continuous, and simultaneous space and ground-based observations, as well as on mutual data comparison with other missions of similar quality.

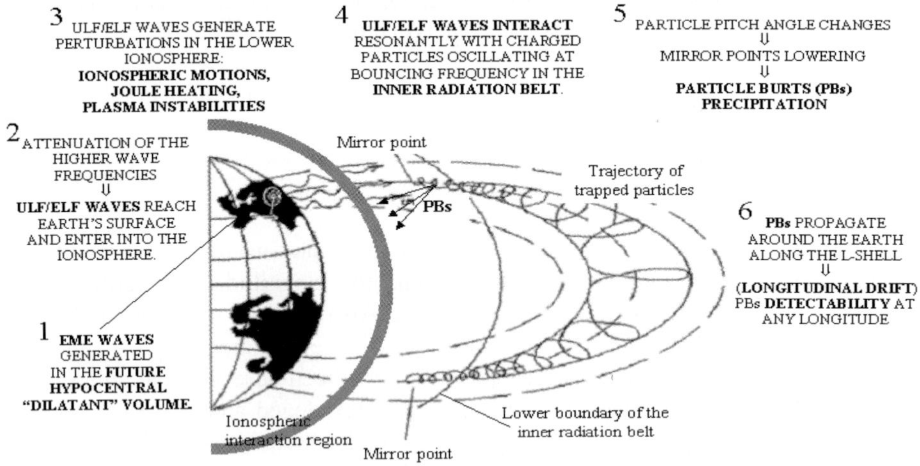


**Fig. 1.** Seismic EME-waves generation, propagation and interaction with ionospheric plasma and magnetospheric trapped particles.

## Science background

Preseismic ground mechanical, electric and electromagnetic fields have been observed as well as ULF-HF EME-waves in the ionosphere and magnetosphere over seismic regions. At present it doesn't exist an exhaustive physical model to justify these seismo-associated phenomena. For a review on the subject see Hayakawa (1999); Sgrigna and Malvezzi (2003); Aleksandrin et al., (2003). A qualitative representation of the phenomenology, as suggested by several authors, is shown in figure 1. Observation of EME-waves generated by anthropogenic activities at the Earth's surface revealed that sources of these waves are power line harmonic radiation, VLF and HF transmitters (Parrot and Zaslavski, 1996).

## ESPERIA main features

1. ORBIT CHARACTERISTICS
   - Ground track repetition with accuracy of 10 km
   - Revisit time ≤ 24 h; geosynchronous orbit: 14 orbits / day
   - Altitude=813km; inclination=11°.5, eccentricity=0; Orbit period:110'
   - Orbit knowledge and time resolution ≈ 100 m and 1s, respectively
   - Field of view: ± 38.5°
   - Maximum oscillation around the magnetic equator: ± 23°
2. SPACECRAFT
   - Platform MITA with Nadir pointing
   - FEEP thrusters applied to the platform (constant altitude)
3. MISSION DURATION ≥ 2 years
4. PAYLOAD INSTRUMENTS
   - **Electric Field Analyser (EFA)**
     - frequency range: ~DC ÷ 10 MHz
     - accuracy: 300 nV/m
     - dynamic range: 120 dB
   - **Magnetic Field Analyser (MAFA)**

     FLUX – GATE:
     - frequency range: ~DC ÷ 10 Hz
     - accuracy: a few (6-8) pT; resolution: 24 bit

     SEARCH – COIL:
     - frequency range: ~10 Hz ÷ 100 kHz
     - sensitivity: $10^{-2}$ pT /(Hz)½ (at 1 kHz)
   - **Langmuir Probe & Retarding Potential Analyser (LP&RPA)**

     LP : electron temperature: 300÷15000K and density: $10^{2}$÷$10^{7}$cm$^{-3}$

     RPA: ionic temperature: 300÷10000K and density: $10^{2}$÷$10^{7}$cm$^{-3}$
   - **Particle Detector Analyser (PDA)**
     - Energy range: 300 KeV ÷ 2GeV
     - Pitch angle precision < 4° with particle identification
     - Geometry: 5 silicon strip telescope + 1 calorimeter<br>1 silicon strip telescope + 1 calorimeter

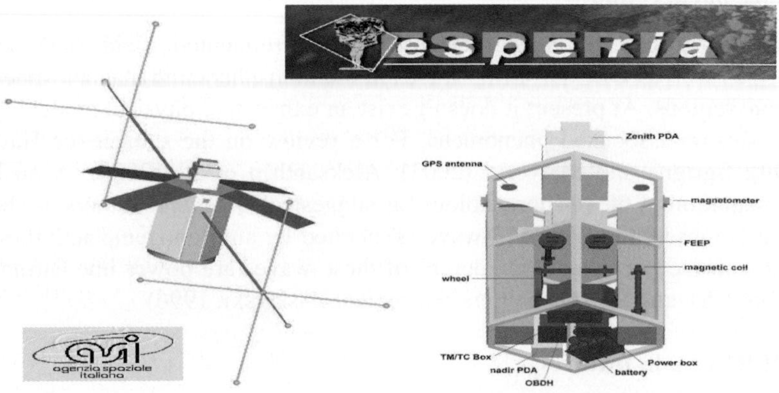


**Fig. 2.** Overview of the ESPERIA equatorial satellite.

An overview of the satellite is given in figure 2. At the topside of the satellite are located PDA, LP, and RPA instruments. Each MAFA sensor system is at one different end of the two primary expanding booms (of ≈5m), so that search-coils are separated by about 10m from flux-gate sensors and each sensor system is at a distance of about 5m from the satellite body. Electric probes with preamplifiers inside are located at the end of secondary booms (of ≈2m), so that each electric sensor system is about 2m from the nearest magnetic sensor system. Instead of magnetic torques, the attitude control is implemented by 3 reactions wheels (of known EM spectra). In this way, magnetic disturbances are less than 2pT at a distance of 5m (deployed booms/magnetic probes accommodation), that is magnetic disturbances are less than instrumental sensitivity.

All probes will be switched off during the desaturation of reaction wheels or when (for short periods) FEEP will be active. To build a "quiet" spacecraft we also planned a surface plating and a special solar panels design in order to ensure the equipotentiality of external satellite surfaces. This requires uniform electrical and optical characteristics of the external surfaces of the satellite and equal to that of areas of electric EFA probes. The velocity vector of the satellite is perpendicular to the direction of primary booms.

## The ARINA Experiment

The ARINA experiment consists of a proton-electron telescope to be installed on board the polar *LEO* Russian satellite RESURS-DK1, which launch is scheduled for mid-2004. The orbit will be elliptic with altitude ranging from 300 km to 600 km and inclination 70.4°. The duration of the mission will be ≥ 3 years. The scientific objective of the experiment is to detect fluxes of high-energy charged particles (3÷100 MeV), from the inner radiation belt and correlate them with seismic activity. So, ARINA may be considered as a preliminary test of the ESPERIA project.

**Relation with other missions and science teams**

ARINA and DEMETER (Parrot, 2002) are two simultaneous polar missions which allow comparisons of particle data. ESPERIA may be considered as a second generation of the DEMETER concept and can profit by the information from previous missions. But the equatorial character of the ESPERIA mission and some augmentation or changes in its payload allow to extend the original scientific objectives of this mission for geo-electric and magnetic field mapping, for studying sun activity and cosmic rays, and investigating structure and dynamics of the magnetospheric cavity. In this sense, four proposals for adding experiments to be carried out on board of ESPERIA have been suggested by the scientific community. They are listed below:

1. Atmospheric and ionospheric structure and dynamics (pressure and temperature profile, ionospheric electron density profile, TEC). A blackjack limbsounding and reflections GPS receiver for occultation measurement is requested[1].
2. Geomagnetic equatorial measurements. ESPERIA may be considered as an equatorial complement to polar missions as SWARM. A scalar magnetometer and an attitude star imager are requested[2].
3. Luminous emissions (Sprites, Blue Jets, etc.) during thunderstorm activity. An Imaging Camera is requested[3].
4. Equatorial electrojet[4].

# 3 Conclusions and Outlook

The ESPERIA space mission has been proposed and the ARINA experiment is in progress for defining the near-Earth electromagnetic, plasma, and particle environment, and for investigating perturbations and instabilities caused by seismicity and anthropogenic EME in the ionosphere-magnetosphere transition region. There is a fruitful collaboration between ESPERIA, ARINA, and DEMETER science teams. The ESPERIA equatorial mission could also be used as a complement to polar Earth's magnetic field mapping missions, like SWARM, and to study relevant phenomena caused by external sources (sun and cosmic rays) or generated inside the magnetospheric cavity.

---

[1] J. LaBrecque (NASA/SENH), A. Farrington and A. Mannucci (JPL), *Personal communication*, 2003.

[2] E. Christensen and F. Primdhal (DSRI/DTU) and P. Taylor (NASA/GSFC), *Personal communication*, 2003.

[3] E. Christensen (DSRI), *Personal communication*, 2003.

[4] R. Singh (IITK), *Personal communication*, 2003.

## References

Aleksandrin S Yu, et al. (2003) High-Energy Charged Particle Bursts in the near-Earth space as earthquake Precursors. Annales Geophysicae *21*: 597-602.

Hayakawa M (1999) Atmospheric and Ionospheric Electromagnetic Phenomena Associated with Earthquakes. TERRAPUB, Tokyo.

Parrot M, and Zaslavski Y (1996) Physical mechanisms of man-made influences on the magnetosphere. Surveys in Geophysics *17*: 67-100.

Parrot M (2002) The micro-satellite DEMETER. Journal Geodyn *33*: 535-541.

Picozza P (PAMELA/ARINA Collaboration) (2003) The PAMELA mission. http://wizard.roma2.infn.it/pamela/index.htm .

Sgrigna V (ESPERIA collaboration) (2001) ESPERIA Phase A Report, ASI Program for Small Scientific Missions.

Sgrigna V, Malvezzi V (2003) Preseismic creep strains revealed by ground tilt measurements in central Italy on the occasion of the 1997 Umbria-Marche Apennines earthquake sequence. Pure Appl Geophys *160*: 1493-1515.

# Status of the CHAMP ME Data Processing

Martin Rother[1], Sungchan Choi[1], Wolfgang Mai[1], Hermann Lühr[1] and David Cooke[2]

[1] GeoForschungsZentrum Potsdam
[2] Air Force Research Laboratory, Hanscom, MA, USA

**Summary.** The routine processing of the CHAMP data has started in May 2001. Since then calibrated data, (science quality) Level-2 products, are made available to the community through our data centre ISDC. The gained experience with the behaviour of the instruments prompted some small adjustments of the processing codes. Also feedbacks from the users were considered when appropriate. Here we will report on the availability and quality of the data on an instrument-by-instrument base. Within the Magnetic and Electric field (ME) group we process the scalar and vector magnetic field data, evaluate the star camera readings and the ion drift-meter measurements. For a complete cross-calibration between the scalar and vector magnetometers two dedicated attitude manoeuvres have been performed. A modified processing of the star camera readings has reduced the attitude noise during periods of single head solutions. As a rather recent product, we make the local electron density available. It is planned to estimate also the electron temperature. Both quantities are derived from the Planar Langmuir Probe.

**Key words:** magnetometer, cross-calibration, calibration parameter history, star camera, common reference frame, jumps.

## 1 Long Time Behaviour of Sensor Parameter

The main parameters of the vector magnetometers (FGM), scale factors, offsets and misalignment angle of the sensor, are checked routinely every 15 days in a calibration procedure, which is based on the measurements of the absolute scalar magnetometer. Fig. 1 and Fig. 2 are summarising the obtained

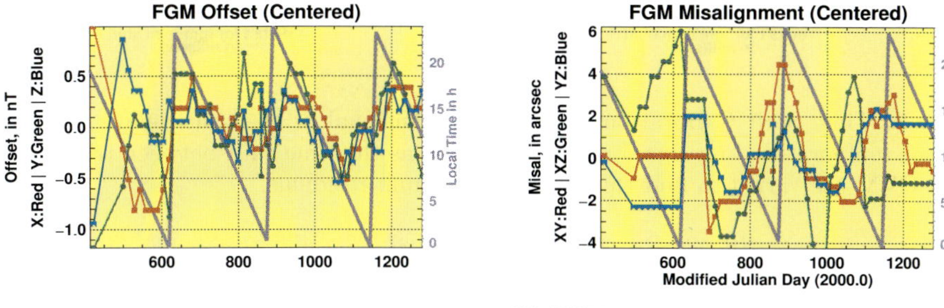


**Fig. 1.** Offset and Misalignment FGM-1. Saw tooth curve shows the local time of the orbit plane.

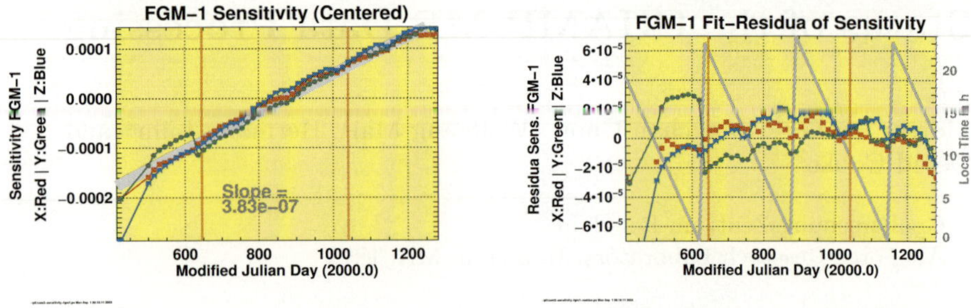


**Fig. 2.** Sensitivity FGM-1

sensor-parameters sets for FGM-1. Offset and Misalignment in Fig. 1 are in good correlation with the Local Time. The vertical lines in Fig. 2 are indicating attitude manoeuvres where the satellite was turned by 90° about the yaw axis to check the sensitivity value of the Y-axis. At the first manoeuvre the jump in the Y-component sensitivity shows the result of this additional information while the second manoeuvre forces no significant improvement. The slope of all three axis (the bias is removed in this graph) seems to be very similar for the considered period (May 2000 - Jul 2003) where the public data are available), the value of the slope is very much in the expected range. Even the residuals to the fitted line show a signal correlated with the local time. The reason for the correlation is probably the local time dependent temperature variation.

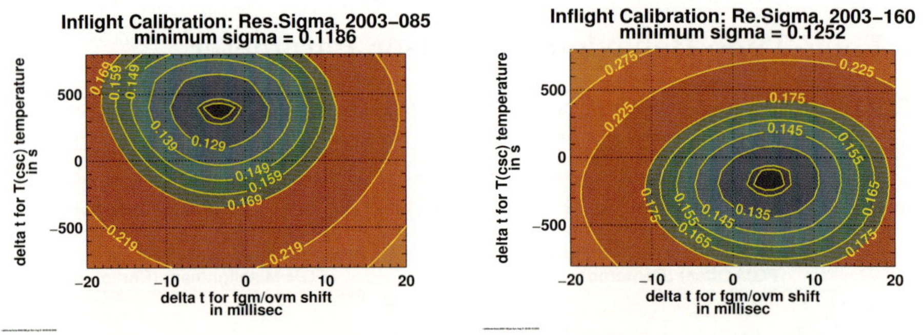


**Fig. 3.** Time Shift Calibrations. In a two-dimensional fit the time delay between sensor temperature and signal response, as well as the time shift between the FGM and OVM are varied to find the point of minimum residuals.

## 2 Time Shift Corrections

Fig. 3 shows two Single Time Delay Calibrations for FGM-1, scanning both the phase shift of the sensor temperature signal and the delta t between FGM and OVM signal. The latter exhibits and shows the unexpected variability of these shifts. The time shift value of the T(CSC) signal is absolute, the other is relative to the a priori setting.

The right part of Fig. 4 shows an overview of the History of Time Shifts of the temperature signal and the time difference between OVM and FGM-1. This variability is consequently also visible in the selected processing parameter values, which are usually valid for a 15 day period. Here is shown a longer period from start of the publicly available data set.

The more recent period and calibration run for the Temp(CSC) and OVM/FGM-1, displayed in the left part of Fig. 4, illustrates the some times quite rapid change of the calibrated time delay. The standard procedure is to check both parameters simultaneously in a five-day raster.

## 3 Electron Density (PLP) Data

Data from the Planar Langmuir Probe is available starting from 2001-05-26 as a public data product with a resolution in time of 15s. The format is a simple ASCII listing with a simple header. Because of the diamagnetic effect,

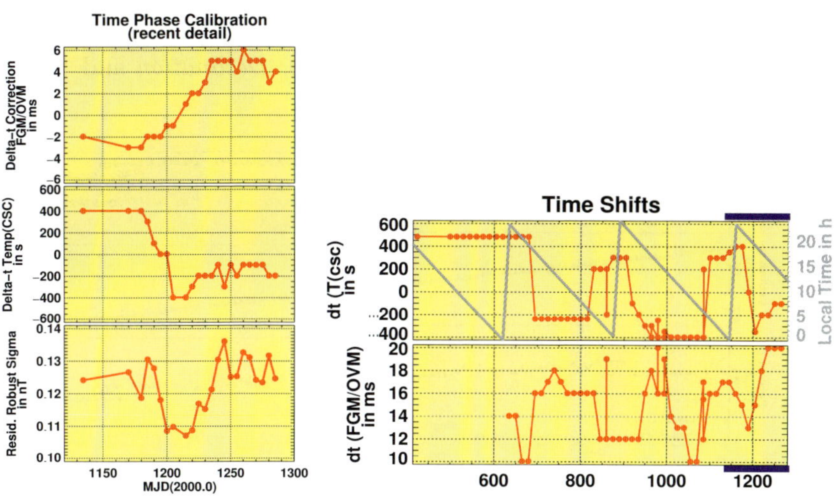


**Fig. 4.** History of used time-shifts (right panel) and a time-shift history detail of a recent calibration attempt (left panel). The time interval of the detailed 2-D fitting attempt is marked in the right panel with blue horizontal bars.

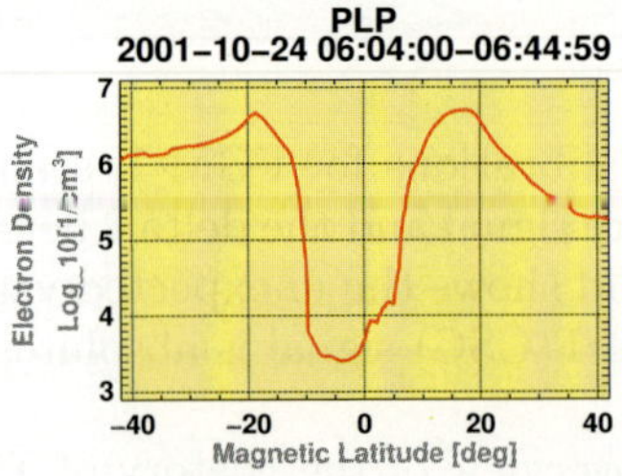


**Fig. 5.** Electron density example.

it is strongly recommended to involve this PLP data in the interpretation of the magnetic field readings. For first results see [1].

The example in the Fig. 5, showing the Electron Density on logarithmic scale versus magnetic latitude, visualises the measured variability of about three decades close to the magnetic equator.

# 4 Star Camera, Attitude and Transformations

The knowledge of the spacecraft (S/C) attitude is an important parameter for the interpretation of the high resolution data of Fluxgate Magnetometers (FGM), which can be fully used only, if the attitude is known to an arc-second accuracy. For this purpose two star cameras are mounted together with the FGM sensors on an optical bench in the middle of the boom. For the convenience of the attitude control system on CHAMP, a quartet of quaternions are describing the attitude in the S/C frame. The quaternions can equivalently be presented as Euler angles of a 3-2-3 rotation matrix, where the angles describes the *Right Ascension*, *Declination* and *Rotation* about the boresight direction.

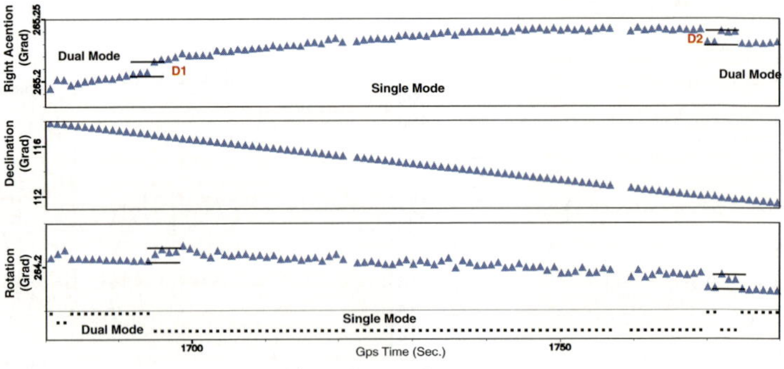


**Fig. 6.** Dual/single mode cross-over event.

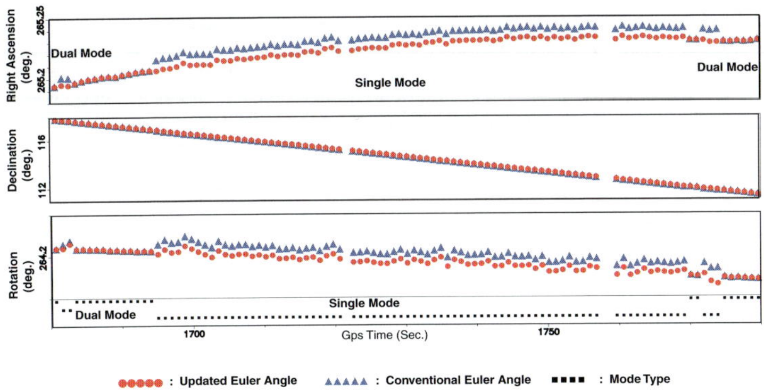


**Fig. 7.** Adjusted dual/single mode cross-over event.

## 5 Star Camera Preprocessing

The first processing step is to perform a 3-1-3 rotation from the S/C frame back into the camera head system (DC_CHU1 and 2) by using the direction cosine (DC) matrix. To calculate the DC matrix in the Common Reference frame (DC_CR_DUAL) from the associated two matrices, we perform the Dual vector operation. If attitude data is available from one head only, the transformation into the Common Reference frame is using the matrix M1 for DC_CHU1 and M2 for DC_CHU2 respectively. The required M1 and M2 can be computed as:

$$M1 = \mathrm{DC\_CR\_DUAL} \cdot \mathrm{DC\_CHU1}^T$$
$$M2 = \mathrm{DC\_CR\_DUAL} \cdot \mathrm{DC\_CHU2}^T.$$

Finally, for the level-2 format, the attitude data in CR frame are produced after polynomial filtering over 20 s (usually of degree 3) and resampled to the time stamps of the FGM readings.

## 6 Star Camera Data Continuity

As shown in Fig. 6, in the cross-over area (e.g. at second 1695 in upper Fig.) from the Dual to the Single mode and (at second 1770) from the Single to the Dual mode, a remarkable data discontinuity is observed, which means that the obtained matrices M1 and M2 are not appropriate in general. The up-shifted Single Mode data in Fig. 6 caused unexpected jumps in the vector magnetometer data in MFA frame of about 30 nT in the X component and of about 5 nT in the Y component (dashed line in Fig. 8). To correct the Single Mode Data in order to get continuity in these areas, we at first rotate the attitude into the single head frame and determine there the differences (D1 and D2 in Fig. 6) at the interfaces of Dual and Single mode in following way:

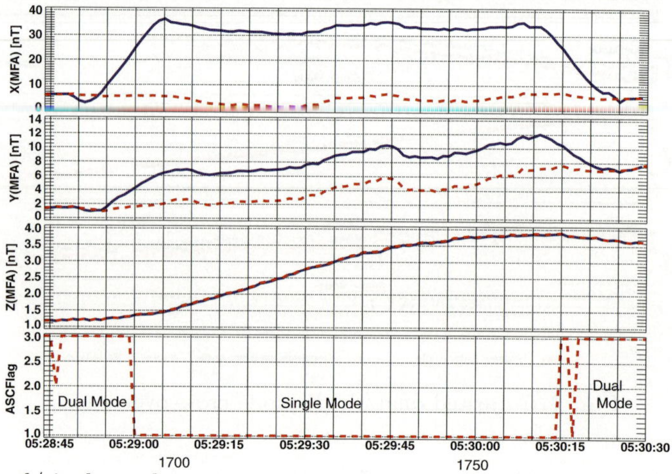


**Fig. 8.** Dual/single mode cross over event for magnetometer data, in MFA frame.

$$D1 = \mathrm{DC1}^T \cdot \mathrm{DC\_CHU1}$$

where

$$DC1 = \mathrm{M1}^T \cdot \mathrm{DC\_CR}$$

$$DC\_CHU1 = \mathrm{M1}^T \cdot \mathrm{DC1\_CR}$$

DC1 is the Dual Head and DC_CHU1 the Single Head matrix in the camera frame. The next step is to shift the uncertain Roll angle by the following method: Hereby DC_CR is Dual Head Data in camera direction and

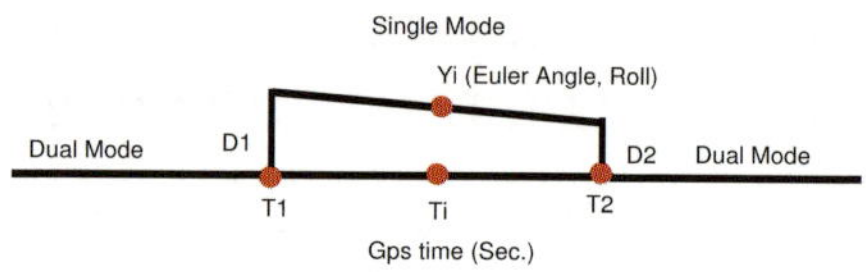


$$\mathrm{Yi} = (\mathrm{Ti} \cdot \mathrm{slope}) + \mathrm{D1}, \text{ where slope} = (\mathrm{D2} - \mathrm{D1})/(\mathrm{T2} - \mathrm{T1})$$

As a result the ASC data is continuous in the cross-over areas between two Modes (see dots in Fig. 6), which makes the updated MFA data free of jumps (dashed line in Fig. 8).

*Acknowledgement.* The operational support of the CHAMP mission by the German Aerospace Center (DLR) and the financial support for the data processing by the Federal ministry of Education and Research (BMBF) are gratefully acknowledged.

# References

[1] Lühr H, Rother M, Maus S, Mai W and Cooke D (2003) The diamagnetic effect of the equatorial Appleton anomaly: Its characteristics and impact on geomagnetic field modeling. Geophys Res Lett *30(17)*: 1906, doi:10.1029/2003GL017407.

# Neutral Atmosphere and Ionosphere

Retrieval Algorithms, RO Data Analysis, Data validation

# Atmospheric and Ocean Sensing with GNSS

Thomas P. Yunck and George A. Hajj

Jet Propulsion Laboratory, California Institute of Technology

**Summary.** The 1980s and 1990s saw the Global Positioning System (GPS) transform space geodesy from an elite national enterprise to one open to the individual researcher. By adapting the tools from that endeavor we are learning to probe the atmosphere and the ocean surface in novel ways, including ground-based sensing of atmospheric moisture; space-based profiling of atmospheric refractivity by active limb sounding; and global ocean altimetry with reflected signals. Ground-based GPS moisture sensing is already being tested for weather prediction. Limb sounding is less mature but offers a variety of attractions, including high accuracy, stability, and resolution; all-weather operation, and low cost. GPS "reflectometry" is least advanced but shows promise for a number of niche applications.

**Key words:** GPS, GNSS, occultation, limb-sounding, bi-static radar, reflectometry

## 1 Introduction

GPS atmospheric sounding first arose as a necessary element of ground-based GPS geodesy. It had been recognized from the earliest days that the effects of atmospheric moisture would ultimately limit the accuracy of space geodesy [e.g., Davis et al., 1985]. A good deal of attention was therefore given to possible calibration techniques. One of the most effective strategies involved modeling the zenith delay as a random walk and estimating a delay correction at every time step – typically every few minutes [Lichten and Border, 1987].

Although zenith delay estimates were at first treated as "nuisance" parameters, it was soon found that with reliable surface pressure data to calibrate the "dry" delay, these estimates provided measurements of the wet delay as accurate as any known [Elgered et al., 2003]. The wet delay solutions can be readily converted to estimates of precipitable water (PW) for use in weather and climate modeling. Alternatively, the delay estimates can be assimilated directly into weather models [Ha et al., 2003; Ridal and Gustafsson, 2003]. It quickly became evident that GPS networks could provide a windfall of atmospheric moisture data – one of the most critical quantities for numerical weather prediction (NWP).

## 2 Atmospheric Sounding from Space

In the late 1980s a few groups began to consider the possibility of atmospheric limb sounding from Earth orbit by active radio occultation, using GNSS signals as sources. This was inspired by the success of planetary radio occultation over the

**Fig. 1.** Observing geometry for atmospheric limb sounding by GPS radio occultation.

previous 25 years, beginning with the 1964 Mariner IV mission to Mars [Fjeldbo, 1964; Kliore et al., 1964]. Although Earth radio occultation was suggested in the 1960s, the first practical GNSS technique did not appear until Yunck et al. [1988]. The first GPS occultation flight experiment was GPS/MET, led by the University Corporation for Atmospheric Research (UCAR) in Boulder [Rocken et al., 1997].

Figure 1 illustrates the GNSS occultation geometry. A low Earth orbiter (LEO) tracks GNSS signals as they rise and set through the atmosphere, measuring the changing carrier phase at both L-band frequencies. The atmosphere acts as a lens, bending and retarding the signals, inducing an additional path delay that leads to additional accumulated phase and a Doppler shift. Depending on the observing geometry, it may take a minute or more for the observed signal to pass from the top of the atmosphere (~100 km altitude) to the surface, or vice versa. Since the changing geometry is dominated by the rapid motion of the LEO, most soundings occur within ±45° of the forward and reverse velocity directions.

Radio occultation differs fundamentally from ground-based zenith delay estimation. Because we are looking at (usually) a single ray traversing the atmosphere horizontally, we obtain extremely high vertical resolution – of the order of 100 m from the surface to the stratosphere. Moreover, the technique yields a diversity of atmospheric parameters distributed evenly around the globe, enabling a broader variety of applications [Anthes et al., 2000]. In addition, from the precisely known observing geometry we can compute the effective absolute height of each measurement to better than 10 m, enabling the computation of precise pressure gradients, contour maps, and such derived products as geostrophic wind fields in the troposphere and stratosphere. This ability to recover precise "geopotential heights" is unique among spaceborne sensors.

## 2.1 Performance

A geodetic receiver can measure the instantaneous carrier phase rate to a few mm/sec. With GNSS-based precise LEO orbit determination, we can isolate the atmospheric Doppler shift to better than 0.1 mm/sec. At 70 km altitude the excess atmosphere path delay is negligible; at the surface, it can reach 2.5 km, with rates exceeding 100 m/sec, or nearly a million times the measurement precision. The standard occultation analysis assumptions – local spherical symmetry, hydrostatic equilibrium, mixing ratios of atmospheric constituents – while imperfect, are good in most instances and result in relatively small refractivity errors.

Extensive studies [Kursinski et al., 1997; Hajj et al., 2002] suggest that because the major errors tend to be random or quasi-random, averaging of multiple profiles for long-term climate studies could yield an effective temperature error of less than 0.1 K. That assessment includes expected long-term bias variations; that is, the averaged profiles may be absolutely accurate, independent of the instrument used or the time flown, to better than 0.1 K over a sizable altitude range, an improvement by a factor of anywhere from 10 to 40 over current global techniques.

## 2.2 Climate Signal Detection

Perhaps the greatest concern in Earth science today is global climate change. A central question is whether Earth has entered a period of accelerated warming and, if so, to what extent this may result from human activity [e.g., Mann et al., 2003]. To discern the effects of various drivers we require precise data from all levels of the atmosphere, something that at present is extremely difficult to obtain globally.

Human-induced global warming may be in the range of 0.1-0.3 K per decade at the surface [Houghton et al., 1990], and even subtler (or reversed) in upper strata. To identify such faint signals we require great precision and long-term stability in our sensors – a level of 0.1 K/decade or better. Current spaceborne sensors fall well short of this. The sensor suite is dominated by passive radiometers, primarily IR and microwave, including MODIS, MISR, AIRS/AMSU, MLS, and AVIRIS. Although great ingenuity is employed in the design and operation of these instruments to maintain precise calibration, on the scale of a decade or more this is extremely difficult to achieve. Sensor aging and replacement with new designs undermine stability. Over a decade or more, the best that can reasonably be achieved with these systems is ~1 K absolute stability.

In an open solicitation released in 2002, NASA declared: "Perhaps the greatest roadblock to our understanding of climate variability and change is the lack of robust and unbiased long-term global observations." They went on to say that for climate monitoring "the focus is on...construction of consistent datasets from multi-instrument, multi-platform, and...multi-year observations with careful attention to calibration...over the lifetime of the measurement" [NASA, 2002].

With GNSS occultation each profile is largely self-calibrating and virtually independent of the particular instrument used or when in its lifetime the profile is acquired. One can argue that the only evolution will be further improvement in the already exceptional accuracy as retrieval techniques and instrument characteristics are refined, and GNSS signal strength is boosted. Equally important, GNSS occultation will provide absolute calibration data that will normalize and stabilize atmospheric products from sensors now in place.

### 2.3 Early Experimental Results

Establishing average temperature accuracies of 0.1 K or better over an extended altitude range is difficult at best. The primary standards to date have been of two types: radiosonde data and climate analyses generated by the US National Center for Environmental Predictions (NCEP) and by the European Center for Medium-Range Weather Forecasting (ECMWF), derived by assimilating radiosonde and space data into numerical models. In neither case can we expect better than about 1 K accuracy from the comparison data.

Figure 2 summarizes more than 4000 CHAMP refractivity profiles from May and June 2001, compared with ECMWF daily analyses [Beyerle et al., 2003]. The GFZ team has used two different techniques: standard "geometric optics" as described above (blue), and a "canonical transform" technique (red) developed by Gorbunov [2002], which offers several practical advantages. The solid curves are for the whole earth while the dashed curves are for the northern hemisphere, for which, owing to more abundant data, the analyses are expected to be somewhat better. Mean offsets (left) and sigmas (r) correspond roughly to temperature mean offsets of 0.5 K and sigmas of 1-2 K, above 2-3 km. See also Hajj et al. [2003].

### 2.4 CHAMP/SAC-C Comparisons

Analysis prospects improved markedly when the Argentine SAC-C spacecraft carrying a second BlackJack occultation receiver was launched into a quite different orbit in November 2000. For the first time we could directly compare data from two occultation instruments, provided they occasionally observed the same region at about the same time. Such coincidences are rare but not unknown. Of more than 60,000 soundings acquired from 10 July 2001 to 9 June 2002, sixty pairs occurred within 200 km and 30 min of one another. Of those, fewer than 30 pairs were within 100 km and 30 min of one another. As these coincidences are not exact there are in fact real differences of ~0.1K between pairs.

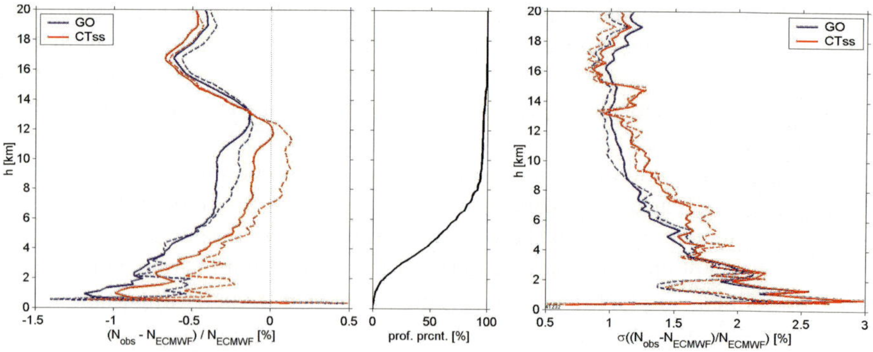


**Fig. 2.** Summary of 4,043 CHAMP refractivity profiles made with geometric optics (blue) and canonical transform techniques, differenced with ECMWF analyses. Standard deviations (right) and mean offsets (left) are similar to those of GPS/MET. (Beyerle et al., 2003).

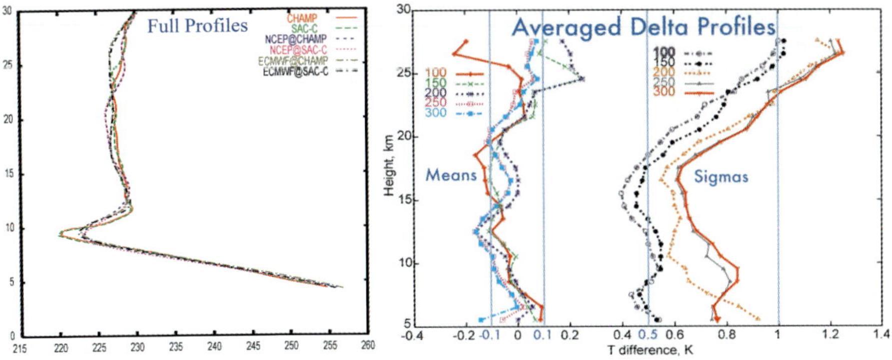


**Fig. 3.** Coincident profile pair from CHAMP and SAC-C plotted against NCEP and ECMWF analyses at each profile location (left); and the average differences of several dozen coincident pairs at five different maximum separations (right). (Hajj et al., 2003)

Figure 3 compares CHAMP and SAC-C occultations. The left panel shows a typical coincident pair (solid red and green) plotted together with standard NCEP and ECMWF analyses for the locations and times of the two profiles. What stands out is the closeness of the occultation profiles despite their differences in space, time, and observing angle. Both differ noticeably from the analyses, which are similar to one another but lack fine vertical detail. As the weather analyses involve a fair degree of smoothing, they cannot reliably capture such sharp features as the tropopause at ~9.5 km shown in the occultations. That two occultations acquired from different angles agree so exactly in this detail is compelling.

Figure 3 also shows the mean differences and sigmas for all occultation pairs coincident within 30 min, for maximum separations ranging from 100 to 300 km. Between 5 and 25 km altitude the average differences fall within or near 0.1 K. Compare this with Fig. 2 showing differences with the weather analyses, in which the mean offsets are ~5 times greater. For the closest pairs, the sigmas (which represent the combined errors of two profiles) are near 0.5 K at 20 km and below. These increase for wider separations, presumably because of actual differences between the sampled locations. For further discussion see Hajj et al. [2003].

These results are far from conclusive; there could well be common biases or trends in the occultation profiles not revealed in the comparisons. Indeed, we know that at higher altitudes where the bending is small and the retrieval is initialized (above 30 km), and also near the surface where the observations can become corrupted by complex atmospheric effects, biases can creep in. The causes of such biases, however, are well understood and refined techniques are now in development that promise to reduce them considerably.

Perhaps the greatest appeal of GNSS occultation is its potential for improving weather forecasts. As yet, with the still meager volumes of high quality data, little has been accomplished in the way of data assimilation and impact studies. But what has been done is encouraging. Preliminary assimilation studies have been reported by Zou et al. [2003], Healy et al. [2003], Kuo et al. [2003], and Aoyama et al. [2003]. Despite the sparseness of the data sets, all four report improvements in forecasts of various kinds – some going out 4 or 5 days – when occultation data were included. Healy, who typically assimilated only 40 profiles at once, comments that the "results are very encouraging [and] support the case for assimilating RO measurements operationally."

Despite these successes GNSS occultation science is still in its infancy. The gulf between the tens of daily profiles available today and a future of tens of thousands is beyond our power to bridge by intuition. In the box below we tabulate seven key attractions – the seven cardinal virtues – of GNSS limb sounding. Soon, this simple trick of radio metrology may offer a key to one of the most vexing problems in Earth science: discerning the faint signatures of global climate change. Its potential for comprehensive real-time monitoring of the global troposphere has equally profound implications for weather prediction.

**Principal Attractions of GNSS Occultation**

- Offers order-of-magnitude improvement in accuracy and long-term stability;
- Operates undiminished in all weather, day or night;
- Offers near-uniform global coverage from the stratopause to the surface;
- Provides unrivalled vertical resolution, to better than 100 m near the surface;
- Yields absolute geopotential heights to <10 m, permitting derivation of global pressure contours and non-equatorial geostrophic wind fields;
- Exploits different physical principles from other atmospheric sensors, remote or in situ, presenting an independent comparison and calibration standard;
- Requires only a palm-sized digital module costing less than 1 % of the tens to hundreds of M$ of many of today's passive sensors.

## 3. GNSS Surface Reflection

Traditional radar altimetry (e.g., Jason-1) observes vertically, obtaining one nadir height at a time. Bistatic GNSS, by contrast, can track a dozen or more reflections from many angles at once with one LEO receiver. This offers the prospect of higher temporal and spatial resolution for discerning finer scale, short-lived features, such as mesoscale eddies [Treuhaft et al., 2003], which play an important role in the transport of momentum, heat, salt, nutrients, and chemicals within the ocean. Hajj and Zuffada [2003] show that eight orbiters acquiring GPS and Galileo reflections could provide global 3-cm ocean heights in 1 day at 200-km spatial scales, or sub-decimeter heights in 4 days over 25-50 km scales.

While the technical feasibility of bistatic GNSS is clear, it remains to be seen whether the technique will offer practical advantages over alternative approaches, such as wide-swath altimetry [e.g., Rodriguez, 2001] or simply flying multiple nadir altimeters [Raney, 2001]. Here we consider some basic requirements and configurations for a practical bistatic GNSS system.

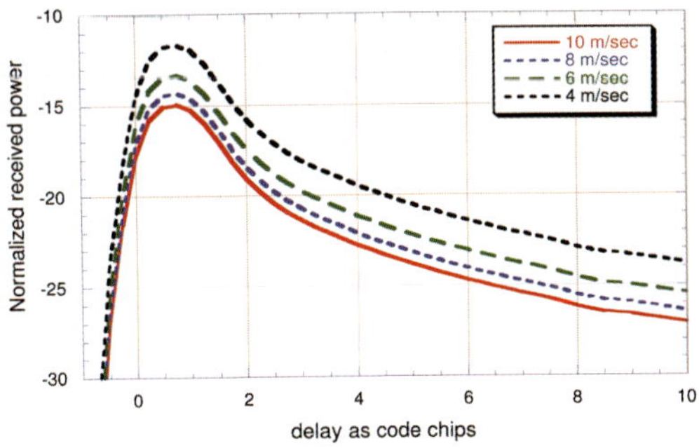


**Fig. 4.** Hypothetical cross-correlations of a GPS reflection with a model signal at many different lags, for four different surface wind speeds.

### 3.1 Basic Principles

Because the ocean is rough, a GNSS signal reflects to an orbiting receiver from an extended portion of the surface. A given wavefront will arrive first from the specular reflection point; decreasing energy will then come in from surrounding points, with greater delays. By cross-correlating the received signal against a model signal at many different delays, one can precisely map the returned energy vs delay. The function rises abruptly then decays gradually as weaker outlying reflections trail in, as illustrated in Fig. 4. The detailed shape of this function contains a good deal of information about the ocean surface.

The sharp leading edge of the correlation function permits precise determination of the arrival time – and hence total delay – of the reflection from the specular point; from that delay, along with other geometrical modeling information, one can derive an estimate of the ocean height at that point. Detailed analysis of the shape of the correlation tail can reveal information about the surface roughness, significant wave height, and wind speed.

Among the challenges in making such measurements are the weakness of the reflected signals at orbital altitude and the rapid signal de-coherence due to ocean roughness. The latter implies that under most ocean conditions and viewing geometries we cannot acquire the continuous GNSS precise phase observable.

### 3.2 Occultation / Reflection Synergy

There are, however, reasons for optimism. For a given delay measurement precision, the greatest sensitivity to ocean height occurs at nadir. Reflected signal quality, however, is poorest at nadir, owing both to lower returned signal strength and greater decoherence. Moreover, for a given solid angle the number of reflections to be seen is least at nadir. At the horizon, the average number of reflections per unit solid angle is fully two orders of magnitude greater than at nadir.

Figure 5 illustrates the distinct geometries for nadir and near-limb reflection sensing. The Rayleigh criterion tells us that for GNSS nadir reflections the ocean scatters coherently only for wave heights $h < 2$ cm; near the horizon, however, this

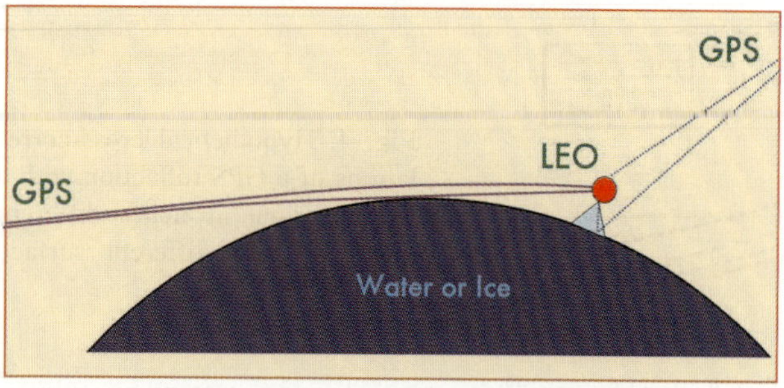


**Fig. 5.** The differing geometries of reflections acquired near nadir and at the horizon.

occurs for h <1 m. The former condition is almost never satisfied while the latter is satisfied much of the time, offering the prospect of precise off-nadir altimetry.

As it turns out, this is more than just a prospect. One of the bigger surprises of CHAMP and SAC-C atmospheric analysis was the discovery of glancing reflections at the horizon in a high percentage of occultation profiles.

The CHAMP science team at GeoForschungsZentrum (GFZ) in Potsdam has done a thorough study of these reflections, with promising results [Beyerle et al, 2002]. Figure 6 shows all occultations acquired by CHAMP over a 4-week period in the spring of 2001 (blue dots) together with all detected reflections (red circles). The first thing we notice is the relative absence of reflections in the tropics. This is presumed to owe to the combination of dense tropical moisture, which inhibits the penetration of all signals; the attenuation from the PN code offset; and the less frequent penetration of tropical occultations sufficiently close to the surface to yield the required small delay offsets. With coming refinements in "open-loop" signal acquisition [Sokolovskiy, 2002] and onboard modeling of the reflections, this shortfall should be alleviated.

Also evident in Fig. 6 is the presence at high latitudes of reflections in almost equal densities from water and ice, offering the possibility of cryosphere as well as ocean sensing. In fact, at high latitudes nearly all occultations occurring over water or ice are accompanied by a detectable reflection. We are thus seeing an unexpected convergence of GNSS occultation and reflectometry.

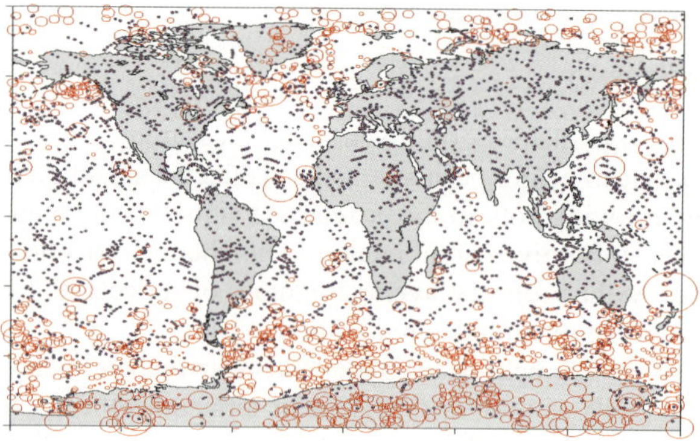

**Fig. 6.** Occultations acquired by CHAMP over 4 weeks in 2001 (blue dots) and accompanying reflections (red circles). Circle size indicates relative signal strength. (From Beyerle et al., 2002)

Beyerle et al. show that the reflected Doppler shift depends on the height of the reflecting surface, its slope (when ice), and the moisture density of the atmosphere, and argue that in different circumstances the signal may be used to estimate each of these. It is, however, too early to tell whether these serendipitous signals will provide information of real science value.

## 4 Future Prospects

The question arises, how many sensors are enough? Much is expected for atmospheric science from the first six or eight. For ocean reflections, we've seen that a dozen or so could be of value. A hint of an upper limit on useful numbers comes from a science panel convened by NASA in 1997 in Easton, Pennsylvania. Their charter was to assess the scope of Earth remote sensing data needed in the next 10-15 years to address NASA's Earth science agenda. The panel endorsed an occultation constellation providing global profiles with an average horizontal spacing of 50 km, twice per day, to initialize numerical weather models. That converts to about 400,000 profiles per day. Achieving such numbers with a 60-satellite GNSS constellation will require 250-300 LEO sensors. A constellation of six or eight occultation receivers will no doubt be incomparably superior to none. But in view of the rich history of GNSS-based science we have little doubt that the dividends will grow steadily with larger numbers, even into the hundreds.

*Acknowledgment.* Work described in this paper was performed in part by the Jet Propulsion Laboratory under contract with the National Aeronautics and Space Administration.

## References

Anthes RA, Rocken C and Kuo Y-H (2000) Applications of COSMIC to meteorology and climate, in Lee et al., 115-156.

Aoyama Y, Ozawa E and Tada H (2003) Data assimilation using temperature retrieved from CHAMP occultation data, Proc. Int. Workshop on GPS Met., Tsukuba, Japan.

Beyerle G et al. (2003) The radio occultation experiment aboard CHAMP (Part II): Advanced retrieval techniques in atmospheric sounding and GPS reflectometry, Proc. Int. Workshop on GPS Meteorology, Tsukuba, Japan.

Beyerle G et al. (2002) GPS radio occultations with CHAMP: A radio holographic analysis of GPS signal propagation in the troposphere and surface reflections, J Geophys Res *107(D24)*: 10.1029/2001JD001402.

Davis JL et al. (1985) Geodesy by radio interferometry: Effects of atmospheric modeling errors on estimates of baseline length, Radio Sci *20*: 1593-1607.

Elgered G et al. (2003) Analysis of atmospheric parameters derived from ground-based GPS observations, Proc. Int. Workshop on GPS Meteorology, Tsukuba, Japan.

Fjeldbo G (1964) Bistatic-radar methods for studying planetary ionospheres and surfaces, Sci. Rpt. #2, NsG-377, SU-SEL-64-025, Stanford Elect. Lab, Stanford, CA.

Gorbunov ME (2002) Canonical transform method for processing radio occultation data in the lower troposphere, Radio Sci *37(5)*: 10.1029/2000RS002592.

Ha S-Y, Kuo Y-H and Lim G-H (2003) Assimilation of GPS slant wet delay data and its impact on the short-range NWP, Proc. Int. Workshop GPS Met., Tsukuba, Japan.

Hajj GA et al. (2003) CHAMP and SAC-C atmospheric occultation results and intercomparisons, J Geophys Res, in press.

Hajj GA et al. (2002) A technical description of atmospheric sounding by GPS occultation, J. Atmos. and Solar-Terrestrial Physics *64*: 451-469.

Hajj GA and Zuffada C (2003) Theoretical description of a bistatic system for ocean altimetry using the GPS signal, Radio Science, to appear.

Healy S, Jupp A and Marquardt C (2003) A forecast impact trial with CHAMP radio occultation measurements. Abstract EAE03-A-08710, EGS-EUG-AGU Joint Assembly, Nice, France, April 2003.

Houghton JT, Jenkins GJ and Ephraums JJ, eds. (1990) Climate Change, The IPCC Scientific Assessment, Univ. of Cambridge, Cambridge, England.

Kliore AJ, Hamilton TW, and Cain DL (1964) Determination of some physical properties of the atmosphere of Mars from changes in the Doppler signal of a spacecraft on an earth occultation trajectory, Technical Rept. *32-674*, JPL, Pasadena, CA.

Kuo Y-H, Wee T-K and Sokolovskiy S (2003) Analysis and assimilation of radio occultation data, Proc. Int. Workshop on GPS Meteorology, Tsukuba, Japan.

Kursinski ER et al. (1997) Observing Earth's atmosphere with radio occultation measurements, J Geophys Res *102(D19)*: 23429-23465.

Lee, LC, Rocken C and Kursinski R, eds. (2000), Applications of Constellation Observing System for Meteorology, Ionosphere & Climate, Springer-Verlag, ISBN 962-430-135-2.

Lichten SM and Border JS (1987) Strategies for high-precision Global Positioning System orbit determination, J Geophys Res *92:* 12751-12762.

Mann M et al. (2003) On past temperatures and anomalous late-20$^{th}$ century warmth, EOS, Trans. AGU *84(27)*: 256-257.

NASA (2002) Cooperative Agreement Notice CAN-02-OES-01, Earth Science Research, Education and Applications Solutions Network (REASoN).

Raney RK and Porter DL (2001) WITTEX: An innovative multi-satellite radar altimeter constellation: A summary statement for the high-resolution ocean topography science working group, http://www.deos.tudelft.nl/gamble/docs/wittex.pdf.

Ridal M and Gustafsson N (2003) Assimilation of ground-based GPS data within the European COST-716 action, Proc. Int. Workshop on GPS Met., Tsukuba, Japan.

Rocken, C et al. (1997) Verification of GPS/MET data in the neutral atmosphere, J Geophys Res *102*: 29,849-29,866.

Rodriguez E, Pollard BD and Martin JM (2001) Wide-swath altimetry using radar interferometry, IEEE Trans. Geosci. and Rem. Sensing.

Sokolovskiy S (2001) Tracking tropospheric radio occultation signals from low Earth orbit Radio Sci *36(3)* doi:10.1029/1999RS002305.

Treuhaft RN et al. (2003) Monitoring coastal eddy evolution with GPS altimetry, Proc. Int. Workshop on GPS Meteorology, Tsukuba, Japan.

Yunck TP, Lindal GF and Liu C-H (1988) The role of GPS in precise Earth observation, Proc. IEEE Position Loc. and Nav. Symp. (PLANS 88).

Zou X, Liu H and Shao H (2003) Further investigations on the assimilation of GPS occultation measurements, Proc. Int. Workshop on GPS Met., Tsukuba, Japan.

# Amplitude Variations in CHAMP Radio Occultation Signal as an Indicator of the Ionospheric Activity

Alexander Pavelyev[1], Jens Wickert[2], Christoph Reigber[2], Torsten Schmidt[2], Yuei-An Liou[3], Chen-Young Huang[3], Stanislav Matyugov[1], Dmitrii Pavelyev[1]

[1] Institute of Radio Engineering and Electronics of Russian Academy of Sciences, (IRE RAS), Fryazino, Vvedenskogo sq. 1, 141120 Moscow region, Russia, fax: +7 095 203 84141, *pvlv@ms.ire.rssi.ru*

[2] GeoForschungsZentrum Potsdam, Telegrafenberg, 14473 Potsdam, Germany, Department Geodesy and Remote Sensing, phone: +49- 331-288-1758; fax: +49-331-288-1732, *wickert@gfz-potsdam.de*

[3] Center for Space and Remote Sensing Research, National Central University, Jung-Li, 320, Taiwan, fax: +886 3 4254908, *yueian@csrsr.ncu.edu.tw*

**Summary.** We showed that the amplitude of GPS occultation signal is important indicator of the ionospheric activity. Amplitude is more sensitive to small-scale ionospheric disturbances than the phase of the radio occultation (RO) signals. Local mechanism of strong ionospheric influence on the amplitude and phase of RO GPS signals is described. Critical points (tangent points) in the ionosphere, where the gradient of the electron density is perpendicular to the RO ray trajectory, strongly influence on the amplitude and phase of RO signals, and introduce multi-RO ionospheric effect in the experimental RO data. Positions of the critical points depend on the structure of the ionospheric disturbances. Analytical model for description of multi-RO ionospheric effect is introduced. Model accounts for the horizontal gradients in the ionosphere and gives analytical expressions for the phase path excess and refraction attenuation of the radio wave propagating through the disturbed ionosphere. Analytical model and analysis of the CHAMP RO data indicated that the centers of strong ionospheric influence on RO signals can exist, for example, in the sporadic E-layers inclined by 3-6 degrees relative to the local horizontal direction. In this case one can observe simultaneously with atmospheric RO the appending ROs in the ionospheric layers. Multi- RO effect can be a cause of the ionospheric interference in the communication and RO signals. Multi- RO effect can be used to study the structure of the ionospheric disturbances using the amplitude variations in RO signals. Multi- RO effect allows introducing a classification of the ionospheric influence on RO signals using the amplitude data. This indicates a possibility for separating the regular and random parts in the ionospheric contribution in the RO signals.

**Key words:** ionosphere, electron density, sporadic E-layer, radio occultation

## 1 Introduction

High-precision radio signals of global positioning satellite system (GPS) are effective tool for global RO studies of the atmosphere and ionosphere of the Earth. The radio signals emitted at two GPS frequencies $f_1$=1575.42 and $f_2$=1227.63 MHz by a GPS radio navigational satellite and received by a small satellite installed on a low earth orbit (LEO) are used for the RO investigations (Melbourne et al., 1994, Kursinski et al., 1997, Hajj and Romans, 1998, Steiner et al.,1999, Wickert et al., 2004). During a RO experiment the ray linking LEO receiver to a GPS satellite (marked by point L and G in Figure 1, respectively) immerses sequentially into the ionosphere and atmosphere. The main contribution to the amplitude and phase data of the RO signal is introduced by relatively small area along the ray GTL with center at the tangent point T (Figure 1). In this area the direction angle of the ray trajectory is near zero relative to the local gradient of the refractivity.

The amplitude and phase variations of the RO signal are functions of minimal ray height *h* above the Earth surface (Fig. 1) and satellites positions and velocities. These functions can be utilized to determine the height dependence of the refraction angle if precise orbital data are given (Melbourne et al., 1994, Hocke et al., 1999). Then Abel transform is used to obtain the vertical profiles of the electron density in the ionosphere and refractivity in the atmosphere (Vorob′ev et al., 1999, Steiner et al., 1999, Igarashi et al., 2000, Wickert, 2001, Gorbunov, 2002). Amplitude data can be used also to obtain independently the vertical gradients of the electron density in the ionosphere and refractivity in the atmosphere (Pavelyev et al., 2002, Liou et al., 2002).

Physical changes in the near-Earth space environment in response to variations in solar radiation, solar plasma ejection, and the electromagnetic status of the interplanetary medium produce disturbances in the ionosphere. The disturbed ionosphere changes the amplitude and phase of RO signal. Strong amplitude and phase frequency dependent variations of the RO signals are often observed in the heights interval 40-100 km. Emerging of strong oscillations in the amplitude and phase of the RO signal may be connected with developing critical (tangent) points at the ionospheric part of the ray trajectory. The tangent points 1, 2 (Fig. 1) can originate because influence of the regular horizontal gradients in the E- and F-layer. The regular horizontal gradients can exist in the ionosphere at the boundary day-night

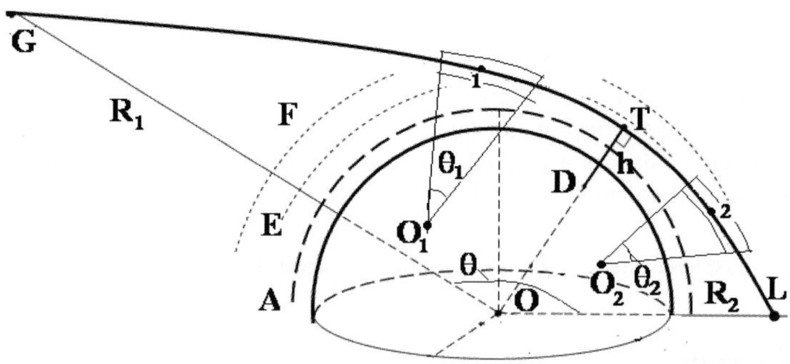


**Fig. 1.** RO measurements in the atmosphere can be disturbed by ionospheric contribution because propagation of the RO signal through parts 1 and 2 of the ionosphere. The additional tangent points 1 and (or) 2 in the ionosphere can be a cause of oscillations in the amplitude and phase of the RO signal.

owing to diurnal motion of the sub- solar point. Also the horizontal gradients can emerge because difference in the electron density concentration between the polar and equatorial regions.

The aim of this paper consists in demonstrating multi-RO effect by means of analysis of the CHAMP RO experimental data (Wickert et al., 2004), and introducing classification of the ionospheric influence on the amplitude of RO signals.

## 2 Multi-RO effect

The most important case for practice is one- ray propagation with several tangent points along the RO ray. In this case we can combine the contributions of different tangent points disposed in different locally spherical symmetric media. The centers of local spherical symmetry associated with a ray path in the ionosphere and atmosphere are located at different points $O_1$, $O_2$, O (Fig. 1) corresponding to the three parts of the ray trajectory G1T (the ionosphere between transmitter and atmosphere), 1T2 (atmospheric part), and T2L (the ionosphere between the receiver and atmosphere). The phase path excess $\Phi$ corresponding to the ray G1T2L (Fig. 1) can be presented as a sum:

$$\Phi=\Phi_1+\Phi_2+\Phi_3-[(R_1^2-p_s^2)^{1/2}+(R_2^2-p_s^2)^{1/2}] \qquad (1)$$

where $\Phi_1$, $\Phi_2$, $\Phi_3$ are the phase changes corresponding to three parts of the ray trajectory, $R_1$, $R_2$ are the distances OL, OG, respectively (Fig. 1), $p_s$ is the impact parameter corresponding to propagation in free space along the path GL (Fig. 1). The refraction attenuation $X_L$ at the path G1T2L is equal to

$$X_L = X_1X_2X_3, \qquad (2)$$

where $X_1$ is the refraction attenuation at the ray path G1, $X_2$ is the refraction attenuation at the ray path in the atmosphere, $X_3$ is the refraction attenuation pertinent to the ray path 2L (Fig. 1).

Equations (1), (2) can be easily generalized for the case of arbitrary M locally spherical symmetric medium disposed along the RO signal trajectory. The phase and amplitude of the GPS occultation signal is a superposition of the phase changes and result of multiplication of the refraction attenuation variations, respectively, occurred in each medium. In any spherical symmetric medium RO event can be developed and effects the amplitude and phase of GPS occultation signal. The corresponding variations in the amplitude and phase will be associated with influence of the atmospheric tangent point T on the straight line OT (Fig. 1). For example, the amplitude and phase variations, associated with an inclined sporadic E-layer disposed at the heights 95-110 km in the mesosphere, will be projected along the trajectory GTL (Fig. 1) to the height $h_p$ in the interval 40-90 km. The height $h_p$ depends on the inclination of the ionization layer relative to the local vertical direction. In result one can observe strong amplitude and phase variations in the height interval 40-90 km, which is unusual both for atmospheric and ionospheric contributions expected for the spherically symmetric atmosphere and ionosphere. The inclination of the sporadic E-layer $\delta$ respective to the local horizontal direction and its displacement d can be evaluated using relationships

$$\delta=(2\Delta h/a)^{1/2},\ d=(2\Delta ha)^{1/2} \qquad (3)$$

where $\Delta h$ is the apparent deflection in height of the ionospheric response as seen in the RO signal, and a is the Earth radius. The second equation (3) gives the displacement of the layer measured as a horizontal distance d between the atmospheric and ionospheric tangent points.

The influence of the two parts of the ionosphere in the RO experiments can be significant if the two necessary and sufficient conditions are fulfilled: 1) the corresponding part of the ionosphere contains a tangent point on the ray trajectory; 2) there is a refractivity layer with sharp vertical gradient in the essential vicinity of the tangent point.

The importance of the contribution of the critical points to the phase and amplitude variations of the RO signal has been noted early by Igarashi et al., 2001. As they showed, the influence of the ionospheric layers depends on the length $L_c$ of coherent interaction with RO signal. The length L of coherent interaction with the locally spherical symmetric layered ionospheric irregularity having the vertical correlation radius $l_c$ depends on the elevation angle $\varepsilon$ between the local horizon and ray trajectory: $L\approx l_c/\sin\varepsilon$. In the vicinity of the tangent point of the ray trajectory the angle $\varepsilon$ is near to zero. Due to this effect the contribution of the ionospheric irregularities to the phase and amplitude variations is diminished by the factors $W_{p,a}$, respectively, in comparison with the amplitude and phase changes caused by a layered structure having the vertical size $L_d$ at the critical point T. According to Igarashi et al., 2001, one may obtain the next relations for the factors $W_{p,a}$:

$$W_p\approx\Delta N_d L_c \sin\varepsilon/(l_c\Delta N_f)M;\ W_a\approx(l_c/L_d)^2 W_p;\ L_c\approx 2(2r_c L_f)^{1/2};\ M=(2D/l_c)^{1/2} \qquad (4)$$

where $r_c$ is the distance from the critical point to the center of the local spherical symmetry, $\Delta N_d$, $\Delta N_f$ are, respectively, the refractivity perturbations near the tangent point T (Fig. 1) and in the ionospheric irregularity, $L_f$ is the size of the Fresnel zone, $L_d$ is the vertical correlation radius of the layered irregularity near the tangent point, D is the vertical size of the part of the ionosphere having the elevation angle $\varepsilon$. Factor M (4) describes a summation effect of the contributions of the $2D/l_c$ independent ionospheric irregularities. The amplitude factor $W_a$ includes the ratio $(l_c/L_d)^2$ because the amplitude depends on the second derivative of the main part of the phase excess on the impact parameter (Pavelyev, 1998). Now the factor $W_{p,a}$, may be estimated. For $L_f\approx 0.7$ km, $r_c=6400$ km, $L\approx 200$ km, $\sin\varepsilon\approx 0.25$; $l_c\sim 10$ km, $L_d\approx 0.5$ km, $D\sim 200$ km, $M\sim 6.4$; $W_p\approx 32\Delta N_d/\Delta N_f$, $W_a\approx 12800\Delta N_d/\Delta N_f$. This estimation demonstrates: (1) high sensitivity of the amplitude (in comparison with the phase) to variations of the refractivity near the critical point, and (2) relatively small level of the contribution of the ionospheric irregularities (about 40-50 db below the contribution of the atmospheric critical point T) during quiet conditions in the ionosphere.

To estimate dependence of the length of coherent interaction of a critical point $L_c$ on the RO ray on the horizontal $\Lambda_h$ and the vertical wavelength $\Lambda_v$ of wave structure in the spatial refractivity field (caused, for example, by internal wave in the atmosphere or in the ionosphere). In the case of spherical symmetry the main contribution to the RO signal is introduced by relatively small essential area with center at the ray tangent point T (Fig. 1). At this point the direction angle $\alpha$ of the ray trajectory relative to the local vertical is zero and radio waves propagate nearly along the layer (Igarashi et al., 2001). Because the ray inside the essential zone is nearly perpendicular to the vertical gradient of the refractivity dN(r)/dr it is evident that $L_c = |\xi(p)n(r)/[dN(r)/dr]|$, where $\xi(p)$ is the refraction angle, p is the ray impact parameter, n(r) is the refraction index at point T (Fig. 1). Using analytical expressions for $\xi(p)$ and N(r) given in the radio- physical model (Pavelyev et al., 1996, Pavelyev, 1998), one can estimate the value $L_c = (2\pi rH)^{1/2}/M_d$, for the case of the standard atmosphere with effective height H, where r is the distance from the tangent point T to the center of spherical symmetry O (Fig. 1), $M_d = M'_r(r)/n(r)$, $M(r) = n(r)r$ is the modified refraction index in the atmosphere. For the case of internal wave model gives $L_c = (r\Lambda_v)^{1/2}/M_d$. One obtains for H=8 km, $\Lambda_v$= 1 km, r=6400 km, $L_c$= 580 km, and $L_c$ = 80 km, respectively. Thus for internal waves with vertical period $\Lambda_v$ in the interval 1–16 km the length $L_c$ is changing in the range 80 – 320 km. One can estimate the lower limit of the length of the essential area $L_c$ and the value $\alpha$ at its boundary. The angle $\alpha$ is connected with $L_c$: $\alpha = 0.5\, L_c/r$. For coherent interaction of the wave with the layered structure at T the difference $\alpha - \xi(\alpha) = \alpha(1 - d\xi/d\alpha) = \alpha M_d$ must not exceed the angular size of the Fresnel zone at the point T (Fig. 1), and one obtains $\alpha \le (\lambda X/L_c)^{1/2}/M_d$. At the boundary of the essential area both values are equal, and one can find $L_c$ and $\alpha$: $\alpha = 2^{-1/3}(\lambda X/r)^{1/3} M_d^{-2/3}$, $L_c = 2^{2/3} r^{2/3} \lambda^{1/3} X^{1/3} M_d^{-2/3}$. For X=1, $\lambda$=20 cm, $M_d \approx 1$, r=6400 km, $L_c \approx 32$ km. The Fresnel's zone radius $\Delta h$ corresponding to $L_c$ is equal to $\Delta h$ = 40 m. Thus the RO method is sensitive to the wave structures with horizontal and vertical wavelength greater than 32 km and 40 m, respectively. It follows from this analysis, that the lower diffraction limit of the intrinsic horizontal resolution of the RO method depends on the wavelength of radio waves $\lambda$, radius of curvature of the layered structure in the atmosphere r, vertical gradient of the modified refraction index $M_d$, and does not depend on the vertical wavelength of the internal wave $\Lambda_v$. When $M_d \approx 0$, the length of coherent interaction can be high, this corresponds to the phenomenon of the wave guide propagation and strong diffraction effect. In usual conditions the physical optics formulas for the intrinsic horizontal resolution $L_c = (r\Lambda_v)^{1/2}/M_d$ and vertical resolution $\Delta h = [\lambda X/(4M_d)]^{1/2}(r\Lambda_v)^{1/4}$ are valid, which indicates dependence of $L_c$ and $\Delta h$ on the vertical wavelength $\Lambda_v$ of internal wave as ~ $\Lambda_v^{1/2}$ and $\Lambda_v^{1/4}$, respectively. One can obtain expression for the critical value of the vertical wavelength $\Lambda_{vc}$ when the coherent length $L_c$ found from the physical optics and diffraction limit coincide $\Lambda_{vc} = 2^{4/3} r^{1/3} \lambda^{2/3} X^{2/3} M_d^{2/3}$ $\approx 160$ m for $X = M_d = 1$, $\lambda$=20 cm, r=6400 km. This analysis makes the local character of the RO method clear, and indicates the values of the physically realizable resolution in the RO investigation of the internal waves in the atmosphere and ionosphere.

## 3 Classification of ionospheric influence on the amplitude and phase of CHAMP RO signal

In the case of CHAMP RO experiment the quiet ionospheric conditions have come to light in the form of smooth dependence of the amplitude in the height interval 40-100 km

$$\sigma^2 = \langle (A(t) - \langle A \rangle)^2 \rangle / \langle A \rangle^2 \quad (5)$$

where <A> is the average amplitude of RO signal corresponding to propagation in the free space. Examples of quiet and noisy ionospheric conditions observed during CHAMP RO experiment is shown in Fig. 2. The amplitude curve have relatively high number of intersections with level 1, M=447, indicating noise origin of fluctuations, and a low level of the amplitude variations $rms=(\sigma^2)^{1/2}=0.0107$. The phase path excesses $\Phi_1(t)$, $\Phi_2(t)$ at the frequencies $f_1$, $f_2$ are shown in Fig. 2 by curves F1, F2. The curves F1, F2 have been obtained after subtracting of the second power Lagrange polynomial approximating a phase trend connecting with the

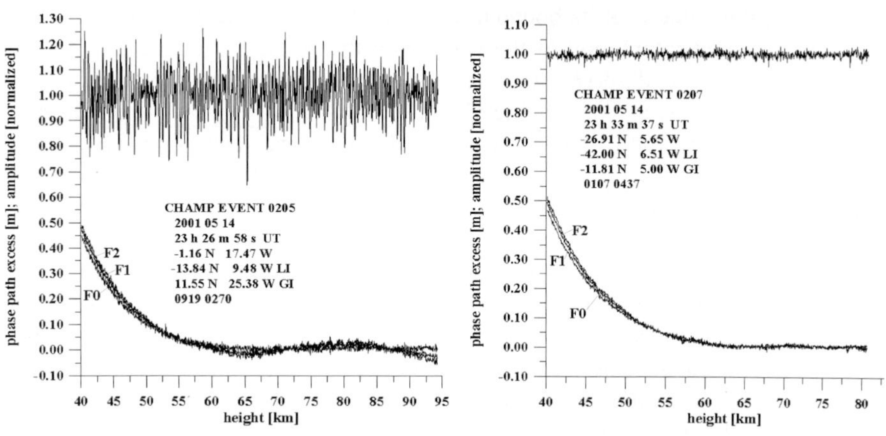


**Fig. 2.** Right panel: the amplitude and phase of CHAMP RO signal (event No. 0207, May, 14, 2001). The insert indicates the time of event, coordinates of the atmospheric tangent point T, coordinates of the intersection of the ionosphere at the height 280 km from the site of LEO satellite (letters "LI") and GPS satellite (letters "GI"). The rms of the amplitude variations $rms=(\sigma^2)^{1/2}$ multiplied by 10000 is shown and the number M of the intersections of level A=1 by the amplitude curve A is given. Left panel: the same as in the right panel for noisy CHAMP event (No. 0060, May 14, 2001) with high amplitude variations (rms=0.0919, M=270) in the equatorial region at local evening.

F-layer influence. The curve F0 relates to the combined frequency F0, obtained from the ionospheric correction formula:

$$\Phi(p)=[f_1^2\,\Phi_1(t)-f_2^2\Phi_2(t)]/[\,f_1^2-f_2^2] \qquad (6)$$

The quiet ionospheric conditions have been observed during 81% of CHAMP events of data base analyzed (Doy 134-160, 2001; Doy 96/97 2001). An example of CHAMP event with high amplitude variations (rms=0.0919, M=270) is shown in Fig. 2 (left panel). This event can be classified as a typical case, relevant to the noisy ionospheric contribution caused by intense ionospheric irregularities in the equatorial region after sunset at 22 h LT. The main part of the amplitude variations is limited to the interval 0.75≤A≤1.25. The phase variations are relatively small and indicated only slow changes in the TEC which may be connected with moving long-scale bubbles in the fountain region of the ionosphere. The isolated quasi-regular event with influence of the sporadic E-layer is shown in Fig. 3 (left panel). The fine structures corresponding to an inclined sporadic E-layer have been projected to the heights 75-80 km. Owing to the projection rule the phase and amplitude vertical distributions corresponds to an inclined sporadic E-layers dis posed at the height about 93-103 km in the evening ionosphere. Using the data shown in Fig. 3 one can find from (3) the position of the ionospheric tangent point. For Δh≈20 km, one can obtain from (3) δ≈2°, d≈250 km. The sporadic E-layer declined by 2° and disposed at the distance 250 km away from the atmospheric tangent point.

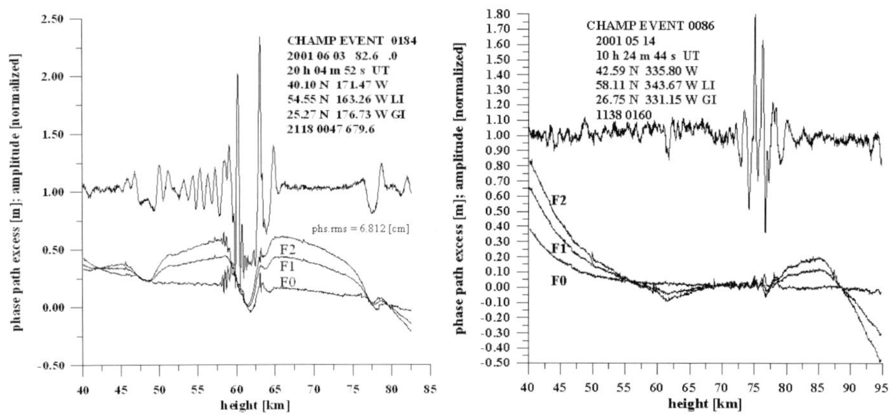


**Fig. 3.** Left panel: the same as in Fig. 2 for the CHAMP RO event No. 0086 (May 14, 2001, mid-latitude day-time ionosphere) with isolated quasi-regular amplitude and phase variations connected with inclined by 2° sporadic E-layer, which influence has been projected to the height interval 75-80 km. Right panel: the same as in Fig. 2 for CHAMP RO event No. 0184 (June 05, 2001). The amplitude and phase data indicate diffraction of radio waves on sharp gradients of the electron density in an inclined by 4° sporadic E- layer of the ionosphere.

The strong ionospheric event with diffraction structures in the RO signal is demonstrated in Fig. 3 (right panel). The diffraction pattern connected with intersection of the caustic boundary is seen at the heights 60 and 62.5 km.

The electron density distribution and its gradient in the inclined ionospheric layers can be restored by appropriate inversion algorithms for type 2 and 3 of the ionospheric disturbances

## 4 Example of recovering the refractivity perturbations in the inclined ionospheric layer from the CHAMP RO signal

Below we will analyze the data of CHAMP RO event provided on May 20, 2001 in the equatorial region. In this event the ionospheric impact can be seen as an influence of a weak inclined sporadic E-layer. The results of the refractivity restoration from the phase and amplitude data are presented in Fig. 4. The refractivity profile (rough top curve in Fig. 4 (right panel)) indicates quasi-regular structures associated with ionospheric influence at the heights between 51 and 55 km. The smooth curve in Fig. 4 (right panel) shows the regular exponential diminishing of the refractivity connected with neutral gas from 0.78 [N-unit] at h=40 km up to 0.02 [N-unit] at h=85 km. Rms of the perturbations of the refractivity is 0.024 [N-unit]. In Fig. 4 (left panel) the smoothed perturbations in the refractivity and its gradient recovered on the amplitude (the second and third top curves) and phase data (top curve) are shown. The time interval for smoothing was equal to 0.38 seconds (19 samples). For comparison the two bottom curves in

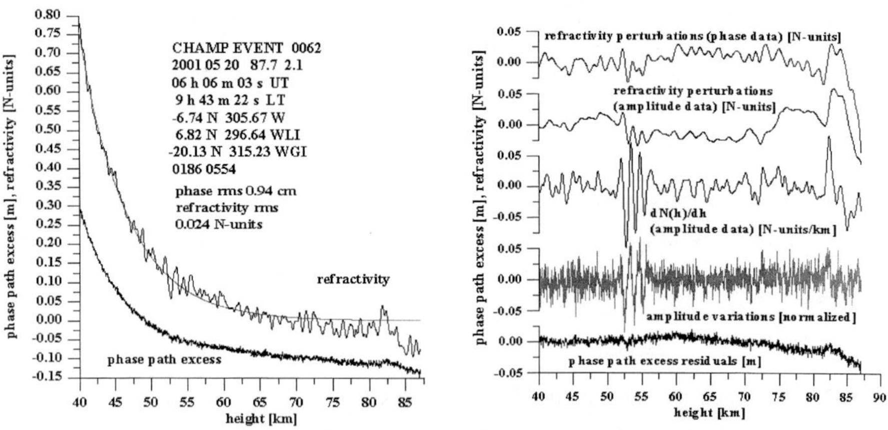


**Fig. 4.** Left panel: Abel's retrievals of the refractivity distribution (smooth and rough top curves) in comparison with the phase path excess at frequency F0 (curve in bottom, displaced by 20 cm). Right panel: results of restoration of the refractivity perturbations on the phase and amplitude data (upper two curves), vertical gradient of the refractivity (middle curve) in comparison with the amplitude variations and the phase path residuals (bottom curves).

Fig. 4 (left panel) show the amplitude and phase path excess variations. Some correspondence is seen between the refractivity perturbations obtained from the phase and amplitude data (e.g., sharp maximum in Fig. 4 (left panel) in the height interval 70-85 km. Small-scale structure of the refractivity perturbations and its gradient is better seen in the amplitude than in phase retrievals. For example, the amplitude retrievals revealed fine structure in the height interval 51-55 km, which is associated with corresponding feature in the amplitude variations (Fig. 4 (left panel), the second and third bottom curves). In this feature the vertical gradient of the refractivity changes in the interval ±0.07 [N-units/km], with corresponding variations in the vertical gradient of the electron density $\pm 4 \cdot 10^3$ [el/(cm$^3$km)] in the inclined by $5^{\circ}$ sporadic E-layer. This reveals high sensitivity of the amplitude data to small-scale variations in the refractivity. The phase data are more appropriate for retrieval of large-scale variations, however they are needed in the preliminary ionospheric correction.

## Conclusions

Multi-RO effect is detected in the CHAMP RO data. The effect is observed as strong quasi-regular variations in the amplitude of RO signal in the height interval 40-80 km. Effect is connected with emerging additional tangent points in the inclined ionospheric layers with sharp gradients of the electron density. Analysis of CHAMP RO data indicated importance of the amplitude channels for classification of the ionospheric influence on the RO signals. Preliminary analysis revealed five types of the ionospheric influence on the CHAMP RO data: (1) quiet events; (2) isolated quasi-regular flashes (connected with influence of inclined sporadic E-layers) in the amplitude and phase of RO signal; (3) events with quasi-periodical changes of the amplitude and phase of RO signal (possible source – wave structures in the electron density); (4) diffractive events with clearly seen diffraction pattern in the amplitude and phase of RO signal; (5) events with noisy contribution of the ionospheric disturbances to the amplitude of the RO signal. Analysis of the CHAMP RO data indicated a possibility to evaluate the electron density distribution and its gradient in the inclined ionospheric layers. This reveals a possibility of simultaneous observations of the ionospheric and atmospheric structures by the RO method.

*Acknowledgments.* We are grateful to GFZ Potsdam for access to the CHAMP RO data and financial support. Work has been partly supported by National Science Council of Taiwan, R.O.C., grants NSC 92-2811-M008-001, NSC 91-2111-M008-029, Office of Naval Research (ONR) of USA, grant N00014-00-0528, and Russian Fund of Basic Research, grant No. 03-02-17414.

## References

Gorbunov ME (2002) Ionospheric correction and statistical optimization of radio occultation data. Radio Sci *37(5)*: 1084, 2000RS002370.

Hajj GA and Romans LJ (1998) Ionospheric electron density profiles obtained with the the Global Positioning System: Results from GPS/MET experiment. Radio Sci *33(1)*: 175-190.

Hocke K, Pavelyev A, Yakovlev O, Barthes L, and Jakowski N (1999) RO data analysis by radio holographic method. JASTP *61*: 1169-1177.

Kursinski ER, Hajj GA, Schofield JT, Linfield RP, and Hardy KR (1997) Observing Earth's atmosphere with radio occultation measurements using the Global Positioning System. J Geophys Res *102*: 23,429-23,465.

Igarashi K, Pavelyev A, Hocke K, Pavelyev D, Wickert J (2001) Observation of wave structures in the upper atmosphere by means of radio holographic analysis of the RO data. Adv Space Res *27(6-7)*: 1321-1327.

Igarashi K, Pavelyev A, Hocke K, Pavelyev D, Kucherjavenkov IA, Matugov S, Zakharov A, and Yakovlev O (2000) Radio holographic principle for observing natural processes in the atmosphere and retrieving meteorological parameters from RO data. Earth Planets Space *52*: 968-875.

Liou Y-A, Pavelyev AG, Huang C-Y, Igarashi K, and Hocke K (2002) Simultaneous observation of the vertical gradients of refractivity in the atmosphere and electron density in the lower ionosphere by radio occultation amplitude method. Geophys Res Lett *29(19)*: doi 10.1029/2002GL015155.

Melbourne WG, Davis ES, Duncan CB, Hajj GA, Hardy KR, Kursinski ER, Meehan TK, Young LE, and Yunck TP (1994) The Application of Spaceborne GPS to Atmospheric Limb Sounding and Global Change Monitoring. JPL Publication 94-18, 147 pp.

Pavelyev A (1998) On the possibility of radio holographic investigation on communication link satellite-to-satellite. Journal of Communications Technology and Electronics, *43(8)*: 939-944.

Pavelyev A, Volkov AV, Zakharov AI, Krytikh SA, and Kucherjavenkov AI (1996) Bistatic radar as a tool for earth investigation using small satellites. Acta Astronautica *39*: 721-730.

Pavelyev AG, Liou YA, Huang C Y, Reigber C, Wickert J, Igarashi K, Hocke K (2002) Radio holographic method for the study of the ionosphere, atmosphere and terrestrial surface using GPS occultation signals. GPS Solutions *(No. 6)*: 101-108.

Steiner AK, Kirchengast G, and Landreiter HP (1999) Inversion, error analysis, and validation of GPS/MET occultation data. Ann Geophys *17*: 122-138.

Vorob'ev VV, Gurvich AS, Kan V, Sokolovskiy SV, Fedorova OV, and Shmakov AV (1999) Structure of the Ionosphere from the Radio-Occultation GPS-"Microlab-1' Satellite Data: Preliminary Results. Earth Observations and Remote Sensing *15*: 609-622.

Wickert J et al. (2004) The radio occultation experiment aboard CHAMP: Operational data analysis and validation of atmospheric profiles. J Meteorol Soc Japan *82(1B)*, Special issue 'Application of GPS Remote Sensing to Meteorology and Related Fields': 381-395.

Wickert J et al. (2001) Atmosphere sounding by GPS radio occultation: First results from CHAMP. Geophys Res Lett *28*: 3263-3266.

# About the Potential of GPS Radio Occultation Measurements for Exploring the Ionosphere

Norbert Jakowski[1], Konstantin Tsybulya[1], Stanimir M. Stankov[1], and Andreas Wehrenpfennig[2]

[1] Deutsches Zentrum für Luft- und Raumfahrt, Institut für Kommunikation und Navigation, Kalkhorstweg 53, D-17235 Neustrelitz, Deutschland, *Norbert.Jakowski@dlr.de*
[2] Fachhochschule Neubrandenburg, Brodaer Str. 2, D-17033 Neubrandenburg, Deutschland

**Summary.** The GPS radio occultation technique onboard LEO satellites such as CHAMP is a rather simple and relatively inexpensive tool for profiling the ionospheric electron density from satellite orbit heights down to the bottomside. The paper addresses capabilities of the ionospheric radio occultation (IRO) technique for globally monitoring the ionosphere on a routine basis to derive value added data products and study various ionospheric processes including perturbations. A model assisted retrieval technique, operational data processing and validation of vertical electron density profiles are also discussed. These profiles may not only be used to validate ionospheric models, they provide also a good data basis for developing new models of key ionospheric parameters such as the critical frequency foF2, the peak height hmF2 and the scale height Hs. Such models are helpful to improve retrieval procedures and tomographic reconstruction techniques. Due to the operational data processing capabilities the data products may contribute to space weather monitoring.

**Key words:** Ionosphere, GPS, radio occultation, electron density, total electron content

## 1 Introduction

Low Earth Orbiting (LEO) satellites are capable of monitoring the ionospheric ionization on a global scale. This has been demonstrated already by several satellites such as Microlab-1 with the GPS/MET experiment (e.g. Hajj and Romans, 1998, Schreiner et al., 1999), Oerstedt, CHAMP (Reigber et al., 2000, Jakowski et al., 2002) or SAC-C. In this paper we review the capabilities of Ionospheric Radio Occultation (IRO) measurements onboard GPS equipped LEO satellites for probing the ionosphere. This is illustrated by presenting results obtained from analysis of CHAMP data providing electron density profiles.

## 2 Observations and Data Processing

After receiving the CHAMP GPS data in the DLR Remote Sensing Data Center Neustrelitz they are immediately processed by an operational data processing system (Wehrenpfennig et al., 2001). For retrieving the vertical electron density profiles (EDP) from IRO measurements a model assisted technique is applied (Jakowski et al., 2002). According to the requirements of potential users in science and space weather applications the deduced EDPs are available within 3 hours after data dump. Subsequently, the computed ionospheric radio occultation (IRO) data products are submitted to the international science community via the Information and Science Data Center (ISDC) of GFZ Potsdam. From about 200 IRO measurements onboard CHAMP, about 150 EDPs are successfully retrieved every day. More than 80000 profiles have been collected so far, forming a powerful data base for more detailed studies. In parallel with these first studies, validation work has to be continued to gain more detailed knowledge on the accuracy and reliability of IRO data ( Jakowski et al., 2004a).

## 3 Global Monitoring of Ionospheric Key Parameters

LEO satellites in near polar orbits may provide all parameters on a global scale. This is a big advantage compared with other techniques that provide only local (ionosondes) or regional (incoherent scatter) information of the vertical electron density structure. Key ionospheric parameters are the peak electron density NmF2 and the corresponding height hmF2. NmF2 is correlated with the critical frequency (foF2) via the relation NmF2 = $0.0124\times(\text{foF2})^2$ in SI units. Fig. 2 clearly indicates the strong latitudinal dependence of the ionospheric ionization and the so-called seasonal anomaly in the peak density as well as in the ionospheric total electron content (ITEC). ITEC is the integrated electron density profile calculated up to 1000 km height taking into account that the densities above the CHAMP altitude contribute via the adjusted model used in the retrieval technique.

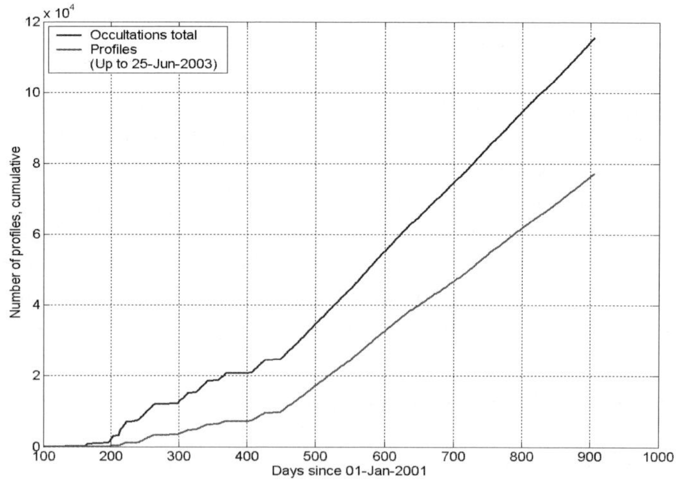


**Fig. 1.** Cumulative number of observations (upper curve) and EDP retrievals (lower curve) from 11 April 2001 till 25 June 2003.

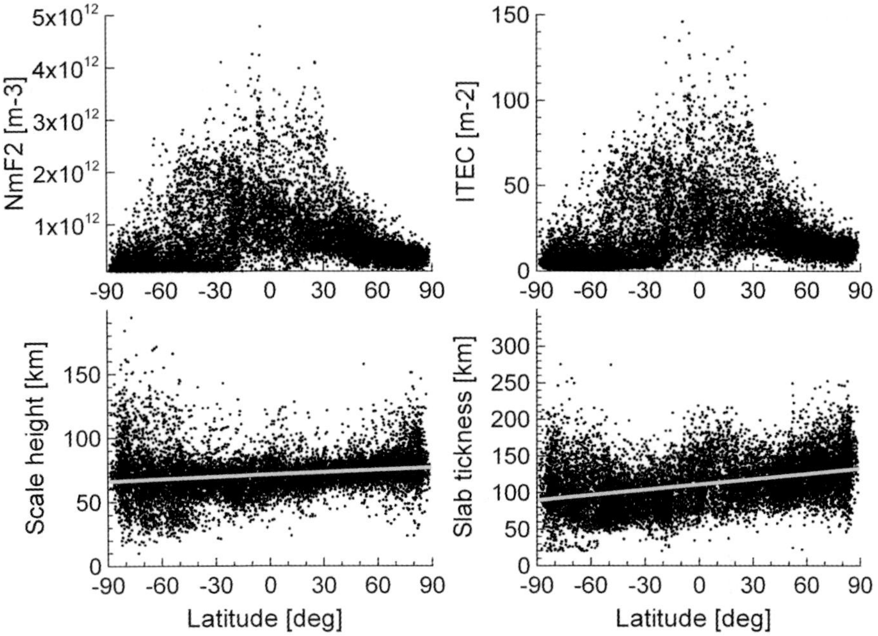


**Fig. 2.** Latitudinal dependence of the peak density NmF2 (top left), ITEC (top right), Hs(425) (bottom left), and bottomside equivalent slab thickness $\tau_b$ (bottom right) during north summer (DOY:121-222). Local time and longitude dependencies are ignored.

Whereas seasons were separated in Fig. 2, local time and longitude dependencies are not considered. It is interesting that, independently from season, the northern crest is much more pronounced than the southern one. This observation needs further investigation for possible local time and longitude dependencies.

The knowledge of the vertical electron density distribution enables us to extract the plasma scale height Hs describing the EDP shape as a function of the plasma temperature in diffusive equilibrium conditions. To get a first-order estimation of the plasma scale height, we extracted the scale height at 425 km height being aware that a certain influence of the adjusted topside ionospheric model still exists. More accurate estimations of the scale height will be obtained at a later stage after more comprehensive validation. Nevertheless, these preliminary estimations will help us to improve the retrieval process. The scale height at 425 km altitude, Hs(425), is plotted in Fig. 2 in comparison with the equivalent slab thickness $\tau_b$ = TEC(hmF2)/NmF2 of the bottom side electron density profile. What is apparently seen is a slight increase of the scale height and the slab thickness towards the summer hemisphere. This fact can easily be explained by enhanced plasma temperatures due to higher solar energy input at the summer hemisphere. The similar behaviour of the bottomside slab thickness $\tau_b$ and the topside scale height is probably the reason why it is possible to extrapolate bottomside scale height to the topside for estimating the EDP above the F2 layer height (Belehaki and Tsagouri, 2002).

## 4 Ionospheric Perturbations

Ionospheric perturbations are far from being understood in detail. IRO measurements are able to provide valuable information on the vertical distribution of ionization perturbations driven by highly dynamical forces such as neutral winds and electric fields. The left panel in Fig. 3 indicates pronounced wavelike structures in the IRO TEC data. Because the retrieval algorithm greatly smoothes out the structures, we prefer studying these phenomena by analyzing the TEC data derived from differential GPS carrier phases. The right panel in Fig. 3 shows the global distribution of wavelike structures similar to the ones shown in the left panel. The strength of these structures is evaluated by an arbitrary scale indicating high dynamics at high scale values. Considering the observations during northern summer (May–July 2002), we find a strong enhancement at southern high latitudes i.e. during winter. In addition, enhanced traveling ionospheric disturbances (TID) activity is also observed close to the geomagnetic equator as expected.

The observed pronounced activity in winter at high latitudes indicates eventually a tracing of atmospheric gravity waves (AGW's) coming from lower atmosphere layers. Lower temperatures during winter may modify the atmospheric filter function in such a way that AGWs may penetrate upwards more favorably than under summer conditions.

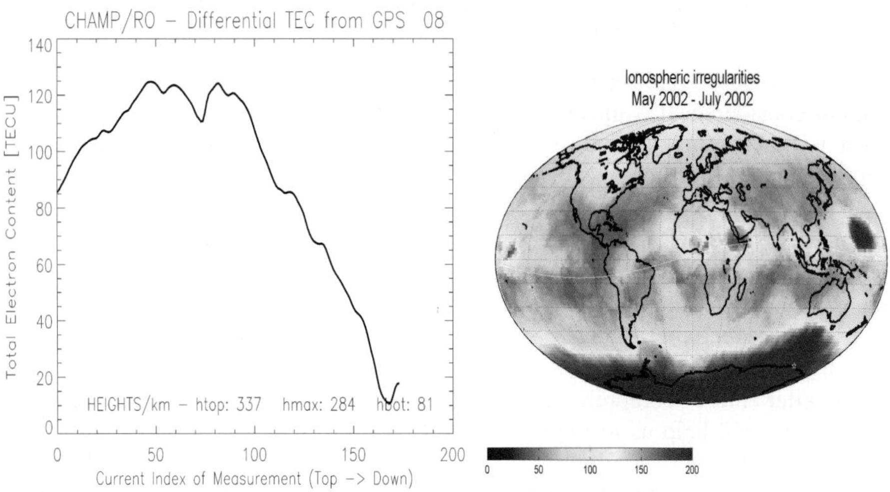


**Fig. 3.** Left panel: Relative TEC data (not calibrated) as observed during an individual occultation event (here with GPS satellite PRN 8). Right panel: Map of the global TID distribution during northern summer, May- July 2002.

## 5 Development and validation of models

The continuously growing number of observation data is an excellent basis for evaluating and developing ionospheric models (Jakowski et al., 2004b). By averaging electron density profiles under certain geophysical conditions such as season, local time, latitude or longitude, one gets information about the ionospheric climatological behaviour as described in models. To support such activities, the computation of value added products is foreseen. Fig. 4 gives an impression of such a product that includes averaged electron density profiles as a function of geographic latitude at 15°E (coincidence radius 10°). In nighttime conditions, there is an indication of a strong latitudinal dependence not only in the ionization as expected from Fig. 2, but also in the profile shape. These preliminary results shall not be overestimated but what can be seen clearly, is a strong geographical asymmetry of the profile shape at 15°E indicating a significant northward shift of the geomagnetic equator and the corresponding low latitude crest. Getting global information on vertical EDPs is an advantage of the IRO method. If validation is good, such studies may indeed lead to new findings in ionospheric physics, e.g. the global development of vertical redistribution of storm-time ionospheric plasma may be studied in a new more effective way.

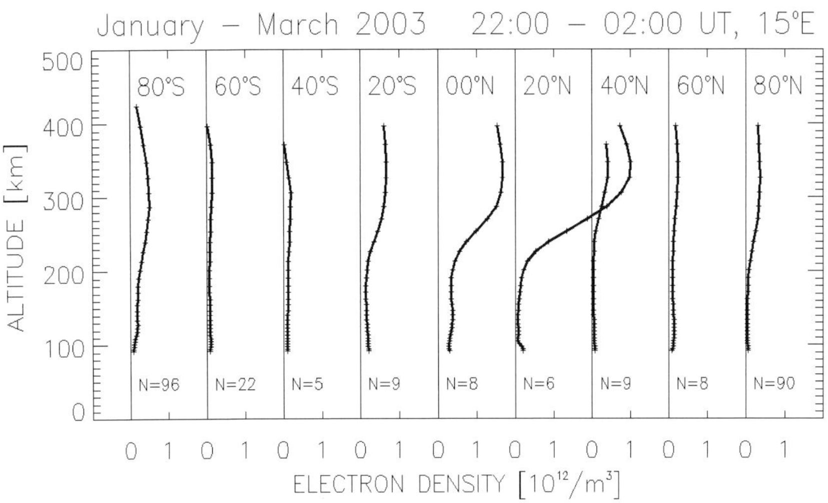


**Fig. 4.** Averaged night-time EDPs obtained from January–March 2003 along the 15°E meridian choosing a coincidence radius of 10 degrees.

# 6 Conclusions

It has been shown that the reception of GNSS signals onboard LEO satellites provides a powerful tool for sounding the vertical structure of the ionosphere on a global scale (for CHAMP - about 150 EDPs per satellite and day). Ionospheric parameters such as NmF2 (foF2), hmF2, and ITEC may easily be monitored globally. The same is valid for the EDP shape parameters scale height and slab thickness. Although the validation work is not yet finished, it can be concluded that CHAMP IRO data clearly indicate the winter anomaly effect of the F2 layer at high solar activity conditions. Since the vertical structure of TIDs may be detected and analysed, there is a good potential for new findings in our understanding of the perturbation mechanisms. Value added data products contribute for developing and testing ionospheric models. Furthermore, it is evident that IRO observations provide valuable information for ionospheric tomography.

# References

Belehaki A, Tsagouri I (2002) Investigation of the relative bottomside/topside contribution to the Total Electron Content estimates. Annals of Geophysics *45(1)*: 73-86.

Hajj GA, Romans LJ (1998) Ionospheric electron density profiles obtained with the Global Positioning System: Results from the GPS/MET experiment. Radio Sci *33*: 175-190.

Jakowski N (1996) TEC Monitoring by Using Satellite Positioning Systems. In: Modern Ionospheric Science, eds. Kohl H, Rüster R, Schlegel K, EGS, Katlenburg-Lindau, ProduServ GmbH Verlagsservice, Berlin, pp 371-390.

Jakowski N, Wehrenpfennig A, Heise S, Reigber Ch, Lühr H, Grunwaldt L, Meehan T (2002) GPS Radio Occultation Measurements of the Ionosphere from CHAMP: Early Results. Geophys Res Lett *29*: No. 10, 10.1029/2001GL014364.

Jakowski N, Tsybulya K, Mielich J, Belehaki A, Altadill D, Jodogne JC, Zolesi B (2004a) Validation of GPS Radio Occultation Measurements on CHAMP by Vertical Sounding Observations in Europe. This issue.

Jakowski N, Tsybulya K, Radicella S M, Cueto M, Herraiz M (2004b) Comparison of electron density profiles from CHAMP data with NeQuick model. This issue.

Reigber Ch, Lühr H, Schwintzer P (2000) CHAMP mission status and perspectives. Suppl. to EOS, Transactions, AGU *81*: 48, F307.

Schreiner W S, Sokolovskiy S V, Rocken C (1999) Analysis and validation of GPS/MET radio occultation data in the ionosphere. Radio Sci *34*: 949-966.

Wehrenpfennig A, Jakowski N, Wickert J (2001) A Dynamically Configurable System for Operational Processing of Space Weather Data. Phys Chem Earth *26*: 601-604.

# Validation of GPS Ionospheric Radio Occultation results onboard CHAMP by Vertical Sounding Observations in Europe

Norbert Jakowski[1], Konstantin Tsybulya[1], Jens Mielich[2], Anna Belehaki[3], David Altadill[4], Jean-Claude Jodogne[5], and Bruno Zolesi[6]

[1] Deutsches Zentrum für Luft- und Raumfahrt (DLR) / Institut für Kommunikation und Navigation, Kalkhorstweg 53, D-17235 Neustrelitz, Germany, *Norbert.Jakowski@dlr.de*
[2] Institute of Atmospheric Physics, University of Rostock, Juliusruh
[3] Institute for Space Applications and Remote Sensing, National Observatory of Athens, Metaxa and Vas. Pavlou str., Palaia Penteli 15236, Greece
[4] Observatori de l'Ebre, 43520 Roquetes, Tarragona, Spain
[5] Royal Meteorological Institute, Avenue Circulaire 3, B-1180 Brussels, Belgium
[6] Istituto Nazionale di Geofisica e Vulcanologia, Rome, Italy

**Summary.** Ionospheric radio occultation (IRO) measurements have a big potential for monitoring the ionospheric behavior on global scale for now- and forecasting the ionospheric impact on radio systems. In this article we validate the retrieved vertical electron density profiles (EDPs) derived from IRO measurements onboard CHAMP by using vertical sounding measurements at five European vertical sounding stations - Athens, Dourbes, Juliusruh, Rome and Tortosa. Since first IRO measurement onboard CHAMP in April 2001, more than 70000 electron density profiles have been retrieved by a model assisted technique so far. The comparison of IRO retrieved EDPs with ionosonde profiles obtained from the above mentioned stations will be discussed.

**Key words:** Ionosphere, radio occultation, ionosonde, electron density, validation

## 1 Introduction

Vertical electron density profiles (EDPs) from GPS ionospheric radio occultation (IRO) measurements onboard of Low Earth Orbiting (LEO) satellites present a rather new data type (e.g. Jakowski et al, 2002). In view of further application for ionospheric monitoring and research these data need validation to investigate their capacities and restrictions.

In this paper we report preliminary validation results of EDP retrievals based on GPS IRO measurements from CHAMP (Reigber et al., 2000). For this purpose EDPs acquired by vertical sounding (VS) at 5 European ionosonde stations (Athens, Dourbes, Juliusruh, Rome and Tortosa) have been compared with coinciding CHAMP EDPs. To support these activities, a special validation group has been formed last year within the European COST 271 action (www.cost271.rl.ac.uk).

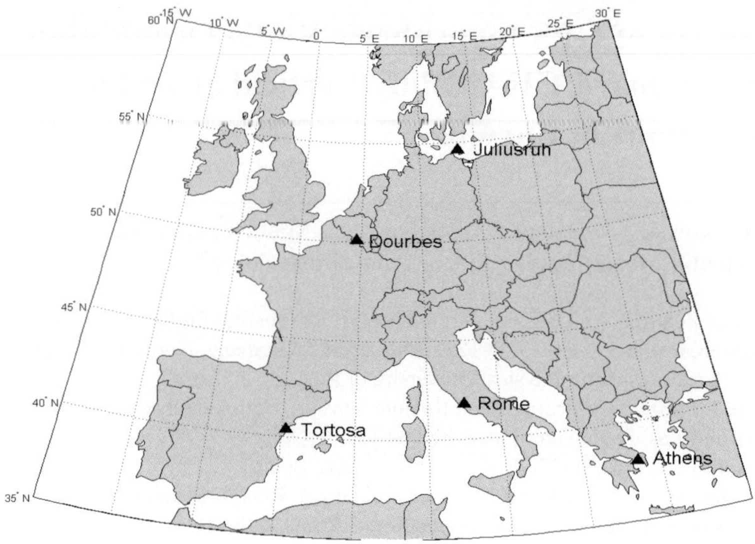


**Fig. 1.** Geographic positions of vertical sounding stations used for IRO EDP validation.

## 2 Data Analysis and Results

The data base of the CHAMP IRO electron density profiles includes more than 70,000 with a sampling period between April, 2001 up to now. However, only 470 of them were found to coincide with the considered ionosonde observations. Coincidence conditions are set spatially within 8°, and temporally within a 15 min time window. The main validation point is based on a statistical analysis over all samples. Therefore, their mean deviation from each other (systematic bias) and corresponding RMS are calculated for electron densities (Ne) and for plasma frequencies (fp) derived from both methods. The two characteristics are handled separately due to their non-linear dependence on each other.

The comparisons were performed in 95–400 km altitude range, in steps of 5 km. Results of the statistical analysis are shown in Fig. 2. The centered line in each panel represents the mean bias surrounded by two lines indicating $1\sigma$ deviation. The comparisons indicate a good general agreement over the whole height range, with a systematic positive bias for the IRO profiles of 0.4–0.5 MHz fp and $0.7 \times 10^{11}$ $m^{-3}$ Ne, respectively. RMS dispersion is uniform over almost all altitude ranges (about 1 MHz fp or $1–2.5 \times 10^{11} m^{-3}$ Ne).

Fig. 3 illustrates typical deviations of VS and coinciding IRO profiles. Sometimes topside profiles may deviate strongly whereas bottomside profiles are almost identical. This deviation can be attributed either to numerical models used in the ionosonde software (e.g. Huang and Reinisch, 2001) or to the model used to estimate the upper ionosphere density in IRO retrieval algorithms (e.g. Jakowski et al., 2002), but considering Fig. 2, these deviations are not statistically significant.

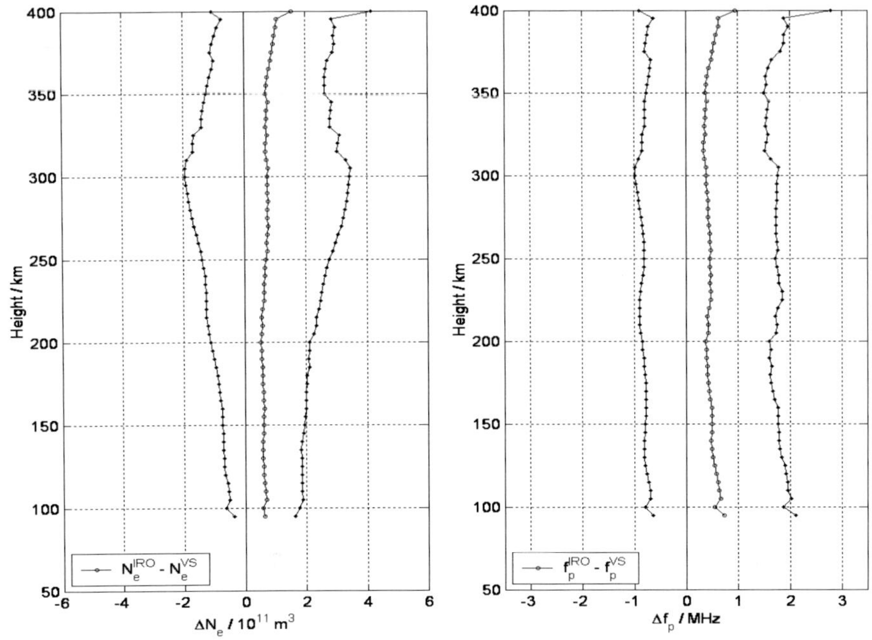


**Fig. 2.** Statistical comparison of IRO and coinciding VS EDPs (474 profiles, Athens, Dourbes, Juliusruh, Rome, Tortosa, Jan. 2002–Mar. 2003). Left: electron density. Right: plasma frequency.

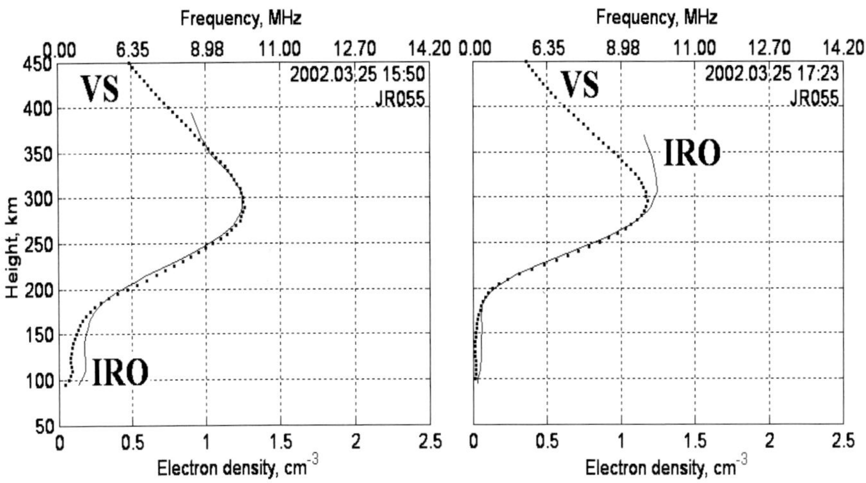


**Fig. 3.** IRO and VS EDP comparison for Juliusruh, Mar. 25, 2002. Very good agreement (left panel) and typical topside deviation (right panel).

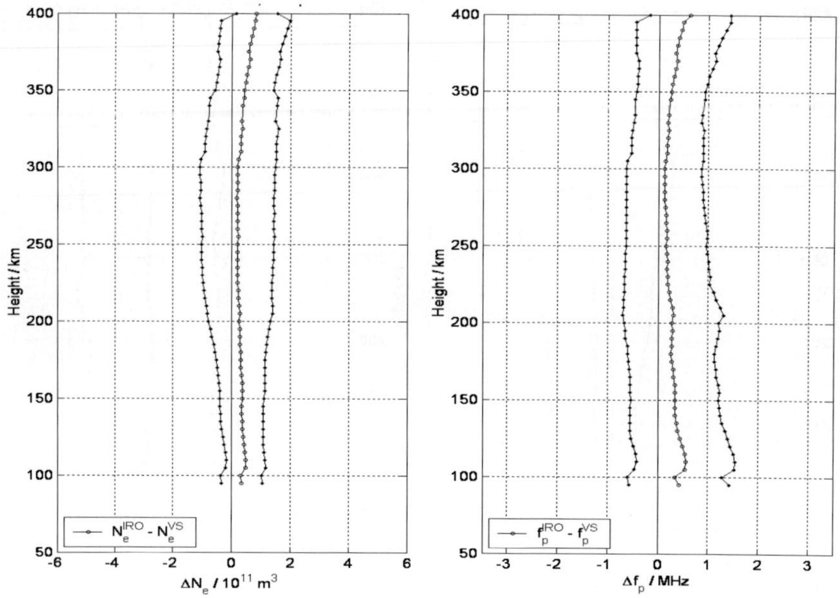


**Fig. 4.** Statistical comparison of IRO and VS profiles from Juliusruh (Jan.2002–Mar.2003, 261 profiles). Left: electron density. Right: plasma frequency.

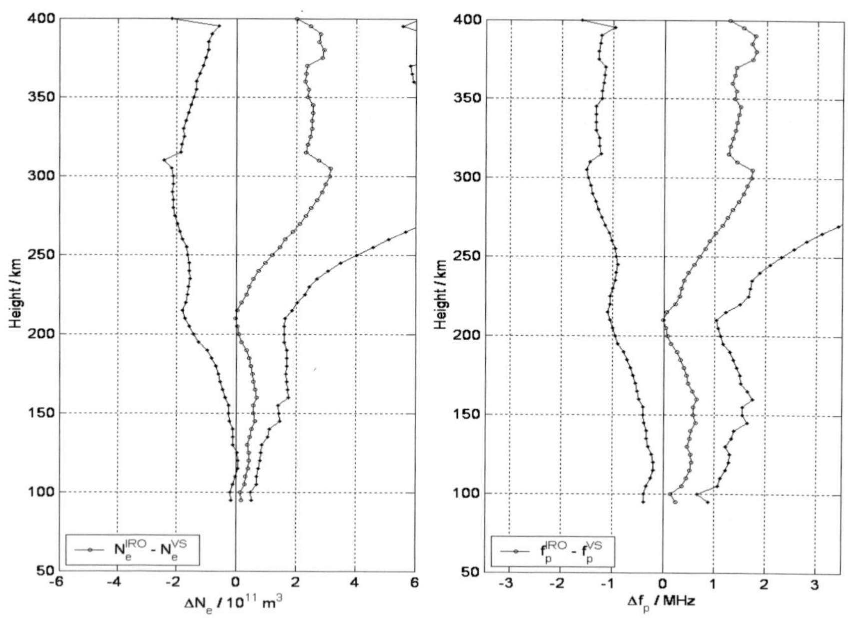


**Fig. 5.** Same as Fig. 4 but for Dourbes station (48 profiles).

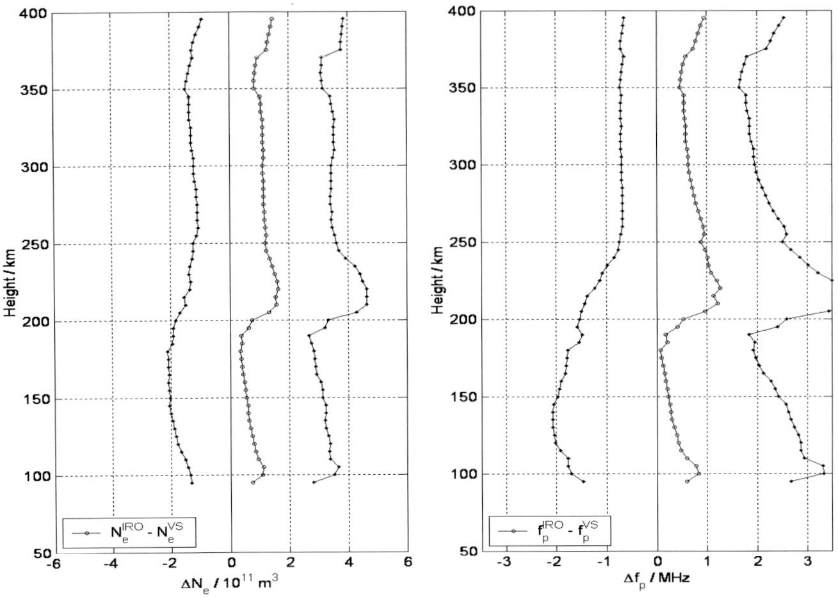


**Fig. 6.** Same as Fig. 4 but for Tortosa station (50 profiles).

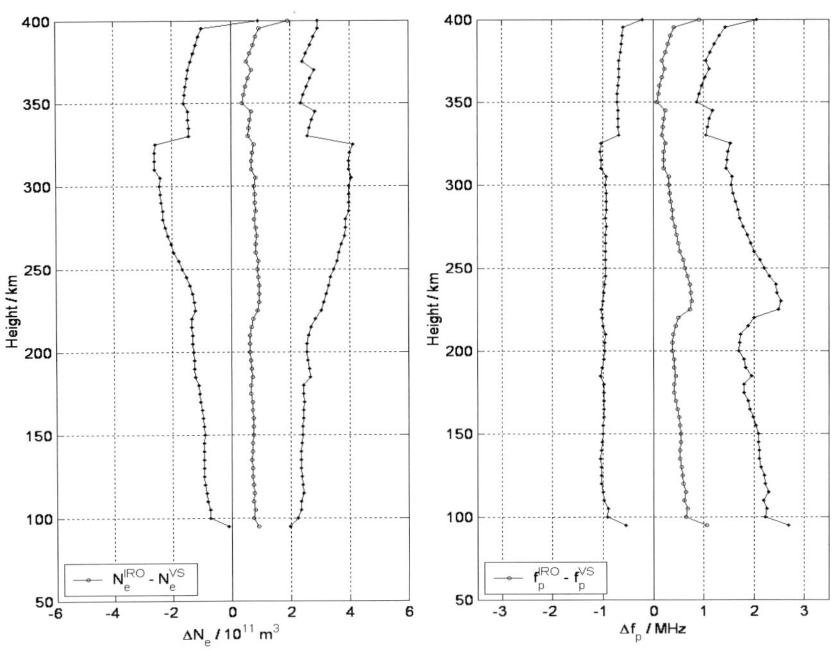


**Fig. 7.** Same as Fig. 4 but for Athens station (93 profiles).

Figs. 4–7 show comparison of the profiles for some ground stations separately. The difference between them is somewhat surprising – from excellent agreement with Juliusruh profiles to greater deviations from Athens and Tortosa data to even greater difference of 2–3 MHz with Dourbes station. Notice that the greatest difference with Dourbes data is at the level of the F2 maximum and higher.

## 3 Conclusions

Although the validation work is far from being finished, we believe that this report can contribute to get a certain understanding of the relationships between averaged IRO derived vertical electron density profiles and local profiles obtained from ground based ionosondes. One has to be aware that these are different data types that cannot fit perfect.

All samples clearly indicate a slight overestimation of the IRO derived data that should be in the order of about 0.5 MHz or even less throughout the entire vertical electron density profile. Though results of comparisons between radio-occultation method results and the classic technique of vertical sounding are rather ambiguous, there are definite signs that discrepancies really are about 1 MHz or even less.

Reasons of high discrepancies of IRO data with some ionosonde station data need more discussion in future work. Comprehensive validation of IRO data has to be confirmed on global scale in particular at high (e.g. Stolle et al., 2003) and low latitudes using a more extended data basis.

*Acknowledgments.* The authors are very grateful to all colleagues from the international CHAMP team who keep CHAMP in operation.

## References

Huang X, Reinisch BW (2001) Vertical total electron content from ionograms in real time. Radio Science *36(2)*: 335-342.

Jakowski N, Wehrenpfennig A, Heise S, Reigber Ch, Lühr H, Grunwaldt L, Meehan T (2002) GPS Radio Occultation Measurements of the Ionosphere from CHAMP: Early Results. Geophys Res Lett *29*: No. 10, 10.1029/2001GL014364.

Reigber Ch, Lühr H, Schwintzer P (2000) CHAMP mission status and perspectives. Suppl to EOS, Transactions, AGU *81*: 48, F307.

Stolle C, Jakowski N, Schlegel K, Rietveld M (2003) Comparison of high latitude electron density profiles obtained with the GPS radio occultation technique and EISCAT measurements. Subm. to Ann Geophys.

# Ionospheric Tomography with GPS Data from CHAMP and SAC-C

Miquel García-Fernández[1], Angela Aragón[1], Manuel Hernandez-Pajares[1], Jose Miguel Juan[1], Jaume Sanz[1], and Victor Rios[2]

[1] gAGE/UPC, Mod C3 - C. N.- Jordi Girona 1-3, 08009 Barcelona, Spain
[2] Physics Department, Univ. Nacional de Tucuman, Argentina

**Summary.** Abel inversion offers a straightforward way to obtain the vertical distribution of electron density with low computational load. Nevertheless the treatment of the electron density above the LEO orbit must not be neglected, specially for satellites with very low orbit such as CHAMP. This work extends previous results obtained by inverting real GPS data from LEO data, coming from satellites such as CHAMP or SAC-C. In this work, the topside ionosphere is modelled using positive elevation data. To overcome the spherical symmetry assumption, occultations are processed with the aid of Vertical Total Electron Content, estimated from ground GPS data or models. The resulting electron density profiles are compared with external real data consisting basically on basic parameters or true-height vertical profiles obtained from ionosonde measurements.

**Key words:** GPS, LEO, Electron density, Occultations, Abel transform, Separability hypothesis

## 1 Introduction

As it is known, Abel inversion techniques are used to obtain high vertical resolution electron density profiles ([3],[4],[9]) which performance may vary between 10% and 20% for foF2 estimations. The assumption of spherical symmetry is not realistic in general, and in particular for occultations where high Vertical Total Electron Content (VTEC) gradients take place. For Low Earth Orbiters (LEO) such as GPS/MET or SAC-C (with nominal orbits of 700km approximately), an exponential extrapolation may be enough to account for the electron content above the LEO [7]. Nevertheless, a more accurate modeling for LEOs at very low orbit such as CHAMP (approx. 400km) is required in order to avoid this assumption leading to incorrect vertical profiles. Therefore, other approaches should be considered. For instance, a modelling of the topside ionosphere with an external model [8]. This work proposes to extend the performance study of two modifications on the classical approach in order to, first, overcome the assumption of spherical symmetry by using horizontal VTEC gradients and, second, model the electron content above the LEO satellite, which becomes crucial for LEOs such as CHAMP, by using LEO GPS observations associated to positive elevation.

The classical approach of Abel inversion assumes spherical symmetry (i.e. electron density only dependent on height) and neglects the electron content above the LEO receiver. Considering that each ray of a radio occultation defines a spherical shell of the atmosphere, the Slant Total Electron Content (STEC) seen by the LEO can be modelled using a discrete form as shown in Eq. 1. Therefore the electron density can be computed in a recursive way starting from the outer ray (the one with greatest distance between ray and earth surface)

$$STEC(p_i) = \sum_{j=1}^{i} 2 \cdot l_{ij} \cdot N_e(p_j) \tag{1}$$

where $l_{ij}$ and $N_e(p_j)$ stand for the ray path length crossing the j-th layer when the i-th observation is carried out and the electron density value at the j-th layer, respectively. In order to overcome the limitations of the assumption of spherical symmetry, an alternative formulation was introduced in [7] and further developed in [2] stating the *separability hypothesis* expressed in Eq. 2.

$$N_e(LT, LAT, H) = VTEC(LT, LAT) \cdot F(H) \tag{2}$$

The vertical profile can be expressed as the product between a VTEC function depending on the latitude, longitude and universal time and a height (H) dependent shape function F, the unknown to be determined through the iterative process of Abel inversion. Moreover, an extra unknown is added to the STEC in order to account for the electron content above the LEO, $F_p$. Therefore, taking into account the STEC expression and the upper ionosphere and plasmasphere contribution, the $L_I \equiv L_1 - L_2$ (GPS geometric free combination) can be expressed as:

$$\begin{aligned} L_I(p_i) &= b_I + \alpha \cdot l_p \cdot VTEC(LT_{ip}, LAT_{ip}) \cdot F_p + \\ &\alpha \cdot \textstyle\sum_{j=1}^{i} l_{ij} \cdot [VTEC(LT_{ij}, LAT_{ij}) + VTEC(LT'_{ij}, LAT'_{ij})] \cdot F(p_j) \end{aligned} \tag{3}$$

where $\alpha = 1.05 m_{L_I} / 10^{17} e/m^2$, $LT_{ip}$ and $LAT_{ip}$ are the slab plasmaspheric pearce point coordinates and $l_p$ the corresponding length, $VTEC(LT_{ij}, LAT_{ij})$ and $VTEC(LT'_{ij}, LAT'_{ij})$ are the VTECs at the two locations of the j-th spherical layer illuminated by the i-th ray. The information about the VTEC gradients has been obtained by means of a data driven model provided by the Technical University of Catalonia (UPC) in IONEX format ([1] and [6], using the GPS ground receiver network of the Interational GPS Service (IGS). In the data pre-processing a cycle-slip detection is performed, nevertheless a carrier phase ambiguity estimation is required. The extra unknown $b_I$ accounts for the carrier phase ambiguities and clock biases. If the occultation shows large number of cycle slips in the negative elevation observations, it is not possible to estimate all ambiguities and the inverted profile is likely to be unrealistic. Therefore, a manual check of the profiles is required as well as a previous step to validation with ionosondes.

## 2 Results with Separability Hypothesis

### 2.1 Scenario

The study performed comprises a period of 16 days, from the 1st to the 16th of November 2002. Abel inverted estimations of electron densities and heights are carried out for CHAMP and SAC-C data for these days. In order to evaluate them, the occultations have been checked with the measured parameters provided by ionosondes that are mainly placed in mid latitudes (the ionosonde data, basically autoscaled for the period of this dataset, has been obtained from the SPIDR web server). As a general procedure for comparison, it has been considered that for a single occultation and ionosonde the valid comparisons were those made with the ionosonde measurements comprising an interval of 1 hour centered at the epoch that the occultation took place. Moreover, the maximum co-location distance between an ionosonde and the occultation has been set to 2000km.

The separability hypothesis requires a constant slab thickness during the occultation ($\tau = VTEC/NmF2$, with NmF2 as the electron density peak over the ionosonde location), that is a proportional relationship between the NmF2 and VTEC. Therefore, the dispersion of the slab thickness as a function of the local time has been studied for checking the consistency of the measured values of peak density of the ionosondes and VTEC computed from the GPS ground data. A statistical filter of $\tau$ (bias and standard deviation, $\sigma$, for local time) has been designed and outliers (values outside $3\sigma$) have been discarded. This filter basically rejects extremely large/low values of slabthickness and peaks in its temporal variation. Around 10% of measurements are discarded by this filter. Note that the shape function at the peak of the profile is in fact the inverse of the slab thickness. When the slab thickness does not show a constant behaviour, the VTEC will not be able to describe the variations of the electron density, leading to an increase of the error in the estimation.

### 2.2 foF2 estimation

Table 1 summarizes the performance regarding the comparison of critical frequencies of the F2 layer (i.e. foF2) provided by the ionosondes with spherical symmetry and separability hypothesis Abel transforms of CHAMP and SAC-C data. These results are distinguished through three moments along the day: Day, Dawn/Dusk (D&D) and Night. This distinction is necessary because the D&D period is characterized for a high variability in the profile (both in density and height). For both satellites, the results show that the performance with respect to ionosonde measurements significantly improve under the separability assumption in all cases (about 40%). It is remarkable that under both assumptions, the worst results are obtained at night time, when the foF2 maximum decreases. During this period, as well as during D&D, the

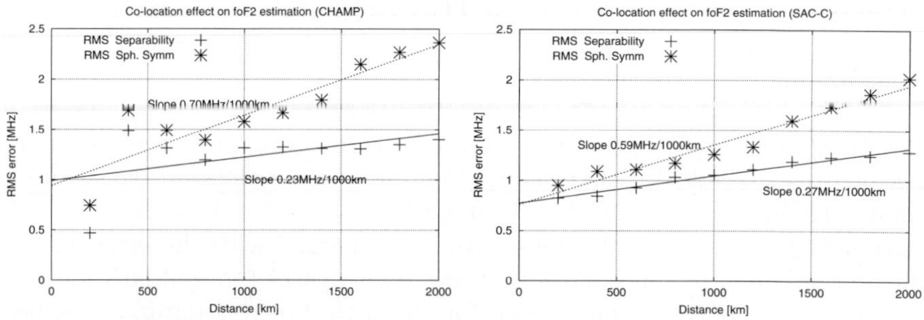


**Fig. 1.** Effect of the collocation distance to the absolute RMS error of foF2 estimation with respect to Ionosonde measurements: CHAMP vs Ionosondes (left) and SAC-C vs Ionosondes (right). The large absolute error at small distance (in the left plot) is explained by a comparison with high error, while the number of comparisons is still low. As distance increases, the number of comparisons increases as well and the effect of this occultation is mitigated and masked by the effect of the co-location distance.

**Table 1.** CHAMP and SAC-C results for foF2 estimation

| | | nr. compar. | Abs. Err. [MHz] Separability | Rel. Err. [%] Separability | Rel. Err. [%] Classic Abel |
|---|---|---|---|---|---|
| CHAMP | Day | 1966 | 1.40 | 14.4 | 24.3 |
| vs | Dawn/dusk | 189 | 1.14 | 21.5 | 42.0 |
| Ionos. | Night | 879 | 1.55 | 28.4 | 54.6 |
| SAC-C | Day | 5868 | 1.29 | 12.5 | 19.6 |
| vs | Dawn/dusk | 529 | 0.97 | 18.1 | 27.6 |
| Ionos. | Night | 1946 | 1.29 | 25.4 | 41.4 |
| CHAMP | Day | 286 | | 13.6 | 22.0 |
| vs | Dawn/dusk | 26 | | 23.6 | 52.0 |
| SAC-C | Night | 143 | | 23.2 | 39.6 |

slab thickness variations can be coped with to a certain extend by the separability hypothesis, thus providing better results than from spherical symmetry. Notice that the relative error for CHAMP is greater than for SAC-C at any of the day time periods. This is due to the difficulty of modelling the ionosphere above the LEO. Moreover, the increase of distance between the ionosonde and the occultation becomes a source of error, as shown in Fig. 1, being more important in the case of classical Abel inversion.

### 2.3 Upper ionosphere and plasmasphere estimation

In order to account for the electron density above the LEO, the proposed method includes an extra unknown of the shape function ($F_p$) as shown in

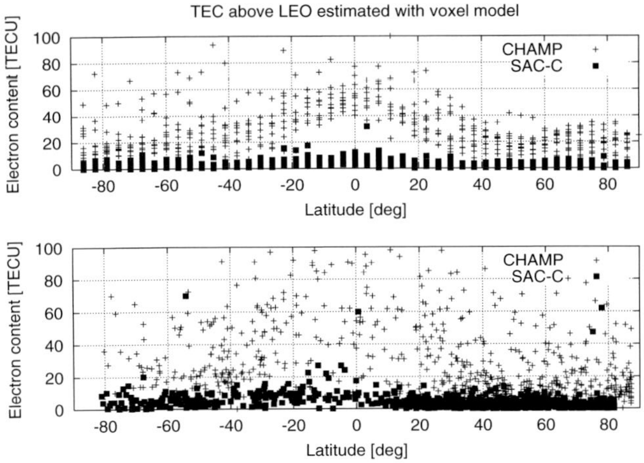


**Fig. 2.** Electron content estimation above LEO orbit: comparison between voxel model (top) and the independent estimates performed in each occultation (bottom).

Eq. 3. To check the validity of the approach, the upper ionosphere and plasmasphere estimation has been compared to a voxel model [5]. Fig. 2 shows the electron content over CHAMP and SAC-C for the 29th of October 2002 estimated with the voxel model using all Precise Orbit Determination (POD) antenna data simultaneously (with an elevation mask of $10^o$, upper plot), and computed as an extra unknown with no aprioris where occultations have been processed separately (bottom plot). It can be seen that the estimations using both approaches are compatible.

## 3 Summary and Conclusions

This study has been focused on the implementation of shape functions using the CHAMP and SAC-C data set which take into account the height dependency in the electron density expression. This approach is an improvement over the classical spherical symmetry assumption. From comparing with ionosondes and intercomparison between CHAMP and SAC-C, it has been shown that the error due to co-location is significantly reduced by estimating the electron density profile using shape functions, as a consequence, the frequency estimation performance is better (average value 30%). With the proposed method, the upper ionosphere contribution does not affect significantly the electron density estimates.

*Acknowledgement.* We are grateful to the IGS, UCAR and the GeoForschungsZentrum Potsdam for providing the ground GPS data, the SAC-C data and the CHAMP observations respectively. The ionosonde data have been extracted from *SPIDR*, http://spidr.ngdc.noaa.gov. This work has been partially supported by the *Generalitat de Catalunya* under the fellowship 2000FI00395 and the Spanish projects TIC-2000-0104-P4-03 and TIC-2001-2356-C02-02.

## References

1. Feltens J, Schaer S (1998) IGS products for the ionosphere. Proc IGS Analysis Center Workshop, ESA/ESOC Darmstadt, Germany, pp. 225–232.
2. García-Fernández M, Hernandez-Pajares M, Juan JM, Sanz J (2003) Improvement of ionospheric electron density estimation with GPSMET occultations using Abel inversion and VTEC information. J Geophys Res *108(A9)*: 1338.
3. Hardy K, Hajj GA, Kursinski E, and Ibañez-Meier R (1993) Accuracies of atmospheric profiles obtained from GPS occultations. Proc ION GPS-93 Conference, pp 1545–1556.
4. Hajj GA, Romans LJ (1998) Ionospheric electron density profiles obtained with the Global Positioning System: Results from the GPS/MET experiment. Radio Science *33(1)*: 175–190.
5. Hernandez-Pajares M, Juan JM, Sanz J (1998) Global observation of the ionospheric electronic response to solar events using ground and LEO GPS data. J Geophys Res *103(A9)*: 20,789–20,796.
6. Hernandez-Pajares M, Juan JM, Sanz J (1999) New approaches in global ionospheric determination using ground GPS data. J Atmos Solar Terr Phys *61*: 1237–1247.
7. Hernandez-Pajares M, Juan JM, Sanz J (2000) Improving the Abel inversion by adding ground data LEO radio occultations in the ionospheric sounding. Geophys Res Lett *27(16)*: 2743–2746.
8. Jakowski N, Wehrenpfennig A, Heise S, Reigber C, Lühr H, Grunwaldt L, and Meehan TK (2002) GPS radio occultation measurements of the ionosphere from CHAMP: Early results. Geophys Res Lett *29(10)*: 1457, doi:10.1029/2001GL014364.
9. Schreiner WS, Sokolovskiy SV, Rocken C, Hunt DC (1999) Analysis and validation of GPS/MET radio occultation data in the ionosphere. Radio Science *34(4)*: 949–966.

# Topside Plasma Scale Height Modelling Based on CHAMP Measurements: First Results

Stanimir M. Stankov and Norbert Jakowski

Deutsches Zentrum für Luft- und Raumfahrt (DLR) - Institut für Kommunikation und Navigation (IKN), Kalkhorstweg 53, D-17235 Neustrelitz, Deutschland

**Summary**. Presented are first results in retrieving, analysing, and modelling the topside plasma scale height by using CHAMP ionospheric radio occultation observations. The plasma scale height value in the region situated immediately above the ionospheric F2-layer density peak, is very important for the TEC calculation and plasma density reconstruction procedures based on GPS radio occultation measurements. Based on the year-long time series data accumulated so far, obtained were latitudinal, diurnal, and seasonal variations of the topside plasma scale height. Considering the growing CHAMP measurement data base, it is possible to develop a new empirical model to be used for improving the process of electron density profile retrieval by delivering an improved initial guess of the topside electron density profile.

**Key words:** Scale Height, Occultation Method, Empirical Modelling

## 1 Introduction

One of the most important characteristics of the ionosphere-plasmasphere system is the plasma scale height, defined as $H_P=kT_P/m_ig$ , where $T_P=0.5(T_i+T_e)$ is the plasma temperature, $T_i$ and $T_e$ – ion and electron temperatures, k – Boltzmann's constant, $m_i$ – ion mass, and g – acceleration due to gravity. Considering its dependence on the plasma temperature and composition, obviously $H_P$ varies with altitude. The initial assessment of the topside ionosphere plasma scale height and its intrinsic connection with the topside electron density profile (EDP) is of crucial importance for the reliability of various GPS TEC calculation and plasma density reconstruction techniques (Heise et al., 2002; Stankov et al., 2003). The CHAMP satellite (Reigber et al., 2003), orbiting the Earth in the altitude region from 450 km down to about 300 km, provides excellent opportunities for observation of the topside ionosphere on a global scale, including the plasma scale height.

Based on CHAMP ionospheric radio occultation (IRO) measurements (Jakowski et al., 2002), presented here are first results in retrieving, analysing, and modelling the topside plasma scale height value. Latitudinal, diurnal, and seasonal variations of the plasma scale height have been already obtained. Considering the steady growing CHAMP measurement data base, it will be possible to develop a new empirical model of the plasma scale height to be used in several applications such as IRO-based EDP retrieval by delivering an improved initial guess or various studies of the ionosphere-plasmasphere composition and dynamics.

## 2 Scale Height Retrieval

The GPS ionospheric radio occultation (IRO) measurements, carried out onboard CHAMP, started on 11 April 2001. The dual frequency carrier phases of the GPS signals are used to compute the total electron content (TEC) along the 1 Hz sampled occultation ray paths. To enable consideration of the horizontal gradients, particularly during ionospheric storms and/or near the crest region, a tomographic approach (Jakowski et al., 2002) is utilised for retrieving the vertical electron density profile instead of the widely used Abel inversion technique.

An occultation event is defined by a series of TEC measurements along ray paths traversing the ionosphere with tangential heights of these ray paths decreasing to the bottom of the ionosphere. Constructed is a system of linear equations for the electron density in the different shells. If the electron density distribution above the top shell (traversed by the first ray of the occultation event) is available a priori (first guess), e.g. from a model, then the electron density in this top shell can be deduced. Hence, starting from the top shell, the electron density in each shell below can be successively determined. Finally, the solution of the above-mentioned system provides the IRO-based vertical electron density profile.

Considering the CHAMP decaying orbit from 454 km at the beginning of the mission down to approximately 300 km at the end of the projected 5 year lifetime, special care is required when determining the upper boundary condition (at the CHAMP altitude). This is due to the fact that the above-lying ionosphere and plasmasphere can contribute with up to estimated 50% of the TEC. To overcome this problem, the inversion is assisted by an adaptive electron density model of the topside ionosphere and plasmasphere (TIP):

$$N_e(h) = N_mF_2 \exp(0.5(1 - z - \exp(-z))) + N_{P0} \exp(-h / H_{PP})$$

where h is the altitude, $N_e(h)$ is the electron density, $z = (h - h_mF_2) / H_{PI}$ , $N_mF_2$ and $h_mF_2$ are the peak electron density and height, $H_{PI}$ is the plasma scale height in the topside ionosphere, $H_{PP}$ is the plasma scale height (fixed at a constant value of 10000 km) in the plasmasphere, and $N_{P0}$ is the electron density at the plasmasphere basis. In the occultation retrieval, the free parameters $N_mF_2$, $h_mF_2$ and $H_{PI}$ are adjusted iteratively, starting from some plausible values. The iteration process deliver a 'smooth' transition from TIP electron densities to the values computed from IRO data. This procedure yields directly the plasma scale height at the upper boundary of the retrieved electron density profile. A comparison between measurements of $N_mF_2$ and $h_mF_2$ deduced from IRO-based profiles and from ground ionosondes shows standard deviations of 18% and 13% respectively.

## 3 Scale Height Behaviour

Using observations from the period April 2002 – March 2003, an important insight into the plasma scale height behaviour has been acquired for various latitude, local time, and season conditions. Three seasons are considered - winter, equinox, and summer, defined as 91 day periods centred on the 356, 81 and 264, 173 day of year, with respect to the opposite seasons in both hemispheres. Day-time and night-time conditions are investigated using data from variable local time ranges (windows) depending on season: larger 10-hour windows are used for extracting summer day-time and winter night-time values. Diurnal variations are obtained using 1-hour windows but results are not yet reliable due to scarcity of data in some periods. For easier investigation of the temporal variations, 3 main geomagnetic-latitude regions are defined (both hemispheres) - low (0°-30°), middle (30°-60°), and high (60°-90°). In this study we have used only high-quality values of the retrieved scale height (internal consistency check applied). To facilitate the analysis, plotted (Figures 1-3) are the median values (solid circles) together with the number of observations on which the corresponding median is based (vertical bars).

### 3.1 Latitudinal variations

To better understand the latitudinal behaviour of the topside plasma scale height, the observations have been sorted into 10°-wide latitude bins with respect also to the above-defined night-time / day-time conditions and the 3 seasons (Fig. 1).

During night-time, the plasma scale height tends to increase at higher latitudes. The increase is most obvious in summer when the median values range between 60 km and 74 km. In winter, the maximum seems to occur at about ±60° latitude.

During day-time, the latitudinal increase is generally preserved but with a few exceptions. In equinox, the values increase from 63 km up to 70 km. In summer, the scale height is higher – it averages 65 km over the equator and increases up to the record 74 km over the poles. An interesting case is the winter day-time distribution – the median values in the equatorial region are around 68 km and are of the same magnitude as the median values over the poles. At the same time, a pronounced decrease is observed in the 45°-55° latitude band where the scale heights drop even below the 50 km mark. This phenomenon is probably due to the ion trough and/or serious violation of the diffusive equilibrium conditions, e.g. when strong vertical plasma fluxes occur (Jakowski et al., 1981).

### 3.2 Diurnal variations

Due to correlation with the plasma temperature, the plasma scale height is expected to be higher during day and lower during night. Although this is true for summer-time conditions (Fig. 2, top right), the diurnal behaviour is actually quite complex. For example, during winter (Fig. 2, top left), in addition to the mid-day peak, two other maxima appear - one in the morning and one in the evening hours.

The data base is limited in some time intervals, so definitive conclusions is too early to make. However, there is an indication of possible correlation between the plasma scale height and the equivalent slab thickness ($TEC/N_mF_2$) measurements. For comparison, median values of the mid-latitude winter-time and summer-time scale height are plotted (Fig. 2, upper panels) together with the corresponding slab thickness estimates over an ionosonde station in the same latitude band (Fig. 2, bottom panels). The correlation is obvious in the winter morning and evening hours, despite the fact that the scale height observations are made during high solar activity and the ionosonde observations are gathered during low solar activity. Substantial differences are detected between the diurnal behaviour of the scale height at different latitudes but additional data are required for a proper analysis.

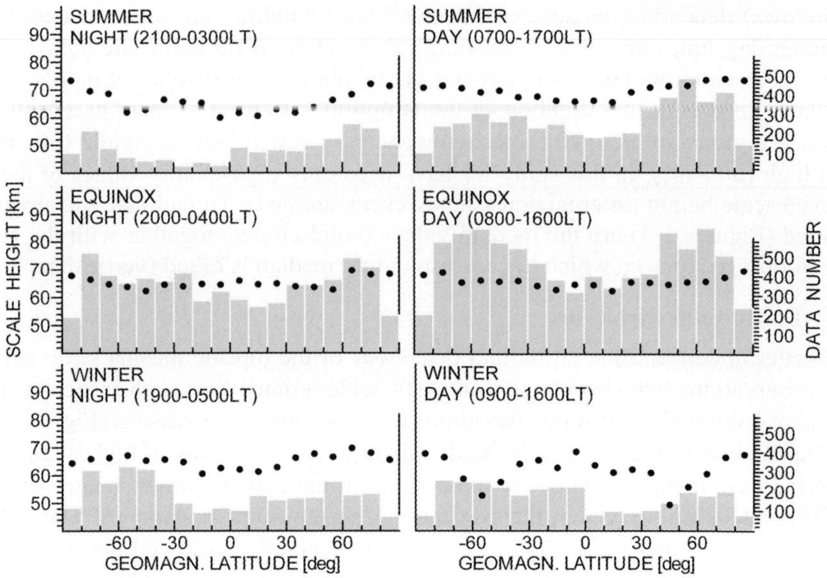


**Fig. 1.** Topside plasma scale height deduced from IRO measurements – observed latitudinal variations in winter (bottom panels), equinox (middle panels), and summer (top panels), during night-time (left) and day-time (right) conditions.

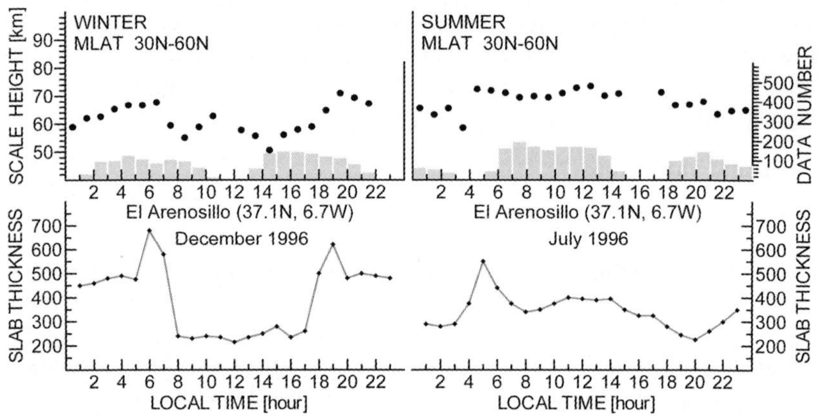


**Fig. 2.** Comparison between diurnal winter- and summer-time variations of the IRO-deduced plasma scale height (top panels) and the equivalent slab thickness (bottom panels) calculations at the El Arenosillo (37.1°N,6.7°W) ionosonde station.

### 3.3 Seasonal (annual) variations

Based on IRO observations from a full year, preliminary estimations of the seasonal scale height behaviour at high solar activity have been obtained. Given here are results (Fig. 3) from the Northern hemisphere observations only.

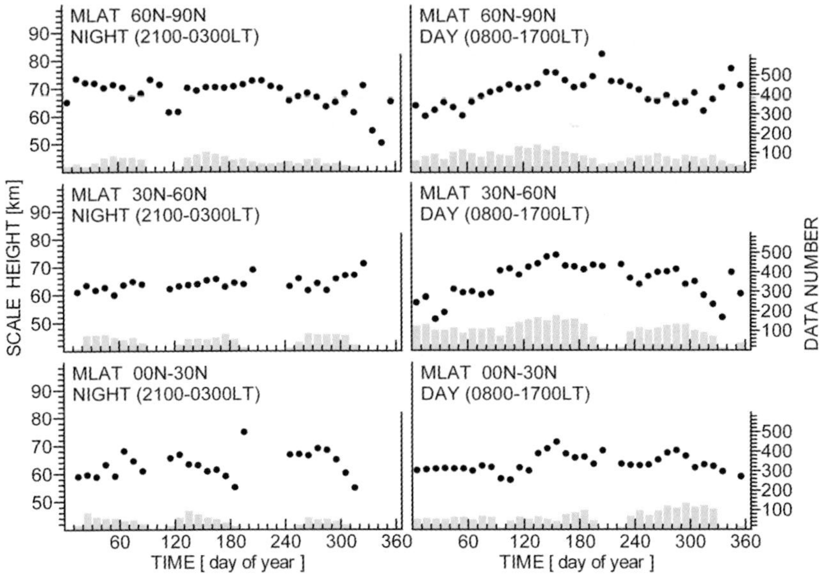


**Fig. 3.** Topside plasma scale height deduced from IRO measurements – observed annual variations at low (bottom panels), middle (middle panels), and high (top panels) latitudes, during night-time (left) and day-time (right) conditions.

Night-time, the seasonal differences in the median values are relatively small. However, the latitudinal increase is obvious – the median, calculated over a whole year, increases from about 64 km at low latitudes up to about 71 km at the high latitudes. Day-time, the seasonal differences are larger and much more evident. Again, the latitudinal increase is preserved. Strongest seasonal differences are observed at middle latitudes where the winter-time minimum is at 51 km and the summer-time maximum is at 74 km, a quite impressive increase of about 45%.

## 4 Empirical modelling of the plasma scale height

Considering that the plasma scale height is important for ionosphere studies and many other applications, it is clear that an empirical model would be quite helpful and such a model is currently being developed. In this model, the plasma scale height is approximated by a multi-variable polynomial delivering the scale height values with respect to geomagnetic latitude, local time, and season. The method of least-squares fit is applied for determining the polynomial coefficients.

Based on the year-long measurements and the analysis presented in the previous parts, a preliminary version of the model has been already prepared and some exemplary model calculations are given in Fig. 4. The model is regularly being upgraded in step with the growing IRO measurement database. It is envisaged that longitude and solar activity dependence will be included as well.

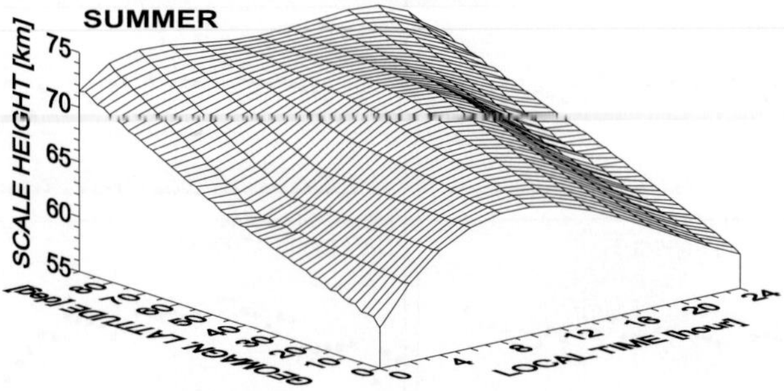


**Fig. 4.** Exemplary model calculations of the plasma scale height for summer conditions: latitudinal vs. diurnal variations.

## 5 Conclusions

Presented were first results from retrieving, analysing, and modelling the topside plasma scale height using IRO measurements onboard CHAMP. It has been shown that it is possible to utilize these observations for acquiring a valuable knowledge of the plasma scale height behaviour. Foremost, it has been proved that the topside plasma scale height depends strongly on the ionosphere-plasmasphere temperature, composition and dynamics. It is clear that the scale height generally increases in poleward directions, particularly during equinox and summer. Strong seasonal differences are observed at middle latitudes where day-time median values in summer are about 45% higher than in winter. Diurnal variations appear to be very complex and additional data is needed. No significant hemispheric differences are detected. A new empirical model is being developed, which will be capable of delivering the scale height with respect to latitude, local time, and season. Model results can be directly implemented into the density profile retrieval procedure by delivering an improved initial guess.

## References

Heise S, Jakowski N, Wehrenpfennig A, Reigber C, Lühr H (2002) Sounding of the topside ionosphere/plasmasphere based on GPS measurements from CHAMP: Initial results. Geophys Res Lett *29(14)*: doi 10.1029/2002GL014738.

Jakowski N, Bettac HD, Lazo B, Lois L (1981) Seasonal variations of the columnar electron content of the ionosphere observed in Havana from July 1974 to April 1975. J Atmos Terr Phys *43(1)*: 7-11, doi 10.1016/0021-9169(81)90003-9.

Jakowski N, Wehrenpfennig A, Heise S, Reigber C, Lühr H, Grunwaldt L, Meehan T (2002) GPS radio occultation measurements of the ionosphere from CHAMP: Early results. Geophys Res Lett *29(10)*: doi 10.1029/2001GL014364.

Reigber C, Lühr H, Schwintzer P (Eds.) (2002) First CHAMP Mission Results for Gravity, Magnetic and Atmospheric Studies. Springer-Verlag Berlin Heidelberg New York.

Stankov SM, Jakowski N, Heise S, Muhtarov P, Kutiev I, Warnant R (2003) A new method for reconstruction of the vertical electron density distribution in the upper ionosphere and plasmasphere. J Geophys Res *108(A5)*: 1164, doi 10.1029/2002JA009570.

# Differential Code Bias of GPS Receivers in Low Earth Orbit: An Assessment for CHAMP and SAC-C

Stefan Heise[1], Claudia Stolle[2], Stefan Schlüter[3], and Norbert Jakowski[3]

[1] GeoForschungsZentrum Potsdam (GFZ), Dep. 1, Potsdam, Germany
*heise@gfz-potsdam.de*
[2] Institute for Meteorology, University of Leipzig, Germany
[3] DLR, IKN, Neustrelitz, Germany

**Summary.** The knowledge of the transmitter and receiver differential code biases (DCB) plays a key role for the calibration of GPS based measurements of total electron content (TEC). To estimate the DCB of the CHAMP receiver concerning the zenith looking antenna a model assisted technique has been developed which takes advantage of the known GPS biases and comparatively low TEC above Low Earth Orbiter (LEO) altitudes in polar regions and during nighttime intervals. For assistance the Parameterized Ionospheric Model (PIM) is used. Applying this method we derived a reliable bias solution for the CHAMP receiver varying within a spread RMS below 1 TECU for the years 2001 and 2002. The SAC-C receiver DCB (zenith antenna) could be derived comparatively stable. In this paper we describe the model assisted calibration technique and present differential code bias estimations for CHAMP and SAC-C. The model assisted approach has been used to estimate the CHAMP occultation DCB as well. This bias could also be assessed by simultaneous GPS observations of the zenith and occultation antenna for selected intervals in 2001. DCB estimations from both approaches are shown.

**Key words:** Differential Code Bias, GPS, CHAMP, SAC-C, Ionosphere

## 1 Introduction

The dispersive propagation properties of the ionosphere for radio waves enable ionospheric remote sensing by using differential GPS navigation signals. Ground based GPS measurements are already well established as a powerful tool for operational ionospheric monitoring. In this context, pseudorange-leveled differential carrier phases are widely used to derive TEC information along the signal ray paths. The so derived TEC is not absolute since it still includes the receiver and satellite differential code biases.

Starting from the GPS/MET mission, which demonstrated that space based GPS measurements are a valuable data source for remote sensing of the Earth's atmosphere (e.g. [7]), the number of GPS equipped LEO missions like CHAMP and SAC-C is growing continuously. Such satellite missions provide a unique GPS data base for ionospheric investigations on global scale.

While the DCBs of the GPS satellites and ground based receivers are routinely estimated in a reliable manner (e.g. [4]), there is no well-established method so far to assess this bias for GPS receivers on board LEO satellites. On the other hand, the calibration of space based TEC measurements is of growing interest in terms of ionospheric tomography and data assimilation (e.g. [3], [6]). In this paper we illustrate a model assisted technique which has been applied to assess the DCBs of the GPS receivers on board the LEO satellites CHAMP and SAC-C. We present results concerning the zenith looking navigation antennas of both satellites and the occultation antenna of CHAMP.

## 2 Retrieval technique

GPS ionospheric sounding is based on measurements of the carrier and code phases (i.e. pseudoranges). Both, differential carrier (L1-L2) as well as differential code phases (P2-P1) provide information on the TEC along the ray path which is assumed to be identical for both GPS frequencies. Pseudorange-leveled carrier phase differences are widely used to resolve the ambiguities of the comparatively low-noise differential carrier phases. The resulting relative TEC ($TEC_{rel}$) still includes satellite and receiver differential code biases. A detailed description on the estimation of TEC as well as GPS ground receiver and satellite DCBs is given e.g. by [5].

### 2.1 Model assisted LEO receiver DCB estimation

The relation between $TEC_{rel}$ and the absolute link related TEC measure $TEC_{abs}$ is given by

$$TEC_{abs} = TEC_{rel} + DCB_{GPS} + DCB_{REC} \tag{1}$$

where $DCB_{GPS}$ and $DCB_{REC}$ denote the DCBs of the involved GPS satellite and receiver, expressed in TEC units (1 TECU = $10^{16}\mathrm{m}^{-2}$). Since the GPS DCBs are known from ground station network bias solutions, equation (1) enables the assessment of the receiver DCBs if an adequate estimation of $TEC_{abs}$ is available. Assuming a sufficient average agreement between the real ionization state and an ionospheric model on selected radio links with expected small TEC values (e.g. high latitude nighttime measurements with high elevation angles), the receiver DCB can be assessed by

$$DCB_{REC} \approx 1/n \sum_{i=1}^{n} (TEC_{mod}(i) - TEC_{rel}(i) - DCB_{GPS}) \tag{2}$$

where $TEC_{mod}$ denotes the modeled TEC according to the selected $TEC_{rel}$ ray paths. This approach assumes a stationary receiver DCB at least in a temporal range of one day which is equally supposed for DCBs of ground receivers and GPS satellites. Since the receiver DCB is mainly influenced by the hardware temperature, this assumption seems to be valid if the LEO internal temperature handling is stable like on CHAMP. To calculate the modeled TEC we used PIM [1] which includes also the Gallagher plasmasphere model [2]. The practical applicability of equation (2) is restricted by the required mean agreement between the model and the real ionization state along the selected radio links. Especially in case of CHAMP this assumption is not always well fulfilled due to the low orbit altitude of about 400 km. Furthermore, the DCB assessment is sensitive to the selection criteria. In order to reduce the model dependence of the receiver DCB estimation we followed an approach which does not demand an average agreement between model and real ionization state but only assumes that all $TEC_{abs}$ exceed a certain minimum value ($TEC_{min}$) which is estimated by the model according to the observation ray paths. By this assumption the receiver DCB can be derived using a simple iterative procedure which gradually increases a knowingly underestimated initial receiver DCB. This initial DCB value can be derived e.g. from equation (2) if we replace $TEC_{mod}(i)$ with $TEC_{min}$. For the iteration increment we used the median value of all positive differences $TEC_{min} - TEC_i$ where $TEC_i$ denotes $TEC_{abs}$ after the i$^{th}$ iteration step according to equation (1). To reduce the influence of possible inaccuracies or outliers of $TEC_{rel}$ we stopped the iteration when positive differences $TEC_{min} - TEC_i$ occurred for less than three different transmitter receiver combinations. At first, receiver DCBs have been estimated on a daily basis but finally they were calculated as sliding averages over 5 days.

### 2.2 Simultaneous navigation and occultation measurements

Due to higher ionization along occultation ray paths the model assisted estimation of DCBs concerning occultation antennas is more difficult than in the case of zenith looking navigation antennas. If the satellite is equipped with both kinds of antennas and the zenith antenna bias ($DCB_{top}$) is known, simultaneous $TEC_{rel}$ measurements ($TEC_{top}$ and $TEC_{occ}$) from the same GPS satellite can be used to assess the occultation DCB ($DCB_{occ}$) according to $DCB_{occ} = 1/n \sum_{i=1}^{n}(TEC_{top}(i) - TEC_{occ}(i) + DCB_{top})$. This procedure requires occultation antenna measurements with positive elevation angles.

## 3 Results

The algorithm described in section 2.1 has been used to assess the DCBs concerning the zenith looking navigation antennas of CHAMP and SAC-C. As Fig. 1 shows, the DCB estimation has been considerably stable for both satellites during 2002 but the DCB estimates for SAC-C are generally more stable than for CHAMP. This difference is considered as a consequence of

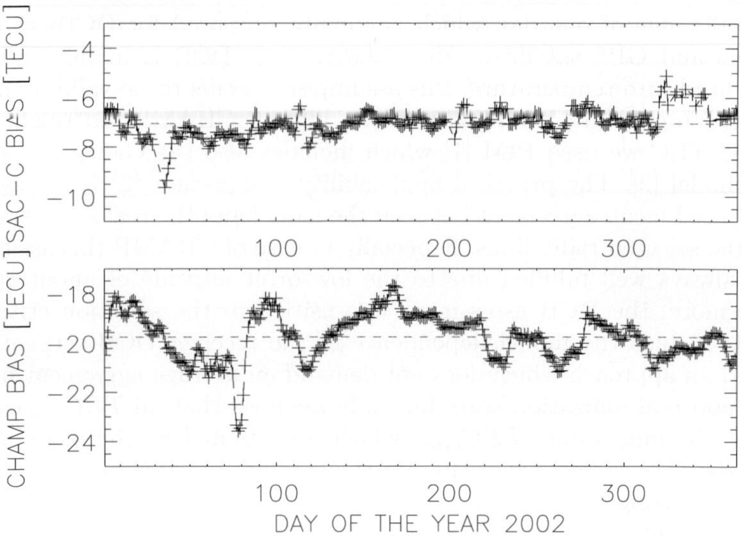


**Fig. 1.** DCB estimation for 2002 concerning zenith antennas of SAC-C (upper panel) and CHAMP (lower panel).

the different orbit altitudes of both satellites (about 400 and 700 km for CHAMP and SAC-C respectively). One should expect that the model based estimation of $TEC_{min}$ (see section 2.1) is more reliable and stable for SAC-C due to the lower ionization level which holds lower error sources. In case of SAC-C the DCB is varying around -7.0 TECU within a spread RMS of only 0.5 TECU. For CHAMP we calculated for 2002 a mean DCB of -19.8 TECU varying within a spread RMS of 0.9 TECU. The CHAMP results for 2001 (not shown here) look quite similar with a DCB of -20.0 TECU and a RMS of 0.7 TECU.

Furthermore, the same technique has been applied to assess the CHAMP occultation DCB for a long period in 2002 (Fig. 2 upper panel). Here the DCB is varying around 13.1 TECU within a spread RMS of 4.0 TECU. In a second step we replaced PIM by results of a 3-d electron density assimilation based on CHAMP zenith antenna TEC [3]. These can be considered as an improved PIM output. The corresponding results (Fig.2 lower panel) look similar but show generally less fluctuations (14.4 TECU mean, 2.7 TECU RMS). Fig. 2 reveals that the CHAMP occultation DCB estimation has been quite stable during the summer months of 2002 but shows considerable fluctuations around the autumn equinox. This could be explained by generally increased $TEC_{min}$ values which contain as well increased error sources for the DCB estimation. Since $TEC_{min}$ values usually occur in the polar regions they increase if these regions are equally exposed to the ionizing radiation of the sun.

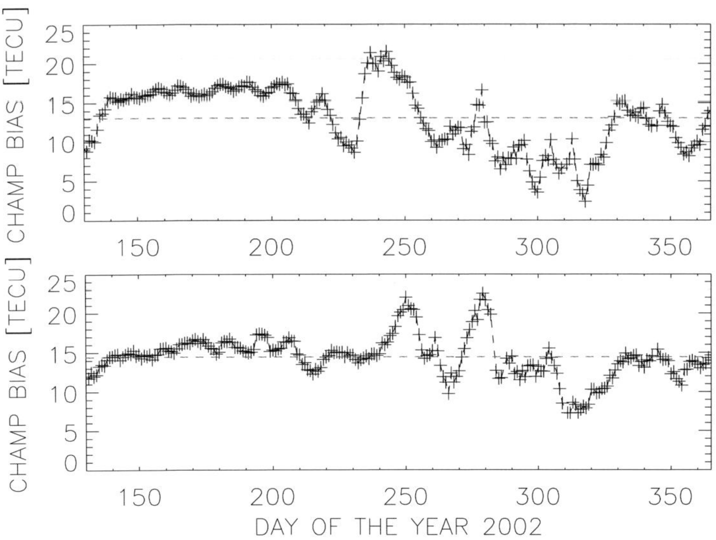


**Fig. 2.** Model assisted estimation of CHAMP occultation DCB for 2002, day of the year 130 until 365. Upper panel: using PIM. Lower panel: using topside assimilation.

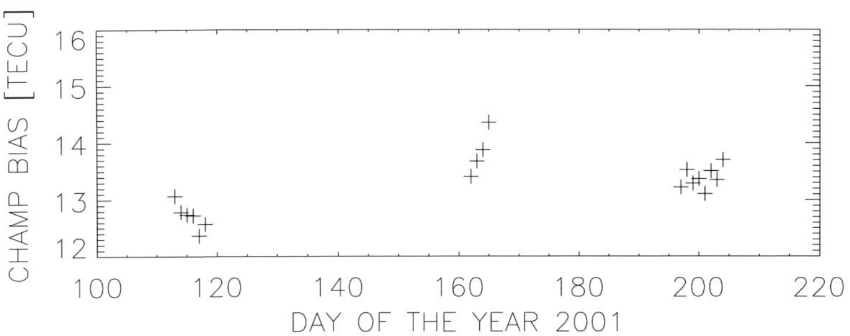


**Fig. 3.** Estimation of CHAMP occultation DCB based on simultaneous observations from occultation and zenith antenna for special periods in 2001.

The CHAMP occultation DCB has also been retrieved by simultaneous navigation and occultation TEC measurements. Results are accessible only for some short periods in 2001 (Fig. 3) due to the restricted availability of positive elevated occultation measurements. The bias solution is stable over the considered time intervals with a mean value of 13.3 TECU.

## 4 Conclusion

A model (PIM) assisted technique has been developed to estimate the differential code bias (DCB) of GPS receivers in low earth orbit. Adequate DCB estimations concerning the navigation antennas of CHAMP and SAC-C have been derived using this method. The model assisted approach has also been applied to estimate the CHAMP occultation DCB. Results are promising for a long period in 2002 but show special uncertainties during equinox times. Therefore, 3-d electron density assimilation results based on CHAMP zenith antenna TEC have been used in place of PIM to assess the CHAMP occultation DCB. The resulting bias estimation occurs more stable and reliable. For specific periods in 2001 the CHAMP occultation DCB was estimated by synchronous observations from zenith and occultation antenna. Results show considerable stability during all selected periods.

*Acknowledgments.* The authors are grateful to the teams of CHAMP and SAC-C and to the data providers GFZ/ISDC and JPL for the excellent GPS data base.

## References

1. Daniell RE, Brown LD, Anderson DN, Fox MW, Doherty PH, Decker DT, Sojka JJ, Schunk RW (1995) Parameterized ionospheric model: A global ionospheric parameterization based on first principles models. Radio Sci *30*: 1499–1510.
2. Gallagher DL, Craven PD, Comfort RH (1988) An empirical model of the Earth's plasmasphere. Adv Space Res *8(8)*: (8)15–(8)24.
3. Heise S, Jakowski N, Cooke D, Reigber Ch (2004) Ionosphere/plasmasphere imaging based on GPS navigation measurements from CHAMP and SAC-C. This issue.
4. Sardón E, Zarraoa N (1997) Estimation of the total electron content using GPS data: How stable are the differential satellite and receiver instrumental biases? Radio Sci *32(5)*: 1899–1910.
5. Sardón E, Rius A, Zarraoa N (1994) Estimation of the transmitter and receiver differential biases and the ionospheric total electron content from Global Positioning System observations. Radio Sci *29(3)*: 577–586.
6. Stolle C, Schlüter S, Jacobi Ch, Jakowski N, Heise S, Raabe A (2004) Three-dimensional monitoring of the polar ionosphere with ground- and space-based GPS. This issue.
7. Ware R, Exner M, Feng D, Gorbunov M, Hardy K, Melbourne W, Rocken C, Schreiner W, Sokolovsky S, Solheim F, Zou X, Anthes AR, Businger S, Trenberth K (1996) GPS soundings of the atmosphere from low earth orbit: Preliminary results. Bull Amer Meteor Soc *77*: 19–40.

# Ionosphere/Plasmasphere Imaging Based on GPS Navigation Measurements from CHAMP and SAC-C

Stefan Heise[1], Norbert Jakowski[2], and David Cooke[3]

[1] GeoForschungsZentrum Potsdam (GFZ), Dep. 1, Potsdam, Germany
*heise@gfz-potsdam.de*
[2] DLR, IKN, Neustrelitz, Germany
[3] Air Force Research Laboratory, Hanscom AFB, MA, USA

**Summary.** Both the CHAMP and SAC-C satellites permanently track several GPS satellites using dedicated zenith looking antennas for precise orbit determination. These 0.1 Hz sampled dual frequency navigation measurements provide valuable information on the ionization state of the upper ionosphere and plasmasphere up to GPS altitudes on a global scale. After preprocessing and calibration, link related TEC measurements are derived from the GPS navigation observations. Three-dimensional electron density information is retrieved by assimilating these TEC data into the Parameterized Ionospheric Model (PIM). In specific periods CHAMP and SAC-C circle the earth in nearly the same orbit plane due to their different orbit properties. We focus on these constellations which are of special interest for the combined assimilation of TEC measurements from both satellites. Reconstruction results for selected assimilation examples are presented by means of two dimensional slices along the respective CHAMP/SAC-C orbit plane. Using electron density measurements from the CHAMP Langmuir Probe we validate our reconstruction results along the CHAMP path and discuss the impact of the TEC measurements from SAC-C on the reconstruction result at CHAMP orbit altitudes.

**Key words:** Ionosphere, Plasmasphere, CHAMP, SAC-C, GPS

## 1 Introduction

Dual frequency GPS observations are well proved and widely used for ionospheric monitoring. Information on the link related total electron content (TEC) can easily be derived from differential GPS phase measurements due to the frequency dependent ionospheric impact on radio signals. Beside the well established GPS ground receiver networks, the installation of GPS receivers on board of LEO (Low Earth Orbiting) satellites like CHAMP and SAC-C provides a valuable data source for ionospheric remote sensing on a global scale. Both satellites, CHAMP and SAC-C, were launched in 2000 and carry on board a TRSR2 space receiver ("Black Jack") for GPS occultation and navigation measurements. For our investigations we used the 0.1 Hz sampled zenith antenna navigation measurements from both satellites. In the following we briefly describe the observation scenario and data processing for the reconstruction of topside ionosphere and plasmasphere electron density distributions by assimilation of the link related TEC measurements from CHAMP and SAC-C. We present assimilation and validation results and discuss comparisons between reconstructions based on TEC measurements from only one and from both satellites.

## 2 Observation scenario and data processing

Both CHAMP and SAC-C orbit the earth in nearly polar orbits (inclination: 87.3° and 98.2° respectively). The orbit time differs slightly (about 93 and 97 minutes) due to the different altitudes of CHAMP (about 400 km) and SAC-C (about 700 km). But while the SAC-C orbit is sun synchronous and therefore fixed in local time, the CHAMP orbit plane moves slowly through all local time sectors. Thus, there are periods when both satellites circle the earth in nearly the same orbit plane but at different altitudes. These periods are of special interest for the combined assimilation of CHAMP/SAC-C TEC data since they promise the strongest interactions between measurements from both satellites. In this paper we focus on periods when angles between both orbit planes were less than 15°. Fig. 1 shows the angle between the CHAMP and SAC-C orbit plane for 2002. According to this we selected the following days of 2002 for our investigations: 30-47, 163-176 and 293-309.

The data processing is similar to that which has been applied for the retrieval of 3-d electron density information using only CHAMP GPS data (e.g.[4]). It basically consists of 3 parts: preprocessing, calibration and assimilation. During the preprocessing we use an algorithm as described by [1] to save the GPS data quality concerning outliers and cycle slips. Furthermore, we derive relative TEC from pseudorange-leveled differential carrier phases and allocate orbit information to each measured GPS transmitter - LEO receiver link. To derive absolute TEC the differential code biases (DCB) of the transmitter and receiver involved respectively are needed. For our calibration we use transmitter DCB estimations from ground receiver network solutions (e.g. [7]) and assess the CHAMP and SAC-C receiver DCB by using a model assisted technique described by [5]. Three-dimensional electron density information is retrieved by assimilating the integral TEC measurements into the

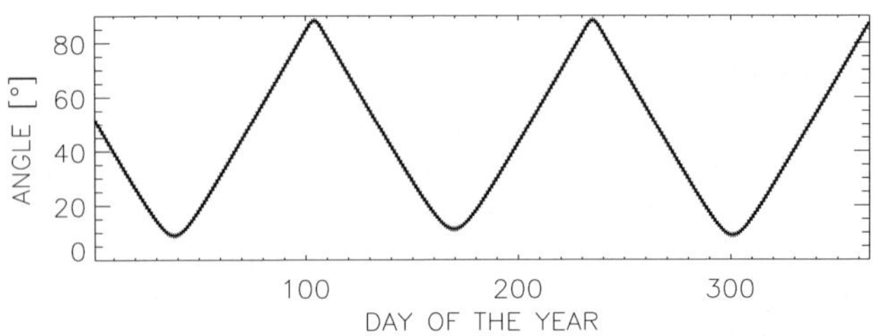


**Fig. 1.** Angle between CHAMP and SAC-C orbit plane for 2002.

Parameterized Ionospheric Model (PIM, [2]) which includes the Gallagher model [3] of the plasmasphere. Here we use the assimilation system originally designed for CHAMP TEC measurements (e.g. [4]) which has been upgraded for the processing of multi satellite TEC information. For each reconstruction we apply about 7000 CHAMP/SAC-C TEC measurements from 93 minute time intervals (one CHAMP revolution). During the assimilation we adapt the initial model assumption (PIM) to the link related TEC information by applying multiplicative corrections according to the widely used Multiplicative Algebraic Reconstruction Technique (MART, e.g. [6]). For this purpose, PIM is discretized on a global voxel structure which covers the ionosphere and plasmasphere up to GPS altitudes. The validity of the resulting reconstruction depends of course on the data coverage and the initial model. Nevertheless, we obtain a physically reasonable reconstruction which represents at least an improvement over the initial model assumption.

## 3 Results

The assimilation result is a global 3-dimensional electron density distribution, covering the region between CHAMP and GPS orbit altitudes. The strongest assimilation impact is obviously found in the region near the CHAMP/SAC-C orbit plane while other areas of the voxel structure give the pure initial model, completely unaffected by the assimilation process. Thus, to present assimilation results we consider 2-dimensional slices with respect to the common CHAMP/SAC-C orbit plane. The left panel of Fig. 2 shows such a slice for a selected assimilation example and gives an impression of the reconstructed electron density distribution near the orbit plane (from around 22:30 LT in the East to 10:30 LT in the West). Beside the clear difference between day- and nighttime ionosphere and plasmasphere, Fig. 2 (left panel) reveals a considerably increased electron density in the nighttime northern hemisphere plasmapause region. This is also clearly visible in the right panel of Fig. 2 which shows the percentage deviation of the assimilation result from the initial model assumption. The remarkable deviation in the plasmapause region can be considered as an assimilation correction of the initial (PIM) plasmapause position. In this context it is interesting to compare the assimilation result shown in Fig. 2 with the corresponding reconstruction based on CHAMP TEC data only. This is shown in Fig. 3 (left panel) and reveals significant differences to the combined CHAMP/SAC-C reconstruction in the plasmasphere region. Fig. 3 (right panel) indicates that this difference is not caused by an absence of data coverage (given in link related TEC measurements per voxel). Thus, we should expect that TEC data from SAC-C hold additional information on the plasmaspheric electron density distribution which are not covered by the CHAMP TEC measurements. This can be explained by the different orbit altitudes. While the integral TEC measurements from CHAMP are obviously dominated by the rather high electron densities of the lower topside

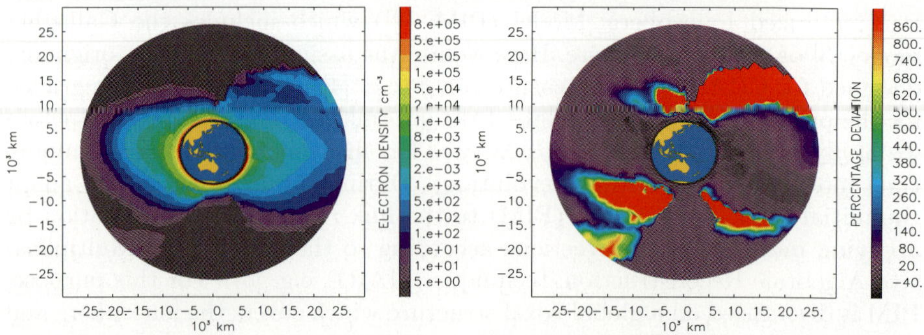


**Fig. 2.** Meridional slices in orbit plane (along 233° E, local time 22:30), assimilation example for October 27, 2002, beginning of assimilation: 06:13 UT, duration: 93 minutes. Left panel: Reconstruction of ionospheric/plasmaspheric electron density distribution after assimilation of TEC data from CHAMP and SAC-C (full CHAMP revolution). Right panel: Percentage deviation of assimilation result from initial model (PIM).

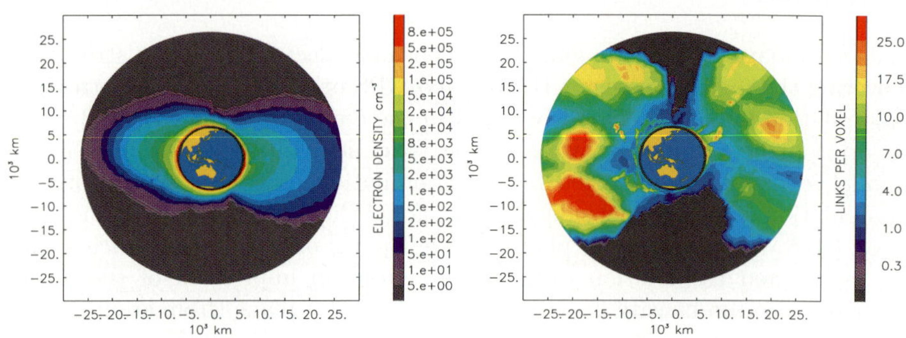


**Fig. 3.** Meridional slices in orbit plane, same assimilation example as shown in Fig. 2 but using only CHAMP TEC data. Left panel: Reconstruction of ionospheric/plasmaspheric electron density distribution. Right panel: Corresponding data coverage.

ionosphere this is not the case for SAC-C which circles the earth far above the ionospheric electron peak densities.

The in situ electron density measurements from the CHAMP Langmuir Probe (LP) provide a unique data source to validate the assimilation results along the CHAMP path and to investigate the impact of the SAC-C TEC data on the reconstruction result at CHAMP orbit altitudes. For this purpose Fig. 4 (left panel) shows comparisons between LP on the one hand and CHAMP/SAC-C assimilation, pure CHAMP assimilation and PIM on the other hand. The comparison between PIM, assimilation and LP reveals a significant improvement over the initial model by the assimilation. While PIM coincides with LP electron densities within an RMS spread of only $52.4\text{x}10^{10}\text{m}^{-3}$

the assimilation results agree within 37.7x$10^{10}$m$^{-3}$ and 34.3x$10^{10}$m$^{-3}$ for CHAMP/SAC-C and pure CHAMP assimilation respectively. Furthermore, Fig. 4 shows a noticeable improvement of the pure CHAMP assimilation by additional use of SAC-C TEC data. Showing a similar RMS agreement, the mean absolute deviation of the pure CHAMP reconstruction is -7.2x$10^{10}$m$^{-3}$ while the combined CHAMP/SAC-C assimilation shows a mean deviation of only -3.6x$10^{10}$m$^{-3}$ from the LP electron densities. Beside the comparisons along the CHAMP path, pure SAC-C reconstructions can be compared with corresponding CHAMP results and PIM to validate pure CHAMP assimilation results at SAC-C altitudes. Fig. 4 (right panel) reveals a much better agreement between SAC-C and CHAMP results than between SAC-C and PIM (12.4x$10^{10}$m$^{-3}$ RMS and 3.7x$10^{10}$m$^{-3}$ mean absolute deviation in comparison to 16.9x$10^{10}$m$^{-3}$ RMS and 6.1x$10^{10}$m$^{-3}$ mean absolute deviation). Thus, the assimilation of CHAMP TEC data significantly improves on the performance of the initial model at SAC-C orbit altitudes.

## 4 Conclusion

The GPS technique on board LEO satellites is a powerful tool for imaging large scale structures of the ionosphere and plasmasphere on a global scale. The assimilation of link related TEC data from CHAMP and SAC-C provides an estimation of the electron density distributions of the upper ionosphere and plasmasphere. Validation with Langmuir Probe measurements indicate a noticeable improvement over pure CHAMP reconstruction results by the additional assimilation of link related TEC measurements from SAC-C.

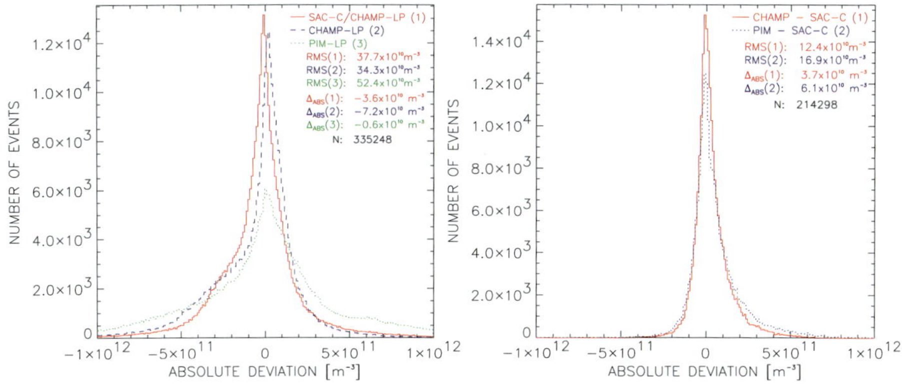


**Fig. 4.** Electron density comparisons for 2002, days of the year 36-47, 166-176, 293-309. Left panel: Absolute deviation between CHAMP Langmuir Probe (LP) and: CHAMP/SAC-C assimilation, pure CHAMP assimilation and PIM. Right panel: Absolute deviation between pure CHAMP and SAC-C assimilation and between PIM and SAC-C assimilation along SAC-C path.

The comparison of pure SAC-C reconstructions with corresponding CHAMP results and PIM shows significant improvements over the initial model at SAC-C orbit altitudes produced by the assimilation of CHAMP TEC data. Further improvement of assimilation results can be expected from additional link related TEC measurements (e.g. GRACE) and improved initial model assumptions.

*Acknowledgement.* The authors are grateful to the teams of CHAMP and SAC-C and to the data providers GFZ/ISDC and JPL.

## References

1. Blewitt G (1990) An automatic editing algorithm for GPS data. Geophys Res Letters *17(3)*: 199–202.
2. Daniell RE, Brown LD, Anderson DN, Fox MW, Doherty PH, Decker DT, Sojka JJ, Schunk RW (1995) Parameterized ionospheric model: A global ionospheric parameterization based on first principles models. Radio Sci *30*: 1499–1510.
3. Gallagher DL, Craven PD, Comfort RH (1988) An empirical model of the Earth's plasmasphere. Adv Space Res *8(8)*: (8)15–(8)24.
4. Heise S, Jakowski N, Wehrenpfennig A, Reigber Ch, Lühr H (2002) Sounding of the topside ionosphere/plasmasphere based on GPS measurements from CHAMP: Initial results. Geophys Res Lett *29(14)*: 10.1029/2002GL014738.
5. Heise S, Stolle C, Schlüter S, Jakowski N (2004) Differential code bias of GPS receivers in low Earth orbit: An assessment for CHAMP and SAC-C. (this issue).
6. Kunitsyn VE, Tereshchenko ED (2003) Ionospheric Tomography. Springer, Berlin, Heidelberg, New York.
7. Sardón E, Rius A, Zarraoa N (1994) Estimation of the transmitter and receiver differential biases and the ionospheric total electron content from Global Positioning System observations. Radio Sci *29(3)*: 577–586.

# Three-Dimensional Monitoring of the Polar Ionosphere with Ground- and Space-Based GPS

Claudia Stolle[1], Stefan Schlüter[2], Christoph Jacobi[1], Norbert Jakowski[2], Stefan Heise[3], and Armin Raabe[1]

[1] Institute for Meteorology, University of Leipzig *stolle@uni-leipzig.de*
[2] DLR, Institute of Communications and Navigation *stefan.schlueter@dlr.de*
[3] GeoForschungsZentrum Potsdam *heise@gfz-potsdam.de*

**Summary.** Integrated electron density measurements along GPS signal ray paths provide a useful data source for 3-dimensional ionospheric reconstruction. Since a substantial data amount is required for the application of the inverse problem, ground-based inversions are restricted to regions with high ground receiver density. The additional incorporation of space-based GPS data from LEO satellites allows the monitoring in areas less densely covered by ground receivers. We present a case study of 3-dimensional reconstruction for Nov. 2001 in the North polar region. Input data are calibrated TEC derived from IGS and CHAMP measurements. The results show increased electron densities near the geomagnetic pole. The comparison of the reconstruction results with measurements of the Langmuir Probe onboard CHAMP shows a good agreement along the CHAMP orbit. The large-scale structure of the ionospheric observations are closely related to the present geomagnetic conditions.

**Key words:** 3D-Tomography, Ionosphere, GPS, CHAMP, Radio-Occultation

## 1 Introduction

The impact of free ionospheric electrons on L-band radio wave propagation makes GPS an interesting tool for ionospheric investigations. GPS services like the International GPS Service (IGS) maintain a network of globally distributed GPS receivers. Ionospheric GPS based measurements provide information on the Total Electron Content (TEC) which is electron density integrated along the signal ray path. The latter two circumstances allow 3-dimensional reconstructions of the electron content by combining several ionospheric GPS measurements through a tomographic inversion algorithm. The need for a substantial amount of records inside the area of interest limits the ground-based calculations to regions of high receiver densities.
In the last decade, the development of space-based GPS measurements onboard Low Earth Orbiter (LEO) satellites has opened a variety of improved possibilities for ionospheric research. An advantageous characteristic of this kind of investigation is global availability. The location of the data solely depends on the location of the LEO and the GPS satellites. Therefore, hardly

accessible regions are also covered by data, e.g., deserts, oceans, or poles. In the case of 3-dimensional monitoring of electron density distributions, this means that the area of interest can be shifted to regions that are not necessarily densely covered by ground-based receivers. In this paper we present a reconstruction study over the North polar region using IGS data as well as CHAMP occultation data. The example is situated during the night of Oct. $31^{st}$, 2001 and has been chosen due to CHAMP data availability and interesting magnetic observations. This case study should serve as an analysis of the potential of electron density reconstructions under disturbed conditions. It is shown that the use of LEO data is helpful for reconstructions in areas with low ground-based receiver density.

## 2 Method and data

The applied reconstruction method is one of the widely used Multiplicative Algebraic Reconstruction Techniques (MART) and it is initialised with a combined IRI/GPCM model. The tomographic procedure has been presented in detail by [1] with application in the European sector, also including remarks on validation. However, some modifications have been undertaken for this study, which should be mentioned. The reconstruction area covers the entire polar cap from 50° northward, with a grid size of 4° in latitude and 10° in longitude. Height ranges between 80 km and 1000 km in steps of 20 km. The lower resolution compared to European applications has been chosen with respect to the more sparse data and to the wider region. In Europe data from $\approx$ 60 IGS stations are available, while in the entire North polar region only $\approx$ 25 IGS stations are in use. The iteration stop criterion is set when real and simulated data reach small differences in its mean or if the improvement of this difference is very small.
In Fig.1a, an example of the data coverage over the North polar region is presented. The ray lengths in the height range of 80-1000 km are projected on the Earth's surface. The often separately appearing rays are ground-based measurement paths. The black columns in fact consist of many parallel CHAMP received signal rays.
Before entering the tomographic procedure, TEC data have been corrected for instrumental biases of the satellite and ground-based receiver ([2]), or of the CHAMP receiver in occultation mode ([3]), respectively. The occultation bias has been extrapolated to 13.7 TECU on the base of the available bias results for 2001 using simultaneous zenith and occultation measurements. The increased noise level of carrier and code phase of low looking zenith antenna data has been already mentioned in [4]. Since in our case only carrier phase enters directly the calculations, the maximal additional noise error of 0.02 TECU is well within the error band of our procedure, which is $\approx$ 1 TECU.
The night of Nov. $1^{st}$, 2001 is accompanied by interesting magnetic observations. For this date, Kp=5 and Dst=90 reveal only moderate perturbations

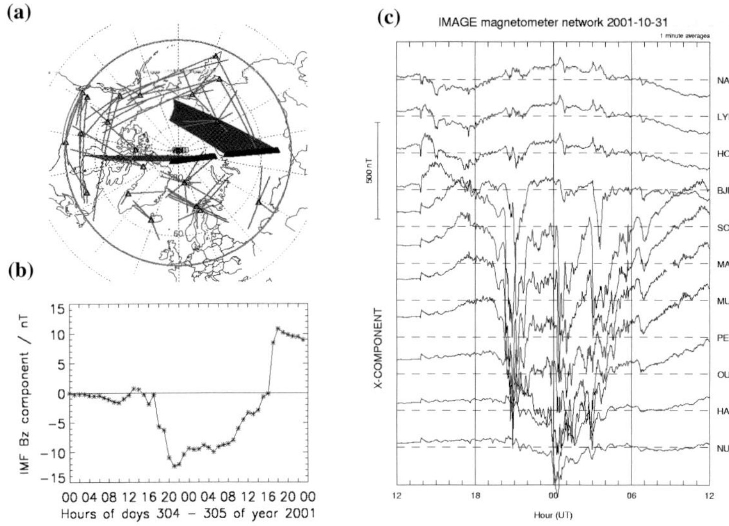


**Fig. 1.** (a): GPS ray data coverage over the North polar region on Nov. $1^{st}$, 2001; (b): Z-component of the IMF plotted over time for days 304-305 of year 2001; Data comes from the ACE2 satellite. (c): X-component of quasi-longitudinal aligned ground magnetometer data from stations of the IMAGE network (Ny Ålesund: 78.92°N → Nurmijärvi: 60.5°N).

in the global geomagnetic field. In contrast, magnetic observations concerning North polar cap regions show significant deviations from the moderate state. In Fig.1b a strong negative Z-component (southward) of the Interplanetary Magnetic Field (IMF) indicates open geomagnetic field lines, which, inter alia, may favour solar wind plasma intrusions into the magnetosphere. Linked to southward IMF, strong electric fields might develop in the ionosphere. Corresponding to this theory, high cross polar cap potentials up to 100 kV between midnight and 0100 UT of this day (not shown here) have been monitored. Fig.1c shows the X-component of the geomagnetic field measured along a chain of ground magnetometers in Europe from the IMAGE network (http://www.geo.fmi.fi/image/). The diagram shows signatures of a well isolated substorm. At first sight, magnetic disturbances are typically restricted to auroral latitudes. A sudden decrease of the geomagnetic field is firstly observed around 2100 UT of Oct. $31^{st}$, 2001. A second, well pronounced sudden decrease is seen just a few minutes after midnight and finally a third, weaker one, around 0300 UT of Nov. $1^{st}$, 2001.

Near the moment of strongest decrease of the geomagnetic field, around 0100 UT of Nov. $1^{st}$, 2001, CHAMP crossed the Northern polar region and 3-dimensional ionospheric reconstruction could be performed.

## 3 Tomographic results

Three-dimensional monitoring has been undertaken for the time interval centered at 01 UT Nov. $1^{st}$, 2001. The results are expressed with averages over time segments of about ten minutes. This is due to the sampling time of about five minutes during one ionospheric GPS occultation observed onboard CHAMP. Two occultations are presented here (see Fig.1a).
In Fig.2a vertical TEC, re-integrated from reconstructed electron density distributions, is represented. Local noon is situated on the top of the figure representing expected higher electron densities compared with the night side. Near the geomagnetic pole (near Thule in Northwest Greenland) high ionisation exceeding 40 TECU, can be recognized. At the same time, a *tongue* of high values connects the day side with this local maximum. This tongue is bounded by lower values at local dawn and dusk.
The thorough interpretation of the physical background of this phenomenon would exceed the scope of this paper. Nevertheless, the local maximum seems to be a joint result of substorm activity and advected plasma from the dayside. The assumptions of magnetic activity effects is clearly supported by the magnetic data shown in Fig.1c as described in section 2. An advection pattern across the polar cap can arise for southward directed IMF. In this case of open magnetic field lines, magnetospheric electric fields may map into the ionosphere. This is supported by the high cross polar cap potential up to 100 kV (see section 2). Southward magnetic field and dawn-dusk directed ionospheric electric field forces large scale antisunward plasma convection via the $E \times B$ drift ([5]).
It is interesting, that reconstructions for the CHAMP paths around 20:20 UT and 23:20 UT of Oct. $31^{st}$, 2001 did not reveal such a high ionospheric

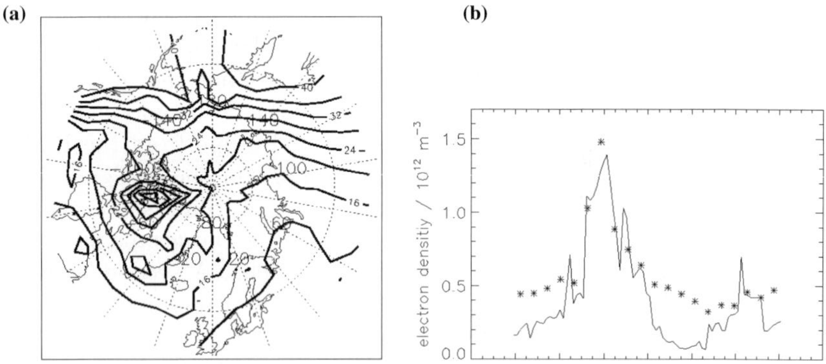


**Fig. 2.** (a): vertical integrated TEC retrieved from reconstruction results for 01 UT of Nov. $1^{st}$, 2001. (b): electron densities from the same reconstruction at 410 km height (asterisks) and those from the Langmuir Probe (solid line) onboard CHAMP in a dawn-dusk cross section.

signature near the geomagnetic pole, even though geomagnetic data suggest that and data availability was similar.
When comparing the data distribution in Fig.1a and the result in Fig.2a an obvious relation between the local maximum near the geomagnetic pole and dense CHAMP data input is seen. (Note, that the projections of the CHAMP rays of the concerned occultation hide the high vertical availability of occultation measurements in this case.) In highly under-determined inverse problems, like this one, the question of mis-reconstructions or artefacts arises.
Fig.2b shows the reconstructed electron densities at 410 km altitude together with the Langmuir Probe electron densities measured onboard CHAMP. Reconstructed densities are aligned along a longitudinal cut at 85°W and 105°E. The CHAMP path follows nearly continously 76°W and, after passing its nearest point to the geographic pole 91°E longitude, respectively. Fig.2b indicates a good correspondence between the reconstruction results and Langmuir Probe data. Especially in the local maximum and therefore for locations of high data coverage a good agreement is found. This conformity encourages us to interprete the result as a physically originated one rather than mis-reconstructions. Poor agreement is seen toward the dusk side of the globe. The ensemble of the occultation rays introduced here is widely horizontally spread, resulting in a reduced vertical distribution. Therefore, vertical smoothing during the tomographic procedure strongly influences the result and peak height densities may be displaced to the real height distribution.
It is not possible to access independent data to validate the entire tomographic array. To get an impression of the validity of the reconstruction results over the whole North polar region, polar vertical TEC-maps using the same GPS ground stations (see http://www.kn.nz.dlr.de/space_weather/npolcap.html) have been subtracted from re-integrated TEC of Fig.2a (not shown here). Differences range from 4 TECU to -20 TECU and with few deviating points down to -27 TECU around 50°– 60°N at local noon. Deviations within +/- 4 TECU are distributed over large areas on the dawn and dusk side connected by a narrow band across the geomagnetic pole. The mean deviation was calculated by -6 TECU and spatial gradients were moderate. A slight underestimation from tomographic TEC is reasonable since TEC-maps include the plasmasphere. Nevertheless, the moderate spatial gradients in deviations to TEC-maps and the above mentioned conclusions from the comparison to the Langmuir probe indicate that the large-scale electron density distribution is well represented despite the sparse input data.

## 4 Conclusion

The combination of ground- and space-based GPS TEC measurements allows 3-dimensional reconstruction over areas with low ground receiver density. We

presented a case study at 0100 UT on Nov. $1^{st}$, 2001 over the North polar region. To this aim, data from IGS and the CHAMP satellite have been used. A new approach is the inclusion of calibrated LEO TEC.
The reconstruction revealed a local maximum of electron density near the geomagnetic pole. The local maximum is connected with local noon by a tongue of increased densities. Magnetic observations suggest that the high values are a joint result of substorm activity and convected ionospheric plasma from the day side of the globe. The reconstruction result has been compared with electron densities measured by the Langmuir Probe onboard CHAMP and showed good agreement especially where good data coverage were available. Moderate deviations from vertical TEC-maps indicate the potential of the reconstruction to rebuild the ionospheric large-scale behaviour.
The results of this study encourage the inclusion of occultation TEC data from further LEO satellites and promises better reconstructions in interesting areas like polar and equatorial regions that are less covered by GPS ground receiver.

*Acknowledgments.* The authors are grateful to the CHAMP team, coordinated by the GFZ Potsdam, for the given support and GPS data. We thank all members of the IGS community for providing their data service. Thanks also for the provision of magnetic observations, Langmuir Probe data and useful discussions to H. Lühr, P. Ritter and W. Mai. This work is supported by DFG under grant No RA 569/5-2.

## References

1. Schlüter S, Stolle C, Jakowski N, Jacobi Ch (2003) Monitoring of 3-dimensional ionospheric electron density distributions based on GPS Measurements. In: Reigber Ch, Lühr H, Schwintzer P (Eds): First CHAMP Mission Results for Gravity, Magnetic and Atmospheric Studies, Springer, Berlin: 521–527.
2. Sardón E, Rius A, Zarraoa N (1994) Estimation of the transmitter and receiver differential biases and the ionospheric total electron content from Global Positioning System observations. Radio Sci *29(3)*: 577–586.
3. Heise S, Stolle C, Schlüter S, Jakowski N (2004) Differential code bias of GPS receivers in low Earth orbit: an assessment for CHAMP and SAC-C. This issue.
4. Montenbruck O, Kroes R (2003) In-flight performance analysis of the CHAMP BlackJack GPS Receiver. GPS Solutions *7*: 74–86.
5. Weimer DR (1995) Models of high-latitude electric potentials derived with a least error fit of spherical harmonic coefficients. J Geophys Res *100*: 19,595–19,607.

# Comparison of Electron Density Profiles from CHAMP Data with NeQuick Model

Norbert Jakowski[1], Konstantin Tsybulya[1], Sandro M Radicella[2], Marta Cueto[3], Miguel Herraiz[3]

[1] Deutsches Zentrum für Luft und Raumfahrt, e.V. (DLR), Institut für Kommunikation und Navigation, Kalkhorstweg 53, Neustrelitz, Germany, *Norbert.Jakowski@dlr.de*
[2] Aeronomy and Radiopropagation Laboratory, Abdus Salam ICTP, 34100 Trieste, Italy.
[3] Department of Physics of the Earth, Astronomy and Astrophysics I, Universidad Complutense de Madrid, Avenida Complutense s/n 28040, Madrid, Spain.

**Summary.** Vertical electron density profiles (EDP) derived from ionospheric radio occultation (IRO) measurements onboard the German CHAMP satellite are calculated by the Institute of Communications and Navigation of DLR on a regular basis since 11 April 2001. In order to validate the ionospheric radio occultation data obtained from this satellite, any systematic comparison with independent measurements but also with results from ionospheric models should help to get a better feeling about the quality of the data. On the other hand, if the IRO data quality is sufficient, the data may contribute to evaluate the accuracy of the ionospheric models. In this study we compare CHAMP EDP data derived in 2002/03 with the NeQuick model, which is one of the three electron density models developed at the Abdus Salam ICTP in Trieste (Italy) and the Institute for Meteorology and Geophysics in Graz (Austria). We discuss results of this comparison, showing changes in bias and deviation with latitude and local time. The best agreement between both types of data was found above a height of 300 km.

**Keywords:** Ionosphere, radio occultation, NeQuick, electron density, validation

## 1 Introduction

Low Earth Orbiting satellites carrying a dual frequency GPS receiver onboard offer a unique chance to monitor the actual state of the global ionosphere on a continuous basis. No other profiling technique unifies profiling through the entire F2-layer with global coverage.

The German CHAMP satellite provided first ionospheric radio occultation (IRO) measurements on 11 April 2001. Preliminary IRO retrieval and validation results are discussed in Jakowski et al. (2002). In this paper we discuss a comparison of up to about 50000 CHAMP measurements with corresponding model values derived from the ionosperic model NeQuick referenced in the following as NeQ.

The NeQ model is one of the three electron density models developed at the Abdus Salam ICTP in Trieste and the Institute for Meteorology and Geophysics in

Graz (Hochegger et al., 2000; Radicella and Leitinger, 2001). All these models are based on the DGR "profiler" concept further developed by Radicella and Zhang (1995) and provide electron density as a function of solar activity (10.7 cm flux), month, Universal Time, geographic latitude, longitude and height. The "quick calculation" model NeQ uses a simple formulation for the topside F-layer, which is essentially a semi-Epstein layer with a thickness parameter that increases linearly with height. In its general form the model is driven by the ITU-R (former CCIR) global coefficients that describe the f0F2 and M3000 ionospheric characteristics. The version of the NeQ model used in this study was the one available on February 4, 2003.

In order to come up with significant conclusions on the comparison between the profiles retrieved by the IRO technique and the NeQ model, statistical studies were performed taking into account about 50000 electron density profiles. Mean values of electron density and plasma frequency differences between NeQ and IRO profiles (called biases) at fixed heights have been calculated from 95–400 km height with a resolution of 5 km. Because the relationship between plasma frequency (fp) and electron density (Ne) is not linear, both were calculated separately.

The dispersion of these differences (RMS) have also been calculated to indicate the variability between the model and the observations.

## 2 Data basis

The CHAMP data are automatically processed in the DLR/Institute of Communications and Navigation by an operational data processing system (Wehrenpfennig et al., 2001). The computed data products are made available to the international science community via the Information and Science Data Center (ISDC) of GFZ Potsdam.

For retrieving vertical electron density profiles, a tomographic approach is established that uses spherically layered voxels with constant electron density (Jakowski, 1999, Jakowski et al., 2002). The IRO measurements onboard CHAMP provide more than 150 vertical EDPs per day (see Jakowski et al., this issue). The database for this comparative study (about 50000 profiles, period of May 2002 – April 2003) has been processed automatically without additional checking.

The CHAMP EDP retrievals have been validated from the CHAMP orbit down to the E-layer altitude by comparison with vertical sounding data. Jakowski et al. (this issue) report corresponding results for the European region, which indicate a systematic positive bias of the IRO data in the order of less than 0.6 MHz plasma frequency (fp) or $8 \times 10^{10}$ m$^{-3}$ electron density (Ne) respectively and a standard deviation of less than 1.0 MHz ($1.3 \times 10^{11}$ m$^{-3}$) throughout the entire profile.

When comparing and discussing the IRO data with NeQ values in the following section, these validation results can be used as a valuable reference.

## 3 Results and Discussion

Keeping in mind the validation results of IRO derived EDP for mid-latitudes, the comparison with NeQ data might provide valuable information on the quality of the NeQ model if deviations exceed the measured bias and the standard deviation range.

Fig. 1 shows the bias and standard deviation between NeQ and IRO data for all measurements under consideration. The centered line in each panel represents the mean bias, surrounded by two lines indicating 1σ deviation. The comparison indicates quite good agreement in the F2 layer altitude range above 300 km height. Ignoring the E-layer, the maximum deviation (underestimation) of –1 MHz fp ($1.4x10^{11}$ $m^{-3}$ Ne) is observed in the 200–250 km height range. To study the latitudinal and local time dependence of the difference, the data set has been divided into three parts representing high ($|\phi| > 60°$), medium ($30 \leq |\phi| \leq 60°$) and low latitude ($|\phi| < 30°$) ranges. These data sets are subdivided in 4 groups corresponding to different local times (night, morning, day, evening) as it is shown in Fig. 2. Ignoring the E-layer, the mean deviation of the plasma frequency fp (electron density Ne) is generally less than 1.2 MHz ($1x10^{11}$ $m^{-3}$) at high latitudes and less than 1.8 MHz ($3x10^{11}$ $m^{-3}$) at mid latitudes under daytime conditions in the 220–230 km height range. At low latitudes the bias reaches about 1.0 MHz ($2.5x10^{11}$ $m^{-3}$) that is most pronounced at all day-time conditions in the altitude range 200–280 km.

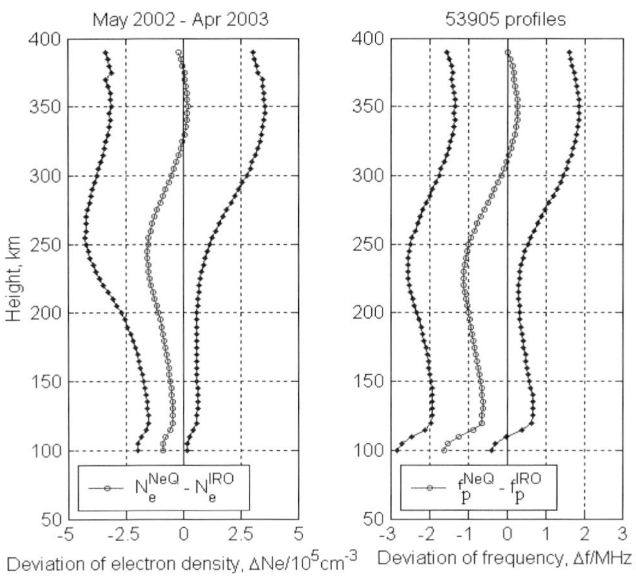


**Fig. 1.** Comparison of CHAMP IRO data with corresponding NeQ EDP, according to the difference NeQ–IRO (May 2002 – April 2003, 53905 profiles). Left – electron density Ne, right – plasma frequency fp.

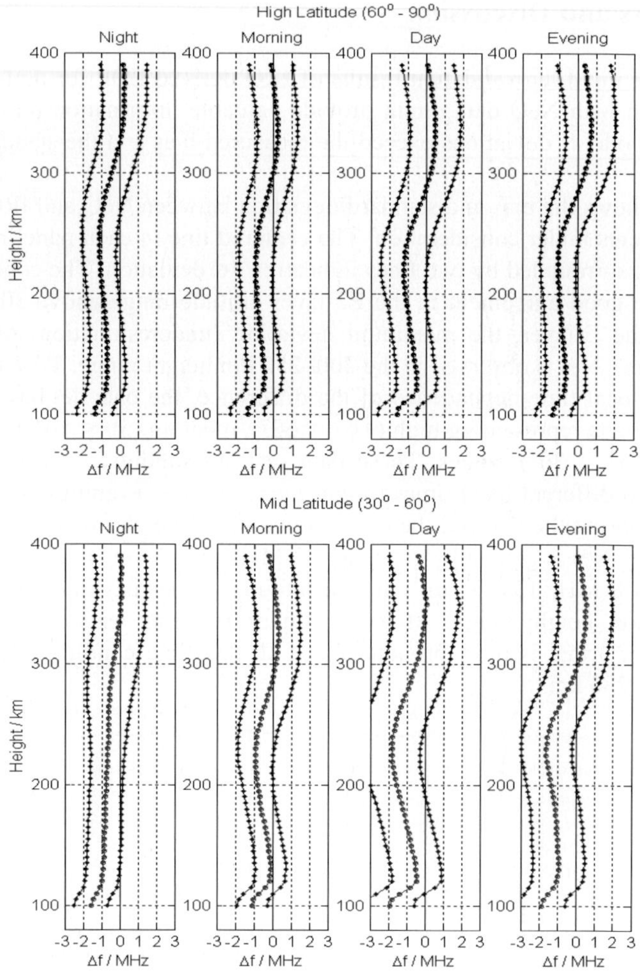


**Fig. 2a, b.** EDP comparison (NeQ–IRO) at *high* (upper panel) and *mid* (lower panel) latitude sectors (May 2002 – April 2003).

**Table 1.** Maximum electron density and plasma frequency biases, RMS values and corresponding heights for high, mid and low latitudes (according to NeQ – IRO).

| Latitude | Electron Density | | | Plasma Frequency | | |
|---|---|---|---|---|---|---|
| | Bias (x1011 m-3) | RMS (x1011 m-3) | Height (km) | Bias (MHz) | RMS (MHz) | Height (km) |
| High | < 1.5 | < 2.5 | 240 | < 1.0 | < 1.5 | 230 |
| Mid | < 3.5 | < 4.0 | 240 | < 1.6 | < 2.0 | 240 |
| Low | < 2.5 | < 6.0 | 270 | < 1.0 | < 2.3 | 260 |

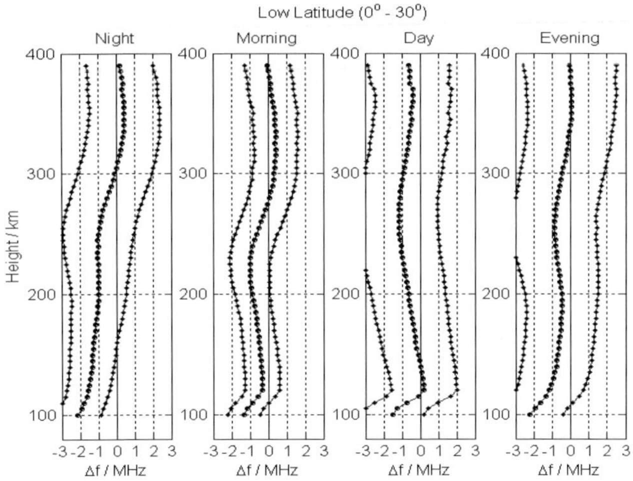


**Fig. 2c.** Same as 2a, b, but for *low* latitude.

It is difficult to comment the low latitude results because entire IRO derived Ne profiles have not yet been validated in a systematic way for this latitude range. We are aware of the fact that the spherical symmetry assumption in IRO retrievals is violated in particular in the crest region due to strong horizontal gradients of the electron density distribution and therefore errors cannot be excluded. However, it has to be mentioned that the biases are smaller than those obtained for mid-latitudes. The enhanced absolute dispersion is surely due to the high ionisation at low latitudes.

Considering the fact that the bias is only slightly positive (< 0.4 MHz) throughout the entire profile at least in mid-latitudes (Jakowski et al., this issue), the underestimation of the electron density in the height range of about 200–300 km is a clear finding. Due to the persistence of this feature at all latitude ranges we assume that NeQ generally underestimates the ionospheric ionisation in this height range under high solar activity. In order to check the consistency of the results, computations have been made separately for the Northern and Southern hemispheres. Comparing bias and dispersion values of corresponding latitude ranges, no substantial difference can be found. Excluding the E-layer range, the percentage accuracy (Δfp/fp) is generally better than 24%. Although the Δfp/fp may exceed 100% in the E-layer range, it is surprising that the agreement is best at low latitudes (deviation < 16 % ) where the ionospheric conditions for retrieving the EDPs from IRO measurements are not optimal.

The maximum average deviations of the electron density and plasma frequency are summarized in Table 1 together with the corresponding altitudes. In any case, further improvements, e.g. the improvement of the topside ionosphere model, are planned to enable more accurate observations.

## 4 Summary and Conclusions

We have reported preliminary results obtained by analyzing GPS radio occultations through the ionosphere measured onboard the CHAMP satellite in the period May 2002–April 2003. The retrieval results obtained by applying a model assisted retrieval technique are promising. Because the radio occultation technique is rather new, a systematic validation of data products in particular with ionosonde and incoherent scatter data is going on. Generally speaking, the comparison with ionospheric models is helpful for improving their performance. The comparison of IRO data with corresponding NeQ data indicates a systematic underestimation of the electron density in the 200–300 km height range of NeQ compared with the IRO data. Future versions of NeQ (here we used the version of February 4, 2003) should take into account this fact. It should be underlined that the IRO vertical density profiles analyzed here are the direct output of the automatically working IRO processing system without further reviewing. Although the number of IRO retrieval outliers is rather small (< 1%), their removal would certainly improve the agreement between the NeQ model data and the CHAMP IRO measurements.

*Acknowledgments.* The authors are very grateful to all colleagues from the international CHAMP team who keep CHAMP in operation.

## References

Hochegger G, Nava B, Radicella S, Leitinger R (2000) A family of ionospheric models for different uses. Phys Chem Earth (C) *25(4)*: 307-310.

Jakowski N (1999) Capabilities of radio occultation measurements onboard LEO satellites for ionospheric monitoring and research. Proc 4th COST 251 Workshop 'The Impact of the Upper Atmosphere on Terrestrial and Earth-Space Communications' (ed. Vernon A), 22-25 March, Funchal, Madeira, Portugal, pp 116-121.

Jakowski N, Wehrenpfennig A, Heise S, Reigber Ch, Lühr H, Grunwaldt L, Meehan TK (2002) GPS radio occultation measurements of the ionosphere from CHAMP: Early results. Geophys Res Lett *29(10)*: 95(1)-95(4).

Radicella S, Zhang M-L (1995) The improved DGR analytical model of electron density height profile and total electron content in the ionosphere. Annali di Geofisica *38(1)*: March.

Radicella S, Leitinger R (2001) The evolution of the DGR approach to model electron density profiles. Adv Space Res *27(1)*: 35-40.

Wehrenpfennig A, Jakowski N, Wickert J (2001) A Dynamically Configurable System for Operational Processing of Space Weather Data. Phys Chem Earth *26*: 601-604.

Jakowski N, Tsybulya K, Stankov S, Wehrenpfennig A (2004) About the Potential of GPS Radio Occultation Measurements for Exploring the Ionosphere. This issue.

# Model for Short-term Atmospheric Density Variations

Mark Zijlstra[1], Stephan Theil[2], and Silvia Scheithauer[2]

[1] Faculty of Aerospace Engineering, Delft University of Technology, Kluyverweg 1, 2629 HS Delft, The Netherlands, *MarkZijlstra@gmx.net*

[2] Center of Applied Space Technology and Microgravity, University of Bremen, Am Fallturm, D-28359 Bremen, Germany, *theil@zarm.uni-bremen.de*, *scheithauer@zarm.uni-bremen.de*

**Summary.** Rising accuracy requirements for attitude and translation control systems of satellites require better models for the expected disturbance forces. A major non-conservative disturbance force in Low Earth Orbits (LEO) is the atmospheric drag. With current atmospheric models an estimation of the drag force can be calculated, however these atmospheric density models, like the NRLMSISE-00[3] model, do not include short-term density variations. These density variations can no longer be disregarded. This paper deals with the development of a model that does include these short-term density variations.

**Key words:** short-term density variations, model, CHAMP accelerometer data

## 1 Introduction

The accuracy requirements for attitude and translation control are rising continuously for current and future scientific satellites. Missions like Gravity Probe-B, STEP, GOCE, MICROSCOPE and precision pointing missions like HYPER have to take into account disturbance forces and torques which were negligible for former satellites.

For control system design, as well as for its verification by simulation, an accurate model of the disturbances is essential. On one hand disturbance forces are a precondition for the design, on the other hand they are used for verification and performance analysis. For Low Earth Orbits, the Earth atmosphere exerts one of the main non-conservative disturbance forces. Especially the short-term density variations, which a satellite undergoes in its orbit, are of major importance for the design of a satellite control system.

Current atmospheric models, like the NRLMSISE-00[3] model (Naval Research Laboratory, Mass Spectrometer, Incoherent Scatter Radar Extended model 2001) do not include these short-term density variations. In this paper we derive a model for the short-term density variations of the upper atmosphere from the CHAMP[2] accelerometer data.

## 2 Preparations

For the purpose of obtaining realistic density time series, use is made of the drag force measurements of the CHAMP satellite, which is equipped with a highly accurate accelerometer. Because the atmospheric density is the most dominant non-conservative disturbance force on this satellite, the standard formula for the drag force (formula 1) can be used to estimate the local atmospheric density

$$F_d = \frac{1}{2}\rho * C_D * A * V_i^2 = m_{sc} * a_{sc} \tag{1}$$

where $F_d$ is the drag force, $\rho$ is the atmospheric density, $C_D$ is the experimental drag coefficient (regarded as fitting parameter), $A$ is the surface of the spacecraft perpendicular to the velocity, and $V_i$ the velocity of the atmosphere relative to the satellite. Further $m_{sc}$ and $a_{sc}$ are respectively the mass and the non-conservative acceleration of the spacecraft in flight direction. This calculation is made for all the data between July $1^{st}$ 2001 and January $31^{st}$ 2002, and will be referred to as the 'CHAMP density time series'.

A matching time series is calculated with the NRLMSISE-00 atmospheric model, which will be referred to as the 'NRLMSISE-00 time series'.

## 3 Theory

Comparison of the CHAMP density time series with the NRLMSISE-00 density time series reveals that the NRLMSISE-00 model is missing the characteristic high frequency content of the real density (see figure 1). A model will be made (called the Density Variation Model or DVM) to compensate for this. The DVM does not aim at predicting the density, but aims at reproducing this typical high frequency content found in real density time series. This high frequency content is the result of the density fluctuations on short-term, and on small spatial distances. The frequency spectrum may possibly also reveal some underlying characteristics. It will for example be possible to find a cell structure in the thermosphere (as is common in the low atmosphere) if these cells have a particular size-distribution.

More approaches have been tried to create a model that can produce density time series that contain a similar high frequency content.

### 3.1 Frequency domain noise isolation

The DVM must be able to produce time series that are in shape comparable to the residual of a true density time series. This can also be done by creating a signal that has a Power Spectral Density (PSD) which is equal to the difference of the PSDs of the CHAMP density time series and the NRLMSISE-00 density time series.

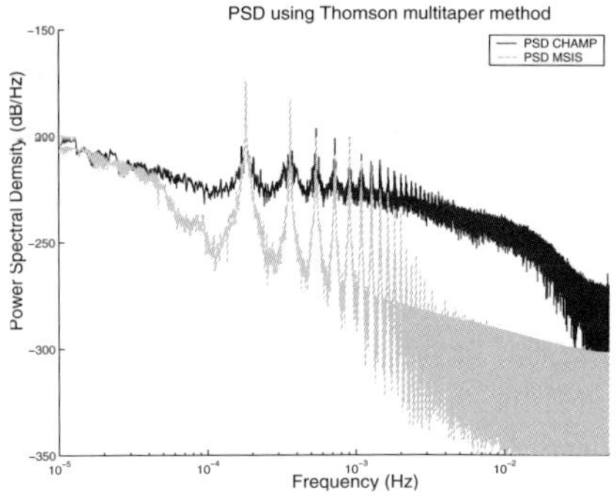


**Fig. 1.** PSDs for density time series of January 2002, using Thomson multitaper method.

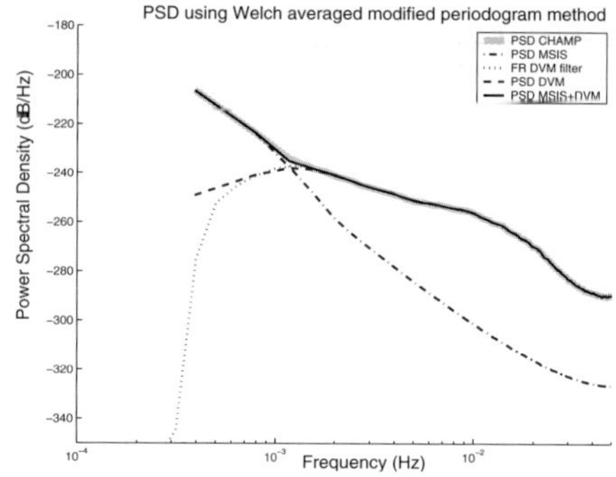


**Fig. 2.** PSDs for density time series of January 2002, using Welch's averaged modified periodogram method.

By creating a time series of white Gaussian noise, and leading it through a filter, a time series can be made for which its power spectral density equals this difference. The filtered time series is the DVM.

The sum of the DVM time series and the NRLMSISE-00 density time series therefore will have the same PSD as the PSD of the CHAMP density time series, and can be used as a more realistic density time series for the purpose of testing drag free control algorithms.

The desired frequency response of the DVM-filter is exactly the difference between the PSD of the CHAMP derived density measurements and the PSD of the NRLMSISE-00. To create a smooth PSD estimation for both time series a Welch's averaged modified periodogram method [1] is used, which averages the PSD over the neighboring frequency bands. One must keep in mind that this method performs badly in the neighborhood of spectral peeks. Next the desired frequency response (the difference between the two Welch PSDs) is interpolated onto a dense evenly spaced grid. The filter coefficients are obtained by applying an inverse fast Fourier transform to the grid and multiplying it by a Hamming window.

## 4 Implementation

The DVM is constructed the way described in the previous paragraph, using the frequency domain noise isolation method.

Figure 1 contains a PSD plot of the CHAMP and NRLMSISE-00 density time series, calculated with the Thomson multitaper method[1] (MATLAB function 'pmtm'). Because these PSDs have an irregular character, smoother PSDs are calculated using the Welch's averaged modified periodogram method, which can be found in figure 2.

Some distinct properties of these plots are:

- The orbit frequency is the first peak on the left at a frequency of $10^{-3.75}Hz$ corresponding to an orbital period of about 1.5 hours (see figure 1). With a sampling frequency of $0.1Hz$, the Nyquist frequency is $0.05Hz$ and is positioned outside this figure.
- On the far right of both figures, all the noise bands make a turn to follow a horizontal line. This feature is directly connected to the interpolation method (interpolation error) used to match the coordinates of the CHAMP satellite, to the accelerometer measurements. This interpolation had to be done, because the sampling rate of the accelerometers of CHAMP is much higher than the sampling rate of the calculated coordinates. The PSD becomes unreliable in the neighborhood of the sampling frequency of the coordinates and will therefore be disregarded from there on.
- The discrepancy between the PSDs of the NRLMSISE-00 and the CHAMP derived density is clearly not evenly distributed over the spectrum. Figure 2 clearly indicates that the frequency content of the NRLMSISE-00 density time series (PSD MSIS) only matches the PSD of the CHAMP density time series (PSD CHAMP) for the low frequencies (a couple of times the orbit frequency). The NRLMSISE-00 model's PSD lacks on power at higher frequencies.
- Figure 2 also contains the difference between the smoothened PSDs of the NRLMSISE-00 and the CHAMP derived density ('PSD DVM'), which forms the wanted frequency response for the DVM model.

Now a digital filter is made that has an equal frequency response as the wanted Frequency Response (figure 2, 'PSD DVM') using the MATLAB 'fir2' command. This digital filter can now be used to filter white Gaussian noise, in order to obtain the wanted DVM time series. The 'FR DVM filter' line in figure 2 gives the frequency response of the DVM filter. Keep in mind that the frequencies below $10^{-3.3}Hz$ are repressed for reasons that will later on be discussed. The filter is of the form

$$H(z) = \frac{b(1) + b(2) * z^{-1} + \ldots + b(m+1) * z^{-m}}{a} \tag{2}$$

where $H(z)$ is the transfer function, $b(i)$ are the filter coefficients with $i = 1 \ldots m+1$ and $a = 1$ denominator coefficient. The difference between the frequency response of the DVM filter and the wanted frequency response ('PSD DVM') is very small, which validates the filter.

Looking at the time series found this way, it became clear that the DVM time series had multiple very slow cyclical signals with a high amplitude in it. Of course the small difference between the PSDs of the NRLMSISE-00 model and the CHAMP derived densities for low frequencies was the reason.

The frequency cut-off visible for the NRLMSISE-00 model time series is obviously caused by the finite length of the polynomial series used in the model. This cut-off frequency can roughly be estimated by looking again to figure 2. It is clear that the frequency content of the NRLMSISE-00 model and the real situation really starts to differ around $10^{-3}Hz$.

For this reason the digital filter has been altered in such a way that the low frequency signals below $10^{-3.3}Hz$ (about 1/hour) are filtered out. The frequency response of the filter clearly indicates this (see figure 2, dotted line). An example of the resulting time series can be found in figure 3.

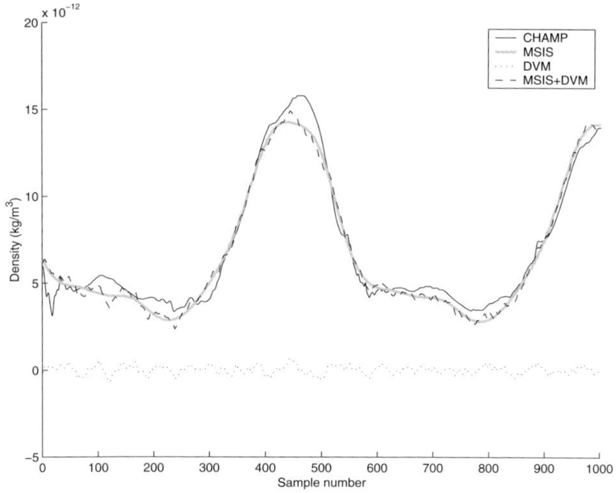


**Fig. 3.** CHAMP, NRLMSISE and DVM time series (10 sec per sample).

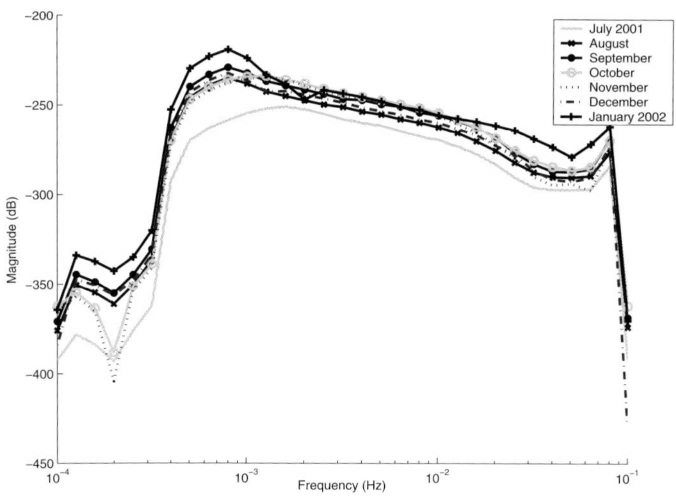


**Fig. 4.** Frequency responses of monthly DVMs.

To validate the DVM, the DVM modeling algorithm is applied to different time periods of data. The resulting DVMs can be seen in figure 4.

The DVMs all look alike, although one typical feature cannot be ignored. The level of 'noise' in the DVM time series seems to increase with time. This is the result of the slowly declining altitude of the CHAMP satellite over time. It can be concluded that the shape of the PSDs does not change with altitude, but that a quantitative correction is needed which is linearly related with the mean density. Nevertheless the frequency responses can be normalized with the mean atmospheric density sensed over that month.

## 5 Conclusion & Outlook

The Density Variation Model (DVM) was developed as an add-on to atmospheric standard models in order to model the high frequency density variations of the upper atmosphere.

The Density Variation Model performed as it was intended, and is therefore practically applicable for the purpose of testing drag-free control algorithms.

Of course every model can be improved, and a possible direction for this improvement can be found in making the model dependent on variables like the local solar time, altitude, latitude and the solar activity.

## References

1. Hayes M (1996) Statistical Digital Signal Processing and Modeling. John Wiley & Sons.
2. The CHAMP Mission. GeoForschungZentrum Potsdam http://gfz-potsdam.de/champ
3. NRLMSISE-00 Model (2001) National Aeronautics and Space Administration ftp://nssdcftp.gsfc.nasa.gov/models/atmospheric/msis/nrlmsise00/ nrlmsis00_c_version

# Atmospheric Profiling with CHAMP: Status of the Operational Data Analysis, Validation of the Recent Data Products and Future Prospects

Jens Wickert, Torsten Schmidt, Georg Beyerle, Grzegorz Michalak, Rolf König, Julia Kaschenz, and Christoph Reigber

GeoForschungsZentrum Potsdam (GFZ), Department 1, Geodesy & Remote Sensing, Potsdam, Germany, *jens.wickert@gfz-potsdam.de*

**Summary.** Atmospheric sounding with the German CHAMP (CHAllenging Minisatellite Payload) satellite is successfully performed since February 2001. In total ~145,000 precise globally distributed vertical profiles of refractivity, temperature and water vapor were provided as of April 2004. The operational occultation infrastructure from GFZ allows for the demonstration of Near-Real Time (NRT) data analysis since February 2003. An average delay of ~5 hours between each measurement and provision of corresponding analysis results is continuously reached. A comparison with more than 10,000 radiosonde measurements shows nearly bias-free refractivity and temperature between ~7 and ~30 km. The standard deviation is ~1 % and ~2 K, respectively. Data of the SAC-C (Satelite de Aplicaciones Cientificas-C) occultation mission are used to prepare for multi-satellite capability of the operational data analysis system. Future prospects of the provision of occultation analysis results at GFZ are given.

**Key words:** GPS, radio occultation, CHAMP, SAC-C, atmospheric profiling

## 1 Introduction

Atmospheric profiling aboard the German CHAMP [1] satellite was activated in February 2001 [2]. The experiment brought significant progress [5] for the innovative GPS (Global Positioning System) radio occultation technique [3] in relation to the pioneering GPS/MET (GPS/METeorology) mission [4]. The main advantages of this calibration-free sounding method are global coverage, high vertical resolution and all-weather-capability combined with high accuracy. These properties allow for various applications in atmospheric/ionospheric research, weather forecast and climate change detection [3, 6, 7].

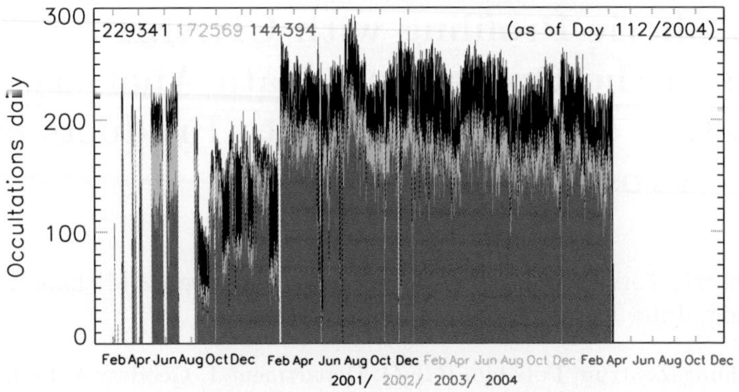


**Fig. 1.** Number of daily CHAMP occultations (duration > 20 s) as of April 21, 2004. The total height of the columns corresponds to the number of daily measurements. The black and light grey color indicate occultations with calibration failure and insufficient data quality, respectively. The height of the dark grey columns corresponds to the number of vertical atmospheric profiles provided to the CHAMP data center at GFZ.

## 2 Operational data analysis

The occultation measurements aboard CHAMP are performed by the state-of-the-art GPS receiver 'BlackJack' provided by the Jet Propulsion Laboratory (JPL). The operational GFZ occultation ground infrastructure allows for demonstration of NRT data analysis and provision. It was installed within the German GPS Atmosphere Sounding Project (GASP, *www.gfz-potsdam.de/gasp*). Details on the infrastructure can be found in [5, 8]. The current NRT scenario is described by [9]. An average delay of ~5 hours between each measurement and provision of corresponding analysis results is continuously reached since February 2003. This demonstration is an important milestone for the future assimilation of GPS radio occultation data in numerical weather models. A positive impact of the CHAMP data on global weather forecasts was already shown by [10]. The results are encouraging while awaiting the future multi-satellite occultation missions as e.g. COSMIC [11] or ACE+ [12]. The usefulness of CHAMP data for the possible detection of global climate trends was demonstrated by [13].

## 3 Status of the experiment

Occultation measurements were performed during 1,043 days since February 2001 as of April 21, 2004; giving a total of 229,341 recorded events (~220 daily). The calibration process was successful for ~75,2 %, thus 172,569 atmospheric excess phases are available (see Fig. 1). Vertical profiles of atmospheric parameters were derived for 144,394 occultations (~63.0 %). The yield of the profiles in relation to the number of measurements is currently investigated in more detail within the framework of the Radio Occultation Sensor Evaluation activity (ROSE), jointly initiated by GFZ, JPL and UCAR (University Corporation for Atmospheric Research) [14, 15]. ROSE is aimed to evaluate and optimize the quality of CHAMP's analysis results and to improve the involved occultation processing systems. A similar study (starting at the atmospheric excess phase level) is performed by GFZ and the University of Graz [16].

## 4 Validation of vertical atmospheric profiles

Early CHAMP analysis results were validated by [2, 17], here we focus on the recent product version (004). The next generation of data products (005), generated by using more advanced retrieval techniques will be provided as of spring 2005. A comparison of more than 35,000 dry temperature and refractivity profiles with ECMWF (European Centre for Medium-Range Weather Forecasts) between 10 and 35 km shows a temperature bias of <0.5 K and a mean deviation of the refractivity of <0.5 %. A height dependent standard deviation of ~1 K (~0.6 %) at 10 km and 2 K (~1 %) at 30 km is observed in the temperature and refractivity comparison, respectively [5]. A statistical comparison of ~10,000 CHAMP measurements with corresponding radio sonde data confirms these good agreements (see Fig. 2). CHAMP data were also compared with ENVISAT/MIPAS (Michelson Interferometer for Passive Atmospheric Sounding) within this height interval [18]. A negative refractivity bias in relation to meteorological analyses and radio sonde data is found in the lower troposphere, which is characterized in more detail in [19, 20]. The application of advanced analysis techniques [19] can significantly reduce

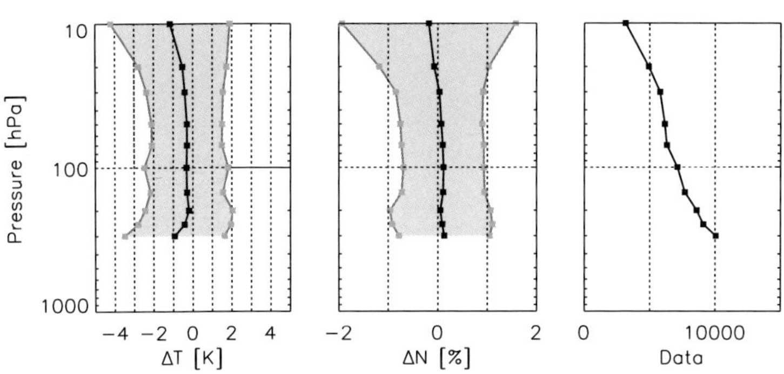


**Fig. 2.** Statistical comparison of CHAMP dry temperatures (left) and refractivities (middle) with radiosonde (RS) data (CHAMP-RS) for ~10,000 soundings ($\Delta$ t<3 h, $\Delta$ d<300 km). The height dependent number of compared data is shown in the right panel. The decreasing number with increasing height is caused by missing radiosonde data.

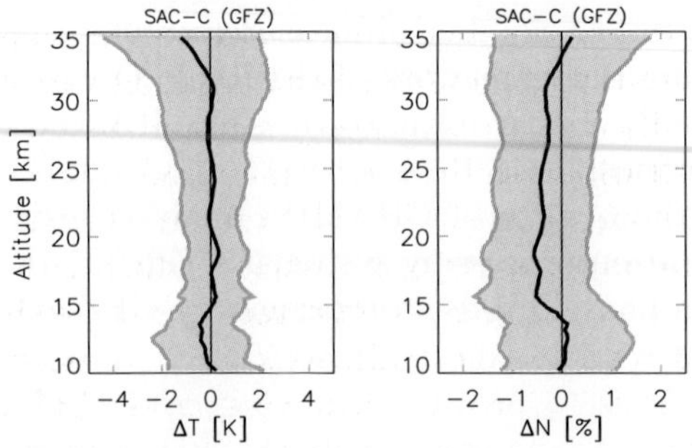


**Fig. 3.** Statistical comparison (SAC-C - ECMWF) between ~1,200 vertical temperature (left) and refractivity (right) profiles from SAC-C and corresponding analysis data (between March 12 and 20, 2002).

this bias. However further work is needed to implement these techniques to the operational occultation data analysis.

## 5 Initial results of operational SAC-C data analysis

GPS radio occultation measurements are also performed by the Argentinean SAC-C satellite (launched on November 21, 2000 [21]). The data are provided by JPL (*http://genesis.jpl.nasa.gov*) and were used to test the multi-satellite capability of the GFZ occultation processing system. 56,051 measurements, recorded during 272 days in 2001 and 2002 were analyzed using the operational occultation analysis system at GFZ. The 3D accuracy of the SAC-C orbits is ~10 cm [22]. A preliminary validation study, comparing ~1,200 vertical temperature and refractivity profiles with corresponding ECMWF data (see Fig. 3) between 10 and 35 km shows nearly identical results as for CHAMP (see Chap. 4).

## 6 Conclusions and outlook

After more than two and a half years of GPS radio occultation with CHAMP ~230,000 occultation measurements were performed. As the mission is expected to last until 2007, the first unique long-term data set of GPS occultation measurements is anticipated. The data and analysis results are available at the Information system and Data Center (ISDC) at GFZ (*http://isdc.gfz-potsdam.de/champ*). A NRT occultation analysis is continuously demonstrated since February 2003. Currently an average delay of ~5 hours between each measurement aboard CHAMP and data product provision is reached. Recent information on the status of the experiment, the operational data analysis at GFZ and current validation results can be obtained via WWW (*http://www.gfz-potsdam.de/gasp*). The multi-satellite capability of the GFZ occultation analysis system was demonstrated using data of the SAC-C satellite. Validation results indicate that upper troposphere and stratospheric temperatures agree well with analyzes and radio sonde observations. Mean temperature deviations are below 1 K and standard deviations fall within the 1-2 K range. The mean refractivity deviation is <0.5 %. A negative refractivity bias in the lower troposphere is observed. It can be significantly reduced by using advanced retrieval methods, which are foreseen for the lower troposphere data analysis to provide next version of CHAMP data products. The CHAMP data and analysis results are in use by more than 40 scientists and research groups all over the world. They are exploited to prepare and improve analysis centers for future occultation missions, to demonstrate improvement of global weather forecasts, to show the ability to detect climate trends, and to improve other satellite based atmosphere sounders. The successful CHAMP radio occultation experiment is an important milestone towards operational constellations for continuous and precise monitoring of the Earth's atmosphere using the GPS radio occultation technique.

*Acknowledgement.* We are grateful to all scientists, engineers and technicians, which made the success of the CHAMP mission possible. We thank our colleagues from JPL for providing the GPS data from the SAC-C experiment, K. Schöllhammer for the radio sonde data preparation, U. Foelsche and G.B. Larsen for constructive review of the manuscript.

## References

1. Reigber Ch, et al. (2002) CHAMP home page, (*www.gfz-potsdam.de/champ*)
2. Wickert J, Reigber Ch, Beyerle G, König R, Marquardt C, Schmidt T, Grunwaldt L, Galas R, Meehan TK, Melbourne WG, and Hocke K (2001) Atmosphere sounding by GPS radio occultation: First results from CHAMP. Geophys Res Lett *28*: 3263–3266.
3. Kursinski ER, Hajj GA, Hardy KR, Schofield JT, and Linfield R (1997) Observing Earth's atmosphere with radio occultation measurements using the Global Positioning System. J Geophys Res *102*: 23429–23465.
4. Ware R et al. (1996) GPS Sounding of the atmosphere from low Earth orbit: Preliminary results. Bull Am Meteorol Soc *77*: 19-40.
5. Wickert J, Schmidt T, Beyerle G, König R, Reigber Ch, and Jakowski N (2004) The radio occultation experiment aboard CHAMP: Operational data analysis and validation of vertical atmospheric profiles, J Meteorol Soc Jpn *82(1B)*: 381–395.
6. Hajj GA, Lee LC, Pi X, Romans LJ , Schreiner WS, Straus PR, and Wang Ch (2000) COSMIC GPS Ionospheric Sensing and Space Weather. Terrestrial, Atmospheric and Oceanic Science *11*: 235–272.
7. Kuo YH, Sokolovskiy SV, Anthes RA, and Vandenberghe F (2000) Assimilation of GPS radio occultation data for numerical weather predicition. Terrestrial, Atmospheric and Oceanic Science *11*: 157–186.

8. Wickert J (2002) The CHAMP radio occultation experiment: Algorithms, Processing system, and First results (in German). Scientific Technical Report 02/07, GFZ Potsdam, Germany.
9. Schmidt T, Wickert J, Beyerle G, König R, Galas R, and Reigber Ch (2004) The CHAMP Atmospheric Processing System for Radio Occultation Measurements. This issue.
10. Healy SB, Jupp AM, and Marquardt C (2003) A forecast impact trial with CHAMP radio occultation measurements. Abstract EAE03-A-08710, EGS-EUG-AGU Joint Assembly, Nice, France, April 2003.
11. Rocken C, Kuo Y-H, Schreiner WS, Hunt D, Sokolovskiy S, and McCormick Ch (2000) COSMIC System description. Terrestrial, Atmospheric and Oceanic Science *11*: 21–52.
12. Hoeg P, and Kirchengast G (2002) ACE+ Atmosphere and Climate Explorer based on GPS, GALILEO, and LEO-LEO radio occultation (ESA Earth Explorer Opportunity Mission, Proposal). Wissenschaftl. Bericht 14, IGAM, University of Graz, Austria.
13. Schmidt T, Wickert J, Beyerle G, König, and Reigber Ch (2004) Tropopause parameters derived from GPS radio occultations with CHAMP. This issue.
14. Ao CO, Schreiner WB, and Wickert J (2003) First Report on the CHAMP Radio Occultation Intercomparison Study. JPL Publication 03-016, Pasadena, U.S.
15. Wickert J, Ao CO, Schreiner WB, Beyerle G, Schmidt T, König, Reigber Ch, Hajj GA, Iijima BA, Hunt D, Sokolovskiy S, and Rocken C (2003) GPS radio occultation with CHAMP: First comparisons of analysis results from GFZ, JPL and UCAR. Abstract EAE03-A-03096, EGS-EUG-AGU Joint Assembly, Nice, France, April 2003.
16. Wickert J, Gobiet A, Beyerle G, Steiner AK, Kirchengast G, Foelsche U, and Schmidt T (2004) GPS radio occultation with CHAMP: Comparison of inversion results from GFZ Potsdam and IGAM Graz. This issue.
17. Marquardt C, Schöllhammer K, Beyerle G, Schmidt T, Wickert J, and Reigber Ch (2002) Validation and Data Quality of CHAMP Radio Occultation Data. In: First CHAMP mission results for gravity, magnetic and atmospheric studies, edited by Reigber C, Lühr H, and Schwintzer P, Springer-Verlag, Berlin: 384–396
18. Wang DY et al. (2004) Comparisons of MIPAS/ENVISAT and GPS-RO/CHAMP temperature profiles. This issue.
19. Beyerle G, Wickert J, Schmidt T, König R, and Reigber Ch (2004) An analysis of the lower tropospheric refractivity bias by heuristic sliding spectral methods. This issue.
20. Beyerle G, Wickert J, Schmidt T and Reigber Ch (2004) Atmospheric sounding by GNSS radio occultation: An analysis of the negative refractivity bias using CHAMP observations, J Geophys Res *109 (D1)*: doi: 10.1029/2003JD003922.
21. Hajj GA, Ao CO, Iijima BA, Kuang D, Kursinski ER, Mannucci AJ, Meehan TK, Romans LJ, de la Torre Juarez M, and Yunck TP (2003) CHAMP and SAC-C atmospheric occultation results and intercomparisons. J Geophys Res, submitted.
22. König R, Michalak G, Neumayer KH, Schmidt R, Zhu SY, Meixner H, and Reigber Ch (2004) Recent developments in CHAMP orbit determination at GFZ. This issue.

# Simulated Temperature and Water Vapor Retrieval from Bending Angles and Refractivity Measurements using an Optimal Estimation Approach

Axel von Engeln[1] and Gerald Nedoluha[2]

[1] University of Bremen, Institute of Environmental Physics, Otto-Hahn-Allee 1, D-28359 Bremen, Germany *engeln@uni-bremen.de*

[2] Naval Research Laboratory, Remote Sensing Division, Code 7227, 4555 Overlook Ave SW, Washington DC 20375, USA *nedoluha@nrl.navy.mil*

**Summary.** We present an Optimal Estimation processing of simulated radio occultation data. Temperature and water vapor profiles are retrieved from either bending angle or refractivity measurements. The advantages of one over the other are assessed. A comparison with idealized calculations shows the potential of ray tracer calculations.

**Key words:** radio occultation, optimal estimation, ray tracer, assimilation

## 1 Introduction

Radio occultation measurements can be converted into a profile of bending angles. This can be further processed to yield a refractivity profile by applying an Abel transform [2]. A direct processing of the refractivity profiles allows to derive temperature information in altitude regions where water vapor is insignificant; most of the early retrievals were focusing on this approach. At lower tropospheric altitudes either temperature or water vapor can be determined with this direct approach, assuming that the other quantity is well known [6].

More recently the measurement assimilation by variational systems has been addressed, e.g. [8, 5, 7, 9], allowing for the simultaneous determination of temperature and water vapor profiles by incorporating a priori information. Either bending angle or refractivity profiles can be used, where a forward model simulates the measurement, for example based on a simple 1-dimensional atmosphere, or a ray tracer. A ray tracer allows to properly address horizontal gradients along the integration path, which are especially important in the lower troposphere. But the ray tracer approach requires large computer resources and a 1-dimensional processing might be more desirable.

Within this study, temperature and water vapor profiles are retrieved from quasi-realistic simulated measurements using an Optimal Estimation approach to quantify the potential of radio occultation data especially in the lower troposphere [10]. Optimal Estimation represents a simplified variational processing. Additionally, these simulations are compared to idealized retrieval calculations in order to visualize the potential of ray tracer algorithms.

A simulation approach was used, since this allows us to prescribe the true state of the atmosphere. This is especially important if only small differences are present between the bending angle or refractivity retrieval approach, or between different retrieval setups. Within this study we shall use simulated measurements made from a single LEO satellite for one particular day in 2001 (May 19, 2001). In total 550 GPS occultation events were found for the simulated LEO satellite, and a subset of 110, chosen to provide a wide range of latitude measurements, was used in this study.

## 2 Setup

Measurements were generated using 2 forward models:

- The End-to-end GNSS Occultation Performance Simulator (EGOPS) is used to derive quasi-realistic simulations of bending angle and refractivity profiles [3, 4]. A 3-dimensional ECMWF field is input to a ray tracing process. The ECMWF horizontal resolution is about 0.351°, the vertical varies from about 20 m at the surface to about 250 m at 1 km altitude, and about 1 km to 3 km in the stratosphere. EGOPS includes a 3-dimensional ionosphere model, and a full error specification following the future radio occultation GRAS receiver [1].
- A 1-dimensional forward model, based on the Abel integral for bending angles and for refractivity on the work of [11] is used within the Optimal Estimation approach. This forward model is also used for the measurement generation in idealized calculations, thus neglecting horizontal inhomogeneities, ionosphere, and all non-Gaussian error distribution (e.g. biases). One dimensional ECMWF profiles following the tangent point are used here.

Measurement errors for bending angle and refractivity are given in Table 1, along with the vertical sampling [1]. These errors are used to fill the different error covariance matrices, thus assuming a normal distribution. The setup for the Optimal Estimation Method is briefly outlined in Table 2. We assume that there are no correlations in the a priori data and between different measurement levels. The a priori is taken from an ECMWF global field of the previous day for quasi-realistic and idealized calculations, the assumed error is in general agreement with a NWP model short range forecast capability [8].

**Table 1.** Sampling grid used for the simulated measurements, along with the corresponding bending angle (BA) and refractivity (REF) errors assumed.

| Height | Sampling $dz$ [km] | BA Error [$\mu$rad] | REF Error [1] |
|---|---|---|---|
| $00 \le z \le 25$ | 0.25 | 4.0 | 0.05 |
| $25 \le z \le 40$ | 0.50 | 2.8 | 0.036 |
| $40 \le z \le 60$ | 1.00 | 2.0 | 0.027 |

**Table 2.** Vertical retrieval grid and a priori errors used in the Optimal Estimation.

| Species | Height [km] | Sampling $dz$ [km] | A priori error |
|---|---|---|---|
| Temperature | $00 \le z \le 20$ | 0.25 | 2.5 K |
| | $20 \le z \le 30$ | 0.5 | 3.6 K |
| | $30 \le z \le 40$ | 1.0 | 5.8 K |
| | $40 \le z \le 60$ | 2.5 | 9.0 K |
| | $60 \le z \le 100$ | 10.0 | 15.6 K |
| Water Vapor | $00 \le z \le 20$ | 0.25 | 40 % |

## 3 Results

Figure 1 visualizes the retrieval results for temperature and water vapor from quasi-realistic simulations. Standard deviations with respect to the true profile (as determined within the ray tracing process at the tangent point) and improvements (as defined in [8]) over the a priori information are given for both species. A total of 100 occultations is shown, 10 occultations failed a quality check, showing strong adjustments in water vapor to compensate for temperature deviations. These erroneous adjustments are a general feature of this radio occultation processing using the Optimal Estimation Method and can also be found in idealized retrievals.

The actual standard deviation of the temperature a priori data is around 2 K to 4 K. Water vapor is provided with a standard deviation of about 50 % to about 200 % in the lowest 10 km. These values deviate from the assumed a priori errors just as in a retrieval from real measurements. The improvement plot for temperature shows that the retrieval from refractivity measurement works slightly better in the altitude range of 15 km to 35 km, yielding improvements over the a priori data of about 85 % to 45 %. Above, bending angle measurements are slightly better, since the integration of bending angles through all atmospheric layers above the tangent point provides additional information in an altitude region where the sensitivity for temperature retrieval decreases.

Improvements in temperature at lower altitudes are only found down to about 7 km, where water vapor is becoming more important. Retrieved temperatures below can even lead to a decrease of the retrieval quality over the

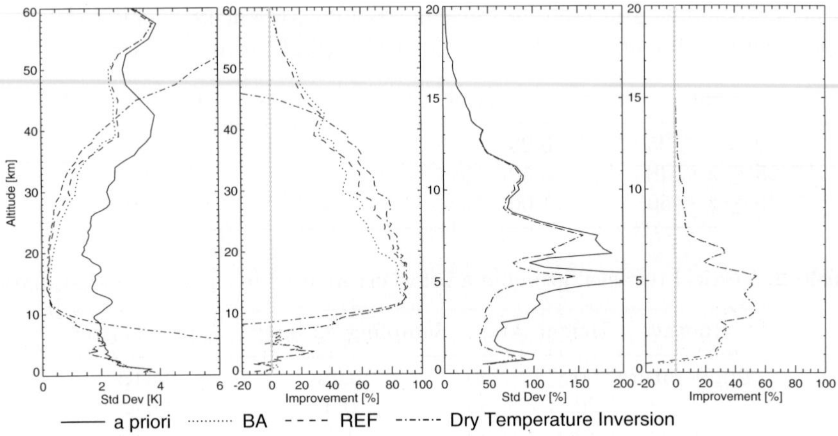


**Fig. 1.** Standard deviation and improvements of temperature (left) and water vapor (right) wrt true profile, using either bending angles (BA) or refractivity (REF) measurements. The dry temperature inversion, as given by EGOPS is also shown for temperature. Note: different altitude ranges.

a priori data. A more rigorous quality processing is necessary to filter out retrievals that lead to erroneous temperature results in the lower troposphere.

The direct processing to obtain a dry temperature profile yields higher improvements over the altitude range of 15 km to 42 km, but a priori data supports the temperature retrieval of the Optimal Estimation Method above and below, leading to better results. An optimization of the assumed retrieval errors will actually lead to similar results for direct and Optimal Estimation retrievals [12].

Water vapor improvements vary between about 20 % and 40 % for an altitude range of 1 km to 7 km, results are very similar for bending angle and refractivity measurements. This similarity does not hold for all locations on the Earth and is a feature of the horizontal ECMWF resolution and the averaging over all latitudes. Negative improvements are only found below 1 km, where the retrieved profile is less accurate than the a priori. But only a few occultations penetrate this layer, thus results are insufficient to draw conclusions.

A separation of the improvement by latitude band is shown in Fig. 2 for the refractivity retrievals. Although the number of occultation within each latitude band is small, some dominant features can also be reproduced for even smaller datasets.

The temperature retrieval in the lower troposphere is severely affected by the presence of water vapor. Low latitude temperature retrievals show improvements up to only about 10 km, while mid and high latitudes reach further down. This feature is independent of the retrieval simulation. Water vapor can be retrieved from quasi-realistic simulations with an improvement

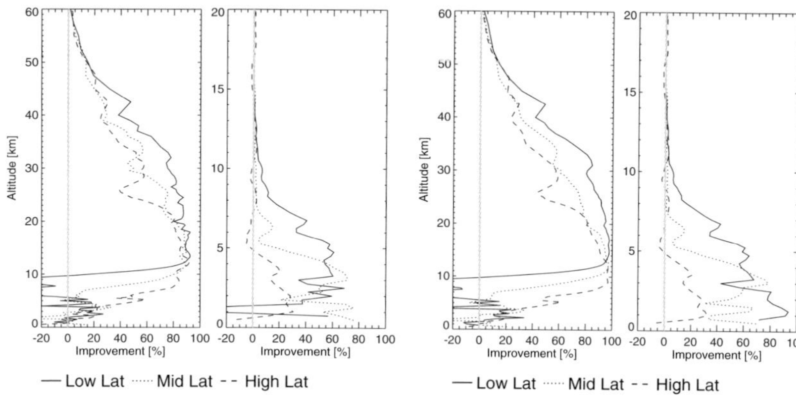


**Fig. 2.** Improvements of temperature (1. and 3. plot) and water vapor (2. and 4. plot) for quasi-realistic simulations (left) and for idealized simulations (right). Note: different altitude ranges.

of about 20 % for high latitudes and 50 % for mid and low latitudes within the lowest 4 km. Our idealized simulations, which do not include horizontal variability, show higher improvements in the lower troposphere, especially for low latitude occultations.

The advantages and disadvantages of bending angle assimilation using a ray tracer over refractivity assimilation are discussed in [13]. A comparison of quasi-realistic with the idealized retrievals shows the impact of horizontal inhomogeneities. Our calculations show that the major potential of a ray tracing algorithm is found for low and mid latitude water vapor retrieval.

## 4 Summary

Quasi-realistic simulations of radio occultation measurements have been used in an Optimal Estimation algorithm to assess the potential of radio occultation data assimilation. A total of 110 measurements were simulated, using high resolution ECMWF atmospheric fields. Profiles of temperature and water vapor were retrieved.

A comparison of refractivity over bending angle assimilation shows generally small differences, refractivity shows slightly better results in the lower stratosphere, while bending angles are slightly better in the upper stratosphere, lower mesosphere. The two measurement assimilation methods show almost no differences in the lower troposphere, neither for the water vapor, nor for the temperature profile retrieval. Refractivity allows faster CPU processing and seems therefor in most cases advantageous over bending angle assimilation.

*Acknowledgement.* This work was supported by an Internal Government Study from the Integrated Program Office of the National Polar-orbiting Operational Environmental Satellite System (NPOESS) and by the Office of Naval Research (ONR). A. von Engeln was partly funded by the German Federal Ministry of Education and Research (BMBF), within the AFO2000 project UTH-MOS, grant 07ATC04. The authors wish to thank Dr. N. Kreitz (ECMWF, Reading, UK) for support with the ECMWF data extraction.

## References

1. ESA/EUMETSAT (1998) The GRAS instrument on METOP. ESA/EUMETSAT Rep. (ESA No. VR/3021/PI, EUM No. EPS/MIS/IN/9), 38 p., ESA/ESTEC, Noordwijk, Netherlands.
2. Fjeldbo G, Kliore A, and Eshleman V (1971) The neutral atmosphere of Venus as studied with the Mariner v radio occultation experiments. Astron J *76*: 123–140.
3. Kirchengast G (1998) End-to-end GNSS Occultation Performance Simulator (EGOPS) overview and exemplary applications. Wissenschaftl. Ber. No. 2/1998, 138 pp, Inst. for Meteorol. and Geophys., Univ. of Graz, Austria.
4. Kirchengast G, Fritzer J, and Ramsauer J (2002) End-to-end GNSS occultation performance simulator version 4 (EGOPS4) software user manual. Techn. Rep. ESA/ESTEC-3/2002, 472 pp., Inst. for Geophys., Astrophys. and Meteorol., Univ. of Graz, Austria.
5. Kuo Y, Sokolovskiy S, Anthes R, and Vandenberghe V (2000) Assimilation of GPS radio occultation data for Numerical Weather Prediction. Terrestial, Atmospheric and Oceanic Science *11(1)*: 157–186.
6. Kursinski E, Hajj G, Schofield J, Linfield R, and Hardy K (1997) Observing earth's atmosphere with radio occultation measurements using GPS. J Geophys Res *102*: 23,429–23,465.
7. Liu H, Zou X, Shao H, Anthes R, Chang J, Tseng J, and Wang B (2001) Impact of 837 GPS/MET bending angles on assimilation and forecasts for the period June 20-30, 1995. J Geophys Res *106(D23)*: 31,771–31,786.
8. Palmer P, Barnett J, Eyre J, and Healy S (2000) A non-linear optimal estimation inverse method for radio occultation measurements of temperature, humidity, and surface pressure. J Geophys Res *105(D13)*: 17,513–17,526.
9. Poli P, Joiner J, and Kursinski E (2002) 1DVAR analysis of temperature and humidity using GPS radio occultation refractivity data. J Geophys Res *107(D20)*: ACL 14-1 to ACL 14-20.
10. Rodgers C (2000) Inverse Methods for Atmospheric Sounding: Theory and Practise. Series on Atmospheric, Oceanic and Planetary Physics, *vol 2*, World Scientific Publ (February 2000).
11. Smith E and Weintraub S (1953) The constants in the equation for atmospheric refractive index at radio frequencies. Proceedings of the IRE *41*: 1035–1037.
12. von Engeln A and Nedoluha G (2003) Retrieval of temperature and water vapor profiles from radio occultation refractivity and bending angle measurements using an optimal estimation approach: A simulation study. J Geophys Res, (submitted).
13. Zou X, Wang B, Liu H, Anthes R, Matsumura T, and Zhu Y (2000) Use of GPS/MET refraction angles in three-dimensional variational analysis. Q J R Meteorol Soc *126*: 3013–3040.

# An Analysis of the Lower Tropospheric Refractivity Bias by Heuristic Sliding Spectral Methods

Georg Beyerle, Jens Wickert, Torsten Schmidt, Rolf König, and Christoph Reigber

GeoForschungsZentrum Potsdam (GFZ), Dept. 1: Geodesy and Remote Sensing, Potsdam, Germany, *gbeyerle@gfz-potsdam.de*

**Summary.** The canonical transform (CT) and full spectrum inversion (FSI) method together with their heuristic sliding spectral modifications are validated using end-to-end simulation data and one week of CHAMP observations. In general, we observe a pronounced correlation between small refractivity biases and enhanced penetration altitudes. Processing of simulated occultation data shows that the heuristic methods exhibit smaller retrieval errors vindicating the assertion that the sliding spectral approaches react less sensitive to receiver tracking errors. The corresponding mean retrievals errors found in the CHAMP data analysis, however, are consistent within 0.5%; differences are observed with respect to penetration altitudes and the retrieval errors' standard deviations.

**Key words:** Remote sensing, GPS radio occultation, canonical transform, full spectrum inversion, refractivity bias

## 1 Introduction

Since February 2001 the GPS radio occultation instrument aboard CHAMP (CHAllenging Minisatellite Payload) [12, 13] continuously monitors the atmospheric refractive index $n$ in the tropo- and stratosphere. At mid and low latitudes water vapor may contribute to refractivity $N = (n-1)\cdot 10^6$ by up to 30%. Inhomogeneous distributions of atmospheric water vapor translate into strong vertical gradients of the refractivity profile allowing for the occurrence of multipath signal propagation. Bending angle retrievals based on single ray propagation, thus, are bound to fail at low latitudes.

To solve the problem of calculating bending angle profiles within multipath zones the canonical transform (CT) method [4, 5] and the full spectrum inversion (FSI) technique [9] have been introduced. In the upper troposphere and lower stratosphere good agreements between CT/FSI retrievals of CHAMP observations and meteorological model analyses are found. In the lower troposphere, however, validation studies consistently report on a negative refractivity bias of more than one percent [18, 11, 8, 3]. This bias was first noticed and discussed by Rocken et al. [14] within the GPS/MET data validation study. The origin of the lower tropospheric refractivity bias is commonly attributed to critical refraction and receiver tracking errors [5, 15, 1, 2, 3]. Here we report on bending angle retrievals using the CT and FSI methods, as well as their heuristic sliding spectral modifications. The four retrieval techniques are applied to two data sets: 30 occultation profiles derived from a forward model simulation and about 1100 CHAMP observations recorded during the time period 1–7 January 2003.

## 2 Wave optics retrieval methods

The canonical transform (CT) method was developed to solve the problem of calculating bending angle $\alpha(p)$ as a function of impact parameter $p$ within regions of multipath signal propagation [4, 5]. First, the signal observed by the GPS receiver along the LEO trajectory is backpropagated to an auxiliary line using the Kirchhoff diffraction integral [7]. Then, interfering rays are disentangled by transforming the backpropagated field to impact parameter space. The bending angle $\alpha(p)$ follows from the transformed signal's phase. For a detailed description see [4, 5].

End-to-end simulation studies including the receiver signal tracking process suggest that bending angles derived by the CT method are sensitive with respect to receiver-induced phase errors. A heuristic modification of the CT technique, the canonical transform / sliding spectral (CTss) method, seems to be more robust against receiver-induced tracking errors at the expense of a reduced vertical resolution [3]. In the CTss approach a radio occultation signal is subjected to a CT in a sliding aperture. That aperture position which provides maximal signal amplitude for a given impact parameter is used for estimating the bending angle [6]. For details on the CTss approach see [3].

Jensen et al. [9] found an alternative solution to the multipath problem. The full spectrum inversion (FSI) method takes advantage of the fact that for circular orbits the Doppler angular frequency occurs at most once during an occultation. I.e. the arrival time of a particular ray is a single-valued function of the signal's Doppler frequency. Deviations of the LEO orbit from the circular shape are taken care of by adding a phase correction to the observed signal [9],

$$u(\theta) \equiv a(\theta)\, \exp[i\,\phi(\theta)]\ , \tag{1}$$

and by using the angle $\theta = \angle(\mathbf{r}_G, \mathbf{r}_L)$ spanned by the GPS's and LEO's position vectors, $\mathbf{r}_G$ and $\mathbf{r}_L$ instead of time $t$. Jensen et al. [9] show that the relation between $\theta$ and its frequency $\omega$ can be obtained from the phase $\Phi$ of the signal's $u(\theta)$ Fourier transform, $\mathcal{F}[u(\theta)] \equiv A(\omega)\, \exp(i\,\Phi(\omega))$, and is found to be

$$\frac{d\Phi(\omega)}{d\omega} = -\tilde{\theta} \equiv -\theta(\omega)\ . \tag{2}$$

Impact parameter $p(\omega)$ is given by [9]

$$\omega = k\,p + k\,\left.\frac{dr_L}{d\theta}\right|_{\theta=\tilde{\theta}} \sqrt{1-\left(\frac{p}{r_L(\tilde{\theta})}\right)^2} + k\,\left.\frac{dr_G}{d\theta}\right|_{\theta=\tilde{\theta}} \sqrt{1-\left(\frac{p}{r_G(\tilde{\theta})}\right)^2} \tag{3}$$

where $r_G \equiv |\mathbf{r}_G|$, $r_L \equiv |\mathbf{r}_L|$ and $k = 2\pi/\lambda$ denotes the wave vector ($\lambda =$ 19.03 cm). Finally, the bending angle follows from

$$\alpha(p(\omega)) = \theta(\omega) - \arccos\left(\frac{p(\omega)}{r_G}\right) - \arccos\left(\frac{p(\omega)}{r_L}\right) \tag{4}$$

Details are given in [9].

The CTss method originates from the sliding spectral technique [16, 6] which might be regarded as a heuristic modification of the FSI technique (in the following denoted by FTss). Again we argue that the noise component of $\Phi(\omega)$ is amplified by the derivative in Eqn. 2; in the sliding spectral FTss approach this equation is replaced by an algorithm to determine the relationship between $\omega$ and $\theta$ from the Fourier transform of $u(\theta)$ in a sliding rectangular window of width $\Delta\theta$, i.e.

$$U(\omega, \theta) \equiv \mathcal{F}[w(\theta - \theta')\, u(\theta')] \tag{5}$$

where $w(\theta - \theta') = 1$ for $\theta - \Delta\theta/2 < \theta' < \theta + \Delta\theta/2$ and zero elsewhere, $\Delta\theta \approx 1$ mrad. For each $\omega$ the value of $\theta$, that maximizes $|U(\omega, \theta)|$, is regarded as the angle $\tilde{\theta}$ that corresponds to $\omega$, i.e.

$$\max_\theta(|U(\omega, \theta)|) = |U(\omega, \theta = \tilde{\theta})| \tag{6}$$

where the subscript indicates that the maximum is determined in $\theta$ space for a given $\omega$. Again, from $\tilde{\theta}$ the impact parameter $p = p(\omega)$ is calculated using Eqn. 3 and the bending angle $\alpha$ follows from Eqn. 4 [16].

## 3 Analysis and discussion

The four retrieval methods, CT and FSI, together with the corresponding sliding spectral techniques, are applied to a set of 30 simulated bending angle profiles. The occultation profiles are calculated using a Multiple Phase Screen forward model [10, 17], the corresponding refractivity profiles are obtained from radio sonde observations recorded on the Atlantic ocean between 66°N and 37°S in October/November 1996. First, the simulated signal is sampled at 1 kHz and −20 dB Gaussian noise is added (corresponding to a carrier-power-to-noise density ratio $C/N_0 = 47$ dB Hz at the beginning of an occultation). Then, carrier signal tracking is simulated using a receiver model described by Ao et al. [1] and a second order phase-locked loop with 25 Hz bandwidth. C/A code and data bit modulation is not taken into account. From bending angles refractivity profiles are obtained using Abel inversion.

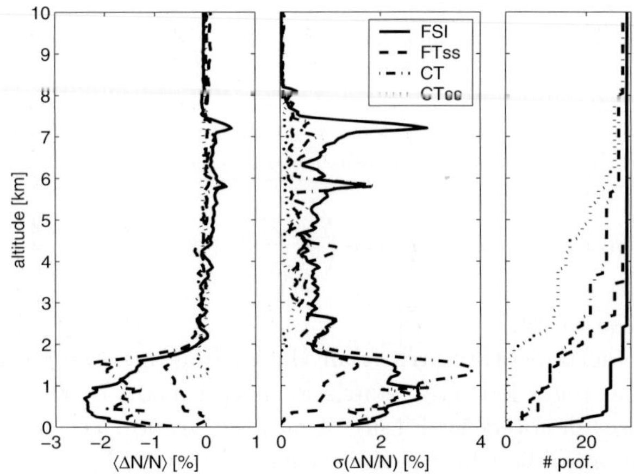


**Fig. 1.** Results obtained by four retrieval methods on the basis of 30 simulated occultations. Left and middle panel show the mean fractional refractivity error and its 1 $\sigma$ standard deviation, respectively. Right: the number of retrieved values.

The retrieval error in terms of the fractional refractivity deviation $(N - N_0)/N_0$ is plotted in Fig. 1. Here, $N$ and $N_0$ denote the retrieved and true refractivity, respectively. The mean and standard deviation of $(N - N_0)/N_0$ are shown in the left and middle panel. The number of data points as a function of altitude is plotted in the right panel. Here, the altitude, at which the number of processed profiles is reduced to 50%, is denoted as penetration altitude $z^\star$. Profile segments that are affected by critical refraction (dN/dz $< -160$ km$^{-1}$) are excluded from the analysis. Above about 8 km good agreement between the four retrievals is observed; the FTss result exhibits a small positive bias, though. Below 8 km, in particular below 2 km, receiver-induced tracking errors lead to significant biases reaching 2.5% in the case of the FSI solution. We note, however, that the FSI profiles have the smallest penetration altitude with 11 of 30 FSI profiles reaching the surface. Both heuristic methods exhibit a smaller bias compared to the CT and FSI result.

Retrieval results obtained from the processing of about 1122 CHAMP occultation profiles recorded during 1–7 January 2003 are shown in Fig. 2. The retrieved refractivity profiles are compared with corresponding ECMWF analyses. FSI and CT biases and standard deviations agree within 0.5%. Below 2 km altitude the CTss solution yields the smallest bias and standard deviation at the expense of an increase in penetration altitude.

## 4 Conclusions

Four bending angle retrieval methods, the CT and FSI methods, together with their heuristic modifications, CTss and FTss, are validated using end-

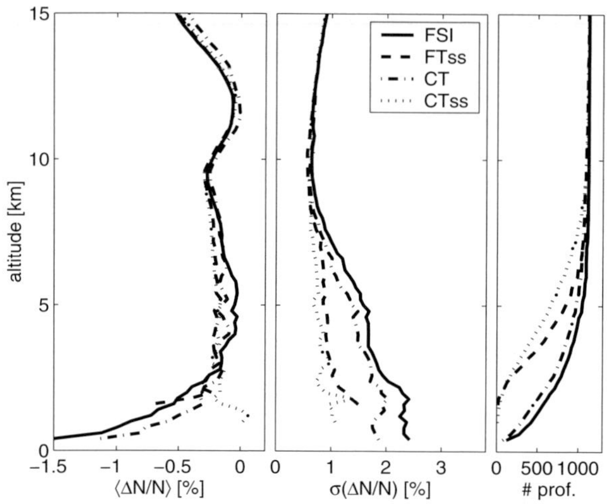


**Fig. 2.** Same as Fig. 1 however showing retrieval errors obtained from 1122 CHAMP observations covering the time period 1–7 January 2003.

to-end simulation data and one week of CHAMP observations. The results show a pronounced correlation between small refractivity biases and large penetration altitudes. In general, the heuristic methods exhibit smaller retrieval errors in the simulation study. The gain in accuracy, however, comes at the price of a reduced vertical resolution and loss of refractivity data in the lower troposphere.

*Acknowledgement.* We thank R. Weller, Alfred Wegener Institute for Polar and Marine Research, Bremerhaven, Germany for high-resolution radio sonde data. We are grateful to two anonymous reviewers whose comments and corrections strengthened and improved the manuscript significantly. The European Centre for Medium-Range Weather Forecasts provided meteorological analysis fields.

## References

1. Ao CO, Meehan TK, Hajj GA, Mannucci AJ, and Beyerle G (2003) Lower-troposphere refractivity bias in GPS occultation retrievals. J Geophys Res *108(D18)*: doi:10.1029/2002JD003216.
2. Beyerle G, Gorbunov ME, and Ao CO (2003) Simulation studies of GPS radio occultation measurements. Radio Sci *38(5)*: doi:10.1029/2002RS002800.
3. Beyerle G, Wickert J, Schmidt T, and Reigber Ch (2003) Atmospheric sounding by GNSS radio occultation: An analysis of the negative refractivity bias using CHAMP observations. J Geophys Res *108*: doi:10.1029/2003JD003922.
4. Gorbunov ME (2002) Canonical transform method for processing GPS radio occultation data in lower troposphere. Radio Sci *37*: doi:10.1029/2000RS002592.

5. Gorbunov ME (2002) Radio-holographic analysis of Microlab-1 radio occultation data in the lower troposphere. J Geophys Res *107(D12)*: doi:10.1029/2001JD000889
6. Gorbunov ME (2002) Radioholographic analysis of radio occultation data in multipath zones. Radio Sci *37(1)*: doi:10.1029/2000RS002577.
7. Gorbunov ME, Gurvich AS, and Bengtsson L (1996) Advanced algorithms of inversion of GPS/MET satellite data and their application to reconstruction of temperature and humidity. Report 211, Max-Planck-Institut für Meteorologie, Germany, Hamburg.
8. Hajj GA, Ao CO, Iijima BA, Kuang D, Kursinski ER, Mannucci AJ, Meehan TK, Romans LJ, de la Torre Juarez M, and Yunck TP (2003) CHAMP and SAC-C atmospheric occultation results and intercomparisons. J Geophys Res, submitted.
9. Jensen AS, Lohmann M, Benzon H-H, and Nielsen A (2003) Full spectrum inversion of radio occultation signals. Radio Sci *38(3)*: doi:10.1029/2002RS002763.
10. Karayel ET and Hinson DP (1997) Sub-Fresnel-scale vertical resolution in atmospheric profiles from radio occultation. Radio Sci *32(2)*: 411–423.
11. Marquardt C, Schöllhammer K, Beyerle G, Schmidt T, Wickert J, and Reigber C (2003) Validation and data quality of CHAMP radio occultation data. In: Reigber C, Lühr H, and Schwintzer P, eds, First CHAMP mission results for gravity, magnetic and atmospheric studies, Springer–Verlag, Berlin: 384–396.
12. Reigber Ch, Lühr H, and Schwintzer P (2000) CHAMP mission status and perspectives. Suppl to EOS Transactions, AGU *81(48)*: F307.
13. Reigber Ch, Lühr H, and Schwintzer P (2002) CHAMP mission status. Adv Space Res *30(2)*: 129–134.
14. Rocken C, Anthes R, Exner M, Hunt D, Sokolovskiy S, Ware R, Gorbunov M, Schreiner W, Feng D, Herman B, Kuo Y-H, and Zou X (1997) Analysis and validation of GPS/MET data in the neutral atmosphere. J Geophys Res *102(D25)*: 29,849–29,866.
15. Sokolovskiy S (2003) Effect of superrefraction on inversions of radio occultation signals in the lower troposphere. Radio Sci *38(3)*: doi:10.1029/2002RS002728.
16. Sokolovskiy SV (2001) Modeling and inverting radio occultation signals in the moist troposphere. Radio Sci *36(3)*: 441–458.
17. Sokolovskiy SV (2001) Tracking tropospheric radio occultation signals from low Earth orbit. Radio Sci *36(3)*: 483–498.
18. Wickert J, Schmidt T, Beyerle G, König R, Reigber Ch, and Jakowski N (2003) The radio occultation experiment aboard CHAMP: Operational data processing and validation of atmospheric parameters. J Meteorol Soc Jpn *82(1B)*: 381–395.

# Diffractive Integrals for Bistatic Remote Sensing Using GPS Signals

Alexander Pavelyev[1], Jens Wickert[2], Yuei-An Liou[3]

[1] Institute of Radio Engineering and Electronics of Russian Academy of Sciences (IRE RAS), Fryazino, Vvedenskogo sq. 1, 141120 Moscow region, Russia, Fax +7 095 203 84141, *pvlv@ms.ire.rssi.ru*

[2] GeoForschungsZentrum Potsdam (GFZ), Department Geodesy and Remote Sensing, Telegrafenberg, 14473 Potsdam, Germany

[3] Center for Space and Remote Sensing Research, National Central University, Jung-Li, 320, Taiwan, Fax +886 3 4254908, *yueian@csrsr.ncu.edu.tw*

**Summary**. Diffractive integrals use a reference signal to derive the field distribution from the radio holograms. The phase of reference signal coincides with phase of the Green function and can be found by solving the scalar wave equation. This property of the reference signal demonstrates a fundamental nature of the radio holographic focused synthetic aperture principle (RFSA) specifically in its application to the radio occultation (RO) data analysis. The RFSA method has been validated by direct observations of multi-beam propagation in the atmosphere and weak reflections from the terrestrial surface, its vertical resolution has been estimated early as about 70 m. To compare the Canonical Transform (CT), Back Propagation (BP) and RFSA methods a general inverse operator (GIO) is introduced. CT as a partial case of GIO method can resolve physical rays in multi-path situations under an assumption of the global spherical symmetry of the atmosphere and ionosphere. RFSA method can account for the multi-path in the case when the global spherical symmetry is absent by using the appropriate model of the refractivity and has a promise to be effective for operational data analysis.

**Key words**: diffractive integrals, bistatic remote sensing, radio occultation

## 1 Introduction

Diffractive integrals connect the electromagnetic field on some interface or curve in space with the fields in a medium between the transmitter and receiver. Zverev's transform has been introduced early in [1] to connect the angular spectrum of the field measured in some plane with the radio fields of the direct and back-propagating waves. A reference signal constructed using a Green function of a diffraction boundary problem in a 3-D medium has been employed in [2] to detect the spatial structure of the Saturn's rings, and spatial resolution about of 1/10 – 1/100 of the Fresnel's zone size has been obtained. A scalar diffractive 2-D integral has been introduced in [3] to heighten the vertical resolution in the RO experiments. RFSA method has been derived in [4]-[7] to obtain radio images of the atmosphere and terrestrial surface and to retrieve the vertical profiles of the physical parameters in the atmosphere and mesosphere. Vertical resolution ~70 m has been demonstrated [6]. Systematic RFSA investigation of the reflected signals in the GPS/MET data has been carried out in [8] to measure the humidity in the boundary layer of the atmosphere. Recently the CT method has been derived [9] using Fourier integral operators to find dependence of the refraction angle on impact parameter for each physical ray in multi-path conditions. Full spectrum inversion (FSI) method established a novel connection between the derivative of the phase of a physical ray on the instantaneous frequency in the full Fourier spectrum of RO signal and the time of intersection of the physical ray with orbital trajectory of LEO satellite [10].

The aim of this contribution consists in short description of the features of the RFSA method, illustration of its capability to compress the angular spectrum of RO signal and achieve high vertical resolution. The second aim is to introduce a GIO method to obtain the CT and BP transforms and to establish their limitations.

## 2 Green function and reference signal for RFSA method

For the RO case the vector diffractive integral has a form [7]

$$\boldsymbol{E}(P_i) = A_o\phi(\boldsymbol{r},t) - i\, k_o\, (4\pi)^{-1}\int_V \boldsymbol{E}\cdot\nabla(\ln\varepsilon)\,\nabla\phi\, dv \qquad (1)$$

$$\Delta\phi + k^2\phi = -4\pi\delta(\boldsymbol{r} - \boldsymbol{r}'),\ k = k_o n \qquad (2)$$

where $A_o$ is the amplitude of a source $G$ (Fig. 1), $\varepsilon$ is the electric permittivity, $\boldsymbol{E}(P_i)$ is the electric field at observation point $P_i(\boldsymbol{r}')$ in terms of the values of $\boldsymbol{E}$ inside the inhomogeneous part of the volume $V$, $\phi(\boldsymbol{r}-\boldsymbol{r}')$ is the Green function, $\boldsymbol{r}$ determines the co-ordinates of the integration element $dv$, $\delta(\boldsymbol{r}-\boldsymbol{r}')$ is a delta-function, $k_o$ is the free space wave number, and $n$ is the refraction index. At the GPS frequencies the volume integral in (1) can be neglected and the field is proportional to the Green function.

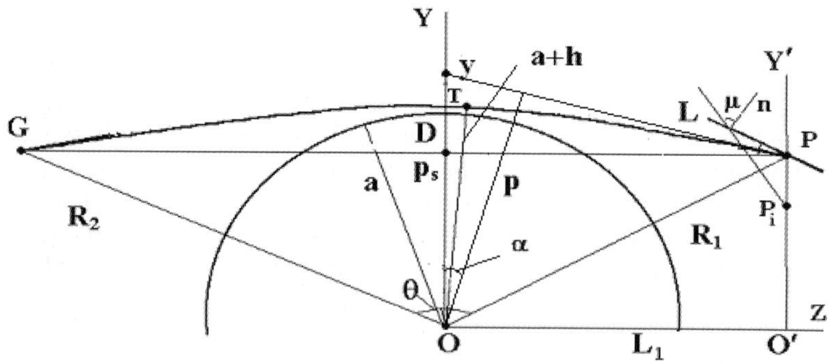


**Fig. 1**. Scheme of RO measurements. RFSA method reconstructs the field in the plane *ZOY* (e.g. rays *P'E'*, *PE*) and thus obtains the radio images of the atmosphere.

The record of the complex radio signal $\phi_e(r,t)$ is the Green function's envelope, that contains the temporal dependencies of the amplitude $A(t)$ and phase path excess $\psi(t)=kS_e(t)$, $\phi_e(\mathbf{r},t)=A(t)exp[-i\psi(t)]$ [4]. The reference signal $E_m(t)=A_m^{-1}(t)exp[i\psi_m(t)]$ must have maximum coherence with the main part of RO signal relevant to the main ray *GTP* (Fig. 1). To achieve this aim the appropriate model of the refractivity in the RO region can be used to determine the temporal dependence of $\psi_m(t)$. An example of a model describing multi-path effect is given in [5]. A temporal convolution of the reference and RO signals and Fourier analysis has been applied in [4], [6], [7] to obtain the compressed angular spectrum. The size of the focused synthetic aperture $L$ and the vertical resolution of the RFSA method $\delta h(\theta)$ are given by equations [7]:

$$\delta h(\theta) = \lambda R_1/L;\ L = 2Tv \tag{3}$$

where $T$ is the time interval of the focused aperture synthesis, $v$ is practically equal to the orbital speed of LEO satellite. For numerical estimation we can let $L$=20 km, $v$=8 km/s, $T$=2.5 s, $\lambda$=20 cm, $R_1$= 7000 km, then we obtain from (3) $\delta h(\theta)$=70 m. Thus the RFSA method is justified as a method, which use the phase of the Green function for determination of the impact parameters and refraction angles of physical rays in RO experiments [4]. Example of radio images of the atmosphere and Earth's surface is shown in Fig. 2. For construction of the reference signal an exponential model of the refractivity has been used: $N(h) = N_o\ exp(-h/H)$, $N_o$=300 (N-units), $H$= 6.3 km. This model corresponds to single-beam propagation. The images relate to RO event No. 0392 (February 05, 1997; 13 h 54 m 42 s UT; 55.6° N, 139.2° E). The radio images of the stratosphere in the height interval 24-30 km (Fig. 2, left panel) contain mainly sharp peaks having vertical width ~70 m at the half power level. It corresponds to an angular resolution of about 17-23 microradians, and spatial compression of the RO signal ~ 1/10 of the Fresnel zone size. The radio brightness distribution in the boundary layer at a height of 0-2 km is shown

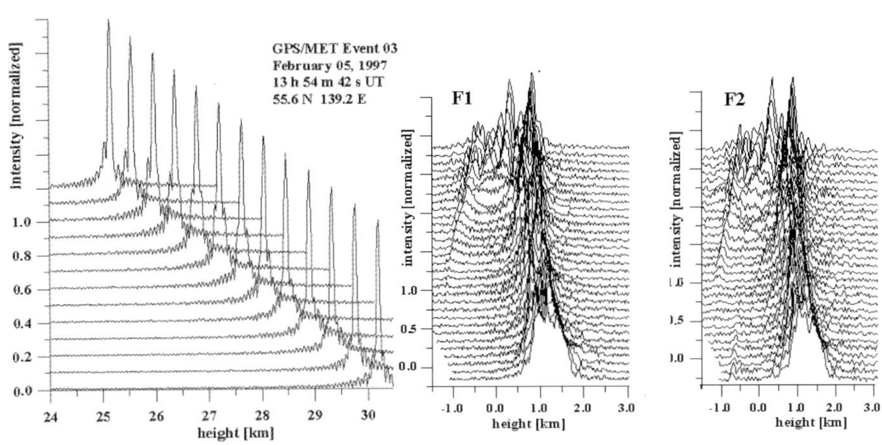


**Fig. 2.** Left panel: radio images of the stratosphere. Middle and right panels: radio images of the boundary layer of the troposphere at frequencies F1 and F2. The radio images are given in sequence with time interval 0.32 s.

in Fig. 2 (middle and right panels). Negative-height values correspond to the signals reflected from the Earth's surface. The main peak corresponds to a radio occultation signal propagating along the path GTP (Fig. 1). In the boundary layer of the troposphere the spatial compression effect in the main peak is diminished because influence of multi-path propagation. However the weak reflected signal is seen with better resolution. The weak reflected signal is slowly moving toward the main tropospheric signal. The behavior of the reflected signal is similar at both GPS frequencies F1 and F2 thus indicating the minimal level of the possible systematic receiver's errors. It follows from Fig. 2 that the vertical resolution of the RFSA method depends on the coherence of the reference signal with main RO signal and can be about of 70-100 meters.

## 3 GIO, RFSA, BP and CT methods

Zverev's equation [1] can be applied to determine the field $E(y,z)$

$$E(y,z) = k_o/2\pi \int d\alpha\, A(\alpha) \exp\{ik_o [\Phi(\alpha,y,z) - \Phi_j(\alpha)]\}, \tag{4}$$

$$\Phi(\alpha,y,z) = z\cos\alpha + y\sin\alpha, \quad \Phi_j(\alpha) = \Phi(\alpha,y_j,z_j)$$

where $A(\alpha)$ is the angular spectrum of the field, $\Phi(\alpha)$, $\Phi_j(\alpha)$, $y_j,z_j$ are the phase function, initial phase, and initial co-ordinates of the *j-th* physical ray. The centre of co-ordinate system $y$, $z$ coincides with point $O$. The axis $OY$ is perpendicular to direction $GP$ (Fig. 1). We introduce a GIO transform $I(p)$:

$$I(p) = k_o/2\pi \int d\eta B(p,\eta) \exp\{ik_o [f(p,\eta) - d(\eta)]\}\, I_1(\eta), \tag{5}$$

$$I_1(\eta) = \int ds \exp(-ik_o \eta s)\, R(s)\, E(y(s), z(s))$$

where $I_1(\eta)$ is the internal operator, $R(s)$ is the reference signal, $B(p,\eta)$ is the amplitude function, $d(\eta)$ is the phase correction function, $f(p,\eta)$ is the impact function, $p$ is the parameter, having different physical interpretation depending on $f(p,\eta)$, $s$ is the path of integration along the orbital trajectory of LEO satellite. The GIO transform can be considered as a generalization of the Egorov's Fourier integral operator [11]. The reference signal $R(s)$ is included in GIO with aim to compress the angular spectrum and to account for the form of the orbital trajectory. The functions $f(p,\eta)$, $B(p,\eta)$, $d(\eta)$ can be arbitrary and their physical meaning can be revealed using Zverev's presentation for the field (4). The RFSA method uses only the operator $I_1(\eta)$ with the reference signal $E_m(t(s))$ instead of $R(s)$ in (5). To integrate along the real orbit we introduce the new co-ordinate system $y'$, $z'$ with axis $PY'$ oriented along the tangent to the orbital trajectory of LEO satellite at angle $\gamma$ relative to the co-ordinate system $y$, $z$ (Fig. 1). The coordinate $y'$ is reckoned from point $P$ (Fig. 1) and connected with $y$, $z$ by equations

$$y = y_p + y'\cos\gamma + z'\sin\gamma; \quad z = z_p + z'\cos\gamma - y'\sin\gamma \tag{6}$$

where $y_p$, $z_p$ are the co-ordinates of the point $P$ in the co-ordinate system $ZOY$ (Fig. 1). For real orbit the integration variable $y'$ can be used instead of $s$ in the operator

$I_1(\eta)$. After substitution (4), (6) into (5) and changing the order of the integral operators one can perform integration on $y'$ letting $z'=0$, $R(s)=1$.

$$I_1(\eta) = k_o/2\pi \int dy' \exp\{-ik_o y'[\eta - \sin(\alpha-\gamma)]\} = \delta(\eta - \sin(\alpha-\gamma)) \tag{7}$$

The second integration on $\eta$ gives

$$E(p) = k_o/2\pi \int d\alpha\, A(\alpha) \exp\{ik_o [f(p,\eta(\alpha)) - d(\eta) + z_p \cos\alpha + y_p \sin\alpha - \Phi_j(\alpha))]\}\, B(p,\eta(\alpha)),\ \eta(\alpha) = \sin(\alpha-\gamma). \tag{8}$$

One can chose the phase correction function $d(\eta)$ in (5) to simplify equation (8)

$$d(\eta) = z_p \cos(\gamma + \sin^{-1}\eta) + y_p \sin(\gamma + \sin^{-1}\eta). \tag{9}$$

The BP case can be obtained from (8) by choosing $B(p,\eta)=1$ and $f(p,\eta)=p\sin(\gamma+\sin^{-1}\eta)$ in the inverse operator (5). In this case the right part of the first equation in (8) coincides with (4), if $z=0$, and corresponds to the field distribution along $OY$. Parameter $p$ has a geometrical sense of the co-ordinate $y$ (Fig. 1). The second partial case is $f(p,\eta)=p(\gamma+\sin^{-1}\eta)$. The case $\gamma=0$ corresponds to CT method [9]. The stationary phase method connects the direction angle $\alpha_j$ of the $j$-$th$ physical ray and parameter $p$

$$P = -z_j \sin\alpha_j + y_j \cos\alpha_j \tag{10}$$

Equation (10) defines $p$ as the distance between the $j$-$th$ physical ray and the centre of the co-ordinate system - point $O$ (Fig. 1). If the centre of global spherical symmetry of the medium coincides with point $O$, $p$ is the impact parameter of the $j$-$th$ ray. When the modified refraction index $M(r)$ is a monotonic function, only one physical ray can correspond to the impact parameter $p$. A possibility of the multi-path effect corresponding to monotonic $M(r)$ profiles has been shown early in [5]. In this case the GIO operator and CT method [9] disentangle the multi-path rays expressing the ray direction angle $\alpha$ as a single-valued function of the impact parameter $p$. The ray direction angle $\alpha$ can be determined by differentiating the phase of the field $E(p)$ [9]: $\alpha = d\arg E(p)/dp$. Note, that in this case the BP method may be a subject of multi-path distortion. In reality only the centres of the local spherical symmetry exist for different parts of the ray trajectories in the ionosphere and atmosphere. In this case the phase of the field transformed by the GIO operator, CT and BP methods may contain distortion connected with horizontal gradients in multi-path situation.

The RFSA method can account for the horizontal gradients using appropriate models of in the ionosphere and atmosphere and can work in multi-path conditions.

## 4 Conclusion

The phase of the reference signal used in the RFSA method, developed early with aim to heighten vertical resolution in analysis of RO data, coincides with the phase of the Green function of the scalar wave equation corresponding to 3-D inhomogeneous medium. This substantiated the radio holographic focused synthetic aperture principle for the atmospheric investigation using RO data. The CT and BP method are the partial cases of the GIO transform and may be a subject of multi-path distortion when the global spherical symmetry is absent. The RFSA method can, in principle, solve the multi-path problem.

*Acknowledgments.* We are grateful to UCAR for access to GPS/MET data. We are grateful to Office of Naval Research (ONR) of USA for support under the grant N00014-00-0528. Work has been supported by Russian Fund of Basic Research, grant No. 03-02-17414.

## References

1. Zverev VA (1975) Radio-optics. Soviet Radio, Moscow.
2. Marouf EA and Tyler GL (1982) Microwave edge diffraction by features in Saturn's rings: Observations with Voyager 1. Science *217*: 243-245.
3. Gorbunov ME, Gurvich AS, Bengtsson L (1996) Advanced algorithms of inversion of GPS-MET satellite data and their application to reconstruction of temperature and humidity. Max-Planck-Inst Meteor, Rep 211, ISSN 0937-1060.
4. Hocke K, Pavelyev A, Yakovlev O, Barthes L and Jakowski N (1999) Radio occultation data analysis by radio holographic method. J Atmos Solar-Terr Phys *61:* 1169-1177.
5. Pavelyev AG (1998) On the possibility of radio holographic investigation on communication link satellite-to-satellite. J Comm Technology and Electronics *43(8)*: 126-131.
6. Igarashi K, Pavelyev A, Hocke K, Pavelyev D, and Wickert J (2001) Observation of wave structures in the upper atmosphere by means of radio holographic analysis of the RO data. Adv Space Res *27(6-7)*: 1321-1327.
7. Pavelyev AG, Liou YA, Huang CY, Reigber C, Wickert J, Igarashi K, Hocke K (2002) Radio holographic method for the study of the ionosphere, atmosphere and terrestrial surface from space using GPS occultation signals. GPS Solutions *(No. 6)*: 101-108.
8. Beyerle G, Hocke K, Wickert J, Schmidt T, Marquardt C, and Reigber Ch (2002) GPS ROs with CHAMP: A radio holographic analysis of GPS signal propagation in the troposphere and surface reflections. J Geophys Res *107(D24)*: 10.1029/2001JD001402.
9. Gorbunov M E (2002) Canonical transform method for processing radio occultation data in the lower troposphere, Radio Sci *37(5)*: 10.1029/2000RS002592.
10. Jensen AS, Lohmann MS, Benzon HH and Nielsen A S (2003) Full spectrum inversion of radio occultation signal, Radio Sci *38(3)*: 10.1029/2002RS002763.
11. Yu V and Egorov (1985) Lectures on partial differential equations. Additional chapters, Moscow State University Press, Moscow (in Russian).

# Canonical Transform Methods for Analysis of Radio Occultations

Michael E. Gorbunov[1] and Kent B. Lauritsen[2]

[1] Institute of Atmospheric Physics, Moscow, Russia *gorbunov@dkrz.de*
[2] Danish Meteorological Institute, 2100 Copenhagen, Denmark *kbl@dmi.dk*

**Summary.** We discuss the application of canonical transform (CT) methods for analyzing radio occultations with multipath behavior. In the present work we discuss a mapping to the representation of approximate impact parameter. This method generalizes the full spectrum inversion (FSI) method for the case of non-circular orbits. The method is based on mapping the field by a Fourier integral operator that maps directly from the measured time-dependent field to the impact parameter representation without first doing a back propagation. Furthermore, our method allows for simple, asymptotic direct modelling of wave propagation. We perform processing of simulated radio occultations and show that our method provides the same accuracy and resolution as the standard CT method.

**Key words:** Canonical transform, Fourier integral operators, multipath

## 1 Introduction

The processing of radio occultation data involves the reconstruction of the geometric optical rays from measurements of wave fields, possibly with multipath behavior. The standard retrieval methods use the connection between ray direction and Doppler frequency and can therefore only be applied in single-ray areas. Transforming radio occultation measurements from the space to the impact parameter representation was recently shown to be a powerful retrieval method for signals with multipath behavior [1, 2] (for a discussion of backpropagation and radio-holographic methods see, e.g., Ref. [1]). The transformation is given by a Fourier integral operator (FIO) whose phase function is the generating function of the geometric optical canonical transform (CT). If each value of the impact parameter, $p$, occurs not more than once, this transform allows for disentangling multiple rays. The CT method provides high accuracy and resolution (about 50 m) of refraction angle profiles.

Another inversion method capable of handling multipath behavior is the full-spectrum inversion (FSI) method [3]. The FSI method also transforms the complex field to the impact parameter representation. This is accomplished by a Fourier transform of the complete record of the measured, complex field multiplied by a reference signal.

A synthesis of the CT and FSI inversion methods was discussed in Ref. [4]. It is possible to derive a general formula for the FIO applied directly to radio occultation data measured along a generic low-Earth orbiter (LEO) trajectory without the back-propagation (BP) procedure. This FIO is referred to as being of type 2 (FIO2). Along these lines one can develop a CT algorithm of type 2 (denoted CT2) that in analogy to the FSI algorithm effectively amounts to a Fourier transform of the measured field [5]. Furthermore, FIOs can also be used for efficient direct modelling of wave propagation [6].

The rest of this paper is organized as follows: In Sec. 2 we introduce Fourier integral operators of type 2, in Sec. 3 we present the CT2 algorithm, and in Sec. 4 we present simulation results. In Sec. 5 we summarize and conclude.

## 2 Fourier integral operators of type 2

We consider the complex field $u(y) = A(y) \exp(ik\Psi(y))$ recorded along the observation trajectory parameterized with some generic coordinate $y$, which can be e.g. time, arc length, or satellite-to-satellite angle $\theta$ [3]. The generic FIO2 operator that maps the field from the $(y, \eta)$ to the $(p, \xi)$ representation ($\eta$ and $\xi$ are the canonical momenta associated to $y$ and $p$) reads [5]:

$$\hat{\Phi}_2 u(p) = \sqrt{\frac{-ik}{2\pi}} \int a_2(p, y) \exp(ikS_2(p, y)) u(y) dy \,. \tag{1}$$

The functions $a_2$ and $S_2$ are denoted amplitude and phase functions. The phase space coordinates are linked by the following canonical transform:

$$\eta = -\partial S_2(p, y)/\partial y, \qquad \xi = \partial S_2(p, y)/\partial p. \tag{2}$$

The transformed field is expressed as $w(p) \equiv \hat{\Phi}_2 u(p) \equiv A'(p) \exp(ik\Psi'(p))$, where $A'(p)$ and $\Psi'(p)$ can be found using the stationary phase method.

The differential equations (2) define the phase function from the known canonical transform $(y, \eta) \to (p, \xi)$. The new momentum $\xi$ is a known function of impact parameter and bending angle. In particlar, it can be chosen to be equal to bending angle $\epsilon$ [5, 7]. The bending angle $\epsilon$ can thus be found from the momentum $\xi$ and impact parameter $p$. For a circular occultation geometry the FIO2 reduces to the Fourier transform.

Thus, for the reconstruction of the refraction angle profile from the measurements of the wave field along the orbit we apply the FIO2 operator (1), which produces a function $\hat{\Phi}_2 u(p) = A'(p) \exp(ik\Psi'(p))$ of impact parameter $p$. The derivative of its eikonal $\Psi'(p)$ with a negative sign is equal to the refraction angle $\epsilon = \epsilon(p, y_s(p))$, where $y_s(p)$ is the coordinate of the trajectory point, where the ray with impact parameter $p$ was observed [$y_s(p)$ is the stationary point of the oscillating integral (1)].

The amplitude function $a_2(p, y)$ of the FIO does not play any role in the computation of refraction angles. For example, we could set $a_2(p, y) \equiv 1$.

However, the correct definition of the amplitude of the transformed wave field is necessary for the retrieval of atmospheric absorption. For the derivation of the amplitude function $a_2(p, y)$, we use energy conservation. For a generic coordinate $y$, we must introduce a measure $\mu(p, y)$ and replace $dy$ with $\mu dy$. The expression for the amplitude function reads [5]:

$$a_2(p, y) = \sqrt{\mu \left| \frac{\partial^2 S_2(p, y)}{\partial p \, \partial y} \right|}, \tag{3}$$

with the measure derived from energy conservation:

$$\mu(p, y) = \left( \sqrt{r_L^2 - p^2} \sqrt{r_G^2 - p^2} \right)^{1/2} \times \left| \dot{\theta} - \frac{\dot{r}_G}{r_G} \frac{p}{\sqrt{r_G^2 - p^2}} - \frac{\dot{r}_L}{r_L} \frac{p}{\sqrt{r_L^2 - p^2}} \right|^{1/2}. \tag{4}$$

## 3 Canonical transform method of type 2: approximate impact parameter representation

The FIO2 defined by the phase and amplitude functions $S_2$ and $a_2$ solves the problem of the extraction of refraction angles from measurements of the complex field along a satellite trajectory, directly without back propagation. However, this operator cannot be implemented as a Fourier transform and therefore its numerical implementation, especially for high frequencies such as 10-30 GHz, is slow. Here we shall describe an approximation that allows for writing the FIO2 operator in the form of a Fourier transform [5].

Consider the measured complex field $u(t) = A(t) \exp(ik\Psi(t))$ and corresponding momentum $\sigma = d\Psi/dt$. We use an FIO2 associated with the canonical transform from the $(t, \sigma)$ to the $(p, \xi)$-representation. The impact parameter $p$ is a known function of $t, \sigma$: $p = p(t, \sigma)$. Instead of exact impact parameter we introduce its approximation $\tilde{p}$ (with momentum $\tilde{\xi}$):

$$\tilde{p}(t, \sigma) = p_0(t) + \frac{\partial p_0}{\partial \sigma} (\sigma - \sigma_0(t)) = f(t) + \frac{\partial p_0}{\partial \sigma} \sigma, \tag{5}$$

$$f(t) = p_0(t) - \frac{\partial p_0}{\partial \sigma} \sigma_0(t), \tag{6}$$

where $\sigma_0(t)$ is a smooth model of Doppler frequency, $p_0(t) = p(\sigma_0(t), t)$, and $\partial p_0/\partial \sigma = \partial p/\partial \sigma|_{\sigma=\sigma_0(t)}$. We compute $\sigma_0(t)$ by differentiation of the eikonal with a strong smoothing over approximately 2 s time interval. We parameterize the trajectory with the coordinate $Y = Y(t)$, where we use the notation $Y$ in order to distinguish between this specific choice of the trajectory coordinate and the generic coordinate $y$. For brevity we use the notation $u(Y)$ instead of $u(t(Y))$. For the coordinate $Y$ and the corresponding momentum $\eta$ we use the following definitions:

$$dY = \left(\frac{\partial p_0}{\partial \sigma}\right)^{-1} dt = \frac{\partial \sigma}{\partial p_0} dt, \qquad \eta = \frac{\partial p_0}{\partial \sigma} \sigma. \tag{7}$$

Then, we can write the linear canonical transform: $\tilde{p} = f(Y) + \eta, \xi = -Y$, where $f(Y)$ is short notation for $f(t(Y))$. The generating function of this canonical transform follows from the differential equation: $dS_2 = \xi\, d\tilde{p} - \eta\, dY = -Y\, d\tilde{p} - (\tilde{p} - f(Y))\, dY$, with the result: $S_2(\tilde{p}, Y) = -\tilde{p}Y + \int_0^Y f(Y')\, dY'$.

Because $\left|\partial^2 S_2/\partial\tilde{p}\partial Y\right| = 1$, the amplitude function (3) equals $\sqrt{\mu}$ and it can be approximated by: $a_2(\tilde{p}, Y) = (\sqrt{r_L^2 - \tilde{p}^2}\sqrt{r_G^2 - \tilde{p}^2})^{1/2}$ [5]. In addition, the amplitude function $a_2(\tilde{p}, Y)$ can be replaced with $a_2(\tilde{p}, Y_s(\tilde{p}))$, where $Y_s(\tilde{p})$ is the stationary point, and factored out from within the integral. The resulting operator can be written as the composition of adding a model, $ik \int f(Y) dY$, to the phase, the Fourier transform, and an amplitude factor:

$$\hat{\Phi}_2 u(\tilde{p}) = \sqrt{\frac{-ik}{2\pi}} a_2(\tilde{p}, Y_s(\tilde{p})) \int \exp\left(-ik\tilde{p}Y\right) \exp\left(ik \int_0^Y f(Y')\, dY'\right) u(Y)\, dY. \tag{8}$$

The function $Y_s(\tilde{p})$ equals $-\tilde{\xi}$, where the momentum $\tilde{\xi}$ is the derivative of the eikonal of the integral term in (8). This operator maps the wave field to the representation of the approximate impact parameter $\tilde{p}$. Practically, the difference between $p$ and $\tilde{p}$ is small and can be neglected.

The FIO2 mapping (8) generalizes the FSI method [3]: FSI uses a composition of a phase model with a Fourier transform with respect to the angle $\theta$. Our definition of the coordinate $Y$ takes into account the generic occultation geometry (and for circular orbits $Y = \theta$). We also use a more accurate derivation of the amplitude function $a_2$. We refer to this CT inversion technique based on the FIO of the second type as the CT2 algorithm.

## 4 Numerical simulations

Next, we compare the performance of the CT2 inversion technique introduced above and the standard composition of BP and CT techniques. We model a spherically symmetric atmosphere using a high-resolution tropical radio sonde profile of refractivity. We simulate radio occultation signals using multiple phase screens (MPS) and asymptotic, direct modeling (A) (see [5, 6]) for the standard GPS frequencies.

The simulated radio occultation signals were sampled at realistic rates and processed by the CT and CT2 inversion algorithms and the results of the reconstruction of refraction angle profile were compared with the exact geometric optical solution. The comparison presented in Fig. 1 indicates a very good agreement between both algorithms and the geometric optical

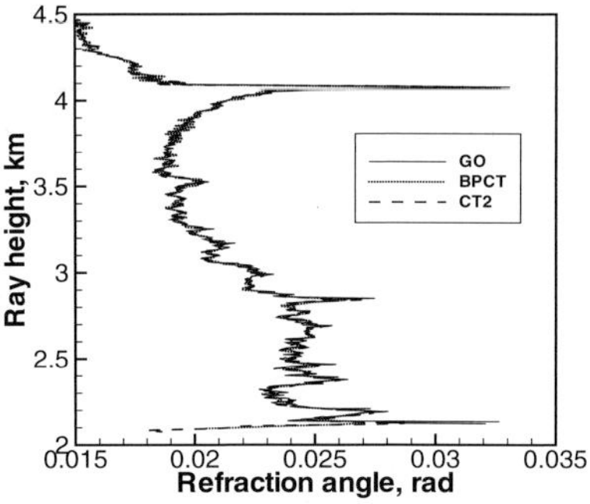


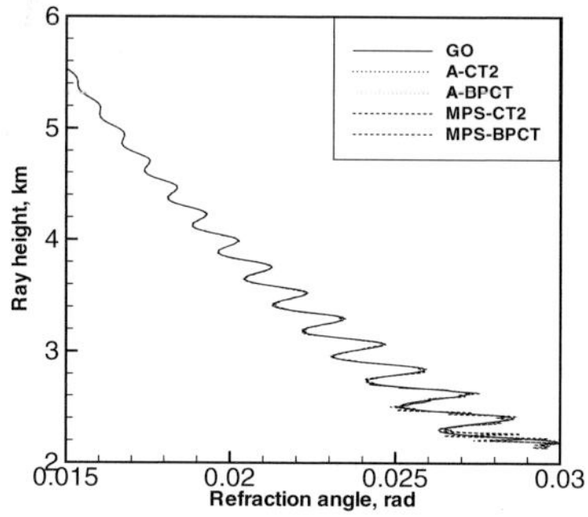


**Fig. 1.** (Left) Comparison of different modifications of the CT technique. Refraction angle profiles as functions of ray height (impact parameter minus Earth's curvature radius): (1) reference geometric optical solution (GO, solid line), (2) standard composition of BP and CT (BPCT, dotted line), and (3) CT2 algorithm (dashed line).

**Fig. 2.** (Right) Refraction angle profiles from geometric optical ray tracing and from simulated radio occultation data: (1) reference GO profile (GO, solid line), (2) asymptotic simulation processed by CT2 (A-CT2, dotted line), (3) asymptotic simulation processed by BP+CT (A-BPCT, dotted line), (4) MPS simulation processed by CT2 (MPS-CT2, dashed line), and (5) MPS simulation processed by BP+CT (MPS-BPCT, dashed line).

solution. All the differences are in small scales below 50 m, which cannot be effectively resolved for the GPS frequencies, due to diffraction inside the atmosphere.

We also performed simulations with a simple spherically-symmetrical phantom $n(z) = 1 + N_0 \exp\left(-\frac{z}{H}\right)\left[1 + \alpha\cos\left(\frac{2\pi z}{h}\right)\exp\left(-\frac{z^2}{L^2}\right)\right]$, where $z$ is the height above the Earth's surface, $N_0 = 300 \times 10^{-6}$ is the characteristic refractivity at the Earth's surface (300 N-units), $H = 7.5$ km is the characteristic vertical scale of refractivity field, $\alpha = 0.003$ is the relative magnitude of the perturbation, $h = 0.3$ km is the period of the perturbation, $L = 3.0$ km is the characteristic height of the perturbation area.

Figure 2 shows the geometric optical refraction angle profile and the results of the inversion of the simulated data. We present four combinations of the two simulation techniques: (1) the FIO asymptotic solution (A) and (2) multiple phase screens (MPS); and the two inversion techniques: (1) CT2 and (2) the standard combination of BP and CT. The figure shows good agreement between the GO solution and the retrieved refractivity profile. The strongest deviations of retrieved refraction angles from the reference GO profile are observed for processing MPS simulations in the lowest 200 m. This can be accounted for by the diffraction on the Earth's surface.

## 5 Conclusions

Different techniques for processing radio occultations based on Fourier integral operators have a common point: they utilize an FIO that maps the measured wave field to the impact parameter representation. The original CT method was based on first applying a back propagation of the measured signal. In contrast to this, the FSI and CT2 methods are applied directly to the measured field. The advantage of such methods is that they do not assume stationarity of the GPS satellite. Furthermore, these methods allow for efficient numerical implementations based on FFTs by mapping to an approximate impact parameter representation.

Another application of FIOs is the direct modeling [5, 6]. The asymptotic solution of the direct problem uses the mapping of the geometric optical solution in the impact parameter representation to the standard coordinate representation. A method based on the inverse CT2 is very efficient numerically because it can be implemented as the composition of the geometric optical solution and a single Fourier transform. This is important for direct modeling and for processing radio occultation data at high frequencies (10-30 GHz), where the computation of diffractive integrals may be numerically slow.

*Acknowledgement.* M. E. G. has been supported by the Russian Foundation for Fundamental research (grants nos. 01-05-64269 and 03-05-64366); K. B. L. has been supported by the GRAS Meteorology SAF.

## References

1. Gorbunov ME (2002) Canonical transform method for processing radio occultation data in the lower troposphere. Radio Sci *37(5)*: 1076, doi:10.1029/2000RS002592, 9-1–9-10.
2. Gorbunov ME (2002) Radio-holographic analysis of Microlab-1 radio occultation data in the lower troposphere. J Geophys Res *107(D12)*: doi:10.1029/2001JD000889, 7-1–7-10.
3. Jensen AS, Lohmann MS, Benzon H-H, and Nielsen AS (2003) Full spectrum inversion of radio occultation signals. Radio Sci *38(3)*: 1040, doi:10.1029/2002RS002763, 6-1–6-15.
4. Gorbunov ME and Lauritsen KB (2002) Canonical transform methods for radio occultation data. Scientific Report 02-10, Danish Meteorological Institute, Copenhagen, Denmark, 2002; http://www.dmi.dk/f+u/publikation/vidrap/2002/Sr02-10.pdf.
5. Gorbunov ME and Lauritsen KB (2003) Analysis of wave fields by Fourier integral operators and their application for radio occultations. submitted to Radio Science.
6. Gorbunov ME (2003) An asymptotic method of modeling radio occultations. J Atmos Terr Phys *65(16-18)*: 1361-1367, doi: 10.1016/j.jastp.2003.09.001.
7. Jensen AS, Lohmann MS, Nielsen AS, and Benzon H-H (2003) Geometrical optics phase matching of radio occultation signals. submitted to Radio Science.

# GPS Radio Occultation with CHAMP: Comparison of Atmospheric Profiles from GFZ Potsdam and IGAM Graz

Jens Wickert[1], Andreas Gobiet[2], Georg Beyerle[1], Andrea K. Steiner[2], Gottfried Kirchengast[2], Ulrich Foelsche[2], and Torsten Schmidt[1]

[1] GeoForschungsZentrum Potsdam (GFZ) Department Geodesy and Remote Sensing, Telegrafenberg, 14473 Potsdam, Germany, *jens.wickert@gfz-potsdam.de*
[2] Institute for Geophysics, Astrophysics and Meteorology (IGAM), University of Graz, Austria

**Summary.** A major application of the CHAMP occultation data is the preparation of processing systems for future occultation missions. The University of Graz (IGAM) uses data and analysis results to prepare for the ACE+ multi-satellite occultation mission, which is currently foreseen to be launched in 2008. We compare vertical profiles of refractivity, derived by GFZ and IGAM, with ECMWF data (European Centre for Medium-Range Weather Forecasts) and discuss the deviations. Good agreement is observed in the upper troposphere and stratosphere. The magnitude of the observed refractivity bias in the lower troposphere depends significantly on the quality control criterion. The bias can be nearly eliminated by the application of the Full Spectrum Inversion analysis method down to ~1 km above the Earth's surface.

**Key words:** GPS, radio occultation, remote sensing, IGAM Graz

## 1 Introduction

Significant progress for the innovative GPS (Global Positioning System) occultation technique was achieved within the atmospheric profiling experiment onboard the German CHAMP (CHAllenging Minisatellite Payload) satellite [1, 2]. Results of the operational data analysis at GFZ are provided via the CHAMP data center (*http://isdc.gfz-potsdam.de/champ*) and are used by scientist all over the world to improve global weather analyzes, to demonstrate the ability to detect climatic trends and also to develop and improve their occultation analysis systems, e.g., to prepare for the upcoming GPS occultation missions [3]. Comparison studies help to improve the data analysis and to optimize the exploitation of the CHAMP data. One example for such activity is ROSE (Radio Occultation Sensor Evaluation), jointly initiated by GFZ, JPL and UCAR [4, 5].

IGAM Graz uses CHAMP data within the CHAMPCLIM project [6] and to prepare for the ACE+ multi-satellite occultation mission. ACE+ was selected by ESA (European Space Agency) in May 2002 as top priority future

EEOM (Earth Explorer Opportunity Mission) and is currently in phase A until mid 2004. After confirmation the operational phase is foreseen to start in 2008 [7, 8]. ACE+ will provide more than 5,000 occultations daily.

We present first results of a comparison study for GFZ and IGAM inversion results. Both centers use the operationally derived atmospheric excess phase data from GFZ.

## 2 Data analysis and quality control

The CHAMP data are continuously processed by the operational occultation analysis system at GFZ, detailed descriptions can be found in [1, 9]. The standard products (current ATM version 004) are analyzed using the geometrical optics (GO) approach. Profiles are excluded, if at least one data point above 8 km shows deviations of >10 % in relation to ECMWF. The profiles are cut off in the lower troposphere, if the difference to the analyzes begins to exceed 10 % below 8 km.

The inversion software (geometrical optics) of IGAM is described by [10, 11, 12]. IGAM uses analyzes for quality control as well, refractivity profiles deviating from ECMWF more than 10 % between 5 and 35 km are sorted out entirely. Additionally, the profiles are cut off in the lower troposphere when severe impact parameter/bending angle ambiguities occur, one indication for the occurrence of multipath effects, which make the geometrical optics assumption invalid.

A 7-day period (January 1-7, 2003) was selected for the comparison. The standard analysis results from GFZ form the data set GFZ GO (1,253 profiles). The IGAM set is named as IGAM GO (1,200 profiles). In addition, results of advanced (wave optics based) data analysis were provided by GFZ. The Full Spectrum Inversion technique (FSI, [13]) was used for the processing (GFZ FSI, 1,234 profiles). An internal cut-off criterion without using external data was applied. The profiles were cut-off, when the smoothed FSI amplitude falls below half of its maximum value. A resulting set of 1,147 coincidencing profiles (IN ALL) was used for the comparisons.

## 3 Results

Whereas the CHAMP results in the upper troposphere and stratosphere exhibit excellent accuracy (e.g. [1]), the lower troposphere data suffer from a negative refractivity bias in relation to independent meteorological data [14]. Therefore we consider these altitude intervals separately. We focus on comparisons of refractivity profiles with interpolated data from 6-hourly operational meteorological analyzes from ECMWF.

### 3.1 Upper troposphere and stratosphere

The comparison of the GFZ and IGAM refractivity data with ECMWF is shown in Fig. 1. In general both data sets show good agreement. The GFZ profiles exhibit slightly negative bias of <0.5 K, similar bias is observed in the IGAM retrieval. The standard deviation is ~0.7 % at 10 km for both, GFZ and IGAM, beginning at 15 km to be slightly larger for the IGAM retrievals and becoming ~1.2 % for GFZ and ~1.5 % for IGAM at 30 km. The inversion results are comparable in this height interval, despite of the fact, that different methods for the optimization of the bending angles using the MSISE-90 [15] climatology are applied. GFZ uses the approach by [16], assuming 20 % error of MSIS and no vertical error correlations. IGAM applies statistical optimization [17] with 15 % MSIS error and vertical error correlation length of 1 km for the observations and 6 km for the background.

The comparison of temperatures (not shown here) shows significantly different results for both analysis centers. Whereas the GFZ temperatures agree well with the analyzes (bias <0.5 K, STD ~1 K at 10 km, ~2 K at 35 km), the IGAM retrievals begin to exhibit significant warm bias above 15 km, reaching ~5 K at 35 km with standard deviation of ~10 K. We relate these differences to the initialization of the temperature for the integration of the hydrostatic equation. GFZ uses ECMWF data at 43 km. IGAM integrates up to 120 km, where the air pressure is assumed to be zero.

Our results indicate that the CHAMP refractivity seems to be more appropriate than temperature to build up climatologies for stratospheric altitudes, because it depends less on additional assumptions or external data, as also was concluded by [18].

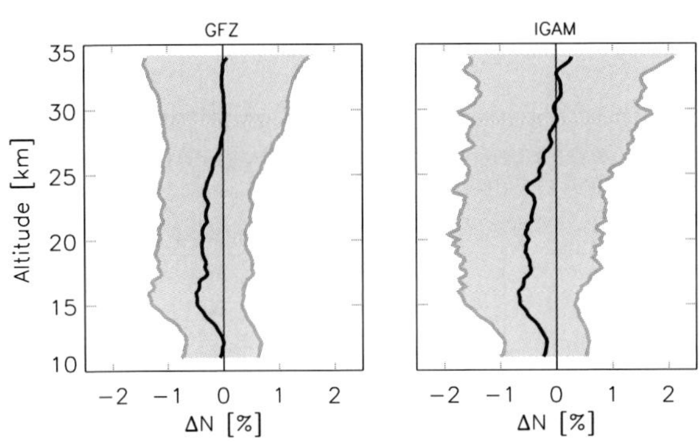


**Fig. 1.** Statistical comparison of 1,147 vertical CHAMP refractivity profiles (GO) from GFZ (left) and IGAM (right) with corresponding ECMWF data (between January 1 and 7, 2003).

### 3.2 Lower troposphere

The lower troposphere comparison of the refractivity is shown in Fig. 2. The various cut off criteria (see Sec. 2) result in different distributions of data points per height for each data set, e.g. the total number reduces to 50 % at different altitudes: 5.0 km (IGAM GO), 4.3 km (GFZ FSI) and 3.2 km (GFZ GO), respectively. The GFZ GO set shows a negative bias of 1 % at 2 km and ∼1.5 % at 1 km, the standard deviation is ∼2 and ∼2.5 % respectively. IGAM GO shows reduced bias and standard deviation characteristics, which we relate to the more rigorous cut-off criterion (see Sec. 2). This obviously removes more data, which are suspect to be influenced by multi-path effects, but also nearly halve the data points in the lower troposphere. The lowest bias exhibits the GFZ FSI data set. Almost bias-free behavior can be observed down to ∼1 km, with a standard deviation up to ∼1 %. Below, a negative bias up to ∼1 % appears. The slightly negative bias of ∼[illegible] % disappears, which is observed already at 10 km in the GFZ GO and nea[illegible] constant down from 10 km almost to the Earth's surface. This is achieved even with having more data available as the IGAM GO set.

In a next step we perform the comparison only at altitudes, where data points from all 3 sets exist. The result (Fig. 3) is completely different from that shown in Fig. 2 and underlines the importance of the cut-off criterion for the lower troposphere data quality and validation. The deviations to ECMWF are nearly identical for all 3 data sets. Almost no bias occurs from 10 km down to 1 km, the standard deviation is nearly constant at ∼1 %. Most noticeable is the slightly negative bias (∼0.3) in the IGAM GO and GFZ GO retrievals, which is eliminated in the GFZ FSI profiles.

## 4 Conclusions

First results of a study comparing CHAMP occultation analysis results from GFZ and IGAM are presented. The refractivity comparison shows good agree-

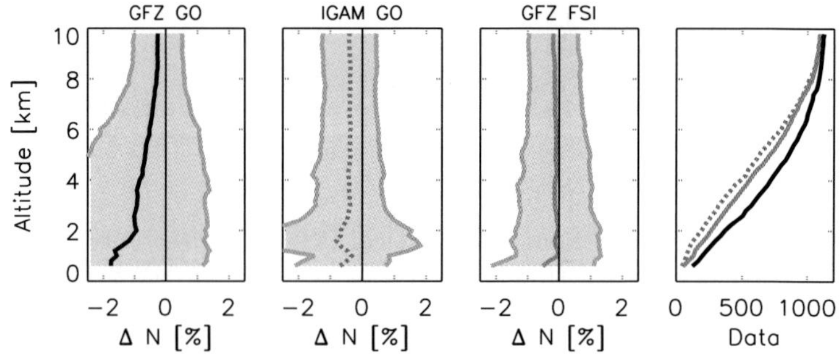


**Fig. 2.** Statistical comparison of 1,147 vertical CHAMP refractivity profiles (data sets: GFZ GO, IGAM GO and GFZ FSI) with corresponding ECMWF data.

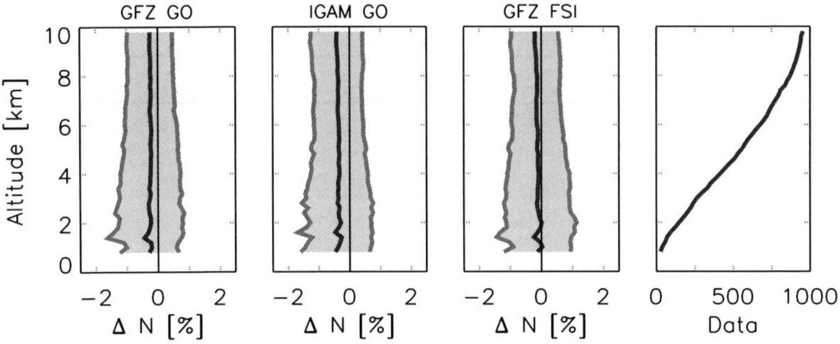


**Fig. 3.** Statistical comparison of the data sets: GFZ GO, IGAM GO and GFZ FSI with ECMWF (only altitudes were compared, where data from all 3 sets were available).

ment with ECMWF and almost identical results for the upper troposphere and stratosphere, even though different methods for the bending angle optimization were used. The temperature is sensitive to the initialization of the hydrostatic equation, differences in the derived temperatures are already observed at altitudes above 15 km. To carefully compare the data in the lower troposphere, the cut off and quality check criteria and the resulting data points vs. height have to be included to the comparison. The simple IGAM cut off criteria successfully removes problematic data in the lower troposphere, but reduces the data significantly compared to the GFZ GO retrievals. The application of the FSI method is very promising a) for the implementation to operational data analysis systems b) to significantly reduce the negative refractivity bias and c) to provide a cut-off criterion, which can be derived from the data analysis (magnitude of the FSI amplitude) and not from comparison with external meteorological data.

*Acknowledgement.* We are grateful to the CHAMP team led by Ch. Reigber for making the satellite mission including the occultation experiment successful. J. Wickert thanks both institutes, GFZ and IGAM, for giving the unique opportunity for a longer research stay at IGAM, during which the main work of this paper was done.

## References

1. Wickert J, Schmidt T, Beyerle G, König R, Reigber Ch, and Jakowski N (2004) The radio occultation experiment aboard CHAMP: Operational data analysis and validation of vertical atmospheric profiles. J Meteorol Soc Jpn *82(1B)*: 381–395.
2. Hajj GA, Ao CO, Iijima BA, Kuang D, Kursinski ER, Mannucci AJ, Meehan TK, Romans LJ, de la Torre Juarez M, and Yunck TP (2003) CHAMP and

SAC-C atmospheric occultation results and intercomparisons, J Geophys Res, submitted.

3. Wickert J, Schmidt T, Beyerle G, Michalak G, König R, and Reigber Ch (2004) Atmospheric profiling with CHAMP. Status of the operational data analysis, validation of the recent data products and future prospects. This issue.
4. Ao CO, Schreiner WB, and Wickert J (2003) First Report on the CHAMP Radio Occultation Intercomparison Study, JPL Publication 03-016, Pasadena, U.S.
5. Wickert J, Ao CO, Schreiner WB, Beyerle G, Schmidt T, König R, Reigber Ch, Hajj GA, Iijima BA, Hunt D, Sokolovskiy S, and Rocken C (2003) GPS radio occultation with CHAMP: First comparisons of analysis results from GFZ, JPL and UCAR. EGS-EUG-AGU Joint Assembly, Nice, France, April 2003.
6. Foelsche U, Kirchengast G, Gobiet A, Steiner AK, Löscher A, Wickert J, and Schmidt T (2004) The CHAMPCLIM project: An overview. This issue.
7. Hoeg P, and Kirchengast G (2002) ACE+ Atmosphere and Climate Explorer based on GPS, GALILEO, and LEO-LEO radio occultation (ESA Earth Explorer Opportunity Mission, Proposal). Wissenschaftl. Bericht 14, IGAM, University of Graz, Austria.
8. Kirchengast G, and Hoeg P (2003) The ACE+ Mission: An Atmosphere and Climate Explorer based on GPS, GALILEO, and LEO-LEO Radio Occultation. Proc. 1st Intl. Workshop on Occultations for Probing Atmosphere and Climate, Sep. 16-20, 2002, Graz, Austria, Springer Verlag.
9. Wickert J (2002) CHAMP GPS occultation data and operational analysis results. Technical Note, available via (*http://www.gfz-potsdam.de/gasp*).
10. Steiner AK, Kirchengast G, and Ladreiter HP (1999) Inversion, error analysis, and validation of GPS/MET occultation data. Ann Geophysicae *17*: 122–138.
11. Gobiet A, Kirchengast G, Wickert J, Retscher C, Wang DY, and Hauchecorne A (2004) Evaluation of Stratospheric Radio Occultation Retrieval Using Data from CHAMP, MIPAS, GOMOS, and ECMWF Analysis Fields. This issue.
12. Gobiet A, and Kirchengast G (2003) Advancement of GNSS Radio Occultation Retrieval in the Upper Stratosphere. Proc. OPAC-1, Springer Series, in print.
13. Jensen AS, Lohmann M, Benzon HH, and Nielsen A (2003) Full Spectrum inversion of radio occultation signals. Radio Sci *38(3)*: 10.1029/2002RS002763.
14. Beyerle G, Wickert J, Schmidt T and Reigber Ch (2003) Atmospheric sounding by GNSS radio occultation: An analysis of the negative refractivity bias using CHAMP observations. J Geophys Res: doi 10.1029/2003JD003922.
15. Hedin AE (1991) Extension of the MSIS thermosphere model into the middle and lower atmosphere. J Geophys Res *96*: 1159–1172.
16. Sokolovskiy S, and Hunt D (1996) Statistical optimization approach for GPS/MET data inversion. URSI GPS/MET workshop, Union Radio Sci Int, Tucson, Ariz.
17. Healy SB (2001) Smoothing radio occultation bending angles above 40 km. Ann Geophys *19*: 459–468.
18. Vedel H, and Stendel M (2003) On the direct use of GNSS RO Refractivity Measurements for Climate Monitoring. Proc 4th Oersted Science Meeting, Copenhagen: 275–278.

# Evaluation of Stratospheric Radio Occultation Retrieval Using Data from CHAMP, MIPAS, GOMOS, and ECMWF Analysis Fields

Andreas Gobiet[1], Gottfried Kirchengast[1], Jens Wickert[2], Christian Retscher[1], Ding-Yi Wang[3], and Alain Hauchecorne[4]

[1] Institute for Geophysics, Astrophysics, and Meteorology (IGAM), University of Graz, Austria, *andreas.gobiet@uni-graz.at*
[2] Department Geodesy and Remote Sensing, GeoForschungsZentrum Potsdam (GFZ), Germany
[3] Institut für Meteorologie und Klimaforschung (IMK), Forschungszentrum Karlsruhe/ Universität Karlsruhe, Germany
[4] Service d'Aéronomie du Centre national de la recherche scientifique (CNRS), Verrières-le-Buisson, France

**Summary.** Two radio occultation (RO) retrieval schemes designed at IGAM to enhance the performance at high altitudes are presented, applied to CHAMP RO data, and validated against ECMWF analyses, GFZ operational retrieval, ENVSAT/MIPAS-, and ENVIAT/GOMOS-derived temperature profiles. IGAM proposes to include background information into the RO retrieval only at one point at bending angle level in order to be able to track error characteristics of the retrieved product. The results show very good agreement with GFZ retrieval and ECMWF analysis below 15 km and, depending on the background information used, either a significant warm bias or essentially no bias up to 30 km. Compared to MIPAS, the only independent data source, the IGAM/ECMWF retrieval is unbiased up to 40 km.

**Key words:** CHAMP, MIPAS, GOMOS, ECMWF, radio occultation, retrieval, statistical optimization, validation

## 1 Introduction

Current Global Navigation Satellite System (GNSS) radio occultation (RO) retrieval techniques yield excellent results on refractivity, geopotential height (or pressure), and temperature in the upper troposphere and lower stratosphere. Above 30 km, throughout the upper stratosphere, the errors of the retrieved parameters increase significantly. There are two prominent pathways to enhance retrieval quality in the stratosphere: to increase the signal-to-noise ratio by building better GNSS receivers and developing better ionospheric correction algorithms (ionospheric noise is the major error source at high altitudes) and to utilize statistical optimization techniques invoking background information. This paper describes how background information is used at IGAM to improve RO retrieval algorithms in the stratosphere (Sect. 2), presents results retrieved from CHAMP RO data and evaluates these results in comparison with various other data sources (Sect. 3). The interpretation of results is given in Sect. 4.

## 2 IGAM Retrieval Schemes

The basic idea of including background information into RO retrieval is to stabilize two integrals involved: The inverse Abel transform (Eq. 1, $n$: refractive index, $\alpha$: bending angle, $a$: impact parameter) to derive refractivity $N$ ($N=10^6(n-1)$) and the hydrostatic integral (Eq. 2, $p_d$: dry pressure, $g$: acceleration of gravity, $k_1 = 77.60$ K hPa$^{-1}$, $R = 8.3145$ kJ K$^{-1}$ kg$^{-1}$, $M_d = 28.964$ kg kmol$^{-1}$) to derive pressure from refractivity. Dry temperature is proportional to $p_d/N$. Both equations are responsible for downward propagation of errors during the RO retrieval. To keep these errors minimal the concept of statistical optimization was introduced into the field of RO retrieval [1]. It derives the best linear unbiased estimator (BLUE, Eq. 3), $\boldsymbol{\alpha}_{opt}$, from an observed ($\boldsymbol{\alpha}_o$) and a background ($\boldsymbol{\alpha}_b$) bending angle profile under the assumption of unbiased Gaussian errors. **O** and **B** are the observation and background error covariance matrices, respectively.

$$n(a) = \exp\left[\frac{1}{\pi}\int_a^\infty \frac{\alpha(a')}{\sqrt{a'^2 - a^2}}\,\mathrm{d}a'\right] \tag{1}$$

$$p_d(r) = \frac{M_d}{k_1 R}\int_r^\infty g(r')N(r')\,\mathrm{d}r' \tag{2}$$

$$\boldsymbol{\alpha}_{opt} = \boldsymbol{\alpha}_b + \mathbf{B}(\mathbf{B}+\mathbf{O})^{-1}(\boldsymbol{\alpha}_o - \boldsymbol{\alpha}_b) \tag{3}$$

$\boldsymbol{\alpha}_{opt}$ is a fused bending angle profile dominated by background information in the upper part and by the observation in the lower part. Though most recent retrieval schemes initialize the hydrostatic integral with a pressure value derived from a temperature guess at 40 – 50 km, this is not necessary if the refractivity profile derived from Eq. 1 reaches high enough (~120 km). The IGAM retrieval schemes integrate background information only at one point of the retrieval (at bending angle level), so that the results have well defined error characteristics. We implemented statistical optimization in two different ways, both relying on Eq. 3, but using different sources of background information and different ways of pre-processing of this information: IGAM/MSIS uses bending angle profiles extracted from the MSISE-90 climatology [2] and applies best-fit-profile library search and bias correction procedures [3] in order to diminish know biases in the climatology [4]. IGAM/ECMWF uses bending angle profiles derived from operational analyses of the European Center for Medium-Range Weather Forecasts (ECMWF). More details can be found in Tab. 1. Both schemes are geometric optics dry air retrievals. For moist retrieval (not used in this study), IGAM applies the 1D-Var approach [5] below 10 to 15 km.

| | IGAM / MSIS | IGAM / ECMWF |
|---|---|---|
| **Outlier Rejection and Smoothing** | "3σ" outlier rejection on phase delays and smoothing using regularization. | Like IGAM/MSIS |
| **Ionospheric Correction** | Linear combination of bending angles [6]. L2 bending angles < 15 km derived via L1-L2 extrapolation. | Like IGAM/MSIS |
| **Bending Angle Initialization** | Statistical optimization of bending angles 30–120 km. Vert. correlated background (corr. length L=6 km) and observation (L=1 km) errors. Obs. error estimated from obs. profile >60 km. Backg. error: 15%. Backg. information: MSISE-90 best fit-profile, bias corrected [3]. | Like IGAM/MSIS, but co-located bending angle profile derived from ECMWF operational analysis (above ~60 km: MSISE-90) as backg. information. No further pre-processing. |
| **Hydrostat. Integral Init.** | At 120 km pressure = p(MSISE-90). | Like IGAM/MSIS |
| **Quality Control** | Refractivity 5–35 km: $\Delta N/N < 10\%$; Temperature 8–25 km: $\Delta T < 25$ K. Reference: ECMWF analysis. | Like IGAM/MSIS |

**Table 1**: Technical overview of the IGAM RO retrieval schemes.

# 3 Validation

Data from ECMWF analyses, the GFZ operational RO retrieval, MIPAS and GOMOS were used to validate the IGAM retrieval schemes using (dry) temperature profiles as retrieval performance indicator. The validation covers 11 days (~1750 profiles) in 2002 (20.9.-27.9., 11.10.-13.10.) and was performed by calculating error statistics profiles showing the bias and standard deviation between 5 and 40 km.

Comparisons with operational ECMWF analyses (Fig. 1) are a validation for the IGAM/MSIS scheme, but not for the IGAM/ECMWF (background information same as reference). In this case it still helps to interpret the retrieval performance. Wave-like bias patterns of both retrieval schemes between ~10 and 18 km reflect the better vertical resolution of the CHAMP profiles in the tropopause region. Both schemes feature 1-3 K standard deviation up to 25 km. IGAM/MSIS shows a significant warm bias from below 20 km upwards which is due to biases in the background information that could not be accounted for properly in that case and propagate downwards predominately via the hydrostatic integral. This warm bias could be observed in comparisons with GFZ operational retrieval, MIPAS, and GOMOS data as well (not shown) and shows how strong biased background information can influence the retrieved temperatures. However, vali-

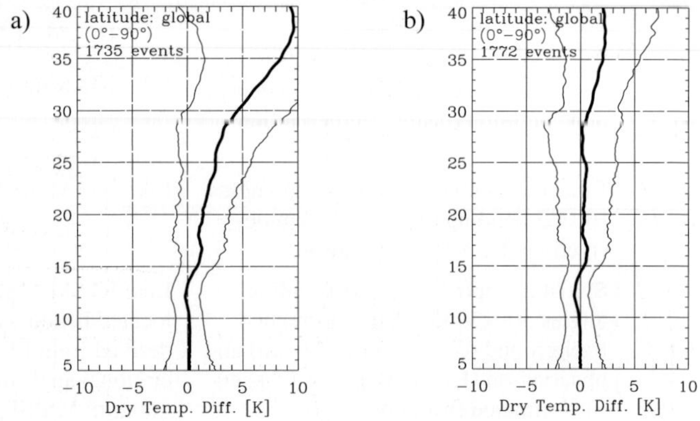


**Fig. 1.** Bias (bold) and standard deviation (thin) of CHAMP dry temperatures compared to ECMWF operational analyses. a) IGAM/MSIS minus ECMWF, b) IGAM/ECMWF minus ECMWF.

dation studies at refractivity level (e.g., [7]) show that IGAM/MSIS- and RO-derived refractivity profiles in general are virtually independent from statistical optimization or high altitude initialization up to 30 – 35 km (no hydrostatic integration needed in this case).

The IGAM/ECMWF retrieved dry temperatures are virtually bias-free up to ~29 km, but warm biased by up to 2 K above which shows that they are not entirely dominated by background information up to above 40 km (Fig. 1b). The transition altitude from observation dominated to background dominated is around 50 km for the CHAMP RO data (not shown).Error statistics for the GFZ (operational retrieval version 4) and IGAM/ ECMWF comparison are shown in Fig. 2a. Below 15 km, the agreement is excellent, the bias and standard deviation is smaller than 0.1 and 0.5 K, respectively. Above 15 km the bias stays below 0.3 K up to 26 km and below 1 K below 29 km, while the standard deviation increases to near 3 K. Above ~29 km, a warm bias qualitatively similar to the bias against ECMWF analyses can be observed, i.e., the GFZ retrieval follows ECMWF more closely than the IGAM/ECMWF retrieval. Both schemes use ECMWF analyses as background information, but in a different way: IGAM/ECMWF to generate the BLUE bending angle profile (GFZ uses MSISE-90-derived bending angle) and GFZ to initialize the hydrostatic integral at 43 km with ECMWF values.

The MIPAS instrument on ESA's ENVISAT is an independent data source for CHAMP retrieval validation. In Fig. 2b, the comparisons of coinciding MIPAS observations (retrieved by IMK) with IGAM/ECMWF is shown. Coincidences were defined to be less than 300 km and 3 h apart. The standard deviation amounts to ~5 K over the entire height range and is basically due to variance in the MIPAS data set. The most salient result is that no significant bias can be observed from 20 to 40 km, though the IGAM/ECMWF and the MIPAS retrieval are entirely independent and both not in agreement with ECMWF analyses above 30 km. From the GOMOS instrument on ENVISAT, temperature profiles are derived using time-

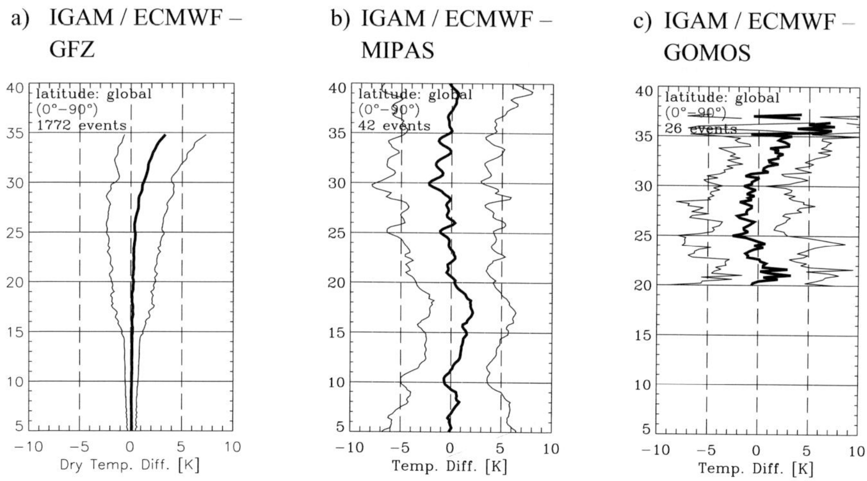


**Fig. 2.** Bias (bold) and standard deviation (thin) of CHAMP dry temperatures (IGAM/ECMWF retrieval) compared to a) GFZ operational CHAMP retrieval, b) coinciding MIPAS measurements, and c) coinciding GOMOS measurements.

delay signals from two fast photometers (retrieval by CNRS). This method is under development and yields only preliminary results so far. In Fig. 2c, IGAM/ECMWF-GOMOS comparison for 26 coinciding occultation events are shown. Roughly, the bias above 30 km shows a similar behavior as in Fig. 1b, which might be due the ECMWF analyses being used for upper-boundary initialization of the GOMOS retrieval.

## 4 Conclusions

The IGAM stratospheric RO retrieval schemes apply statistical optimization based on the best linear unbiased estimator of observed and background bending angle profiles. In order not to give the background information too much weight and to maintain well defined error characteristics, no further background information is involved. Background information is either derived from the MSISE-90 best-fit-profile (bias corrected) or from ECMWF analyses.

Below 15 km, both IGAM retrieval schemes are in very good agreement with ECMWF analyses and strongly conform with the operational GFZ retrieval results (< 0.1 K bias, < 0.5 K std. dev.). Though on refractivity-level no high-altitude bias occurs up to ~35 km [7], the IGAM/MSIS temperatures are warm biased above 15–20 km, which shows that the background bias correction algorithm involved is not fully effective when applied to CHAMP data. The IGAM/MSIS scheme was successfully evaluated in simulation studies using METOP-GRAS receiver specifications [3], the lacking performance applied to CHAMP data is due to worse data quality at high altitudes (higher receiver-noise level, outliers, residual ionospheric noise stemming from small-scale structures in the ionosphere that were not modeled in the simulation study). Since RO retrievals independent from NWP analyses are desirable, this scheme will be further developed to become more robust against noisy data, and better profile-search libraries than the MSISE-90 climatology are envisaged.

The IGAM/ECMWF results show excellent agreement with the GFZ retrieval up to ~26 km (<0.3 K bias, <3 K std. dev.) and a warm bias up to 2 K above that height. Compared to MIPAS, this bias does not occur up to 40 km, which shows that the IGAM/ECMWF retrieval scheme is less dominated by background information in the stratosphere and still yields excellent results.

*Acknowledgments.* The authors gratefully acknowledge valuable scientific discussions with A.K. Steiner and technical assistance by J. Fritzer and J. Ramsauer (IGAM, Univ. of Graz, Austria). A.G. was funded from the START research award of G.K. (Y103-N03) of the Austrian Science Fund. Partial funding was also provided by the CHAMPCLIM contract (ASAP-CO-004/03) of the Austrian Space Agency.

## References

1. Sokolovskiy S, Hunt D (1996) Statistical optimization approach for GPS/Met data inversions. Presentation at URSI GPS/Met Workshop 1996, Tucson, AZ, USA.
2. Hedin AE (1991) Extension of the MSIS thermosphere model into the middle and lower atmosphere. J Geophys Res *96*: 1159–1172
3. Gobiet A, Kirchengast G (2004) Advancement of GNSS Radio Occultation Retrieval in the Upper Stratosphere. Proc. OPAC-1, Sept. 2002, Graz, Austria, accepted.
4. Randel W, Chanin ML, Michaut C (2002) SPARC intercomparison of middle atmosphere climatologies. SPARC Report No. 3, WCRP 116, WMO/TD 11424.
5. Healy SB, Eyre JR (2000) Retrieving temperature, water vapour and surface pressure information from refractive-index profiles derived by radio occultation: a simulation study. Q J R Meteorol Soc *126*: 1661-1683.
6. Vorob'ev VV, Krasnil'nikova TG (1994) Estimation of the accuracy of the atmospheric refractive index recovery from Doppler shift measurements at frequencies used in the NAVSTAR system. Phys Atmos Ocean *29*: 602–609.
7. Wickert J, Gobiet A, Beyerle G, Steiner A, Kirchengast G, Foelsche U, Schmidt T (2004) GPS radio occultation with CHAMP: Comparison of atmospheric profiles from GFZ Potsdam and IGAM Graz. This issue.

# Derivation of the Water Vapor Content from the GNSS Radio Occultation Observations


Francesco Vespe[1], Jens Wickert[2], Catia Benedetto[3], Rosa Pacione[3]

[1] Agenzia Spaziale Italiana / Centro di Geodesia Spaziale 75100 Matera, Italy
[2] GeoForschungsZentrum Potsdam (GFZ), Dept. Geodesy & Remote Sensing, Germany
[3] Telespazio S.p.A./ Centro di Geodesia Spaziale 75100 Matera, Italy



**Summary.** The present work investigates the possibility of retrieving humidity profiles using the bending angle data obtained from GNSS Radio Occultations (Global Navigation Satellite System RO) without data constraint. In particular with the proposed approach, the dry pressure profiles are obtained by fitting the bending angles of the outer troposphere layers (from $h=h_{250K}$ up to the stratopause) using the Hopfield dry atmosphere model. In this model the ground pressure and temperature are the parameters to be estimated. The humidity profiles are extracted by subtracting the contribution due to the dry atmosphere from the measured bending angles. In the first part we discuss the mathematical approach adopted for the derivation of refractivity profiles without using the Abel inversion and water vapor directly from the bending angle. In the second part the results are shown from applying the method on CHAMP data. In particular, we have used refractivity profiles from CHAMP which were derived using a heuristic retrieval algorithm based on the canonical transform.



**Key words:** radio occultation, atmospheric remote sensing, water vapor profiles, inversion techniques, refractivity profiles


## 1 Introduction

The parameters of the neutral atmosphere (pressure "*P*", temperature "*T*" and water vapor pressure "$P_{wet}$") are retrieved from GNSS RO data, after having derived the refractivity from the bending angles, using two equations: the refractivity at microwave wavelengths according to Smith and Weintraub (1953) and the hydrostatic equilibrium law combined with the state equation of ideal gas (Kursinski et al. 2000). In the upper atmosphere water vapor is negligible, so two equations have just two unknowns, P and T: easy to solve for. The atmospheric quantities of interest can be obtained for the lower troposphere by integrating GNSS RO data with external information, e.g. from ECMWF or NCEP re-analysis. Thus, although GNSS RO is a very reliable technique for the measurement of neutral atmosphere key parameters, current retrieval methods for the water vapor content in the lower troposphere depend essentially on conventional observation systems (mainly RAOB, RAdiosonde OBservations). In previous works (Vespe et al. 2002, 2004a) a new method was described: the **B**oundary **Pr**ofiles e**V**aluation (BPV), which is able to derive tropospheric water vapor profiles using GNSS RO refractivities without additional external information. Efforts have been performed recently just to make the GNSS RO a standalone tool for water vapor retrieval (O' Sullivan et al. 2000; de la Torre and Nilsson 2003). A main drawback in the GNSS RO retrieval is the possible existence of horizontal gradients of the refractivity or non-spherical structure in the vicinity of the occultation location which could make the Abel inversion not fully accurate (Kursinski et al. 2000). For this reason it is preferable to work on the processing level of the bending angles; i.e. using the data on which the Abel inversion is not applied. Thus, we are going to describe the new release of the BPV in which the bending angles are used directly for the water vapor retrieval. The new BPV release implies also the reconstruction of the refractivity by applying a sort of variational technique. Thus the second non-negligible result is the possibility to derive refractivity in an alternative way to Abel inversion. The refractivity profiles obtained from CHAMP and GPS/MET data have a negative bias in relation to radiosonde and analysis data (e.g. Ao et al. 2003; Beyerle et al. 2004). We use two versions of CHAMP data products based on Geometrical Optics (GO): version 003 for Recife and Brest sites and 004 (Wickert et al. 2004). In addition we use preliminary results obtained applying a heuristic retrieval algorithm (CTss) based on Canonical Transform (CT) and the sliding spectral (ss) technique to reduce the refractivity bias (Beyerle et al. 2004; Wickert et al. 2004). The use of the wave optics based algorithms reduces the refractivity bias by a factor of ~2 in relation to the GO profiles. Such algorithms will be implemented to the operational CHAMP data analysis as of mid 2004 (product version 005). The new release of the BPV (Vespe et al. 2004b) was applied to the version 003 and 004 GO data, available online at GFZ web site (http://isdc.gfz-potsdam.de/champ/), as well as to the CTss results. Thus a comparison between the two sets of products is another outcome of the present paper.

## 2 Data Analysis

The refractivity $N$ is defined as: $N = 10^6(n-1)$ with the refractive index $n$. For the neutral atmosphere $N$ is the sum of a dry ($N_{dry}$) and a wet component ($N_{wet}$):

$$N(h) = N_{dry}(h) + N_{wet}(h) = \left[ a_1 \frac{P_{dry}(h)}{T(h)} \right]_{N_{dry}} + \left| \frac{a_1 \cdot T(h) \cdot P_{wet}(h) + a_3 \cdot P_{wet}(h)}{T^2(h)} \right|_{N_{wet}} \quad (1)$$

where

| | | | |
|---|---|---|---|
| $N_{dry}$ | dry refractivity | $P_{wet}$ | water vapor pressure (mb) |
| $N_{wet}$ | wet refractivity | T | temperature (K) |
| h | height ( km) | $a_1$ | 77.6 (K/mbar) ± 0.05 |
| P | $P_{dry}+P_{wet}$ total pressure (mb) | $a_3$ | $3.739*10^5$ ($K^2$/mbar) ± $0.012*10^5$ |

Assuming hydrostatic equilibrium and validity of the state equation of the ideal gas, together with Eq. 1 a system of two equations and three unknowns can be formed ($T$, $P_{dry}$ and $P_{wet}$).

Since water vapor is negligible above the tropospheric height $h_{250K}$, $P$ and $T$ can be solved for considering only the dry component. Dry atmosphere approximation does not work in the lower troposphere, where water vapor pressure can be retrieved only if additional external data are available and assimilated as guess values in an iterative procedure. Such data are generally given by meteorological analyses carried out by the ECMWF and the NCEP.

In the first release of our approach (Vespe 2002; Vespe et al. 2004a) we adopted as dry refractivity the Hopfield model (Hf, Hopfield 1969) which relates the dry air refractivity to the fourth power of the height and to the surface values of pressure and temperature $P_0$ and $T_0$. We use the GNSS RO refractivity data of the atmospheric layers between the stratopause and the height, where T=250 K, through which the water vapor content can be considered negligible, for performing the fit. In this fit the surface pressure and temperature of the dry air refractivity (Hf) is estimated. Hf is extrapolated down to the ground with the assumption that the difference of the extrapolated and the observed GNSS RO refractivity is due to the wet part of the atmosphere. Finally Eq. 1 is applied to retrieve the wet pressure. The enhanced BPV is suitable to be applied directly to the bending angles (BA). The necessary steps to derive the water vapor partial pressure profile using the GNSS RO BA data are described in detail by Vespe et al. (2004b). They are briefly summarized here:

1. Fit the BA above $h_{250K}$ up to the stratopause retrieved by GNSS RO.
2. From the estimated parameters $T_0$ and $P_0$ the dry BA is computed down to the ground by extrapolating the Hf model. In Fig. 1 the measured and dry BA are compared for selected sites.
3. A Taylor expansion is applied to the BA just to built refractivity profiles using the dry Hf as starting profile. The difference between total and dry refractivity gives the wet contribution as in (1).
4. When the refractivity is built, the first release of the BPV method is applied for the water vapor retrieval (Vespe et al. 2004a). Results are shown in Fig. 3.

## 3 Results and Discussion

The method proposed seems to be a promising alternative to the Abel inversion for retrieving refractivity profiles from the BA. The advantages of our approach is evident in the plots of Fig. 2. The profiles retrieved with the described technique are systematically very close to the RAOB data. In the previous work (Vespe et al. 2004b) the refractivity profiles seemed to be slightly positively biased if compared with CHAMP GO data for occultations reaching the Earth's surface.

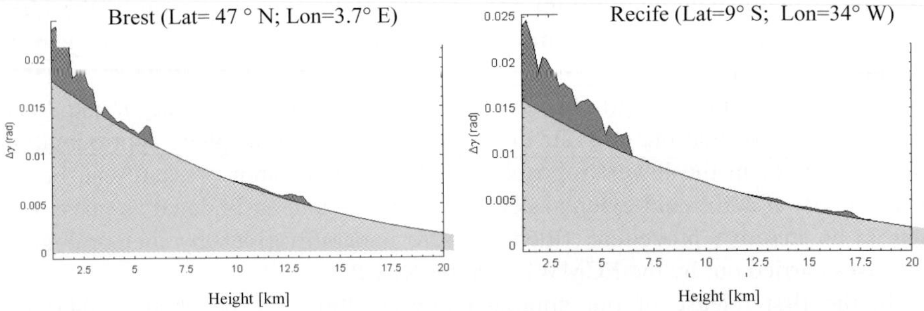


**Fig. 1.** The plots show the results of the fits of the BA with the dry Hf model (light gray). The differences with the measured BA (dark gray) are supposed to be due to atmospheric water vapor content. The equatorial site (Recife, right plot), as expected, show a larger difference than the European site.

Such positive bias not only vanishes with the new CHAMP data (CTss), their agreement is so close that they are practically identical (Fig. 2). We relate this to the CTss algorithms introduced, as described by Beyerle et al. (2004). The discrepancies of the refractivity with RAOB data were really huge for MANGALORE (Vespe et al. 2004b). The use of the CTss profiles has completely diminished these differences. Furthermore some humps, evident in the RAOB refractivity profile, were clearly detected by BPV (Recife profiles) using the GO based CHAMP data. Such humps are detected now also by using the new products. The main drawback of the GO profiles is the cut off when the refractivity deviates by more than 10% from ECMWF for quality check reasons (Wickert et al. 2004). The application of this criterion sometimes seems to be not wise because we have experienced, on the contrary, a good agreement of the cut GNSS RO data with RAOB observations. Combined with the BPV method these data can help to detect local atmospheric structures. In the Recife refractivity profile for example (see Fig. 2) the agreement with RAOB data is good also with GO data down to the ground (Vespe et al. 2004b). So it would be preferable to introduce only a warning flag. These preliminary results seem to confirm that the BPV applied to the BA could in part overcome the severe problem of the negative refractivity bias over the equatorial regions as previously announced (Vespe et al. 2003). We would emphasize that such comments are only weak suppositions for the small number of analysed profiles. A more extensive statistical investigation is planned to confirm them. The water vapor retrievals are shown in Fig. 3. Our results after applying the BPV on the CTss CHAMP data do not confirm the previous ones (Vespe et al. 2004b). The BPV applied to the GO profiles results in an overestimation of the water vapor content. The results in Fig. 3 on the contrary outline an underestimate. Such discrepancy must be carefully analyzed because it can be due to the efficiency of our approach in detecting irregularities and super-refractivity events but also on account of some inaccuracies in it. The reconstruction of the

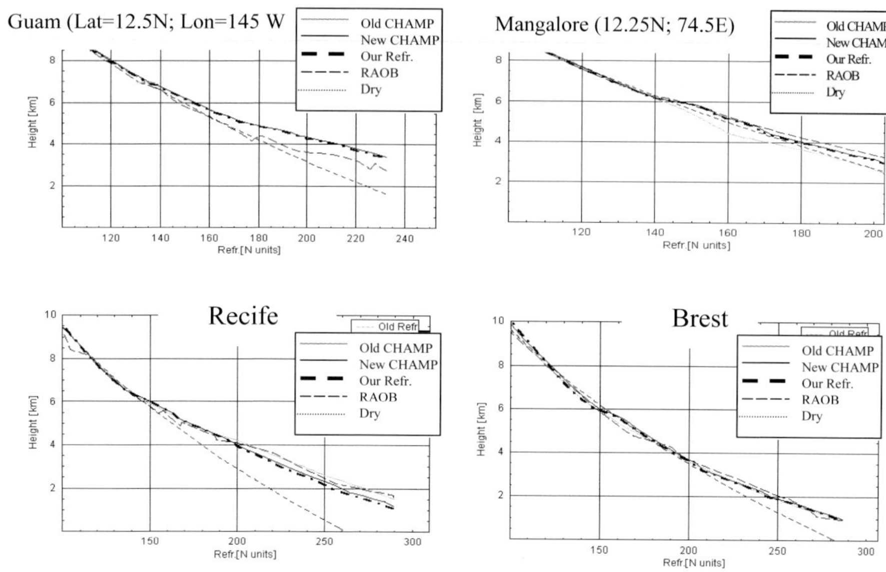


**Fig. 2.** The figures show the refractivity obtained applying our approach (dash-dot thick line) and compared with RAOB (dash line), CHAMP CTss data (black line), CHAMP GO (old 003 gray line, new 004 black) and dry model (small dashed).

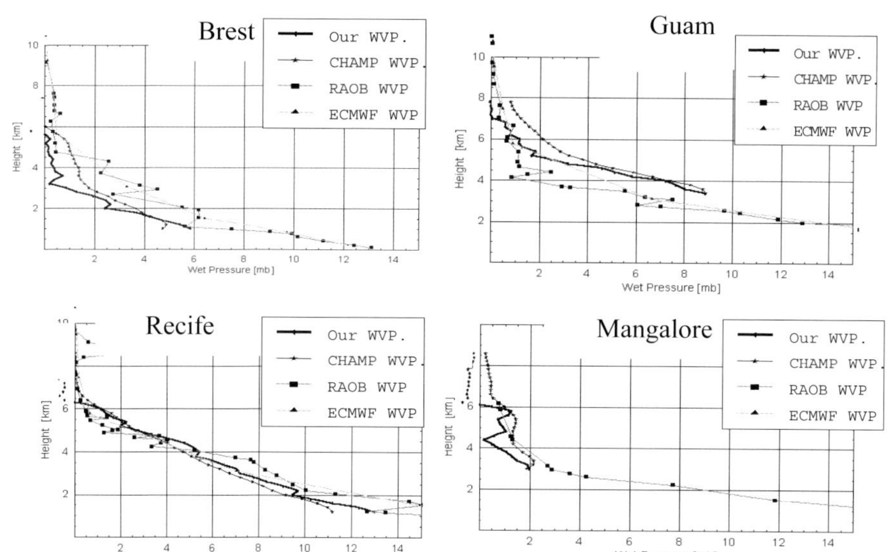


**Fig. 3**. The water vapor profiles retrieved by using the BPV approach, compared with CTss CHAMP, ECMWF and RAOB data.

temperature profiles in the lower atmosphere is one of the pending solution in the proposed procedure. The linear interpolation of the temperatures between the surface values as given by the Hf model (h=0) and the height where $T_{dry}$ is <250 K, sounds not so rigorous. It is worthwhile to be investigated if our approach improves its performance in case of applying the same algorithms described by O'Sullivan et al. (2000). The extension of the BPV method to the BA has confirmed the promising results achieved in previous studies on refractivity (Vespe et al. 2002 and 2004a).
The results of the present work, together with efforts of other groups (O'Sullivan et al. 2000; de la Torre and Nilson 2003), are encouraging to make the GNSS RO a standalone technique for supporting weather forecasts and climate research.

## References


Ao C O, Hajj G A, Leroy S S, Meehan T K, de la Torre Juarez M, Iijima B A, Mannucci J (2003) Backpropagation processing of GPS radio occultation data. In: Reigber Ch, Lühr H and Schwintzer P (Eds): First CHAMP Mission Results for Gravity, Magnetic and Atmospheric Studies, Springer-Verlag, 415-422.

Beyerle G, Wickert J, Schmidt T, Reigber Ch (2004) Atmospheric sounding by GNSS radio occultation: An analysis of the negative refractivity bias using CHAMP observations. J Geophys Res *109(D1)*: doi: 10.1029/2003JD003922.

de la Torre Juàrez M, Nilsson M (2003) On the Detection of Water Vapor Profiles and Thin Moisture Layers from Atmospheric Radio Occultations, J Geophys Res, in print.

Hopfield H S (1969) Two-Quartic tropospheric refractivity profile for correcting satellite data. J Geophys Res *74(18)*: 4487-4499.

Kursinski R E, Hajj G A, Leroy S S, Herman B, (2000) The GPS Radio Occultation Technique. TAO *11(1)*: 53-114.

O' Sullivan D B, Herman B M, Feng D E, Flittner D E, Ward D M (2003) Retrieval of Water Vapor Profiles from GPS/MET Radio Occultations. Bull Am Meteor Soc *81(5)*: 1031-1040.

Smith E K, Weintraub S (1953) The constants in the equation for atmospheric refractive index at radio frequencies. Proc IRE *41*: 1035-1037.

Vespe F (2002) Water Vapor Retrieval by GNSS Radio Occultation Technique with no external information. Proc COSMIC Science Workshop, Boulder, August 2002.

Vespe F, Benedetto C, Pacione R (2004a) The use of refractivity retrieved by radio occultation technique for the derivation of atmospheric water vapor content. Phys Chem Earth, in print.

Vespe F, Benedetto C, Pacione R (2004b) GNSS Radio Occultation: from the bending angles to the atmospheric profiles. Proc URSI workshop on Atmospheric Remote Sensing by GNSS, Oct. 2003, Matera Italy, in print.

Wickert J, Schmidt T, Beyerle G, König R, Reigber Ch, Jakowski N (2004) The radio occultation experiment aboard CHAMP: Operational data analysis and validation of vertical atmospheric profiles. J Meteorology Soc Jap *82(1B)*, Special issue 'Application of GPS Remote Sensing to Meteorology and Related Fields': 381-395.

# Processing of CHAMP Radio Occultation Data Using GRAS SAF Software

Georg Bergeton Larsen, Kent Bækgaard Lauritsen, Frans Rubek, Martin Bjært Sørensen

Danish Meteorological Institute (DMI), Lyngbyvej 100, DK-2100 Copenhagen, Denmark, *gbl@dmi.dk*

**Summary.** The developments of an operational system for processing of radio occultations from the GRAS (Global navigation satellite system Receiver for Atmospheric Sounding) receiver onboard the future Metop satellite is the primary scope of the EUMETSAT project: GRAS Satellite Application Facility (SAF). The planned GRAS SAF data products and processing system is discussed together with a preliminary product validation based on statistical analysis of retrieved refractivity and temperature using radio occultation data from the German CHAMP (CHAllenging Minisatellite Payload) satellite. The CHAMP data from the first 8 weeks of 2003 have been processed using the GRAS SAF retrieval system and we present the results as compared to the ECMWF. The mean difference in refractivity is approx. 0.5% with standard deviations less than 1% between 8-25 km and approximately 2% both close to the surface and at 35 km.

**Key words:** GPS, radio occultation, CHAMP

## 1 Introduction

The GRAS SAF started in 1999 with participation from the Danish Meteorological Institute (DMI), Institut d'Estudis Espacials de Catalunya (Spain), and Met Office (UK). The host institute is the DMI and this will also be the physical location of the operational GRAS SAF data facility. The first Metop satellite, in a series of three polar satellites carrying the GRAS receiver, will be launched in the second half of 2005. This will be the space segment of the EUMETSAT polar system (EPS), which the GRAS SAF is a part of. The GRAS receiver measures the signals from the Global Positioning System (GPS) satellites. It will measure setting and rising occultations, and also reference GPS signals for navigation and precise orbit determination [1].

The GRAS SAF will receive Level 0, Level 1a (phases and amplitudes) and Level 1b data (bending angles, impact parameters) from the EPS Core Ground Segment (CGS), process the data into Level 2 products (refractivity, temperature, humidity), and distribute these to Numerical Weather Prediction (NWP) centres and climate research users. The operational data products will be available from the second half of 2006.The GRAS SAF data products are vertical profiles of refractivity, temperature, pressure, and humidity. They come in two types: Near-

Real Time (NRT) products, disseminated less than 3 hours after sounding, and as offline products (reprocessed to benefit from improved orbit determination), disseminated/available less than 30 days after sounding. All data products will be archived at the GRAS SAF Archive at DMI. Archived products will be available through EUMETSATs Unified Meteorological Archive and Retrieval Facility (UMARF).

The use of GPS radio occultations for accurate measuring, with high vertical resolution, of the Earth atmosphere was first demonstrated by the proof of concept mission GPS/MET in 1995 [2]. Since then the field of remote sensing using GPS occultations has evolved rapidly and more research satellite missions have followed. The successful CHAMP mission provides continuously GPS occultation measurements since mid 2001 resulting in approximately 150 atmospheric profiles per day [3]. We have here used data from CHAMP to provide a first validation of the GRAS SAF processing system. In chapter 2 we briefly introduce the GRAS SAF products and processing, the validation results using CHAMP data are presented in chapter 3, with conclusions in chapter 4.

## 2 Data products and processing system

The GRAS SAF data products will be derived from the measurements of the GRAS receiver onboard Metop. The first steps of the radio occultation data processing will take place at the EPS CGS at the EUMETSAT headquarters in Darmstadt, Germany. This includes data collection from Metop as well as ground based measurements in real time of the GPS satellites for precise orbit determination and clock differencing [4]. The near real time processing (within 2 hrs and 15 min) up to and including bending angles as function of impact parameter is the responsibility of the CGS. The GRAS SAF is responsible for the near real time processing (3 hrs) and dissemination of atmosphere products, including: refractivity, temperature and humidity as function of height [5], with accuracies as specified by the GRAS SAF user requirements [6].

The GRAS SAF processing software consists of inversion routines to obtain the refractivity, dry pressure and temperature, and the 1D-var routine [7, 8] for deriving temperature and humidity. The details of the processing system are described in the architectural design [9]. The GRAS SAF also includes continuously reprocessing (offline data products) of all data products when improved orbit and reference GPS data are available. The NRT Metop orbit determination is subdivided into segments of 10 minutes each and relies on the predicted orbits for the GPS satellites. The Metop orbit will be reprocessed by CGS to adjust for orbit manoeuvres and precise GPS orbits. The offline data products contain furthermore the reprocessed excess phases and bending angles. The GRAS SAF also develops dedicated software deliverables for NWP centres to facilitate the direct data assimilation of both bending angles and refractivity using the 3/4D-var techniques. The list of GRAS SAF products (data and software) is shown in Table 1.

| GRAS SAF Products List | Description of product |
|---|---|
| **NRT Data Products** | |
| Refractivity profile | Refractive index of the atmosphere in units of $(n-1)10^6$ |
| Temperature profile | Temperature as function of height at tangent point |
| Pressure profile | Pressure as function of height at tangent point |
| Specific humidity profile | Specific humidity as function of height at tangent point |
| Surface pressure | Pressure estimate at surface level |
| Error covariance matrix | Error covariance matrix as average for all profiles |
| **Offline Data Products** | |
| Bending angle profile | Bending angle as function of impact parameter |
| Refractivity profile | Refractive index of the atmosphere in units of $(n-1)10^6$ |
| Temperature profile | Temperature as function of height at tangent point |
| Pressure profile | Pressure as function of height at tangent point |
| Specific humidity profile | Specific humidity as function of height at tangent point |
| Error covariance matrix | Error covariance matrix as average for all profiles |
| Global map of temperature | Global map based on monthly averages of temperature profiles |
| Global map of specific humidity | Global map based on monthly averages of humidity profiles |
| Global map of geopotential height | Global map of monthly averages of geopotential height profiles |
| **Software Products** | |
| 1D-var software | Software module used to generate NRT temperature, pressure and humidity products from refractivity and a user's background profile |
| 3/4D-var assimilation software | Forward operators and their adjoints to allow for 3/4D-var data assimilation of GRAS SAF and bending angle products into existing NWP models |
| Pre-processing tools | Pre-processing tools to assist the data assimilation of GRAS SAF and bending angle products |

**Table 1.** GRAS SAF data and software products.

The refractivity is derived using geometrical optics [5, 10], and is the input to the 1D-var algorithm to derive the atmosphere parameters of temperature and humidity. We use a background error covariance matrix used for operational processing of ATOVS data, from the Met Office, with 43 pressure levels (surface to 0.1 hPa). The measurement error-covariance matrix is defined as follows: the standard deviation is 2% at the surface, falling linearly to the constant value 0.2% from 12 km and up. In addition, the minimum (absolute) error is fixed at 0.02 N-units. This corresponds to an increase in the relative error above 25 km, reaching approx 2% at 40 km. The correlations are assumed to fall off exponentially with a correlation length of 3.0 km. These error magnitudes are based on results from simulation experiments [10, 11] combined with analysis of CHAMP data. The background field is obtained from the ECMWF (European Center for Medium-range Weather Forecasts). The field resolution is 1° with values for temperature and humidity given at 60 model levels interpolated to the 43 fixed pressure levels of the background error covariance matrix.

## 3. Validation results using CHAMP data

CHAMP data from the first 8 weeks of 2003 starting 1 January and continuously until 25 February 2003 (56 days) has been selected for the validation analysis. We do not include data below the height corresponding to the time when the fly-wheeling flag is set for the first time, i.e. when the signal noise is so large that the closed-loop tracking is lost.

With the quality settings we are using, we obtain that about 84% of the CHAMP profiles are accepted by the 1D-var algorithm resulting in a total of 7280 profiles. The observed refractivity, derived from bending angles using the Abel transform, is compared against the ECMWF values in Fig. 1. The mean difference is approximately 0.5% and standard deviations are less than 1% between 8-25 km, and approximately 2% both close to the surface and at 35 km. These results compare well with the CHAMP results by other groups [3, 12]. The small bulge of negative bias in refractivity seen close to 15 km could be caused by e.g. coarse resolution in the ECMWF model (also observed by [3] for an even larger sample of CHAMP occultations), but it requires more analysis and validation of other periods to investigate possible seasonal effects.

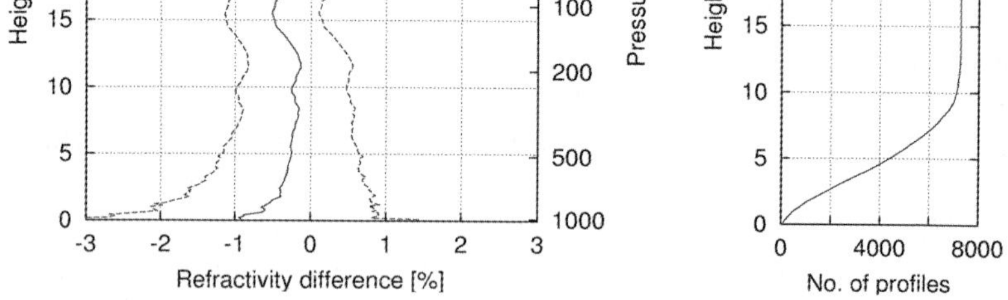


**Fig. 1.** Left panel shows the relative difference between the observed refractivity and the analysis data from ECMWF. The solid line shows the mean difference and the dashed lines indicate +/- one standard deviation away from the mean. Right panel shows the number of measurements used in the statistics as a function of height.

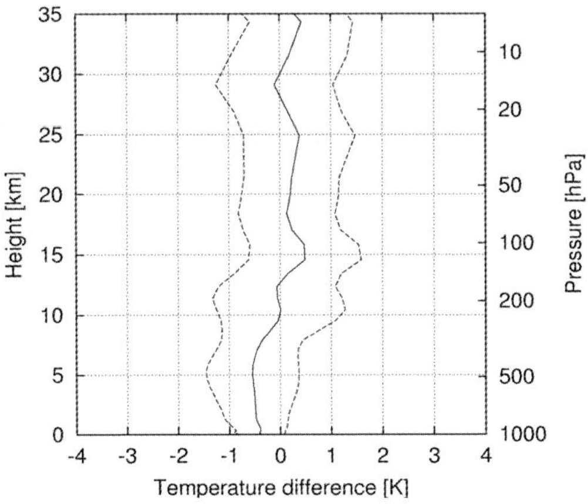


**Fig. 2.** Temperature difference between the derived 1D-var temperature and the ECMWF. The solid line shows the mean difference and the dashed lines indicate +/- one standard deviation away from the mean. The number of measurements as a function of height is identical to that used in Fig.1.

In order to measure the quality of the 1D-var solution we construct the difference of the 1D-var solution and the ECMWF background field at the given location.

In Fig. 2 we show the difference for the 1D-var temperature profile. The mean difference is less than 0.5 K and the standard deviation is approximately 1 K between the surface and 35 km. As the 1D-var solution and the background are not independent, the interpretation of the statistics in Fig. 2 is not straightforward and hence the apparent deviations cannot directly be used as error estimates. The deviations become smaller close to the surface as the measurements decrease in number near the surface, which gives more weight to the model in the solution.

The fact that we obtain temperature profiles which still include deviations in the order of 1K suggest that we have found a reasonable balance between the background errors and the measurement errors used in the 1D-var algorithm.

## 4. Conclusions

We have presented a brief overview of the GRAS SAF data products which will be produced on the basis of the future GRAS radio occultation mission onboard the Metop satellites. Refractivity and temperature profiles derived using CHAMP data are compared against analysis data from ECMWF. The mean refractivity difference shows high accuracy of 0.5% to 1% in the range between 8-25 km, with a standard deviation about 2% or more close to the surface or at 35 km. This is an upper limit on the accuracy on the retrieval and consistent with the assumed errors in the 1D-var algorithm. The 1D-var temperature compared against ECMWF shows consistent deviations of less than 1K, however this cannot directly be interpreted as an accuracy estimate. A comparison against a completely independent measurement is difficult to achieve, as e.g. the radiosonde measurements that are suited for direct comparison already has been used as one of the inputs to the ECMWF model. Future GRAS SAF validation studies will be carried out using the CHAMP radio occultation measurements and will study effects such as seasonal variation, residual ionosphere contributions, and skew profiles using high vertical resolution ECMWF data.

## References

1. Christensen J, Bonnedal M, Lindgren T, Stricker N, Loiselet M, and Menard Y (2002) EUMETSAT Meteorological satellite Conference, Dublin, Ireland, Proceedings, p.147.
2. Rocken, C, Anthes R, Exner M, Hunt D, Sokolovskiy S, Ware R, Gorbunov M, Schreiner W, Feng D, Herman B, Kuo Y-H, and Zou X (1997) Analysis and validation of GPS/MET data in the neutral atmosphere, J Geoph Res *102:* 29,849-29,860.
3. Wickert J, Schmidt T, Beyerle G, König R, Reigber C, and Jakowski N (2004) The radio occultation experiment aboard CHAMP: Operational data analysis and validation of vertical atmospheric profiles. Special issue of J Meteorol Soc Japan *82(1B)*: 381-395.
4. Luntama J-P, Wilson JJW (2003) GRAS Level 1 Product Generation Function Specification, Issue 6 Rev 3, EPS/SYS/SPE/990010.
5. Rubek F, Larsen GB, Lauritsen KB, Sørensen MB, Offiler D, Healy S, Higgins M, Aparicio JM, Rius A (2003) GRAS SAF Science Plan.
Ref. SAF/GRAS/DMI/ALG/SP/031, Nov 2003.
6. Offiler D, Larsen GB, Benzon H-H, Lauritsen KB, Rubek F, Healy S, Higgins M, Aparicio JM, Rius A (2001) GRAS SAF User Requirement Document.
Ref. SAF/GRAS/METOFFICE/RQ/URD/001, Nov 2001.
7. GRAS SAF: Processing Radio Occultation Data with the 1D-var Technique: An Overview and User Guide. Ref: SAF/GRAS/UKMO/ALG/1DV/001, 2001.
8. Healy SB and Eyre JR (2000) Retrieving temperature, water vapour and surface pressure information from refractive-index profiles derived by radio occultation: Assimilation study. Q J R Meteorol Soc *126*: 1661–1683.
9. Sørensen MB, Aparicio JM, Larsen GB, Lauritsen KB, Rubek F, Rius A (2003) GRAS SAF Architectural Design Document. Ref. SAF/GRAS/IEEC/DPR/ADD/025, Nov 2003.
10. Høeg P, Larsen GB, Benzon H-H, Grove-Rasmussen J, Syndergaard S, Mortensen M D, Christensen J, and Schultz K (1998) GPS atmosphere profiling methods and error assessments. Scientific Report 98-7, Danish Meteorological Institute, Copenhagen Denmark.
11. Steiner AK and Kirchengast G (2003b) Empirical error characteristics of radio occultation data for data assimilation systems. paper presented at EGS-AGU-EUG Joint Assembly 2003, Nice, France, April 7–11.
12. Kuo Y-H, Wee T-K, Sokolovskiy S, Rocken C, Schreiner W, Hunt D, and Anthes A (2004) Inversion and error estimation of GPS radio occultation data. Special issue of J Meteorol Soc Japan *82(1B)*: 507-531.

# Gravity Wave "Portrait" Reconstructed by Radio Holographic Analysis of the Amplitude of GPS Radio Occultation Signals

Yuei-An Liou[1], Jens Wickert[2], Alexander Pavelyev[3], Christoph Reigber[2], Torsten Schmidt[2], Chen-Young Huang[1], Shen Yan[1]

[1] Center for Space and Remote Sensing Research, National Central University, Jung-Li, 320, Taiwan, *yueian@csrsr.ncu.edu.tw*

[2] GeoForschungsZentrum Potsdam (GFZ), Department Geodesy and Remote Sensing, Telegrafenberg, 14473 Potsdam, Germany, *wickert@gfz-potsdam.de*

[3] Institute of Radio Engineering and Electronics of Russian Academy of Sciences, (IRE RAS), Fryazino, Vvedenskogo sq. 1, 141120 Moscow region, Russia, *pvlv@ms.ire.rssi.ru*

**Summary.** The analysis of the amplitude of the GPS/MET and CHAMP radio occultation (RO) events revealed clusters (quasi-regular structures) with a vertical size of about 10 km and an interior vertical period of ~0.8 to 2 km in the tropopause and lower stratosphere. The height interval of the clusters changes from 10 to 40 km. Restored from the RO amplitude data, variations of the vertical temperature gradient in clusters dT/dh are from -8 to -9 K/km up to 6 to 8 K/km. These variations can be associated with the influence of internal waves (gravity waves; GWs) propagating through the atmosphere and mesosphere. We found a height dependence of the GW phase and amplitude (the GW "portrait"), using for example the amplitude data corresponding to the GPS/MET and CHAMP RO events. For a GPS/MET event we estimated the horizontal wind speed perturbations which are in fairly good agreement with radiosonde data. Between 10 and 40 km, the horizontal wind speed perturbations v(h) are changing in the range v from ~±1 to ±9 m/s with vertical gradients dv/dh ~±0.5 to ±15 m/(s km). The height dependence of the GW vertical wavelength was derived by differentiation of the GW phase. The analysis of this dependence led to the estimation of the GW intrinsic phase speed which changed for the considered events in the interval from 1.5 to 5 m/s. The analysis of RO data shows that the RO signal amplitude contains valuable information for studying the GW activity in the atmosphere.

**Key words**: gravity waves, radio occultation, amplitude, vertical gradient of temperature

## 1 Introduction

GWs play a decisive role in transporting energy and momentum, in contributing turbulence and mixing, and in affecting the atmospheric circulation and temperature regime [1]. For theoretical studies of the GW phenomena it is important to use experimental data, which can reproduce the phase and amplitude dependence of the GWs on height. Radiosonde and rocketsonde GW measurements, radar observations and lidar studies are limited to ground-based sites [1,2], mainly over specific land parts of the Northern Hemisphere, whereas the RO method can be applied to measure the GW parameters on a global scale. Atmospheric studies using the RO technology are based on the analysis of the phase and amplitude of radio waves after propagating through the atmosphere. The radio signals emitted at two GPS frequencies (F1 = 1575.42 and F2 = 1227.6 MHz) by the radio navigational satellites and received by a small satellite on a low earth orbit (LEO) have been used for the RO investigations [3]. The analysis of temperature variations derived from RO phase data furnishes an opportunity to investigate the global morphology of the GW activity in the stratosphere and to measure GW characteristics in the atmosphere as shown early in [2,4]. However, these papers were concerned mainly with GW statistical parameters. The analysis of the RO signal amplitude is a promising tool for the atmospheric research [5,6].

The aim of this contribution is to demonstrate the possibility of direct observations of the quasi-regular structure of atmospheric and ionospheric waves using the amplitude of RO signals.

## 2 Wave trains in the RO data

The scheme of the RO experiment is shown in Fig. 1. Point O is the center of spherical symmetry of the atmosphere. Radio waves emitted by the GPS satellite (point G) are propagating to the receiver onboard the LEO satellite (point L) along the ray GTL where T is the tangent point in the atmosphere. The point T denotes the minimum ray distance from the Earth surface h. The vicinity of the point T introduces the major contribution to the changes of the RO signal. Recordings of the RO signal along the LEO trajectory contain the amplitudes A1(t), A2(t) and the phases of the radio field at frequencies F1, F2 as functions of time. The time interval for RO measurements $\tau$ depends on the orientation between the vertical direction at the point T and the occultation ray path. The time $\tau$ reaches its minimum (~30 s) when the orbital planes of the LEO and GPS satellites are parallel. Thus the RO experiments record practically simultaneously the impact of the GWs on the RO signal because the GW frequencies are usually well below $1/\tau \sim 0.03\ s^{-1}$. The amplitude RO method is appropriate for observation of the internal wave in the atmosphere and ionosphere [5,6]. The horizontal resolution of the RO method for GWs is changing in the range from 100 to 170 km [4]. The amplitudes of the RO signals are shown in Figure 2, panel a) for the GPS/MET event 0316 (July 10, 1995) and CHAMP event (May 21, 2001). Results of the simulations of the amplitude dependence on height are shown in Fig. 2 a) by the curves M0 and M. For calculating M0 we used the refractivity model $N(h) = N_o \exp(-h/H)$ with $N_0 = 340$ (N-units) and H = 6.4 km. We applied the same method to obtain the curves M in Fig. 2 a). Here the refractivity model was a sum of damped complex exponentials $N(h) = \mathrm{Re}[N_j \exp(-\alpha_j h)]$ with real and complex $N_j$ and $\alpha_j$.

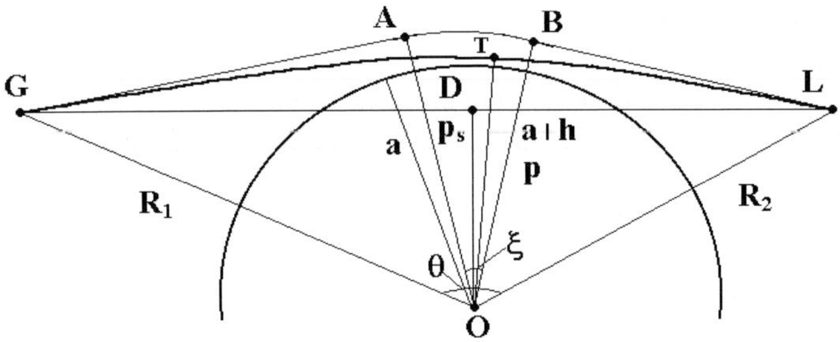


**Fig. 1.** Radio occultation ray path for the spherical symmetric case.

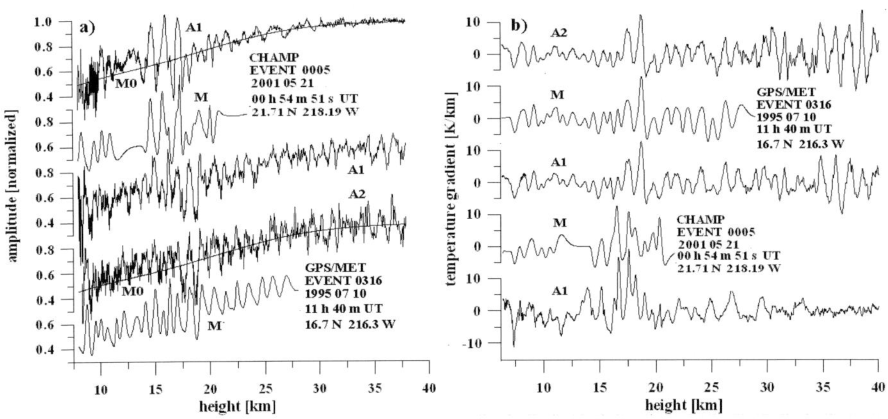


**Fig. 2.** Panel a): Wave train in the amplitudes A1, A2 (GPS/MET event) and A1 (CHAMP event). Curves M0, M correspond to results of the amplitude evaluation using the standard and wave models of the refractivity. Panel b): Vertical gradient dt/dh retrieved from the amplitude RO data.

In order to recover the amplitude variations, which concur with the experimental data, the parameters $N_j$ and $\alpha_j$ were determined. To obtain the vertical profiles of temperature and its gradient we use the approach given in [6]. Wave trains with intense amplitude variations can be found between 8 and 40 km. Their internal vertical wavelength ranges from 0.8 to 2 km. The results of the simulation (Fig. 2, panel a), curves M0, M) are in good agreement with the experimental data, which corresponds to the reduced level of systematic errors in the amplitude data and sufficient accuracy of the model used for simulation. Vertical gradients of temperature retrieved from the amplitude data are presented in Fig. 2, panel b). The quasi-regular wave structure with a vertical wavelength $\lambda_h \sim 0.8 - 2$ km is clearly perceptible both in the experimental and model data. The intensity and spatial period of the wave structure are associated to the phase speed and the horizontal wind perturbations of the GWs.

## 3 "Portrait" of gravity waves

The GW dispersion relationship [1] is valid if the intrinsic frequency of the GW is greater than the inertial frequency f but below the buoyancy frequency $\omega_b$:

$$v_i = |c - U\cos\varphi|: \lambda_h = 2\pi v_i/\omega_b \qquad (1)$$

where $\lambda_h$ is the vertical wavelength of the GW, $v_i$ is the intrinsic phase speed, U is the background wind speed, c is the ground-based GW horizontal phase speed, and $\varphi$ is the azimuth angle between the background wind and the GW propagation vectors, which can be measured by an observer moving with the background wind velocity. The GW polarization relation [1] has the following form:

$$v(h) = \mathrm{Re}[ig/(T_b\omega_b)t(h)],\ \omega_b^2 = g/T_b\Gamma,\ \Gamma = \partial T_b/\partial h + 9.8°/\mathrm{km} \qquad (2)$$

$$dv(h)/dh = d\mathrm{Re}[ig/(T_b\omega_b)t(h)]/dh \approx \mathrm{Re}[ig/(T_b\omega_b)dt(h)/dh]$$

$$dt(h)/dh = \mathrm{Re}\{a_t(h)\exp[i\Phi_t(h)]\};\ dv(h)/dh = \mathrm{Re}\{a(h)\exp[i\Phi(h)]\} \qquad (3)$$

where t(h) is the complex amplitude of temperature variations, v(h) is the GW horizontal wind perturbation, g is the gravity acceleration, $\Gamma$ is the adiabatic lapse rate, and $T_b$ is the background temperature. The function dv(h)/dh can be derived from (2) by differentiating if $T_b(h)$ and $\omega_b(h)$ are slowly changing at the vertical scales $\sim\lambda_h$. Factors $T_b$ and $\omega_b$ are known from the model of the atmosphere in the region. To get dv(h)/dh, the analytic presentation of the real signal dt(h)/dh (3) can be implemented where $a_t(h)$ and $\Phi_t(h)$ are the amplitude and phase of the vertical gradient of the temperature. The functions a(h), $\Phi$(h) present together a GW "portrait' which is shown in Fig. 3. The phase curve of the GW "portrait" depends linearly, on average, on the height h. The amplitude changes in the interval 0.5 to 16 m/(s km). In Fig. 3 the altitudes with high (17-19 km, 29-31 km, 35-38 km) and low (32-34 km) GW activity are clearly visible. By differentiating the phase $\Phi$(h), $\lambda_h$ can be derived as function of height and, using relation (1), the GW intrinsic phase speed $v_i$ can be estimated. As seen in Fig. 4, left panel, the value $v_i(h)$ changes in the range 1.5 to 5 m/s, corresponding to quiet conditions in the atmos-

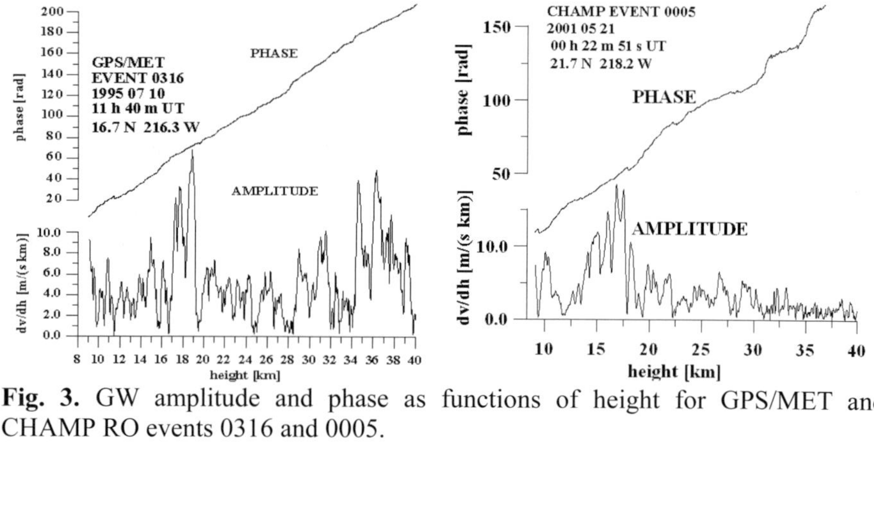


**Fig. 3.** GW amplitude and phase as functions of height for GPS/MET and CHAMP RO events 0316 and 0005.

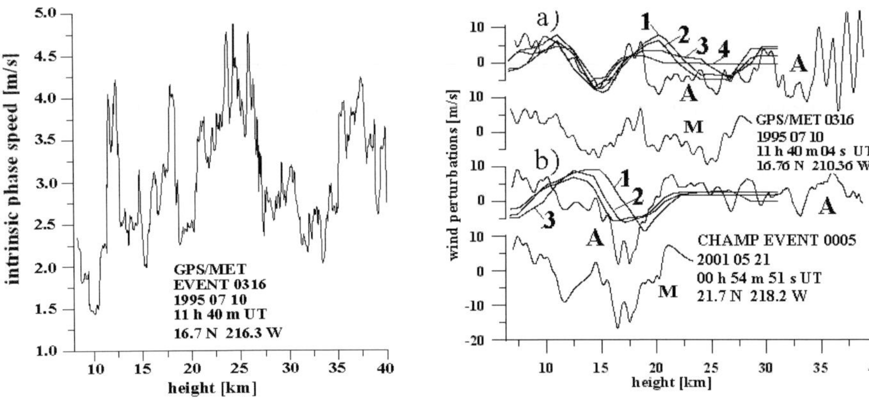


**Fig. 4.** Left panel: the GW intrinsic phase speed $v_i(h)$. Right panel: horizontal wind speed perturbations v(h) associated with GW influence for GPS/MET (part a) and CHAMP (part b) RO events.

phere. The integration of the vertical gradient dv(h)/dh reveals the horizontal wind perturbations v(h) which are depicted in Fig. 4, right panel, for GPS/MET and CHAMP RO events. Curve A in Fig. 4, part a) was obtained by integrating the average wind speed gradient $dv/dh = [dv_1(h)/dh + dv_2(h)/dh]/2$ restored on the GPS/MET RO amplitude data. Curve A in Fig. 4, part b) was obtained by integrating the wind speed gradient dv/dh corresponding to the bottom curve A1 in Fig. 2 b). The curves M (Fig. 4, part a), b)) describe the simulation results. The curves 1-4 in Fig. 4, part a) indicate the radio sondes (RS) data corresponding to two stations in Taiwan: Hualian (1,4) (24.0° N, 238.4° W) and Taipei (2,3) (25.0° N, 238.5° W), obtained on July 15, 1995 at 00 h UT (1,2) and 12 h UT (3,4), respectively. The curves 1-3 in Fig. 4, part b) data correspond to the same stations and indicate the RS data, obtained on May 30, 2001 at 00 and 12 h UT (1,2) (Taipei) and 12 h UT (3) (Hualian), respectively. The difference between the Taiwan and the GPS/MET RO region latitudes for event 0316 is about 8°, the corresponding value for longitudes is about 28°. The difference between the Taiwan and the CHAMP RO region latitudes for event 0005 is about 3°, the corresponding value for longitudes is about 20°. The time difference between the RO events and RS observations was chosen in accordance with the average background westward wind velocity of about 6 to 10 m/s between 8 and 30 km. The RS wind perturbations (curves 1-4 in Fig. 4, part a), curves 1-3 in Fig. 4, part b)) were obtained by subtracting the polynomial approximation of the fifth power from the experimental vertical profiles of the horizontal wind speed. As can be seen in Fig. 4, part a), the RS data (1-4) are in fairly good agreement with the results obtained from the amplitude of RO signals and simulation (curves A, M in Fig. 4, part a), respectively). Some discrepancy of about ±1-3 m/s exists in the height interval 19 to 30 km. The discrepancy between the CHAMP (curve A in Fig. 4, part b) and RS data (curves 1-3 in Fig. 4, part b)) is larger (~±3 to 5 m/s). This difference may correspond to a current state of inversion accuracy. Note that the difference between RS and experimental data may be connected with the instability of the receiver gain and the transmitter power, and increasing low-frequency noise (containing trends and bias) caused by the integration during the inversion process. Note that RS data do not reveal high-spatial frequencies as observed in the RO results. Apparently this is due to smoothing effects of the RS measurements. Also, RS data can have some distortion in the measured GW vertical periods, if the intrinsic phase speed of GWs is comparable with the vertical component of the RS velocity.

## 4 Conclusion

The amplitude of RO signals can be used to retrieve the GW "portrait". The GW "portrait" contains amplitude and phase of the GW as functions of height. The analytic form of the GW presentation is convenient for the analysis of experimental data and can be implemented for the determination of the GW intrinsic phase speed and horizontal wind speed perturbations associated with the GW influence. The radiosonde wind measurements rhyme satisfactorily with the wind speed perturbations retrieved from the GPS/MET RO data. The discrepancy is about 1 to 2 m/s for heights below 18 km and ~2 to 4 m/s for altitudes between 20 and 30 km. Thus the RO method appears to be a promising tool to measure the regular characteristics of the GWs on a global scale.

*Acknowledgments.* We are grateful to GFZ Potsdam for access to the CHAMP RO data and financial support. The work was supported by the National Science Council of Taiwan, R.O.C., grant NSC 91-2111-M008-029, the Office of Naval Research (ONR) of USA, grant N00014-00-0528, and the Russian Fund of Basic Research, grant No. 03-02-17414.

## References

1. Friits DC and Alexander MJ (2003) Gravity wave dynamics and effects in the atmosphere. Rev Geophys *41*: 3-1-3-64.
2. Steiner A K and Kirchengast G (2000) Gravity wave spectra from GPS/MET occultation observations. J Atmos Ocean Tech *17*: 495-503.
3. Kursinski ER, Hajj GA, Schofield JT, Linfield RP and Hardy KR (1997) Observing Earth's atmosphere with radio occultation measurements using the Global Positioning System. J Geophys Res *102*: 23,429-23,465.
4. Tsuda T and Hocke K (2002) Vertical wave number spectrum of temperature fluctuations in the stratosphere using GPS occultation data. J Meteor Soc Japan *80*: 1-13.
5. Pavelyev A, Igarashi K, Reigber Ch, Hocke K, Wickert J, Beyerle G, Matyugov S, Kucherjavenkov A, Pavelyev D, and Yakovlev O (2002) First application of radioholographic method to wave observations in the upper atmosphere. Radio Science *37(3)*: 1043, DOI: 10.1029/2000RS002501.
6. Liou Y-A, Pavelyev AG, Huang C-Y, Igarashi K, Hocke K, and Yan SK (2003) Analytic method for observation of the gravity waves using radio occultation data. Geophys Res Lett *30(20)*: DOI: 10.1029/2003GL017818.

# Global Analysis of Stratospheric Gravity Wave Activity Using CHAMP Radio Occultation Temperatures

Christoph Jacobi, Madineni Venkat Ratnam, and Gerd Tetzlaff

Institute for Meteorology, University of Leipzig, *jacobi@uni-leipzig.de*

**Summary.** Global analyses of $E_p$ as a proxy for gravity wave activity in the stratosphere obtained using GPS radio occultation (RO) data from the CHAMP satellite are presented. Large $E_p$ values are noticed at tropical latitudes, but also at midlatitudes during winter. A correspondence between tropical gravity wave activity and tropical deep convection is found. $E_p$ values are also large during the sudden stratospheric warming that occurred in the southern hemisphere during September-October 2002.

**Key words**: Gravity waves, radio occultation, stratosphere

## 1 Introduction

In the Earth's atmosphere, gravity waves exert a major influence on the large-scale circulation and structure of the middle and upper atmosphere. Taking this into account, it is essential that gravity waves and their dissipation have to be modelled or parameterised, respectively, with great accuracy and these models should be tested by experimental observations. This requires the need for measuring systems that can observe gravity waves on a global scale.

During the past two decades, considerable effort has been devoted in characterizing gravity waves using ground-based systems as radar, lidar and radiosondes. Unfortunately, although these techniques can provide observations with good temporal and vertical resolutions, the ground-based instruments are sparse and unevenly distributed. Satellite observations are able to provide global coverage ([1], [2]), but are of poor spatial resolution and are not suitable to retain the spectral properties of gravity waves. However, they can give a quantitative picture of wave activity. Recently, using GPS/MET satellite radio occultation (RO) observations, [3] provided a global analysis of stratospheric gravity wave activity with special emphasis on winter months. However, radio occultation data from the GPS/MET experiment are limited to some campaigns and therefore to short time intervals compared with the climatological time scale. Therefore, in the present study, gravity wave activity on a global scale, observed by the CHAMP/GPS satellite, is presented with higher statistical significance.

Most of the sources for the generation of these gravity waves lie in the troposphere ([4]). In the tropics, it is generally thought that gravity waves are mostly generated by cumulus convection ([5], [6]). Therefore the possible connection of CHAMP satellite derived gravity wave activity with convection will be analysed.

## 2 Potential energy as a proxy for gravity waves

Under linear gravity wave theory, the energy spectrum of gravity waves can be separated in an energy density $E_0$ and a horizontal wavenumber and frequency part. While the later two cannot be derived from temperature profiles, $E_0$ is chosen as a measure of gravity wave activity. $E_0$ is defined as (see, e.g., [3], [7]):

$$E_0 = \frac{1}{2}\left[\overline{u'^2} + \overline{v'^2} + \overline{w'^2} + \left(\frac{g}{N}\right)^2 \overline{\left(\frac{T'}{\overline{T}}\right)^2}\right] = E_k + E_p \qquad (1)$$

where $E_k$ and $E_p$ are the kinetic and potential energy per unit mass, u', v' and w' are the perturbation components of the zonal, meridional and vertical wind, respectively, g is the acceleration due to gravity, N is the Brunt-Väisälä frequency, and $\overline{T}$ and T' are the mean and perturbation temperatures. According to the linear theory of gravity waves, the ratio of kinetic to potential energy becomes constant, therefore it is possible to estimate $E_0$ from temperature observations only ([3]). However, since presenting $E_0$ after having applied linear theory would not give additional information, and because we have no wind measurements here to prove a relationship between $E_0$ and $E_p$, in the following we discuss results of $E_p$ calculation alone. This calculation mainly depends on the estimation of the temperature fluctuation. For this, the procedure adopted from [3], i.e., calculating temperature fluctuation by high-pass filter with a cut-off at 10 km is closely followed here. Shifting a 10-km window through the respective vertical temperature profiles allows the calculation of $E_p$ for different altitudes while, however, one has to take into account that $E_p$ values referring to a specific height includes information from a 10 km height interval.

When interpreting potential energy from radio occultation data, one has to keep in mind that horizontal soundings generally see reduced temperature variances, which is especially true for short horizontal wavelengths [8,9]. From theoretical estimations, [3] found an up to 21% decrease in the amplitudes of radio occultation temperatures. Since the reduction of measured variance also depends on the viewing angle relative to the propagation direction of the gravity waves, which is not known, a simple straightforward correction procedure is not available. Therefore, one has to keep in mind that the following results may contain the effect of $E_p$ reduction, which may be different for different heights or regions.

## 3 Global distribution of potential energy

Zonally averaged seasonal means (Nov-Feb, Mar-Apr, May-Aug, Sep-Oct) of $E_p$ are shown as height-latitude cross-sections in Fig. 1. Data measured during May 2001 through Jan 2003 is used. Large $E_p$ values at low latitudes up to a height of 25 km is visible from the figure. For the lowermost heights one has to consider a possible influence of the tropopause, however, this should not be the case with the data near 25 km height, so that we may conclude that the high values of $E_p$ are at least qualitatively realistic.

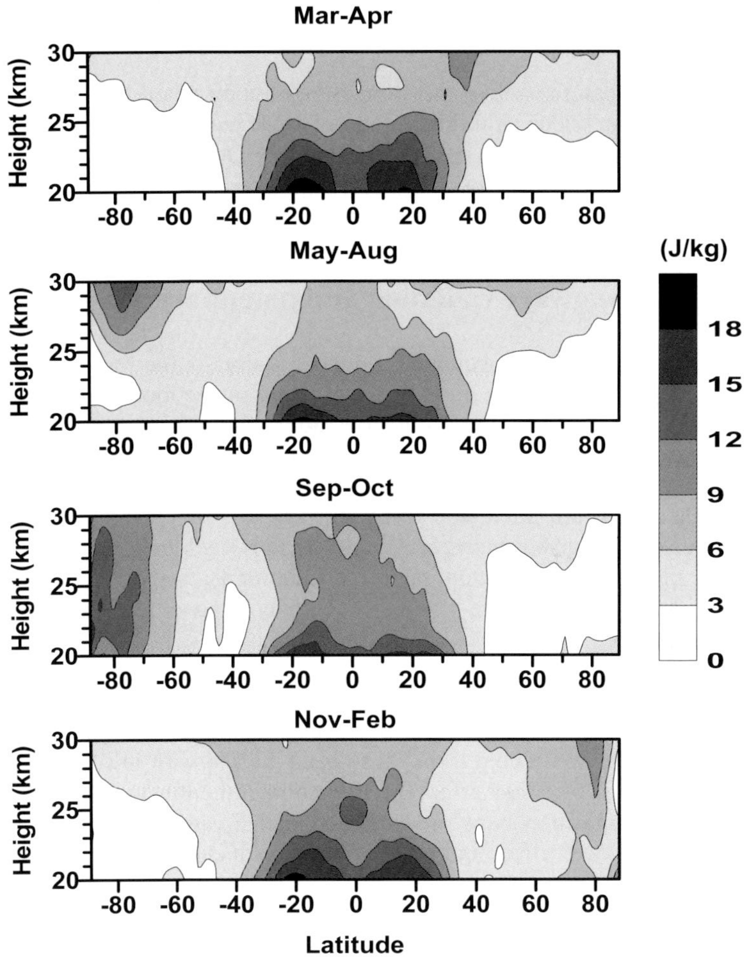


**Fig. 1.** Height-latitude cross-sections of $E_p$ in different seasons during May 2001 through Jan 2003.

Another source of uncertainty especially in the tropics is a possible contribution of Kelvin or Rossby waves to the analysed $E_p$. With the approach chosen, i.e., analysing each temperature profile separately, we are not able to distinguish between gravity waves and large-scale waves if they have similar vertical wavelength. The region of large $E_p$ in the lower stratosphere extends to $30^0$ in both hemispheres in all seasons. Midlatitude $E_p$ values are nearly independent of season in both hemispheres, and increase with height. At higher latitudes, in winter, $E_p$ values are larger in both hemispheres. During March-April, the latitudinal distribution of $E_p$ values is nearly symmetric with respect to the equator, which is the case with the entire height range considered. Another interesting feature visible in Fig. 1 is that large values of $E_p$ are found at southern hemisphere polar latitudes during southern hemisphere spring (Sep-Oct), probably in connection with the major stratospheric warming in 2002.

Many of the features above mentioned are already noted by [3]. Fundamental differences are the wider latitudinal range of large equatorial values and the large values of $E_p$ at southern polar latitudes during Sep-Oct. Further investigations will be necessary to reveal the possible reason for this significant enhancement and wider latitudinal range.

## 4 Outgoing longwave radiation and potential energy

In the tropics, convection is believed to be the major source for gravity waves. In order to study the coherence of $E_p$ with tropical convection, outgoing long wave radiation (OLR) is used as a proxy for the latter. As an example, departures of $E_p$ and OLR from their zonal means in September 2002 is shown in Fig. 2a (top panel). The data refer to equatorial values with latitudes up to 10°N or S, while the CHAMP data are again calculated as means over 20-25 km height.

An inverse relationship between $E_p$ and OLR is visible. This is also shown in the scatter plot in Fig. 2b (bottom panel), presenting $E_p$. vs. OLR in all seasons together. High OLR values are connected with cooling of the lowermost stratosphere due to deep convection. Thus, the correlation seen in Fig. 2b indicates a connection between equatorial gravity waves and convection. Hence it may be confirmed from the data that convection is a primary source for the observed gravity wave activity in the tropics. Still, however, it has to be taken into account that of course a particularly a cold tropopause is connected with low OLR values, and the tropopause structure may affect the $E_p$ results in the lower stratosphere. Therefore the clear connection between the calculated $E_p$ and OLR possibly includes both the gravity wave effect and some residual influence of the sharp tropopause in the vertical profile.

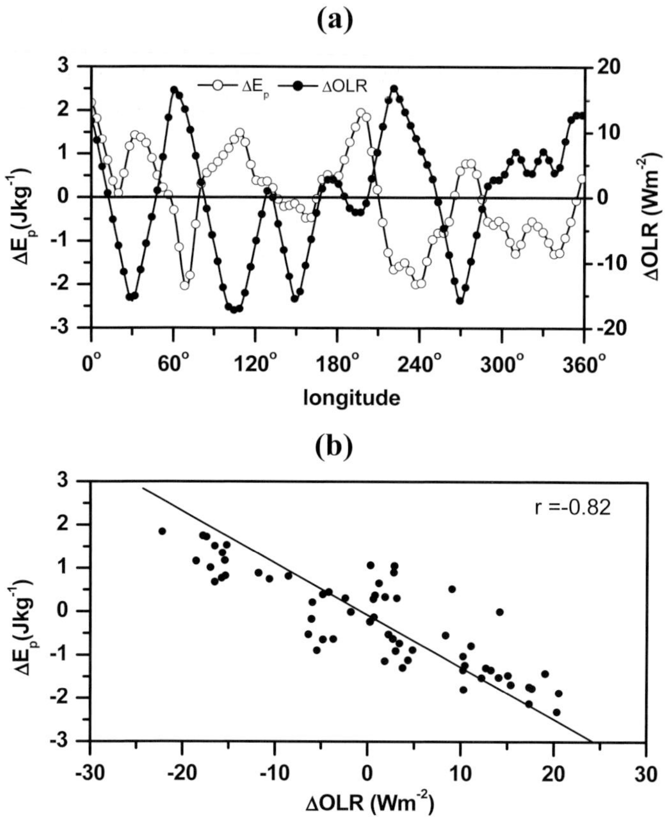


**Fig. 2.** **(a)** Longitudinal variation of low-latitude $E_p$ and OLR deviations from the zonal mean in September 2002, **(b)** $E_p$ vs. OLR at 20-25 km near the equator.

## 5 Conclusions

We have presented a first global picture of potential energy derived from CHAMP RO data for heights of 20-30 km. We restricted ourselves to that height range to avoid effects due to a sharp tropopause in the tropics and to ionospheric residuals at higher altitudes. Since the tropopause is situated at lower heights at middle and high latitudes, for these latitudes $E_p$ can be calculated at lower heights also, but here we did not do that in order to ensure the comparableness between latitudes, and leave the more detailed analysis of high latitude $E_p$ to a later study.

As expected, large $E_p$ values are found near the equator, as has already been shown by [3], but the latitudinal range of this tropical maximum is wider than in [3]. Larger values are also found in winter higher latitudes. During southern hemisphere spring, in 2002 very large values have been measured, probably due to the stratospheric warming event then. At tropical latitudes, gravity waves are clearly owing to deep convection, which can be shown by a comparison of OLR and $E_p$.

CHAMP RO provide a tool for effectively producing maps of potential energy and thus gravity wave activity. Of course, not all relevant parameters are available from the vertical profiles, so information about horizontal wavenumber and frequency is not available, so that the momentum flux cannot be calculated. This means that the use of CHAMP-derived gravity wave maps as input for numerical models is limited. However, information on seasonal variability, vertical distribution and, in future, interannual variability will be available, so that CHAMP GPS measurements will significantly increase our understanding of middle atmosphere dynamics.

*Acknowledgements:* We wish to thank GFZ Potsdam for providing CHAMP/GPS data and NOAA for providing OLR data through the CDC. Our sincere thanks to W.N. Chen for providing Taiwan radiosonde observations. This research was funded by the DFG under grant JA 836/4-2.

## References

1. Wu DL, Waters JW (1996) Satellite observations if atmospheric variances: A possible indication of gravity waves. Geophys Res Lett *23*: 2631-2634.
2. McLandress C, Alexander MJ, Wu DL (2000) Microwave limb sounder observations of gravity waves in the stratosphere: A climatology and interpretation. J Geophys Res *105*: 11,947-11,967.
3. Tsuda T, Nishida M, Rocken C, Ware RH (2000) A global morphology of gravity wave activity in the stratosphere revealed by the GPS occultation data (GPS/MET). J Geophys Res *105*: 7257-7273.
4. Fritts DC, Nastrom GD (1992) Sources of mesoscale variability of gravity waves II: frontal, convective and jet stream excitation. J Atmos Sci *49*: 111-127.
5. Alexander MJ, Holton J (1997) A model study of zonal forcing in the equatorial stratosphere by convectively induced gravity waves. J Atmos Sci *54*: 408-419.
6. Alexander MJ, Beres JH, Pfister L (2000) Tropical stratospheric gravity wave activity and relationship to clouds. J Geophys Res *105*: 22,299-22,309.
7. Allen SJ, Vincent RA (1995) Gravity wave activity in lower atmosphere: Seasonal and latitudinal variations. J Geophys Res *100*: 1327-1350.
8. Lange M, Jacobi Ch (2003) Analysis of gravity waves from radio occultation measurements. In: Reigber Ch, Lühr H, Schwintzer P (eds), First CHAMP Mission Results for Gravity, Magnetic and Atmospheric Studies, Springer, Berlin: 479-484.
9. Preusse P, Dörnbrack A, Eckermann SD, Riese M, Schaeler B, Bacmeister JT, Broutman D, Grossmann KU (2002) Space-based measurements of stratospheric mountain waves by CRISTA 1. Sensitivity, analysis method, and a case study. J Geophys Res *107*: doi: 10.1029/2001JD000699.

# Tropical Tropopause Characteristics from CHAMP

Torsten Schmidt, Jens Wickert, Georg Beyerle, and Christoph Reigber

GeoForschungsZentrum Potsdam (GFZ), Department 1, Telegrafenberg A17, D-14473 Potsdam, Germany, *tschmidt@gfz-potsdam.de*

**Summary.** In this paper an overview of the temperature structure in the tropical upper troposphere and lower stratosphere (UTLS) region is given using Global Positioning System (GPS) radio occultation (RO) data from the German CHAMP (CHAllenging Minisatellite Payload) satellite mission. Several climatologies for tropopause parameters based on radiosonde data and model analyzes have been published in recent years. Both the data sources suffer either to less global coverage or low vertical resolution. This fault can be overcome by the GPS RO technique due to its global coverage, high vertical resolution, all-weather viewing, and long-term stability. CHAMP RO data are available since 2001 with up to 200 high resolution temperature profiles per day. Using CHAMP RO data during May 2001-September 2003, the structure and variability of the tropical tropopause is presented.

**Key words:** CHAMP, GPS, radio occultation, tropopause, UTLS region, GFZ Potsdam

## 1 Introduction

The tropopause separates the troposphere and stratosphere which have fundamental different characteristics, for example, chemical composition and static stability. Thus, the identification of the tropopause is of importance for describing climate variability and change, and for understanding the troposphere-stratosphere exchange [1, 2]. According to the World Meteorological Organization (WMO), the tropopause is defined as the lowest level at which the temperature lapse-rate is less than 2 K/km and the lapse-rate average between this level and the next 2 km does not exceed 2 K/km [3]. This is the so-called lapse-rate tropopause (LRT). Beside this definition in the tropics the potential temperature and the minimum temperature in a vertical temperature profile (cold-point) is used to locate the tropical tropopause.

In the past, several studies have been performed leading to tropopause climatologies from different data sources [4, 5, 6]. The most important data source for the determination of tropopause parameters are radiosonde data whereas model analyzes (ECMWF or NCEP) suffer from low vertical resolution and attendant biases. Despite of good vertical resolution of radiosonde measurements, a global coverage is impossible. On the other hand, investigations of long-term radiosonde data have to take into account for the inhomogeneous time series due to sensor changes. This effect can be overcome by the

GPS RO technique with its (1) high vertical resolution (less than 1 km), (2) long-term stability, (3) all-weather capability, and (4) global coverage [7, 8, 9].

The proof-of-concept GPS RO experiment GPS/MET (GPS/Meteorology) performed between 1995 and 1997 has demonstrated for the first time the potential of GPS-based limb sounding from LEO satellites for deriving atmospheric temperature and water vapor profiles [8]. Further missions with radio occultation experiments followed (Ørsted, SAC-C, and CHAMP), but the German geoscience satellite CHAMP is the only one measuring continuously atmospheric profiles in an operational manner since mid-2001 [10, 11].

First investigations of the thermal structure and variability in the tropical UTLS region based on GPS RO measurements were performed by [12, 13] on the basis of the GPS/MET data from 1995-1997 with focus to the so-called 'prime-times' (June-July 1995, December-February 1996/97) [9]. First applications of CHAMP RO data to the UTLS region with a 1-year data set were published in [14]. Here we present a thermal tropical tropopause statistics based on GPS RO measurements aboard CHAMP for the May 2001-September 2003 period.

## 2 The tropical tropopause

### 2.1 Validation

The temperature in the UTLS region derived from CHAMP is in excellent agreement with radiosonde measurements serving as an independent data source for CHAMP RO data validation. As shown by [15], the temperature bias between CHAMP and radiosonde data is less than 0.5 K between 300-30 hPa for more than 10,000 RO that meet the condition that the radiosonde was launched within a distance of less than 300 km and with a time delay less than 3 hours from the CHAMP measurement.

### 2.2 Time-averaged and spatial structure

Averaged tropopause statistics were computed from individual high-resolution CHAMP temperature profiles. For the identification of the LRT, the WMO definition for the thermal tropopause was used [3]. In addition to the LRT parameters, the cold-point tropopause (CPT) characteristics and the 100-hPa level features are also included here. The tropical tropopause has not a sharp boundary. As discussed, e.g., in [5, 16, 17] a tropical transition layer (TTL) is introduced in which the interaction of convection, the stratospheric wave driven circulation and horizontal transport processes determine the troposphere-stratosphere exchange. [17] suggest that the CPT forms the upper boundary of the TTL. The 100-hPa level that has also been used for the identification of the tropical tropopause can only serve as a proxy. On the

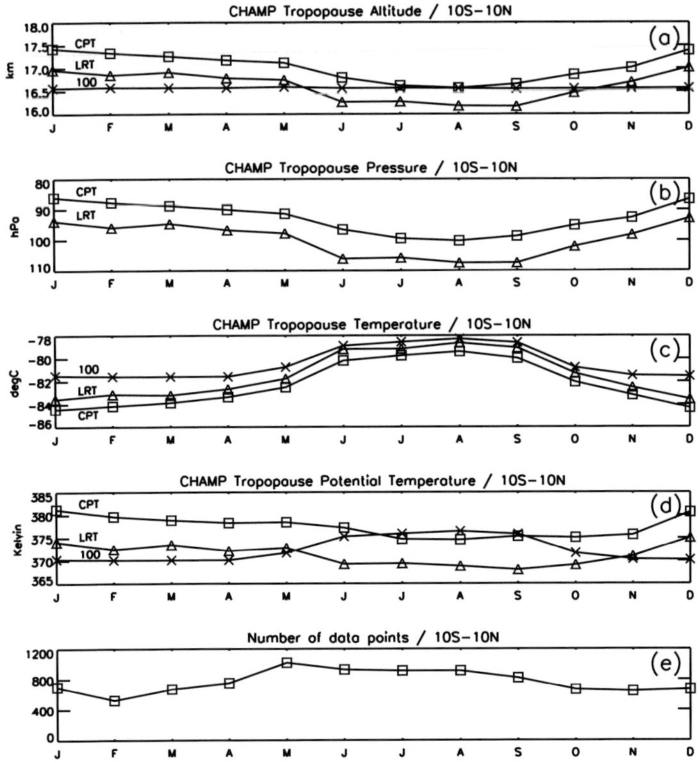


**Fig. 1.** Monthly means of CHAMP tropical tropopause parameters for the period May 2001-September 2003 (LRT: lapse-rate tropopause, CPT: cold-point tropopause, 100: 100-hPa level).

basis of the individual CHAMP RO measurements monthly averages of the tropopause parameters were calculated.

Fig. 1 shows the monthly-average tropopause parameters for the equatorial zone (10S-10N). The tropopause is highest and coldest during the northern hemisphere winter months, and lowest and warmest during the northern summer. A discussion of this annual cycle can be found in, e.g., [18, 19]. In the CHAMP data the LRT altitude has its maximum at 17.0 km in December and January and the minimum in August and September (16.2 km). During all months the CPT is about 300-500 m higher than the LRT. The 100-hPa level is at 16.6 km all time long. In the northern hemisphere winter the 100-hPa level is in the troposphere and during the southern hemisphere winter months it is located in the stratosphere. Fig. 1b shows the annual cycle of LRT and CPT pressure supporting the interpretation for LRT and CPT altitude: minimum of LRT pressure in December-January (93-95 hPa) and maximum in August-September (107 hPa). The CPT pressure is about 5-10 hPa lower than the LRT pressure. Fig. 1c shows annual and zonal tropopause temperature values from CHAMP RO data. In the equatorial zone the tropopause temperature is lowest in the northern hemispheric winter months and highest during northern summer. Monthly averaged CPT temperatures reaching -84°C in December-January and increasing to -79°C in August. LRT values are about 1 K higher than CPT temperatures. The potential temperature is an important value because the tropical tropopause corresponds to an isentropic surface with a potential temperature of on average about 380 K [1]. Typical values of potential temperature in the equatorial zone found in the CHAMP RO measurements are between 368-374 K for LRT and 374-381 K for the CPT. The number of CHAMP RO temperature profiles that are the basis for these investigations can be found in (Fig. 1e).

The spatial structure of the tropical LRT derived from CHAMP RO measurements for the northern hemispheric winter months is shown in Fig. 2. The highest LRT altitudes during December-February are >17 km in the tropical western Pacific region and over South America. A smaller maximum altitude is located over Central Africa and the western Indian Ocean (Fig.

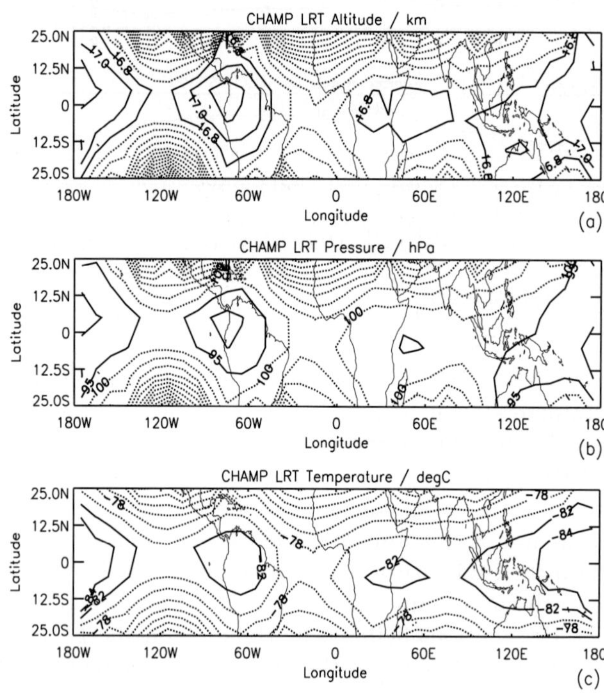


**Fig. 2.** LRT altitude (a), pressure (b), and temperature (c) for the northern hemisphere winter months (December-February). Contour intervals are (a) 0.2 km, and values above 16.8 km are plotted with solid lines; (b) 5 hPa, and values below 95 hPa are plotted with solid lines; (c) 2 K, and values below -82°C are plotted with solid lines.

2a). The associated pressure patterns are shown in Fig. 2b. The coldest LRT temperatures (Fig. 2c) are less than -82°C correlated with the maximum LRT altitudes. During the northern hemisphere summer (not shown here) the area of highest tropopause (>17 km) is moving northward to the south-Asian monsoon region. The temperature in that region is less than -80°C as also in the western Pacific area but here with LRT altitudes reaching only 16.2-16.4 km.

The basis for the contour plots in Fig. 2 are mean tropopause parameters representing an area of ±5° in latitude and longitude. The 6 latitudes (25°S, 15°S, 5°S, 5°N, 15°N, and 25°N) and 35 longitudes (175°W, 165°W, 155°W, ... , 155°E, 165°E, and 175°E) are the centers of the bins. Each of this areas contain on average about 40 CHAMP measurements. Thus, already this single satellite constellation leads to a global coverage of the tropical region for seasonal plots of climate parameters where no interpolation is necessary. With increasing numbers of RO experiments in the next future the spatial information density (not only in the UTLS region) of atmospheric climate change parameters will increase rapidly.

## 3 Conclusions

Tropical tropopause parameters on the basis of CHAMP RO measurements for the period May 2001-September 2003 have been discussed. Because of accuracy, high vertical resolution, and globally distributed temperature data in the tropopause region the relatively new RO technique is suitable for global monitoring of the UTLS as an important part of the atmosphere. Changes in tropopause parameters can be used for the detection of climate change as already shown by [2]. The CHAMP RO experiment generates the first long-term RO data set, other satellite missions will follow (GRACE, TerraSAR-X, METOP, COSMIC) establishing the RO technique for global temperature monitoring in the UTLS region.

## References

1. Holton JR, Haynes PH, McIntyre ME, Douglass AR, Rood RB, and Pfister L (1995) Stratosphere-troposphere exchange. Rev Geophysics *33*: 403–439.
2. Sausen R and Santer BD (2003) Use of changes in tropopause height to detect influences on climate. Meteorol Z *12(3)*: 131–136.
3. WMO (1957) Definition of the tropopause. WMO Bull *6*.
4. Hoinka PH (1998) Statistics of the global tropopause pressure. Mon Weather Rev *126*: 3303–3325.
5. Highwood EJ and Hoskins BJ (1998) The tropical tropopause. QJR Meteorol Soc *124*: 1579–1604.

6. Seidel DJ, Ross RJ, Angell JK, and Reid GC (2001) Climatological characteristics of the tropical tropopause as revealed by radiosondes. J Geophys Res *106(D8)*: 7857–7878.
7. Melbourne WG, Davis ES, Hajj GA, Hardy KR, Kursinski ER, Meehan TK, and Young LE (1994) The application of spaceborne GPS to atmospheric limb sounding and global change monitoring. JPL Publication *94-18*, Jet Propulsion Laboratory, Pasadena, California.
8. Ware R, Exner M, Feng D, Gorbunov M, Hardy K, Melbourne W, Rocken C, Schreiner W, Sokolovsky S, Solheim F, Zou X, Anthes R, Businger S, and Trenberth K (1996) GPS soundings of the atmosphere from low earth orbit: Preliminary results. Bull Amer Meteor Soc *77*: 19–40.
9. Rocken C, Anthes R, Exner M, Hunt D, Sokolovskiy S, Ware R, Gorbunov M, Schreiner W, Feng D, Herman B, Kuo YH, and Zou X (1997) Analysis and validation of GPS/MET data in the neutral atmosphere. J Geophys Res *102(D25)*: 29,849–29,866.
10. Wickert J, Reigber C, Beyerle G, König R, Marquardt C, Schmidt T, Grunwaldt L, Galas R, Meehan TK, Melbourne WG, and Hocke C (2001) Atmosphere sounding by GPS radio occultation: First results from CHAMP. Geophys Res Lett *28*: 3263–3266.
11. Wickert J, Schmidt T, Beyerle G, König R, Reigber C, and Jakowski N (2004) The radio occultation experiment aboard CHAMP: Operational data analysis and validation of atmospheric profiles. J Meteorol Soc Japan *82(1B)*: 381–395.
12. Nishida M, Shimizu A, Tsuda T, Rocken C, and Ware RH (2000) Seasonal and longitudinal variations in the tropical tropopause observed with the GPS occultation technique (GPS/MET). J Meteorol Soc Japan *78*: 691–700.
13. Randel WJ, Wu F, and Rios WR (2003) Thermal variability of the tropical tropopause region derived from GPS/MET observations. J Geophys Res *108(D1)*: 4020, doi:10.1029/2002JD002595.
14. Schmidt T, Wickert J, Marquardt C, Beyerle G, Reigber C, Galas R, and König (2004) GPS radio occultation with CHAMP: An innovative remote sensing method of the atmosphere. Adv Space Res, in print.
15. Wickert J, Schmidt T, Beyerle G, Michalak G, König R, Kaschenz J, and Reigber C (2004) Atmospheric profileing with CHAMP: Status of the operational data analysis, validation of the recent data products and future prospects. This issue.
16. Holton JR and Gettelman A (2001) Horizontal transport and the dehydration of the stratosphere. Geophys Res Lett *28*: 2799–2802.
17. Gettelman A and de Forster PM (2002) A climatology of the tropical tropopause layer. J Meteorol Soc Japan *80*: 911–924.
18. Reid GC and Gage KS (1996) The tropical tropopause over the western Pacific: Wave driving, convection, and the annual cycle. J Geophys Res *101(D16)*: 21,233–21,241.
19. Yulaeva E, Holton JR, and Wallace JM (1994) On the cause of the annual cycle in tropical lower-stratospheric temperatures. J Atmos Sci *51*: 169–174.

# Comparisons of MIPAS/ENVISAT and GPS-RO/CHAMP Temperatures

Ding-Yi Wang[1], Jens Wickert[2], Gabriele P. Stiller[1], Thomas von Clarmann[1], Georg Beyerle[2], Torsten Schmidt[2], Manuel López-Puertas[3], Bernd Funke[3], Sergio Gil-López[3], Norbert Glatthor[1], Udo Grabowski[1], Michael Höpfner[1], Sylvia Kellmann[1], Michael Kiefer[1], Andrea Linden[1], Gizaw Mengistu Tsidu[1], Mathias Milz[1], Tilman Steck[1], and Herbert Fischer[1]

[1] Forschungszentrum Karlsruhe GmbH und Universität Karlsruhe, Institut für Meteorologie und Klimaforschung (IMK), Postfach 3640, 76021 Karlsruhe, Germany *ding-yi.wang@imk.fzk.de*

[2] GeoForschungsZentrum Potsdam (GFZ) Department 1: Geodesy and Remote Sensing, Telegrafenberg, 14473 Potsdam, Germany

[3] Instituto de Astrofísica de Andalucía, CSIC, Apartado, Postal 3004, Granada, Spain

**Summary.** The temperatures retrieved from MIPAS/ENVISAT limb mid-infrared emission and CHAMP GPS radio occultation measurements are compared at altitudes between 8 - 30 km during the stratospheric major sudden warming in the southern hemisphere winter of 2002. The mean differences between the correlative measurements of the two instruments are less than ~1 K with rms deviations of ~3-5 K. The MIPAS temperatures are slightly higher than those of GPS-RO around 30 km. Possible explanation is discussed.

**Key words:** MIPAS, CHAMP, GPS radio occultation, temperature

## 1 Introduction

The Michelson Interferometer for Passive Atmospheric Sounding (MIPAS) instrument onboard the ENVISAT measures stratospheric temperature and volume mixing ratio (VMR) of various gas species by limb-observing mid-infrared emissions. The MIPAS temperatures produced by the IMK (Institut für Meteorologie und Klimaforschung) data processor (for details see [1],[2], and [5]) are compared with retrievals of other MIPAS data processors based on synthetic spectra [1], and with a number of other satellite measurements [6]. These comparisons show reasonable agreements within their individual specifications.

The Global Positioning System (GPS) receiver onboard the German satellite CHAMP (CHAllenging Mini-Satellite Payload) and Argentinean satellite SAC-C (Satelite de Aplicaciones Cientificas-C) have provided radio occultation (RO) temperature measurements since 2001. Comparison between GPS-RO/SAC-C and MIPAS temperatures show good consistency [3]. The GPS-RO/CHAMP temperatures processed at the GeoForschungsZentrum Potsdam (GFZ) show agreement with corresponding ECMWF (European Centre for Medium-Range Weather Forecasts), with a mean difference less than 1 K and a standard deviation of 1.5 K or less above the tropopause [8]. This study compares the MIPAS and GPS-RO temperatures produced in turn by IMK and GFZ. The comparisons will benefit both observation systems, in order to evaluate the methods, to assess their effectiveness, and to produce definite statements about the exploitability of these data.

## 2 Data Description and Analysis Methods

The MIPAS temperatures used for this study are IMK product version V1.0. They are taken from 24 July and the 8 days (see Figure 2) of the September/October in 2002. The retrievals are calculated based on the operational ESA level-1B data (i.e. calibrated and geo-located radiance spectra). The ECMWF temperatures are used as initial guess, but not as *a priori* constraint in the sense of optimal estimation. Local thermodynamic equilibrium (LTE) is assumed, since the non-LTE effects are not significant in the region below 70 km except for the lower mesosphere in the polar winter regions [2]. The retrieval is performed between 6 and 70 km on a 1-km grid below 44 km and 2-km above. The observations provide global coverage with 14.4 orbit per day. The sampling rate is $\sim$ 500 km along-track and $\sim$ 2800 km across-track at the equator. Some measurements with severe cloud contamination were rejected. The number of available measurements for each day varies from several tens to low hundreds.

GPS-RO/CHAMP temperatures are taken from GFZ product version 4. CHAMP provides $\sim$230 globally distributed vertical profiles per day in the region of 0-50 km. The vertical resolution ranges from 0.5 km in the lower troposphere to 1.5 km in the stratosphere and the resolution along the ray path is around a few hundred km. The GPS-RO measurements are used to derive the bending angle of each ray by assuming a spherical symmetry of the atmosphere. This bending angle can be inverted to a profile of the refractive index. The atmospheric density, temperature, and pressure are obtained from the known relationship between density and refractivity, the hydrostatic equilibrium, and the equation of state for an ideal gas. When water vapor is present, additional information is required to determine the humidity and density from refractivity profiles. In the GFZ data processor, the ECMWF temperature profile is used to initialize hydrostatic equation at 43 km and to derive humidity profiles from the calculated refractivity in an iterative procedure (for details see [7] and [8]).

For the MIPAS and GPS-RO temperature comparisons presented here, we have used a measurement coincidence criteria of 5 degrees latitude, 10 degrees longitude, and 6 hours. We have experimented with other reasonable coincidence criteria and found the overall results to be not overly sensitive to the adopted values. Total 844 MIPAS profiles are available for comparison. The correlative profiles are interpolated to a common altitude grid as that used by the MIPAS-IMK data. No resolution or averaging kernel is used in this preliminary study.

## 3 Comparison Results

Detailed profile-by-profile comparisons (not shown here for brevity) are performed for the MIPAS and GPS-RO measurements between 8 and 30 km at solstice and polar winter conditions. The cut-off altitudes are due to the GPS-RO data. Below 8 km the presence of water vapour results in "cold-biased" dry-temperature profiles. Above 30 km, the influence of the ECMWF initialization using ECMWF dominates the derived temperatures. The latitude coverage of the measurements is global for the 9 days.

We assembled sets of comparisons to reduce influences of geophysical variations in temperature in order to judge whether any biases are present in the two observation systems. The mean residuals and root-mean-squared deviations are averaged on daily basis over all available overlapped profiles. They are presented in Figure 1 for the solstitial and polar winter time measurements. The daily averages of the residuals are in the range between 1 and 3 K with rms deviations of 5-8 K, with larger values occurred for the sudden warming period. The 8-day mean differences for September/October period are less than 1 K below 27 km with the rms deviations of ~4 K, indicating good agreement between the two data sets. However, the 8-day averages are seen to slightly increase with increasing height between 25 and 30 km. Around 30 km, the 8-day mean differences reach a maximum of ~ 2 K, with the MIPAS temperatures higher than GPS-RO/CHAMP.

We further examine latitudinal behaviour of the differences. The daily zonal means of the residuals between MIPAS and GPS-RO temperatures are displayed in Figure 2 for MIPAS downleg (corresponding to daytime) observations during the period of September/October of 2002. They are calculated by sorting available correlative profiles at each leg and individual heights into latitude bins of a 30° width. For the eight days, the daily zonal mean differences are less than $\pm(0-2)$K in wide latitude and altitude regions, with significant latitude and day-to-day variations. They have relatively small magnitudes in the latitudes between 30°S and 90°N. On the contrary, the mean differences tend to increase at higher latitudes south of 60°S and reach maximum around the south pole, where the MIPAS temperatures are generally lower than GPS-RO at the low altitudes 15 km, but higher above. The discrepancies between MIPAS and GPS-RO temperatures are significantly increased on 26 September, peaked around 20-25 km with a maximum magnitude of 4-6 K. On that day, the polar vortex just splat from zonal wavenumber one

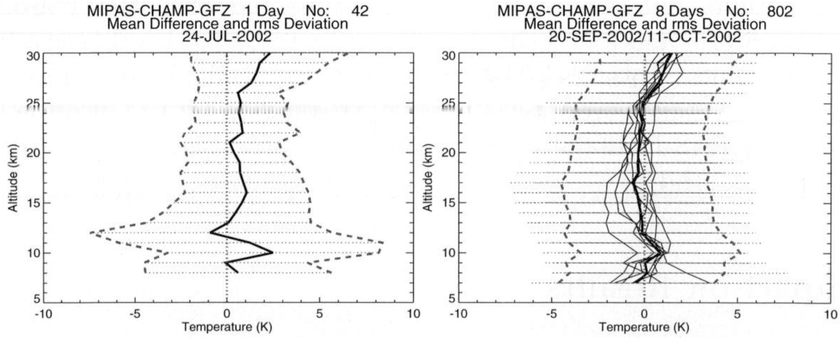


**Fig. 1.** Mean differences (solid) and root-mean-squared deviations (dotted) of MIPAS and GPS-RO/CHAMP temperatures (in Kelvin). The data are averaged on daily basis (thin line) over all available correlative profiles for 24 July (left) and 8 days of September/October, 2002 (right). The thick lines denote overall means for all days available during the time period.

to two pattern during the unprecedented Antarctic winter stratospheric sudden warming. The increased deviations between the MIPAS and GPS-RO measurements suggest strong influences of the enhanced wave activities.

## 4 Concluding Remarks

A comparison of ∼850 profiles shows generally good agreement between the MIPAS and GPS-RO/CHAMP temperatures. The overall mean differences are estimated at 1 K with rms deviations of 3-4 K. Around 30 km, the MIPAS temperatures tend to be higher than GPS-RO/CHAMP, with a maximum mean difference of ∼2 K at 30 km. Larger discrepancies of 4-6 K are also observed at high latitudes of 60°S and around 25 km. In order to understand the consistency and discrepancy, we compared (not shown here) the MIPAS and GPS-RO temperatures with the ECMWF data. The GPS-RO temperatures show better agreement with the ECMWF with the mean differences less than 1 K throughout the entire altitude region of 8 - 30 km. In contrast, the mean differences between the MIPAS and ECMWF temperatures are less than 1-3 K, with an apparent tendency for the MIPAS temperatures to be higher than the ECMWF at upper heights around 30 km, in particular at the high latitude region between 60°S-90°S.

The tendency for the MIPAS temperatures higher than GPS-RO and ECMWF around 30 km could be ascribed with the known cold bias in the ECMWF temperatures between 30 and 45km [4]. It is known that the current GPS-RO/CHAMP temperatures at these heights depend on the ECMWF initialization[7]. If one compares the GFZ retrievals with the ECMWF at 43 km, zero bias and standard deviation (pure ECMWF) are expected. This feature is not associated with the measurement principles, but simply due to

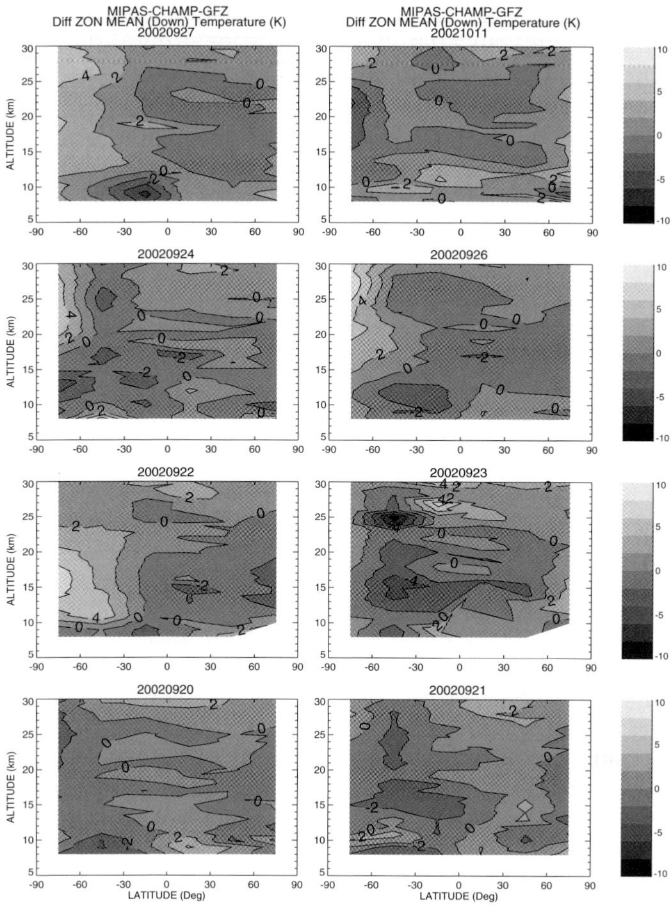


**Fig. 2.** Daily zonally averaged temperature differences (in Kelvin) of correaltive MIPAS and GPS-RO/CHAMP measurements for the eight days of September/October, 2002. MIPAS observations are in descending node (daytime). The contour interval is 2 K.

the very straightforward manner of using the initialization information in the current retrieval scheme. In contrast, the MIPAS-IMK retrieval is performed for all measurements (spectral microwindows and tangent altitudes) simultaneously in a global fit sense with a Tikhonov-type first order smoothing constraint. The retrieved profiles are not biased towards the initial guess, i.e. the ECMWF field, but just made smooth[1]. Since the two algorithms use the background information in different ways, comparison of the derived temperatures is a bit difficult.

The CHAMP temperature and water vapor profiles also can be derived by using optimal estimation methods from the observed refractivity. The implementation of appropriate algorithms is currently under investigation and it is planned to be implemented at the operational data analysis level. We note that the refractivity, beside the bending angle the key observable of the GPS radio occultation method, can be derived without ancillary information. Therefore it can be regarded as independent data set, which can be assimilated to meteorological analyses. This is a new quality of satellite based remote sensing data.

*Acknowledgement.* ECMWF data has been provided via the NILU database. The research work of IMK-IAA MIPAS group has been funded by EC via contract No. EVG1-CT-1999-00015 (AMIL2DA), BMBF via project No. 07 ATF 43/44 (KODY-ACS), 07 ATF 53 (SACADA), and 01 SF 9953 (HGF-VF), and ESA via contract No. 15530/01/NL/SF (INFLIC).

## References

1. v Clarmann T et al (2003a) A blind test retrieval experiment for limb emission Spectrometry. J Geophys Res *(108)*, D23, 4746, doi:10.1029/2003JD003835.
2. v Clarmann T et al (2003b) Retrieval of temperature and tangent altitude pointing from limb emission spectra recorded from space by the Michelson interferometer for passive atmosphere (MIPAS). J Geophys Res *108*, D23, 4736, doi:10.1029/2003JD003602.
3. Jiang JH et al. (2004) Comparison of GPS/SAC-C and MIPAS/ENVISAT temperature profiles and its implementation for MLS-AURA observations. *this issue.*
4. Randel W, Chanin M-L, and Michaut C (2002)SPARC Intercomparison of Middle Atmosphere Climatologies. WCRP-116, WMO/TD-No-1142, SPARC Report No. 3.
5. Stiller GP, Steck T, Milz M, von Clarmann T, Grabowski U, and Fisher H (2003) Approach to the cross-validation of MIPAS and CHAMP temperature and water vapour profiles. In: First CHAMP Mission Results for Gravity, Magnetic and Atmospheric Studies, Ch Reigber, Lühr H, Schwintzer P Eds, Springer: 550–556.
6. Wang DY et al. (2004) Comparisons of MIPAS-Observed Temperature Profiles with Other Satellite Measurements. Proc SPIE (in print).
7. Wickert J et al. (2001) Atmosphere sounding by GPS radio occultation: First results from CHAMP. Geophys Res Lett *28(17)*: 3263–3266.
8. Wickert J, Schmidt T, Beyerle G, König R, Reigber Ch, and Jakowski N (2004) The radio occultation experiment aboard CHAMP: Operational data analysis and validation of vertical atmospheric profiles. J Meteor Soc of Japan *82(1B)*: 381–395.

# Comparison of GPS/SAC-C and MIPAS/ENVISAT Temperature Profiles and Its Possible Implementation for EOS MLS Observations

Jonathan H. Jiang[1], Ding-Yi Wang[2], Larry L. Romans[1], Chi O. Ao[1], Michael J. Schwartz[1], Gabriele P. Stiller[2], Thomas von Clarmann[2], Manuel López-Puertas[3], Bernd Funke[3], Sergio Gil-López[3], Norbert Glatthor[2], Udo Grabowski[2], Michael Höpfner[2], Sylvia Kellmann[2], Michael Kiefer[2], Andrea Linden[2], Gizaw Mengistu Tsidu[2], Mathias Milz[2], Tilman Steck[2], and Herbert Fischer[2]

[1] Jet Propulsion Laboratory, California Institute of Technology, California, USA
[2] Forschungszentrum Karlsruhe GmbH und Universität Karlsruhe, Institut für Meteorologie und Klimaforschung (IMK), Karlsruhe, Germany
[3] Instituto de Astrofísica de Andalucía, CSIC, Apartado, Granada, Spain

**Summary.** This analysis presents comparisons of the atmospheric temperatures retrieved from GPS/SAC-C radio occultation observations using the JPL retrieval software, and from MIPAS/ENVISAT infrared spectrum measurements using the IMK data processor. Both individual profiles and zonal means of the atmospheric temperature at different seasons and geo-locations show reasonable agreement. For the temperatures at altitudes between 8-30 km, the mean differences between the correlative measurements are estimated at less than 2 K with rms deviations less than 5 K. A similar cross comparison technique can be used to help validate the observed temperatures from the new EOS MLS instrument, to be launched in 2004.

**Key words:** GPS/SAC-C, MIPAS, EOS MLS, temperature.

## 1 Introduction

Global Positioning System (GPS) radio occultations are active limb sounding measurements of the Earth's atmosphere, with the advantages of global coverage, high vertical resolution, self-calibration, and capability to operate under all-weather conditions. By placing a GPS receiver on a low earth orbiter (LEO), the phase delays of GPS carrier signals induced by the intervening medium can be accurately measured as the GPS-LEO satellite link descends through the atmosphere. Under the assumption of geometric optics and local spherical symmetry of the atmosphere, the phase delay measurements can be directly inverted to yield the index of refraction profile with vertical resolution that varies from about 0.5 km in the lower troposphere to about 1 km in the lower stratosphere [*Kursinski et al.*, 1997; *Hajj et al.*, 2002]. Coupled with the hydrostatic equation, the index of refraction profile can be converted unambiguously into a temperature profile above ~5 km, where the water vapor contribution is small. This concept of GPS occultation was successfully demonstrated with the GPS/MET (GPS Meteorology) mission in 1995 [*Ware et al.*, 1996]. The recently launched, polar orbiting, Argentinian satellite SAC-C (Satelite de Aplicaciones Cientificas-C) carries a new-generation occultation-enabled GPS receiver, the "BlackJack" supplied by the Jet Propulsion Laboratory (JPL). GPS/SAC-C has been collecting occultation data nearly continuously since July 2001, with a typical throughput of about 200 soundings per day. The data are routinely processed at JPL, and are publicly available.

The Michelson Interferometer for Passive Atmospheric Sounding (MIPAS) on board the ENVISAT satellite provides vertical profiles of stratospheric temperature and volume mixing ratio (VMR) of various gas species by limb-observing mid-infrared emissions [*Fischer and Oelhaf*, 1996]. The IMK data processor provides simultaneous retrieval of temperature and line-of-sight parameters, as well as pressure from measured spectra and the spacecraft ephemerids. Details of the retrieval scheme and its accuracy have been discussed by *Clarmann et al.* [2003] and *Stiller et al.* [2003]. The IMK MIPAS temperatures have been compared with a number of other satellite observations [e.g. *Wang et al.* 2003], including GPS/CHAMP (Challenging Mini-Satellite Payload for Geoscientific Research and Application), HALOE (Halogen Occultation Experiment), and SABER (Sounding of the Atmosphere using Broadband Emission Radiometry), as well as with UKMO (United Kingdom Meteorological Office) stratospheric assimilated data. Both individual profiles and zonal means of the atmospheric temperature at different seasons and geo-locations show reasonable agreement. The overall mean differences are on the order 1 K with root-mean-square (rms) deviations of ~5 K.

| Time | MIPAS Total | SAC-C Total | MIPAS/SAC-C |
|---|---|---|---|
| 24-JUL-2002 | 204 | 186 | 41/45 |
| 20-SEP-2002 | 480 | 183 | 117/136 |
| 21-SEP-2002 | 384 | 175 | 82/96 |
| 22-SEP-2002 | 412 | 174 | 96/107 |
| 23-SEP-2002 | 214 | 135 | 16/16 |
| 24-SEP-2002 | 213 | 180 | 52/58 |
| 26-SEP-2002 | 475 | 141 | 101/114 |
| 27-SEP-2002 | 568 | 166 | 109/129 |
| 11-OCT-2002 | 292 | 166 | 61/66 |

**Table 1.** Number of coincidence profiles used for comparison. Coincidence Criteria: latitude and longitude differences are less than 5° and 10°, respectively, the time differences are less than 6 hours.

In this study, we will compare temperatures measured by MIPAS and GPS/SAC-C. This analysis benefits not only the two different limb-viewing observation systems, but also the future cross-comparison with temperature measurements from the soon-to-be launched EOS MLS instrument.

## 2 Data Description and Comparison Methods

The GPS/SAC-C temperature data are produced by the JPL retrieval software. The data provide global coverage with more than one hundred temperature profiles per day (see the 3rd column of Table 1). The retrieval uses Abel inversion under the assumption of hydrostatic equilibrium, spherical symmetry of the atmosphere about the ray tangent point, a refractivity expression and the equation of state [*Hajj et al.*, 2002]. Water vapour is negligible except in the lowest 5 km of tropical atmosphere where NCEP analysis is used to provide water vapour profile. The NCEP temperature at 30 km is used as an initial guess in the JPL retrieval. For brevity, we here-after call the GPS/SAC-C data simply as SAC-C data or SAC-C measurements.

The MIPAS temperature data used in this study are version V1.0 of the IMK data products. The MIPAS limb-viewing observations from ENVISAT provide global coverage with 14.4 orbits per day. At a given latitude over the course of a day, about 30 longitudinal points are sampled, 15 each in the ascending and descending modes. The sampling in the two legs corresponds to two different local times, each of which remains approximately constant for successive orbits, changing by only 20 minutes per day. The V1.0 retrieval is performed between 6 and 70 km with a grid spacing of 1 km below 44 km and 2 km above. Local thermodynamic equilibrium (LTE) is assumed in the retrievals, since the non-LTE effects are not significant in the region below 70 km except for the lower mesosphere in the polar winter regions. Some measurements were rejected due to severe cloud contamination. The number of available measurements for each day is usually a few hundreds (see the 2nd column of Table 1).

The temperature data used for this comparison analysis are taken from different geo-locations for two periods in 2002 when the MIPAS data are available. Due to characteristics of the sampling scenarios for the two datasets, it is difficult to achieve excellent spatial and temporal coincidence for individual pairings of MIPAS profiles with correlative SAC-C measurements. As a first order approximation, individual paired-profile comparisons are conducted for those measurements with latitude and longitude differences smaller than $5^o$ and $10^o$ , respectively. The time differences between the paired profiles are limited to $<$ 6 hours. The paired profiles are then interpolated to a common altitude grid as that used by the MIPAS data. No averaging kernel is used in this study.

## 3 Results

Individual temperature profile comparisons are performed for the MIPAS and SAC-C measurements between 5-30 km. The total MIPAS, SAC-C sampling profiles and the numbers of MIPAS/SAC-C correlative profiles used for this study are listed in Table 1. One MIPAS profile may have multiple SAC-C coincidences due to the sampling characteristics and our coincidence criteria.

Note that by the time this study was conducted, MIPAS measurements were only available for July 24 and 15 days in September-October (8 days used in this study). During the later period, an unprecedented Antarctic winter stratospheric sudden warming event occurred [*Manney et al.*, 2004].

For brevity, plots of detailed profile-by-profile comparisons are not displayed here. The mean differences and the rms deviations averaged over all available paired profiles are presented in Figure $1_{\text{left}}$ for July 2002 and Figure $1_{\text{right}}$ for the September/October periods. In general, the MIPAS and SAC-C temperature profiles agree well, with overall mean differences $<2$ K and the rms deviations $<5$ K or more throughout troposphere and lower-stratosphere. For the July 24 profile, the mean MIPAS temperatures are slightly warmer ($\sim$1 to 2 K) than the SAC-C temperatures. Both data sets show good agreement above 20 km, although the mean MIPAS temperatures are somewhat cooler by $\sim$1 K near the tropopause ($\sim$12-18 km).

In order to examine the latitudinal behavior of the two data sets, we compare the daily zonal means of these temperature differences. To determine the zonal means, all available longitudinal data at individual heights are sorted into latitude-bins of a specified width (10° in latitude are used for this study, in order to have enough data points available in each bin). The daily means of available data points are computed separately for each leg. Figure 2 show the daily zonal mean temperature differences between the correlative MIPAS (descending mode) and SAC-C measurements. In general, there are random differences of 2-4 K in wide latitude and altitude regions. The noticeable large differences are about 6-10 K at high latitude of 60°S around 20-25 km during 22-27 September, 2002, when the MIPAS temperature measurements reported a major polar stratospheric warming event [*Manney et al.*, 2004]. The cause of this MIPAS warm-bias over SAC-C, however, is not clear at present.

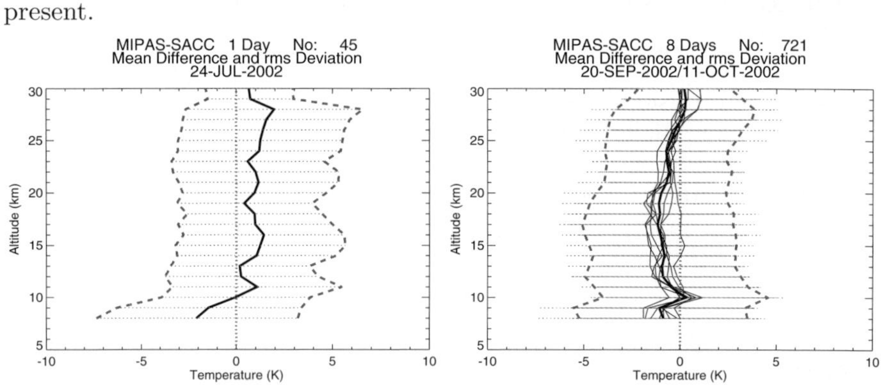


**Fig. 1.** Mean differences (solid) and rms deviations (dotted) of MIPAS and SAC-C temperatures (K). The light-lines are data averaged over all available paired profiles on a single day, dark solid-lines are averaged over all the measurement days: 1 day for July 24 (left), 8 days for September-October (right).

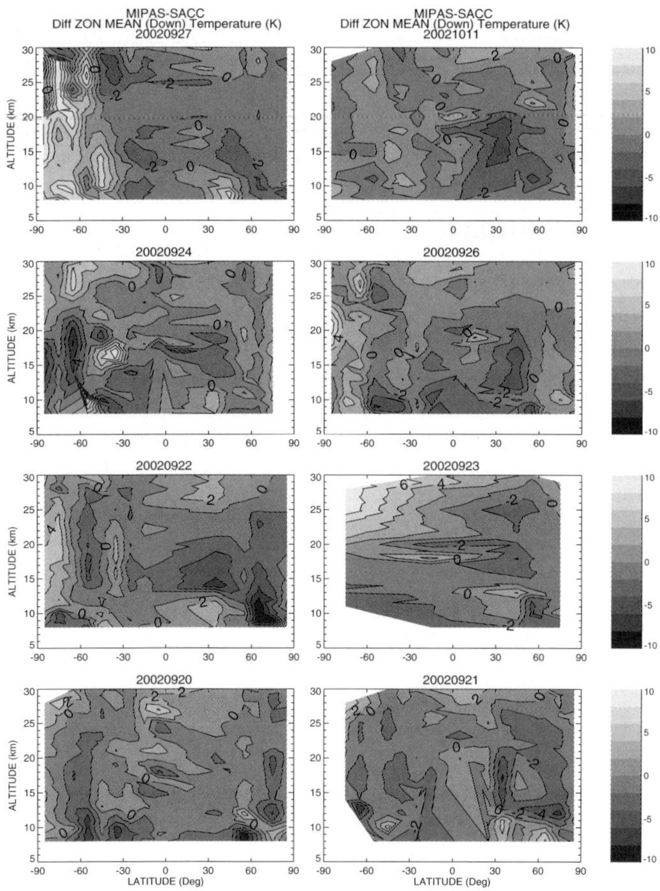


**Fig. 2.** Daily zonal averaged temperature differences (K) of correlative MIPAS and SAC-C measurements. MIPAS observations are taken from the descending mode. The temperatures are contoured by 2 K interval.

# 4 Future comparison with EOS MLS

The Earth Observing System (EOS) Microwave Limb Sounder (MLS) is a 'second-generation' MLS experiment [*Waters et al.*, 1999] to be launched on-board the NASA EOS Aura satellite in 2004. EOS MLS measures atmospheric composition, temperature, cloud ice, and pressure from observations of thermal emission at microwave wavelengths as the instrument field-of-view is scanned through the atmospheric limb. For temperature measurement, EOS MLS has a vertical and global coverage similar to that of MIPAS. The standard MLS temperature product is expected to have ~2 K precision or better from 500 to 0.001 hPa (some smoothing will be applied between 10-0.001 hPa in both vertical and along-track direction). The cross-comparison technique that is used in this study provides a useful comparison and validation tool for the temperature measurements from EOS MLS. In addition, cross-comparison and validations of temperatures, water vapor, ozone and other species measured by MLS and MIPAS will be especially valuable for studies related to stratosphere ozone depletion, greenhouse gases transformation, troposphere-stratosphere exchange, and radiative forcing of climate change.

## 5 Concluding Remarks

We compared ∼770 correlative profiles of MIPAS and SAC-C temperatures between 5-30 km. Both, individual profiles and zonal means of the temperature at two different seasons and various geolocations show reasonable agreement. The overall mean differences are estimated to be less than ∼2 K with rms deviation of less than 5 K. No apparent height dependence exists. Similar technique can be applied for a cross-comparison between MIPAS and EOS MLS instruments.

*Acknowledgement.* We thank the two anonymous reviewers for helpful comments and suggestions. This study was supported by the Institut für Meteorologie und Klimaforschung of Germany, and the Jet Propulsion Laboratory, California Institute of Technology, under contract with National Aeronautics and Space Administration of U.S.A. J. H. Jiang is especially grateful for the support and guidance from Drs. D. L. Wu and E. R. Kursinski for the GPS related study in 1999-2000.

## References

1. v Clarmann T, et al. (2003) Remote sensing of the middle atmosphere with MIPAS. In: Remote Sensing of Clouds and the Atmosphere VII, Schafer K and Lado-Bordowsky O, eds., Proc. SPIE, *4882*: 172–183.
2. Fischer H and Oelhaf H (1996) Remote sensing of vertical profiles of atmospheric trace constituents with MIPAS limb emission spectrometers. Appl Opt *35(16)*: 2787–2796.
3. Hajj GA, Kursinski ER, Romans LJ, Bertiger WI, and Leroy SS (2002) A technical description of atmospheric sounding by GPS occultation. J Atmos Solar-Terr Phys *64(4)*: 451–469.
4. Kursinski ER, Hajj GA, Schofield JT, Linfield RP, and Hardy KR (1997) Observing Earth's atmosphere with radio occultation measurements using the Global Positioning System. J Geophys Res *102(D19)*: 23,429–23,465.
5. Manney GL et al. (2004) Simulations of Dynamics and Transport During the September 2002 Antarctic Major Warming. J Atmos Sci: (accepted).
6. Stiller GP et al. (2003) Early IMK/IAA MIPAS/ENVISAT results. In: Remote sensing of Clouds and the Atmosphere VII, Schafer K and Lado-Bordowsky O, eds., Proc. SPIE *4882*: 184–193.
7. Wang DY et al. (2003) Comparisons of MIPAS-Observed Temperature Profiles with Other Satellite. Proc. SPIE (accepted).
8. Ware R et al. (1996) GPS sounding of the atmosphere from low Earth orbit - preliminary results. Bull Am Meteorol Soc *77*: 19–40.
9. Waters JW et al. (1999) The UARS and EOS Microwave Limb Sounder Experiments. J Atmos Sci *56*: 194–218.

# Structure and Variability of the Tropopause Obtained from CHAMP Radio Occultation Temperature Profiles

Madineni Venkat Ratnam, Gerd Tetzlaff, and Christoph Jacobi

Institute for Meteorology, University of Leipzig, *vratnam@uni-leipzig.de*

**Summary.** The global structure and variability of the tropopause observed using CHAMP/GPS radio occultation observations from May 2001 through September 2003 are presented. Tropopause temperature and height observed by CHAMP/GPS has been compared with radiosonde observations at a sub-tropical site. At polar latitudes, the tropopause sharpness found to be highest in summer and lowest in winter with slight differences between hemispheres.

**Key words**: Tropopause, radio occultations

## 1 Introduction

Using tropopause characteristics as indicator for climate variability has focused international interest on this region of the atmosphere ([1]). However, due to the coarse vertical resolution of ECMWF or other analysis data, and the uneven distribution of radiosondes with gaps especially over Oceans, at low latitudes and in the polar regions, the comprehensive investigation of tropopause characteristics is difficult. A new and promising tool to analyze the tropopause characteristics and variability are GPS radio occultation data, which are characterized by high vertical resolution and global sampling. The latter is particularly advantageous in the tropics, where radiosonde measurements are sparse. Moreover, radio occultation observations are calibration-free, therefore have excellent long-term stability, and they are insensitive to clouds and rain.

Recently, using GPS/MET observations, the variability of the tropical tropopause region was studied by [2,3]. They showed the high accuracy of GPS/MET retrievals especially in tropics. It is also reported that sub-seasonal variability of tropopause temperature and height appears to be related to wave-like fluctuations (such as gravity and Kelvin waves) and significant correlations are also found between GPS/MET observations and outgoing long wave radiation data. However, these studies are restricted to tropical latitudes.

Using CHAMP measurements, extended analysis of the tropopause region is possible using the advantages of this system. These are a larger number of occultations compared to GPS/MET due to improved GPS receiver technique (JPL's state-of-the-art 'BlackJack' flight receiver) and optimized occultation infrastructure that allows continuous atmospheric sounding independent of the Anti-Spoofing mode of the GPS system.

For the present study we use level 3 version 004 data from May 2001 to September 2003 that are produced by GFZ Potsdam [4]. Besides this data, temperatures observed with radiosondes are also collected to compare with CHAMP/GPS data.

## 2 Comparison with standard radiosonde data

Before analyzing tropopause characteristics by radio occultation data, the observed temperature profiles need to be compared with other well-established techniques. We use standard pressure-level radiosonde temperatures (10-30 km) over a sub-tropical latitude (Pan Chiao, Taiwan, 25°N, 121°E) to compare with CHAMP/GPS data. Notwithstanding the global coverage of CHAMP occultations, very close coincidences with radiosonde ascents are very rare. Therefore differences of ± 2° latitude, ± 20° longitude and ± 2 hours have been accepted for coincidences of ground based and satellite derived profiles. It is true that comparison with one radiosonde is not a validation. Furthermore, the radiosonde data are not high-resolution results, however, Pan Chiao radiosonde ascents of similar resolution have been used to study tropopause characteristics [5]. Fig. 1 therefore gives an impression, which structures are visible in the CHAMP GPS data in comparison with the standard analyses of radiosondes. To give an impression of the representation of the tropopause in both datasets, an example of vertical temperature profiles is shown in Fig. 1, left panel.

The results for all seasons (with a total of 59 profiles meeting the above men-

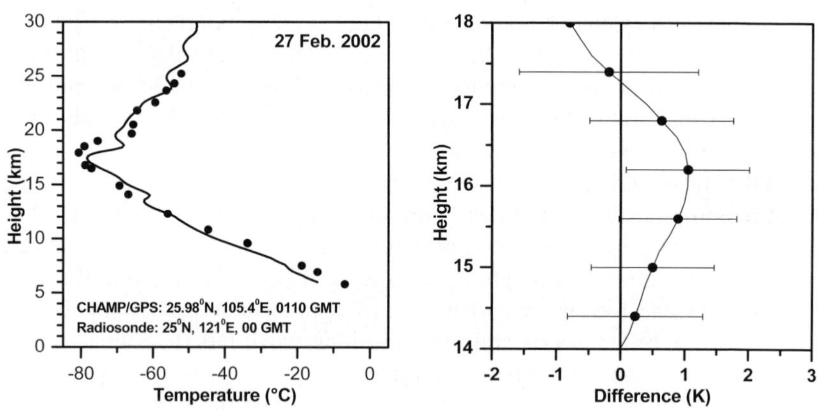


**Fig. 1.** Example of a CHAMP radio occultation temperature profile together with radiosonde data (left panel). In the right panel the mean differences radiosonde-CHAMP are plotted for 59 profiles of nearest coincidence.

tioned criteria) are shown in the right panel of Fig. 1. The differences observed with CHAMP-radiosonde measurements are plotted between 14 and 18 km during May 2001 to June 2002. The "error bars" show the standard deviation. The comparison of the CHAMP profiles with radiosonde shows excellent agreement. Near the sub-tropical tropopause (around 16 km), the mean deviation is about 0.5-1 K, with colder CHAMP data, which also has been observed by [6] using ECMWF analysis. This is possible due to a better vertical resolution of radio occultation measurements in comparison to the analyses that were available on standard pressure levels (radiosonde). This comparison reveals the high accuracy of CHAMP/GPS measurements, especially at a sharp tropopause, which is more often seen at tropical latitudes.

## 3 Tropical and midlatitude tropopause

To demonstrate the potential of CHAMP tropopause analysis, Fig. 2 shows the cold point tropopause temperatures in December 2001 between 50°S and 50°N. The coldest temperatures are found near the equator. Minimum temperatures are lower than 185 K, but non-zonal structures are well visible with the coldest tropopause over the Indonesian sector.

Fig. 3 shows the tropopause height and temperatures near the equator (±10° latitude) obtained using CHAMP observations during May 2001 to Dec. 2002. Tropopause height and temperature show a clear annual cycle with peaks during Northern Hemisphere winter and summer, respectively. The annual mean equato-

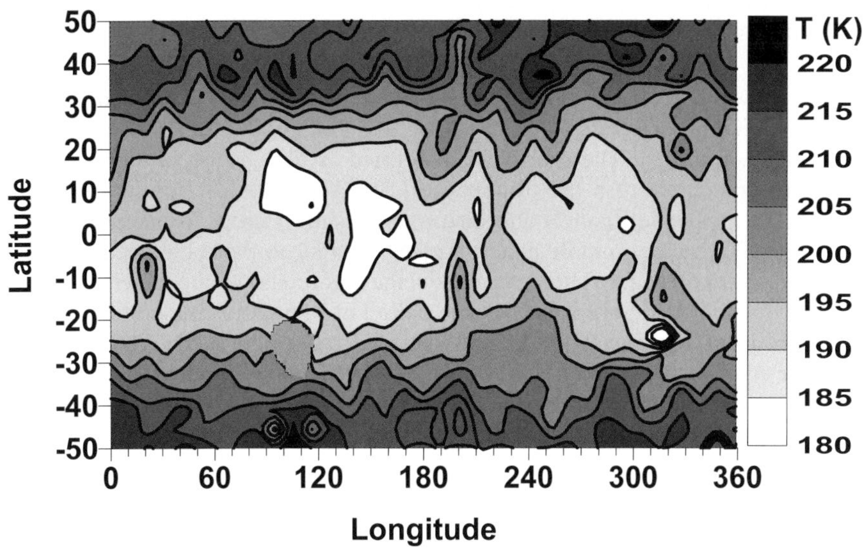


**Fig. 2.** Latitude-longitude plot of the cold-point tropopause temperature in December 2001.

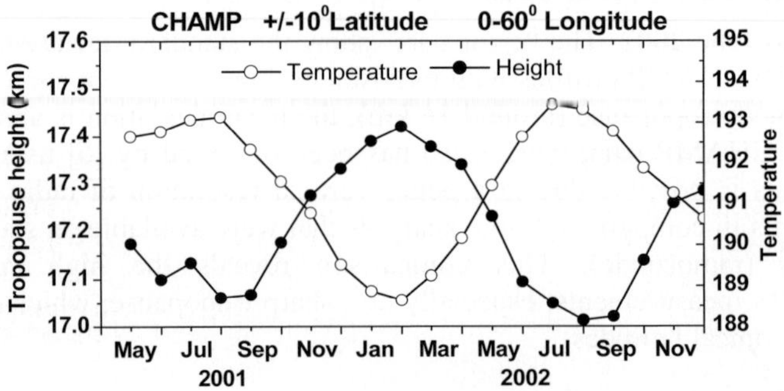


**Fig. 3.** Time series of the monthly mean zonal mean tropical tropopause height and temperature. Data between 10°N and 10°S are used.

rial tropopause height and temperature is 17.2 km and 191.5K, respectively. Using 30 years (1961-1990) of data from a radiosonde network, [7] reported corresponding values of 16.9 km and 197.7 K. While the height is comparable with the CHAMP data, the radiosonde temperature is 6 K higher, probably due to the lower vertical resolution of standard radiosonde analyses. However, one has to take into account that here we compare a rather short CHAMP data set with a climatological statistics, so that conclusions should be drawn with caution. For more detailed analyses of the tropical tropopause region the reader is referred to [8].

## 4 High-latitude tropopause

Fig. 4 shows the comparison of the radiosonde and CHAMP observed polar mean tropopause sharpness over both the Artic and Antarctic. Radiosonde results have been taken from [9]. The sharpness is defined as the change of the vertical temperature gradient across the tropopause [10]. The CHAMP results over Central Antarctica (upper left panel) and also over Eastern Europe/Western Siberia (70-80N; 40-100E, upper middle panel) are very similar to those obtained with the radiosonde network data. The error bar in the figure shows the standard deviation obtained while averaging over the time period of May 2001-Sep. 2003.

There exist significant differences between radiosonde and CHAMP observations over of Alaska-Canada (70-80N, 200-260E) in some months, but above all over Eastern Europe/Asian higher midlatitudes (55-60N; 40-100E). The reason for these observed discrepancies could be partly attributed to the different times of observations. We have to take into account that CHAMP observations data still base upon few years. Possible differences therefore should be further investigated using additional observations from CHAMP and other GPS radio occultation missions in future.

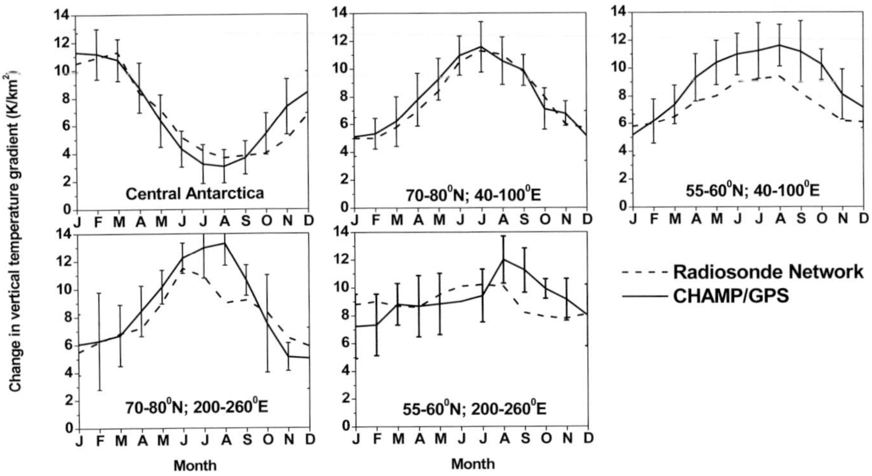


**Fig. 4.** Comparison CHAMP/GPS observed tropopause sharpness with data observed by the radiosonde network (taken from [6]). Note the different time interval that data are referring to (radiosondes 1989-1993, CHAMP 2001-2003).

# 5 Conclusions

Using CHAMP GPS radio occultation temperature profiles it is possible to obtain a global picture of the tropopause temperature and height. Since the temperature profiles are available with high vertical resolution, the results, particularly for the tropics, are superior to those obtained from standard radiosonde analyses. In addition, the global coverage of GPS profiles allows the construction of global fields without limitations due to insufficient data coverage.

Monitoring tropopause parameters has proven to be an appropriate mean for detecting climate variability and man-made influence on climate [1]. The potential for improving the quality of tropopause monitoring through GPS radio occultation makes these satellite measurements an important tool for further climate research.

*Acknowledgements:* We wish to thank GFZ Potsdam and the CHAMP team for providing CHAMP/GPS data. This study was supported by DFG under grant JA 836/4-2.

## References


1. Sausen R, Santer BD (2003) Use of changes of tropopause height to detect human influences on climate. Meteorol Z, NF *12*: 131-136.
2. Randel WJ, Wu F, Rios WR (2003) Thermal variability of the tropical tropopause region derived from GPS/MET observations. J Geophys Res 108: 10.1029/2002JD002595.
3. Nishida M, Shimizu A, Tsuda T, Rocken C, Ware RH (2000) Seasonal and longitudinal variations in the tropical tropopause observed with the GPS occultation technique (GPS/MET). J Meteorol Soc Japan *78*: 691-700.
4. Wickert J, Schmidt T, Beyerle G, König R, Reigber C, Jakowski N (2004) The radio occultation experiment aboard CHAMP: Operational data analysis and validation of atmospheric profiles. J Meteorol Soc Japan, Special issue 'Application of GPS Remote Sensing to Meteorology and Related Fields', *82(1B)*: 381-395.
5. Thulasiraman TS, Nee JB, Chen WN, Chen JH (1999) Temporal characteristics of tropopause and lower stratosphere over Taiwan during 1990-1995. J Atmos Solar-Terr Phys *61*: 1299-1306.
6. Wickert J, Reigber C, Beyerle G, König R, Marquardt C, Schmidt T, Grunwaldt L, Galas R, Meehan TK, Melbourne WG, Hocke K (2001) Atmospheric soundings by GPS radio occultation: First results from CHAMP. Geophys Res Lett *28*: 3263–3266.
7. Seidel JD, Ross RJ, Angell JK, Reid GC (2001) Climatological characteristics of the tropical tropopause as revealed by radiosondes. J Geophy Res *106*: 7857–7878.
8. Schmidt T, Wickert J, Beyerle G, König R, Reigber C (2004) Tropopause parameters derived from CHAMP. This issue.
9. Zängl G, Hoinka KP (2001) The tropopause in the polar regions. J Clim *14*: 3117-3139.
10. Wirth V (2000) Thermal versus dynamical tropopause in upper tropospheric balanced flow anomalies. Quart J Roy Meteor Soc *126*: 299-317.

# An Assessment of an Ionospheric GPS Data Assimilation Process

Matthew Angling

Centre for RF Propagation and Atmospheric Research, QinetiQ, St.Andrew's Road, Malvern, Worcestershire, WR14 3PS, UK. *mjangling@QinetiQ.com*

**Summary.** A data assimilation process that employs best linear unbiased estimation has been applied to slant total electron content data from the IGS network and CHAMP. The data has been assimilated into a three dimensional electron density grid for a test day in June, 2003. The accuracy of these assimilations has been assessed through the use of independent GPS data.

**Key words:** Ionosphere, ionospheric data assimilation, slant total electron content

## 1 Introduction

Comprehensive, global and timely specifications of the earth's atmosphere (particularly the refractivity of the troposphere and ionosphere) are required to ensure the effective operation, planning and management of many radio frequency systems. Although many ground based techniques have been developed to measure atmospheric refractivity, methods using the Global Positioning System (GPS) are being increasingly investigated. These include measurements from both ground based receivers and from space based radio occultation (RO) receivers. RO involves monitoring transmissions from GPS satellites using receivers on Low Earth Orbiting (LEO) satellites [*Hajj and Romans*, 1998].

The Electron Density Assimilative Model (EDAM) has been developed to exploit GPS measurements of slant total electron content (TEC) [*Angling and Cannon*, 2002]. The problem is mathematically under-determined because the amount of information that can be extracted from GPS TEC measurements is low compared to the required resolution of the electron density field under investigation. Therefore it is necessary to utilise *a priori* information about the state of the ionosphere. EDAM employs data assimilation techniques to incorporate GPS TEC data into a background model of the environment. Such techniques are well suited to both sparse and under-determined data sets, where the aim is to combine measurement data with a background model in an optimal way [*Rodgers*, 2000]. Since both the observations and the background model contain errors it is not possible to find the true state of the environment. Nevertheless, a good statistical estimate of the state may be found. One additional advantage of a data assimilation approach is that it overcomes the limitations of the traditional Abel Transform method of inverting RO data: spherical symmetry is not assumed, occultations do not have to be close to the orbital plane of the receiver, and both the transmitter and receiver need not be outside of the atmosphere.

The emphasis of EDAM's development has been to produce a practical approach that enables the efficient assimilation of many measurements on a single processor PC. Furthermore, the intention has been to develop a scalable system that can be applied to a wide range of data densities. This paper describes the algorithm employed by EDAM and presents a test assimilation of data from the International GPS service (IGS) and the CHAMP satellite. The quality of the assimilation will be assessed through a comparison with independent slant TEC data.

## 2 The Electron Density Assimilative Model

The following sections describe the best linear unbiased estimation (BLUE) technique that has been used in EDAM to directly modify electron density grids produced by an ionospheric model.

### 2.1 Observations

The observation vector (**y**) consists of a series of *p* GPS slant TEC measurements. The TEC is calculated from both the L1/L2 phase and pseudorange differences. The phase ambiguity can then be found by minimising the squared difference between the phase and pseudorange TECs. For the purposes of this work the remaining differential code biases (DCB) have been removed by using the values calculated by the Centre for Orbit Determination in Europe (CODE) in Switzerland for the GPS satellites and IGS ground stations. A DCB of 9ns has been estimated for the CHAMP RO receiver [*Heise et al.*, 2004]. Associated with each slant TEC is the time, date, and positions of both the receiver and the GPS satellite. The errors associated with each slant TEC are assumed to be independent and variances are all assumed to be equal. Therefore the observation error covariance matrix **R** is diagonal (with dimensions $p \times p$) and has the same value in each diagonal element.

### 2.2 Background model

The Parameterised Ionospheric Model (PIM), which contains the Gallagher Plasmasphere model, has been used to produce EDAM's background model. Values of electron density are generated on a 4° × 4° latitude-longitude grid and at 60 altitudes ranging from 90 km to 24000 km. The vertical resolution ranges from 2 km in the E region to 2000 km in the plasmasphere. From this grid, those voxels that are intersected by, or are close to voxels intercepted by, rays from the GPS to the LEO are rasterised to form the background model ($\mathbf{x}_b$, containing *n* elements).

The background model error covariance matrix, **B**, is block diagonal with dimensions $n \times n$. For ground based measurements, the background error variance at each height, (i.e. the values which comprise the diagonal elements of the background error covariance matrix) is assumed to be proportional to the square of the electron density in the background model at that height. For RO measurements the diagonal elements are assumed to be proportional to electron density to the power of 1.5 . This acts to maximise the absolute errors in the F region, where it is expected that the ionosphere is most variable. The higher power is required for the ground based measurements because these measurements contain less vertical information and must, therefore, be more strongly constrained. The error covariances are estimated from the error variances, the vertical separation, and an estimate of the vertical scale height in the ionosphere. Only matrix elements within ±4 altitude levels of the diagonal elements of **B** have their covariances calculated. All other elements are assumed to be zero. For points that are separated in latitude or longitude, the covariance is further scaled by the horizontal ionospheric scale length. A scale length of four degrees has been assumed.

### 2.3 Observation operator

The observation operator, **H**, describes the way in which the rays from the GPS satellite pass through the background model. The operator consists of a $p \times n$ matrix, where each row is associated with one point in a modelled observation. Such a point (e.g. the $i^{th}$) is a measure of the slant TEC along the ray and can be modelled by the sum of the electron density in the $j^{th}$ voxel ($x_j$) multiplied by the ray length within the $j^{th}$ voxel ($H_{ij}$):

$$TEC_i = \sum_{j=1}^{n} H_{ij} x_j \Rightarrow \mathbf{TEC} = \mathbf{H}\mathbf{x}_\mathbf{b} \tag{1}$$

### 2.4 Analysis

The BLUE technique produces an analysis ($\mathbf{x_a}$) based on the background model ($\mathbf{x_b}$), which is modified linearly by differences between the observation vector ($\mathbf{y}$) and the background state acted on by the observation operator [*Rodgers*, 2000]:

$$\begin{aligned} \mathbf{x_a} &= \mathbf{x_b} + \mathbf{K}(\mathbf{y} - \mathbf{H}\mathbf{x_b}) \\ \mathbf{K} &= \mathbf{B}\mathbf{H}^\mathrm{T}\left(\mathbf{H}\mathbf{B}\mathbf{H}^\mathrm{T} + \mathbf{R}\right)^{-1} \end{aligned} \tag{2}$$

where the weight matrix, **K**, is dependent on the background and observation error covariance matrices.

### 2.5 Time evolution

During each assimilation period (i.e. 15 minutes) the ionosphere is assumed to be unchanging and each measurement is assimilated in turn. This reduces the memory requirements for the process, since for each measurement only a sub-set of the full background model need be used.

After all the measurements from one assimilation period have been assimilated it is necessary to evolve the analysis in time to provide the background model for the next period of assimilation. For the results reported here, a simple persistence model (in geomagnetic latitude and local time) has been adopted for the ionospheric evolution.

## 3 Test assimilation

EDAM has been used to assimilate data from the IGS network and from the CHAMP RO payload for day 165 of 2003 ($14^{th}$ June). Only IGS data available in hourly RINEX files, and for which a CODE DCB is available has been used, resulting in the assimilation of approximately 75 stations distributed globally (Figure 1). Four stations (GODE, PERT, SUTH and ZIMM) have been selected to provide a test set against which to test the EDAM analysis. These stations are not included in the assimilation itself.

Data is assimilated in periods of 15 minutes. At the end of each assimilation period, slant TEC data from each of the four test stations is compared to slant TECs generated by integrating through the EDAM analysis and through the output of PIM (run for the appropriate time).

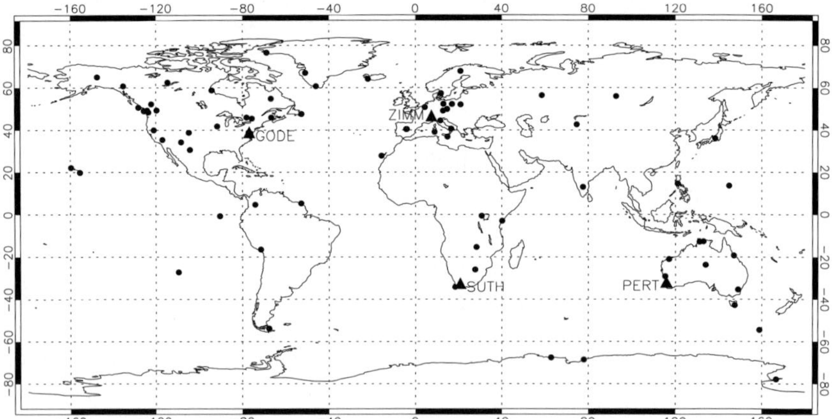


**Fig. 1.** Map of IGS stations used in the assimilation (circles) and used for validation (triangles).

Histograms of the slant 96 TEC differences generated for the test day show the reduction in TEC difference when EDAM is used in preference to PIM (Figures 1 to 4). Furthermore, the RMS slant TEC differences show an improvement when using EDAM (Table 1). For example, the RMS slant TEC error at the ZIMM IGS station is reduced from 11.0 to 4.1 TECu.

Due to the sparse nature of the CHAMP measurements and the fixed locations of the test points, the inclusion of CHAMP data in the assimilation does not have a large impact on the results. The impact of RO on ionospheric assimilation is likely to increase as the number of satellites increases. Future studies will assess the impact of radio occultation constellations on EDAM assimilations.

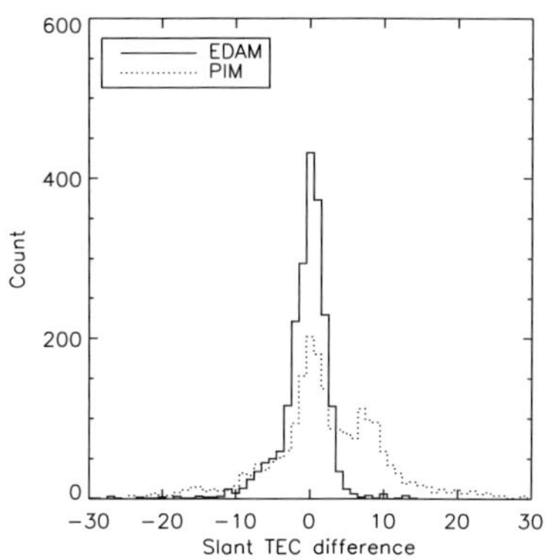


**Fig. 2.** Slant TEC difference between the EDAM analysis (solid line) and PIM (dotted line) and the GODE IGS station.

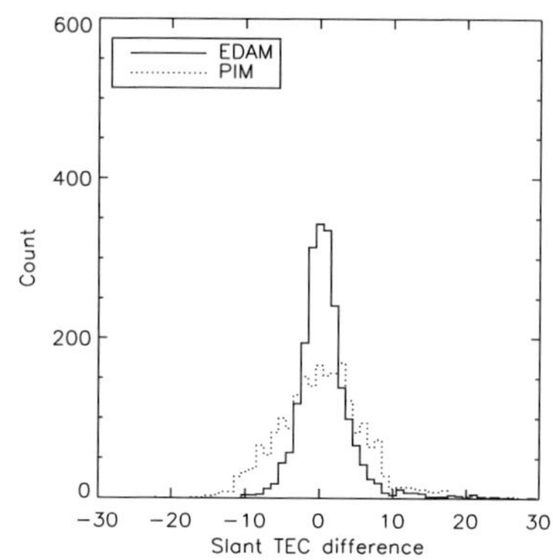


**Fig. 3.** Slant TEC difference between the EDAM analysis (solid line) and PIM (dotted line) and the PERT IGS station.

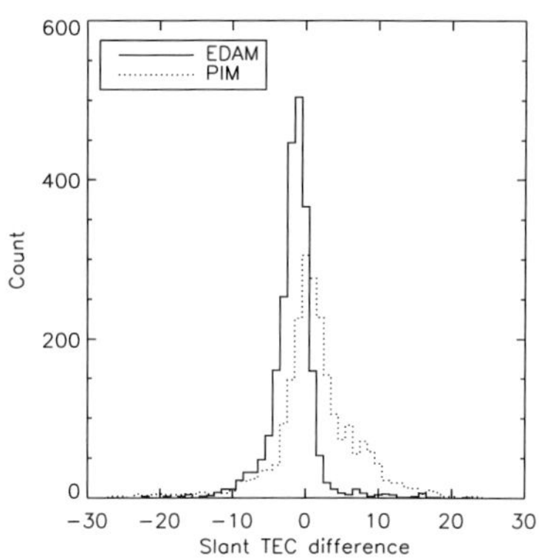


**Fig. 4.** Slant TEC difference between the EDAM analysis (solid line) and PIM (dotted line) and the SUTH IGS station.

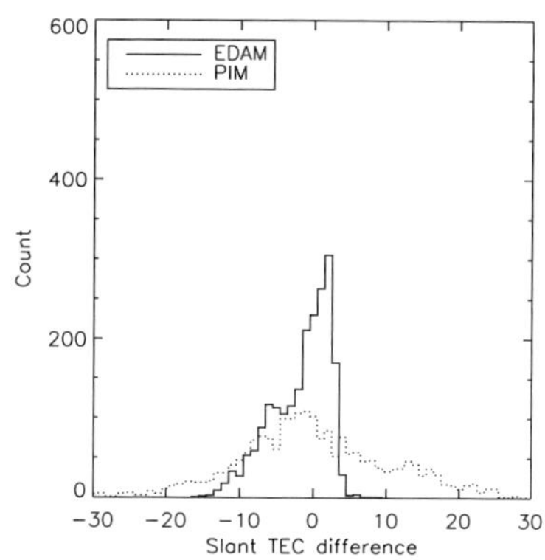


**Fig. 5.** Slant TEC difference between the EDAM analysis (solid line) and PIM (dotted line) and the ZIMM IGS station.

| | PIM | EDAM |
|---|---|---|
| GODE | 8.5 | 4.0 |
| PERT | 6.1 | 4.6 |
| SUTH | 6.1 | 3.5 |
| ZIMM | 11.0 | 4.1 |

**Table 1.** RMS slant TEC differences for EDAM and PIM. Values are in TEC units.

## 4 Conclusions and future work

EDAM has applied data assimilation techniques to ground and spaced based GPS slant TEC measurements of the ionosphere. EDAM shows the potential to provide a consistent method of combining measurements and a background model, thus allowing a more accurate specification of ionospheric electron density.

The effectiveness of data assimilation is highly dependent on the accuracy of the relevant error covariance matrices (both observation errors and background errors) and further work is required on their specification. In particular, different values for the latitudinal and longitudinal horizontal scale lengths should be developed that can be varied for different locations.

*Acknowledgements.* This work was carried out under funding from the United Kingdom Ministry of Defence Operational Environments Corporate Research Programme. CHAMP data was supplied by the CHAMP Information System and Data Centre. IGS data was obtained from the IGS Operational Data Centre at JPL. Differential code biases where obtained from the Centre for Orbit Determination in Europe.

## References

Hajj GA, Romans LJ (1998) Ionospheric electron density profiles obtained with the Global Positioning System: Results from the GPS/MET experiment. Radio Science *33(1)*: 175-190.

Angling MJ, Cannon PS (2002) Assimilation of radio occultation measurements into background ionospheric models. Presented at Ionospheric Effects Symposium, Alexandria, USA.

Rodgers CD (2000) Inverse methods for atmospheric sounding: theory and practice. In: Atmospheric, Oceanic and Planetary Physics. ed. Taylor FW, Vol. 2, World Scientific Publishing, Singapore.

Heise S, Stolle C, Schlüter S, Jakowski N (2004) Differential Code Bias of GPS Receivers in Low Earth Orbit: An Assessment for CHAMP and SAC-C. This issue.

# The Continuous Wavelet Transform, a Valuable Analysis Tool to Detect Atmospheric and Ionospheric Signatures in GPS Radio Occultation Phase Delay Data

Achim Helm, Georg Beyerle, Stefan Heise, Torsten Schmidt, and Jens Wickert

GeoForschungsZentrum Potsdam, Dept. 1 'Geodesy & Remote Sensing', Telegrafenberg, D-14473 Potsdam, Germany, *helm@gfz-potsdam.de*

**Summary.** GPS radio occultation phase delay (PD) data contain valuable information about atmosphere and ionosphere and its structure, based on the recorded changes of the direct GPS signal between LEO and GPS satellite. PD data contain signatures of surface reflections mainly above water and snow/ice covered regions which have successfully been detected and analyzed with the radio holographic (RH) method recently. Reflection signatures in the PD data can also be detected and analyzed with the continuous wavelet transform (CWT) in a strait forward manner, e.g. without requiring any additional information/forward model or reference field. The use of the CWT as an additional tool to analyze PD data is described and the assets and drawbacks in comparison to RH are discussed. Two years of consistent CHAMP PD data have been analyzed with CWT. The signature of surface reflections can be detected successfully. The CWT based method also reveals weak signatures during times when the direct GPS signal is not influenced by the atmosphere and travels only through the ionosphere. Different classes of ionospheric signatures can be isolated. The geographical distribution of each signature class reveals different patterns. Such observations may contribute to investigations of ionospheric structures like sporadic E or spread F layer.

**Key words:** CHAMP, GPS, Radio occultation, Continuous wavelet transform, CWT, Atmosphere, Ionosphere, Reflections

## 1 Introduction

GPS radio occultation PD data contain valuable information about the atmosphere (e.g., [2]) and ionosphere (e.g., [3], [4]) and its structure, based on the recorded changes of the direct GPS signal between LEO and GPS satellite. [1] already applied a wavelet based method on a simulated occultation signal. This study evaluates qualitatively which signatures beside the direct occultation signal can be retrieved from about 2 years of CHAMP PD data using the CWT [5].

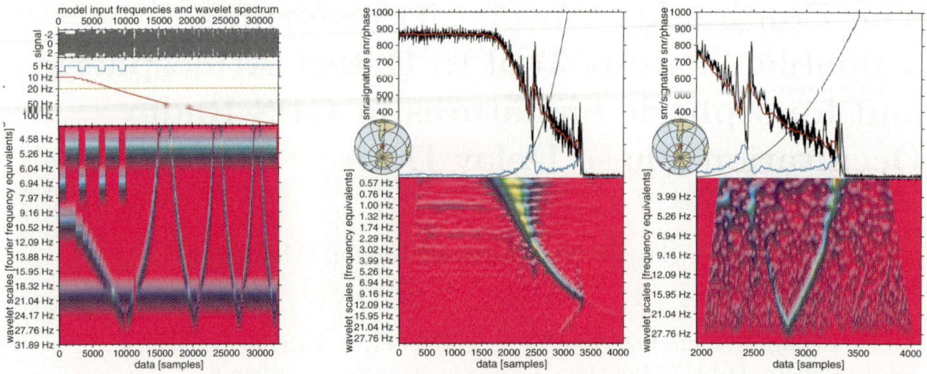


**Fig. 1.** The wavelet spectrum presents low energy by pink, higher energy by blue/green/yellow. The x-axis represents time given in data-samples. The y-axis starts at high frequencies and goes up to lower frequencies. Left: Wavelet spectrum of a model signal composed of different varying frequencies. Middle and right: The amplitude/snr of the PD data is plotted in black, the phase in blue. A running mean of the amplitude is printed in red, its standard deviation in turquoise. A star on the globe marks the location of the occultation. Middle: Wavelet spectrum of CHAMP PD data. Right: Type ATMO-A signature, after subtracting the direct signal.

## 2 Method

The CWT of a discrete sequence is defined as the convolution of the sequence with a scaled and translated version of a function, called the mother wavelet. By varying the wavelet scale and translating along the localized time index, one can construct a two-dimensional picture showing both the amplitude of any features versus the scale and how this amplitude varies with time. The result represents the wavelet energy spectrum. The wavelet function must have zero mean and must be localized in time and frequency. In this study the Morlet wavelet is used. It is constructed of a plane wave and modulated by a Gaussian. The time series must have equal time spacing. The relationship between wavelet scale and Fourier frequency depends on the used wavelet function and can be derived analytically for a particular wavelet function. In the wavelet spectrum signals with constant frequency appear as horizontal lines or as line segments when the signal occurs only sporadically. Frequencies higher than the Nyquist frequency (in this study 25 Hz) are aliased back into the spectrum (Fig. 1, left panel).

CHAMP calibrated atmospheric phase delay data (CH-AI-2-PD) [6] are interpreted as complex time series with a sampling rate of 50 Hz. The signal-to-noise ratio for the L1 Coarse/Aquisition data is used as amplitude and the associated atmospheric phase delay is used as phase. A running mean phase is calculated and subtracted in order to restrain the energy of the direct occultation signal (Fig. 1, middle and left panel). The wavelet energy

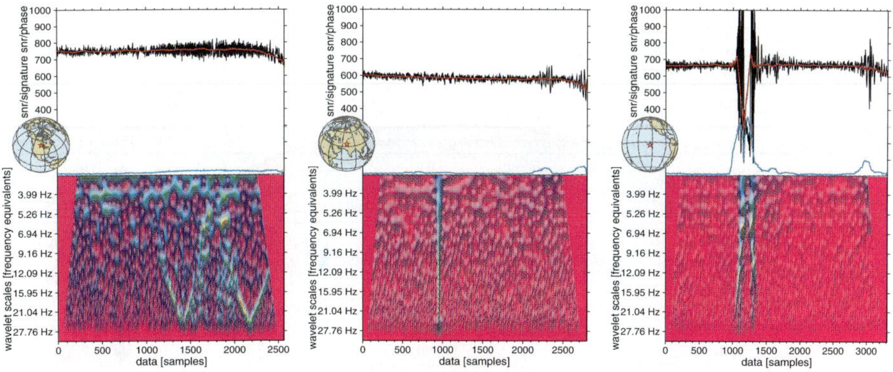


**Fig. 2.** Three different types of ionospheric signatures can be identified with CWT-CS. From left to right: type IONO-A, IONO-B and IONO-C

spectrum is calculated in a 2-dimensional grid and represents the normalized energy at each wavelet scale and discrete time sample. In order to screen the large PD data set a classification scheme (CWT-CS) was developed which automatically analyzes the 2-D wavelet spectrum for structured energy patterns. CWT-CS classifies the energy patterns into 4 characteristic types of signatures which could be regularly observed in the PD data. The signature is classified as atmospheric type ATMO-A (Fig. 1, right panel) in case the energy pattern is observed when the phase delay starts to be influenced by the neutral atmosphere, else as ionospheric type IONO-A, -B or -C (Fig. 2).

## 3 Data Analysis and Discussion

106,160 CHAMP GPS radio occultation events are used from the time period between 14 May 2001 and 22 August 2003. CWT-CS detected 26,870 atmospheric signatures and 14,010 ionospheric signatures which could be separated into 3 different classes, 6,140 of type IONO-A, 2,887 of type IONO-B and 4,983 of type IONO-C. Each signature class reveals a different global distribution pattern (Fig. 3). Most ATMO-A events occur over Antarctica's inland ice and over the shelf ice between 330°–150°E and 160°–270°E longitude respectively. Events also cluster at the area of the Bering Sea and the equatorial current belt systems. During a shorter data period (26 August 2001 – 1 February 2002) the CWT-CS detected atmospheric signatures are compared to the results obtained with RH. RH uses a reference model which only allows those rays which undergo a reflection at Earth's surface [2]. From 27,546 recorded CHAMP occultation events RH identifies 6,471 reflection signatures while CWT-CS detects 4,934 ATMO-A signatures. As can be seen in Fig. 3 and [2], CWT-CS can detect strong reflections from ice- and snow covered areas, but unlike RH CWT-CS also registers a large number of

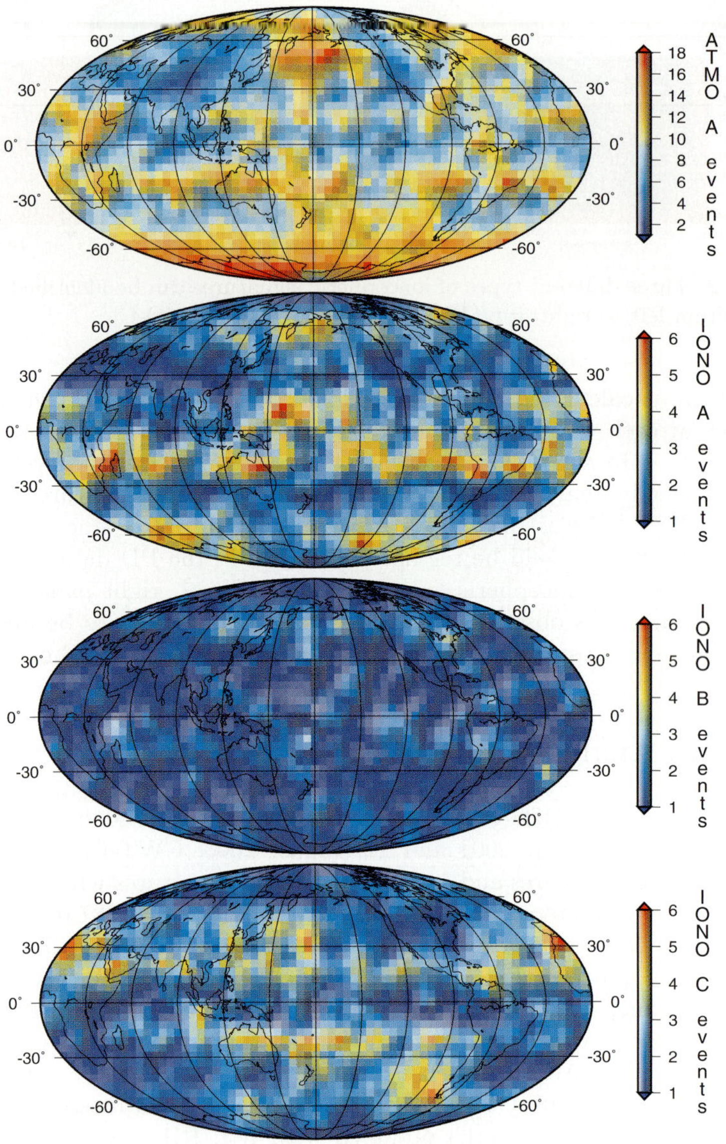


**Fig. 3.** Global distribution of CWT-CS detected signature events using CHAMP PD data between 14 May 2001 and 22 August 2003. A Mollweide projection with a geographical grid of 30° x 30° is used, with the central meridian located at the left border. The events are counted within a global grid of 3° x 3° resolution.

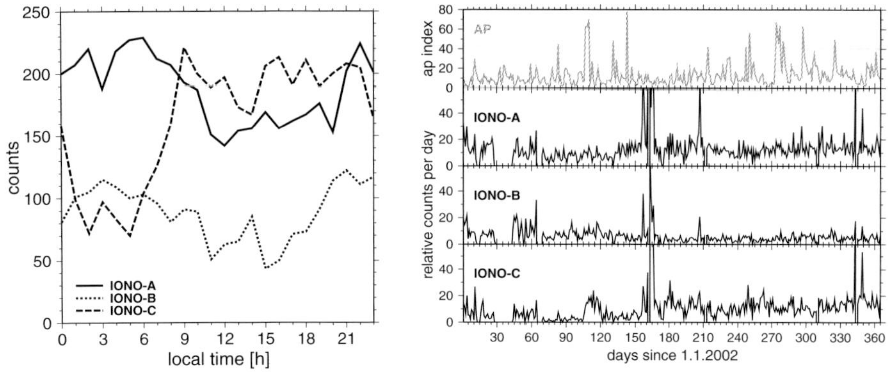


**Fig. 4.** Left: Ionospheric class distribution relative to local time (LT) between 14 May 2001 and 22 August 2003, data excluded from polar regions (above ±65° latitude). Right: Daily occurrence of each ionospheric signature in relation to the daily number of registered PD-data in comparison to the planetary equivalent daily amplitude (Ap) index during 2002.

events at locations on continents where reflections are not expected. IONO-A events can be frequently observed within ± 30° latitude with a strong local pattern. Additionally IONO-A events concentrate at the Bering Sea and locations around -60° latitude. IONO-B events do not show such pronounced local pattern. IONO-C events are mainly distributed between ± 15°–60° latitude, in the southern hemisphere roughly limited to the region of the southern pacific ocean. The number of IONO-A and IONO-B events decreases during local-day-time (10:00–20:00 LT), while IONO-C events rapidly drop after local-midnight (01:00–06:00 LT, Fig. 4, left). During 2002 some signals/peaks can be observed in the relative number of all ionospheric signatures (Fig. 4, right), e.g. between day of the year (DOY) 40–65, 155–175 and 340–365. The comparison with magnetic field disturbances represented by the Ap index shows no direct correlation. The highest variability is observed in the IONO-C data set. Between DOY 70–105 only a low number of IONO-C events can be counted daily. From DOY 105–125 a strong increase can be observed while the Ap index has a high amplitude for several days. A similar longer lasting high amplitude of the Ap index (DOY 270–285) does not result in a higher occurrence of IONO-C events.

Taking into account the PD signal form and the height (point of closest approach) of about 90–130 km, IONO-C events can be partly connected to sporadic E. Due to the integral character of PD data, parts of IONO-C events may also be connected to spread F and the equatorial fountain effect because of the geographical pattern and the local time distribution of IONO-C signatures. During IONO-A events longer-lasting amplitude fluctuations can be observed in the PD data. In the CWT spectrum an energy signature can be observed with a varying frequency shift to the direct PD signal, similar to ATMO-A.

## 4 Conclusions and Outlook

In comparison to CWT-CS RH can better identify surface reflection events. CWT-CS detects 4 different types of signatures in 2 years of CHAMP PD data. All show different geographical distribution patterns. Parts of ATMO-A signatures can be identified as surface reflections and the energy signature can be interpreted as frequency shift of the reflected signal. Parts of IONO-C signatures can be connected to ionospheric structures and may contribute to further investigations of sporadic E or spread F. Additionally, CWT-CS can potentially be used to watch for short and weak noise signals (e.g. 1 Hz in Fig. 1, middle panel, sample 500–1500) in the PD data. Nevertheless CWT-CS offers an adequate tool for analyzing PD data in case no suitable reference model is available or no background information is known, but interpretation of the results is more difficult.

*Acknowledgement.* The authors are very grateful to all the colleagues of the CHAMP team for the given support and data. The National Geophysical Data Center kindly provided the Ap index data.

## References

1. Benzon H and Lauritsen K (2002) Radioholographic and wavelet analyses of radio occultation data. Geophys Res Abstr: pp. A–05682.
2. Beyerle G, Hocke K, Wickert J, Schmidt T, and Reigber Ch (2002) GPS radio occultations with CHAMP: A radio holographic analysis of GPS signal propagation in the troposphere and surface reflections. J Geophys Res *07(D24)*: doi:10.1029/2001JD001402.
3. Hocke K, Igarashi K, Nakamura M, Wilkinson P, Wu J, Pavelyev A, and Wickert J (2001) Global sounding of sporadic E layers by the GPS/MET radio occultation experiment. J Atmos Solar-Terr Phys *63(18)*: 1973–1980.
4. Pavelyev A, Wickert J, Liou Y, Igarashi K, Hocke K, and Huang C (2003) Vertical gradients of refractivity in the mesosphere and atmosphere retrieved from GPS/MET and CHAMP radio occultation data. In: Reigber Ch, Lühr H, and Schwintzer P, eds, First CHAMP Mission results for Gravity, Magnetic and Atmospheric Studies, Springer: 500–507.
5. Torrence Ch and Compo GP (1998) A practical guide to wavelet analysis. Bull Amer Meteor Soc *79*: 61–78.
6. Wickert J, Galas R, Beyerle G, König R, and Reigber Ch (2001) GPS ground station data for CHAMP radio occultation measurements. Phys Chem Earth (A) *26(6-8)*: 503–511.

# The CHAMP Atmospheric Processing System for Radio Occultation Measurements

Torsten Schmidt, Jens Wickert, Georg Beyerle, Rolf König, Roman Galas, and Christoph Reigber

GeoForschungsZentrum Potsdam (GFZ), Department 1, Telegrafenberg A17, D-14473 Potsdam, Germany, *tschmidt@gfz-potsdam.de*

**Summary.** In this paper a description of the CHAMP atmospheric processing system for radio occultation data at GFZ Potsdam is given. The generation of radio occultation products, as e.g. atmospheric excess phases, vertical profiles of refractivity, temperature or water vapour is a complex process. Besides the scientific challenge the design and installation of an automatic data processing system is also of great importance. This system must be able to process the different input data from external data sources, coordinates the different data streams and scientific software modules, and feeds the results into the data centre automatically. Caused by different user demands the CHAMP Atmospheric Processor is divided into two parts: A rapid processing mode makes radio occultation analysis results available on average five hours after measurements. In the standard processing mode quality checked profiles of atmospheric parameters are available with a latency of about two days.

**Key words:** CHAMP, GPS, radio occultation, CHAMP Atmospheric Processor, near-real time, GFZ Potsdam

## 1 Introduction

The GPS radio occultation (RO) technique is an excellent method for global and continuous monitoring of the Earth's atmosphere [1, 2, 3]. Data assimilation studies using radio occultation data have already shown that this new type of remote sensing data improves the accuracy of global and regional weather analyses and predictions [4].

The use of analysis results obtained from radio occultation data, as e.g. atmospheric excess phases, by weather prediction centres requires an operational data processing system generating and delivering data products automatically within a certain time limit. Such a continuous Near-Real Time (NRT) data processing system, the CHAMP Atmospheric Processor (CAP), is operating at GFZ Potsdam with radio occultation data from the German CHAMP satellite [5].

Because of different user demands with respect to the availability of CHAMP RO products two data processing modes were introduced: a rapid mode and a standard mode. Currently in the rapid processing mode the daily averaged time delay between occultation measurements and availability of analysis results at the Information System and Data Centre (ISDC) [6] is about 5 hours. In the standard processing mode quality checked profiles of atmospheric parameters are available with a latency of about 2 days.

The realization of CAP follows the principle of separating the processing system into a scientific and controlling part leading to an independent and more flexible software development [7]. Basic considerations and requirements for an automatic processing system with respect to software and hardware can also be found in [7]. This paper is focussed on the controlling components of the CAP that generate a continuous data flow between the scientific software modules and the data centre. The CAP is designed to be easily extendable by additional scientific modules or input data. Thus, it also allows for an extension to other single- or multi-satellite radio occultation missions, as e.g. SAC-C, GRACE, TerraSAR-X.

## 2 Infrastructure and input data

The generation of radio occultation products requires a complex infrastructure that was developed and installed for CHAMP within the GPS Atmosphere Sounding Project (GASP) beginning in 2000 [8, 9].

The main components are (Fig. 1):

- the Black-Jack GPS receiver (provided by NASA/JPL) onboard CHAMP,
- two downlink stations receiving CHAMP data (Neustrelitz, Germany and Ny-Aalesund, Spitsbergen) in cooperation with DLR,
- the fiducial low latency and high rate GPS ground network operated in cooperation between GFZ and JPL [10],
- an orbit processing facility for determination of precise orbits for the GPS satellites and CHAMP [11],
- the CHAMP Atmospheric Processor generating CHAMP radio occultation data products (atmospheric excess phases, profiles of bending angle, refractivity, temperature, and humidity), and
- the CHAMP Information System and Data Centre (ISDC) archiving and distributing CHAMP data and analysis results [6].

## 3 The CHAMP Atmospheric Processor

The goal of the CAP is to coordinate the different data streams and to start the various scientific applications automatically when all input data for an application are available. The CAP is a modular structured and dynamically configurable software package consisting of a scientific and controlling part

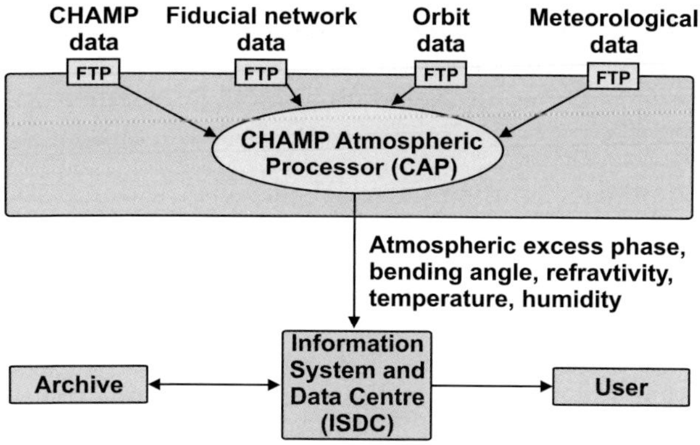


**Fig. 1.** Infrastructure and input data for GPS radio occultation data processing from CHAMP.

(Fig. 2). The scientific software modules (e.g. DDIFF, ATMO, WVP) calculate the atmospheric excess phases and vertical atmospheric profiles [12]. The controlling components ensure the continuous data flow of all input data through the scientific analysis modules and provide the interface for feeding the radio occultation products into the data centre (ISDC).

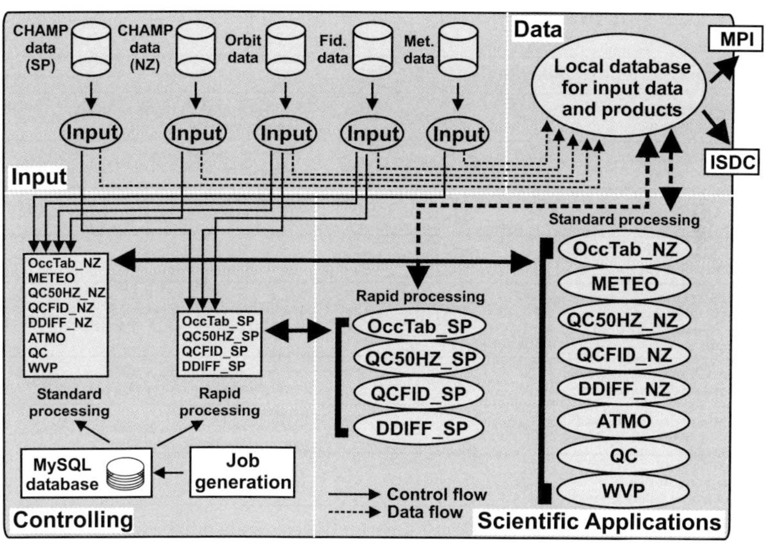


**Fig. 2.** Schematic view of the CHAMP Atmospheric Processor at GFZ Potsdam.

CAP runs on multiple UltraSparc II/III processor machines and is a combination of different subroutines and scripts written in C++, Fortran, IDL, Perl, and C-Shell.

# 4 Standard and rapid processing

The processing of radio occultation data at GFZ Potsdam is divided into two parts (Fig. 2):

- the standard processing using CHAMP data via the dump station Neustrelitz, Germany and Rapid Science GPS and CHAMP Orbits (RSO) [13],
- the rapid processing using CHAMP data via the polar dump station at Ny-Aalesund, Spitsbergen and Ultra rapid Science Orbits (USO) [13].

These two modes are necessary due to demands on product availability. The availability of radio occultation data products, i.e. the time delay between measurement and delivery of products to the data centre, depends on two factors: (1) the moment at which all input data for the respective applications (Fig. 2) are available and (2) the duration that CAP needs to process the input data and generate products. Tab. 1 gives an overview of input and output data for the different applications including the availability of the data with respect to the time of the occultation measurements for the standard processing mode.

The delivery of calibrated atmospheric excess phases to weather service centres in a certain time window depends on the availability of CHAMP occultation data, on the precise orbits of GPS satellites and CHAMP, and on the time needed for data processing within this time window. The first

| Input | Delay [h] | Scientific Applications | Output | Delay [h] |
|---|---|---|---|---|
| - CHAMP 50 Hz data (dump files from NZ) | 0.5-12 | OccTab_NZ | - hourly CHAMP 50 Hz data (RINEX)<br>- occultation table | 0.5-12 |
| - daily ECMWF analysis<br>- occultation table | 24-48<br>0.5-12 | METEO | - meteorol. data at each occultation point | 28-52 |
| - hourly CHAMP 50 Hz data | 0.5-12 | QC50HZ_NZ | - quality checked hourly CHAMP 50 Hz data | 0.5-12 |
| - hourly fiducial network data (1 Hz) | 0.5-1.5 | QCFID_NZ | - quality checked hourly fiducial network data | 0.5-1.5 |
| - occultation table<br>- hourly CHAMP data<br>- hourly fiducial data (1 Hz)<br>- orbit data | 0.5-12<br>0.5-12<br>0.5-1.5<br>12-36 | DDIFF_NZ | - calibrated atmospheric excess phases | 16-40 |
| - met. data at occultation points<br>- atmospheric excess phases | 28-52<br>16-40 | ATMO | - vertical atmospheric profiles (refractivity, dry temperature) | 29-53 |
| - atmospheric excess phases<br>- vertical atmospheric profiles | 29-53 | QC | quality checked<br>- atmospheric excess phases<br>- vertical temperature profiles | 29-53 |
| - quality checked temperature profiles<br>- met. data at occultation points | 30-54 | WVP | - vertical water vapour profiles | 31-55 |

**Table 1.** Input and output data including data availability with respect to the occultation measurement for the standard processing mode.

requirement is fulfilled due to the installation of a polar dump station. This enables a contact to CHAMP about every 90 minutes. Since April 2002 the GPS and CHAMP satellite orbits are available every 3 hours.

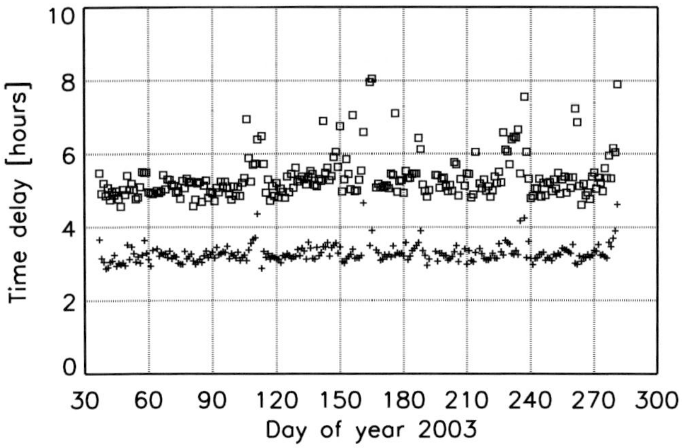


**Fig. 3.** Daily averaged time delay between CHAMP occultation measurement and product delivery to MPI (squares: average, crosses: minimum).

These USOs cover a 14 hour (CHAMP) and a 24 hour (GPS satellites) time window. The accuracy is comparable to the standard RSOs [12].

A pilot project in cooperation with the Max Planck Institute for Meteorology (MPI) demonstrates for the first time the delivery of CHAMP radio occultation products (atmospheric excess phases) in the rapid processing mode since the beginning of 2003. The averaged time delay between CHAMP measurements and data product delivery is about 5 hours. For single products a time delay less than 3 hours is reached (Fig. 3). The operational use of CHAMP radio occultation products by weather service centres (DWD, Met Office, ECMWF) is in preparation.

## 5 Summary and outlook

The CAP enables an operational data processing and the delivery of analysis results to the data centre. Since February 2001 more than 116.000 CHAMP radio occultation products, as calibrated atmospheric excess phases, refractivity, temperature and water vapour profiles, have been generated (about 170 daily, as of September 2003).

Due to the modular structure CAP is suitable for easy extension to other single- or multi-satellite radio occultation experiments.

In the standard processing mode quality checked products (refractivity, temperature and water vapour profiles) are available with a latency of about 2 days. A near-polar receiving station at Spitsbergen and a 3-hourly CHAMP and GPS orbit production cycle made it possible to implement for the first time a NRT processing at GFZ Potsdam. Since the beginning of 2003 atmospheric excess phases with a average time delay of about 5 hours between measurement and availability of the analysis results are generated continuously. Weather service centres (Met Office, ECMWF, DWD) will soon follow the current cooperation with the MPI and use the rapid processing products.

*Acknowledgement.* This study was carried out in the GASP (GPS Atmosphere Sounding Project) strategy fonds program under the grant of the German Federal Ministry of Education and Research (BMBF) no. 01SF9922/2.

## References

1. Melbourne WG, Davis ES, Hajj GA, Hardy KR, Kursinski ER, Meehan TK, and Young LE (1994) The application of spaceborne GPS to atmospheric limb sounding and global change monitoring. JPL Publication, *94-18*, Jet Propulsion Laboratory, Pasadena, California.
2. Kursinski ER, Hajj GA, Hardy KR, Schofield JT, and Linfield R (1997) Observing Earth's atmosphere with radio occultation measurements using the Global Positioning System. J Geophys Res *102*: 23,429–23,465.
3. Anthes RA, Rocken C, and Kuo YH (2000) Applications of COSMIC to meteorology and climate. Terrestrial, Atmospheric and Oceanic Sciences *11*: 115–156.
4. Kuo YH, Sokolovsky S, Anthes RA, and Vandenberghe F (2000) Assimilation of GPS radio occultation data for numerical weather prediction. Terrestrial, Atmospheric and Oceanic Sciences, *11*: 157–186.
5. Wickert J, Reigber C, Beyerle G, König R, Marquardt C, Schmidt T, Grunwaldt L, Galas R, Meehan TK, Melbourne WG, and Hocke K (2001) Atmosphere sounding by GPS radio occultation: First results from CHAMP. Geophys Res Lett *28*: 3263–3266.
6. www.isdc.gfz-potsdam.de/champ
7. Wehrenpfennig A, Jakowski N, and Wickert J (2001) A Dynamically Configurable System for Operational Processing of Space Weather Data. Phys Chem Earth (C) *26*: 601–604.
8. Reigber C et al. (1998) GPS Atmosphere Sounding: An innovative approach for the recovery of atmospheric parameters. HGF Strategy fund proposal, Potsdam, Germany.
9. www.gfz-potsdam.de/champ, www.gfz-potsdam.de/gasp
10. Galas R, Wickert J, and Burghardt W (2001) High Rate low latency GPS ground tracking network for CHAMP. Phys Chem Earth (A) *26*: 649–652.
11. König R, Zhu SY, Reigber C, Neumayer KH, Meixner H, Galas R, Baustert G, and Schwintzer P (2002) CHAMP Rapid Orbit Determination for GPS Atmospheric Limb sounding. Adv Space Res *30(2)*: 289–293.
12. Wickert J, Schmidt T, Beyerle G, König R, Reigber C, and Jakowski N (2003) The radio occultation experiment aboard CHAMP: Operational data analysis and validation of atmospheric profiles. J Meteorol Soc Japan *82(1B)*: 381–395.
13. König R, Michalak G, Neumayer KH, Schmidt R, Zhu SY, Meixner H, and Reigber C (2003) Recent developments in CHAMP orbit determination at GFZ. This issue.

# Potential Contribution of CHAMP Occultation to Pressure Field Improvement for Gravity Recovery

Shengjie Ge and C. K. Shum

Laboratory for Space Geodesy and Remote Sensing, The Ohio State University, Columbus, Ohio 43210, USA, *ge.18@osu.edu, ckshum@osu.edu*

**Summary.** This paper studies one aspect of use of the CHAMP GPS occultation data, namely the improvement of the atmospheric pressure field, particularly over Antarctica. Previous studies indicate that pressure differences between ECMWF and ground truth data reach 5.18 hPa RMS over Antarctica [Ge et al., 2003]. In this study, comparisons of pressure profiles (January–March 2003) from data (CHAMP occultation and radiosonde) and models (ECMWF and NCEP), indicate large discrepancies over different regions, notably over southern polar region. Global pressure differences between CHAMP and radiosonde and model outputs reach 4 hPa RMS at 1 km above MSL. We found a positive bias in CHAMP data (CHAMP measures larger pressure values) when comparing with both radiosonde and ECMWF. Analysis shows the lack of adequate penetration of CHAMP occultation data in the planetary boundary layer particularly in the tropical region (only ~10% signal is within 1 km above MSL), as compared to ~80% penetration in Arctic and Antarctica. However, CHAMP provides improved data coverage in temporal, spatial and vertical resolution globally. We conclude that the CHAMP occultation data could potentially improve the surface pressure modeling to benefit temporal gravity recovery, in particular over data sparse region such as Antarctica.

**Key words:** GPS occultation, surface pressure, time variable gravity

## 1 Introduction

Atmospheric pressure from current operational weather analysis products is not adequate to compute atmospheric loading for accurate static and temporal gravity field recovery using CHAMP and GRACE data [Ge et al., 2003], in particular for GRACE data as it is significantly more accurate than the CHAMP data. Studies indicate that the discrepancy between the model and *in situ* observation is inadequate for GRACE data processing, especially in the data sparse region such as Antarctica [Ge et al., 2003]. In addition, Han et al. [2003] show that satellite sampling of atmosphere causes temporal aliasing of the signal, which corrupts high-frequency errors in GRACE monthly gravity field solutions. At present, atmospheric mass variations are modeled for GRACE using 6-hourly, $0^{\circ}.5$ ECMWF model outputs. An improved spatial and temporal resolution would reduce the aliasing error. In principle, GPS occultation data has improved spatial and vertical resolutions for measuring global atmospheric pressure field, shown in Fig. 1 comparing with meteorological and radiosonde stations over Antarctica.

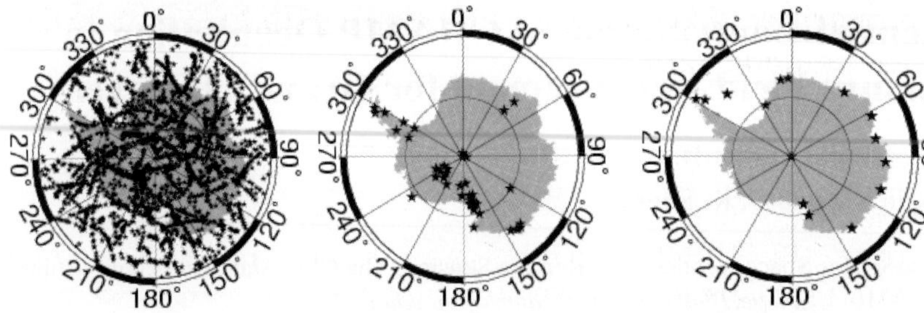


**Fig. 1.** Coverage map for CHAMP occultation data, January–March, 2003 (left), Automatic Weather Stations (middle) and Radiosonde Stations in Antarctica (right).

## 2 GPS Radio Occultation Technique

The principles and applications of GPS occultation have been extensively discussed in the literature [e.g. Wickert, 2002]. Overall, signal tracking and penetration are among the most critical limitations for accurate retrieval of surface pressure and its separation from water vapor and temperature.

Due primarily to the complexity of water vapor and signal tracking problem in the Planetary Boundary Layer (PBL), radar signals often cannot penetrate down to the surface. Fig. 2 illustrates the distribution of CHAMP occultation penetration over three regions. In the tropical region (30°N–30°S), which is warm and moist, only ~10% of signals penetrate down to 1 km above the MSL (Fig. 2b), while in the cold and dry Arctic oceanic area (Fig. 2a), ~80% of the profiles reach within 1 km above the MSL due to lack of water vapor and rapid changing small scale features. Before removing the topographic inconsistency, only ~50% profiles reach 1 km MSL in Antarctica (Fig. 2c), which is a high elevation, cold and dry region. After referencing the occultation profiles to the ECMWF topography, Ge and Shum [2003] show that the Antarctica signal penetration is similar to the Arctic region (~80% signals penetrate to within 1 km above the MSL).

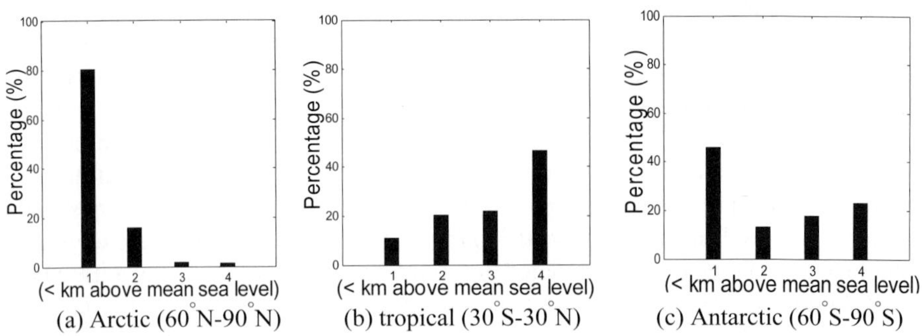


**Fig. 2.** Histograms of CHAMP occultation penetration depth (Jan-March, 2003, referencing to the MSL) over Arctic (~80% penetration), tropical (~10%) and Antarctic (~50%) region.

## 3 Comparison of GPS Derived Pressure with ECMWF, NCEP and Radiosonde data

### 3.1 GPS Occultation Datasets

The period chosen in this study is from Jan–Mar 2003 (or day of year–doy 001–093, 2003). The CHAMP wet profiles were obtained from the COSMIC website (http://www.cosmic.ucar.edu). doy 016, 028, 044, 053 are excluded from this analysis because the data are unavailable at the time. ECMWF and NCEP profiles were obtained by interpolating along the occultation path. If only CHAMP and ECMWF data are considered, there are a total of 13,422 matched profiles. When all 4 types of data are considered, there are only 3,891 matched profiles.

### 3.2 Method

To compare the profiles, CHAMP, ECMWF, NCEP and radiosonde pressure profiles are interpolated between 1 km and 30 km altitude above the MSL in 1 km steps. The comparison is conducted for the data only when all 4 types of data exist so that there would be no extrapolation. We divide the study areas into 5 regions from $30^{\circ}$ to $60^{\circ}$ latitude zonal bands: southern polar (SP), southern mid-latitude (SM), tropical (TP), northern mid-latitude (NM) and northern polar (NP) regions (Fig. 3). The mean and RMS are computed for each pair of data in the analysis period. Large amount of profiles in other datasets have to be edited, simply because radiosonde measurements only exist over land area, and there are very few data match-ups in the southern hemisphere (Fig. 3, right panels). Without losing generality, we compute and compare the statistics of 13,422 and 3,981 matched CHAMP and ECMWF profiles (reduced to 3,981 in order to match radiosonde profiles). In this study, we only discuss results for the reduced dataset.

### 3.3 Results

The differences between 4 types of pressure profiles are shown in Fig. 3. We compare CHAMP with ECMWF, NCEP and radiosonde pressure profiles in Fig. 3a, 3b and 3c, respectively. It is obvious from left panels (mean differences) of Fig. 3a, 3b and 3c that CHAMP pressure is positively biased or larger than other three data types. If one relates pressure to temperature through hydrostatic equation and the universal gas law, CHAMP derived temperature is colder than models and radiosonde measurements, which agrees with other analysis [e.g., Leroy, 1997]. The bias is shown to have larger discrepancy at tropopause (TP and SM), above Planetary Boundary Layer (PBL) (TP, SM) and near ground (SP) in Fig. 3a. It is nearly 3 hPa at 1 km altitude for the Antarctic region (SP). Larger differences for NM and SP region at 1 km altitude are shown in the middle panel (RMS) in Fig. 3a. With the exception of the tropical region, the 1 km altitude cases for other regions are generally larger than 3 hPa RMS. It is surprisingly to find that tropical region has the smallest RMS. Similar to the previous study [Ge et al., 2003], the comparison of ECMWF and CHAMP pressure in the SP region is found to be the worst both in terms of bias and RMS.

In Fig. 3b, the large bias between NCEP and CHAMP pressure profiles also occurs at tropopause (TP and SM). SP is still the area has the largest bias at 1 km altitude above MSL. It is expected that the bias would be even larger at the surface. The RMS figures (middle panels) show similar pattern for the ECMWF/CHAMP comparison. Tropical region (middle panels) still has the smallest RMS. The similarity of Fig. 3a and Fig. 3b implies that ECMWF agrees relatively well with NCEP as shown in Fig. 3d. The bias and RMS for comparisons in all regions are smaller than 2 hPa. The largest bias occurs at around 4 km altitude cases. Although the RMS is relatively small compared to other cases, the SP region has the largest discrepancy between the two models.

Radiosonde is an invaluable independent data source in this study. Fig. 3c, 3e and 3f compare the radiosonde with CHAMP, ECMWF and NCEP. Fig. 3c (left panel) could probably confirm that the bias between CHAMP and ECMWF and between CHAMP and NCEP comes from CHAMP pressure profile. Fig. 3c (middle panel, RMS differences) also shows that the largest RMS is near the surface, with the TP region has the smallest discrepancy. However, it is interesting to note that there are several pikes in the TP curve. Comparing with Fig. 3a and 3b, we find no similar patterns. This seems to imply that the pikes could be originated from radiosonde measurements. One possibility is that the radiosonde measurements are less accurate in these 3 locations, since usually occultation has the best performance in these areas. Similar pikes are also found in Fig. 3e and 3f for the TP region case. One should note that there also could be errors in the profiles because of quality control or interpolation errors, since CHAMP occultation cannot exactly match radiosonde both in location and time.

Fig. 3e and 3f show that there are large mean differences between the models and radiosonde measurements in the SP region. This is reasonable because of lack of data in the southern polar region which causes poor model performance. The differences are probably larger near the surface.

## 4 Conclusion

In conclusion, we found that there exists a positive bias in CHAMP derived pressure comparing with model (ECMWF and NCEP) and in situ data (radiosonde). The bias is larger in the SP region than any other regions globally. The RMS difference between the CHAMP pressure and other data sets is as large as 4 hPa globally at 1 km above MSL. It is surprising that the best comparison occurs at the TP region which is usually expected worst because of large water vapor signal and poor penetration of radar signals to the surface.

The comparison of ECMWF and NCEP agrees better than any other comparisons both in bias and RMS. This reflects the two models are consistent to a certain degree. However, SP region is still the area where the largest discrepancy exists.

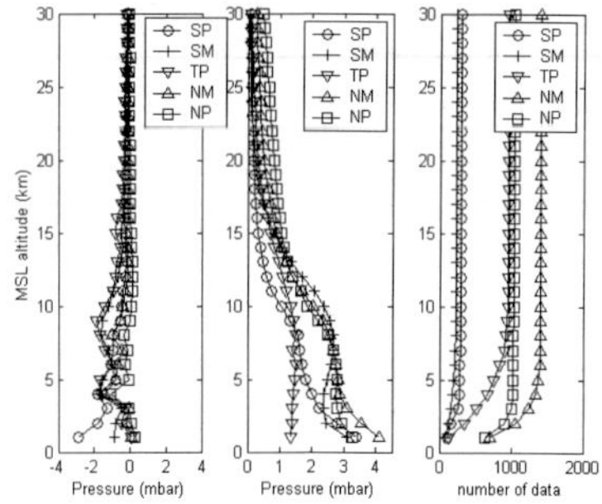


(a) $P_{ECMWF}$-$P_{CHAMP}$

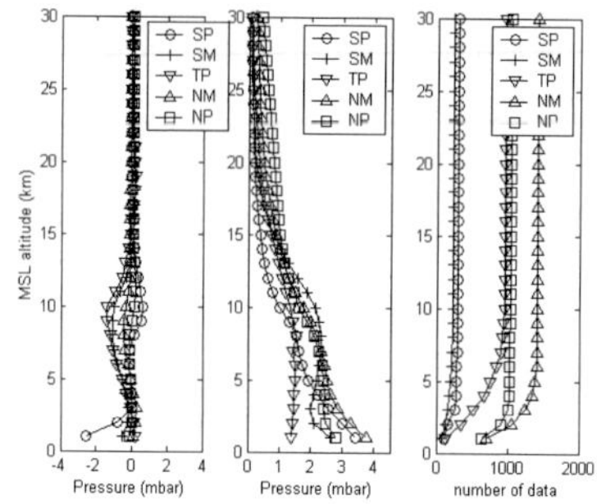


(b) $P_{NCEP}$-$P_{CHAMP}$

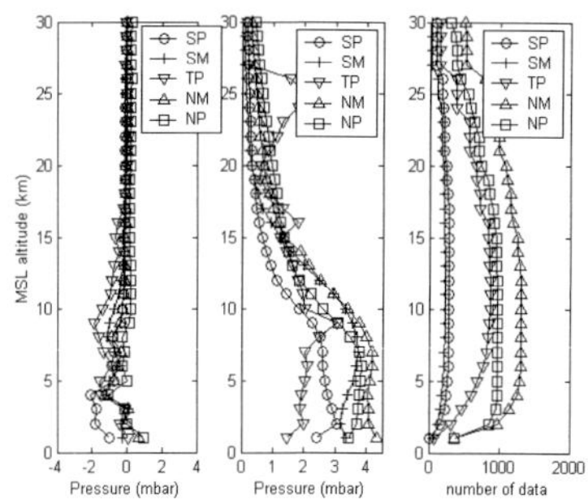


(c) $P_{SON}$-$P_{CHAMP}$

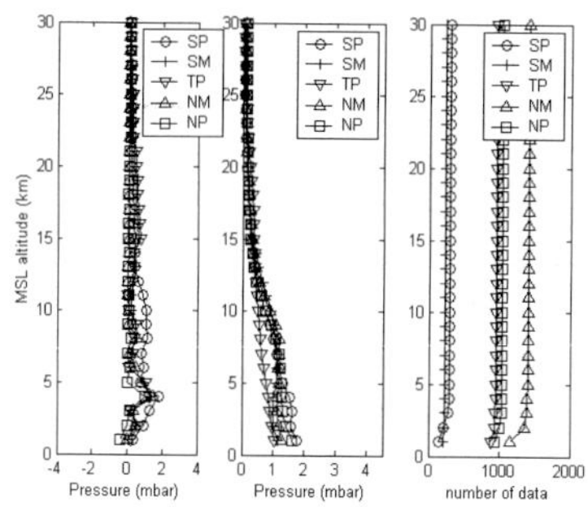


(d) $P_{NCEP}$-$P_{ECMWF}$

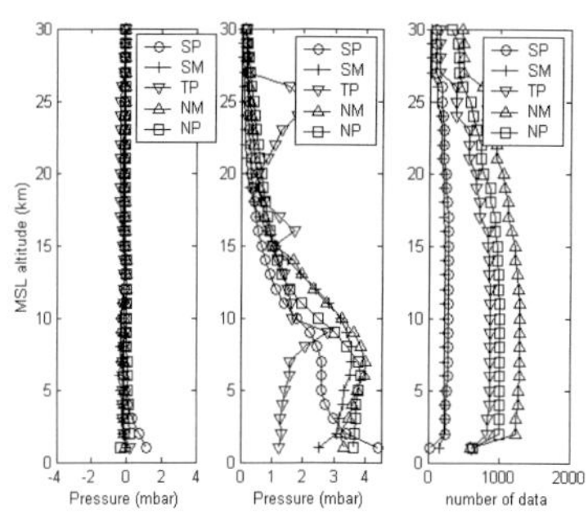


(e) $P_{ECMWF}$-$P_{SON}$

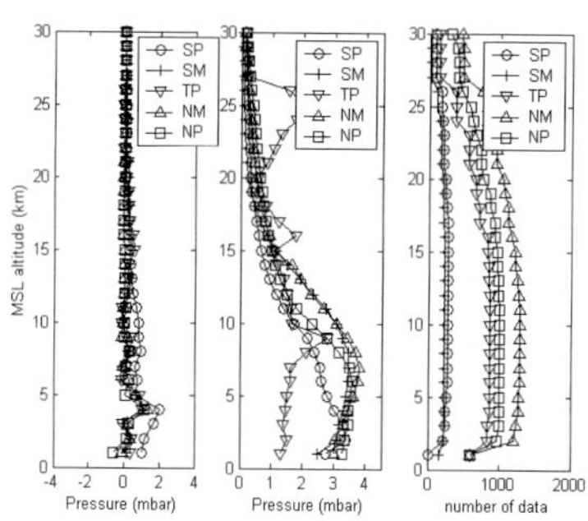


(f) $P_{NCEP}$-$P_{SON}$

**Fig. 3.** Intercomparison of CHAMP derived pressure profile ($P_{CHAMP}$), ECMWF pressure profile (($P_{ECMWF}$), NCEP pressure profile ($P_{NCEP}$) and radiosonde pressure profile ($P_{SON}$). SP—southern polar region (60°S–90°S), SM—southern mid-latitude region (30°S–60°S), TP—tropical region (30°S–30°N), NM—northern mid-latitude region (30°N–60°N), NP—northern polar region (60°N–90°N). For each figure: left panel—mean of pressure difference, middle panel—standard deviation of pressure difference, right panel—number of values in each level used for the comparison.

This once again confirms our conclusion that models have worse performance in this area. Radiosonde, as an independent measurement, provides us another alternative to evaluate the occultation data. Through the comparison of CHAMP pressure with radiosonde pressure, we conclude that CHAMP pressure is positively biased. Biases or errors exist in one or both measurements. One possibility is the poor vertical resolution from the radiosonde could cause larger differences in the tropospheric region. Poor temporal and spatial sampling could contribute to the differences. Meanwhile, the radiosonde data does not agree well with the models (NCEP and ECMWF) in data sparse regions such as SP. After "correcting" the CHAMP occultation profiles to reference to the ECMWF topography over Antarctica, the signal penetration improves to 80%, similar to the penetration in the Arctic region. The cause of the CHAMP bias is at present unknown and the understanding of its origin and eliminating it would provide a validated CHAMP occultation data product. Then, the CHAMP (along with other) GPS occultation observations, with better vertical resolution and global coverage, could improve global pressure field modeling and in particular, over Antarctica.

Future works include using techniques such as 1DVAR to potentially enhance accuracy and penetration of signals. The use of a finer resolution (3 hr sampling, 50 km or finer) mesoscale model in a 4DVAR scheme to assimilated GPS occultation measurements [Kuo et al. 2002] is anticipated to further improve pressure field accuracy and reduce temporal aliasing for gravity field mission data.

*Acknowledgement.* This research is supported by grants from NASA's Interdisciplinary Science program and from the University of Texas under a prime contract from NASA. We acknowledge GFZ for providing CHAMP data and UCAR for providing retrieved CHAMP data products for this study.

## References

Ge S, Shum C, Wickert J, Reigber C (2003) GPS Occultation: Potential A New Data Source for Improvement of Antarctic Pressure Field, Wuhan University Journal of Natural Science *8(28)*: 636-648.

Ge S, Shum C (2003) Improvement of Antarctic Pressure Field Using GPS Occultation Data, International Union of Geodesy and Geophysics Symposia 2003, Sapporo, Japan, June 30-July 11.

Han S-C, Jekeli C, Shum C (2003) Time-variable effects of ocean tides, atmosphere, and continental water mass on monthly mean GRACE gravity field, J Geophys Res, in review.

Kuo Y-H, Wee T-K, Bromwich D (2002) Potential Impact of COSMIC GPS Radio Occultation Data on Regional Weather Analysis and Prediction over the Antarctic, Radio Occultation Science Workshop, NCAR Foothills Laboratory, Boulder, Colorado, August 21-23.

Leroy S (1997) Measurement of geopotential heights by GPS occultation, J Geophys Res *102*: D6: 6971-6986.

Wickert J (2002) The CHAMP radio occultation experiment: Algorithms, Processing system, and First results, PhD thesis (in German), University of Graz, Scientific Technical Report 07/02, GFZ Potsdam, Germany.

# Analysis of Gravity Wave Variability from SAC-C and CHAMP Occultation Profiles between June 2001 and March 2003

Alejandro de la Torre[1], Toshitaka Tsuda[2], Ho Fang Tsai[2], George Hajj[3] and Jens Wickert[4]

[1] Departamento de Fisica, FCEN, Universidad de Buenos Aires, Pab I, Ciudad Universitaria, 1428 Buenos Aires, Argentina *delatorr@df.uba.ar*

[2] Radio Space Center for Space and Atmosphere, Kyoto University, Uji, Japan (HF. Tsai now at National Space Program Office (NSPO), Hsinchu City, Taiwan) *tsuda@kurasc.kyoto-u.ac.jp, tsai@kurasc.kyoto-u.ac.jp*

[3] Jet Propulsion Laboratory, California Institute of Technology, Pasadena, CA, USA *George.Hajj@jpl.nasa.gov*

[4] GeoForschungsZentrum Potsdam, Dept. Geodesy and Remote Sensing, Potsdam, Germany *jens.wickert@gfz-potsdam.de*

**Summary.** The gravity wave activity on a global scale between 10 and 27 km was analyzed by using temperature profiles retrieved from GPS occultation events detected with SAC-C and CHAMP satellites. Examples of its variability with altitude, longitude, latitude and time are presented for different wavelength ranges. A particular attention is given to gravity wave generation in the Andes Range region, in connection with the forcing of the mean wind by the mountains, and in the equatorial region, in possible connection with convective activity. A clear signature, possibly related to mountain wave generation, is seen above Andes mountains, between 30 and 40S. It is also appreciated the global enhanced wave activity in equatorial regions during the winter hemisphere.

**Key words:** Radio occultation, Andes Mountains, Gravity wave activity

## 1 Introduction

Gravity wave sources have great influence on local meteorology, as well as on the energy and momentum exchange between different altitudes, playing a crucial role in controlling the large-scale circulation of the Earth's higher troposphere and middle atmosphere. The development of improved schemes for the parameterization of small-scale gravity wave drag is of crucial importance in determining the circulation of the middle atmosphere. The effects of gravity waves which are not properly parameterized could give rise to simulated winds excessively weak, strong or even wrongly directed. The main tropospheric sources of gravity waves are topographic forcing, convective and frontal activity in the tropics, and at middle latitudes, wind shear, geostrophic adjustment, thunderstorms and typhoons (see e.g. [8] and references therein). Topographic and deep convection events are of particular relevance near the tropics. In these two cases, the relationship between tropospheric sources and stratospheric wave activity may sometimes be quite obvious, and the interactions between gravity waves are easy to identify (e.g. absorption at critical levels, reflection and breaking). In other cases, the observation of gravity waves mainly related to nonorographic and nonconvective sources are clearly insufficient to produce adjusted parameterizations in the numerical models.

GPS radio occultation for remote sensing of the Earth's atmosphere was pioneered by the U.S. American GPS/MET (GPS/Meteorology) experiment. The LEO (LOW EARTH ORBIT) satellites CHAMP and SAC-C, launched in 2000, carry a new generation of Global Positioning System (GPS) receivers to perform radio occultation soundings of the ionosphere and neutral atmosphere. Up to date, occultations in the order of $10^5$ have been gathered (see e.g. [2],[12]).

Useful information of the gravity wave activity in the upper troposphere and stratosphere is provided by horizontal wind and temperature variances (e.g. [1]). [11] have extracted mesoscale temperature perturbations from vertical wavelengths ranging from 2 to 10 km, using temperature profiles obtained by the GPS/MET experiment. They calculated the global potential energy, assumed to be caused by atmospheric gravity waves at 20-45 km height, during NH (Northern Hemispheric) winter. The highest values were found around the equator with considerable longitudinal variations. Longitudinal energy variations in the same height range in midlatitudes yielded higher energy values over the continents than over the oceans. Latitude variations largely concentrated around the equator at 20-30 km, in addition to local enhancements at midlatitudes, were observed in winter months in both hemispheres. Using the same observations, [7] found enhanced gravity wave activity of the lower stratosphere at a height of 22-28 km, associated to areas of increased tropospheric water vapor pressure at 4-6 km, a measure of tropical convection activity. Recently, [8] studied stratospheric gravity wave fluctuations at midlatitudes, observing significant asymmetries in their longitudinal dependences between both hemispheres, well correlated in SH (Southern Hemispheric) with the topographic morphology. In section 2, we present the data of temperature profiles retrieved by the GPS occultation experiments aboard SAC-C and CHAMP satellites and analyze examples of gravity wave activity as a function of altitude, latitude and longitude above Andes Mountains and in equatorial regions. In section 3, some conclusions are drawn.

## 2 SAC-C and CHAMP GPS Occultation Data

We analyze the gravity wave activity using temperature profiles retrieved by the GPS occultation experiments on board SAC-C and CHAMP satellites. A total of 14,876 globally distributed temperature profiles were successfully retrieved from occultation observations made by both satellites between June 2001 and March 2003. No wind data may be obtained from this technique. Nevertheless, from linear theory of gravity waves, the ratio of kinetic to potential energy is constant (e.g., [6]). This has been confirmed and extensively discussed from experimental evidence under different atmospheric conditions and wavelength ranges (e.g. [4]). Thus, it is possible to estimate the wave energy variability by the calculation of the potential energy or the temperature variance alone, from temperature profiles. We will use below the relative temperature variance to detect wave energy activity. We examine its global variability with altitude, during two months in which considerable wave activity has been observed near to the mid-latitude Andes and in equatorial latitudes, respectively: August and December, 2001. We subdivided both hemispheres in independent, adjacent cells of 5 per 10 degrees in longitude and latitude respectively, making a total of 1,296 cells. The occultation events are rather sparse at equatorial latitudes, in comparison with middle and high latitudes. A compromise between latitudinal, longitudinal and temporal resolution should be specified, in order to get information about seasonal and geographical variability of wave activity. With an average number of $T$ (Temperature) profiles per cell and month near to 4, the estimated statistical uncertainty in the potential energy in each cell is $4^{-1/2}$ times the standard deviation. Satellite missions scheduled in the near future are expected to increase in an order of magnitude the number of available occultation events per day, so decreasing accordingly the associated statistical error. In addition to the statistical error, the standard GPS retrieval techniques which are usually implemented suffer from a high sensitivity to measurement noise (see e.g. [10]). Compared to the standard approach, the variational framework could provide a more detailed understanding of the signal-to-noise issue. In the data set considered here and processed at JPL (Jet Propulsion Laboratory), bias in the temperature fluctuations greater than 3 K are frequently observed above and below 29 and 8 km respectively, and sometimes even within this range (see e.g. [10]). Nevertheless, as our main interest here is the upper troposphere and lower stratosphere regions, we show below results as a function of altitude within 10-27 km only.

## 3 Analysis Results

In recent GPS/MET stratospheric studies, normalized temperature fluctuations $(T'/T_f)^2$ were used, where $T'$ is the difference between the retrieved temperature $T$ and its lowpass filtered profile $T_f$ ([11], [8]).

The temperature profiles, after a spline process selecting a step of .2 km height, were high-pass filtered with a cutoff at 3.5 km (see below). The filter applied is non-recursive, and to avoid Gibbs effects, a Kaiser window was used (see e.g. [3]). Vertical fluctuations with wave lengths above and below 3.5 km are separately considered. Near to the Andes, long vertical wavelengths are expected, associated with orographic waves [5]. In equatorial regions, shorter vertical wavelengths are expected, and the contribution of Kelvin and Rossby-gravity waves are supposed to be quite reduced below the cutoff selected at 3.5 km. In Figure 1, a tomographic longitude-height view of mean $(T'/T_f)^2$ during August 2001 is shown, for occultation events registered at latitudes between 40 and 30S. Empty cells are represented by vertical white bands. A clear enhancement near to the Andes is observed from 10 km height up to the middle stratosphere, in the region limited by longitudes 65 and 70W. A band-pass filter between 3.5 and 9 km has been applied here. Nevertheless

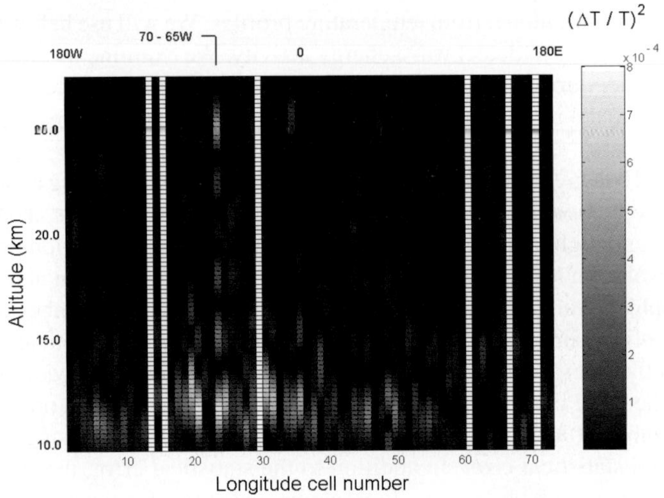


**Fig. 1.** Longitude-height view of mean relative temperature variance, $(T'/T_f)^2$, between 30 and 40S, during August 2001

still some differences between real and apparent vertical wavelengths should be expected. The generation of mountain waves is mainly related to three factors: mean forcing winds, orographic height and atmospheric stability. Since a detailed stability analysis lies outside the scope of this contribution, we recall that the highest Andes mountains are situated between 65-70W and 30-40S. An enhanced wave activity is observed just there in Figure 1. At other latitude bands to the North and South of Central Andes Range, the signal is clearly lower, as it may be seen in Figure 2. Here we show, in a latitude-longitude plot, the integrated relative temperature variance, in the lower stratosphere, between 19 and 26 km height. Here the same band-pass filter of 3.5-9 km was applied. The prominent enhancement already shown in Figure 1, is seen here, pointed out within the oval. At midlatitudes, this is the cell exhibiting the largest wave activity in both hemispheres, between 19 and 26 km height. Checking the variability of mean monthly zonal wind at 850 and 700 mbar between June 2001 and March 2003 from NCEP data (not shown here), high mean values are observed during August 2001 and August, September and October 2002. In all these cases, clear signals of enhanced wave activity similar to that present in Figures 1 and 2 may be appreciated. In Figure 2 it may be seen the enhanced wave activity at equatorial regions, where contributions of diverse origin, mainly including convective activity, are expected. As a last example, in Figure 3, wave activity for shorter vertical wavelengths are shown, during December 2001 in the lower stratosphere between 20 and 25 km height. The contribution of Kelvin and Rossby-gravity waves are supposed to be quite reduced below vertical wavelengths of 3.5 km. Note in the global distribution of wave activity the enhancements around the equator, mainly above Brazil, India and Indonesia. According to previous results from GPS/MET data ([11]) the

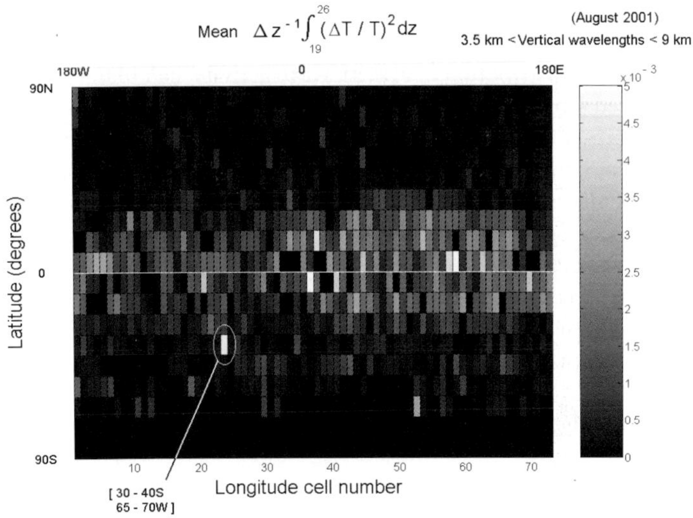


**Fig. 2.** Latitude-longitude distribution of integrated relative variance in the lower stratosphere, between 19 and 26 km height, during August 2001

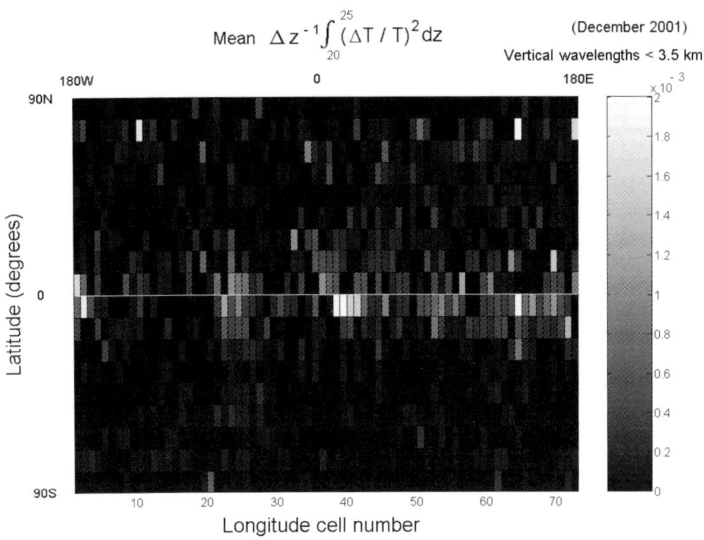


**Fig. 3.** Latitude-longitude distribution of integrated relative variance in the lower stratosphere, between 20 and 25 km height, during December 2001

wave activity during the winter hemisphere is clearly higher at middle and high latitudes.

## 4 Conclusions

Examples of global distribution of gravity wave activity, calculated from temperature profiles retrieved from GPS occultation events detected with SAC-C and CHAMP satellites, are presented here. Its variability with altitude, longitude and latitude is presented in tomographic plots, for two wavelength ranges. For long vertical wavelengths, a particular attention is given to gravity wave generation in the Andes Range region during August 2001, in connection with the forcing of the mean wind by the mountains and its detection up to the middle stratosphere. This enhancement is also shown by plotting the latitude-longitude wave energy distribution. For short wavelengths, the global wave activity is illustrated during December 2001. The expected enhancements in the winter hemisphere may be clearly seen here.

## References

[1] Alexander MJ (1998) Interpretations of observed climatological patterns in stratospheric gravity wave variance. J Geophys Res *103*: 8627–8640.

[2] Ao CO, Hajj GA, Meehan TK, Leroy SS, Kursinski ER, de la Torre Juarez M, Iijima BA, and Mannucci AJ (2003) Backpropagation processing of GPS radio occultation data. in: First CHAMP Mission Results for Gravity, Magnetic and Atmospheric Studies, Springer-Verlag, Berlin Heidelberg New York: 415–422.

[3] Brigham EO (1973) The Fast Fourier Transform. 246 pp, Prentice-Hall, New Jersey.

[4] de la Torre A, Alexander P, and Giraldez A (1999) The kinetic to potential energy ratio and spectral separability from high-resolution balloon soundings near the Andes mountains. Geophys Res Lett *26*: 1413–1416.

[5] Eckermann SD and Preusse P (1999) Global measurements of stratospheric mountain waves from space. Science *286*: 1534–1537.

[6] Fritts DC and VanZandt TE (1993): Spectral estimates of gravity wave energy and momentum fluxes. Part I: Energy dissipation, acceleration and constraints. J Atmos Sci *50*: 3685–3694.

[7] Hocke K and Tsuda T (2001) Gravity waves and ionospheric irregularities over tropical convection zones observed by GPS/MET radio occultation. Geophys Res Lett *28*: 2815–2818.

[8] Hocke K, Tsuda T, and de la Torre A (2002) A study of stratospheric GW fluctuations and sporadic E at midlatitudes with focus on possible orographic effect of Andes. J Geophys Res *107*: D20, ACL5/1-16.

[9] Lott F and H Teitelbaum (1993) Topographic waves generated by a transient wind. J Atmos Sci *50*: 2607–2624.

[10] Marquardt C, Schoellhammer K, Beyerle G, Schmidt T, Wickert J and Reigber C (2003) Validation and data quality of CHAMP radio occultation data. in: First CHAMP Mission Results for Gravity, Magnetic and Atmospheric Studies, Springer-Verlag, Berlin Heidelberg New York: 384–396.

[11] Tsuda T, Nishida M, Rocken C, and Ware RH (2000) A global morphology of gravity wave activity in the stratosphere revealed by the GPS occultation data (GPS/MET). J Geophys Res *105*: 7257–7273.

[12] Wickert J, Reigber C, Beyerle G, König R, Marquardt C, Meehan TK, Melbourne WG and Hocke K (2001) Atmospheric sounding by GPS radio occultation: First results from CHAMP. Geophys Res Lett *28*: 3263–3266.

# The CHAMPCLIM Project: An Overview

Ulrich Foelsche[1], Andreas Gobiet[1], Armin Löscher[1], Gottfried Kirchengast[1], Andrea K. Steiner[1], Jens Wickert[2], and Torsten Schmidt[2]

[1] Institute for Geophysics, Astrophysics, and Meteorology (IGAM), University of Graz, Universitätsplatz 5, 8010 Graz, Austria, *ulrich.foelsche@uni-graz.at*
[2] GeoForschungsZentrum Potsdam (GFZ), Department Geodesy and Remote Sensing, Telegrafenberg, 14473 Potsdam, Germany

**Summary.** The CHAMP radio occultation (RO) data provide the first opportunity to create real RO based climatologies on a longer term. CHAMPCLIM is a joint project of the Institute for Geophysics, Astrophysics, and Meteorology (IGAM) in Graz and the GeoForschungsZentrum (GFZ) in Potsdam. The overall aim of CHAMPCLIM is to ensure that the CHAMP RO data are exploited in the best possible manner, in particular for climate monitoring. The main objectives of the CHAMPCLIM project can be summarized in form of three areas of study as follows: RO data processing advancements for optimizing climate utility, RO data and algorithms validation based on CHAMP/GPS data, and global RO based climatologies for monitoring climate change. Here we show a summary of the current activities and exemplary results.

**Key words:** CHAMP, radio occultation, climate monitoring, CHAMPCLIM

## 1 Introduction

Increasing evidence suggests that the Earth's climate is significantly influenced by human activities (e.g., IPCC 2001). While there is little doubt that the Earth's surface temperature has risen by about 0.6°C during the $20^{th}$ century, the amount and even the existence of temperature trends in the troposphere are still under debate (e.g., Christy and Spencer 2003; Vinnikov and Grody 2003). Additional high quality observations of the atmosphere are therefore particularly required in this context. The Global Navigation Satellite System (GNSS) radio occultation (RO) technique has the potential to substantially contribute to this scientific challenge. For a detailed description of the RO technique see, e.g., the reviews of Kursinski et al. 1997 and Steiner et al. 2001.

With respect to climate studies, one of the most important properties of the RO technique is the expected long-term stability of RO data. It is achieved since precise atomic clocks are the basis for accurate measurements during each single occultation event, independent of whether two events are separated by an hour or by decades. Unlike many traditional satellite data like those from passive microwave sounders, radio occultation data are essentially self-calibrated as the measurement principle is basically counting of wave cycles (including fractional ones).

Sensing of the Earth's atmosphere with the RO method was first successfully demonstrated with the GPS Meteorology (GPS/MET) experiment performing measurement campaigns from April 1995 to March 1997. Analysis and validation of GPS/MET data sets confirmed most of the expected strengths of the technique, like high vertical resolution, high accuracy of retrieved parameters, and insensitivity to clouds (Kursinski et al. 1997; Rocken et al. 1997; Steiner et al. 1999).

The climate monitoring capability of a GNSS occultation observing system has not yet been tested due to the lack of long-term measurements. The CHAMP (Challenging Minisatellite Payload) RO data provide the very first opportunity to create real RO based climatologies. Continuous data are available since 2001 and the mission is expected to last until 2007. Wickert et al. (2004a) give an overview of the CHAMP RO experiment, information on the current status can be found in Wickert et al. (2004b).

CHAMPCLIM is a joint project of the IGAM in Graz and the GFZ in Potsdam. The project started in July 2003, its overall aim is to ensure that the CHAMP RO data are exploited as good as possible, with particular focus on climate monitoring. The three main objectives of CHAMPCLIM and some initial results are described in sections 2-4.

## 2 Radio occultation data processing advancements for optimizing climate utility

The upper stratosphere, where the signal-to-noise ratio of RO data is low and ionospheric effects increasingly influence the signal, is a critical domain for RO based climatologies. The processing advancements part of the study therefore aims at enhancing the RO retrieval algorithms with focus on upper stratospheric retrieval performance (ionospheric correction, statistical optimization). It also deals with better characterization of observation errors, and with the advancement of lower tropospheric retrieval quality.

The inversion of RO data requires some kind of high altitude initialization. So far, an advanced statistical optimization scheme has been developed, which combines observed bending angle profiles with background information derived from the MSISE-90 climatology (Hedin 1991) in a statistically optimal manner (Sokolovskiy and Hunt 1996). The key feature of this scheme is the ability to correct for systematic errors in background information (Gobiet and Kirchengast 2003). Results from a simulation study aiming at validation of the basic IGAM retrieval scheme (without background bias correction) and the advanced IGAM scheme are shown in Figure 1.

The characterization of observation error covariances has been analyzed in simulation studies (Steiner and Kirchengast 2003) and is currently object of studies using measured CHAMP data. The results of these studies aim at advancing the above mentioned statistical optimization and the utility of RO data for use in data assimilation systems.

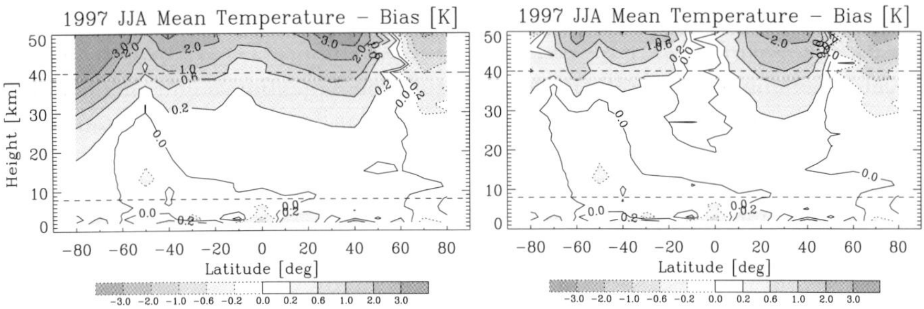


**Fig. 1.** Bias in seasonal mean dry temperature for a typical (northern) summer season based on an ensemble of ~1,000 simulated RO data. Results from the basic IGAM scheme (left panel) and from the enhanced scheme (right panel).

# 3 Radio occultation data and algorithms validation based on CHAMP/GPS data

In this part of the study, atmospheric profiles derived from CHAMP RO data are validated against co-located ECMWF analyses. Figure 2 shows results for an ensemble of 1,753 refractivity profiles as an example. Up to 35 km the relative bias (thick line) is smaller than 1 %, the most prominent feature of the bias profile at low latitudes is due to the fact that the sharp tropical tropopause near 17 km altitude is better resolved by radio-occultations than by the ECMF analyses.

Moreover, CHAMP RO data are validated against data derived from the GOMOS and MIPAS instruments onboard ENVISAT. A detailed discussion of first results of these comparisons can be found in Gobiet et al. (2004) in this issue.

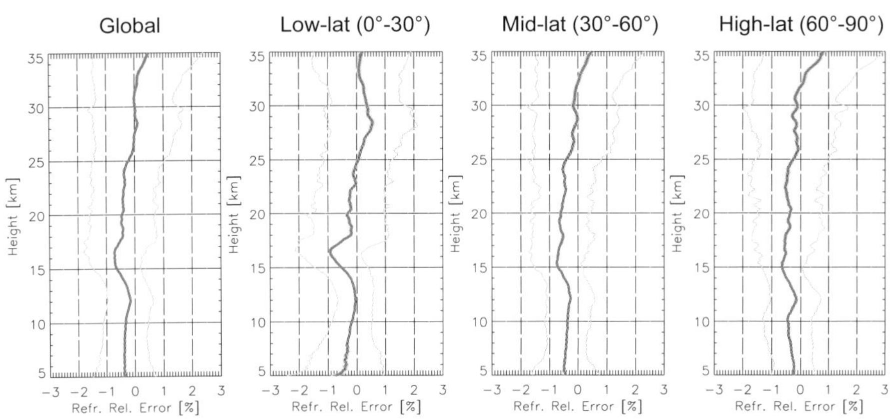


**Fig. 2.** Statistical comparison for a sample of 1,735 CHAMP refractivity profiles (11 days in 2003). Enhanced IGAM retrieval with MSIS initialization compared to ECMWF analysis profiles. Bias (thick black) and standard deviation (grey). Low-lat: 537 events, mid-lat: 672 events, high-lat: 526 events.

Furthermore, the performance of GFZ Potsdam retrieval algorithms is compared with the performance of the enhanced IGAM algorithm. We examine the retrieval of bending angle, refractivity, pressure, geopotential height, temperature, and humidity profiles, respectively, with particular focus on upper stratosphere and lower- to mid troposphere performance. First results of these comparisons are described in Wickert et al. (2004c) in this issue.

## 4 Global RO based climatologies

The potential of RO data for climate monitoring has been shown with simulation studies (e.g., Steiner et al. 2001; Foelsche et al. 2003). CHAMP RO data have already been used to derive the height of the tropical tropopause, a potential indicator for climate change, with high accuracy (Schmidt et al. 2004, this issue). Results for the global tropopause can be found in Ratnam et al. (2004) in this issue.

In a first approach, this part of the CHAMPCLIM study aims at direct (model independent) monitoring of the evolution of climatological refractivity, temperature, geopotential height, and humidity fields. Given the somewhat unfavorable single LEO satellite situation the sampling error has to be considered carefully in this context, as the sampling through occultation events is not dense enough to capture the entire spatio-temporal evolution. Due to the high inclination of the satellite (87.3°), the event density in low latitude regions is particularly small. Comparatively small temperature variations in these bins, however, prevent the sampling error from increasing dramatically. Sampling error studies by Foelsche et al. (2003) suggest that useful temperature climatologies resolving scales > 1000 km can be obtained even with RO data received from a single satellite. In the near future, multi-satellite missions like COSMIC (Lee et al. 2001) and ACE+ (Hoeg and Kirchengast 2002) with ~3,000 and ~4,000 RO profiles per day, respectively, will improve the spatial and temporal sampling significantly.

In a second approach we start to perform 3D-variational assimilation of CHAMP RO-derived refractivity data and ECMWF analysis fields into global climate analyses.

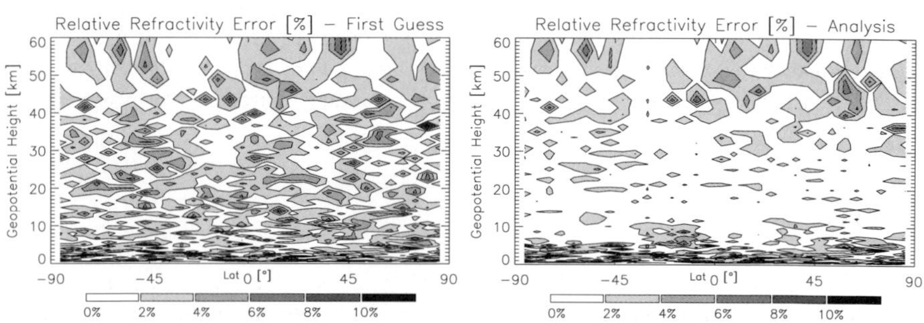


**Fig. 3.** Trial run of the refractivity assimilation system showing the impact of assimilating ~2,000 globally distributed RO profiles into a monthly mean field. Left panel: relative errors of the first guess, right panel: relative errors of the climate analysis.

Figure 3 shows first results of a simulation study to test the assimilation scheme for monthly mean fields (T21L60, ~600 km horizontal resolution at equator). There is an overall improvement of the background field, which is most clearly visible between ~5 km and ~35 km.

The assimilation scheme is tuned for high vertical and moderate horizontal resolution, reflecting the spatial characteristics of RO measurements. The background field (first-guess) was obtained by disturbing the "true" atmospheric mean field with reasonable errors using the error pattern superposition method. Simulated RO measurements have been derived from the "true" atmospheric fields. Realistic errors have been superimposed based on empirical error covariance matrices (Steiner and Kirchengast 2003). About 2,000 globally distributed RO profiles have been assimilated assuming Gaussian error characteristics for the background.

## 5 Summary, conclusions, and outlook

The continuous and ongoing flow of CHAMP radio RO data is the basis for the first RO based climatologies of atmospheric parameters like refractivity and temperature. CHAMPCLIM is a joint project of the Institute for Geophysics, Astrophysics, and Meteorology (IGAM) in Graz and the GeoForschungsZentrum (GFZ) in Potsdam. The overall aim of CHAMPCLIM is to ensure that the CHAMP RO data are exploited in the best possible manner, in particular for climate monitoring. We described the three main objectives of CHAMPCLIM, namely: RO data processing advancements for optimizing climate utility, RO data and algorithms validation based on CHAMP/GPS data, and global RO based climatologies for monitoring climate change. The project started in July 2003, the first results are already encouraging. We are confident to build reliable RO based climatologies, even with RO data from a single receiving satellite. Results of this study will certainly be valuable for multi-satellite RO missions in the near future.

*Acknowledgments.* The authors thank all colleagues in the CHAMPCLIM project for their support. U. Foelsche, A. Gobiet, A. Löscher, and A.K. Steiner were funded for this work from the START research award of G. Kirchengast financed by the Austrian Ministry for Education, Science, and Culture and managed under Program Y103-CHE of the Austrian Science Fund. Part of this study was also funded under contract no. ASAP-CO-004/03 of the Austrian Space Agency, which will be the main funding source of later phases of the CHAMPCLIM project.

## References

Christy JR, Spencer RW (2003) Reliability of satellite data sets, Science *301*: 1046-1047.

Foelsche U, Kirchengast G, Steiner AK (2003) Global climate monitoring based on CHAMP/GPS radio occultation data, in: First CHAMP mission results for gravity, magnetic and atmospheric studies (Reigber et al. eds.), Springer: 397-407.

Gobiet A, Kirchengast G (2003), Advancement of GNSS Radio Occultation Retrieval in the Upper Stratosphere. Proc 1st Int'l Workshop on Occultations for Probing Atmosphere and Climate (OPAC-1), Graz, Austria, September 2002, accepted.

Gobiet A, Kirchengast G, Wickert J, Retscher C, Wang DY, Hauchecorne A (2004) Evaluation of stratospheric radio occultation retrieval using data from CHAMP, MIPAS, GOMOS, and ECMWF analysis fields. This issue.

Hedin AE (1991) Extension of the MSIS thermosphere model into the middle and lower atmosphere. J Geophys Res *96*: 1159–1172.

Hoeg P, Kirchengast G (2002) ACE+ – Atmosphere and Climate Explorer based on GPS, GALILEO, and LEO-LEO radio occultation (ESA Earth Explorer Opportunity Mission Proposal). Wissenschaftl Ber 14, IGAM, University of Graz, Austria, 121 pp.

IPCC (2001) Climate change 2001: The scientific basis. Cambridge Univ Press, 881 pp.

Kursinski ER, Hajj GA, Schofield JT, Linfield RP, Hardy KR (1997) Observing Earth's atmosphere with radio occultation measurements using the Global Positioning System. J Geophys Res *102*: 23,429–23,465.

Lee L-C, Rocken C, Kursinski R (eds) (2001) COSMIC – Applications of constellation observing system for meteorology, ionosphere, and climate. Springer, Hong Kong, Berlin Heidelberg.

Ratnam MV, Tetzlaff G, Jacobi Ch, Structure and variability of the global tropopause. This issue.

Rocken C, Anthes R, Exner M, Hunt D, Sokolovskiy S, Ware R, Gorbunov M, Schreiner W, Feng D, Herman B, Kuo YH, Zou X (1997) Analysis and validation of GPS/MET data in the neutral atmosphere. J Geophys Res *102*: 29,849–29,866.

Schmidt T, Wickert J, Beyerle G, Reigber Ch (2004) Tropical tropopause parameters derived from GPS radio occultation measurements with CHAMP. This issue.

Sokolovskiy S, Hunt D (1996) Statistical optimization approach for GPS/Met data inversions. Presentation at URSI GPS/Met Workshop 1996, Tucson, AZ, USA.

Steiner AK, Kirchengast G, Ladreiter HP (1999) Inversion, error analysis, and validation of GPS/MET occultation data. Ann Geophys *17*: 122–138.

Steiner AK, Kirchengast G, Foelsche U, Kornblueh L, Manzini E, Bengtsson L (2001) GNSS occultation sounding for climate monitoring. Phys Chem Earth *26*: 113–124.

Steiner AK, Kirchengast G (2003) Ensemble-based analysis of errors in atmospheric profiles retrieved from GNSS occultation data. Proc 1st Int'l Workshop on Occultations for Probing Atmosphere and Climate (OPAC-1), Graz, Austria, September 2002, accepted.

Vinnikov KY, Grody NC (2003) Global warming trend of mean tropospheric temperature observed by satellites. Science *302*: 269-272.

Wickert J, Schmidt T, Beyerle G, König R, Reigber Ch, Jakowski N (2004a), The radio occultation experiment aboard CHAMP: Operational data analysis and validation of vertical atmospheric profiles, J Met Soc Japan, Special issue 'Application of GPS Remote Sensing to Meteorology and Related Fields' *82(1B)*: 381-395.

Wickert J, Beyerle G, Schmidt T, Michalak G, König R, Reigber Ch (2004b) Atmospheric profiling with CHAMP: Status of the operational data analysis, validation of the recent data products and future prospects. This issue.

Wickert J, Gobiet A, Beyerle G, Steiner AK, Kirchengast G, Foelsche U, Schmidt T (2004c): GPS radio occultation with CHAMP: Comparison of atmospheric profiles from GFZ Potsdam and IGAM Graz. This issue.

# AUTHOR INDEX

Altadill, D. 447
Angling, M. 585
Ao, C. O. 573
Aragón, A. 453
Asgharzadeh, M. F. 287

Balasis, G. 293, 347
Ballani, L. 311
Balmino, G. 25
Barthelmes, F. 25, 31
Belehaki, A. 447
Benedetto, C. 537
Beutler, G. 89, 157
Beyerle, G. 495, 507, 525, 561, 567, 591, 597
Biancale, R. 25, 77
Bock, H. 89, 157
Braun, A. 108
Bruinsma, S. 77

Casassa, G. 3
Chen, Y. 108
Chernova, T. A. 323
Choi, S. 169, 413
Christiansen, F. 371, 375, 381
Console, R. 407
Conti, L. 407
Cooke, D. 413, 471
Cueto, M. 483

Danilov, S. 211
de la Torre, A. 609
De Lauretis, M. 395
De Santis, A. 317
Denker, H. 59
Dierks, O. 31
Dietrich, R. 181

Everett, M. E. 341

Fengler, M. J. 139, 145
Feuchtinger, M. 127, 151
Fischer, H. 567, 573
Fleming, K. 217
Foelsche, U. 525, 615
Földváry, L. 13
Förste, C. 25, 169
Förster, M. 239
Fox Maule, C. 255
Freeden, W. 139, 145
Fritzsch, B. 211
Frommknecht, B. 13
Funke, B. 567, 573

Galas, R. 597
Galper, A. M. 407
García-Fernández, M. 453
Gaya-Piqué, L. R. 261, 317
Ge, S. 603
Gerlach, C. 13
Gil-López, S. 567, 573
Glatthor, N. 567, 573
Gobiet, A. 525, 531, 615
Golovkov, V. P. 323
Gooding, R. H. 101
Gorbunov, M. E. 519
Grabowski, U. 567, 573
Greiner-Mai, H. 311
Gruber, T, 13
Grunwaldt, L. 175
Gutting, M. 145

Haagmans, R. 95
Haak, V. 249
Hagedoorn, J. 217
Hajj, G. A. 421, 609
Han, S.-C. 108, 193, 230
Harig, S. 211
Hauchecorne, A. 531
Heirtzler, J. 273
Heise, S. 465, 471, 477, 591
Helm, A. 591

Hemant, K. 249
Hernandez-Pajares, M. 453
Herraiz, M. 483
Hinga, M. B. 115
Holme, R. 305, 329
Höpfner, M. 567, 573
Hsu, H. 83
Huang, C.-Y. 431, 549
Hugentobler, U. 89, 157

Ilk, K. H. 53, 127
Imanishi, Y. 31
Iorio, L. 187
Ivins, E. R. 3

Jacobi, C. 477, 555, 579
Jäggi, A. 89, 157
Jakowski, N. 441, 447, 459, 465, 471, 477, 483
James, T. S. 3
Jekeli, C. 108
Jiang, J. H. 573
Jochmann, H. 25
Jodogne, J.-C. 447
Juan, J. M. 453

Kaban, M. K. 199
Kaschenz, J. 495
Keller, W. 205
Kellmann, S. 567, 573
Kiefer, M. 567, 573
Kim, H. R. 261, 279, 287
Kim, J. W. 261
Kirchengast, G. 525, 531, 615
Kis, K. I. 279
Kivman, G. 211
Klokočník, J. 101
Köhler, W. 366
König, R. 25, 65, 71, 175, 495, 507, 597
Koop, R. 223
Koren, W. 395
Korhonen, J. V. 279
Kostelecký, J. 101
Kroner, C. 31
Kuo, C.-Y. 108
Kusche, J. 133, 139, 151
Kuvshinov, A. 353, 359

Larsen, G. B. 543
Lauritsen, K B. 519, 543
Laxon, S. 37
Lehmann, A. 181
Lemoine, J.-M. 25
Lemoine, J. M. 77
Lesur, V. 299
Li, J. 83
Linden, A. 567, 573
Liou, Y.-A. 431, 513, 549
Liu, H. 366
Löcher, A. 53
López-Puertas, M. 567, 573
Löscher, A. 615
Loyer, S. 25, 77
Lu, Y. 47
Lühr, H. 239, 293, 347, 366, 375, 389, 395, 413

Macmillan, S. 299
Magnes, W. 395
Mai, W. 413
Malvezzi, V. 407
Mandea, M. 293
Marquart, G. 223
Martinec, Z. 217, 335
Matyugov, S. 431
Maus, S. 249, 293, 347
Mautz, R. 193, 230
Mayer-Gürr, T. 127, 151
McAdoo, D. 37
McCreadie, H. 335, 401
Meixner, H. 65
Mengistu Tsidu, G. 567, 573
Meurers, B. 31
Michalak, G. 65, 71, 175, 495
Mielich, J. 447
Milz, M. 567, 573
Miquel Torta, J. 317

Moore, P. 19, 163

Nazarova, K. A 267, 273
Nedoluha, G. 501
Neumayer, K.-H. 25, 65, 71
Neumeyer, J. 31
Newell, P. T. 375

Olsen, N. 255, 329, 353, 359

Pacione, R. 537
Parrot, M. 407
Pavelyev, A. 431, 513, 549
Pavelyev, D. 431
Peng, B. 83
Perosanz, F. 25, 77
Peters, T. 13
Petrovic, S. 25
Piancatelli, A. 395
Picozza, P. 407
Poole, S. R. 115
Potts, L. V. 193, 279
Purucker, M. E. 255

Qiang, Z. 19, 163

Raabe, A. 477
Radicella, S. M. 483
Reick, C. 211
Reigber, C. 25, 65, 101, 175, 199, 431, 495, 507, 549, 561, 597
Reinhard, D. 121
Retscher, C. 531
Rignot, E. 3
Rios, V. 453
Roland, M, 59
Roman, L. L. 573
Rothacher, M. 13
Rother, M. 239, 293, 347, 395, 413
Rubek, F. 543
Rummel, R. 13

Sabaka, T. 273
Sanz, J. 453
Schaffrin, B. 230
Scheithauer, S. 489
Schlegel, K. 375
Schlüter, S. 465, 477
Schmidt, R. 65
Schmidt, T 431, 495, 507, 525, 549, 561, 567, 591, 597, 615
Schröter, J. 205, 211
Schwartz, M. J. 573
Schwingenschuh, K. 395
Schwintzer, P. 25, 31, 199
Scrimaglio, R. 407
Seufer, V. 205, 211
Sgrigna, V. 407
Shi, H. 47
Shum, C. K. 108, 193, 230, 603
Sidorenko, D. 211
Sørensen, M. B. 543
Spillantini, P. 407
Staneva, J. 211
Stankov, S. M. 441, 459
Stauning, P. 371, 375, 381
Steck, T. 567, 573
Steigenberger P. 13
Steiner, A. K. 525, 615
Stiller, G. P. 567, 573
Stolle, C. 465, 477
Stromeyer, D. 311
Sun, H.-P. 31
Sutcliffe, P. R. 389
Švehla, D. 13

Tapley B. D. 115
Taylor, P. T. 261, 279, 287
Tetzlaff, G. 555, 579
Thayer, J. P. 375
Theil, S. 489
Thomson A. 299
Tsai, H. F. 609
Tsuda, T. 609
Tsvetkov, Y. 273
Tsybulya, K. 441, 447, 483

Vales, N. 77

van den Ijssel, J. 95
van Loon, J. 133
Velímský, J. 341
Vellante, M. 395
Venkat Ratnam, M. 555, 579
Vespe, F. 537
Villante, U. 395
Virtanen, H. 31
Visser, P. 95
von Clarmann, T. 567, 573
von Engeln, A. 501
von Frese, R. R. B. 193, 261, 279, 287

Wagner, C. A. 37, 101
Wang, D.-Y. 531, 567, 573
Wardinski, I. 305, 311
Watermann, J. 371, 375, 381
Wehrenpfennig, A. 441
Wenzel, M. 205, 211
Wermuth, M. 13
Wesztergom, V. 395
Wickert, J. 431, 495, 507, 513, 525, 531, 537, 549, 561, 567, 597, 591, 609, 615
Wiehl, M. 121, 181
Wittmann, G. 279
Wolf, D. 217
Wu, B. 83
Wu, X. 3
Wünsch, J. 25

Yan, S. 549
Yunck, T. P. 421

Zhang, T.-L. 395
Zhu, S.Y. 65
Zijlstra, M. 489
Zilpimiani, D. 407
Zolesi, B. 447
Zvereva, T. I. 323

# KEYWORD INDEX


3D-tomography 477

Abel transform 453
accelerometer 71, 77, 95, 163, 169
accelerometer calibration parameters 121
accelerometer data 89, 489
accelerometer preprocessing 169
ADMAP 287
aliasing 121
amplitude 549
Andes mountains 609
anomaly continuation 293
Antarctica 261, 287
Arctic 37
assimilation 501
atmosphere 591
atmospheric profiling 495
atmospheric remote sensing 537
attitude 71
attitude preprocessing 169

balloon 273
Baltic Sea 181
bias and scale 163
bi-static radar 421
bistatic remote sensing 513
BlackJack GPS receiver 175
boundary value problem 151
B-spline wavelets 230

calibration 77
calibration parameter history 413
canonical transform 507, 519
cavity modes 389
CHAMP Atmospheric Processor 597
CHAMPCLIM 615
climate monitoring 615
clock bias 83
clock offset 175
common reference frame 413
comparing satellite magnetic anomalies 279
continuous wavelet transform 591
core-mantle boundary 311
core-surface flows 305
correlations 193
cross-calibration 413
crustal 249
crustal magnetization 279
crustal thickness 193, 261, 279
cryosphere 3
cusp region 381
CWT 591
cycle slip 83

database 267
degree variances 59
differential code bias 465
diffractive integrals 513
downward continuation 311
Dst variation 335
dynamic 77
dynamic topography 199

Earth gravity models 187
earthquakes 407
ECMWF 531
efficient and precise orbit determination 157
EIGEN 121
eigenvalue disposal 115
electric fields 239
electrical conductivity of mantle 335
electromagnetic induction 335, 341, 353, 359
electron density 431, 441, 447, 453, 483
elevation-dependent weighting 157
empirical modeling 459

energy balance 53
energy balance approach 133, 151
energy integral 13
EOS MLS 573
equatorial effects 239
equatorial electrojet 401

FAC modelling 371
Fennoscandia 181
FGM data 347
field line resonance 395
field models 293
field-aligned currents 371, 375, 381
fine-scale FAC 381
fluid outer core 311
Fourier integral operators 519
F-region instabilities 347
full spectrum inversion 507

Gauss coefficients 311
Gaussian filter 47
general theory of relativity 187
geoid 37, 108, 205
geoid anomalies 223
geoid change 217
geoid undulations 230
geomagnetic field 293, 407
geomagnetic field modelling 299, 329
geomagnetic jerk 305, 311
geomagnetic spectrum 329
geomagnetism 273, 401
geopotential 108
GFZ Potsdam 525, 597
GIS 249
glacial-isostatic adjustment 217
global 249
global gravity field model 25, 199
GNSS 421
GOMOS 531
GPS 65, 77, 95, 421, 441, 453, 465, 471, 477, 495, 525, 543, 561, 591, 597
GPS clock interpolation 157
GPS occultation 603, 615
GPS radio occultation 507, 567
GPS/SAC-C 573
GRACE 3, 37, 65, 121, 217
gravitational field model conversion 145
gravity 19, 37, 163, 193
gravity anomaly 47
gravity anomaly errors 59
gravity field 13, 31, 101
gravity field modeling 127
gravity field recovery 151
gravity field validation 53
gravity mission 25
gravity variation 31
gravity wave activity 609
gravity waves 549, 555
Greenland 255
Greenland ice sheet 217

high-latitude ionosphere 375
Hudson Bay 193

IGAM Graz 525
inversion techniques 537
ionosonde 447
ionosphere 401, 431, 441, 447, 465, 471, 477, 483, 585, 591
ionospheric data assimilation 585
isostatic gravity anomalies 199
ITG-Champ01E 127

Jacobi integral 53
Joule heating 366
jumps 413

Kalman filtering 211
Kaula's rule 115
kinematic orbit 13, 139

LAGEOS-LAGEOS II orbits 187
Lense-Thirring effect 187
LEO 453

limb-sounding 421
lithosphere 273
lithospheric 267
Lorentz force 169
Low Earth Orbiters 157
low-altitude cusp 375
lumped coefficients 101
lumped harmonics 19

magnetic 249
magnetic anomalies 261
magnetic crustal thickness 255
magnetic field variations 353, 359
magnetic petrology 267
magnetometer 413
magnetosphere 407
magnetosphere/ionosphere interaction 395
magnetospheric currents 299
magnetospheric storm 366
MAGSAT 305, 311
main field 293
main geomagnetic field 317
mantle scaling factor 223
mantle viscosity 223
Maude Rise 287
mean sea surface 205
mid- and low-latitude ionosphere 239
MIPAS 531, 567, 573
model 489
modelling 239
multipath 519
multi-quadrics 230
multiresolution analysis 59
multiscale modelling 145

natural orthogonal components 323
near-real time 597
NeQuick 483
noise 347
non-conservative acceleration 95
nonlinear models 230

occultation 421, 453
occultation method 459
ocean circulation 211
ocean effect 353
ocean model 205
ocean tides 359
oceanographic variations 181
optimal estimation 501
orbit 77
Ørsted 287, 305, 311

Pannonian Basin and South and Central Finland 279
parametrization 121
Pi2 pulsations 389
plasma 407
plasmasphere 239, 471
polar ionosphere 371, 381
polynomial interpolation 151
postglacial rebound 181
precise orbit determination 65, 95
preprocessing 169
pseudo-stochastic orbit modeling 89
PSO orbit 139
pulsations 347

radiation belts 407
radio occultation 431, 441, 447, 477, 483, 495, 501, 513, 525, 531, 537, 543, 549, 555, 561, 579, 591, 597, 609
Rapid Science Orbit 71
ray tracer 501
reduced-dynamic orbit determination 89
reflections 591
reflectometry 421
refractivity bias 507
refractivity profiles 537
regional gravity field recovery 127, 139
regional modeling 317

remote sensing 507, 525
resonances 101
retrieval 531

SAC-C 65, 465, 471, 495
satellite altimetry 205, 211
satellite data 323
satellite geodesy 211
satellite gravity field mission 47
satellite orbit validation 53
satellite-to-satellite-tracking 53, 127
scale height 459
sea ice 37
sealevel rise 3
secular variation 305, 323, 329
separability hypothesis 453
short-term density variations 489
slant total electron content 585
SLR 77
small-scale field-aligned current 366
solar-quiet variations 341
spaceborne GPS receivers 89
spherical cap harmonic analysis 317
spherical harmonics 59, 108
sporadic E-layer 431
star camera 413
statistical assessment 133
statistical optimization 531
stratosphere 273, 555
Superconducting Gravimeter 31
surface pressure 603

temperature 567, 573
temporal gravity field variations 25
temporal variability 19
thermosphere 239, 366
thermospheric winds and composition 239
three-dimensional modellings 359
thrusters 163
Tikhonov-wavelet regularization 139
time-variable gravity 108, 121, 181, 603
topography 193
total electron content 441
tropopause 561, 579

ULF waves 389, 395
UTLS region 561

validation 447, 483, 531
variance components 133
vector magnetometer data 389
vertical gradient of temperature 549

water vapor profiles 537
wavelet coefficients 145
wavelet variances 145
wavelets 145, 347

X3 correction 169